Nanostructured Materials for Type III Photovoltaics

Nanoscience & Nanotechnology Series

Editor-in-chief:
Paul O'Brien CBE FREng FRS, *University of Manchester, UK*

Series editors:
Xiaogang Liu, *National University of Singapore, Singapore*
Ralph Nuzzo, *University of Illinois at Urbana-Champaign, USA*
Joao Rocha, *University of Aveiro, Portugal*

Titles in the series:
1: Nanotubes and Nanowires
2: Fullerenes: Principles and Applications
3: Nanocharacterisation
4: Atom Resolved Surface Reactions: Nanocatalysis
5: Biomimetic Nanoceramics in Clinical Use: From Materials to Applications
6: Nanofluidics: Nanoscience and Nanotechnology
7: Bionanodesign: Following Nature's Touch
8: Nano-Society: Pushing the Boundaries of Technology
9: Polymer-based Nanostructures: Medical Applications
10: Metallic and Molecular Interactions in Nanometer Layers, Pores and Particles: New Findings at the Yoctolitre Level
11: Nanocasting: A Versatile Strategy for Creating Nanostructured Porous Materials
12: Titanate and Titania Nanotubes: Synthesis, Properties and Applications
13: Raman Spectroscopy, Fullerenes and Nanotechnology
14: Nanotechnologies in Food
15: Unravelling Single Cell Genomics: Micro and Nanotools
16: Polymer Nanocomposites by Emulsion and Suspension
17: Phage Nanobiotechnology
18: Nanotubes and Nanowires, 2nd Edition
19: Nanostructured Catalysts: Transition Metal Oxides
20: Fullerenes: Principles and Applications, 2nd Edition
21: Biological Interactions with Surface Charge Biomaterials
22: Nanoporous Gold: From an Ancient Technology to a High-Tech Material
23: Nanoparticles in Anti-Microbial Materials: Use and Characterisation
24: Manipulation of Nanoscale Materials: An Introduction to Nanoarchitectonics
25: Towards Efficient Designing of Safe Nanomaterials: Innovative Merge of Computational Approaches and Experimental Techniques
26: Polymer–Graphene Nanocomposites
27: Carbon Nanotube–Polymer Composites
28: Nanoscience for the Conservation of Works of Art
29: Polymer Nanofibers: Building Blocks for Nanotechnology

30: Artificial Cilia
31: Nanodiamond
32: Nanofabrication and its Application in Renewable Energy
33: Semiconductor Quantum Dots: Organometallic and Inorganic Synthesis
34: Soft Nanoparticles for Biomedical Applications
35: Hierarchical Nanostructures for Energy Devices
36: Microfluidics for Medical Applications
37: Nanocharacterisation, 2nd Edition
38: Thermometry at the Nanoscale: Techniques and Selected Applications
39: Nanoceramics in Clinical Use: From Materials to Applications, 2nd Edition
40: Near-infrared Nanomaterials: Preparation, Bioimaging and Therapy Applications
41: Nanofluidics, 2nd Edition
42: Nanotechnologies in Food, 2nd Edition
43: ZnO Nanostructures: Fabrication and Applications
44: Diatom Nanotechnology: Progress and Emerging Applications
45: Nanostructured Materials for Type III Photovoltaics

How to obtain future titles on publication:
A standing order plan is available for this series. A standing order will bring delivery of each new volume immediately on publication.

For further information please contact:
Book Sales Department, Royal Society of Chemistry, Thomas Graham House, Science Park, Milton Road, Cambridge, CB4 0WF, UK
Telephone: +44 (0)1223 420066, Fax: +44 (0)1223 420247
Email: booksales@rsc.org
Visit our website at www.rsc.org/books

Nanostructured Materials for Type III Photovoltaics

Edited by

Peter Skabara
University of Glasgow, UK
Email: peter.skabara@glasgow.ac.uk

and

Mohammad Azad Malik
University of Manchester, UK
Email: azad.malik@manchester.ac.uk

ROYAL SOCIETY OF CHEMISTRY

THE QUEEN'S AWARDS FOR ENTERPRISE: INTERNATIONAL TRADE 2013

Nanoscience & Nanotechnology Series No. 45

Print ISBN: 978-1-78262-458-5
PDF eISBN: 978-1-78262-674-9
EPUB eISBN: 978-1-78801-250-8
ISSN: 1757-7136

A catalogue record for this book is available from the British Library

The Royal Society of Chemistry is a charity, registered in England and Wales, Number 207890, and a company incorporated in England by Royal Charter (Registered No. RC000524), registered office: Burlington House, Piccadilly, London W1J 0BA, UK, Telephone: +44 (0) 207 4378 6556.

For further information see our web site at www.rsc.org

Printed in the United Kingdom by CPI Group (UK) Ltd, Croydon, CR0 4YY, UK

Preface

The importance of renewable energy has become ever more important with the increasing environmental costs of using fossil fuels and their limited supply. The Sun is the most abundant source of renewable energy. There are two routes for the conversion of sunlight into a usable form of energy: the solar thermal approach, whereby solar energy is converted to heat and then electricity, and the solar photovoltaic route, where semiconductors are used to convert solar radiation directly into electricity. The photovoltaic solar cell has been identified as one of the most promising conversion devices for solar energy because it is both clean and scalable.

Photovoltaic technology has gone through many changes over the last four decades. These changes have now been classified into different generations of photovoltaic technology. Type I photovoltaic technology was based on crystalline silicon; it is expensive, but can give high power conversion efficiencies. Type II photovoltaic technology is based on thin film technology, with films containing polycrystalline silicon, cadmium telluride and copper indium disulphide. It is cheap compared with Type I photovoltaic technology, but offers lower efficiencies. Type III photovoltaic technology, the topic of this book, is based on solution-processed semiconducting nanostructured materials, which are both cheap and efficient. Nanostructured materials include atomic clusters or layered assemblies as thin films in which one of the dimensions is <100 nm. The properties of these materials are remarkably different from the bulk materials. Type III photovoltaic solar cells include organic polymer:fullerene macromolecules, inorganic nanoparticles or hybrids, dye-sensitised solar cells, hybrid polymer and perovskite solar cells.

The materials for Type III solar cells have branched into a series of generic groups. These include: organic 'small molecule' and polymer-conjugated

Nanoscience & Nanotechnology Series No. 45
Nanostructured Materials for Type III Photovoltaics
Edited by Peter Skabara and Mohammad Azad Malik

Published by the Royal Society of Chemistry, www.rsc.org

structures, fullerenes, quantum dots, copper indium gallium selenide nanocrystal films, dyes/TiO_2 for Grätzel cells, hybrid organic/inorganic composites, and perovskites. Although the power conversion efficiencies of organic solar cells are modest compared with other Type III photovoltaic materials, plastic semiconductors provide a cheap route to manufacturing through solution processing and offer flexible devices. However, other types of material are proving to be compatible with this type of processing while providing higher device efficiencies. As a result, the field is experiencing healthy competition between technologies and this is pushing progress at a fast rate. In particular, perovskite solar cells have emerged as a highly disruptive technology, with power conversion efficiencies now just over 22%. Perovskite cells, however, still have to address stability and environmental issues.

With such a diverse range of materials, it is timely to capture the different technologies in a single volume of work. The aim of this book is to give a collective insight into the different parts played by nanostructured materials in Type III solar cells. It aims to be an essential text for those working with any of the devices highlighted here, providing a fundamental understanding and appreciation of the potential and challenges associated with each of these technologies. This book gives an overview of recent developments in the synthesis and use of these nanostructured materials in photovoltaic technology. It is divided into two major sections, although the first chapter critically and importantly deals with the accuracy of photovoltaic cell measurements and is essential reading for those who make and test devices. The next six chapters of the book cover the development and working principles of organic solar cells, while the last seven chapters highlight the progress made in the synthesis, properties and the use of inorganic nanostructured materials in photovoltaic technology.

Peter Skabara and Mohammad Azad Malik

We would like to dedicate this book to Dr Russell Binions of Queen Mary University of London, who died recently at the age of 38, soon after completing his chapter on hybrid solar cells.

Contents

Nanoscience & Nanotechnology Series No. 45
Nanostructured Materials for Type III Photovoltaics
Edited by Peter Skabara and Mohammad Azad Malik

Published by the Royal Society of Chemistry, www.rsc.org

CHAPTER 1

Reliably Measuring the Performance of Emerging Photovoltaic Solar Cells

MATTHEW O. REESE, ASHLEY R. MARSHALL AND GARRY RUMBLES*

National Renewable Energy Laboratory, Golden, Colorado, 80401-3393, USA
*E-mail: Garry.Rumbles@nrel.gov

1.1 Introduction

The most reliable method of determining the power conversion efficiency (PCE) of a new solar cell is to send it to an accredited testing laboratory, such as the National Renewable Energy Laboratory (NREL), the Fraunhofer Institute for Solar Energy Systems or the National Institute of Advanced Industrial Science and Technology. However, this is not always a practical solution and these certification laboratories only determine the efficiencies of solar cells that have already been tested in a researcher's own laboratory and have shown a promisingly high performance. The record efficiency "hero" cells, certified in accredited laboratories, have been tracked and the results have been published semi-annually in *Progress in Photovoltaics: Research and Applications*[2] since 1993.[3] A graphical depiction of these published hero cell performances, and earlier data, is recorded on a chart, the May 2017 version of which is shown in Figure 1.1. The most recent version of this graph can be found online,[1] including an interactive version.[4]

Nanoscience & Nanotechnology Series No. 45
Nanostructured Materials for Type III Photovoltaics
Edited by Peter Skabara and Mohammad Azad Malik

Published by the Royal Society of Chemistry, www.rsc.org

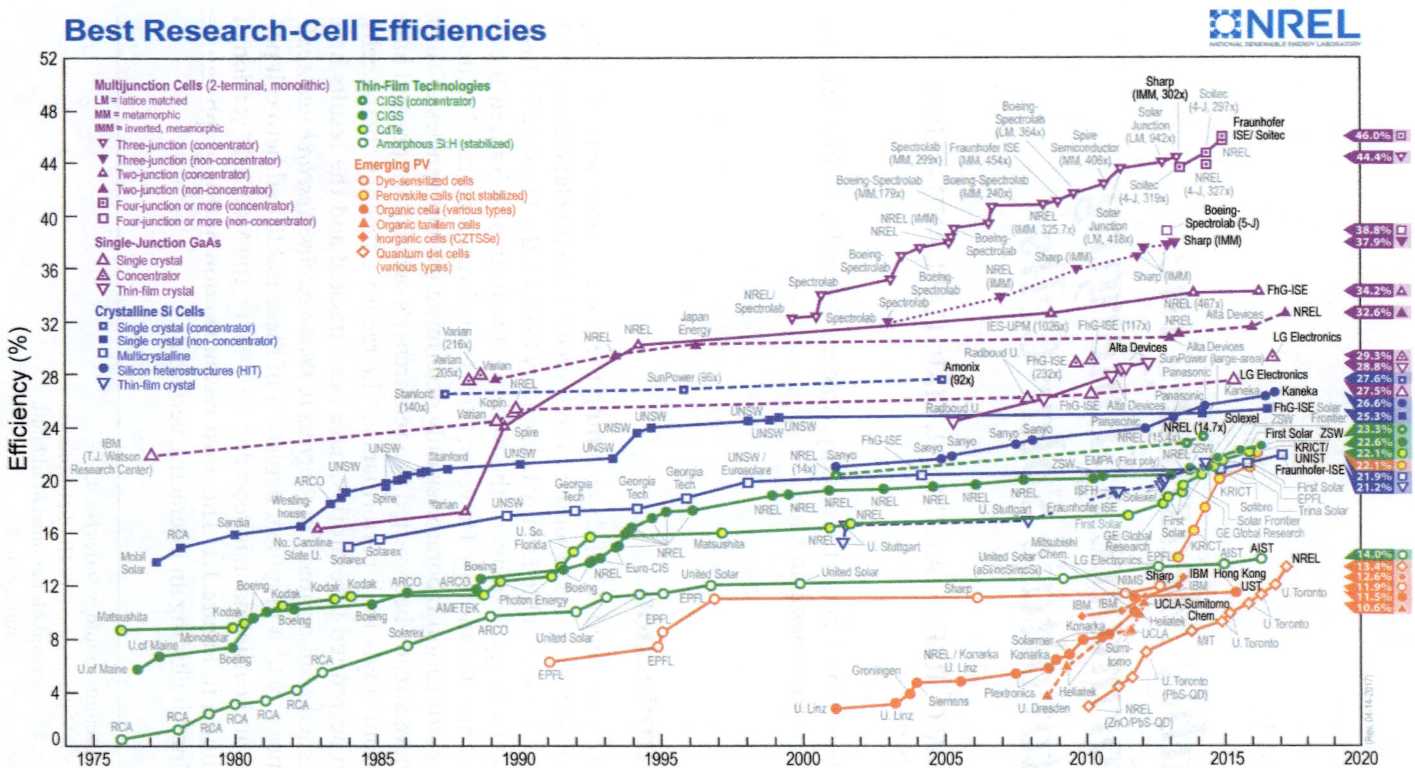

Figure 1.1 Research cell efficiencies from 1976 to 22 May 2017.[1] Certified photovoltaic data reported primarily from the National Renewable Energy Laboratory (NREL), the Fraunhofer Institute for Solar Energy Systems (FhG-ISE) and the National Institute of Advanced Industrial Science and Technology (AIST).

The solar cell systems described in this book fall into the category of emerging photovoltaics and include dye-sensitized solar cells (DSSCs), organic photovoltaic (OPV) cells, quantum dots, perovskite solar cells (PSCs) and the new copper zinc tin sulfide inorganic cells. Although these new photovoltaic technology platforms are identified on both the NREL and IEEE charts, the values are tabulated in *Progress in Photovoltaics: Research and Applications*[2] and on charts as "notable exceptions" because they rarely comply with the full set of stringent tests associated with the certification process. Specifically, the active areas measured for these emerging devices are often less than the required 1 cm^2 and issues associated with device stability during the evaluation process might also be questionable. Regardless, the values cited serve as a valuable target for researchers aiming to improve device performance. The recognition of a researcher's efforts on the chart is a frequent goal of many in the field of solar photovoltaics.

Small devices are commonly made in the world of emerging photovoltaics for several reasons. Small devices are often constructed on a single larger substrate yielding, for example, six or eight discrete pixel-type solar cells that can be constructed from the same set of depositions, thus providing a good range of performance statistics. The amount of active material available is often restricted or expensive and smaller devices reduce the amount of material required to evaluate the performance of the solar cell. A small device also provides an opportunity to reduce the impact of the sheet resistance on the fill factor. Within the ranks of hero cells for emerging photovoltaics, most devices have active areas, apertured or masked, of ~0.05 cm^2, with only the current record DSSCs exceeding the required 1 cm.[2] For example, the current record for PSCs as a "notable exception" is 22.1%, whereas at 1 cm^2 the PSC record is lower at 19.7%.[2]

Why is it important to precisely determine the PCE of a new cell? The primary reason is to ensure that a researcher can identify advances in their own work so that new findings are discovered and can be exploited. It is also necessary for different research groups to be able to repeat published research and obtain similar results. To demonstrate and report a record efficiency, it is necessary to have the device certified by a recognized center. Making a good, reliable efficiency measurement helps to pre-screen devices that are suitable for the certification process, although this can be both costly and time consuming. For systems that have the potential to be scaled up to larger device areas or modules, it is important to know that the starting point is good.

Despite repeated efforts to caution the photovoltaic community of the need to make good device performance measurements and descriptions of the exact methods required to avoid mistakes, erroneous reporting still occurs.[5–9] Some journals have attempted to introduce itemized checklists to prevent this,[10] but it is easy to make the wrong measurements and to inadvertently publish incorrect results. A common method used by researchers at NREL when reading or reviewing a new article is to determine the short circuit current, J_{sc}, predicted from the external quantum efficiency (EQE) data in the article and to compare this with the measured J_{sc} values under one sun illumination.

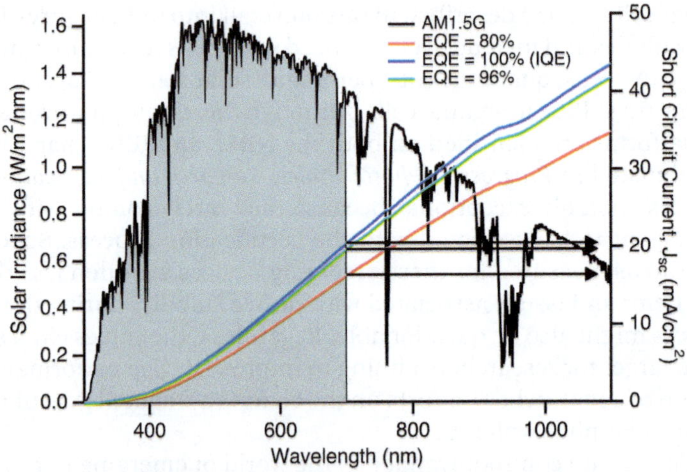

Figure 1.2 Shockley–Queisser current limits. Plots of maximum integrated short circuit current (J_{sc}) as a function of the optical bandgap for ideal EQE scenarios: 100% (blue); 96% (green); and 80% (red). The three arrows show the maximum current associated with a system with an optical bandgap at 700 nm.

The latter are occasionally observed to be far higher due to poor device stability or errors in determining the active device area.[11] These data are often shown to have J_{sc} values that exceed the Shockley–Queisser limit.[12] Figure 1.2 plots the AM1.5 spectrum with three scenarios of short circuit current *versus* the optical bandgap: (1) EQE = 100%, corresponding to the optimum internal quantum efficiency; (2) EQE = 96%, corresponding to a reflectivity of ~4% from the device surface; and (3) EQE = 80%, corresponding to a reasonably good device. A device with an optical bandgap of 700 nm (1.77 eV), as depicted in Figure 1.2, would have maximum short circuit currents under these three situations of 20.5, 19.6 and 16.4 mA cm^{-2}, respectively.

Although the device performance group at NREL is accredited by the American Association for Laboratory Accreditation to ISO-17025 standards, it was necessary to develop additional capabilities to rapidly evaluate the performance of the many new emerging photovoltaic solar cells being constructed prior to identifying hero devices suitable for certification, despite the fact that these hero devices are often certified at one of the other two accredited laboratories. This additional capability is described here, together with the rationale for its design and the protocol used to measure many different types of cells with the goal of obtaining reliable data accurate to ~95%.

To make a good measurement of solar cell efficiency for high-quality science requires the acquisition of the following:

- a well-measured spectrum of the lamp used in the solar simulator;
- a determination of the spectral mismatch factor between the lamp spectrum, the test cell and the reference cell;

- a measurement of the illuminated device area;
- an EQE spectrum of the cell under test; and
- measurements made with care.

1.1.1 Lamp Spectrum

In establishing a new set-up to make efficiency measurements, the most central piece of equipment is generally viewed to be the solar simulator, which can easily cost many tens of thousands of dollars; however, with more care this might not be necessary. When considering a new purchase, and sometimes evaluating others' work, many researchers focus on the class of the simulator as designated by the International Electrotechnical Commission (IEC) standard 60904-9 (there are also country-specific standards such as the American Society for Testing and Materials (ASTM) E 927-05 and the Japanese Industrial Standard (JIS) C 8912). As the class system is intended to designate how easy it will be to make a good measurement, this is not unreasonable. Using an AAA simulator does not guarantee good results, however, just as using a less well-matched simulator does not guarantee bad results. Understanding the measurements and identifying the sources and magnitudes of any errors is the most important way of reducing these to an acceptable level.

We explain here how simulators are categorized into classes. Each letter designator corresponds to a particular quality. The first letter indicates the spectral match to the reference spectrum, the second letter indicates the quality of spatial uniformity and the third letter indicates the quality of both short- and long-term temporal uniformity (Tables 1.1 and 1.2).

Table 1.1 Ideal spectral match for solar simulators elements defined by ASTM, IEC and JIS standards.

Wavelength[a] (nm)	400–500	500–600	600–700	700–800	800–900	900–1000
Total irradiance from 400 to 1000 nm (%)	18.4	19.9	18.4	14.9	12.5	15.9

[a]Spectral distribution not specified for 300–400 or 1100–1400 nm.

Table 1.2 Classes of solar simulator elements showing the spectral, spatial and temporal requirements for the IEC 60904-9-2007, ASTM E927-05 and JIS C8912 classification standards.

Class	Spectral mismatch to all intervals	Spatial non-uniformity (%)	Temporal instability (%)	
			Short-term	Long-term
A	0.75–1.25	2	0.5	2
B	0.6–1.4	5	2	5
C	0.4–2.0	10	10	10

1.1.2 Spectral Quality

The spectral quality of simulators is officially evaluated from 400 to 1100 nm for historical reasons—the field was, and still is, dominated by silicon solar cells and hence all specifications are designed around the needs of that particular technology. There are many aspects of the characterization of solar cells that have the same silicon-based legacy. There are fewer photons in the UV spectrum below 400 nm, especially after transmission through the front sheet used in photovoltaic modules, and silicon is unable to convert a significant number of photons above 1100 nm. To evaluate the spectral quality from 400 to 1100 nm, six bins are generated in which the AM1.5G reference solar spectral irradiance (*e.g.* ASTM G173-03, Figure 1.3) is integrated.

Each bin provides a target percentage of the full irradiance. Class A spectra will have the integrated irradiance of each bin within ±25% (relative) of the target level, class B ±40% and class C −60 or +100%. It is important to note that these are integrated values, so if there is a very strong line in the light source, then it can be balanced by a strong dip (Figure 1.4).

Using this system definition, six light-emitting diodes (LEDs) in the correct bins can provide a class A spectrum, which many might argue looks nothing like the reference spectrum, although it is worth noting that many new LED simulators on the market use closer to 20 LED wavelengths and do provide an excellent match to the solar spectrum.[14] In general, if a sample is not especially sensitive to a specific wavelength, this is not likely to be an issue. However, this may not be true for some devices due to, for example, layer thickness effects, narrow bandwidth absorbers such as quantum dots, the optimization of anti-reflection coatings or the presence of trap states. A light source is characterized by its worst bin. For example, if a light source is within ±10% of 400–900 nm, but has +50% variation from

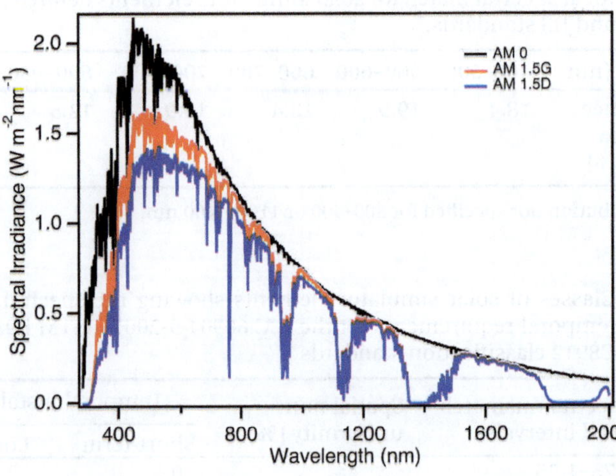

Figure 1.3 AM0, AM1.5G and AM1.5D reference spectra from the ASTM G 173 standard.[13]

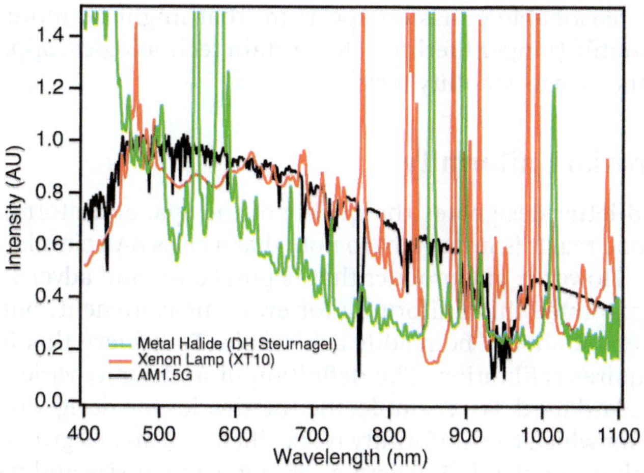

Figure 1.4 Metal halide and xenon lamp spectra, both of which can be found in AAA class solar simulators, relative to the AM1.5G spectrum.

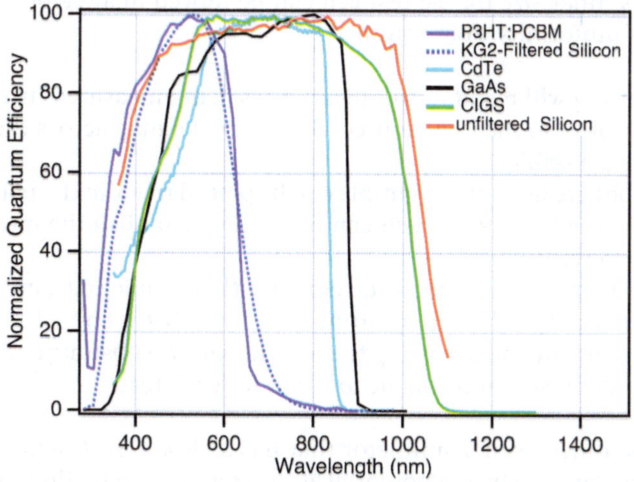

Figure 1.5 Representative quantum efficiency curves for several different photovoltaic technologies illustrating some of the wide variation in spectral responsivity across technologies, all of which require light only up to a wavelength of 1100 nm.

900 to 1100 nm, then this would be designated as a class C spectrum. It is also important to recognize that if a bin is outside the range of class C, it may be designated X or simply not designated. However, there are many technologies that may not require light beyond 900 nm—amorphous silicon, cadmium telluride, DSSCs, OPV cells and PSCs are examples (Figure 1.5)—although PbS quantum dot solar cells do require light up to 1200 nm. Hence it should be recognized that there may be light sources that could

generate a reasonable emission spectrum that might be more affordable or more durable (longer lifetimes) for certain technologies, applications or experiments such as stability tests.

1.1.3 Spatial Uniformity

The second letter designates the quality of the spatial uniformity over the illumination area. It is important to note that a class AAA simulator has ±2% uniformity. However, any product that is purchased and advertised as such does not guarantee this uniformity for every measurement, but the designation means it should be readily achievable. To achieve this level of uniformity requires calibration. The definition of uniformity varies depending on the standard used. For example, the IEC divides the designated illumination area over which the uniformity is specified into an 8×8 grid, whereas the ASTM requires one that is 6×6 with a varying detector size and packing fractions. In a research environment, cells are typically much smaller than the designated area of a particular light source and thus the spatial uniformity should be evaluated with this in mind. A monitor cell that is smaller than, or at least of similar size to, the test cell can be moved in a region to establish the level of uniformity. For example:

- if fixturing will result in samples always being measured in a fixed location relative to the light source, then a spatial map across the substrate area is advisable;
- if probes are used and a sample can be placed freely in the path of a light source, then a monitor cell can be used to establish the most spatially uniform region;
- if a substrate to be tested is designed with multiple cells, increased uniformity can be achieved by lifting the probes between each separate cell measurement and adjusting the location of the substrate such that each cell under test is at the same location for each test.

In this manner, even a simulator that has a class C (±10% uniformity) or poorer spectrum can have a potential 20% error effectively eliminated and, at a minimum, this method should remove relative errors.

Although the simplest way to establish the spatial uniformity of a light source is to take a small photodiode and scan the active (useful) area by hand, several factors should be considered. If this approach is adopted, it is helpful to ensure that the diode remains with a 0° tilt throughout the course of the measurement. If a photodiode is chosen with a package that is impervious to light, no light reaches the active area from the back or the sides. Examples of such currently available photodiodes include the low dark current silicon photodiodes from Hamamatsu: the 1.3 mm × 1.3 mm S1087 photodiode and the larger 2.8 mm × 2.8 mm S1787 photodiode are both available with cut-off filters, although they can also be added separately for use as reference cells. For most laboratory measurements, this

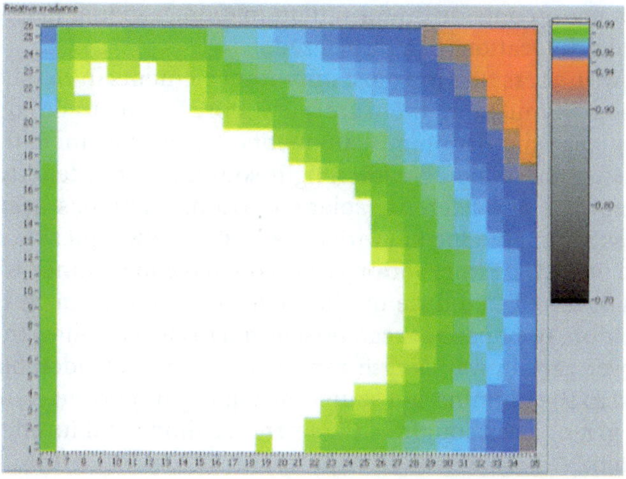

Figure 1.6 Measuring the lamp uniformity of the NREL (×25) system. The plot shows the lamp intensity across an area of 25 cm × 30 cm. A 1 cm × 1 cm grid was used for these measurements.

will easily establish the level of spatial uniformity and associated error. If it is desirable to establish a particularly high-quality spatial map of a light source with a larger area (Figure 1.6)—when for instance, building a custom light source and optics system—a few different methods might be considered. Although spatial uniformity mappers can be purchased, they are expensive and such products are normally only specified for the illumination area of interest. A spatial mapper will also typically have at least one off-area cell that can be used to reduce/eliminate the effects of temporal non-uniformity.

Do-it-yourself approaches can also be made using simple *x*–*y* stages or even by reconfiguring old digital plotters. Again, if this level of effort is desired, it is wise to have a fixed location diode to help remove the effects of temporal non-uniformity. If a single diode is translated, more reliable results may be realized if this is carried out in concert with random *x*–*y* locations than if a sequential scan is used because this partially obviates the effects of cell heating.

1.1.4 Temporal Uniformity

Long-term uniformity correlates with lamp drift, whereas short-term uniformity is a measure of lamp flicker. These values are all dependent on the type of light source, the power supply and the age of the lamp. One of the most effective ways of dealing with this is to incorporate a monitor diode for *in situ* intensity monitoring. After calibrating the monitor diode's output to the reference cell, which could be one and the same, it is easy to establish and then significantly reduce the effects of temporal drift. Even if the *in situ* monitor diode is in a region outside uniform illumination, a spatial map/correlation to individual cell locations can be utilized.

Now that we understand the most important elements in classifying a light source for solar simulation, there are a few more factors to be considered in high-quality solar simulation measurements. The next element to address is spectral mismatch. As noted earlier, light sources do not tend to share the same spectrum as any reference solar spectrum. There has been considerable progress in tunable spectra with the advent of new light sources, such as those based on LEDs. Either various colored LEDs can be tuned to match the spectral class, tuned to reduce or eliminate spectral mismatch in the standard calculation, or phosphors can be selected to closely follow the reference spectrum over certain wavelength ranges. Although considerable attention is often paid to the illumination source, an equally important element that is sometimes given little thought is the reference diode and its effect on spectral mismatch.

1.2 Reference Diodes and Mismatch Factors

The role of the reference diode is to set the intensity for a one sun measurement. To accomplish this, the reference diode must have been calibrated from a known source. This can be done in coordination with one of the accreditation laboratories (*e.g.* NREL, the Fraunhofer Institute for Solar Energy Systems or the National Institute of Advanced Industrial Science and Technology). A calibrated reference diode will have a measured one sun short circuit current and spectral response. Ideally, a commercially obtained cell that has been packaged for enhanced stability is used. Cells intended for calibration measurements will typically include a temperature sensor in the package to enable and promote measurements at 25 °C, which is also part of the standard test conditions. The reference diode should have the same spectral responsivity as the device under test. For example, the Hamamatsu S1087 or the larger S1787 photodiode can both be used with separate color filters, such as the Schott KG and BG series. However, this can be challenging for new material development, such as some of the devices described in this book, as they tend not to be commercially available in a packaged device form and stability can be an issue.

To remove the error of spectral mismatch between the solar simulator lamp, the AM1.5G reference spectrum, the reference diode and the test cell, a spectral mismatch factor can be calculated. The spectral mismatch factor, M, is a convolution of four inputs: the solar irradiance reference spectrum ($E_{\mathrm{AM1.5}}$); the lamp irradiance spectrum (E_{Lamp}); and the EQEs of the reference cell ($\mathrm{QE_{Ref}}$) and the device being tested ($\mathrm{QE_{TestCell}}$).[5]

$$M = \frac{\int_{\lambda_1}^{\lambda_2} E_{\mathrm{AM1.5}}(\lambda) S_{\mathrm{Ref}}(\lambda) \mathrm{d}\lambda}{\int_{\lambda_1}^{\lambda_2} E_{\mathrm{Lamp}}(\lambda) S_{\mathrm{Ref}}(\lambda) \mathrm{d}\lambda} \bullet \frac{\int_{\lambda_1}^{\lambda_2} E_{\mathrm{Lamp}}(\lambda) S_{\mathrm{TestCell}}(\lambda) \mathrm{d}\lambda}{\int_{\lambda_1}^{\lambda_2} E_{\mathrm{AM1.5}}(\lambda) S_{\mathrm{TestCell}}(\lambda) \mathrm{d}\lambda} \tag{1.1}$$

Where $$S_x(\lambda) = \frac{q\lambda}{hc}\mathrm{QE}_x(\lambda) = \text{ spectral responsivity} \quad (1.2)$$

Only relative values are required because each input appears in both the numerator and denominator. It is important to integrate over the larger of the two wavelength ranges for the spectral response of either the reference diode or the test cell. Each integral corresponds to the short circuit current associated with the lamp and the solar spectrum, each measured by both the reference and the test cell.

As an extreme example, Figure 1.7 uses an OPV cell with a filtered (KG2) silicon diode and a halogen lamp spectrum to explain the importance of the

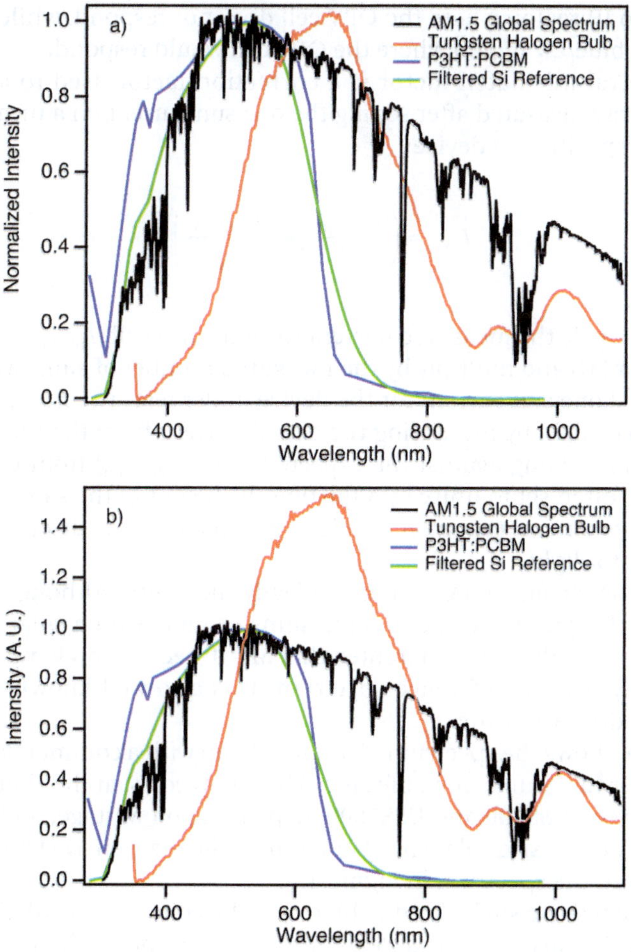

Figure 1.7 Example of four inputs for a spectral mismatch calculation: QE_{ref}, $QE_{TestCell}$, E_{Ref}, E_{Lamp}. (a) All four inputs presented in normalized form. (b) The halogen lamp spectrum is increased to try to generate the one sun current from the reference cell (the filtered Si reference).

mismatch factor and what is being done when setting the one sun intensity with a reference diode. Normalized curves are shown in panel (a). The filtered silicon reference cell has a similar response profile to the OPV cell, although there are some differences toward the red and blue sides of the response. There is, however, a significantly greater mismatch between the tungsten lamp and the two reference spectra. Panel (b) shows roughly what the lamp spectrum might appear to be when set by the filtered silicon reference cell to the one sun intensity relative to AM1.5G. Because the lamp spectrum is deficient in the UV region, the red overall intensity must be significantly increased to generate the one sun current from the reference cell. If a silicon reference cell was used without a filter, then the intensity of the lamp spectrum relative to the AM1.5G spectrum would be lower. This is less appropriate, however, for the OPV cell because much of the intensity is now in the red to IR region where the OPV cell does not respond, while still deficient in the blue/UV region where the OPV cell would respond.

The spectral mismatch factor is a correction factor used to adjust the current density measured after setting the one sun current on a measurement set-up for a specific test device:

$$I_{corr} = \left(\frac{I_{TestCell}}{M}\right) \bullet \left(\frac{I_{Ref,1sun}}{I_{Ref,Meas}}\right) \tag{1.3}$$

Here, we divide the measured current from the test cell ($I_{TestCell}$) by the mismatch factor (M) and multiply by the measured number of suns, which gives the corrected one sun current for the device under test. The number of suns ratio is determined by measuring the actual current from the reference cell ($I_{Ref,\ Meas}$) and ratioing against the expected value ($I_{Ref,1sun}$) from calibration. This ratio is often set to unity by adjusting the height of the sample stage to make these two numbers equal by taking advantage of the divergence of the solar simulator light.

The mismatch factor (M) should, ideally, be unity. Although it is not uncommon for groups to spend large sums of money on a solar simulator light source, this does not guarantee that $M = 1$ because no lamp spectrum looks exactly like the reference spectrum. It is also well known that lamp spectra can drift with time.

Figure 1.8 shows the spectrum of a xenon lamp with a commercial AM1.5G filter set over the course of its lifetime. This lamp costs, at the time of publication, between US\$500 and US\$3000, depending on wattage, and is part of an extremely expensive solar simulator, unlike the inexpensive halogen lamp used to record the spectrum in Figure 1.7.

Notice the relative shift of red to blue over the course of the lifetime of the bulb. This is because there is a clean quartz window on the bulb early in the bulb's life and xenon arc lamps deposit a thin layer of carbon on the window of the bulb as they age. Shorter (blue) wavelengths are transmitted less through this film than longer (red) wavelengths. Although this may appear

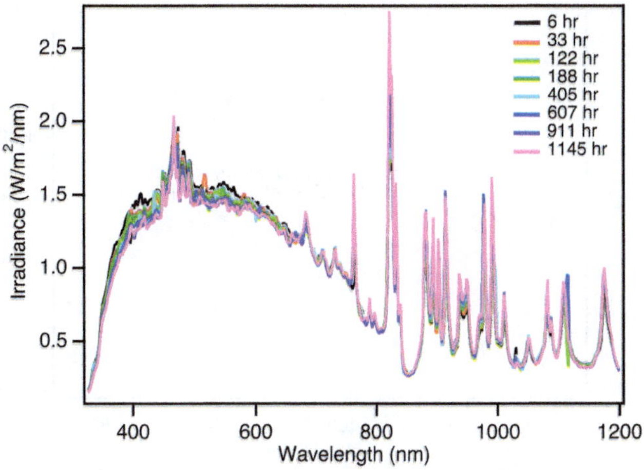

Figure 1.8 Lamp aging of a xenon-based solar simulator. These lamp spectra show how the intensity in the all-important 400–600 nm spectral window decreases and shifts as a function of time, which would require a recalculation of the mismatch factor M if a poor reference cell match was chosen.

to be a relatively minor error, it can have significant effects if there is an improper choice of reference diode.

Figure 1.9 shows the evolution of the spectral mismatch factor using either an unfiltered silicon diode or a KG5-filtered silicon diode and an OPV device.[6] The unfiltered silicon diode begins with $M \sim 1.18$; but in <1000 h it is reduced to ~1.05, with a particularly large decrease in the region between 100 and 200 h. Hence even if spatial and temporal errors are ignored or eliminated, merely having a class A spectrum, as in this example, can result in a drift of up to 13%, which can be very misleading if not taken into consideration. Examining the curve corresponding to the right-hand axis illustrates the effect of using an appropriate choice of filtering. Choosing an appropriately matched reference cell can therefore reduce these spectral errors. In this instance, the mismatch factor ranges between ~1.0275 and ~1.02 with the same lamp spectral shifts.

It is generally good practice to occasionally re-measure the lamp spectrum to track the evolving mismatch factor. It is worth noting, however, that many inexpensive spectrometers are not well calibrated in terms of intensity. Although the wavelength information can be fairly close when examining calibrated spectral lines, the relative intensity at each wavelength can lead to considerable errors if the system is not carefully calibrated for intensity as well as wavelength. A spectral mismatch factor of unity would be achieved if the lamp irradiance spectrum was identical to the solar irradiance spectrum (*e.g.* AM1.5G) or if the spectral response of the reference cell was identical to that of the test cell. Although it is desirable to match both as best as is feasible, it is arguable that it is easier and

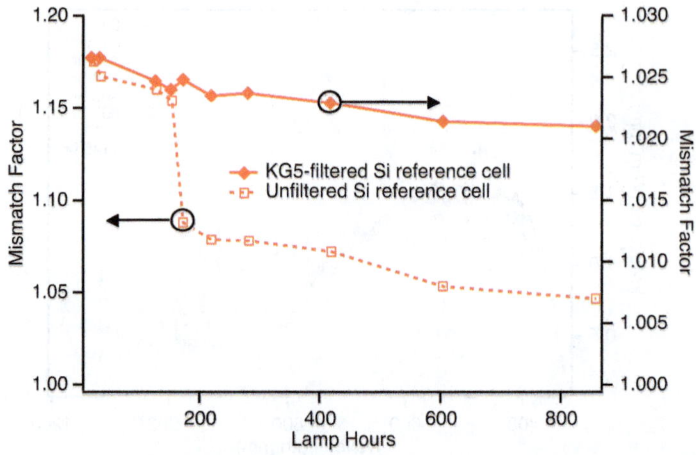

Figure 1.9 Changes in the mismatch factor over time for a xenon lamp. The two plots correspond to the change in the mismatch factor as a function of time using two different silicon reference cells: unfiltered and with a KG5 Schott filter, commonly used for many new solar cells responsive to visible (data reproduced from ref. 6).

less costly to focus on matching to the reference cell's spectral response. For laboratories studying different absorbers, it is prudent to have multiple filter sets to pair with the reference cell. Filter sets that match well with different photovoltaic technologies include some of the Schott KG and BG color glass series.

1.3 Area Measurements

One of the largest sources of error is poor measurements of the illuminated area of a cell under test. This is greatly exacerbated by the propensity of laboratories to make small devices. This happens for several reasons:

- research groups like statistics on relatively small area substrates;
- processes may be highly non-uniform, so small devices may enable research groups to separate dead spots from good regions; and
- higher efficiencies can be reported when using transparent conductors optimized to let more light through at the expense of higher resistance.

This latter point is illustrated in Figure 1.10, where the same spin-cast OPV system of P3HT:PCBM is processed on identical substrates with the same transparent contact and therefore the same sheet resistance; the only major difference is the active area of the two cells. This difference manifests itself primarily as a large loss in fill factor due to the higher resistance pathway of the larger device, with both the short circuit current densities and open circuit voltages remaining almost the same (Table 1.3).

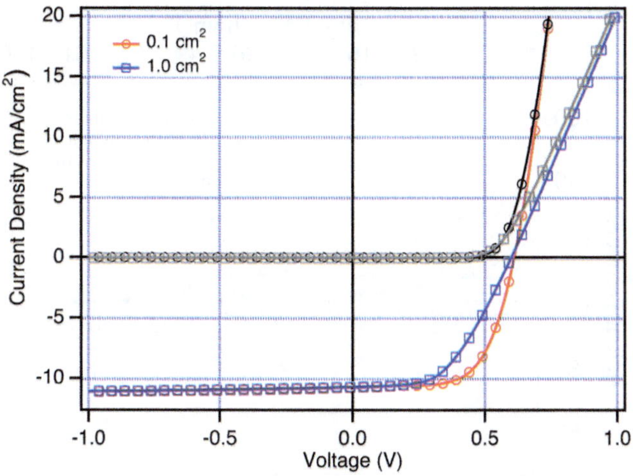

Figure 1.10 Effect of area on efficiency for two OPV devices with different areas. The resistance of the transparent contact associated with the larger device results in a significant decrease in the device fill factor.

Table 1.3 Device performance data for the two device areas (0.1 and 1.0 cm^2) shown in Figure 1.10.

Area (cm^2)	V_{oc} (mV)	J_{sc} (mA cm^{-2})	Fill factor	Efficiency (%)
0.1	609	10.7	0.64	4.14
1.0	599	10.1	0.50	3.03

Although the search for higher efficiencies tends to encourage groups toward devices with a smaller area, the potential for error increases, as illustrated in Figure 1.11. It is not uncommon for devices with areas between 1 and 25 mm^2 to be used. For square devices, an underestimate of the actual side lengths by only 100 μm results in a 21% error for a nominal 1 mm^2 device and >6% for a nominal 10 mm^2 device. Length/area measurements errors can occur for a variety of reasons.

There are several different standard approaches that can be used to define the active area of a laboratory-scale device. Perhaps the most reliable method is to use mesa isolation, where a completed device is etched such that the top contact and the active layer are patterned. In this example, the area of the top contact indisputably defines the area of the device. Doing this generally requires extra processing steps, such as photolithography, and can result in carrier recombination at the edges of the device. If a less reproducible, but cheaper, method than lithography is used, such as a razor blade, measuring a device area precisely becomes far more onerous and prone to error.

An alternative method to mesa isolation is to use a patterned bottom contact and patterned top contact deposited through a shadow mask. The overlap area of the two patterns can then be used to define the area in a cross-bar pattern (Figure 1.12). In addition to avoiding edge recombination, the device

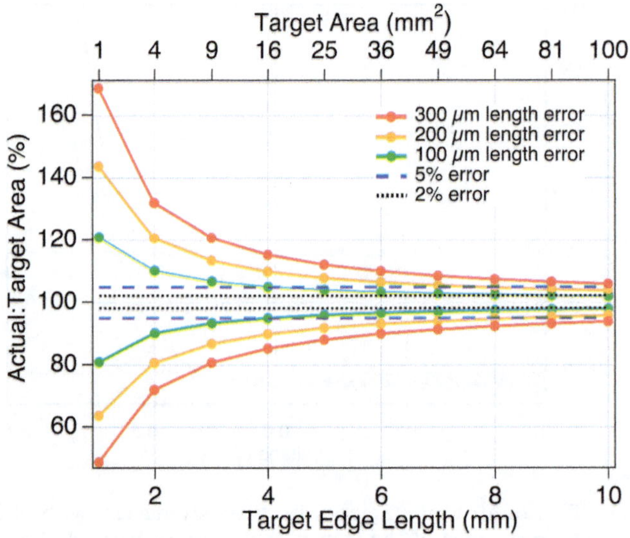

Figure 1.11 Influence of measuring the active (illuminated) area of a device and the uncertainty it can create in the short circuit current and therefore the overall device efficiency. The record-holding emerging photovoltaic devices in Figure 1.1 have reported active areas in the 4–5 mm^2 range, necessitating a dimension measurement better than 100 μm to achieve a 5% precision in device area.

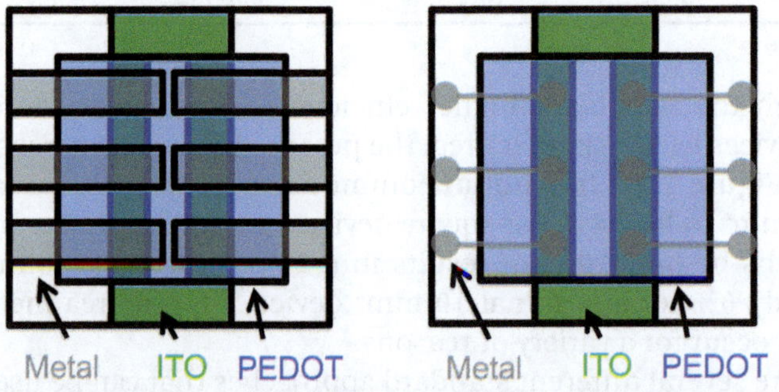

Figure 1.12 (Left-hand panel) "Cross-bar" geometry device. A patterned 2.5 cm × 2.5 cm substrate with six individual pixel cells. Although the overlap of the ITO strip and the metal contacts is about 0.11 mm^2, this is *not* a good measure of the device area and an illumination mask (0.059 cm^2 aperture) is required to define the active area precisely. (Right-hand panel) "Lollipop" geometry device. An alternative overlap lollipop-shaped pattern can reduce the error from a conductive transport layer. Errors from the shadow mask deposition/alignment can still generate issues depending on implementation. A measurement with and without an aperture can determine the magnitude of these errors in both examples.

can now be contacted without the concern of puncturing the top contact. Although the bottom contact may be pre-patterned in a highly reproducible manner through a process such as photolithography, defining dimensions with a shadow mask is less precise. Evaporating through a shadow mask results in much better edge definition than sputtering, but samples are generally rotated to improve uniformity and coverage, which leads to a layer with a variable thickness wider than the opening of a shadow mask. Even without rotation, the thickness of the mask and the locations in which it does not make intimate contact with the sample results in a non-uniform film that is wider than the mask's opening. Although it is easy to either use the specified shadow mask's dimensions or measure the opening with a pair of calipers, this would result in an underestimation of the device area.

Another, typically smaller, error in the other direction occurs from the build-up of deposited films over time. Metal electrode film thicknesses of the order of 100 nm are standard. After 100 such depositions, this can result in a 20 μm reduction in a lateral dimension of the deposition mask (10 μm on each side). It is unlikely that a greater number of depositions would occur, however, without cleaning the mask, or without natural flaking of the mask, returning it to its original dimensions.

The last method of defining the correct active area is a variant of the two previous methods. An unpatterned bottom contact is used in conjunction with an island top contact. Here the same error associated with the deposited area through the shadow mask is generated. In these previous cases, however, there was another potential source of error in the definition of area. In photovoltaic devices, transport, buffer or collection layers are commonly used between the electrode and the active material. These are typically not highly conductive relative to a transparent conductor or metal layer; however, on a length scale of 100s of micrometres they can be efficient. In certain cases, such as an OPV device, an intentionally conductive layer such as PEDOT:PSS is used. Although there is no reason to limit the conductivity of such a layer in a large-scale device, in laboratory-scale devices in which there is a strong desire to limit cross-talk and have well-defined areas suited for the accurate determination of efficiencies, using a low (lateral) conductivity for such layers is a good idea.

It is easy to underestimate the device area by inadvertently exploiting thicker or more conductive layers, which will result in an overstatement of the device's current density. A standard hallmark of this is voltage-dependent collection in the J–V curve (low shunt resistance) (Figure 1.13) and this mechanism has been explored in detail, with an excess efficiency of ~61% demonstrated in small area devices (4 mm^2).[15] These researchers strongly suggested the use of island rather than cross-bar cells.[15] However, it is important to note that although such a device architecture may work for certain thin film stacks, in others it would not fully address the problem. For instance, in cadmium telluride, it is not uncommon to modify the surface with an oxidizing etch that generates a thin layer of tellurium on the surface of an otherwise highly resistive semiconductor. Such a tellurium layer can easily lead to the

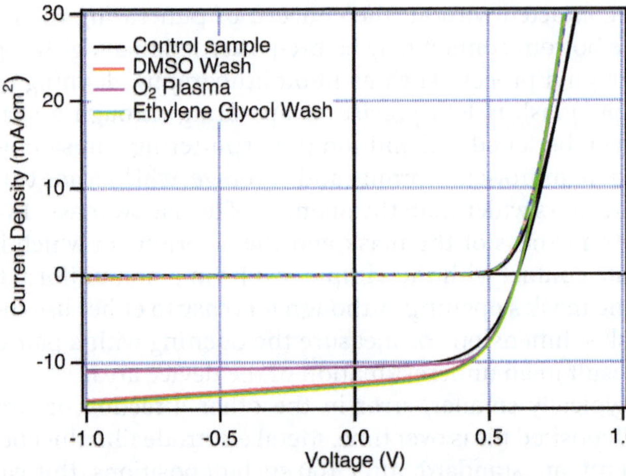

Figure 1.13 Voltage-dependent current collection, indicative of a low shunt resistance, showing a slope that is not flat where the washed samples (DMSO and ethylene glycol) cross the line of zero voltage. Therefore applying more bias will get more current out of the device. This is not true for the control and O_2 plasma curves. PEDOT:PSS (Baytron P VP AI 4083) spun at 6000 rpm.

collection of charge 100 µm, or even further, away from a metal contact. In other systems, such as PSCs, there may be an unpatterned transport layer directly underneath the metal contact and, just as in the case of PEDOT:PSS, this can result in charge collection around even an island geometry. The "lollipop" geometry (Figure 1.12) is a compromise between island and standard overlap patterns and is intended to minimize charge collection outside the area.

Other than mesa isolation, there is only one agreed way in which a cell area can be well defined: an aperture. An aperture can be made of many materials, with the main requirements being that it is opaque and can be used with cells of any architecture. It is placed in contact with the cell being tested (or as close to it as possible in the case of a superstrate or packaged device). In its simplest form, an aperture can be made by applying an opaque tape to a device. Care is required to measure the area because vinyl tapes (such as electrical tape) can stretch, making curved instead of straight lines. A more reliable method is to use a metal mask, which can either be as simple as a drilled/reamed hole in a thin plate or a photo-etched/laser-cut pattern produced in the same manner as a shadow mask. These may be painted black or anodized to remove the influence of reflections.

In any official accreditation measurement of a sample that is not mesa isolated, an aperture is required to establish that there is no outside area collection of current. The mask containing the aperture should ideally extend fully past the edges of the sample to eliminate wave-guiding effects that might occur in a superstrate or packaged device. In a laboratory setting, however, an aperture may be used to establish that the correct short circuit current density (J_{sc}) is measured, although if a sample has the same J_{sc} with and without an aperture, then no aperture is required. If a fixture for measuring a standard patterned sample is used, custom apertures may readily be made that will align with the sample with minimal effort. This can enable a standardized, extremely high-quality measurement to be performed daily. For most technologies, an aperture that is smaller than the active area of the device is usually chosen because this allows the aperture area to be used to calculate J_{sc}. Manually aligning an aperture with only 100 μm tolerance can be challenging, whereas a tolerance of 400 μm is much more reasonable. If an aperture is much smaller than the device area, the shaded cell area in parallel may start to significantly lower the voltage and keeping this area to <40% of the total device area is advised. It is also worth noting that any bus-bar or metallization covered by an aperture does not adversely count against a device's area.

Measurement of a device or aperture area may be accomplished in one of several ways. Calipers, used correctly, can provide sufficient precision and accuracy for measurement of the aperture area. Measuring a cell area without an aperture is generally achieved either with a calibrated microscope stage or using a camera in combination with a calibration standard. With a microscope and camera, the appropriate magnification is chosen, the image saved and image processing software is then used to establish the area. Alternatively, in the case of a calibrated stage, the user may focus on the edge of a device and record the stage location. Many points can be recorded to establish the perimeter of a polygon in the event of non-linear features. A simple examination of samples at high magnification and attempts to measure the area will demonstrate the irregularities and, in some cases, potential difficulties in identifying the edge of a device.

1.4 External Quantum Efficiency

To assess the efficiency of a device, the spectral response must be known as this is needed to decide which filter must be placed on the reference diode. An absolute quantum efficiency is ideal in that it can theoretically be integrated to establish the one sun current of a cell. This value should agree closely with the one sun current from a spectrally corrected *I–V* measurement. Typically, the current density extracted from a spectrally corrected *I–V* curve is assumed to be more accurate, but the numbers should generally agree within a few percentage points. It is important to note that only a relative EQE is required to define the efficiency of a device. The difference between an absolute and relative quantum efficiency measurement is that in an absolute measurement the excitation beam is fully contained in the device's active area. For a relative measurement, the beam's spot is larger than the device, such that the number of incident photons is not known, merely the full spot's relative intensity at each wavelength. This is typically carried out as a lock-in measurement, in which a monochromatic signal is chopped and used to generate an ac signal, which is then amplified for measurement.

Several considerations should be made when performing these measurements. The chopped frequency should be notably different from the line frequency (*i.e.* 60 Hz in the USA, 50 Hz in Europe) and prime numbers are generally good choices for the frequency (*e.g.* 37 Hz). The intensity from the monochromatic signal is typically ≪1% of one sun. Although a white light bias is often not performed, there are material systems for which a white light bias may alter the spectral response. An example of where this may be important is when there is a highly defective layer that may have traps filled by a white light bias. Generally, lock-in measurements of the EQE can take ~5–15 min depending on the system and wavelength range, a time that offers an explanation as to why they are not recorded for every device being tested. In addition, for samples that are particularly unstable under light illumination, this period of time can lead to degradation during the course of the measurement.

Although technically an EQE should be known to choose the appropriate reference diode, scanning the *I–V* curve first in these cases and using a representative EQE from a similar, but stable, device will be reasonably accurate. A recent solution to this problem is the "flash EQE" system developed by the NREL, which enables extremely rapid quantum efficiency curves to be measured (~1 s for each EQE curve). This type of system uses a series of different colored LEDs, each driven at a separate frequency. The signal is then deconvoluted with a Fourier transform to establish the efficiency at each wavelength. Although this is a more limited technique in that the wavelengths are fixed, it has been demonstrated with conventional systems (*e.g.* Si and copper indium gallium selenide solar cells), including tandem cells, to be fairly effective and reproducible.[16,17]

1.5 Making Measurements

Although the points discussed so far in this chapter are generally the most crucial elements that are most commonly poorly performed, it is worth discussing a variety of other aspects of the measurements. The standard test conditions state that a cell should be measured at 25 °C, but cells can easily heat up to 40 °C under full illumination at one sun. Depending on the temperature coefficient, this will lower the performance to varying degrees; the temperature can be better controlled by placing the device on a cooled stage. This is a good solution for substrate cells, but superstrate cells (where light illuminates the cell through the growth substrate) are more challenging to maintain at a fixed temperature. Thermal deviations can be significantly reduced for superstrate cells by fan cooling and by closing the light source's shutter between measurements. To establish the level of heating in a system, a thermocouple or resistive temperature detector may be mounted directly on the sample and a standard test protocol may then be performed and the temperature swings monitored. If no significant light soaking is required, it should be possible to keep even superstrate samples at or below 30 °C.

When there is more concern about temperature effects with simulators that have unused illumination areas away from the test cell, a proxy sample (*i.e.* a cell with a similar absorber) may be set up with a semi-permanent temperature sensor that can be monitored in real time during a measurement. In certified measurements, where it is often difficult to control the device temperature and scan rates are slow, V_{oc} might not actually appear on the final measured *I–V* curve. This is because V_{oc} can be measured independently of the rest of the curve and can exhibit a strong temperature coefficient. A routine can be written to monitor the rise and fall of the open circuit voltage (V_{oc}) while rapidly opening a mechanical shutter.

Probes are the most common method used to couple electrical signals to and from a device, although wires may also be attached directly. If a fixture is made to measure patterned samples, standard connecting methods include pogo pins and integrated circuit test clips. Such a fixture can significantly speed up measurements, especially when coupled to a multiplexor. This may ultimately enable more samples to be measured under many more measurement conditions, such as light soaking or voltage bias holds. The data described in this chapter used such an approach. A fixture can also be fitted with a temperature sensor, an *in situ* light monitor and an aperture. Higher versatility can be achieved with micro-manipulated probes, which are often used in a Kelvin probe configuration to remove contact resistance effects. If a four-wire or remote sensing measurement is performed, then it is important to recognize that the positive and negative probe pairs should be placed close to one another. Typically, they should be within <1 mm, but, more importantly, the resistance between a pair should be kept to <2 Ω. As samples are typically measured with a voltage bias, a non-negligible resistance between one pair of probes means that the measured voltage–current pair may not reflect what is actually seen by the device.

When physically contacting a sample, some device geometries such as the cross-bar (Figure 1.12) may enable probes to be applied with less concern about puncturing a thin film top contact. Gentle pogo pins with rounded tips and low spring forces or gently applied Kelvin probes can typically make contact with island or mesa samples. For more fragile films, placing a soft metal such as indium or gold between the sample and the contact can facilitate the connection. Certain transparent conductors such as SnO_2:F can be difficult to contact; soldering (ultrasonically) indium to the SnO_2:F can greatly ease this process. Placing a bus-bar (*e.g.* an indium solder line) around the border of a transparent electrode can also significantly reduce the series resistance.

To generate *I–V* curves, a voltage bias is applied and the current measured. Although there are many pieces of equipment that can do this, the most common choice in photovoltaic measurement laboratories is to use a source-measure unit (*e.g.* Keithley 2450). The ASTM E 948-05 standard specifies that the current measurement electronics should have a resolution of at least 0.02% of the maximum current (I_{sc}). Hence a sensitivity of 200 nA is required for a J_{sc} of 10 mA cm^{-2} and a 0.1 cm^2 cell. This is readily achieved with a coaxial cable and a source meter; indeed, some source meters can measure down to 0.1 pA (a triaxial cable is required). Although such low level measurements are unlikely to affect the sensitivity of a light curve, a nanoamp or lower sensitivity can be useful in analyses of the dark saturation current density (J_0) from the dark *I–V* curve. Again, the standards specify that the voltage control is within 1% V_{oc}. This is not to say that the step size must be that small, just the control. In practice, this only requires a resolution of a few millivolts, which is not challenging (a 10 bit control is sufficient for 1 mV from 1 V full-scale).

Scans are generally performed as quickly as possible without introducing hysteresis when swept in both directions. Although some technologies are known to exhibit hysteretic behavior (*e.g.* DSSCs and PSCs), it is worth noting that most technologies will display some level of difference between forward ($V < 0$ to $> V_{oc}$) and back ($>V_{oc}$ to <0) scans if a curve is swept quickly enough. With most low-hysteretic technologies, a scan rate of the order of 0.3–1 V s^{-1} will result in minimal hysteresis.

Another approach has been adopted for highly hysteretic systems, such as some PSCs (Figure 1.14a). A quick curve may be swept in both directions and then a maximum power point region identified. After this, a slower sweep around the maximum power point region may be conducted, a voltage is identified and the current settles with time. After the device is not changing by more than a few per cent, a new voltage is applied and the process is repeated. It has become common to report not only the forward and backward scan rates for PSCs, but also a time *versus* maximum efficiency to demonstrate that this is a stable operating point (Figure 1.14b). After establishing a non-hysteretic *I–V* curve, the maximum power point, and hence the efficiency, is calculated by taking current–voltage pairs, multiplying them together and then selecting the highest pair.

There are often stability concerns with new materials that can frustrate good quality measurements of efficiency, but there are several ways of addressing this without having to delve too deeply into an understanding of the degradation mechanisms. If cells are unstable due to the presence of moisture or oxygen, a simulator may be integrated into the inert atmosphere of a glovebox, either through a window or by fully enclosing the system (Figure 1.16). Heat management can become an important issue for a fully enclosed system. Alternatively, various schema can be considered to take a sample out of a glovebox without exposing it to air. One of the more popular methods is to use a UV-curable epoxy and to seal the sample with a glass cover. This has the advantage that it allows many samples to be prepared and removed at the same time. However, it has the disadvantage that it can be messy and there is the potential for chemical interactions of the sealant with the actual test sample. Another approach is to use a re-usable "space suit", in which a KF window blank allows light in, a KF stub is machined to allow electrical connections and the seal is made with a standard KF O-ring and clamp.[19]

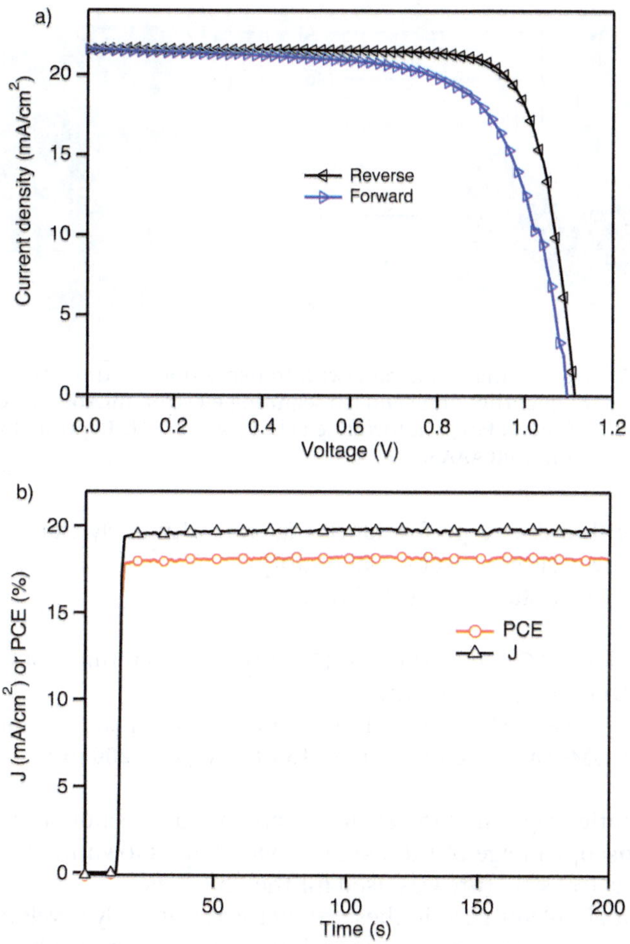

Figure 1.14 Hysteresis in a perovskite solar cell. Device characteristics of a $CH_3NH_3PbI_{3-x}Br_x$-based PSC with a normal cell architecture using TiO_2 and spiro-OMeTAD as the electron and hole transporting materials, respectively. (a) The reverse scan *J*–*V* curve shows a PCE of 19.12% with J_{sc} = 21.6 mA cm^{-2}, V_{oc} = 1.12 V and a fill factor of 0.793. (b) The stabilized PCE of 18.3% was recorded for several minutes under continuous light soaking. Reprinted by permission from Macmillan Publishers Ltd: *Nature Communications* (ref. 18), copyright 2016.

1.6 Example of Two Emerging Photovoltaic Systems

This section describes the characterization of an emerging photovoltaic technology based on colloidal quantum dots:[20] one based on PbS and the other based on $CsPbI_3$. Detailed reports on the optimized performance of these two device systems have been reported elsewhere.[21] We explore the data produced during the development of these devices, which eventually led to a

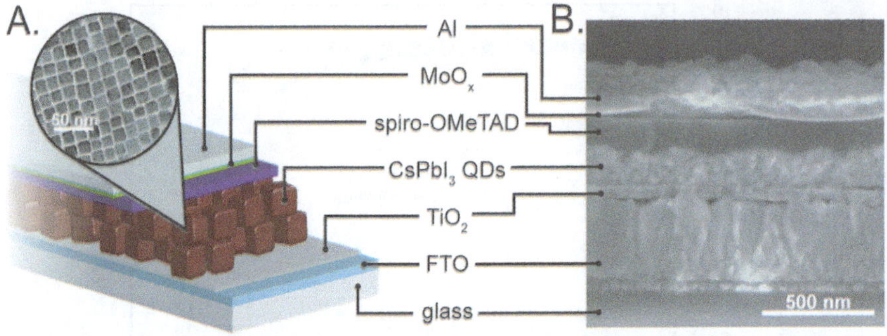

Figure 1.15 (A) Schematic diagram (with transmission electron microscopy image of quantum dots) and (B) scanning electron microscopy cross-section of the $CsPbI_3$ photovoltaic cell. From ref. 21. Reprinted with permission from AAAS.

hero cell with a device performance that was ultimately transferred to the NREL photovoltaic certification laboratory.

The two device stacks are as follows:

1. PbS:glass/FTO (300 nm)/TiO_2 (70 nm)/PbS quantum dots (~400 nm)/ MoO_x (15 nm)/Al (200 nm)
2. $CsPbI_3$:glass/FTO (300 nm)/TiO_2 (70 nm)/$CsPbI_3$ quantum dots (250 nm)/ spiro-OMeTAD (150 nm)/MoO_x (15 nm)/Al (150–200 nm)

A schematic diagram of the structure and a cross-sectional scanning electron microscopy image of the $CsPbI_3$ device[21] are shown in Figure 1.15; an almost identical structure was used for the PbS device.

The solar simulator used in these studies was originally developed for studies of organic photovoltaic devices and has been systematically improved for performance and usability. The new set-up provides multiplexed acquisition of the J–V curves of the six pixel devices on each substrate using the substrate design shown in Figure 1.12. A schematic diagram of the set-up is shown in Figure 1.16. The sample is positioned inside a nitrogen-filled glovebox with the solar simulator housed externally; the solar cells are illuminated from below through an optical window. Similarly, for the EQE measurements, the sample is located inside the glovebox with the light source and electronics located below.

It is first necessary to examine the type of filter to be placed on the reference silicon cell to more closely match its spectral response to that of the test cell; this requires a knowledge of the EQE spectrum of the new device material. Although this spectrum could be estimated from the sample absorption spectrum, it is preferable to determine the EQE spectrum directly. Figure 1.17 shows that the PbS device has a response with a peak EQE of ~80% at 500 nm, which extends into the IR with an optical bandgap at 1100 nm. The $CsPbI_3$ device has a peak EQE at ~450 nm and extends to the optical band edge at 700 nm. No filter was required to match the response of the silicon

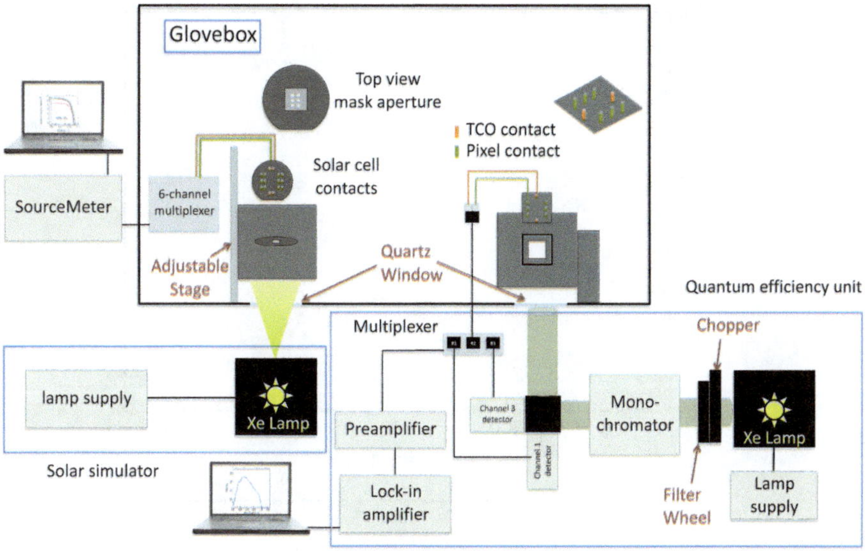

Figure 1.16 Schematic diagram and description of the components of the two systems. Solar simulator, Newport 94043A 450 W 69920 xenon lamp; Phidget multiplexor; Keithley 2450 source meter. Quantum efficiency system: Newport Oriel QEPVSI-b 300 W xenon lamp; 74004 optical chopper; 75151 filter set; 1/8m Oriel 74004cornerstone monochromator; lithium tantalate 70362 pyroelectric detectors (channels 1 and 3); Stanford SR570 low-noise preamplifier; Merlin 70103 lock-in amplifier; Merlin 90029648 multiplexer.

reference cell to that of the test cells when measuring the *J–V* curves for the PbS devices, but the shorter wavelength bandgap of the $CsPbI_3$ devices required a KG2 filter to be placed in front of the silicon reference cell. It should be noted that the unfiltered silicon reference cell is only just suitable for the PbS solar cell because the response extends significantly into the IR. Having completed this task once, it can be reasonably assumed that these filter combinations can be used for future measurements on the same quantum dot systems.

The mismatch factor (M) requires EQE data for not only the solar cell being tested, but also the EQE for the reference cell with an appropriate filter to set the one sun intensity. Figure 1.18 shows four individual plots of lamp irradiance, $E_{\text{Lamp}}(\lambda)$, and the AM1.5G irradiance, $E_{\text{AM1.5}}(\lambda)$, compared with the corresponding product of spectral responsivity for the PbS test cell, $S_{\text{TestCell}}(\lambda)$, and the reference silicon cell, $S_{\text{Ref}}(\lambda)$. Each plot also contains the integrated area under the $E(\lambda)S(\lambda)$ product, the peak value of which corresponds to the integrals defining the mismatch factor. Similar data for the $CsPbI_3$ solar cells are shown in Figure 1.19. The key differences are the extended response of the PbS sample in the 700–1150 nm spectral window and the correspondingly larger integrated short circuit currents determined for the silicon reference

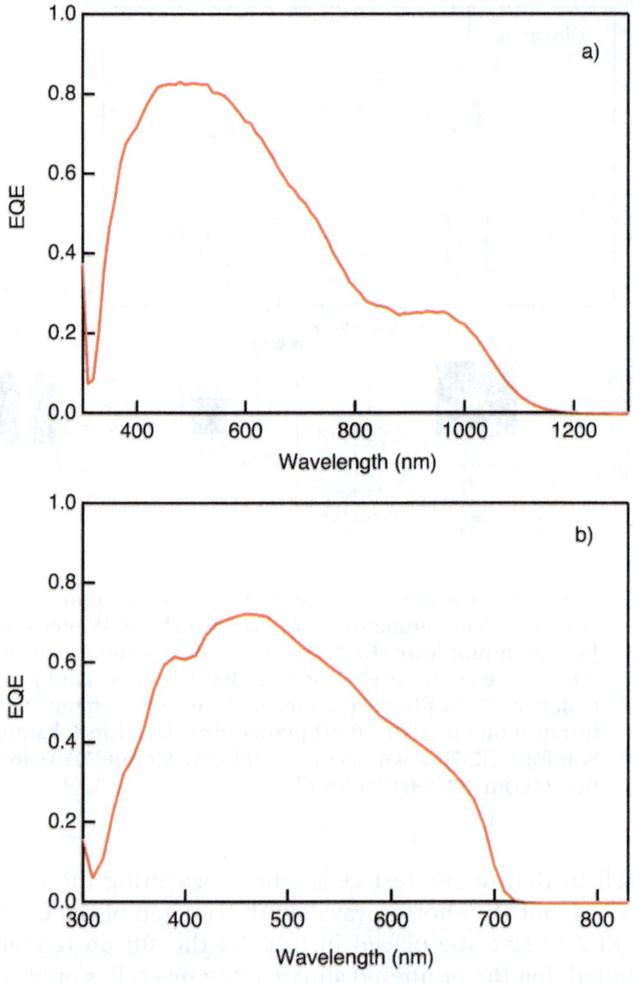

Figure 1.17 EQE spectra of (a) PbS and (b) $CsPbI_3$ quantum dot solar cells.

cell in the absence of the KG2 filter. Relative to determining the *J–V* curves for new test cells, the EQE spectra take longer to record and the relative spectra for individual cells often change very little, resulting in far fewer measurements of this type. As discussed earlier (Figure 1.8), the effects of lamp aging can induce significant changes in the mismatch factor if the reference cell's spectral responsivity, $S_{Ref}(\lambda)$, is not chosen carefully.

Data points in the EQE spectra are often determined at large wavelength intervals. For example, the data shown in Figure 1.17 has values that are taken every 10 nm, which is acceptable providing that the spectra do not exhibit sharp features. It should be noted that the irradiance spectra of both the AM1.5 spectrum and the xenon lamp both exhibit many sharp spectroscopic features. To avoid errors when determining the irradiance product

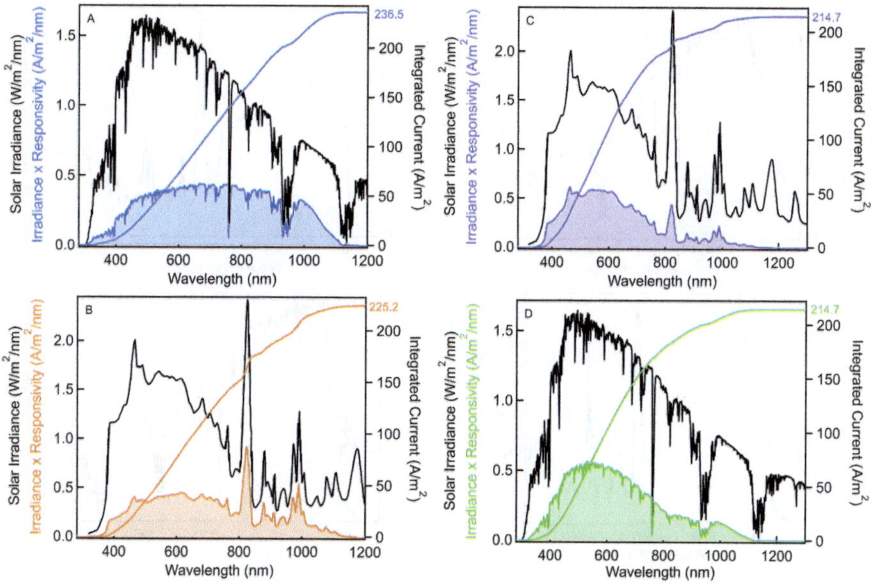

Figure 1.18 Plot of the four key irradiance $E(\lambda)$ values and the product with spectral responsivity, $E_x(\lambda)S_x(\lambda)$, for the PbS quantum dot solar cells. The integrated currents indicated in each plot correspond to the integral of the shaded areas: (A) $E_{AM1.5}(\lambda)S_{Ref}(\lambda)$; (B) $E_{Lamp}(\lambda)S_{Ref}(\lambda)$; (C) $E_{Lamp}(\lambda)S_{Test}(\lambda)$; and (D) $E_{AM1.5}(\lambda)S_{Test}(\lambda)$. These values are used to determine the spectral mismatch factor (M) and to correct for the short circuit currents using eqn (1.1).

with spectral responsivity, $E(\lambda)S(\lambda)$, the number of data points in the EQE spectrum should be increased to match the resolution of the irradiance data by interpolating between known values in the spectral responsivity data. The sharp spectral features in the $E_x(\lambda)S_x(\lambda)$ product shown in Figures 1.18 and 1.19 are evidence of the use of interpolation to retain the sharp features of the irradiance spectra.

The mismatch factor, M, is calculated using eqn (1.1), which can be written as:

$$M = \frac{A}{B} \bullet \frac{C}{D} \tag{1.4}$$

where A and B represent the integrated short circuit currents for the silicon reference cell with AM1.5 and xenon lamp irradiation, respectively. Similarly, C and D correspond to the xenon lamp and AM1.5 irradiation for the new test cell. Using these data, the mismatch factors are:

$$M_{PbS} = \frac{A}{B} \bullet \frac{C}{D} = \frac{236.5}{225.2} \bullet \frac{214.7}{214.7} = 1.05 \tag{1.5}$$

and

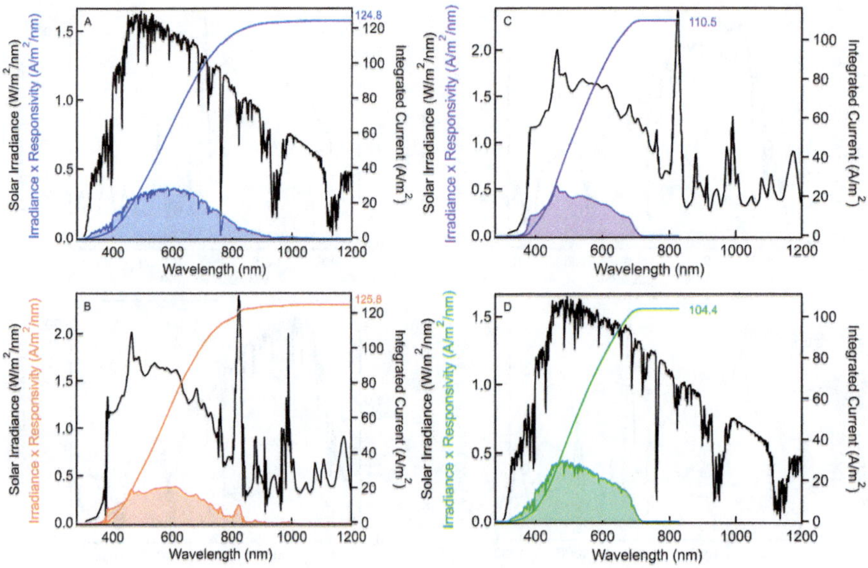

Figure 1.19 A plot of the four key irradiance $E(\lambda)$ values and the product with spectral responsivity, $E_x(\lambda)S_x(\lambda)$, for the $CsPbI_3$ quantum dot solar cells. The integrated currents indicated in each plot correspond to the integral of the shaded areas: (A) $E_{AM1.5}(\lambda)S_{Ref}(\lambda)$; (B) $E_{Lamp}(\lambda)S_{Ref}(\lambda)$; (C) $E_{Lamp}(\lambda)S_{Test}(\lambda)$; and (D) $E_{AM1.5}(\lambda)S_{Test}(\lambda)$. These values are used to determine the spectral mismatch factor (M) and to correct for the short circuit currents using eqn (1.1).

$$M_{CsPbI_3} = \frac{A}{B} \bullet \frac{C}{D} = \frac{124.8}{125.8} \bullet \frac{110.5}{104.4} = 1.05 \tag{1.6}$$

Note that both mismatch factors are >1, which will result in a corrected short circuit current (I_{corr} in eqn (1.3)) that is less than that measured directly using the solar simulator. The 5% correction is only just acceptable and, if the test cell should have exhibited a promising performance, the certification laboratory would probably be required to find a different reference filter or even a different reference cell (*e.g.* GaAs) to reduce this correction to a mismatch factor closer to 1.

Measuring the *J*–*V* curves is the final stage of the process. A silicon reference cell (Hamamatsu S1787-08) of known active area is placed on a simulator stage. Eqn (1.3) (I_{corr}) suggests that the measured current density is used to ratio against what is expected from a one sun AM1.5 spectrum. However, an approach is to adjust the height of the sample mount stage such that the current obtained from the reference cell matches that expected from one sun. Prior to loading the six pixel substrates into the solar simulator mount, a thin, pre-machined stainless-steel mask with six precision apertures (area 0.059 cm^2) defining the active illumination area of each pixel is first placed in the mount. The substrate is then placed on top of the mask with the transparent electrode, FTO in the two examples here, facing down toward the illumination source (a superstrate sample). Metal pogo pins located in a single anodized aluminum block contact with the six metal electrodes with pressure pins, with two additional pins on either end of the transparent conductor 'race-track' electrode (Figure 1.12), which are shorted in this detection configuration. The *J–V* curves for each of the six pixels are measured separately using a source meter. The NREL system has all six devices connected to the source meter *via* a multiplexer, which cycles through 1–6 in sequence. The data for the PbS and $CsPbI_3$ QD devices are shown in Figure 1.20.

Tables 1.4 and 1.5 give the data extracted from each of these curves, along with the average of each of the four characteristic parameters: V_{oc}, J_{sc}, fill factor and efficiency. The open circuit voltage (V_{oc}) and short circuit current (J_{sc}) are obtained by crossing the two axes again using interpolation to identify V_{oc} and applying a zero voltage to identify J_{sc}. The fill factor (FF) is identified from the voltage (V_{max}) and current (J_{max}) at the maximum power point (P_{max}), where $d(JV)/dV = 0$ and $FF = P_{max}/(V_{oc}J_{sc})$. The efficiency is calculated from $V_{oc}J_{sc}FF/P_{in}$, where $P_{in} = 100$ mW cm^{-2}.

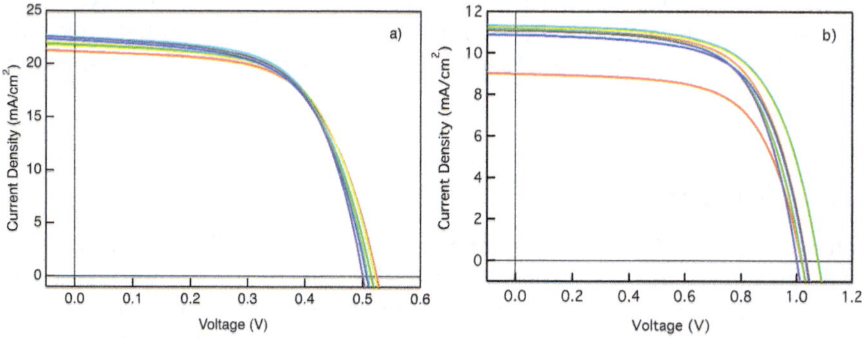

Figure 1.20 *J–V* curves for each of the six pixel devices of (a) PbS quantum dots and (b) $CsPbI_3$ quantum dots.

Table 1.4 *J–V* performance data for the individual pixels of the PbS quantum dot devices.

PbS device	V_{oc} (V)	J_{sc} (mA cm^{-2})[a]	Fill factor	Efficiency (%)[a]	Plot from Figure 1.20a
1A	0.53	21.3	0.617	6.9	Red
1B	0.52	22.1	0.611	7.1	Yellow
1C	0.52	21.89	0.611	6.9	Green
1D	0.51	22.32	0.608	6.9	Blue
1E	0.50	22.55	0.618	7.0	Purple
1F	0.51	22.60	0.614	7.1	Cyan
Average	0.51 ± 0.01	22.1 ± 0.5	0.613 ± 0.004	6.9 ± 0.1	

[a]Data have not been corrected for the spectral mismatch factor.

Table 1.5 JV Performance data for the individual pixels of the $CsPbI_3$ quantum dot devices.

$CsPbI_3$ device	V_{oc} (V)	J_{sc} (mA cm^{-2})[a]	Fill factor	Efficiency (%)[a]	Plot from Figure 1.20b
2A	1.02	9.0	0.646	5.9	Red
2B	1.04	11.2	0.646	7.5	Brown
2C	1.08	11.3	0.638	7.8	Green
2D	1.02	11.9	0.631	7.19	Dark green
2E	1.03	10.9	0.641	7.2	Blue
2F	1.00	11.1	0.646	7.2	Purple
Average	1.03 ± 0.03	10.8 ± 0.9	0.641 ± 0.006	7.1 ± 0.6%	

[a]Data have not been corrected for the spectral mismatch factor.

Therefore the solar PCE of the PbS device is (correcting for the mismatch factor) 6.6 ± 0.6% and the solar PCE of the $CsPbI_3$ is 6.8 ± 0.6%. The short circuit currents measured from integrating the EQE data are 21.5 mA cm^{-2} for the PbS device and 11.1 mA cm^{-2} for the $CsPbI_3$ device. Both results are in reasonable agreement with the currents measured from the *J–V* curves. These two solar cell devices—the best optimized devices—have yielded uncertified efficiencies of 8.3 ± 0.6 and 13.4 ± 0.6%[21] for PbS and $CsPbI_3$, respectively, although these results continue to improve.

1.7 Summary

Measuring the PCE or performance of a new solar cell is a relatively simple procedure, but the process can easily lead to erroneous results that can mislead research directions and foil opportunities to publicize champion solar cell performances. These erroneous results can be minimized by using the following prescribed measurement methodology. The protocol is straightforward:

- measure the EQE spectrum;
- measure a calibrated lamp spectrum from the solar simulator;
- determine the spectral mismatch factor;
- identify the correct reference cell and filter;
- use an aperture test cell to define the illuminated active area;
- measure the *J–V* curves to avoid the effects of any hysteresis;
- take note of sample degradation issues; and
- avoid the temptation to artificially enhance efficiency data.

Obtaining performance data that is 95% reliable can help accurate device optimization and may result in a device being certified at one of accreditation laboratories, with champion efficiencies being published and ultimately referenced on the NREL research cell efficiency chart.

Acknowledgements

We thank Yang Yang (UCLA), Kai Zhu (NREL) and Joey Luther (NREL) for permission to reproduce their data in Figures 1.9, 1.14 and 1.15, respectively. We also thank Tom Moriarty (NREL) and Moritz Riede (Oxford) for the original lamp spectra in Figure 1.4. ARM acknowledges support from the Center for Advanced Solar Photophysics, an Energy Frontier Research Center funded by the US Department of Energy, Office of Science, Office of Basic Energy Sciences. MR acknowledges support from the US Department of Energy, Office of Energy Efficiency and Renewable Energy, Solar Energy Technology Program. GR acknowledges support from the Solar Photochemistry Program of the US Department of Energy, Office of Science, Basic Energy Sciences, Division of Chemical Sciences, Geosciences and Biosciences. All support was under Contract No. DE-AC36-08-GO28308 to NREL.

References

1. NREL Best Research-cell Efficiencies, https://www.nrel.gov/pv/assets/images/efficiency-chart.png.
2. M. A. Green, K. Emery, Y. Hishikawa, W. Warta, E. D. Dunlop, D. H. Levi and A. W. Y. Ho-Baillie, Solar Cell Efficiency Tables (Version 49), *Prog. Photovolt: Res. Appl.*, 2017, **25**(1), 3–13.
3. M. A. Green and K. Emery, Solar Cell Efficiency Tables, *Prog. Photovolt: Res. Appl.*, 1993, **1**(1), 25–29.
4. *Interactive: Record-breaking PV Cells*, http://spectrum.ieee.org/static/interactive-record-breaking-pv-cells.
5. H. J. Snaith, How Should You Measure Your Excitonic Solar Cells? *Energy Environ. Sci.*, 2012, **5**(4), 6513–6520.
6. V. Shrotriya, G. Li, Y. Yao, T. Moriarty, K. Emery and Y. Yang, Accurate Measurement and Characterization of Organic Solar Cells, *Adv. Funct. Mater.*, 2006, **16**(15), 2016–2023.
7. G. Dennler, The Value of Values, *Mater. Today*, 2007, **10**(11), 56.
8. E. J. Luber and J. M. Buriak, Reporting Performance in Organic Photovoltaic Devices, *ACS Nano*, 2013, **7**(6), 4708–4714.
9. H. J. Snaith, The Perils of Solar Cell Efficiency Measurements, *Nat. Photonics*, 2012, **6**(6), 337–340.
10. A Checklist for Photovoltaic Research, *Nat. Mater.*, 2015, **14**(11), 1073.
11. K. A. Emery, Comment on "Roles of Donor and Acceptor Nanodomains in 6% Efficient Thermally Annealed Polymer Photovoltaics" [Appl. Phys. Lett. 90, 163511 (2007)], *Appl. Phys. Lett.*, 2007, **91**(26), 266101.
12. S. Rühle, Tabulated Values of the Shockley-Queisser Limit for Single Junction Solar Cells, *Sol. Energy*, 2016, **130**(C), 139–147.
13. Reference Solar Spectral Irradiance: ASTM G-173, http://rredc.nrel.gov/solar/spectra/am1.5/astmg173/astmg173.html.
14. *LED-based Sun-simulator Design: Technical and Commercial Considerations*, https://www.photonics.com/article.aspx?AID=57253.

15. M.-S. Kim, M.-G. Kang, L. J. Guo and J. Kim, Choice of Electrode Geometry for Accurate Measurement of Organic Photovoltaic Cell Performance, *Appl. Phys. Lett.*, 2008, **92**(13), 133301.
16. D. L. Young, B. Egaas, S. Pinegar and P. Stradins, *A New Real-time Quantum Efficiency Measurement System*, IEEE, 2008, pp. 1–3.
17. G. S. Horner, L. A. Vasilyev, J. Schmidt, J. E. Hudson, M. Arbore, M. Dybiec and E. A. Good, *Flash Quantum Efficiency Measurements for Multi-Junction Solar Cells*, IEEE, 2011, pp. 001716–001720.
18. M. Yang, T. Zhang, P. Schulz, Z. Li, G. Li, D. H. Kim, N. Guo, J. J. Berry, K. Zhu and Y. Zhao, Facile Fabrication of Large-Grain $CH_3NH_3PbI_{3-X}Br_X$ Films for High-Efficiency Solar Cells *via* CH_3NH_3Br-Selective Ostwald Ripening, *Nat. Commun.*, 2016, 7, 12305.
19. M. O. Reese, A. K. Sigdel, J. J. Berry, D. S. Ginley and S. E. Shaheen, A Simple Miniature Controlled-Atmosphere Chamber for Optoelectronic Characterizations, *Sol. Energy Mater. Sol. Cells*, 2010, **94**(7), 1254–1258.
20. B. D. Chernomordik, A. R. Marshall, G. F. Pach, J. M. Luther and M. C. Beard, Quantum Dot Solar Cell Fabrication Protocols, *Chem. Mater.*, 2016, **29**(1), 189–198.
21. A. Swarnkar, A. R. Marshall, E. M. Sanehira, B. D. Chernomordik, D. T. Moore, J. A. Christians, T. Chakrabarti and J. M. Luther, Quantum Dot–Induced Phase Stabilization of A-$CsPbI_3$ Perovskite for High-Efficiency Photovoltaics, *Science*, 2016, **354**(6308), 92–95.

CHAPTER 2

Bulk Heterojunction Organic Solar Cells: Working Principles and Power Conversion Efficiencies

M. C. SCHARBER* AND N. S. SARCIFTCI*

Linz Institute of Organic Solar Cells, Physical Chemistry, Johannes Kepler University Linz, Altenbergerstrasse 69, 4040 Linz, Austria
*E-mail: Serdar.Sariciftci@jku.at; markus_clark.scharber@jku.at

2.1 Introduction

Organic photovoltaic (OPV) cells are promising and relatively new devices for the conversion of light into electricity. The technology currently demonstrates moderate power conversion efficiencies (PCEs) and operational stabilities that do not qualify OPV cells for rapid, large-scale commercialisation. Nonetheless, academic curiosity and the potential of the technology have motivated intense research in this field. Figure 2.1 plots the number of publications per year found on the Web of Science database (Thomson Reuters) when entering the search string Topic = (organic OR polymer) AND Topic = (photovoltaic OR solar$cell$) NOT Topic = (dye-sensitised). The number of published papers on this topic increased almost exponentially from four to >2000 between 1985 and 2015.

Nanoscience & Nanotechnology Series No. 45
Nanostructured Materials for Type III Photovoltaics
Edited by Peter Skabara and Mohammad Azad Malik

Published by the Royal Society of Chemistry, www.rsc.org

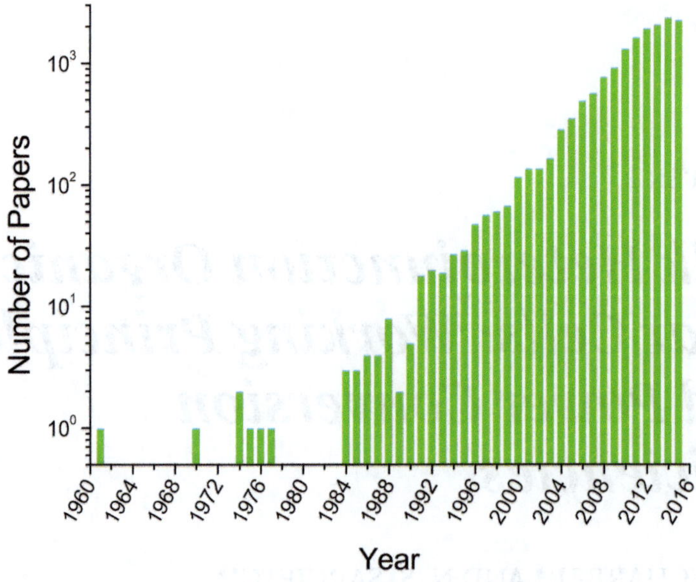

Figure 2.1 Number of publications on OSCs found on the Web of Science database (Thomson Reuters) using the search string Topic = (organic OR polymer) AND Topic = (photovoltaic OR solar$cell) NOT Topic = (dye-sensitized) (retrieved December 2015).

In the same time period, the record PCE of single-junction organic solar cells (OSCs) increased from 1% to ~11% and the PCE of tandem OSCs reached 13.2%.[1,2] The operational lifetime of OSCs has also improved during the last few years. There are many industrial and academic studies reporting lifetimes in the range 5–7 years and lifetimes >20 years have been extrapolated for hermetically sealed OSCs.[3,4]

The properties of OSCs often considered to be advantageous are:

- short energy payback times and a low environmental impact of the manufacturing process;
- significantly lower manufacturing costs compared with conventional inorganic technologies;
- lightweight and flexible photovoltaic modules;
- semi-transparent;
- easy integration into other products;
- new market opportunities, *e.g.* wearable solar cells; and
- manufacturing in a continuous process using state-of-the-art printing and coating tools or vacuum deposition.

Several different life cycle analysis studies have shown that OSCs can have very short energy payback times. Although a state-of-the-art silicon module mounted in the south of Europe needs about 2.5 years to generate the energy required to manufacture the device, it only takes about 2 months for a flexible OSC to reach energy neutrality.[5–7] Kaltenbrunner *et al.*[8] reported a flexible OSC built on a very thin substrate with an output power per weight of 10 W g^{-1}. This is more than an order of magnitude larger than the output power per weight found for a highly efficient crystalline silicon solar cell. Low weight solar cells are important in the space industry because the cost of uplifting 1 kg to an Earth orbit can be very high.

Semi-transparency is essential for the application of solar cells in windows. By choosing the right absorber materials, the window could harvest radiation in the IR region of the solar spectrum while being essentially transparent or distinctively coloured in the visible region.[9,10] This can easily be achieved by tailor-made organic semiconductors. Their mechanical properties and the ability to process flexible thin film OSCs in almost every 2D or 3D shape allow an integration of OPVs into various objects and new products, including cloth, fabric and textiles. Photovoltaic leaves, such as the solar ivy,[11] are good examples of the new possibilities that OSCs are offering.

Several reports have described the roll-to-roll manufacturing of OSCs.[12–14] Based on a pilot line production process, Krebs and co-workers[15] estimated the total manufacturing cost per watt. Even at relatively small volumes, costs <1 € W^{-1} were derived. They found that materials contribute >70% to the total manufacturing costs. Increasing manufacturing volumes should lead to lower material costs – especially related to the organic semiconductor materials – and costs substantially below 1 € W^{-1} appear to be feasible. There is an enormous potential for future applications of OSCs and the achievable performance (efficiency and lifetime) will determine the market penetration of OPV materials.

This chapter focuses on solution-processed OSCs and is organised as follows: the historical development of OPVs; the working principles and the importance of the absorber morphology and materials used in OSCs; different approaches to estimate the ultimate PCE limit of organic single- and multi-junction cells; and different promising concepts for high-efficiency OPVs. The chapter concludes with a summary and an outlook.

2.2 History of Organic Photovoltaic Cells

Pochettino and Atti[16] reported photoconductivity in anthracene in 1906. This discovery triggered systematic investigations of aromatic molecules forming crystals and/or thin films. The dark conductivity and photoconductivity, optical properties, thermo-voltage and Hall effect were measured in several different molecules and polymers.[17] After the discovery of increasing numbers of polycyclic aromatic compounds exhibiting semiconductor-like behaviour, Inokuchi[18] introduced the phrase "organic semiconductor" in 1954. Several reviews were published in the 1960s summarising early work on organic semiconductors.[17,19,20]

The photo-induced generation of free charge carriers is of paramount importance in solar cells. Gudden and Pohl[21] performed systematic studies on the photoconductivity of various materials in the 1920s. They found that efficient charge generation only occurs in materials with a high refractive index. This observation was later explained by the binding energy of the photo-excited electron–hole pair (exciton), which depends on the dielectric constant of the material, the size of the exciton and the effective mass of the charge carriers forming the excited state. These scientists found that a refractive index of about 3 is necessary for the quantitative generation of photo-induced charge at room temperature. Organic semiconductors typically exhibit relatively low refractive indices and it is not surprising that the light-induced charge generation efficiency in these materials is poor. Inspired by photosynthesis in living plants, Kearns and Calvin[22] patented and published the first OSC based on a donor–acceptor heterostructure. They used a bilayer made of magnesium phthalocyanine and air-oxidised tetramethyl-*p*-phenylenediamine and reported an open circuit voltage of 200 mV and a maximum power output for the device of 3×10^{-12} W. They did not give enough detail to calculate the PCE. However, assuming an illumination intensity in the μW cm^{-2} to mW cm^{-2} range, the PCE of their device was $\ll$0.001%.[23] Despite the poor performance and the small number of working examples, the US patent office issued a patent with very broad independent claims. Considering the claims within their broadest possible scope, the patent did cover various OSC designs, including most of the concepts explored today. The first claim of the patent (US3057947, patented 9 October 1962, *Photoelectric Cell Using Organic Materials*) reads:

What is claimed:

1. A photovoltaic cell comprised of an organic semiconductor, having primarily a π-electron system, and an organic material selected from the group consisting of electron acceptors and electrons donors, said semiconductor and organic material being in contact with each other so as to form an electron transferring boundary.

It was several years after the work of Calvin and Kearns that the PCE of OSCs was substantially improved. Tang summarised his OSC work performed at Eastman Kodak in several patents and scientific publications. Optimising the donor–acceptor pair and the device design, PCEs of ~1% were demonstrated in 1986, measured under simulated Air Mass 2 illumination.[24,25] Copper phthalocyanine was applied as the donor compound and a perylene tetracarboxylic derivative as the acceptor material (Figure 2.2).

In 1992, Sariciftci *et al.*[26,28] published the use of conjugated, semiconducting polymers in OPV cells. This report attracted much attention in the scientific community and triggered worldwide activity around OSCs, leading to the enormous increase in scientific reports on OPV cells seen in Figure 2.1.

After the discovery of the photo-induced electron transfer from conjugated polymers onto fullerenes,[26,27] especially C_{60} and C_{70}, different fullerene

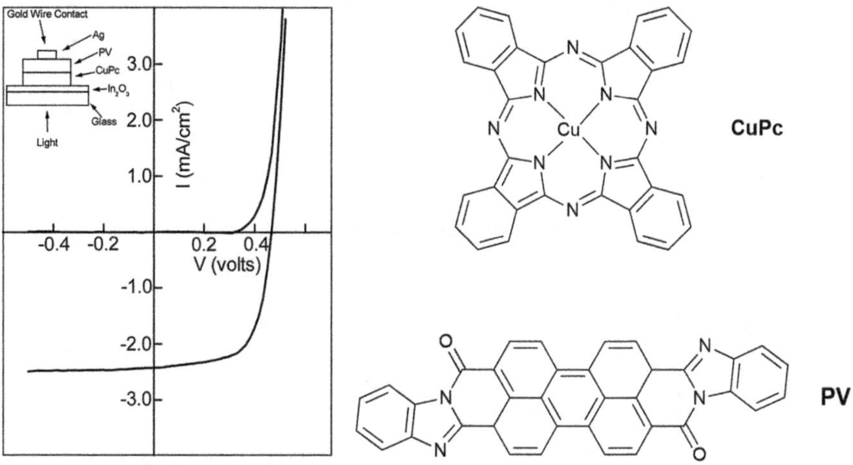

Figure 2.2 Current–voltage curves (measured under illumination and in the dark), device structure and absorber materials used in Tang's 1% efficient OSC. Reproduced with permission from C. W. Tang, *Appl. Phys. Lett.*, 1986, **48**, 183.[25] Copyright (1986) API Publishing LLC.

derivatives were used in OSCs. The seminal patent US5331183 filed in 1992 describes the discovery of conjugated polymer : fullerene PVCs.[28] The bulk heterojunction (BHJ) concept was claimed for the first time in Claim 3 of this patent.

What is claimed

A heterojunction device comprising a. a conjugated polymer which serves as a donor, and adjacent thereto, b. an acceptor material comprising an acceptor selected from the group consisting of fullerenes or fullerene derivatives, polymers comprising fullerenes or fullerene derivatives, organic and or polymeric acceptors having electronegativity in the range to enable a photoinitiated charge separation process defined by the following steps:

Step 1: $D + A \rightarrow {}^{1,3}D^* + A$, (excitation on D);
Step 2: ${}^{1,3}D^* + A \rightarrow {}^{1,3}(D - A)^*$, (excitation delocalised on D – A complex);
Step 3: ${}^{1,3}(D - A)^* \rightarrow {}^{1,3}(D^{\delta+} - A^{\delta-})^*$, (charge transfer initiated);
Step 4: ${}^{1,3}(D^{\delta+} - A^{\delta-})^* \rightarrow {}^{1,3}(D^{+} - A^{-})$, (ion radical pair formed);
Step 5: ${}^{1,3}(D^{+} - A^{-}) \rightarrow D^{+} + A^{-}$ (charge separation);

where donor (D) and acceptor (A) units are either covalently bound (intramolecular), or spatially close but not covalently bonded (intermolecular); "1,3" denotes singlet or triplet excited states, respectively, and where a heterojunction between the conjugated polymer and acceptor material is formed *in situ* by controlled segregation during solidification from a solution containing both the donor and the acceptor moieties.

In 1995, Yu *et al.*[29] reported an OSC based on a conjugated polymer : fullerene derivative absorber mixture with a PCE of ~3%. Although this efficiency

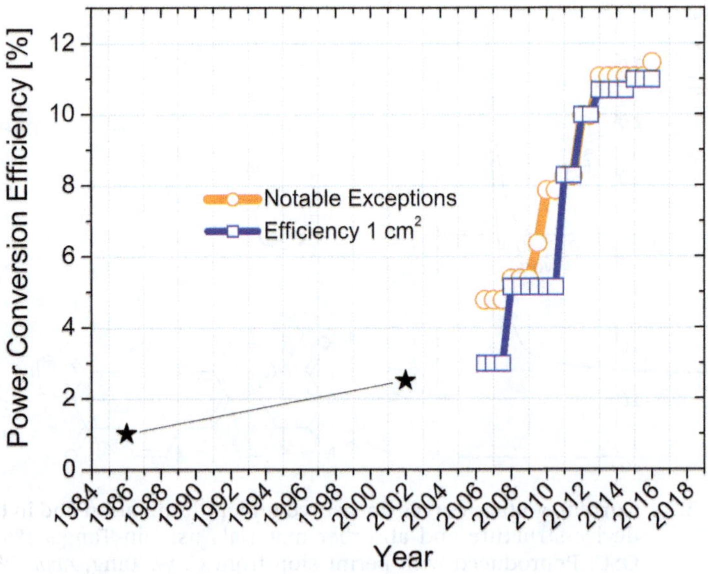

Figure 2.3 Development of the PCE of single-junction OSCs. The squares represent devices with an active area >1 cm^2, whereas the circles represent devices with an active area <1 cm^2. The first two points (1986 and 2001) were taken from original publications. The remaining efficiency numbers were reported in different efficiency tables in *Progress in Photovoltaics*.

was determined under conditions atypical of solar cells (light intensity, spectral distribution), the reported high photon-to-electron conversion efficiency (~30%) of the fullerene-containing devices demonstrated the advantageous properties of the novel acceptor. In 2001, Shaheen *et al.*[30] reported for the first time a certified PCE of an OSC. The Fraunhofer Institute for Solar Energy Systems rated the polymer : fullerene solar cell at 2.5% measured under AM1.5 global illumination. By developing new donor and acceptor materials with optimised energetics, electrical and optical properties, the PCE improved from 2.5% in 2001–11.5% in 2016.[1] Figure 2.3 summarises the development of the PCE of OSCs.

Parallel to the development of single-junction OSCs, multi-junction organic cells have also been explored. Using multiple absorber layers gives a potentially higher conversion efficiency. The first organic tandem solar cell was reported by Hiramoto *et al.* in 1990.[31] Two solar cells made of evaporated small molecules were stacked on top of each other. Forrest and co-workers pioneered the field of vacuum-processed tandem cells starting around the year 2002.[32] Introducing new absorber and transport materials, they demonstrated PCEs up to ~6%.[33] Heliatek GmbH have recently reported several record efficiencies for their multi-junction OSC with an efficiency of 13.2%.[2] It is believed that the PCE of single- and multi-junction OSCs can be improved further. The efficiency limit of OSCs is discussed in Section 4. Further improvements in the PCE are essential for the large-scale commercialisation of OPV cells.

2.3 Working Principle and Device Design

A series of physical processes must occur in an ideal OSC to convert an incident photon into useful electrical energy. These processes include photon absorption in the donor or acceptor material, the formation of a bound electron–hole pair (exciton), diffusion of the exciton to a donor–acceptor interface, exciton dissociation, charge separation and the formation of free charge carriers, charge transport and the collection of the charges at the corresponding electrodes (Figure 2.4). In a real device, several loss processes compete with the charge generation and collection steps. Among others, photons may not be absorbed in the photoactive layer, excitons can recombine before they reach a donor–acceptor interface and generated charge carriers can be trapped at defect sites or recombine before reaching the electrodes.

It is generally accepted that, in state-of-the-art organic semiconductors, a tightly bound electron–hole pair is the primary photo-excitation.[34] As a result of the low dielectric constant of organic materials, there is a strong Coulomb attraction between the electron and the hole and a dissociation of the electron–hole pair into free charges is very unlikely. Considering an oversimplified model of an electron–hole pair separated by 1 nm in a material with a dielectric constant of 3–4, applying Coulomb's law results in an electron-hole binding energy in the range 0.35–0.5 eV. Similar binding energies can be calculated assuming that the electron–hole pair forms a Rydberg-type atom. In this case, the dielectric constant and the effective masses of the bound charges determine the binding energy. In both cases the binding energy substantially exceeds the available thermal energy at room temperature. Therefore a donor–acceptor heterojunction is required for the dissociation

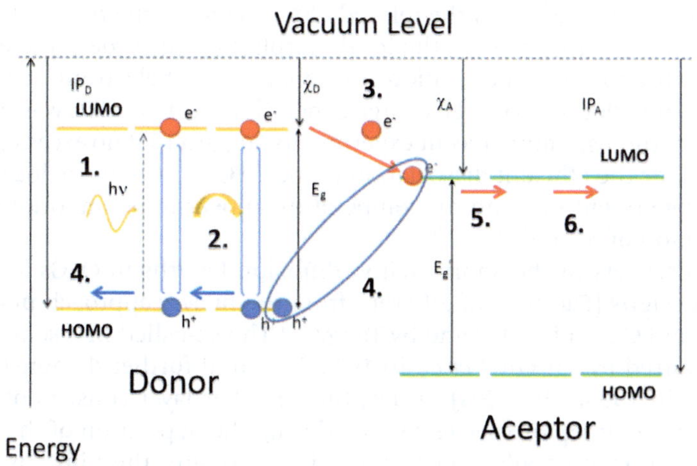

Figure 2.4 Schematic diagram of the donor–acceptor energy levels and the different processes occurring in an OSC. 1, Photon absorption; 2, exciton diffusion; 3, charge transfer; 4, the formation of the CTC; 5, dissociation of the CTC; and 6, charge transport.

of excitons. The difference between the electron affinities or the lowest unoccupied molecular orbital (LUMO) levels of the donor and acceptor material must be larger than the exciton binding energy to support the rapid transfer of an electron from the donor to the acceptor. The same is true for the difference between the highest occupied molecular orbital (HOMO) levels of donor and acceptor moieties to allow efficient hole transfer from the acceptor to the donor.

The exciton is a neutral species and cannot be dissociated by moderate electric fields. Only concentration differences lead to the diffusion of excitons. Recent studies have shown that the exciton diffusion length is in the range 5–10 nm in organic semiconductors.[35] This limits the charge generation to volumes very close to the donor–acceptor interfaces. For efficient exciton dissociation the donor and acceptor species need to be arranged such that essentially all incoming photons are absorbed less than a diffusion length away from the donor–acceptor interface. The detailed donor–acceptor arrangement (morphology) in the photoactive layer is therefore very important and is one of the main research areas in the field of OSCs.[36–39]

After the exciton has been dissociated at the donor–acceptor interface, the following steps leading to free charge carriers are still a matter of debate. It is generally accepted that photo-induced electron transfer between the donor and acceptor molecules leads to the population of a charge transfer complex (CTC) at the donor–acceptor interface. As a result of the proximity of the electron and the hole at this interface, the binding energy of the CTC may be considerable and much higher than the available thermal energy. Several mechanisms and processes have been proposed to explain the efficient charge generation in organic semiconductor blends.[40,41] Dissociation *via* the hot CTC utilising the excess energy available immediately after photoexcitation, dissociation *via* the relaxed CTC, including the role of the charge carrier mobility, the effect of the local morphology (disorder), the electrical field and the role of delocalisation and coherence[42] of electronic states, have all been considered. Although charge generation *via* a hot CTC was proposed in earlier studies,[43] more recent experiments suggest that no excess energy is required for the efficient dissociation of the CTC.[44,45] Charge delocalisation, the change of entropy and the temperature appear to be important for the dissociation of relaxed CTCs.[46]

The discovery of the short exciton diffusion lengths in OSCs led to new device designs (Figure 2.5), different from the bilayer approach pursued by Kearns and Calvin in 1958 and by Tang.[22,23] The so-called BHJ solar cell was first reported by Sariciftci *et al.* in 1992[26,28,47] and further demonstrated by Yu *et al.* in 1995.[25] In a BHJ device, the absorber layer consists of a blend of donor and acceptor species formed during the deposition of the layer on a substrate or in an additional processing step after the film has formed. Often spontaneous phase separation leads to a self-organised nanostructure with typical dimensions ideal for efficient photo-induced charge separation. Absorbed photon-to-electron conversion efficiencies close to 100% have been reported for BHJ devices.[48] A so-called planar heterostructure or graded

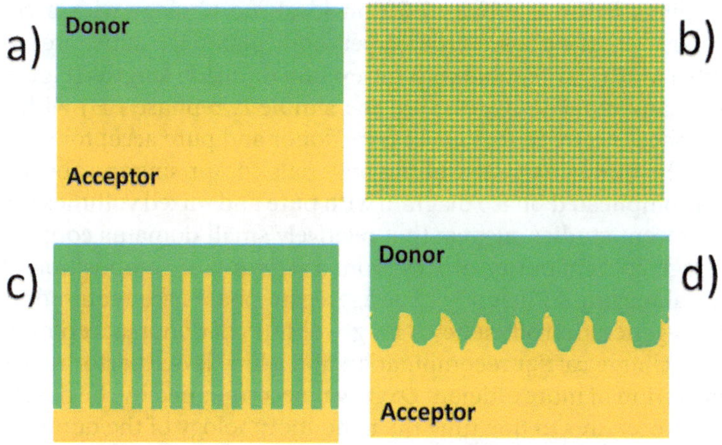

Figure 2.5 Schematic drawing of different donor–acceptor arrangements. (a) Bilayer device, (b) BHJ device, (c) ideal non-morphology and (d) graded bilayer device.

bilayer structure[49] is shown in Figure 2.5d. Here the charge-separating interface is enlarged by forming a donor–acceptor blend between the pristine materials.

BHJ solar cells based on ternary blends of organic semiconductors were found to allow the preparation of high-performance devices. The optical properties of the device can be tuned by using different donors[50] and the electrical properties of the solar cell can be improved. Combining a good absorber with a good charge transport material can lead to a significant increase in the electrical fill factor of the solar cell.[51] The highest reported PCEs for ternary blend devices are comparable with standard donor–acceptor BHJ solar cells and are in the range 9–11%.[51,52] However, as pointed out by Savoie *et al.*,[53] binary and ternary organic photovoltaic devices do have comparable performance potentials. Controlling the molecular arrangement and thermodynamic stability of a three-component system is very challenging.

2.3.1 Importance of the Absorber Morphology

Despite years of research and hundreds of publications on this topic, a consensus has not yet found for the ideal morphology of BHJ solar cells. The precise morphology, the equilibrium phase diagram, the role of vertical phase segregation and the mechanism by which phase separation occurs are not yet fully understood. A very fine dispersion of the acceptor in the donor material will lead to efficient charge generation, but poor charge transport. Ideal charge transport could be achieved by arranging the donor and acceptor in a bilayer configuration. However, charge generation only occurs at the interface between the two materials and will be inherently poor overall.

These findings led to a proposal for an ideal morphology with a comb-like arrangement with the phases ordered perpendicular to the substrate and structures with a width comparable with the exciton diffusion lengths (Figure 2.5c).

Today, many reports suggest that the simple two-phase BHJ with well-defined interfaces between regions of pure donor and pure acceptor volumes do not exist. The finite miscibility of the materials comprising the absorber layer leads to a complicated phase diagram with pure and mixed volumes present in the film. Recent studies suggest that relatively small domains coupled with a high average domain purity, arising from a variation of composition of mixed domains, alongside sufficiently pure aggregate phases are required to obtain high photo-generation efficiency.[54] Large volumes of donor–acceptor mixtures lead to fast charge carrier recombination and a low device performance.

The evolution of more efficient OSCs was accompanied by the development of several procedures to fine tune the nano-morphology of the donor–acceptor blend.[55] The most important are summarised in the following sections.

2.3.1.1 Post-treatment Exposure[56]

The donor–acceptor blend is exposed to solvent vapour or heated to a certain temperature. In both cases, the treatment increases the mobility of the donor and/or acceptor molecules, which leads to a rearrangement of the two components in the film. The type of solvent, the chosen temperature and the exposure time are crucial in both treatments. Thermal and solvent treatments were found to work best for polythiophene : fullerene blends. In particular, poly(3-hexylthiophene) (P3HT):[6,6]-phenyl-C61-butyric acid methyl ester (PCBM) blends represent an illustrative absorber layer for BHJ solar cells. In many cases, using excessive post-treatment exposure leads to a de-mixing of the donor and acceptor components.

2.3.1.2 Solvents and Solvent Mixtures

The importance of choosing the correct solvent was first demonstrated by Shaheen *et al.*[30] Using chlorobenzene instead of dichlorobenzene or toluene resulted in an increase in the PCE of a poly(2-methoxy-5-(3′,7′-dimethyloctyloxy)-1,4-phenylenevinylene): PCBM BHJ by a factor of two. This finding was followed up by a detailed analysis showing that the nano-morphology of the absorber layer depends on the processing solvent used. The impact of different solvents on the device performance of low band gap polymers has been studied. Investigating PCDTBT : phenylC_{71}-butyric acid methyl ester ($PC_{71}BM$)-based cells, Park *et al.*[48] observed the highest efficiency of 6.1% in devices prepared from DCB, which results in significantly smaller nanoscale phase separation. In some cases, solvent mixtures were found to be useful, especially for polymers containing a diketo-pyrrolo-pyrrole unit, which is known for its strong tendency to stack. Using processing solvent mixtures containing chloroform led to a significant improvement in the overall device performance, which could be correlated with an improved morphology of the absorber layer.[57]

2.3.1.3 *Processing Additives*

Octanedithiol, diiodooctane, chloronaphthalene and diphenyl ether are typical processing additives for BHJ solar cells. A few volume per cent are added to the processing solvent. As a result of their high boiling point and selective solubility, additives can strongly influence the formation of the donor–acceptor film. By delaying the solidification of one component (often the fullerene derivative), the second component can form more ordered structures. One of the first examples showing a pronounced additive effect was the low band gap polymer poly[2,6-(4,4-bis(2-ethylhexyl)-4H-cyclopenta[2,1-*b*; 3,4-*b*′]dithiophene)-*alt*-4,7-(2,1,3-benzothiadiazole)] (PCPDTBT), the PCE of which increased from ~3 to ~6% after the addition of a small amount of octanedithiol in the chlorobenzene-based solution.[38] Today, many high-performance materials require the use of processing additives; the use of additives appears to be mandatory to achieve the highest PCEs for the important class of benzodithiophene-based polymers.

2.3.1.4 *Chemical Design*[58]

In addition to the band gap and the energetic position of the electronic levels, the tendency of the donor and acceptor to aggregate and their miscibility are important ingredients for high-performance BHJ devices. Side-chain variations, selective substitution (*e.g.* fluorination) and the correct selection of the building blocks are essential for the design of high-performance materials. A good example of this is the work started by the group of Yu.[59,60] Several modifications were required before PCEs in the range of 10% were achieved.[61] Another good example of the importance of careful molecular design is the class of thiophene : benzothiadiazole-based polymers, which reached efficiencies >11% after the optimisation of the chemical structure and the device processing conditions.[62]

2.3.1.5 *Planar Heterostructures or Graded Junction Solar Cells*

Layer-by-layer solution deposition (with the subsequent spin-coating of a donor and an acceptor material) is one approach leading to a simple processing scheme. Thermal treatment or the selection of a solvent, interacting with the first layer during the deposition of the second layer, leads to an intermixing of the two materials at the interface. Using this method, Ayzner *et al.*[63] demonstrated P3HT : PCBM planar heterojunction solar cells with solar cell efficiencies comparable with BHJ devices. Liu *et al.*[64] reported an OSC produced *via* a sequential processing method with a PCE of 7.59%. Planar heterostructures consisting of materials with high thermal stability and limited miscibility may also show improved stability when operated at elevated temperatures. The more ordered structure at the interface is less prone to inter-diffusion, leading to better thermal stability.

Overall, the nano-morphology of solution-processed OSCs has been a hot topic for many years. The highest PCEs for a given donor–acceptor pair can only be achieved by arranging the photoactive materials carefully at the nanometre scale. However, the ultimate efficiency limit of OSCs is not determined by the morphology of the absorber layer. As discussed in the following sections, the charge separation processes leading to a significant loss in the open circuit voltage of the devices is responsible for the moderate performance of the state-of-the art OPVs available today.

2.3.2 Materials for OPV Cells

Organic semiconductors used for the fabrication of solar cells are formed by sp^2 hybridised carbon atoms along the main chain of the molecules, resulting in a sequence of alternating carbon–carbon single and double bonds. Materials with alternating carbon–carbon single and double bonds are often called conjugated materials. The resulting σ- and π-bonds lead to a semiconductor-like electronic structure with a band gap between the highest fully occupied and the LUMO energy levels. Polyacetylene is the simplest conjugated polymer. Pentacene and copper phthalocyanine are examples of conjugated molecules that show semiconducting properties in the solid state. By changing the arrangement of carbon single and double bonds and by the introduction of heteroatoms such as nitrogen, sulfur or oxygen, the properties of conjugated materials can be modified. This allows easy tuning of the band gap, a property that is especially important for light-emitting diodes and solar cells. Today, quantum chemical calculations can be used to predict the electronic structure fairly precisely. In the course of the Harvard Green Energy Project,[65] the energy levels of thousands of different conjugated molecules were calculated and the PCE potential of each molecule when used in an OPV was estimated. Based on the creativity of organic chemists and the aforementioned calculations, many different organic semiconductors have been synthesised. In particular, by combining electron-rich and electron-poor monomer units, the band gap and the position of the energy levels can be tuned. Essentially, all high-performance conjugated polymers and oligomers are based on this approach. Figure 2.6 gives a few examples of the donors and acceptors used in state-of-the-art OSCs.

In addition to the appropriate alignment of HOMO and LUMO levels in donor–acceptor systems, donors need to be good hole conductors and acceptors need to be good electron conductors. Many of the conjugated molecules, oligomers and polymers do support hole transport, but they are poor electron conductors. This is related to the fact that these molecules can be oxidised, but often not reduced, reversibly. In the best-performing BHJ solar cells, C_{60} or derivatives of the Buckminster fullerene, such as PCBM (Figure 2.6f), indene-C_{60}-bisadduct (ICBA) (Figure 2.6g), C_{60} and other fullerene derivatives are applied as electron acceptors. These molecules show excellent electron transport and have a relatively deep-lying LUMO level, qualifying the material as an acceptor. Their 3D electronic structure and outstanding electron-accepting properties make fullerenes successful components in OSCs. In recent years, non-fullerene acceptors have emerged as an alternative candidate for n-type materials to overcome the poor optical properties of fullerene derivatives (fullerenes have very low absorptivity above 500 nm).

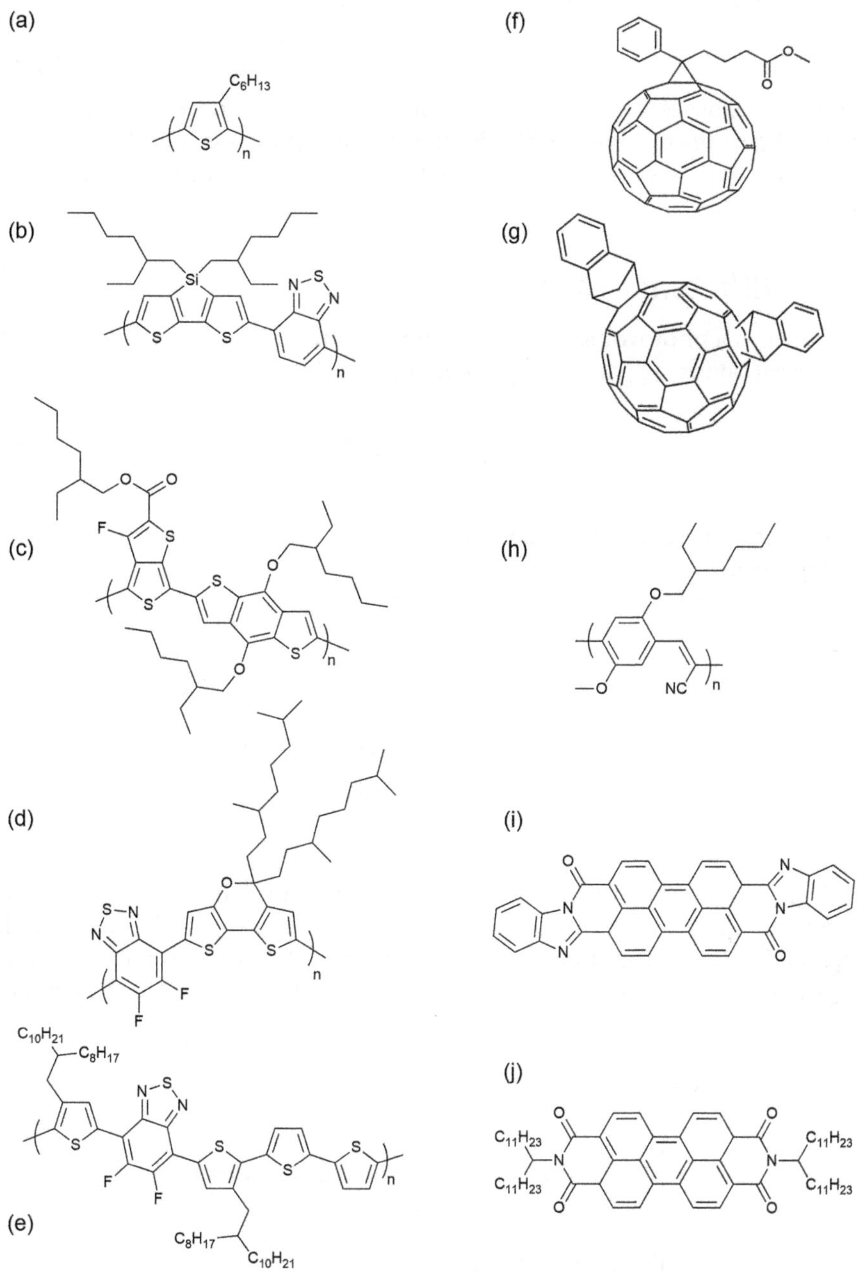

Figure 2.6 Examples of different donors (left-hand column) and acceptors (right-hand column) used in OSCs. (a) Poly-(3-hexylthiophene), (b) poly[(4,4′-bis(2-ethylhexyl)dithieno[3,2-*b*:2′,3′-*d*]silole)-2,6-diyl-*alt*-(2,1,3-benzothiadiazole)-4,7-diyl], (c) poly[[4,8-bis[(2-ethylhexyl)oxy]benzo[1,2-*b*:4,5-*b*′]dithiophene-2,6-diyl][3-fluoro-2-[(2-ethylhexyloxy)carbonyl]thieno[3,4-*b*]thiophenediyl]], (d) poly[2,7-(5,5-bis-(3,7-dimethyloctyl)-5H-dithieno[3,2-*b*:2′,3′-*d*]pyran)-*alt*-4,7-(5,6-difluoro-2,1,3-benzothiadiazole)], (e) (poly[(5,6-difluoro-2,1,3-benzothiadiazol-4,7-diyl)-*alt*-(3,3‴-di(2-octyldodecyl)-2,2′; 5′,2″; 5″,2‴-quaterthiophen-5,5‴-diyl)]), (f) [6,6]-phenyl-C_{61}-butyric acid methyl ester, (g) indene-C_{60}-bisadduct, (h) (poly(2-methoxy-5-ethylhexyloxy-1,4-phenylenecyanovinylene)), (i) perylene dicarboxylic derivative and (j) perylenetetracarboxylic diimide.

Different classes of material with promising PCEs have been reported. Several acceptor polymers,[66] oligomers[67] and groups of perylene diimide compounds[68] have shown efficiencies >8% when combined with high-performance donor polymers.

2.4 Efficiency of OSCs

The PCE of a photovoltaic device is defined as the maximum or the extracted electrical power (P_{el}) divided by the incident solar power (P_{sol}):

$$\eta = \frac{\max(P_{el})}{P_{sol}} \tag{2.1}$$

A typical current–voltage characteristic of a photovoltaic device is shown in Figure 2.7.

The electrical fill factor (FF) is given by:

$$\mathrm{FF} = \frac{V_{max} \cdot I_{max}}{V_{oc} \cdot I_{sc}} \tag{2.2}$$

Using the fill factor, the PCE of a solar cell can be written as:

$$\eta = \frac{V_{oc} \cdot I_{sc} \cdot \mathrm{FF}}{P_{sol}} \tag{2.3}$$

Shockley and Queisser presented a calculation of the ultimate PCE of a photovoltaic device with a single absorber layer in 1961.[69] They applied the detailed balance principle to analyse charge generation and recombination processes in an ideal solar cell. Assuming that (1) all photons with an energy

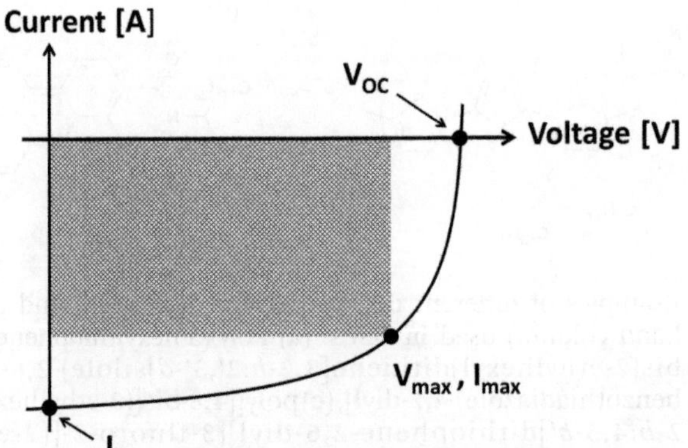

Figure 2.7 Current–voltage curve of a photovoltaic device. The dots indicate the most important points on the curve: the open circuit voltage V_{oc}, the short circuit current I_{sc} and the maximum power point V_{max}, I_{max}.

larger the band gap of the absorber material are absorbed, (2) each absorber photon generates one electron–hole pair, (3) only radiative charge carrier recombination is active and (4) all photo-generated charges are collected at the device electrodes, they calculated the current–voltage curve of an ideal solar cell to be:

$$I(V) = -I_{sc} + I_0\left(\mathrm{Exp}\left(\frac{qV}{k_B T}\right) - 1\right) \tag{2.4}$$

where q is the elementary charge, k_B is the Boltzmann constant, T is the temperature and I_{sc} and I_0 can be calculated by

$$I_{sc} = q\int_{E_g}^{\infty} \phi_{Sol}(E)\cdot \mathrm{d}E \tag{2.5}$$

$$I_0 = q\int_{E_g}^{\infty} \phi_{BB}(E)\cdot \mathrm{d}E \tag{2.6}$$

where E_g is the band gap of the absorber material, ϕ_{Sol} is the photon flux from the Sun and ϕ_{BB} is the photon flux of the surrounding black body radiation. A consequence of allowing only radiative recombination is that an ideal solar cell is also an ideal light-emitting diode. However, an ideal LED is not necessarily an ideal solar cell, *e.g.* conjugated polymers are known for their high fluorescence and electroluminescence quantum yield, but their photo-induced charge generation efficiency is poor due to the large exciton binding energy, leading to low photovoltaic efficiencies.

Using eqn (4), (5) and (6), the characteristics of ideal solar cells can be derived as a function of the band gap (E_g). The cell performance parameters are summarised in Figure 2.8. Calculations were performed assuming a black body radiation source with a surface temperature of 5800 K, an integrated light intensity at the solar cell of 1375 W m^{-2} and 300 K for the ambient temperature. The chosen solar flux corresponds closely to the so-called Air Mass 0 (AM0) spectrum found above the Earth's atmosphere.

Under these conditions, a maximum PCE of ~31% at an optimum band gap of ~1.3 eV is found. For the AM1.5 Global (AM 1.5 G) spectrum, which takes into account typical losses and absorption by the atmosphere, the maximum PCE is ~33% at an optimum gap of 1.34 eV. A detailed analysis[70] shows that, for an ideal solar cell, ~31% of the incoming solar energy is lost because the photon energy is smaller than the band gap energy. About 23% of the incoming solar energy is lost due to the rapid thermalisation of photo-excited electron–hole pairs. Another 12% is assigned to entropic losses and about 1% to radiative losses.

The detailed balance principle was also applied by Rau to derive two basic reciprocity relations between the photovoltaic and electroluminescent actions of light-emitting diodes and solar cells.[71] The first relates the

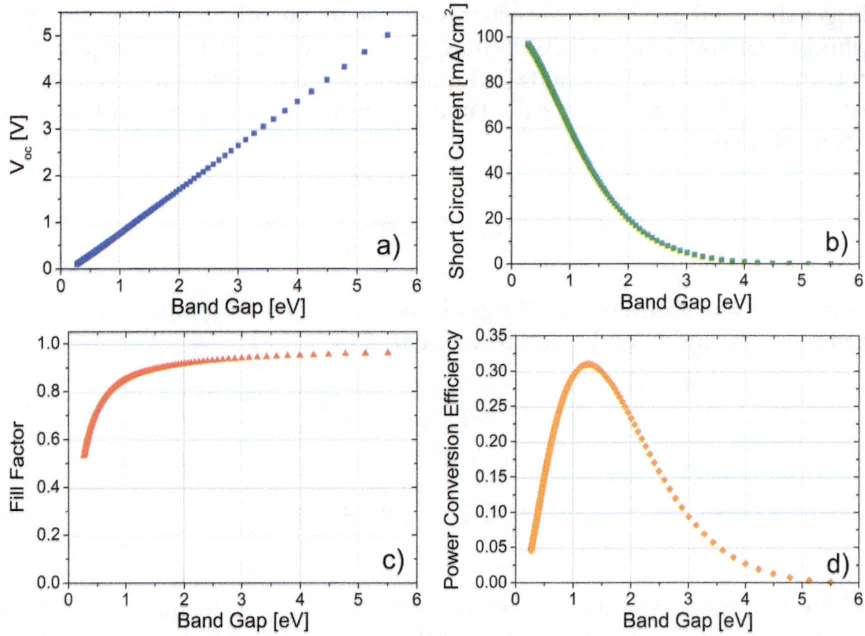

Figure 2.8 (a) Open circuit voltage (V_{oc}), (b) short circuit current density (J_{sc}), (c) fill factor and (d) PCE (η) calculated for an ideal solar cell assuming a black body radiation source with a temperature of 5800 K with an integrated light intensity at the solar cell of 1375 W m^{-2} and an ambient temperature of 300 K.

electroluminescence of solar cells to the photovoltaic external quantum efficiency:

$$\phi_{EL}(E) = \mathrm{EQE}(E) \cdot \phi_{BB}(E) \cdot \left[\exp\left(\frac{qV}{k_B T}\right) - 1 \right] \tag{2.7}$$

The term ϕ_{EL} is the electroluminescence quantum efficiency and the external quantum efficiency (EQE) is the EQE of the solar cell. The second reciprocity relation connects the open circuit voltage of a solar cell and its electroluminescence quantum efficiency. Rau's reciprocity relationship can be used to evaluate the potential of efficiency for existing solar cells. Green analysed different photovoltaic technologies listed in the efficiency table published in *Progress in Photovoltaics*. The resulting summary plot is shown in Figure 2.9, illustrating the importance of the radiative recombination of photo-induced charges in solar cells.[72]

The detailed balance approach can also be used for the estimation of the ultimate PCE of multi-junction solar cells. Combining two cells with different band gaps leads to a maximum efficiency of ~45% under AM1.5 G radiation. Under the same conditions, a triple-junction device can be as efficient as ~51% and a multi-junction cell based on six different absorber layers can have an

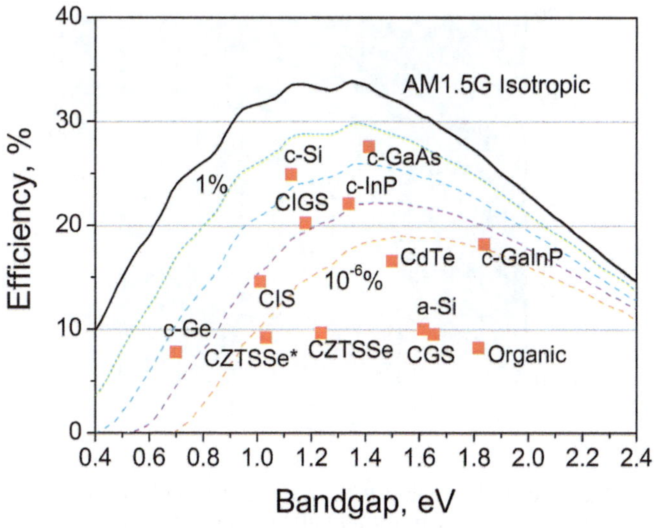

Figure 2.9 Shockley–Queisser limits on the efficiency of conventional single-junction cells and limits for cells of 1, 0.01, 0.0001 and 0.000001% external radiative efficiency (ERE). The ERE is defined as the fraction of the total dark current recombination that results in radiative emission. Confirmed experimental results are also shown (squares). Reproduced with permission from M. A. Green, Radiative efficiency of state-of-the-art photovoltaic cells, *Prog. Photovoltaics Res. Appl.*, John Wiley & Sons.[72] Copyright © 2011 John Wiley & Sons, Ltd.

efficiency up to ~59%. When working with concentrated sunlight, the ultimate PCE is even higher, *e.g.* a multi-junction cell with two different absorbers can have an efficiency of ~60% and fully concentrated AM1.5 G radiation.[73]

2.4.1 CTCs and Detailed Balance Limit for OPV Cells

Strong experimental evidence that a CTC is formed in conjugated polymer : fullerene solar cells was reported for the first time in 2007.[74,75] Initially, the CTC was considered to be an additional state in the already complicated excitation scheme of conjugated polymer : fullerene blends. However, work by Vandewal *et al.*[76] and Veldman *et al.*[77] revealed that the CTC plays an important part in OSCs. The open circuit voltage of OPVs is proportional to the energetic position of the CTC. A detailed analysis of the EQE, the electroluminescence and the current–voltage characteristics of a series of BHJ solar cells was analysed applying Rau's reciprocity relation. All these findings suggest that charge generation and recombination through the CTC are the efficiency-limiting processes in OSCs and can be used for the determination of their ultimate PCE. In 2009, Kirchartz *et al.*[78] determined for the first time the radiative efficiency limit of a state-of-the art BHJ solar cell. Based on detailed analysis of the optical properties of a state-of-the-art polymer : fullerene blend material (PF10TBT : PCBM), they estimated a maximum efficiency >20% for pristine polymer and polymer : fullerene absorbers. As the corresponding devices showed a much lower PCE

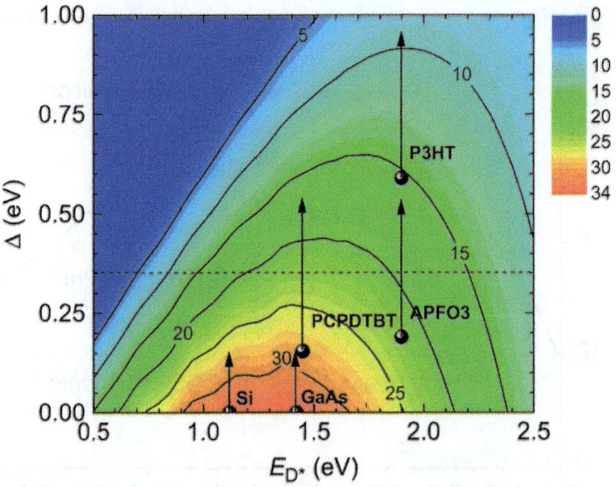

Figure 2.10 Maximum obtainable PCE in the radiative limit as a function of the optical gap of the dominant absorber E_D* and the losses due to charge transfer state formation Δ. Some polymer–fullerene and some inorganic solar cells are indicated. The arrows indicate a further loss in efficiency assuming a non-radiative recombination loss of 0.35 eV (EQE_{EL} ~10^{-6}) for polymer–fullerene solar cells and 0.15 eV (EQE_{EL} ~10^{-2}) for the inorganic solar cells. If the non-radiative recombination is not further reduced for OSCs, only the efficiencies above the dashed line remain accessible.[79] Reprinted from *Semiconductors and Semimetals*, 85, K. Vandewal, K. Tvingstedt and O. Inganäs, Charge Transfer States in Organic Donor–Acceptor Solar Cells, 261–295, Copyright (2011), with permission from Elsevier.

(maximum 4.2%), they analysed the losses leading to this tremendous reduction in PCE. They found: optical losses, charge carrier collection losses due to insufficient mobilities, exciton losses due to the insufficient transport of excitons or due to inefficient exciton dissociation and non-radiative recombination.

By far the most dominant loss was attributed to non-radiative recombination at the donor–acceptor interface. A similar analysis was performed by Vandewal *et al.*[79] They investigated the properties of the CTC in three different organic semiconductor absorbers and determined the radiative efficiency limit (Figure 2.10). They found an ultimate PCE of ~28% for an OSC with the low gap absorber layer PCPDTBT : PC71BM. If the very low radiative recombination quantum yield (~10^{-6}) is included in their calculations, the ultimate efficiency decreases to ~16%. The best measured efficiency for this organic semiconductor blend is close to 6%. This example shows that, under idealised conditions, OSCs could be as efficient as solar cells made of inorganic materials. Today, however, all state-of-the-art absorbers are far from this ideal performance and novel materials will be necessary to approach the Shockely–Queisser limit.

Nayak *et al.*[80] considered the effect of disorder on the PCE of solar cells. They introduced vibronic losses (dynamic disorder), dielectric losses and tail state losses and calculated the efficiency of OSCs following the Shockley–Queisser

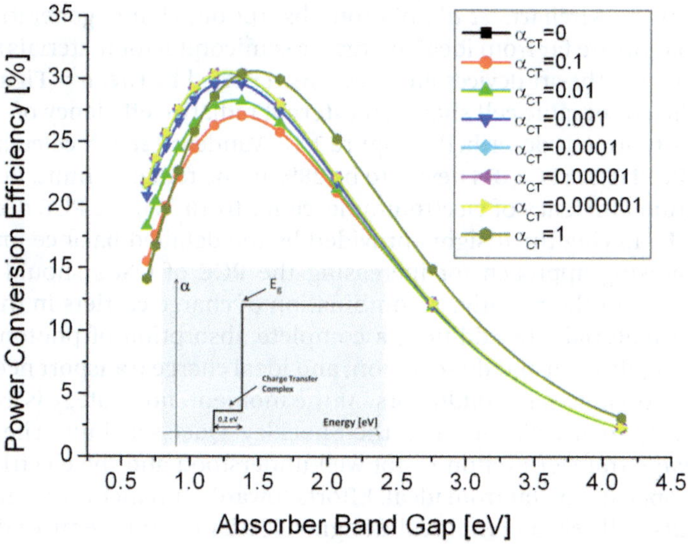

Figure 2.11 PCE of a solar cell with an absorption profile shown in the inset and different absorption strength (α_{CT}) of the charge transfer state. Inset: profile of an ideal absorber with a weak and 0.2 eV broad charge transfer absorption.[83] Reprinted from Progess in Polymer Science, 38, M. C. Scharber and N. S. Sariciftci, Efficiency of bulk-heterojunction organic solar cells, 1929–1940, Copyright (2013), with permission from Elsevier.

approach. When they included all relevant losses (realistic scenario), they derived efficiencies of ~13% under AM1.5 G illumination. The optimum absorber band gap was ~1.7 eV. Nayak *et al.* emphasise the importance of disorder in organic and dye-sensitised solar cells and its disadvantageous effect on the PCE.

Koster *et al.*[81] and Gruber *et al.*[82] pointed out that either a very weakly absorbing or a strongly absorbing charge transfer state leads to the highest PCEs under the radiative efficiency limit. This suggests that the charge transfer state *per se* is not detrimental to high PCEs. A weakly absorbing CTC minimises the V_{oc} losses, whereas a strongly absorbing CTC allows the collection of additional solar photons. In both cases, the ultimate efficiency in the radiative limit is >30%. This situation is illustrated in Figure 2.11. Here we assume a CTC with different absorption strength with a constant spectral width of 0.2 eV. The calculations were performed using a black body radiator with a surface temperature of 5800 K as the light source, a light intensity at the solar cell surface of 1000 W m^{-2} and a solar cell temperature of 300 K.

Similar high efficiencies for BHJ solar cells can be found by applying pure thermodynamic arguments. Giebink *et al.*[84] found a range of the "thermodynamic efficiency limit of exciton solar cells" of 22–27%, demonstrating again the performance potential of OSCs under idealised conditions.

When applying the detailed balance limit to OSCs, theory predicts that efficiencies close to the ultimate limit of ~33% may be achievable. However, as pointed out by Kirchartz *et al.*, photon absorption, charge generation and recombination are far from ideal in organic semiconductor materials and the PCEs of state-of-the-art devices are ~10%. As reported by Green[72] (Figure 2.9), an 8% efficient organic cell shows an external radiative efficiency of ~10^{-7}%. A similar situation is described in Figure 2.10. Vandewal *et al.*[46] calculated the PCE of a PCPDTBT : $PC_{71}BM$ device to be 28% in the radiative limit. Reducing the quantum efficiency of electroluminescence to 10^{-4}%, the PCE decreased to ~16%. Following the insights provided by the detailed balance limit, the most promising approach for increasing the PCE of OSCs should be the enhancement of the radiative recombination of charge carriers in the semiconductor materials. In addition, a complete absorption of photons, loss-free exciton diffusion and dissociation, and ideal charge transport need to be achieved in organic semiconductors. At the moment, no strategy is available to bring BHJ solar cells closer to the Shockley–Queisser limit. The nature of the charge transfer exciton is not well understood and the electrical and optical properties are far from ideal. Efforts towards enhancing the efficiency of BHJ cells still rely on empirical design rules for organic semiconductors, which allow a prediction of the efficiency potential of OSCs.

2.4.2 Empirical Approaches

In 2004, Coakley *et al.*[85] analysed the efficiency potential of BHJ solar cells. They developed a simple model for estimating the ultimate PCE. The calculations were performed based on the following assumptions: all photons with an energy larger than the band gap energy are absorbed; the electrical fill factor is equal to 1; and each photo-generated electron stores the band gap energy minus the losses occurring during charge transfer. With an energy loss of 1 eV due to charge transfer, a maximum efficiency of ~15% with an ideal absorber gap of ~1.75 eV was found.

A more realistic model for the PCE of OSCs was developed after the discovery of a relationship between the open circuit voltage and the energy levels of the donor–acceptor materials.[86] A systematic study of a large set of donor polymers combined with the acceptor molecule PCBM showed that the open circuit voltage of optimised solar cells can be empirically calculated by

$$V_{oc} = \frac{1}{e}\left(E_{HOMO}^{DONOR} - E_{LUMO}^{Acceptor}\right) - 0.3\ V \tag{2.8}$$

Here e is the elementary charge and 0.3 V is a typical loss found in BHJ solar cells. The energies of the donor HOMO level and the acceptor LUMO level are given in electron volts.

Figure 2.12 shows the set of experimental data leading to eqn (10). When using −4.3 eV for the position of the acceptor LUMO level (PCBM), eqn (10) can be deduced from the straight line in Figure 2.12. Assuming a typical average EQE and electrical fill factor, eqn (3), (5) and (8) were used to calculate the PCE as a function of the solar cell band gap and the LUMO position. When the model was first published in 2006, the highest reported EQE and fill factors were ~65%. With a minimum ΔLUMO of 0.3 V (to ensure efficient charge transfer),

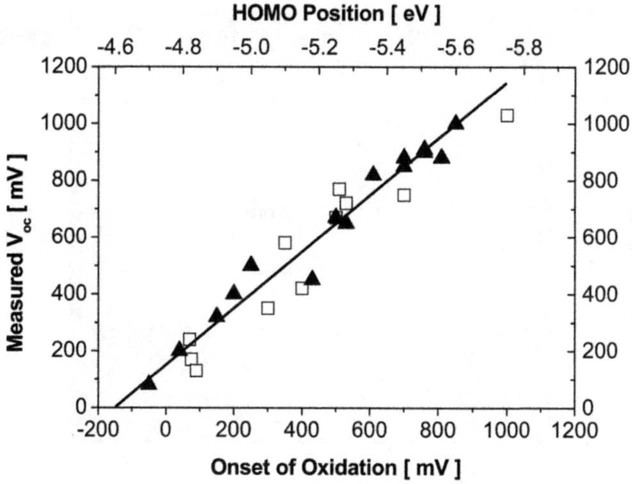

Figure 2.12 Open circuit voltage (V_{oc}) of different BHJ solar cells plotted *versus* the oxidation potential/HOMO position of the donor polymer used in each individual device. The straight line represents a linear fit with a slope of 1. Reproduced with permission from M. C. Scharber, D. Muhlbacher, M. Koppe, P. Denk, C. Waldauf, A. J. Heeger and C. J. Brabec, Design Rules for Donors in Bulk-Heterojunction Solar Cells—Towards 10% Energy-Conversion Efficiency, *Advanced Materials*, John Wiley & Sons.[86] Copyright © 2006 WILEY-VCH Verlag GmbH & Co. KGaA, Weinheim.

a maximum efficiency of 11% was derived and the optimum band gap was found to be 1.45 eV. Figure 2.13 plots the PCE of a BHJ solar cell as a function of the absorber band gap and the LUMO level offset. Devices fabricated today can have fill factors and EQEs up to 75 and 80%, respectively. With a higher EQE and fill factor, an ultimate efficiency up to 16% appears feasible. The model gives clear design rules for high-performance materials and it has been used extensively to evaluate the potential of new organic semiconductors. It has been the basis for computer-based material design approaches, such as the Harvard Clean Energy Project.[65]

The efficiency potential of state-of-the art conjugated polymer : fullerene blends has been evaluated by applying this empirical model.[87] The detailed investigation showed that even the best-performing BHJ solar cells suffer from the same efficiency losses and that, despite its simplicity, the model still applies and describes the efficiency potential of BHJ solar cells very well.

Veldman *et al.*[47] found an alternative empirical equation for the open circuit voltage of OSCs. They measured the HOMO and LUMO levels of donors and acceptors and the open circuit voltage for corresponding solar cells and correlated these with the CTC energy obtained from electroluminescence experiments. They found:

$$qV_{oc} = E_{CT} - 0.47(\pm 0.06)\ \text{eV} \tag{2.9}$$

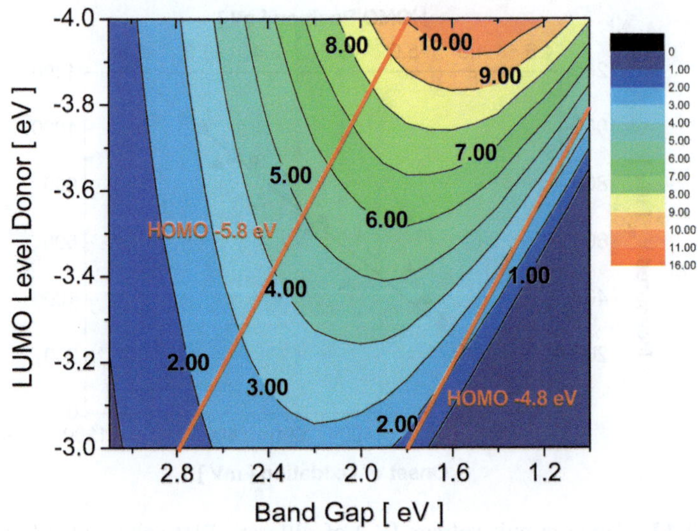

Figure 2.13 Contour plot showing the PCE of a BHJ solar cell with PCBM as acceptor material (LUMO level 4.3 eV). For the calculation, an EQE of 65%, a fill factor of 65% and an open circuit voltage according to eqn (1) were used. Black lines indicate constant HOMO levels of 5.8 and 4.8 eV, respectively. Reproduced with permission from M. C. Scharber, D. Muhlbacher, M. Koppe, P. Denk, C. Waldauf, A. J. Heeger and C. J. Brabec, Design Rules for Donors in Bulk-Heterojunction Solar Cells—Towards 10% Energy-Conversion Efficiency, *Advanced Materials*, John Wiley & Sons.[86] Copyright © 2006 WILEY-VCH Verlag GmbH & Co. KGaA, Weinheim.

In addition, they showed that a relatively small free-energy difference of 0.1 eV between the band gap energy and the energy of the CTC is sufficient for splitting a photo-excited exciton at the donor–acceptor interface. Combining the two findings and assuming a fill factor and an average EQE of 65%, an ultimate PCE of 11% at an optimum band gap of ~1.4 eV is predicted.

Minnaert and Burgelman introduced a more realistic model for the optical absorbance of OSCs by introducing a finite bandwidth of the photo-active layer.[88] Assuming an open circuit voltage given by the donor HOMO minus the acceptor LUMO difference, multiplied by a so-called voltage factor *f* and typical values for the fill factor, they calculated efficiencies in the range 5–15%. Their calculations and results are comparable with those found by Coakley *et al.* and Scharber *et al.*[51,52]

Marks and co-workers discussed the practical efficiency limit of organic photovoltaic cells in a paper published in 2009.[89] They introduced a factor into their model describing the field-dependent exciton dissociation. By applying Green's method, an equation for the electrical fill factor was derived. Using eqn (10) and varying the LUMO offset between the donor and acceptor, a practical efficiency limit of ~14% was found.

Blom and co-workers used a very different approach to estimate the ultimate efficiency of polymer : fullerene BHJ solar cells.[90,91] Applying their advanced drift–diffusion device model, they simulated the current–voltage characteristics of a state-of-the-art poly(3-hexyl-thiophene):PCBM solar cell with an efficiency of 3.5%. The model includes the drift and diffusion of charges, space charge effects and the temperature and field dependence of charge generation. Assuming that an optimised OSC will exhibit similar device parameters, the effect of the LUMO level offset between the donor and the acceptor, the layer thickness and the charge carrier mobility were investigated. They found an ultimate PCE of ~11% for a 100–200 nm thick layer and an optimum band gap of 1.7–1.9 eV.[63] In a more general calculation including the optical properties of organic semiconductors, they found a maximum efficiency of 11.5% in the same band gap range.[64]

These empirical models were predominantly developed for BHJ devices. For bilayer structures, the exciton diffusion length is an additional parameter that needs to be taken into account, whereas in BHJ devices the domains are always considered to be smaller than the diffusion length of excitons and all excitons can reach a donor–acceptor interface. In bilayer devices, there is a competition between complete light absorption (making the donor–acceptor stack optically thick) and the generation of excitons close to the donor–acceptor interface. Rand *et al.*[92] investigated the ultimate PCE in detail. Considering the energetics of the donor–acceptor junction, charge transfer dynamics and the exciton diffusion in the donor and acceptor volume, a maximum PCE of ~12% was derived for long exciton diffusion lengths.

The empirical models discussed here predict maximum PCEs of small-area devices in the range 10–15%. Module efficiencies in this range will be required for large-scale commercial success. Taking into account the fact that the efficiency of commercial thin film modules is ~60–70% of the highest small-area device efficiency, record efficiencies in the range 18–25% are desirable. We now discuss different approaches to obtaining higher PCEs.

2.4.3 Improving Single-junction Devices

A comparison of the record devices made of different absorber materials shows that GaAs or silicon solar cells show much smaller open circuit voltage losses than OSCs (Figure 2.14).

Although high-performance inorganic solar cells show open circuit voltages ~300–400 mV smaller than the band gap, losses in organic cells are in the range 600–800 mV. Polman and Atwater,[95,96] Markvart[97] and Rau *et al.*[98] discussed the different V_{oc} losses using the following equations:

$$qV_{oc}^{rad} = E_g\left(1-\frac{T}{T_{sun}}\right) + k_BT\left[\ln\left(\frac{\gamma(E_g,T_{sun})}{\gamma(E_g,T)}\right) - \ln\left(\frac{\Omega_{emit}}{\Omega_{sun}}\right)\right] \quad (2.10)$$

$$qV_{OC} = qV_{OC}^{rad} - k_BT\ln(QE) \quad (2.11)$$

Eqn (10) describes the ideal open circuit voltage of a solar cell under the radiative limit, where T_{sun} is the temperature of the Sun, T is the temperature

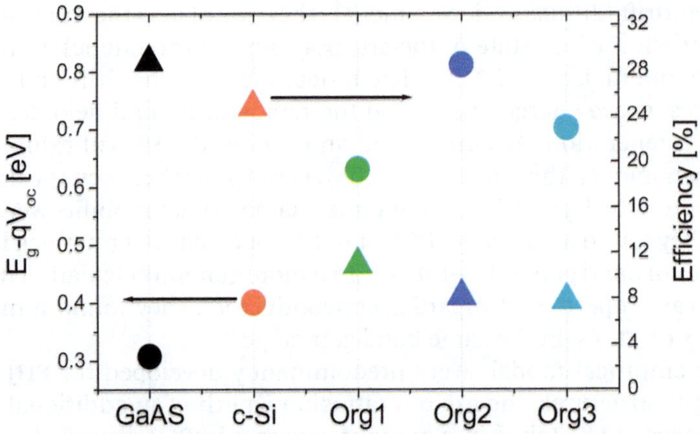

Figure 2.14 PCEs and differences between the semiconductor band gap and the best open circuit voltage for different technologies.[1,93,94] Values for band gap were taken from the published papers or determined using published EQE spectra.

of the solar cell and E_g is the band gap of the semiconductor. $\gamma(E,T)$ is defined as $2k_BT/(c^2h^3)$ $(E^2 + 2k_BTE + 2k_B{}^2T^2)$, where c is the speed of light and h is Planck's constant. Eqn (11) refers to the well-established reciprocity relation correcting for non-radiative losses present in the solar cell.[60]

The first term of eqn (10) on the right-hand side represents the conversion of solar light using a Carnot engine, neglecting spontaneous emission from the semiconductor. The term in the square brackets accounts for entropy-related gains and losses. The first term reflects the entropy decrease due to photon absorption and emission at different temperatures (photo-cooling). The second term reflects the entropy increase due to photon absorption and re-emission through spontaneous recombination. Although the incident solid angle of photons coming directly from the Sun is Ω_{sun} ~6×10^{-5} sr, the outgoing re-emission is directed into a solid angle up to $\Omega_{emit} = 4\pi$ sr. This change in entropy results in an open circuit voltage loss of up to 300 mV. Eqn (11) describes the V_{oc} losses caused by non-radiative recombination of photogenerated charges. QE is the quantum efficiency for radiative recombination and is defined as $R_{rad}/(R_{nrad} + R_{rad})$, where R_{rad} and R_{nrad} are the radiative and non-radiative recombination rates, respectively. For a QE of 0.01, the V_{oc} loss is in the range of 100 mV. In total, all losses can easily account for a V_{oc} reduction of 500 mV or more, especially for devices with a dominant non-radiative recombination.

Although the quantum efficiency is a material property that can be improved, *e.g.* by higher material purity, the first and the second entropy terms can be tackled by optical means. Light management and photonic structures can reflect and redirect emitted light back to the disk of the Sun and trap the incoming light in the semiconductor layer. High-performance inorganic solar cells are all equipped with advanced light-trapping and light management systems. Without surface texturing and anti-reflective coatings, the PCE of a crystalline silicon solar cell would be much lower. Several different light management and light-trapping concepts have been applied to OSCs.[99–105] Reduced optical losses and higher photo-currents have often been demonstrated. However, the reported enhancement in the open circuit voltage or PCE has been moderate. It appears that there is a lot of room for improvement and that a well-designed light management concept applied to an efficient OSC could lead to a significant improvement in performance. However, to have some relevance for future OSC products, the manufacturing should be simple and the light management should not add significant cost to the product, *i.e.* roll-to-roll compatible processes and cost-efficient materials will be required.

Instead of optimising single-junction devices, we could also explore the so-called third-generation concepts suitable for organic semiconductor materials. Green summarised these advanced solar energy conversion concepts in his book *Third Generation Photovoltaics*.[73] The basics and impact of these concepts on OSCs are discussed in the next section.

2.4.4 Third-generation Photovoltaics and OSCs

Third-generation PVC are solar cells that are potentially able to overcome the radiative efficiency limit of Shockley and Queisser for a single-junction device. Several different concepts have been discussed: hot carrier cells; multiple electron–hole pairs per photon device; impurity photovoltaic and multiband cells; and multi-junction solar cells (tandem cells).

In hot carrier cells, photo-excited charges are collected before thermalisation, *i.e.* the excess energy available immediately after photo-excitation is not lost, but is used to achieve a higher open circuit voltage. It is currently unclear how to build such a hot carrier device and there is no evidence that the use of organic semiconductors would be beneficial.

In multiple electron–hole pairs per photon devices, one high-energy photon creates multiple electron–hole-pairs. This could lead to quantum efficiencies >100% and peak efficiencies >40%. Evidence for the generation of multiple electron–hole-pairs per incident photon has been reported in inorganic semiconductors (colloidal nanoparticles) and recently also in organic semiconductors.[106–108] In organic semiconductors, so-called singlet exciton fission has been observed.[109] Singlet fission is a process in which an organic absorber unit in an excited singlet state shares its excitation energy with a neighbouring unit in the ground state and both are converted into triplet excited states. If these triplets are cleaved into electron–hole pairs at a donor–acceptor interface, a quantum yield of 200% could be achieved. Baldo *et al.* reported an OSC with an EQE of 109%.[110] Additional investigations under external magnetic fields showed that singlet fission is active and contributes to the generation of charges. However, the reported PCEs are moderate and substantially more effort is required for the development of highly efficient OSCs with multiple electron–hole pairs per photon generation.

In impurity photovoltaic and multiband cells, the absorption of semiconductors is extended by introducing additional energy levels in the band gap of the semiconductor. Low-energy photons can promote an electron from the valence band to the level in the gap and by a second low energy photon to the conduction band of the semiconductor. Different band structures have been proposed and Lopez *et al.* reported the first evidence of a working multiband solar cell based on a GaN_xAs_{1-x} alloy.[111] A few proposed structures may be realised with organic semiconductors by arranging several different donor and acceptor moieties in a supramolecular structure. It remains to be seen whether this will indeed lead to high-performance solar cells.

Multi-junction or tandem cells are the only third-generation technology that has demonstrated high PCEs. By stacking two or more solar cells with different band gaps on top of each other, the thermalisation losses of photo-excited charge carriers can be reduced and the device absorbs a larger fraction of the incoming solar photons. A multi-junction device made of an InGaP/GaAs/InGaAs stack was rated at 37.9% measured under AM1.5 G conditions. Under concentrated sunlight, the National Renewable Energy Laboratory (Boulder, Colorado, USA) certified an InGaP/GaAs/InGaAs device at an efficiency of 44.4%.[1] Theoretical efficiencies of multi-junction devices are ~45% for a double- and ~51% for a triple-junction device when measured under AM1.5 conditions. Under concentrated sunlight, the maximum efficiencies were ~60 and ~67%, respectively.

Organic tandem cells[112,113] consisting of two cells stacked on top of each other show PCEs in the range 10–13%. Solution-processed devices and devices deposited from the vapour phase have been demonstrated with high efficiencies. Tandem cells with different absorber materials and devices with two identical cells have been reported with efficiencies >10%. Although a tandem cell with two BHJ solar cells made of poly[2,6-(4,4-bis-(2-ethylhexyl)-4H-cyclopenta[2,1-*b*; 3,4-*b′*]-dithiophene)-*alt*-4,7-(2,1,3-difluorobenzothiadiazole) : $PC_{71}BM$ layers showed an efficiency of 10.2%,[114] a stack consisting of one BHJ layer made of P3HT : ICBA and one BHJ layer made of PDTP-DFBT : PCBM was certified at 10.6%.[115] Two BHJ devices with a poly[4,8-bis(5-(2-ethylhexyl) thiophen-2-yl)benzo[1,2-*b*:4,5-*b′*]dithiophene-co-3-fluorothieno[3,4-*b*]thiophene-2-carboxylate] and the acceptor $PC_{71}BM$ absorber stacked on top of each other showed a PCE of 11.3%, whereas the corresponding single-layer device was only 8.6% efficient.[116]

The 8.3% efficient small-molecule-based tandem cell reported by Heliatek in ref. 92 is based on two absorbers with a similar band gap. In this case it was also found that tandem cells with two identical absorbers outperformed the corresponding single-junction device. This indicates that the optimised single layer does not absorb incoming photons effectively and using two absorber layers cures this deficiency. An optically thick single layer is not beneficial in many cases because the available charge carrier mobilities do not support efficient charge extraction. Splitting the absorber into two thinner layers ensures the efficient collection of photo-induced charge.

Several estimates of the ultimate PCE of organic tandem cells have been published. Minnaert *et al.*[87] and Blom *et al.*[89] extended their efficiency models for single-junction solar cells to tandem cells. They calculated maximum efficiencies of 23% (compared with 15% for the single-junction cells) and 14% (compared with 11.5% for the single-junction cells), respectively.

Dennler *et al.*[117] have developed design rules for donors in BHJ tandem cells based on the model used to calculate the contour plot in Figure 2.13. Including the optical effect, assuming that the V_{oc} of the tandem cell is the sum of the individual cells and that the short circuit current is determined by the smaller I_{sc}, the efficiency was calculated as a function of the band gaps of the two absorber layers. The resulting contour plot is shown in Figure 2.15. The maximum efficiency is 15% and the best tandem cell shows a ~40% better performance compared with the best single-junction device. Assuming better performing individual devices (higher EQE and electrical fill factors), the same model delivers higher PCEs. The Brabec group calculated a PCE potential of ~21% for two stacked BHJ solar cells stacked with ideal band gaps.[118]

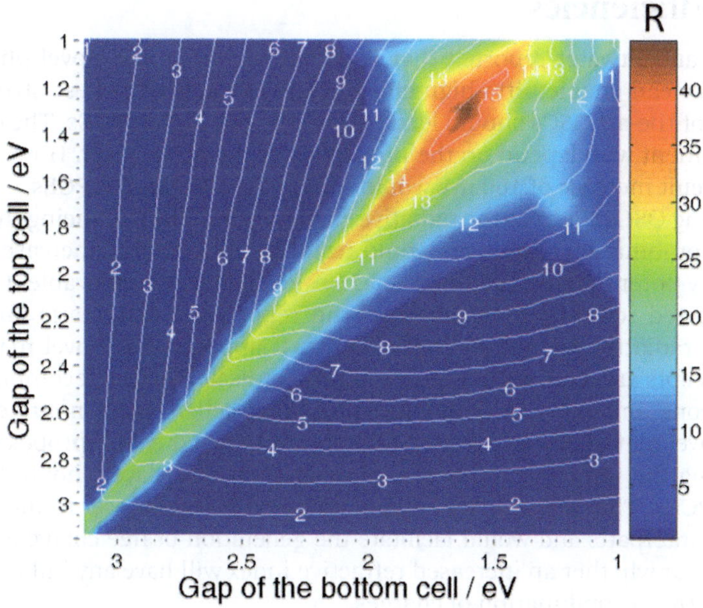

Figure 2.15 Percentage of efficiency increase of a tandem cell over the best single cell (R) for a device consisting of a top sub-cell and a bottom sub-cell based on donors with a LUMO level at −4 eV. The variables are the band gaps of both donors. The lines indicate the efficiency of the tandem cell. Reproduced with permission from G. Dennler, M. C. Scharber, T. Ameri, P. Denk, K. Forberich, C. Waldauf and C. J. Brabec, Design Rules for Donors in Bulk-Heterojunction Tandem Solar Cells–Towards 15% Energy-Conversion Efficiency, *Advanced Materials*, John Wiley & Sons.[117] Copyright © 2008 WILEY-VCH Verlag GmbH & Co. KGaA, Weinheim.

Table 2.1 Comparsion of theoretical solar cell efficiency limits.

	Single-junction cells(%)	Tandem cells (%)	Performance ratio
SQ	33	45	1.4
Minnaert *et al.*	15	23	1.5
Blom *et al.*	11.5	14	1.2
Dennler *et al.*	11	15	1.4

A comparison of single- and tandem devices and the corresponding performance ratio are summarised in Table 2.1.

Organic tandem cells should be 20–50% more efficient than optimised single-junction devices. This makes tandem cells interesting and they might be the most promising approach to close the performance gap between OSCs and state-of-the-art inorganic technologies.

2.5 Summary: Promising Approaches for Higher Efficiencies

Polman and Atwater reported that light management and novel photonic structures can enable ultra-high-efficiency PVCs. Applying these structures to state-of-the-art OSCs should lead to devices with higher PCEs. The overall improvement will depend on the properties of the active layer. However, an open circuit increase of 100 mV would lead to 10–15% higher PCEs.

Bringing OSCs closer to the radiative efficiency limit by enhancing the radiative recombination of charges would result in a substantial increase in the efficiency potential. Instead of 10%, >25% PCEs may be achievable. Despite the expected rewards, there has been no report of a systematic improvement in radiative recombination in OSCs. It appears that novel materials and concepts are required to enforce photon emission on the recombination of electrons and holes in donor–acceptor solar cells. A design strategy for these novel materials has not yet been reported. Koster *et al.*[81] proposed that increasing the refractive index of the organic absorber layer should lead to higher PCEs. This would bring organic semiconductors closer to their inorganic counterparts and would facilitate the generation of free charge carriers. It is unclear whether an increased refractive index will have any influence on the radiative recombination of charges.

Among all third-generation concepts, tandem or multi-junction cells appear to be the most promising approach for OPVs. For the highest efficiencies, novel absorber materials with low band gaps and excellent performance will be required (~1.2 eV). In combination with today's best-performing absorbers with band gaps in the range 1.6–1.7 eV, efficiencies of ~15% could be achieved. Further optimisation of the EQE and the electrical fill factor could lead to efficiencies close to 20%.

Multi-exciton generation appears to be a promising concept in OSCs. Converting a high-energy singlet state into two triplets which, in a second step, are split at a donor–acceptor interface could lead to EQEs >100%. Recent reports describe the observation of multi-exciton generation, but no efficiency enhancement was demonstrated. More research will be required to explore the possibility of using multi-exciton generation in OSCs.

There are several different approaches available to further increase the PCE of OSCs. Closing the efficiency gap between organic and inorganic PVCs will be essential for the large-scale commercialisation of OPV cells. In addition, the durability and the costs of OPV modules need to meet market requirements.

Products for niche markets, such as battery chargers and power supplies for sensors, based on OSC devices may be competitive even today. Having companies define the needs and requirements for new products and giving guidance to the scientific community is another important ingredient for the further development of OSCs.

References

1. M. A. Green, K. Emery, Y. Hishikawa, W. Warta and E. D. Dunlop, Solar Cell Efficiency Tables (Version 47), *Prog. Photovoltaics Res. Appl.*, 2016, **24**(1), 3–11.
2. http://www.heliatek.com.
3. M. Jørgensen, K. Norrman, S. A. Gevorgyan, T. Tromholt, B. Andreasen and F. C. Krebs, Stability of polymer solar cells, *Adv. Mater.*, 2012, **24**(5), 580–612.
4. J. A. Hauch, P. Schilinsky, S. A. Choulis, S. Rajoelson and C. J. Brabec, The impact of water vapor transmission rate on the lifetime of flexible polymer solar cells, *Appl. Phys. Lett.*, 2008, **93**(10), 103306.
5. A. L. Roes, E. A. Alsema, K. Blok and M. K. Patel, Ex-ante environmental and economic evaluation of polymer photovoltaics, *Prog. Photovoltaics Res. Appl.*, 2009, **17**(6), 372–393.
6. R. García-Valverde, J. A. Cherni and A. Urbina, Life cycle analysis of organic photovoltaic technologies, *Prog. Photovoltaics Res. Appl.*, 2010, **18**(7), 535–558.
7. N. Espinosa, R. García-Valverde, A. Urbina, F. Lenzmann, M. Manceau, D. Angmo and F. C. Krebs, Life cycle assessment of ITO-free flexible polymer solar cells prepared by roll-to-roll coating and printing, *Sol. Energy Mater. Sol. Cells*, 2012, **97**, 3–13.
8. M. Kaltenbrunner, M. S. White, E. D. Głowacki, T. Sekitani, T. Someya, N. S. Sariciftci and S. Bauer, Ultrathin and lightweight organic solar cells with high flexibility, *Nat. Commun.*, 2012, **3**, 770.
9. R. R. Lunt and V. Bulovic, Transparent, near-infrared organic photovoltaic solar cells for window and energy-scavenging applications, *Appl. Phys. Lett.*, 2011, **98**(11), 113305.
10. J. Huang, G. Li and Y. Yang, A Semi-transparent Plastic Solar Cell Fabricated by a Lamination Process, *Adv. Mater.*, 2008, **20**(3), 415–419.
11. http://solarivy.com/; www.solarte.de.

12. F. C. Krebs, J. Fyenbo and M. Jørgensen, Product integration of compact roll-to-roll processed polymer solar cell modules: methods and manufacture using flexographic printing, slot-die coating and rotary screen printing, *J. Mater. Chem.*, 2010, **20**(41), 8994.
13. M. M. Voigt, R. C. I. Mackenzie, S. P. King, C. P. Yau, P. Atienzar, J. Dane, P. E. Keivanidis, I. Zadrazil, D. D. C. Bradley and J. Nelson, Gravure printing inverted organic solar cells: The influence of ink properties on film quality and device performance, *Sol. Energy Mater. Sol. Cells*, 2012, **105**, 77–85.
14. H. J. Park, M. Kang, S. H. Ahn and L. J. Guo, A Facile Route to Polymer Solar Cells with Optimum Morphology Readily Applicable to a Roll-to-Roll Process without Sacrificing High Device Performances, *Adv. Mater.*, 2010, **22**(35), E247–E253.
15. B. Azzopardi, C. J. M. Emmott, A. Urbina, F. C. Krebs, J. Mutale and J. Nelson, Economic assessment of solar electricity production from organic-based photovoltaic modules in a domestic environment, *Energy Environ. Sci.*, 2011, **4**(10), 3741.
16. A. Pochettino, R. Atti della and Accademia di Lincei, *Rendiconti della classe di scienze fisiche, matematiche e naturali*, 1906, vol. 35, p. 355.
17. C. Hamann, Organische Halbleiter, *Phys. Status Solidi*, 1965, **12**(2), 483–532.
18. H. Inokuchi, The discovery of organic semiconductors. Its light and shadow, *Org. Electron.*, Apr. 2006, **7**(2), 62–76.
19. N. B. Hannay, *Semiconductors*, Reinhold Publ. Corp., New York, 1959, pp. 634–675.
20. F. Gutmann and A. Netschey, Organic Semiconductors, *Rev. Pure Appl. Chem.*, 1962, **12**, 2–15.
21. B. Gudden and R. Pohl, Neuere Beobachtungen über den Zusammenhang elektrischer und optischer Erscheinungen, *Naturwissenschaften*, 1923, **11**(19), 348–354.
22. M. Calvin and D. R. Kearns, *U.S. Patent* 3057947, 1962.
23. D. Kearns and M. Calvin, Photovoltaic Effect and Photoconductivity in Laminated Organic Systems, *J. Chem. Phys.*, 1958, **29**(4), 950.
24. C. Tang, *U.S. Patent* US4125414, 1978.
25. C. W. Tang, Two-layer organic photovoltaic cell, *Appl. Phys. Lett.*, 1986, **48**(2), 183.
26. N. S. Sariciftci, L. Smilowitz, a. J. Heeger and F. Wudl, Photoinduced Electron Transfer from a Conducting Polymer to Buckminsterfullerene, *Science*, Nov. 1992, **258**(5087), 1474–1476.
27. Y. Wang, Photoconductivity of fullerene-doped polymers, *Nature*, 1992, **356**, 585–587.
28. N. S. Sariciftci and A. J. Heeger, *U.S. Patent* US5331183, 1994.
29. G. Yu, J. Gao, J. C. Hummelen, F. Wudl and A. J. Heeger, Polymer Photovoltaic Cells: Enhanced Efficiencies *via* a Network of Internal Donor-Acceptor Heterojunctions, *Science*, 1995, **270**(5243), 1789–1791.

30. S. E. Shaheen, C. J. Brabec, N. S. Sariciftci, F. Padinger, T. Fromherz and J. C. Hummelen, 2.5% efficient organic plastic solar cells, *Appl. Phys. Lett.*, 2001, **78**(6), 841–843.
31. M. Hiramoto, M. Suezaki and M. Yokoyama, Effect of thin gold interstitial-layer on the photovoltaic properties of tandem organic solar cell, *Chem. Lett.*, 1990, (3), 327–330.
32. A. Yakimov and S. R. Forrest, High photovoltage multiple-heterojunction organic solar cells incorporating interfacial metallic nanoclusters, *Appl. Phys. Lett.*, 2002, **80**(9), 1667–1669.
33. J. Xue, S. Uchida, B. P. Rand and S. R. Forrest, Asymmetric tandem organic photovoltaic cells with hybrid planar-mixed molecular heterojunctions, *Appl. Phys. Lett.*, 2004, **85**(23), 5757–5759.
34. A. J. Heeger, N. S. Sariciftci and E. B. Namdas, *Semiconducting and Metallic Polymers*, Oxford University Press, Oxford UK, 2010, p. 278.
35. O. V. Mikhnenko, H. Azimi, M. Scharber, M. Morana, P. W. M. Blom and M. A. Loi, Exciton diffusion length in narrow bandgap polymers, *Energy Environ. Sci.*, 2012, **5**(5), 6960.
36. X. Yang and J. Loos, Toward High-Performance Polymer Solar Cells: The Importance of Morphology Control, *Macromolecules*, 2007, **40**(5), 1353–1362.
37. F. Padinger, R. S. Rittberger, N. S. Sariciftci, F. Padinger, R. S. Rittberger, R. S. Rittberger, N. S. Sariciftci and N. S. Sariciftci, Effects of postproduction treatment om plastic solar cells, *Adv. Funct. Mater.*, 2003, **13**(1), 85.
38. J. Peet, J. Y. Kim, N. E. Coates, W. L. Ma, D. Moses, A. J. Heeger and G. C. Bazan, Efficiency enhancement in low-bandgap polymer solar cells by processing with alkane dithiols, *Nat. Mater.*, 2007, **6**(7), 497–500.
39. H. Hoppe and N. S. Sariciftci, Organic solar cells: An overview, *J. Mater. Res.*, 2004, **19**(07), 1924–1945.
40. M. A. Loi and A. Troisi, Charge generation mechanism in organic solar cells, *Phys. Chem. Chem. Phys.*, 2014, **16**(38), 20277–20278.
41. F. Gao and O. Inganäs, Charge generation in polymer–fullerene bulk-heterojunction solar cells, *Phys. Chem. Chem. Phys.*, 2014, **16**(38), 20291–20304.
42. L. G. Kaake, D. Moses and A. J. Heeger, Coherence and Uncertainty in Nanostructured Organic Photovoltaics, *J. Phys. Chem. Lett.*, 2013, **4**(14), 2264–2268.
43. T. M. Clarke, A. M. Ballantyne, J. Nelson, D. D. C. Bradley and J. R. Durrant, Free Energy Control of Charge Photogeneration in Polythiophene/Fullerene Solar Cells: The Influence of Thermal Annealing on P3HT/PCBM Blends, *Adv. Funct. Mater.*, 2008, **18**(24), 4029–4035.
44. T. G. J. van der Hofstad, D. Di Nuzzo, M. van den Berg, R. A. J. Janssen and S. C. J. Meskers, Influence of Photon Excess Energy on Charge Carrier Dynamics in a Polymer-Fullerene Solar Cell, *Adv. Energy Mater.*, 2012, **2**(9), 1095–1099.

45. K. Vandewal, S. Albrecht, E. T. Hoke, K. R. Graham, J. Widmer, J. D. Douglas, M. Schubert, W. R. Mateker, J. T. Bloking, G. F. Burkhard, A. Sellinger, J. M. J. Fréchet, A. Amassian, M. K. Riede, M. D. McGehee, D. Neher and A. Salleo, Efficient charge generation by relaxed charge-transfer states at organic interfaces, *Nat. Mater.*, 2014, **13**(1), 63–68.
46. F. Gao, W. Tress, J. Wang and O. Inganäs, Temperature Dependence of Charge Carrier Generation in Organic Photovoltaics, *Phys. Rev. Lett.*, 2015, **114**(12), 128701.
47. N. S. Sariciftci, D. Braun, C. Zhang, V. I. Srdanov, A. J. Heeger, G. Stucky and F. Wudl, Semiconducting polymer-buckminsterfullerene heterojunctions: Diodes, photodiodes, and photovoltaic cells, *Appl. Phys. Lett.*, 1993, **62**(6), 585.
48. S. H. Park, A. Roy, S. Beaupré, S. Cho, N. Coates, J. S. Moon, D. Moses, M. Leclerc, K. Lee and A. J. Heeger, Bulk heterojunction solar cells with internal quantum efficiency approaching 100%, *Nat. Photonics*, 2009, **3**(5), 297–302.
49. D. Meissner and J. Rostalski, Photovoltaics of interconnected networks, *Synth. Met.*, 2001, **121**(1–3), 1551–1552.
50. M. Koppe, H.-J. Egelhaaf, G. Dennler, M. C. Scharber, C. J. Brabec, P. Schilinsky and C. N. Hoth, Near IR sensitization of organic bulk heterojunction solar cells: Towards optimization of the spectral response of organic solar cells, *Adv. Funct. Mater.*, 2010, **20**(2), 338–346.
51. J. Zhang, Y. Zhang, J. Fang, K. Lu, Z. Wang, W. Ma and Z. Wei, Conjugated Polymer–Small Molecule Alloy Leads to High Efficient Ternary Organic Solar Cells, *J. Am. Chem. Soc.*, 2015, **137**(25), 8176–8183.
52. L. Lu, W. Chen, T. Xu and L. Yu, High-performance ternary blend polymer solar cells involving both energy transfer and hole relay processes, *Nat. Commun.*, 2015, **6**, 7327.
53. B. M. Savoie, S. Dunaisky, T. J. Marks and M. A. Ratner, The Scope and Limitations of Ternary Blend Organic Photovoltaics, *Adv. Energy Mater.*, 2015, **5**(3), DOI: 10.1002/aenm.201400891.
54. B. A. Collins, J. R. Tumbleston and H. Ade, Miscibility, Crystallinity, and Phase Development in P3HT/PCBM Solar Cells: Toward an Enlightened Understanding of Device Morphology and Stability, *J. Phys. Chem. Lett.*, 2011, **2**(24), 3135–3145.
55. Y. Huang, E. J. Kramer, A. J. Heeger and G. C. Bazan, Bulk Heterojunction Solar Cells: Morphology and Performance Relationships, *Chem. Rev.*, 2014, **114**(14), 7006–7043.
56. G. Li, V. Shrotriya, J. Huang, Y. Yao, T. Moriarty, K. Emery and Y. Yang, High-efficiency solution processable polymer photovoltaic cells by self-organization of polymer blends, *Nat. Mater.*, 2005, **4**(11), 864–868.
57. M. M. Wienk, M. Turbiez, J. Gilot and R. A. J. Janssen, Narrow-Bandgap Diketo-Pyrrolo-Pyrrole Polymer Solar Cells: The Effect of Processing on the Performance, *Adv. Mater.*, 2008, **20**(13), 2556–2560.
58. Z. B. Henson, K. Müllen and G. C. Bazan, Design strategies for organic semiconductors beyond the molecular formula, *Nat. Chem.*, 2012, **4**(9), 699–704.

59. T. Xu and L. Yu, How to design low bandgap polymers for highly efficient organic solar cells, *Mater. Today*, 2014, **17**(1), 11–15.
60. L. Lu and L. Yu, Understanding Low Bandgap Polymer PTB7 and Optimizing Polymer Solar Cells Based on It, *Adv. Mater.*, 2014, **26**(26), 4413–4430.
61. Z. He, B. Xiao, F. Liu, H. Wu, Y. Yang, S. Xiao, C. Wang, T. P. Russell and Y. Cao, Single-junction polymer solar cells with high efficiency and photovoltage, *Nat. Photonics*, 2015, **9**(3), 174–179.
62. J. Zhao, Y. Li, G. Yang, K. Jiang, H. Lin, H. Ade, W. Ma and H. Yan, Efficient organic solar cells processed from hydrocarbon solvents, *Nat. Energy*, 2016, **1**(2), 15027.
63. A. L. Ayzner, C. J. Tassone, S. H. Tolbert and B. J. Schwartz, Reappraising the Need for Bulk Heterojunctions in Polymer- Fullerene Photovoltaics: The Role of Carrier Transport in All-Solution-Processed P3HT/PCBM Bilayer Solar Cells, *J. Phys. Chem. C*, 2009, **113**(46), 20050–20060.
64. Y. Liu, F. Liu, H. Wang, D. Nordlund, Z. Sun, S. Ferdous and T. P. Russell, Sequential Deposition: Optimization of Solvent Swelling for High-Performance Polymer Solar Cells, *ACS Appl. Mater. Interfaces*, 2015, **7**(1), 653–661.
65. http://cleanenergy.molecularspace.org/.
66. L. Gao, Z.-G. Zhang, L. Xue, J. Min, J. Zhang, Z. Wei and Y. Li, All-Polymer Solar Cells Based on Absorption-Complementary Polymer Donor and Acceptor with High Power Conversion Efficiency of 8.27%, *Adv. Mater.*, 2016, **28**(9), 1884–1890.
67. O. K. Kwon, M. A. Uddin, J.-H. Park, S. K. Park, T. L. Nguyen, H. Y. Woo and S. Y. Park, A High Efficiency Nonfullerene Organic Solar Cell with Optimized Crystalline Organizations, *Adv. Mater.*, 2016, **28**(5), 910–916.
68. Y. Zhong, M. T. Trinh, R. Chen, G. E. Purdum, P. P. Khlyabich, M. Sezen, S. Oh, H. Zhu, B. Fowler, B. Zhang, W. Wang, C.-Y. Nam, M. Y. Sfeir, C. T. Black, M. L. Steigerwald, Y.-L. Loo, F. Ng, X.-Y. Zhu and C. Nuckolls, Molecular helices as electron acceptors in high-performance bulk heterojunction solar cells, *Nat. Commun.*, 2015, **6**, 8242.
69. W. Shockley and H. J. Queisser, Detailed Balance Limit of Efficiency of p–n Junction Solar Cells, *J. Appl. Phys.*, 1961, **32**(3), 510.
70. M. D. Archer and J. R. Bolton, Requirements for ideal performance of photochemical and photovoltaic solar energy converters, *J. Phys. Chem.*, 1990, **94**(21), 8028–8036.
71. U. Rau, Reciprocity relation between photovoltaic quantum efficiency and electroluminescent emission of solar cells, *Phys. Rev. B*, 2007, **76**(8), 085303.
72. M. A. Green, Radiative efficiency of state-of-the-art photovoltaic cells, *Prog. Photovoltaics Res. Appl.*, 2012, **20**(4), 472–476.
73. M. A. Green, *Third Generation Photovoltaics*, Springer-Verlag, Berlin/Heidelberg, 2006, p. 172.

74. J. J. Benson-Smith, L. Goris, K. Vandewal, K. Haenen, J. V. Manca, D. Vanderzande, D. D. C. Bradley and J. Nelson, Formation of a Ground-State Charge-Transfer Complex in Polyfluorene//[6,6]-Phenyl-C61 Butyric Acid Methyl Ester (PCBM) Blend Films and Its Role in the Function of Polymer/PCBM Solar Cells, *Adv. Funct. Mater.*, 2007, **17**(3), 451–457.
75. M. A. Loi, S. Toffanin, M. Muccini, M. Forster, U. Scherf and M. Scharber, Charge transfer excitons in bulk heterojunctions of a polyfluorene copolymer and a fullerene derivative, *Adv. Funct. Mater.*, 2007, **17**(13), 2111–2116.
76. K. Vandewal, K. Tvingstedt, A. Gadisa, O. Inganäs and J. V Manca, On the origin of the open-circuit voltage of polymer-fullerene solar cells, *Nat. Mater.*, 2009, **8**(11), 904–909.
77. D. Veldman, S. C. J. Meskers and R. A. J. Janssen, The Energy of Charge-Transfer States in Electron Donor-Acceptor Blends: Insight into the Energy Losses in Organic Solar Cells, *Adv. Funct. Mater.*, 2009, **19**(12), 1939–1948.
78. T. Kirchartz, K. Taretto and U. Rau, Efficiency limits of organic bulk heterojunction solar cells, *J. Phys. Chem. C*, 2009, **113**(41), 17958–17966.
79. K. Vandewal, K. Tvingstedt, and O. Inganäs, Charge Transfer States in Organic Donor–Acceptor Solar Cells, *Semiconductors*, Elsevier, 2011, vol. 85, pp. 261–295.
80. P. K. Nayak, G. Garcia-Belmonte, A. Kahn, J. Bisquert and D. Cahen, Photovoltaic efficiency limits and material disorder, *Energy Environ. Sci.*, 2012, **5**(3), 6022.
81. L. J. A. Koster, S. E. Shaheen and J. C. Hummelen, Pathways to a New Efficiency Regime for Organic Solar Cells, *Adv. Energy Mater.*, 2012, **2**(10), 1246–1253.
82. M. Gruber, J. Wagner, K. Klein, U. Hörmann, A. Opitz, M. Stutzmann and W. Brütting, Thermodynamic Efficiency Limit of Molecular Donor-Acceptor Solar Cells and its Application to Diindenoperylene/C60-Based Planar Heterojunction Devices, *Adv. Energy Mater.*, 2012, **2**(9), 1100–1108.
83. M. C. Scharber and N. S. Sariciftci, Efficiency of bulk-heterojunction organic solar cells, *Prog. Polym. Sci.*, 2013, **38**(12), 1929–1949.
84. N. C. Giebink, G. P. Wiederrecht, M. R. Wasielewski and S. R. Forrest, Thermodynamic efficiency limit of excitonic solar cells, *Phys. Rev. B*, 2011, **83**(19), 1–6.
85. K. M. Coakley and M. D. McGehee, Conjugated Polymer Photovoltaic Cells, *Chem. Mater.*, 2004, **16**(23), 4533–4542.
86. M. C. Scharber, D. Mühlbacher, M. Koppe, P. Denk, C. Waldauf, A. J. Heeger and C. J. Brabec, Design Rules for Donors in Bulk-Heterojunction Solar Cells—Towards 10% Energy-Conversion Efficiency, *Adv. Mater.*, 2006, **18**(6), 789–794.
87. M. C. Scharber, On the Efficiency Limit of Conjugated Polymer:Fullerene-Based Bulk Heterojunction Solar Cells, *Adv. Mater.*, 2016, **28**(10), 1994–2001.

88. B. Minnaert and M. Burgelman, Efficiency potential of organic bulk heterojunction solar cells, *Prog. Photovoltaics Res. Appl.*, 2007, **15**(8), 741–748.
89. J. D. Servaites, M. A. Ratner and T. J. Marks, Practical efficiency limits in organic photovoltaic cells: Functional dependence of fill factor and external quantum efficiency, *Appl. Phys. Lett.*, 2009, **95**(16), 163302.
90. L. J. A. Koster, E. C. P. Smits, V. D. Mihailetchi and P. W. M. Blom, Device model for the operation of polymer/fullerene bulk heterojunction solar cells, *Phys. Rev. B*, 2005, **72**(8), 085205.
91. J. D. Kotlarski and P. W. M. Blom, Ultimate performance of polymer: fullerene bulk heterojunction tandem solar cells, *Appl. Phys. Lett.*, 2011, **98**(5), 053301.
92. B. P. Rand, D. P. Burk and S. R. Forrest, Offset energies at organic semiconductor heterojunctions and their influence on the open-circuit voltage of thin-film solar cells, *Phys. Rev. B*, 2007, **75**(11), 115327.
93. M. A. Green, K. Emery, Y. Hishikawa, and W. Warta, Solar Cell Efficiency Tables (Version 37), *Prog. Photovoltaics Res. Appl.*, 2011, **19**(1), 84–92.
94. B. P. Rand, J. Genoe, P. Heremans and J. Poortmans, Solar Cells Utilizing Small Molecular Weight Organic Semiconductors, *Prog. Photovolt Res. Appl.*, 2007, **15**, 659–676.
95. A. Polman and H. a. Atwater, Photonic design principles for ultrahigh-efficiency photovoltaics, *Nat. Mater.*, 2012, **11**(3), 174–177.
96. H. A. Atwater and A. Polman, Reply to On the thermodynamics of light trapping in solar cells, *Nat. Mater.*, 2014, **13**(2), 104–105.
97. T. Markvart, The thermodynamics of optical étendue, *J. Opt. A: Pure Appl. Opt.*, 2008, **10**(1), 015008.
98. U. Rau and T. Kirchartz, On the thermodynamics of light trapping in solar cells, *Nat. Mater.*, 2014, **13**(2), 103–104.
99. Z. Hu, J. Zhang and Y. Zhao, Effect of textured electrodes with light-trapping on performance of polymer solar cells, *J. Appl. Phys.*, 2012, **111**(10), 104516.
100. J. N. Munday, D. M. Callahan and H. A. Atwater, Light trapping beyond the 4n2 limit in thin waveguides, *Appl. Phys. Lett.*, 2012, **100**(12), 121121.
101. D. M. Callahan, J. N. Munday and H. A. Atwater, Solar Cell light trapping beyond the ray optic limit, *Nano Lett.*, 2012, **12**(1), 214–218.
102. V. Janković, Y. (Michael) Yang, J. You, L. Dou, Y. Liu, P. Cheung, J. P. Chang and Y. Yang, Active Layer-Incorporated, Spectrally Tuned Au/SiO_2 Core/Shell Nanorod-Based Light Trapping for Organic Photovoltaics, *ACS Nano*, 2013, **7**(5), 3815–3822.
103. L. Song and A. Uddin, Design of high efficiency organic solar cell with light trapping, *Opt. Express*, 2012, **20**(suppl. 5), A606–A621.
104. S. J. Kim, G. Y. Margulis, S. Rim, M. L. Brongersma, M. D. Mcgehee and P. Peumans, *Geometric Light Trapping with a V-trap for Efficient Organic Solar Cells*, 2013, vol. 21, pp. 572–576.
105. V. Andersson, K. Tvingstedt and O. Inganäs, Optical modeling of a folded organic solar cell, *J. Appl. Phys.*, 2008, **103**(9), 094520.

106. A. J. Nozik, M. C. Beard, J. M. Luther, M. Law, R. J. Ellingson and J. C. Johnson, Semiconductor Quantum Dots and Quantum Dot Arrays and Applications of Multiple Exciton Generation to Third-Generation Photovoltaic Solar Cells, *Chem. Rev.*, 2010, **110**(11), 6873–6890.
107. J. Lee, P. Jadhav, P. D. Reusswig, S. R. Yost, N. J. Thompson, D. N. Congreve, E. Hontz, T. Van Voorhis and M. A. Baldo, Singlet Exciton Fission Photovoltaics, *Acc. Chem. Res.*, 2013, **46**(6), 1300–1311.
108. G. Grancini, M. Maiuri, D. Fazzi, a. Petrozza, H.-J. Egelhaaf, D. Brida, G. Cerullo and G. Lanzani, Hot exciton dissociation in polymer solar cells, *Nat. Mater.*, 2013, **12**(1), 29–33.
109. M. B. Smith and J. Michl, Singlet fission, *Chem. Rev.*, 2010, **110**(11), 6891–6936.
110. D. N. Congreve, J. Lee, N. J. Thompson, E. Hontz, S. R. Yost, P. D. Reusswig, M. E. Bahlke, S. Reineke, T. Van Voorhis and M. A. Baldo, External quantum efficiency above 100% in a singlet-exciton-fission-based organic photovoltaic cell, *Science*, 2013, **340**(6130), 334–337.
111. N. López, L. a. Reichertz, K. M. Yu, K. Campman and W. Walukiewicz, Engineering the Electronic Band Structure for Multiband Solar Cells, *Phys. Rev. Lett.*, 2011, **106**(2), 028701.
112. T. Ameri, G. Dennler, C. Lungenschmied and C. J. Brabec, Organic tandem solar cells: A review, *Energy Environ. Sci.*, 2009, **2**(4), 347.
113. T. Ameri, N. Li and C. J. Brabec, Highly efficient organic tandem solar cells: a follow up review, *Energy Environ. Sci.*, 2013, **6**(8), 2390.
114. J. You, C.-C. Chen, Z. Hong, K. Yoshimura, K. Ohya, R. Xu, S. Ye, J. Gao, G. Li and Y. Yang, 10.2% Power Conversion Efficiency Polymer Tandem Solar Cells Consisting of Two Identical Sub-Cells, *Adv. Mater.*, 2013, **25**(29), 3973–3978.
115. J. You, L. Dou, K. Yoshimura, T. Kato, K. Ohya, T. Moriarty, K. Emery, C.-C. Chen, J. Gao, G. Li and Y. Yang, A polymer tandem solar cell with 10.6% power conversion efficiency, *Nat. Commun.*, 2013, **4**, 1446.
116. H. Zhou, Y. Zhang, C.-K. Mai, S. D. Collins, G. C. Bazan, T.-Q. Nguyen and A. J. Heeger, Polymer Homo-Tandem Solar Cells with Best Efficiency of 11.3%, *Adv. Mater.*, 2015, **27**(10), 1767–1773.
117. G. Dennler, M. C. Scharber, T. Ameri, P. Denk, K. Forberich, C. Waldauf and C. J. Brabec, Design rules for donors in bulk-heterojunction tandem solar cells-towards 15% energy-conversion efficiency, *Adv. Mater.*, 2008, **20**(3), 579–583.
118. N. Li, D. Baran, G. D. Spyropoulos, H. Zhang, S. Berny, M. Turbiez, T. Ameri, F. C. Krebs and C. J. Brabec, Environmentally Printing Efficient Organic Tandem Solar Cells with High Fill Factors: A Guideline Towards 20% Power Conversion Efficiency, *Adv. Energy Mater.*, 2014, **4**(11), DOI: 10.1002/aenm.201400084.

CHAPTER 3

High-performance Organic Photovoltaic Donor Polymers

ANDREW WADSWORTH*[a], DERYA BARAN[a,b], JEFFREY GORMAN[a] AND IAIN MCCULLOCH*[a,b]

[a]Department of Chemistry and Centre for Plastic Electronics, Imperial College London, London, SW7 2AZ, UK; [b]Physical Sciences and Engineering Division, KAUST Solar Center (KSC), King Abdullah University of Science and Technology (KAUST), KSC, Thuwal 23955-6900, Saudi Arabia
*E-mail: andrew.wadsworth11@imperial.ac.uk, iain.mcculloch@kaust.edu.sa

3.1 Introduction

With the global demand for energy projected to double by 2040, utilisation of the hugely abundant supply of solar energy must become more commonplace, both to meet our global energy needs and to mitigate the increased consumption of climate-harming fossil fuels.[1] First- and second- generation photovoltaic devices already provide high stability and efficiency; however, their energy-intensive manufacture means initial operational lifetimes of one to four years are needed to counterbalance the emissions used to assemble them. This high-energy fabrication limits cost reduction,[2] but the cost of photovoltaic technologies must become competitive with fossil fuels if their use is to become more widespread. Organic photovoltaic (OPV) modules provide an alternative, novel, low-cost method for harvesting solar energy. These modules use conjugated, plastics-based semiconductors and can be

Nanoscience & Nanotechnology Series No. 45
Nanostructured Materials for Type III Photovoltaics
Edited by Peter Skabara and Mohammad Azad Malik

Published by the Royal Society of Chemistry, www.rsc.org

coated in thinner layers, on flexible substrates, with cheaper roll-to-roll printing methods. It is possible to have operation times as low as a few months before offsetting their energy payback, further reducing the cost of photovoltaic technologies.[3]

3.1.1 Photophysics in OPV Devices

3.1.1.1 Organic Semiconductors and Light Absorption Properties

Organic semiconductors transfer charge through delocalised π-orbitals. The difference between the highest occupied molecular orbital (HOMO, π) and the lowest unoccupied molecular orbital (LUMO, π*) directly affects the energy of the semiconductor bandgap (E_g) (Figure 3.1a). E_g plays an important part in the amount of solar flux harvesting that can be achieved. Importantly, the energy levels of the HOMO and LUMO can be modified by chemical changes to the organic semiconducting materials. Changes to their optical and electrical properties are therefore relatively simple.

Contemporary OPV devices use a blended ~100 nm layer of two different organic semiconductors sandwiched between two electrodes (Figure 3.1b). Organic semiconductors are defined as electron donors and electron acceptors. Organic acceptor molecules have deeper HOMO and LUMO energy levels than their donor counterpart (Figure 3.1a). The difference in LUMO levels (ΔLUMO) plays a pivotal part in charge separation.

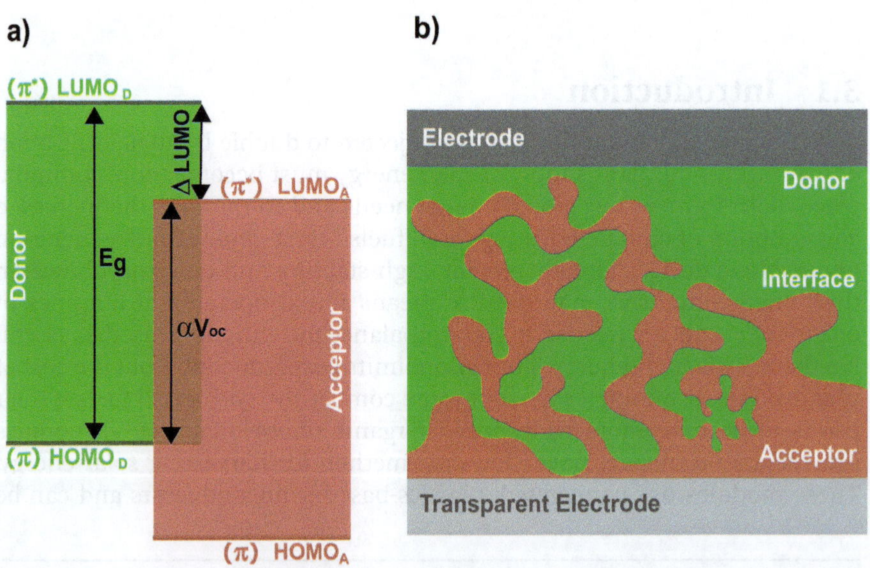

Figure 3.1 Semiconductor system consisting of two organic components with an electron donor (green) and an electron acceptor (red). (a) HOMO and LUMO levels of an organic semiconductors. (b) Blended donor and acceptor active layer, connected to electrodes.

A photoexcitation energy $E \geq E_g$ promotes an electron from the HOMO level to the LUMO level on the donor and leaves behind a positive charge (hole) in the HOMO, a π–π* transition. This π–π* electron transition across E_g occurs *via* a direct E_g in organic semiconductors. First- and second-generation photovoltaics undergo indirect E_g processes. As a result, such devices require micrometre-thick layers, which must be very pure to reduce defect-assisted recombination losses. OPV device layers are in the nanometre range and therefore have less stringent purity requirements, giving thinner and more transparent absorbing layers.[4]

3.1.1.2 Frenkel Excitons and the Photovoltaic Process

Figure 3.2 outlines the photovoltaic process in a donor–acceptor OPV system. In the active layer of the device, a π–π* transition stimulated by photoexcitation causes the formation of an electron–hole pair (step 1). However, the opposing charges of the photogenerated electron and hole have a mutual coulombic attraction. The electron–hole pair attractive species is denoted as an exciton and the energy of this attraction can be described by eqn (1.1):

$$V = \frac{e^2}{4\pi\varepsilon_0\varepsilon_r r_{eh}} \quad (3.1)$$

where e is the electron charge, ε_0 is the vacuum permittivity, ε_r is the dielectric constant and r_{eh} is the electron–hole separation distance. Importantly, the nature of this photoexcited excitonic state differs substantially between OPV

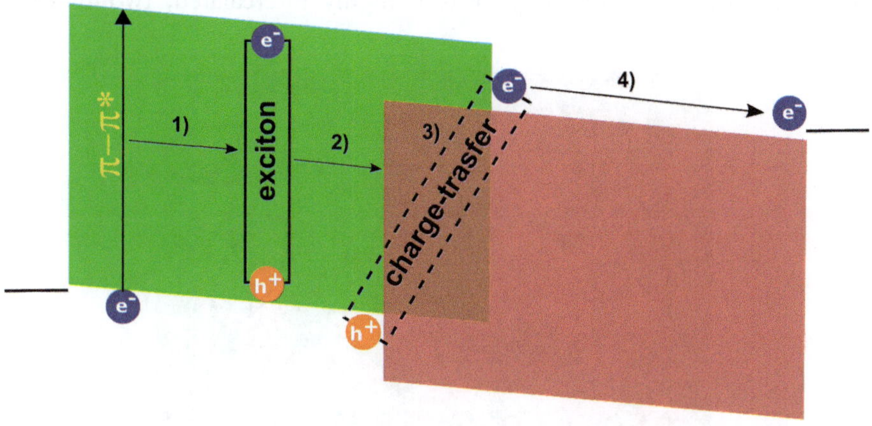

Figure 3.2 Photovoltaic process in OPV devices. Step 1: photoexcitation of ground state electron and exciton formation. Step 2: exciton diffusion to the interface. Step 3: electron transfer and charge transfer state formation. Step 4: diffusion of an electron to the electrode and hole diffusion (not shown).

devices and previous generations of photovoltaic devices.[4] The dielectric constant (ε_r) of organic semiconductors is lower than that of Si and other Group III–V semiconductors. Therefore in OPVs the electron–hole pair remains tightly bound as a Frenkel exciton, where the electron–hole pair is localised on a single molecule. Early OPV devices, which utilised only one organic semiconductor, had poor photocurrents due to the tight binding of the Frenkel exciton.[5] Here, excitons rarely separate into free electrons and holes at room temperature. Tang overcame this by using two different organic semiconductors in a bilayer device.[6] Using a donor and acceptor, the Frenkel excitonic state is overcome and electron transfer to the acceptor can take place.

The attractive force between the hole and electron remains, however, even after the electron is transferred to the acceptor (step 3). The electron and hole are now localised on separate, but neighbouring, donor and acceptor molecules. This bimolecular attractive species is denoted as a charge transfer state.[7,8] The charge transfer state may dissociate into separated charges if the hole–electron pair distance can overcome their mutual coulombic capture radius, r_{eh}, from eqn (1.1).[9] Once the charge transfer state is overcome, the charges are fully separated and can be extracted (step 4) (Figures 3.2 and 3.3).

The low mobilities in organic semiconductors limit the diffusion distance of excitons to approximately 10 nm, many times lower than the active layer thicknesses.[10,11] Therefore, during its lifetime, the exciton must be able to diffuse and reach the donor–acceptor interface.

3.1.1.3 Bulk Heterojunctions and Nanoscale Morphology

The surface area of the donor–acceptor interface was drastically enhanced by blending two semiconductors in the same solution. The resulting solution-processed morphology was a highly intercalated, two-material,

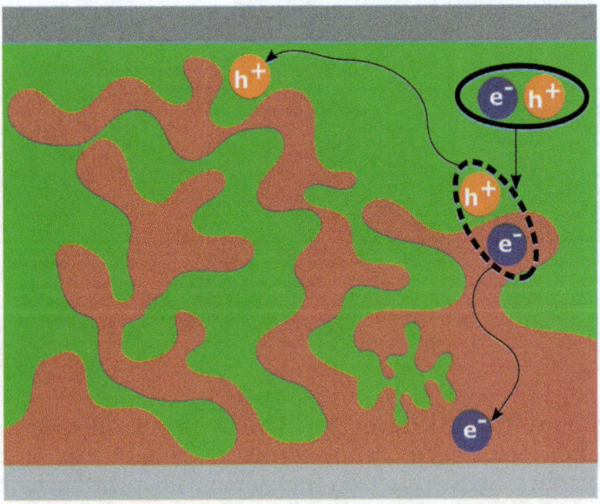

Figure 3.3 Bulk heterojunction with donor (green) and acceptor (red) species showing photovoltaic steps 2–4 described in Figure 3.2.

nanoscale semiconductor layer denoted as a bulk heterojunction (BHJ). The BHJ improved charge separation, the photocurrent and the power output.[12] The highly intercalated BHJ increased the interfacial surface area, which improved the probability of photoabsorption and exciton formation within the exciton diffusion distance (L_D) of a donor–acceptor interface:

$$L_D = \sqrt{\mu t} \tag{3.2}$$

where L_D is the exciton diffusion length, μ is the exciton mobility and t is the exciton lifetime. A downside to the BHJ is that, during the extraction of the separated charge, the probability of a separated hole encountering a separated electron is higher than in the bilayer device. This increases the probability of non-geminate recombination and device efficiency loss.

Even if the donor and acceptor materials have excellent frontier molecular orbital (FMO) energy level alignment, they must form a near-ideal morphology to produce a respectable power output. The complex, nanostructured morphology of an OPV device requires a fine balance of intermixed donor–acceptor domains for exciton dissociation and purer crystalline domains for charge extraction.[13,14] A vast number of solvents, additives, solution concentrations, donor to acceptor mass ratios, deposition techniques and post-deposition annealing treatments must be tested to optimise new materials and generate the highest possible power conversion efficiency (PCE). The generally accepted design of the BHJ is to have to interpenetrating networks of both the donor and acceptor, with widths similar to the exciton diffusion length of ~10 nm.[10,11] Continuous pathways to the corresponding electrode must also be present to extract free charges without recombination

3.1.2 Device Architectures

The organic BHJ layer is sandwiched between two electrodes. Indium-doped tin oxide (ITO) is typically used as the transparent electrode and is coated on a glass or plastic substrate to provide a route for incident light to reach the active layer. The counter electrodes are usually Al or Ag, depending on the device architecture. Single-layer devices are built in either conventional (Figure 3.4a) or inverted (Figure 3.4b) configurations. In a conventional

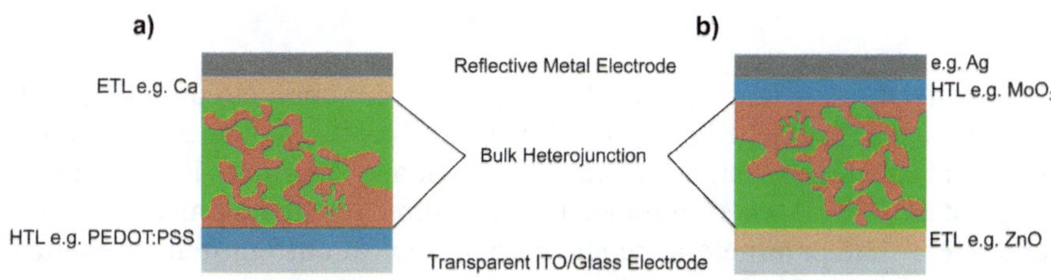

Figure 3.4 (a) Conventional architecture solar cell. (b) Inverted architecture solar cell.

device, the ITO electrode acts as the anode and extracts the holes. A lower work function reflective Al metal cathode extracts the electrons. To improve charge injection between the photoactive organic blend layer and the electrodes, additional semiconducting layers of intermediate energy levels are used. For example, poly(3,4-ethylenedioxythiophene):polystyrene sulfonate (PEDOT:PSS) or MoO_3 are typically used to improve hole injection and electron blocking, with ZnO or Ca for electron extraction. It must be noted that stability issues arise in the conventional device architecture. First, the ITO–PEDOT:PSS interface is unstable as a result of the acidic nature of PEDOT:PSS and water-soluble PEDOT:PSS is very sensitive to humidity and the ambient conditions.[15,16] Second, the Al cathode has a low work function (~4–4.2 eV) and hence is easily doped and degraded by water and oxygen.[17] Inverted devices utilise a more stable, high work function metal such as Ag, with ITO acting as the cathode.

One of the advantages of using organic semiconductors in solar cells is that they have strong absorptions due to their π–π* transitions. However, as the intensity of these transitions varies with different photon energies, they have an absorption maximum and cannot cover the whole visible to near-IR spectrum. Various strategies have matured during the last decade to overcome the absorption limitation of organic materials for BHJ solar cells.[18] For instance, low bandgap polymers converting the near-IR photons that were not previously absorbed have been utilised in ternary organic solar cells.[19,20] The same approach used to add absorbing acceptor molecules rather than fullerenes in a binary blend enhanced absorption in the visible region for higher photocurrents in binary and ternary solar cells.[21–23] Alternatively, multi-junction organic solar cells have been used as a key strategy to increase the open circuit voltage (V_{OC}) or short circuit current (J_{sc}) of single-layer solar cells by using series or parallel connections with efficient interlayers.[24–26]

3.1.3 Organic Photovoltaic Semiconductors

A number of parameters need to be optimised to improve the PCE:

$$\mathrm{PCE} = \frac{P_{\mathrm{out}}}{P_{\mathrm{in}}} = \frac{J_{\mathrm{sc}}\,V_{\mathrm{oc}}\,\mathrm{FF}}{P_{\mathrm{in}}}$$

where

$$\mathrm{FF} = \frac{J_{\mathrm{max}}\,V_{\mathrm{max}}}{J_{\mathrm{sc}}\,V_{\mathrm{oc}}} \tag{3.3}$$

J_{SC} is the short circuit current density, V_{OC} is the open circuit voltage, P_{in} is the power of incident light and FF is the fill factor. J_{max} and V_{max} are the maximum current density and maximum voltage. The judicious chemical design of donor and acceptor molecules is required to enhance all these target parameters. The parameters are shown on a typical current density–voltage (J–V) curve for an OPV device in Figure 3.5.

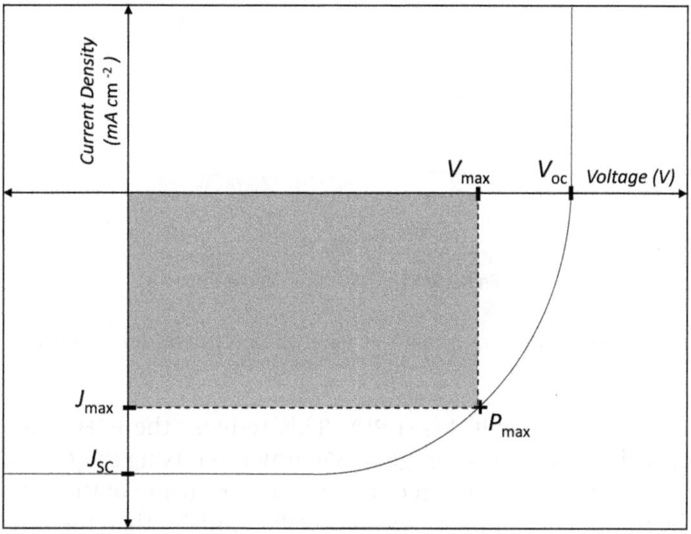

Figure 3.5 *J–V* characteristic curve showing J_{SC}, V_{OC}, J_{max} and V_{max}.

3.1.3.1 Donor Materials

Electron-rich polymers are usually the electron-donating material in the BHJ. Early donor polymers used single aromatic repeat units, typically a phenyl- or thiophene-based repeat unit, to form a homopolymer. These monomers were functionalised with an alkyl or alkoxy side-chain to allow solution processability in common solvents such as chloroform and chlorobenzene. Donor polymers are usually the photon-absorbing material, hence the ability of the donor to absorb an optimum amount of the UV–visible solar spectrum is vital. The E_g of the polymer is inversely proportional to the maximum wavelength the material can absorb, so typically lower E_g donors yield a higher J_{SC}. Much effort has been made to reduce the donor E_g. This can be achieved by chemical tuning of the structure to modify the HOMO and LUMO levels, or by increasing the conjugation length along the backbone. This improves spectral overlap and generally translates to an increased J_{SC} and PCE. Further reduction of the E_g to below 1.9 eV is impractical for homopolymers, however, due to steric hindrance from the neighbouring aromatic units and side-chains inhibiting continuous conjugation and often resulting in a low J_{SC} and PCE.[27,28]

An early example of a donor polymer, poly[2-methoxy-5-(2-ethylhexyloxy)-1,4-phenylenevinylene] (MEH-PPV), utilises a homocoupled vinylphenyl repeat unit with side-chains to improve solubility (Figure 3.6). The wide 2.2 eV E_g of MEH-PPV limits absorption to photons up to ~600 nm.[29] Hence BHJ devices containing MEH-PVV typically produce a low J_{SC} from poor harvesting of the solar spectra. The PCE and J_{SC} were higher, however, for poly(3-hexylthiophene-2,5-diyl) (P3HT). P3HT benefits from a stronger quinodal character and lower steric repulsions of a thiophene homocouple compared with the

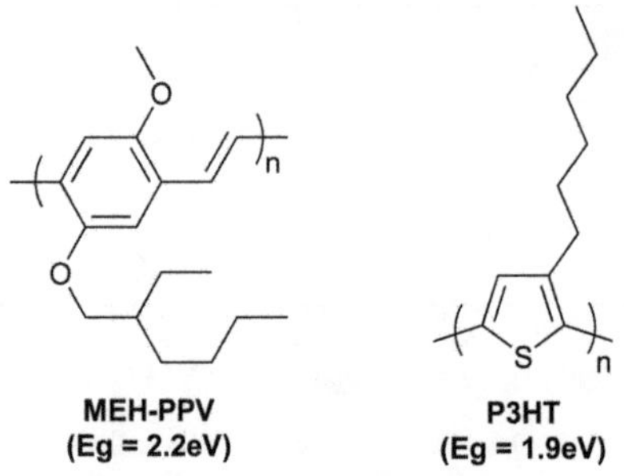

Figure 3.6 Examples of homopolymers used as electron donors in OPV devices.

phenyl-based repeat unit in MEH-PPV. This reduces the P3HT bandgap to 1.9 eV, improving spectral coverage to 650 nm and raising the J_{SC} and PCE.[30] Another important aspect of E_g reduction is the extension of the conjugation length. The thiophene couple is less sterically repulsive than the phenyl couple, the backbone is more planar and conjugation extends further – reducing E_g. The high regioregularity of the **P3HT** backbone enhances self-assembly in the BHJ and improves the PCE.[31]

About 70% of incident sunlight is within the range 350–900 nm, corresponding to an E_g of ~1.5 eV.[32] Further reductions in E_g must be achieved by raising the HOMO, as the retention of the LUMO–LUMO offset (~0.3 eV) is necessary for exciton dissociation. Lowering E_g also reduces the difference between the acceptor's LUMO and the donor's HOMO, which is directly proportional to V_{OC}. V_{OC} is also of vital importance to the output power and therefore a compromise must be made between optimising J_{SC} and V_{OC}.[33] Further bandgap narrowing requires the coupling of electron-deficient and electron-rich co-monomers to instigate intermolecular charge transfer. This reduces the bandgap, with J_{SC} values >20 mA cm^{-2}, which is otherwise unachievable with homopolymers.[34]

The planar conjugated segments along the polymer chain facilitate intramolecular π–π stacking interactions, which improve crystallinity and charge transport across the photoactive layer. The intrinsic disorder of polymer chains, the mixed morphology of the two-component blend and the presence of alkyl chains impede the π–π stacking ability of the polymer. The alkyl chain effect can be exploited and altered to more finely tune the π–π stacking, allowing for some control over the donor and acceptor domain size.[35,36] The alkyl chains also play a vital part in solution processability, where lengthier or branched chains improve solubility and processability, but sterically limit π–π stacking.

3.1.3.2 Acceptor Materials

Functionalised C_{60} fullerenes are the dominant n-type material in OPV blend devices. With the LUMO delocalised over the entire 3D conjugated cage, C_{60} has ideal isotropic transport and π–π stacking ability. Solution processability

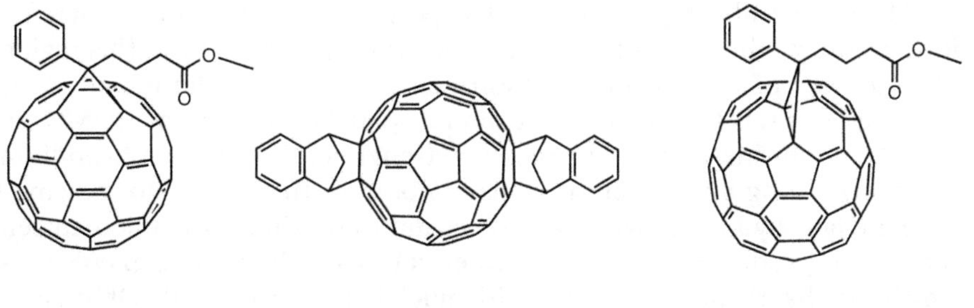

$PC_{60}BM$ **ICBA** **$PC_{70}BM$**

Figure 3.7 Examples of the frequently used fullerene-based acceptor molecules in OPV donor–acceptor blends.

was improved in the seminal work of Wudl in the synthesis of $PC_{60}BM$ by functionalisation with a solubilising ester chain (Figure 3.7).[12,37] Despite the near-ideal transfer properties, the cage-localised LUMO level is difficult to perturb and improve, limiting V_{OC}. The depth of the acceptor's LUMO level dominates V_{OC}, hence much effort has been made to raise the LUMO level. The addition of a second side-chain in bis-$PC_{60}BM$ raises the LUMO by 0.1 eV relative to monoadduct $PC_{60}BM$ and this results in a larger V_{OC} (~100 mV).[38] However, the additional steric bulk inhibits π–π packing and the PCE. Li *et al.* first developed an indene–C_{60} bis-adduct (ICBA), where the addition of two indacene units raised the LUMO energy by 170 mV relative to $PC_{60}BM$.[39,40] ICBA currently represents the near state-of-the-art fullerene blend V_{OC}, with P3HT blends at 840 mV.

Unfortunately, the high symmetry of the C_{60} cage inhibits photon harvesting in the UV–visible range. However, the less symmetrical C_{70} cages allow more optical π–π* transitions in the UV–visible spectrum, leading to an enhanced photocurrent in devices blended with $PC_{70}BM$.[41] The development of a number of non-fullerene, small molecular acceptors has recently been realised; these materials benefit from greater solar harvesting to enhance the photocurrent and more accessible FMOs to improve V_{OC}.[42–45]

3.2 Design Considerations for Donor Polymer Materials

There are several key considerations when designing new high-performance donor polymer materials for OPV applications. The absorption of light, molecular energy levels, charge carrier mobility, blend morphology and solubility must all be taken into account. These properties have an important impact on key device properties such as the J_{SC}, V_{OC} and fill factor, thereby strongly influencing the PCE a device can achieve, as well as affecting the ease and cost of device fabrication.

Fullerene derivatives (*e.g.* $PC_{60}BM$, $PC_{70}BM$ and ICBA) are among the most commonly used acceptors in polymer solar cells and absorb light relatively weakly in the visible region of the solar spectrum (400–900 nm).[46] Therefore it is important that the donor polymer absorbs light strongly in this region. The strength of the polymer's absorption, indicated by the molar extinction coefficient, is largest at the maximum spatial overlap of the HOMO and LUMO, thereby increasing the oscillator strength of the polymer. In addition to having strong absorption, it is also important that the donor polymers have a relatively broad absorption spectrum to maximise the photocurrent that can be produced. This can be achieved by using low bandgap polymers, which can absorb photons of light with much longer wavelengths. It has been shown that extending the absorption profile in this way increases the fraction of photons that can be absorbed and therefore increases the current density that the active layer can produce.[47] Polymers that absorb strongly over a large fraction of the visible region of the solar spectrum can absorb more photons; this leads to the creation of more excitons in the polymer–fullerene blend and therefore larger photocurrents (J_{SC}) can be achieved, leading to higher efficiencies in devices.

The relative energies of the FMOs of the polymer also play a key part in device performance. As can be seen from Figure 1a, the V_{OC} a device can achieve is directly proportional to the difference in energy between the HOMO of the donor polymer and the LUMO of the fullerene acceptor; hence, to maximise the V_{OC}, the donor polymer must have a relatively deep-lying HOMO. The energy of the LUMO is also very important and for facile charge separation at the donor–acceptor interface it is generally accepted that an offset in the LUMO energies (ΔE_{LUMO}) of ~0.3 eV is needed.[48] From the measured electron affinity (closely related to the LUMO energy) values for $PC_{60}BM$ and $PC_{70}BM$ (~3.7 eV),[49,50] it is clear that donor polymers should have a LUMO of about −3.4 eV to facilitate efficient electron transfer to the acceptor. However, because the HOMO and LUMO energies also dictate the bandgap of the polymer, and therefore its absorption spectrum, it is not always possible to maximise the V_{OC} with a deep-lying HOMO and maintain a broad absorption profile for the polymer. Thus a compromise must be reached between V_{OC} and E_g because the ΔE_{LUMO} offset must be maintained for charge separation to occur and the device to function.

For the polymer–fullerene solar cells to be solution-processable, a key advantage of OPV technologies over inorganic photovoltaics, donor polymers must exhibit good solubility in common organic solvents such as chloroform (CF), chlorobenzene (CB) and dichlorobenzene (DCB). Polymers that have good solubility in non-chlorinated solvents such as toluene and xylene are less common, however; moving forwards, this may become important for the industrial viability of this technology. The donor polymers are highly conjugated and therefore have very rigid backbones. In addition, the conjugated polymers are susceptible to strong π–π stacking interactions that lead to a strong tendency to aggregate. As such, these polymers exhibit poor solubility without the inclusion of pendant alkyl or alkoxy chains.

The morphology of the donor–acceptor blend is also of great importance. The excitons that form upon light absorption must reach an interface within their exciton diffusion length (~10 nm) to avoid relaxation back to the ground state before charge separation has occurred.[51] Therefore, to maximise the fraction of absorbed photons that contribute to the device photocurrent, the polymer–fullerene blend must consist of nanoscale domains. Conjugated polymers have a rigid backbone and strong π–π interactions, which can result in the formation of micrometre crystalline domains in the blend. This must be avoided for respectable efficiencies to be achieved; the use of side-chains and the introduction of twists into the polymer backbone are often used to reduce the aggregation tendency of donor polymers and to tune the morphology of the blend. If the donor polymer is too amorphous, then molecular mixing with the fullerene acceptors will be seen in the blend, leading to large recombination losses in the active layer. The morphology of the active layer can also be influenced by the introduction of additives such as 1,8-diiodooctane (DIO) during the processing of the active layer. It has been shown that blend morphology is closely related to the fill factor of a device,[52] thus improving the morphology can lead to great improvements in the PCE.

Losses due to charge recombination can drastically reduce the performance of a solar cell. This can be mitigated to an extent by balancing the hole mobility (μ_h) of the polymer donor with the electron mobility (μ_e) of the fullerene acceptor. $PC_{60}BM$ has a μ_e of ~10^{-3} cm^2 V^{-1} s^{-1},[53] therefore the donor polymers should have a μ_h at least equal to this value to reduce recombination in the active layer of the device. The charge carrier mobility is heavily dependent on the ability of charges to easily hop between polymer chains. As such, the packing exhibited by the polymer and the crystallinity within the donor-rich domains of the BHJ are of great importance, with a high degree of order correlating to greater charge carrier mobility. A high molecular weight (M_w) also leads to improved mobility; although this is not easy to incorporate into the design, it should be considered when optimising the polymerisation conditions.

3.3 High-efficiency Donor Polymers

3.3.1 Introduction to Push–Pull Copolymers

A key feature in designing new high-performance donor polymers is ensuring that a low bandgap is achieved. This is most commonly attained through the use of the push–pull effect exhibited when both electron donor and electron acceptor units are included in the polymer structure, often separated by a π-conjugated spacer. The hybridisation that occurs when the molecular orbitals of the donor and acceptor units interact (Figure 3.8), leads to a narrowing of the bandgap of the polymer. Primarily, the electron-rich donor unit influences the resulting HOMO of the copolymer; the LUMO largely depends on the electron-deficient acceptor unit. In addition to narrowing the bandgap, this also allows for rather facile tuning of the polymer FMOs by modification or substitution of the donor and acceptor units. There are a large number of possible donor and acceptor units that can be utilised in

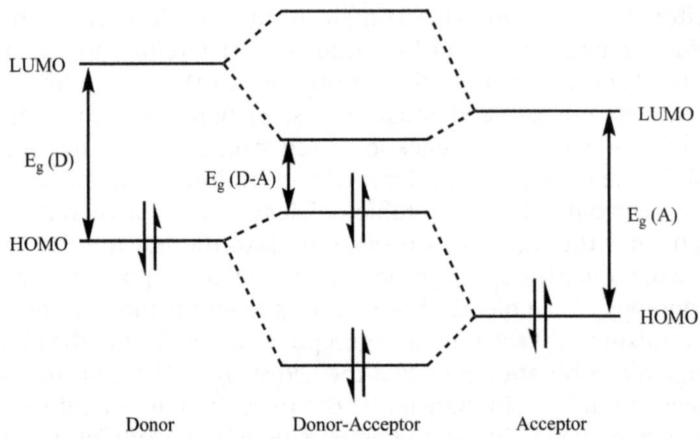

Figure 3.8 Molecular orbital interactions of the donor and acceptor units in a push–pull copolymer.

the copolymers and thus a huge variety of polymer materials are possible. Increasing the conjugation in a polymer is another effective method of reducing its bandgap. This is usually achieved by forcing the polymer backbone into a more planar conformation by reducing the steric hindrance between adjacent units.

Many of the high-performing push–pull donor copolymers are presented in the following sections and their performance is discussed in relation to their structures. The copolymers are classified by the type of donor moiety incorporated into the polymer structure.

3.3.2 Push–Pull Copolymers Based on Thiophenes, Fused Thiophenes and Their Derivatives

Thiophenes are among the simplest donor units used. However, this does not condemn the push–pull copolymers that make use of thiophene units to a modest performance in OPV devices. Thiophene and its analogues are reasonably strong donor moieties. In addition, the reduced steric interactions of the α-hydrogens in neighbouring thiophenes relative to phenyl-terminated donor units allow a greater degree of planarisation in polymers. This leads to the possibility of achieving very low bandgap polymers by increasing the conjugation in the system. Variations of thiophene-based acceptors can be produced by substituting an S atom for other group 16 atoms such as Se and Te. The larger group 16 atoms lead to a decrease in the aromaticity of the donor unit because the bonds are lengthened; this increases the quinoidal character of the donor moiety. As a result, the HOMO is raised and the LUMO is lowered, effectively reducing the bandgap of the polymer. As Figure 3.9 shows, thiophenes can be paired with a wide range of acceptor units, which gives several opportunities to tune the structural and optoelectronic

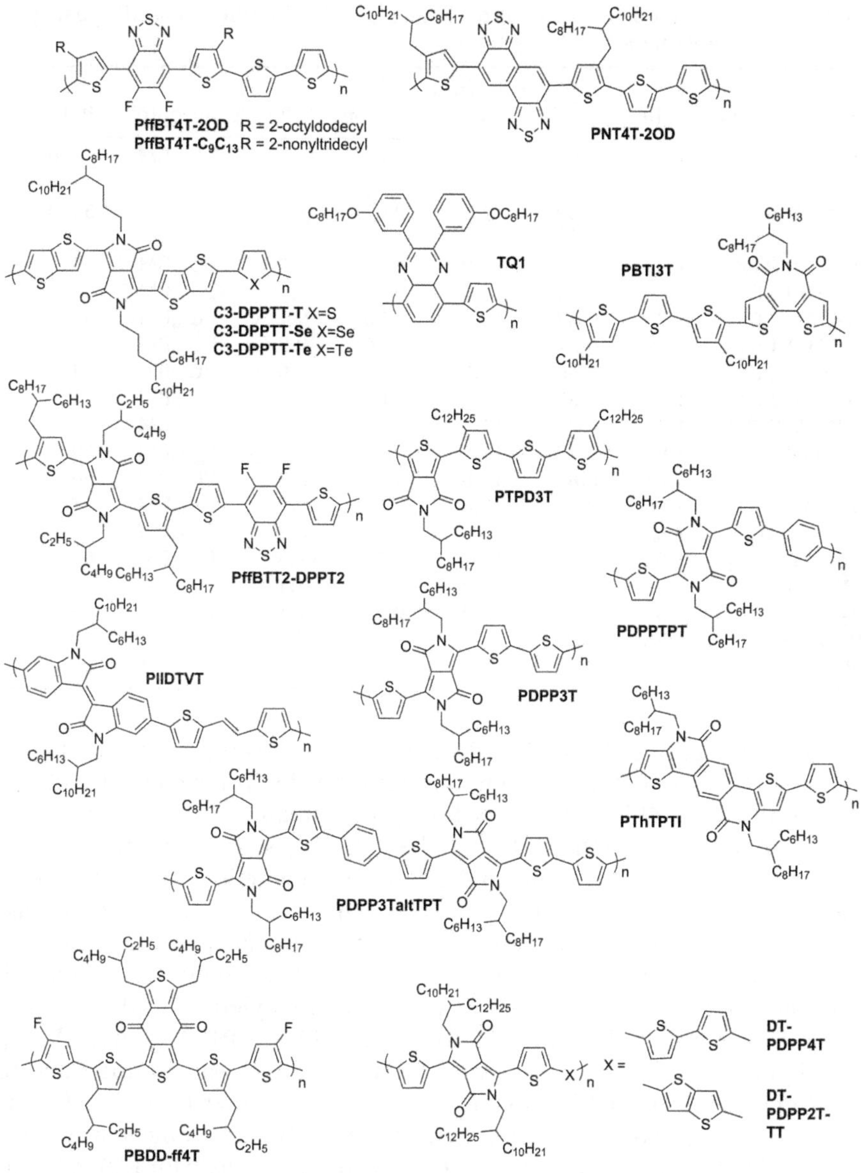

Figure 3.9 Thiophene-based push–pull donor copolymers.

properties of the polymers to better suit them to high-performance devices. Diketopyrrolopyrrole (DPP) and benzothiadiazole (BT) are among the most common acceptor moieties used. Many of these push–pull coploymers are summarised in Table 3.1.

PDPP3T, first reported by Janssen *et al.*,[54,55] was able to achieve a notably low bandgap (1.33 eV) through the choice of monomers utilised. A strong

Table 3.1 *J–V* characteristics and bandgap of thiophene-based push–pull copolymers.

Donor polymer	E_g (eV)	V_{OC} (V)	J_{SC} (mA cm^{-2})	Fill factor	PCE (%)	Acceptor	Ref.
PDPP3T	1.33	0.67	15.4	0.69	7.1	$PC_{70}BM$	54 and 55
PDPPTPT	1.53	0.80	14.0	0.67	7.5	$PC_{70}BM$	54 and 56
PDPP3TaltTPT	1.43	0.75	15.9	0.67	8.0	$PC_{70}BM$	54
DT-PDPP4T	1.43	0.64	16.0	0.69	7.1	$PC_{70}BM$	57
DT-PDPP2T-TT	1.39	0.67	20.1	0.70	9.4	$PC_{70}BM$	58
C3-PDPPTT-T	1.39	0.57	23.5	0.66	8.8	$PC_{70}BM$	34
C3-PDPPTT-Se	1.37	0.56	21.5	0.63	7.6	$PC_{70}BM$	34
C3-PDPPTT-Te	1.32	0.52	21.7	0.63	7.1	$PC_{70}BM$	34
PffBTT2-DPPT2	1.43	0.81	17.3	0.60	8.4	$PC_{70}BM$	59
PTPD3T	1.82	0.79	12.3	0.79	7.7	$PC_{70}BM$	60
PBTI3T	1.81	0.85	12.8	0.76	8.3	$PC_{70}BM$	60
PffBT4T-2OD	1.65	0.77	18.8	0.75	10.9	$PC_{70}BM$	61
PNT4T-2OD	1.63	0.76	19.8	0.68	10.2	$PC_{70}BM$	61
PffBT4T-C_9C_{13}	1.65	0.78	19.8	0.73	11.3	$PC_{70}BM$	62
PIIDTVT	1.61	0.91	13.2	0.59	7.1	$PC_{60}BM$	63
TQ1	1.37	0.91	12.2	0.64	7.1	$PC_{70}BM$	64
PThTPTI	1.78	0.87	13.7	0.66	7.9	$PC_{70}BM$	65
PBDD-ff4T	1.74	0.95	13.2	0.73	9.2	$PC_{70}BM$	66

push–pull effect was afforded by the use of the electron-poor pyrrolo[3,4-*c*]pyrrolo-1,4-dione (DPP) accepting unit and the electron-rich terthiophene (3T) donor unit. This polymer is highly planar due to the lack of steric clashes of protons α to the DPP–thiophene linkages; as such, the extended conjugation that results from this acts to further reduce the bandgap. A reasonably high J_{SC} of 15.4 mA cm^{-2} was achieved using PDPP3T, at least in part due to the narrow bandgap of this polymer. However, the PCE (7.1%) was limited by the modest V_{OC} (0.67 V) that could be achieved. This was addressed by the use of thiophene-phenylene-thiophene (TPT) as the donor unit rather than 3T to produce PDPPTPT. Although the J_{SC} was slightly lower in this case, a higher V_{OC} (0.80 V) was attained due to the deeper lying HOMO in this polymer. Hence a higher PCE of 7.5% was achieved. In this work, the importance of the polymers' molecular weight was also highlighted. The device PCEs reported previously for PDPP3T and PDPPTPT were 4.7 and 5.5%, respectively,[55,56] and the dramatic improvement in the PCE reported more recently was a result of an increase in the molecular weights of the polymers.[54]

PDPP3TaltTPT, an interesting terpolymer making use of DPP, 3T and TPT units, was also reported, where the donor units from both PDPP3T and PDPPTPT were included to tune the optical bandgap and FMOs.[54] The inclusion of both 3T and TPT donor units, *via* a D_1–A–D_2–A strategy, produced a terpolymer that has an intermediate HOMO energy level, relative to PDPP3T and PDPPTPT, and therefore an intermediate bandgap (1.43 eV). The result of this is that PDPP3TaltTPT is able to produce a V_{OC} of 0.75 V in devices, while still maintaining a high J_{SC}. By tuning the FMOs it was possible to reach a compromise of V_{OC} and E_g to produce a terpolymer that achieved an 8.0% PCE in devices, which was an improvement on the parent copolymers from which it was designed.

DT-PDPP4T is both structurally and electronically very similar to PDPP3T, but includes a quaterthiophene (4T) donor unit rather than the 3T unit. The slightly improved PCE of 7.1% was likely to be due to a more favourable morphology in the active layer blend. Transmission electron microscopy showed that DT-PDPP4T formed thinner fibrils in the blend, which has been correlated with higher external quantum efficiencies and hence an improved performance for these thiophene-based DPP polymers.[57] DT-PDPP2T-TT yielded a remarkable PCE of 9.4% in devices. The main reason for this was the ability to form very thick active layers when using this polymer (>300 nm) while still maintaining a high fill factor of 0.70.[58] A consequence of being able to produce such thick active layers is that an impressive J_{SC} of 20.1 mA cm^{-2} could be achieved.

McCulloch *et al.* reported a series of planar copolymers consisting of DPP flanked on either side by thienothiophene (TT) copolymerised with different chalcogenophene units.[34] C3-DPPTT-T, C3-DPPTT-Se and C3-DPPTT-Te were all able to achieve J_{SC} values >20 mA cm^{-2} and the thiophene-containing polymer was able to achieve a notable PCE of 8.8%. It was found that as the size of the chalcogen atom increased (from S to Se to Te), the aromaticity of the chalcogenophene unit decreased, leading to an increase in quinoidal character. This raised the HOMO and lowered the LUMO energy, thereby narrowing the bandgap of the polymers, which was shown by a red shift in the absorption spectra. Another result of this was the decrease in V_{OC} from C3-DPPTT-T to C3-DPPTT-Te (from 0.57 to 0.52 V). The lower fill factor observed in devices using C3-DPPTT-Se and C3-DPPTT-Te can be attributed to an increased tendency for the polymer to aggregate in films. Although the selenophene- and tellurophene-containing copolymers had narrower bandgaps than C3-DPPTT-T, the reduced V_{OC} and fill factor that could be attained in devices led to slightly lower PCEs.

PffBTT2-DPPT2, reported by Yan *et al.*, is another thiophene-based terpolymer that makes use of the DPP acceptor unit.[59] This polymer includes DPP and 5,6-difluoro-2,1,3-benzothiadiazole (ffBT) acceptor units along with bithiophene donor units in a D–A_1–D–A_2 configuration. Despite the narrow bandgap exhibited by this polymer, a V_{OC} of 0.81 V was still obtained. As such, the active layer blend benefits from both a high V_{OC} and J_{SC}, producing a PCE of 8.4% in devices. A result of the electron-withdrawing nature of the ffBT unit was that rather low-lying energy levels were observed for **PffBTT2-DPPT2**. This afforded a deep HOMO level, which produced a donor copolymer that exhibited good stability in air. The interaction of the electronegative fluorine atoms of ffBT with the electron-rich sulphur atoms of the thiophene units acts to planarise the backbone and led to strong aggregation tendencies for the polymer. Thus the inclusion of branched alkyl chains between neighbouring thiophene units was important to control the morphology of the active layer blend. It must be noted that devices using this polymer also exhibited excellent thermal stability, which is likely to be a result of the strong aggregation tendency of this polymer.

Marks *et al.* utilised a different approach to design high-performance, thiophene-containing, push–pull copolymers. Rather than focusing on creating low bandgap materials, the emphasis was instead directed onto obtaining a high fill factor in devices. **PTPD3T** and **PBTI3T** contain a thieno[3,4-*c*]pyrrole-4,6-dione (TPD) and a 2,2′-bithiophene-3,3′-dicarboxamide (BTI) acceptor unit, respectively, along with 3T as the donor in both polymers.[60] The highly planar nature of these polymers and the strategic placement of side-chains to avoid steric torsion allowed a high degree of order to be achieved, while still affording solution-processable polymers. This resulted in high charge carrier mobilities, which allowed outstanding fill factors of 0.79 and 0.76 to be obtained in devices for **PTPD3T** and **PBTI3T**, respectively. The J_{SC} values (~12 mA cm^{-2}) are fairly modest compared with many of the other state-of-the-art donor polymers. This is mainly due to the relatively wide bandgap of these polymers; however, high fill factors compensate for this and allow competitive PCEs (~8%) to be achieved.

Yan *et al.* reported some of the highest performing push–pull copolymers, all of which include 4T donor units. **PffBT4T-2OD** and **PNT4T-2OD**, reported in 2014,[61] include ffBT and naphtha[1,2-*c*:5,6-*c*′]bis[1,2,5]thiadiazole (NT) acceptor units, respectively. The bandgaps of these polymers (~1.6 eV) are not as narrow as many of the other higher performance donor copolymers due to the use of fewer electron-poor acceptor units and the slightly fewer planar polymer backbones compared with many of the thiophene-DPP based copolymers. However, high J_{SC} values (18.8 and 19.8 mA cm^{-2}) and fill factors (0.75 and 0.68) were achieved for **PffBT4T-2OD** and **PNT4T-2OD**, respectively, as a result of the inclusion of the 2-octyldodecyl (2OD) chains on the 4T unit. The 2OD chains afforded a temperature dependence on the aggregation and crystallisation of the donor polymer within the active layer blend, thereby allowing a degree of control over the morphology of the active layer. The result of this was the ability of these polymers to obtain hugely impressive PCEs in devices of 10.9 and 10.2%, respectively. A more recent development was the attainment of **PffBT4T-C$_9$C$_{13}$**, which is structurally very similar to **PffBT4T-2OD**, although the 2OD chains are replaced by 2-nonyltridecyl chains.[62] Although **PffBT4T-C$_9$C$_{13}$** performed slightly worse (9.3%) when using the CB : DIO (97.5 : 2.5) solvent system that had been used for **PffBT4T-2OD**, using a non-chlorinated 1,2,4 trimethylbenzene (TMB), 2.5% 1-phenylnaphthalene (PN) solvent system afforded a PCE of 11.3% for **PffBT4T-C$_9$C$_{13}$** compared with the 9.2% achieved using **PffBT4T-2OD**. The result of the slightly longer alkyl chains and the TMB:PN solvent system was to improve molecular orientation relative to the interfaces, slightly reduce the size of domains in the blend and improve the purity of the polymer domains, leading to an improved PCE.

PIIDTVT, reported by Jung and Jo, makes use of a thienylenevinylene (TVT) donor unit and isoindigo (IID) acceptor unit.[63] A deep-lying HOMO level allows a high V_{OC} (0.91 V) to be obtained, a result of the electron-poor nature of the isoindigo unit. Along with the notable V_{OC}, a reasonable J_{SC} (13.2 mA cm^{-2}) and fill factor (0.59) are also obtained, leading to a PCE of 7.1%.

TQ1 makes use of thiophene (T) and bis-phenylquinoxaline (Q) units to achieve a low bandgap polymer due to the push–pull effect.[64] This polymer achieves a high V_{OC} (0.91 V), as a result of the electron-deficient nature of the quinoxaline unit. Although the J_{SC} achieved is not particularly high (12.2 mA cm^{-2}), a high fill factor is seen (0.64) using 1-chloronaphtalene as an additive, which encouraged small-scale phase separation in the active layer blend. The result of a 7.1% PCE achieved by devices using this polymer is made more impressive by its simple nature, demonstrating a relatively cost-effective polymer that still affords a high PCE.

Ding *et al.*, reported an interesting thiophene-based push–pull copolymer **PThTPTI**, which combines thiophene (T) donor units with a pentacyclic fused lactam system: thieno[2′,3′:5,6]pyrido[3,4-*g*]thieno[3,2-*c*]-isoquinoline-5,11(4H,10H)-dione (TPTI).[65] Despite the moderate bandgap of this polymer, a J_{SC} of ~14 mA cm^{-2} was achieved. This is probably a result of good charge carrier mobilities, which is also suggested by the high fill factor (0.66). The polymer is planar and exhibits partial crystallinity, which leads to good charge carrier mobility. Again, a high V_{OC} was possible due to the deep-lying HOMO of the polymer. The 7.9% PCE that this polymer can achieve is among the highest for moderate bandgap copolymers and it could prove to be an interesting material in tandem solar cells.

PBDD-ff4T incorporates a partially fluorinated 4T donor unit and a benzodithiophene-4,8-dione (BDD) acceptor unit to yield a medium bandgap (1.74 eV) donor polymer.[66] The fluorination of the 4T unit acts to lower the HOMO, narrow the bandgap and lead to the reasonably crystalline nature of the polymer. As a result, a high V_{OC} (0.95 V) and an impressive fill factor (0.73) are possible, leading to a PCE of 9.2%.

3.3.3 Push–Pull Copolymers Based on Cyclopentadithiophene, Dithienosilole, Dithienogermole and Other Bridged Bithiophene Derivatives

Bridged bithiophene-type donor units have also yielded some success in push–pull copolymers. The inclusion of the bridging atom between two ring systems offers some interesting possibilities to tune the structural and optoelectronic properties of the polymers. The bridging atom offers a straightforward opportunity to include side-chains, affording control over the solubility, crystallinity and phase separation exhibited by the polymer. Larger group 14 elements, such as Si and Ge, form longer bonds with the thiophene units when included at the bridging position; this results in longer bonds between the bridging atom and the adjacent carbons, which pushes the bulky alkyl chains further from the π-conjugated backbone of the polymer and therefore allows stronger π–π stacking between polymer chains and increases the crystallinity of the polymer. Variation of the bridging atom between group 14 elements allows some control over the FMOs and E_g of the polymer. The longer bonds between the bridging atom and the thiophenes causes a greater separation of the anti-bonding lobes and consequently lowers the anti-bonding energy, stabilising the HOMO of the polymer. The σ* orbital of the silylene or germylene fragments can also undergo more effective mixing with the π* orbital of the conjugated backbone and this effectively stabilises the LUMO. As Figure 3.10 shows, bridged bithiophene donor moieties are commonly copolymerised with TPD or BT units; many of the highest performing copolymers are summarised in Table 3.2.

F-PCPDTBT, reported by Neher *et al.*,[67] consists of a cyclopenta[2,1-*b*:3,4-*b*′]dithiophene donor unit copolymerised with a monofluorobenzothiadiazole (fBT) unit to yield a relatively low bandgap polymer (1.45 eV). This polymer displays an improved V_{OC} compared with the unflourinated analogue (**PCPDTBT**)[68] as a result of the fluorine atom lowering the level of the HOMO of the polymer. It was also noted that geminate recombination was inhibited by the stabilisation of photogenerated holes away from the donor–acceptor interface. It has been suggested that this is a result of the increased order of the polymer domains in the blend due to the inclusion of the fluorine atom. **PDTSBT-F** is analogous in structure to **F-PCPDTBT**, with a Si bridging atom in place of the carbon atom in the bridged bithiophene unit: silylene-2,2′-bithiophene (DTS).[69] The main effect of this substitution is an increase in the crystallinity of the polymer, which increases the charge carrier mobility and reduces non-geminate losses in the blends. As such, a greater J_{SC} (16.0 mA cm^{-2}) is achieved, despite the very similar bandgap to **F-PCPDTBT**, and an improved PCE of 6.7% was achieved.

Yang *et al.* reported a copolymer using a slightly more unorthodox bridged bithiophene unit, dithieno[3,2-*b*:2′,3′-*d*]pyran (DTP),[70] which was copolymerised with ffBT to form **PDTP-DFBT**. The strongly donating DTP unit, which is a direct result of the electron-donating oxygen atom in the pyran ring, allows for a strong push–pull effect in this copolymer and therefore allows the low bandgap of 1.38 eV to be achieved. Another advantage of the DTP unit is that it displays improved solubility compared with the CPDT and DTS units, allowing for less bulky alkyl chains to be included. The low bandgap achieved by this polymer leads to a relatively high J_{SC} of 18.0 mA cm^{-2} and the inclusion of less sterically bulky solubilising chains is likely to allow for the fill factor of 0.63 to be obtained, which leads to the impressive 7.9% PCE.

The TPD unit has also been used extensively with bridged bithiophenes to create medium to low bandgap donor copolymers. Ding *et al.* first reported **PCPDTTPD-Oc**, a copolymer that incorporates a CPDT donor unit and a TPD acceptor unit, which was able to achieve a PCE of 6.5% in devices.[71] It was reported that the alkyl chain on the TPD unit played an important part in the blend morphology and solubility of the polymer. As a result of the crystalline nature of this polymer, longer (*n*-octyl) chains improved the morphology and therefore gave the optimum performance. This was followed up

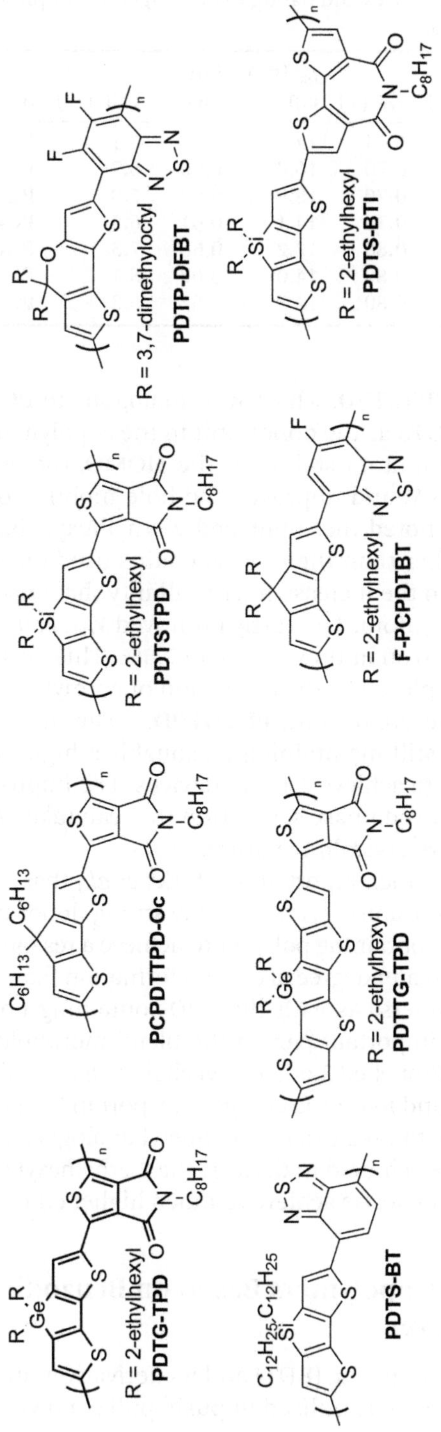

Figure 3.10 Bridged bithiophene based push–pull donor copolymers.

Table 3.2 *J–V* characteristics and bandgap of bridged bithiophene based push–pull copolymers.

Donor polymer	E_g (eV)	V_{OC} (V)	J_{SC} (mA cm^{-2})	Fill factor	PCE (%)	Acceptor	Ref.
F-PCPDTBT	1.45	0.74	14.1	0.58	6.1	$PC_{70}BM$	67 and 68
PDTSBT-F	1.48	0.70	16.0	0.60	6.7	$PC_{70}BM$	69
PDTP-DFBT	1.38	0.70	18.0	0.63	7.9	$PC_{70}BM$	70
PCPDTTPD-Oc	1.59	0.75	14.1	0.61	6.5	$PC_{70}BM$	71
PDTSTPD	1.73	0.88	12.2	0.68	7.3	$PC_{70}BM$	72
PDTGTPD	1.69	0.86	14.0	0.67	8.1	$PC_{70}BM$	73
PDTS-BTI	1.75	0.80	12.8	0.62	6.3	$PC_{70}BM$	74

by the synthesis of **PDTSTPD**, which was analogous to **PCPDTTPD-Oc**, but instead made use of DTS as the donor unit in the copolymer.[72] The inclusion of the bridging silicon atom stabilised the HOMO, thereby increasing the V_{OC} from 0.75 to 0.88 V, and improved the hole mobility of the copolymer. However, it must be noted that branched 2-ethylhexyl chains were needed on the DTS unit, rather than the *n*-hexyl chains used on the CPDT core in **PCPDTTPD-Oc**, due to the increase in crystallinity that arises from the inclusion of the Si bridging atom. The vastly improved V_{OC}, coupled with the high fill factor of 0.68, led to an improved PCE of 7.3%. This push–pull copolymer design was further explored by the utilisation of germylene-2,2′-dithiophene instead of DTS as the donor unit, **PDTGTPD**.[73] The improvement in J_{SC} to 14.0 mA cm^{-2}, while still maintaining reasonably a high fill factor and V_{OC}, allowed this polymer to achieve 8.1% in devices. The improved J_{SC} is likely to be as a result of improved phase separation that can take place in **PDTGTPD** due to its slightly more crystalline nature.

PDTS-BTI is a copolymer, reported by Marks *et al.*, that combines the DTS donor and BTI acceptor units.[74] A reasonably strong donor unit, such as DTS, is necessary for a BTI-containing polymer to achieve a reasonably low bandgap (1.75 eV), allowing a greater degree of exciton formation than is seen with many other BTI-based polymers. As with the TPD-containing polymers, the alkyl chains again play an important part in the blend morphology and therefore device performance. Branched 2-ethylhexyl chains on the imide nitrogen atom disrupt π–π stacking and lower the charge transport in the polymer, and therefore the J_{SC} and fill factor, relative to the *n*-octyl analogue. This is manifested in the PCE that can be achieved in devices; the 2-ethylhexyl analogue achieves 5.5% and the *n*-octyl analogue achieves a much higher efficiency of 6.3%.

3.3.4 Push–Pull Copolymers Based on Benzodithiophene and Its Derivatives

Benzo[1,2-*b*:4,5-*b*′]dithiophene (BDT) and its derivatives are among the most successful donor units to be utilised in push–pull copolymers, with many of the highest performing donor polymers incorporating these moieties into their structures. BDT is very different from many of the typical donor units: it consists of two thiophene units fused directly to one benzene ring without bridging atoms. The lack of bridging atoms means there is no obvious point of attachment for the solubilising alkyl chains and fewer opportunities to tune the energy levels of polymers that incorporate BDT units. It is clear from Figures 3.11 and 3.12 that thieno[3,4-*b*]thiophene is one of the most common acceptor moieties used with BDT; however, other units such as TPD and BT have also been utilised.

PBDTTT-C, reported by Li *et al.*, is a copolymer that incorporates the BDT donor and thieno[3,4-*b*]thiophene moieties into its structure.[75] Among the first high-performance BDT-based copolymers to be reported, PBDTTT-C was able to achieve a PCE of 6.6% in devices due to its reasonably high J_{SC} (14.7 mA cm^{-2}) and fill factor (0.64). However, the V_{OC} that could be achieved with $PC_{70}BM$ was only 0.70 V and limited the efficiency that could be attained using this polymer. As such, a fluorine atom was added to the thieno[3,4-*b*] thiophene unit, PBDTTT-CF.[75] This lowered the HOMO and LUMO of the polymer by ~0.1 eV and allowed a higher V_{OC} (0.76 V) to be achieved. A higher J_{SC} (15.3 mA cm^{-2}) and fill factor (0.67) were achieved due to improved morphology and hole mobility in the polymer, resulting in a PCE of 7.7%. **PTB7**, reported by Yu *et al.*, is very similar in structure to **PBDTTT-CF**, but includes an ester rather than a ketone on the side-chain of the thieno[3,4-*b*]thiophene unit and a branched 2-ethylhexyl chain rather than a straight *n*-heptyl chain.[76] Despite the subtlety of this structural difference, it initially had very little impact on the polymer's performance in devices, also achieving a PCE of 7.4% with very similar J_{SC}, V_{OC} and fill factor values. However, Cao *et al.* fabricated inverted devices using **PTB7**, which also included a poly [(9,9-bis(3′-(*N*,*N*-dimethylamino)propyl)-2,7-fluorene)-alt-2,7-(9,9-dioctylfluorene)] (PFN) interlayer and led to significant improvement in devices.[77] The PFN layer had little effect on the V_{OC}, although it had a dramatic improvement on the J_{SC} (from 14.2 to 17.5 mA cm^{-2}) and a slight improvement in the fill factor (from 0.67 to 0.70), resulting in a hugely impressive PCE of 9.2%. The authors suggested that the improvement in J_{SC} was a result of improved optical absorption made possible by the inverted structure (Table 3.3).

A further improvement was made to the PBDTTT series of polymers by Hou *et al.* with the introduction of alkylthienyl side-chains in the place of alkoxy solubilising chains, **PBDTTT-C-T**.[78] This is a 2D conjugated polymer as a result of the additional thiophene units included on the side-chains; this red shifts and broadens the absorption spectrum slightly and dramatically improves the hole mobility of the polymer. As a result, an improved J_{SC} of 17.5 mA cm^{-2} can be achieved and overall the PCE can be improved to 7.6% despite a decrease in the fill factor to 0.59. Building on this work, Hou *et al.* probed the effect of varying the heteroatom in the side-chain, replacing the alkylthienyl side-chains with alkylfuryl (**PBDTTT-EFF**) and alkylselenophenyl (**PBDTTT-EFS**).[79] As was the case with **PBDTTT-C-T**, these polymers also feature 2D conjugation and, by varying the aromatic units on the side-chain, it is possible to tune the structural and optoelectronic properties of the polymer.

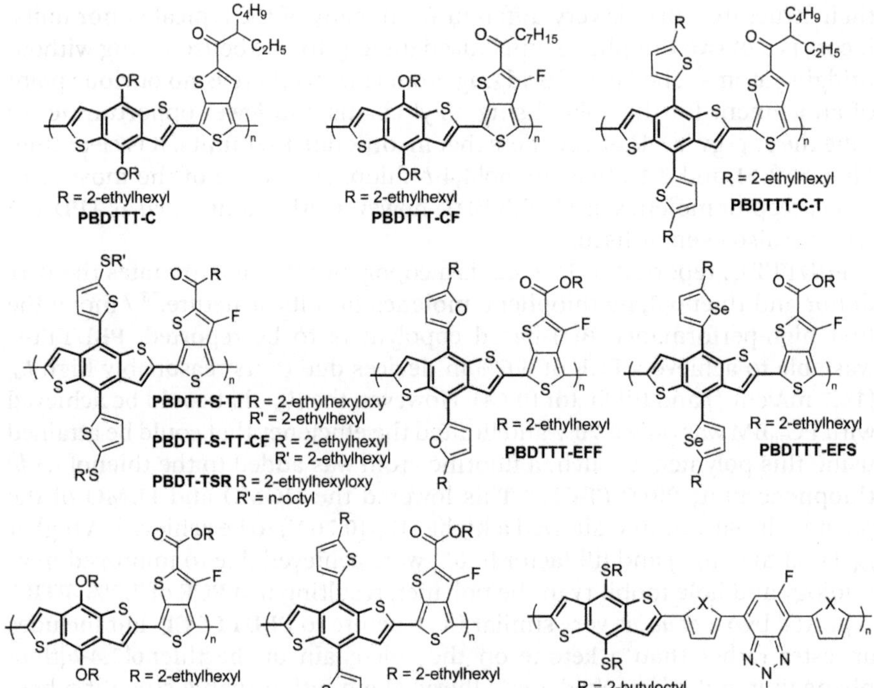

Figure 3.11 Benzodithiophene based push–pull donor copolymers.

The alkylfuryl units in **PBDTTT-EFF** form a smaller dihedral angle with the BDT unit than the alkylselenophenyl units in **PBDTTT-EFS**, which is probably a result of the increased quinoidal character in furan. This had two significant effects on the properties exhibited by the polymers. First, the smaller dihedral angle appeared to facilitate π–π stacking in the polymer, leading to

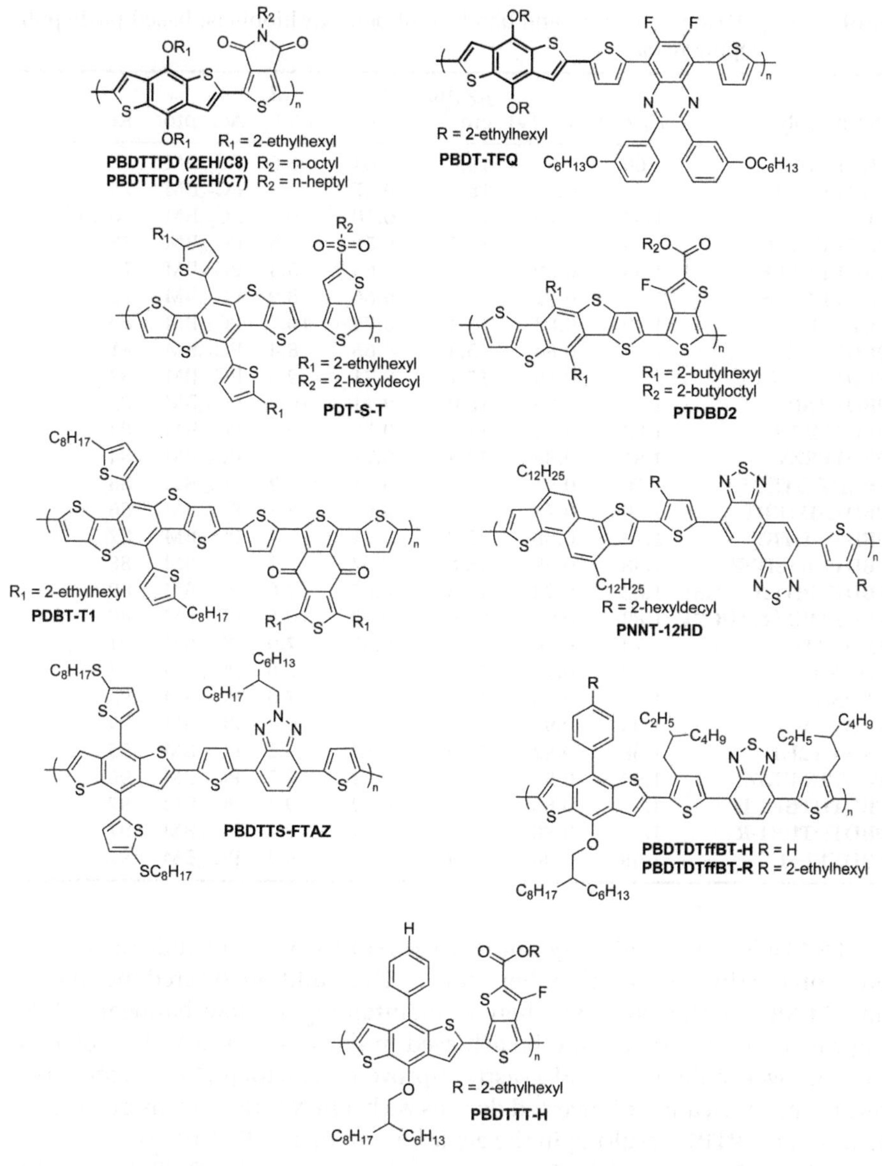

Figure 3.12 Additional benzodithiophene based push–pull donor copolymers.

the formation of domains larger than the nanoscale. Hence a significantly poorer J_{SC} (11.8 *versus* 16.6 mA cm^{-2}) was observed for **PBDTTT-EFF** due to increased recombination in the active layer blend. Second, the increased coplanarity of the alkylfuryl units with the BDT core led to raised HOMO and LUMO levels in **PBDTTT-EFF**, therefore the devices had a lower V_{OC} (0.69 V) than when **PBDTTT-EFS** was used (0.81 V). Overall, this led to much greater PCE of 8.9% to be achieved by **PBDTTT-EFS**.

Table 3.3 *J–V* Characteristics and bandgap of benzodithiophene based push–pull copolymers.

Donor polymer	E_g (eV)	V_{OC} (V)	J_{SC} (mA cm^{-2})	Fill factor	PCE (%)	Acceptor	Ref.
PBDTTT-C	1.60	0.70	14.7	0.64	6.6	$PC_{70}BM$	75
PBDTTT-CF	1.60	0.76	15.2	0.67	7.7	$PC_{70}BM$	75
PTB7	1.84	0.75	17.5	0.70	9.2	$PC_{70}BM$	76 and 77
PBDTTT-C-T	1.58	0.74	17.5	0.59	7.6	$PC_{70}BM$	78
PBDTTT-EFF	1.55	0.69	11.8	0.65	5.3	$PC_{70}BM$	79
PBDTTT-EFS	1.58	0.81	16.6	0.66	8.9	$PC_{70}BM$	79
PBT7-Th	1.59	0.83	17.4	0.74	10.7	$PC_{70}BM$	80
PBDTT-S-TT	1.57	0.84	15.3	0.65	8.4	$PC_{70}BM$	81
PBDTT-S-TT-CF	1.56	0.89	15.1	0.71	9.6	$PC_{70}BM$	82
PBDT-TSR	1.57	0.80	18.0	0.71	10.2	$PC_{70}BM$	83
PBDTFBZO	1.83	0.91	11.8	0.58	6.2	$PC_{70}BM$	84
PBDTFBZS	1.81	0.88	12.4	0.71	7.7	$PC_{70}BM$	84
PBnDT-DTffBT	1.70	0.91	12.9	0.61	7.2	$PC_{60}BM$	85
PBDTt-DTffBT	1.70	0.78	15.4	0.69	8.3	$PC_{70}BM$	86
PBDTP-DTBT	1.70	0.88	12.9	0.71	8.1	$PC_{70}BM$	87
PBDTTT-SeDPP	1.38	0.69	16.8	0.62	7.2	$PC_{70}BM$	88
PBDTTPD (2EH/C8)	1.82	0.93	12.5	0.65	7.6	$PC_{70}BM$	89
PBDTTPD (2EH/C7)	1.82	0.97	12.6	0.70	8.6	$PC_{70}BM$	89
PBDT-TFQ	1.73	0.76	17.9	0.58	7.9	$PC_{70}BM$	91
PDT-S-T	1.59	0.73	16.6	0.64	7.8	$PC_{70}BM$	92
PTDBD2	1.67	0.89	13.0	0.65	7.5	$PC_{70}BM$	93
PDBT-T1	1.85	0.92	14.1	0.75	9.7	$PC_{70}BM$	94
PNNT-12HD	1.68	0.82	15.6	0.64	8.2	$PC_{70}BM$	95
PBDTTS-FTAZ	1.90	0.83	14.0	0.71	8.3	$PC_{70}BM$	96
PBDTDTffBT-H	1.76	0.88	14.9	0.72	9.4	$PC_{70}BM$	97
PBDTDTffBT-R	1.75	0.88	14.0	0.74	9.1	$PC_{70}BM$	97
PBDTTT-H	1.68	0.80	17.4	0.67	9.4	$PC_{70}BM$	97

PTB7-Th is structurally very similar to **PBDTTT-C-T**, but includes a fluorine atom on the thieno[3,4-*b*]thiophene unit.[80] This addition lowered the HOMO and LUMO of the polymer, thereby maintaining its low bandgap while improving the V_{OC} that could be achieved in devices to 0.83 V. A J_{SC} of 17.4 mA cm^{-2} was maintained and a vastly improved fill factor (0.74) was obtained due to the fabrication of inverted devices with a PFN interlayer, as previously reported for **PTB7**, resulting in the achievement of a PCE of 10.7%.

PBDTT-S-TT, reported by Li *et al.*, also makes use of a BDT donor unit with thiophene groups included in the solubilising chains and fluorinated thieno[3, 4-*b*]thiophene acceptor unit.[81] However, one interesting difference compared with many of the polymers discussed here is the inclusion of alkylthiol substituents on the conjugated thiophene side-chains. This had little effect on the structural properties, but did stabilise the HOMO level by ~0.1 eV and red shifted the absorption spectrum of the polymer slightly relative to its analogue with alkylthienyl side-chains. This resulted in a V_{OC} of 0.84 V, among the highest for BDT- and thieno[3, 4-*b*]thiophene-containing copolymers, while maintaining a high J_{SC} (15.3 mA cm^{-2}) and fill factor (0.65), allowing devices to achieve a PCE of 8.4%. In **PBDTT-S-TT-CF** the 2-ethylhexyloxy chain on the TT unit was replaced by a 2-ethylhexyl chain.[82] The notable effect of this was a vast improvement in fill factor (0.71) and a small improvement in V_{OC} (0.89 V), which led to a PCE of 9.6%. Another interesting derivative of **PBDTT-S-TT** is **PBDT-TSR**, reported by Hou *et al.*[83] The only structural change was the use of *n*-octylthio rather than 2-ethylhexylthio side-chains. A more notable difference is that a high degree of regioregularity was ensured. In doing this, a vast improvement in the extinction coefficient of the polymer was observed, leading to the greatly improved J_{SC} (18.0 mA cm^{-2}) and hole mobility, which enhanced the fill factor to 0.71. The overall result of this was a dramatic improvement in the PCE to 10.2%. This highlights the huge benefits of ensuring regioregularity, particularly when using asymmetrical monomers.

BDTFBZO and **BDTFBZS**, reported by Peng *et al.*, are BDT-based copolymers that use monofluorobenzotriazole as the acceptor unit, flanked by either furans or thiophenes, respectively.[84] The nitrogen lone pair in benzotriazole make is less strong as an acceptor unit than benzothiadiazole and benzoxadiazole and, as such, the bandgap of these polymers is relatively wide compared with many of the other BDT copolymers (~1.8 eV) and it has relatively low-lying HOMO and LUMO levels. The wide bandgap limits the photon harvesting possible by these polymers, affording rather modest J_{SC} values of ~12 mA cm^{-2} for both polymers in devices. However, the low-lying HOMO levels led to high V_{OC} values (~0.9 V). The difference in performance between these polymers was the result of a vast difference in the fill factor that can be achieved: **BDTFBZO** could achieve 0.59, whereas **BDTFBZS** achieved 0.71. The main reason for the large difference in fill factor was suggested to be the fact that thiophenes are far better at suppressing recombination than furan units. This resulted in a modest PCE of 6.2% in devices using **BDTFBZO,** but a PCE of 7.7% with **BDTFBZS.**

In addition to the benzotriazole acceptor moiety, the 2,1,3-benzothiadiazole (BT) moiety and its fluorinated analogues have been used successfully with BDT in push–pull copolymers. **PBnDT-DTffBT** makes use of the BDT donor moiety and the 5,6-difluoro-4,7-dithien-2-yl-2,1,3-benzothiadiazole (DTffBT) acceptor unit.[85] A relatively low-lying HOMO was made possible by the inclusion of the fluorine atoms on the BT moiety, therefore a high V_{OC} (0.91 V) could be achieved in devices. A reasonable J_{SC} of 12.9 mA cm^{-2} and fill factor of 0.61 allowed an efficiency of 7.2%. Jiang *et al.* modified this polymer with the inclusion of alkylthienyl side-chains in place of the alkyl side-chains of **PBnDT-DTffBT**; producing **PBDTt-DTffBT.**[86] The marked improvement in PCE to 8.3% can be mainly attributed to the higher J_{SC} and fill factor that are possible as a result of the introduction of 2D conjugation into the polymer. The increased conjugation led to improved hole mobility in the polymer and increased the tendency for it to π–π stack, thereby improving the blend morphology of the active layer. A similar polymer, **PBDTP-DTBT**, has also been reported; this polymer makes use of alkylphenyl solubilising chains on the BDT unit rather than alkylthienyl.[87] We would expect a lower degree of 2D conjugation when alkylphenyl groups are present due to the steric clash of the α protons leading to more twisting. As such, a rather low HOMO was achieved without fluorinating the BT unit, hence a high V_{OC} of 0.88 V was seen. When DIO was used as an additive, it was possible to form an interpenetrating network of nanoscale fibrils and polymer π–π stacks, leading to a reasonable mobility, thus a rather high fill factor of 0.71 was possible. This led to a PCE of 8.1% in devices.

BDT has also been polymerised with a DPP flanked on either side by selenophenes to form the polymer **PBDTTT-SeDPP**.[88] This polymer was able to achieve a rather low bandgap of 1.38 eV due to the large amount of push–pull character in the BDT and DPP units, but also by the use of selenophenes, which reduce the bandgap by increasing the quinoidal character in the donor moiety. In achieving a low bandgap, it is possible for this polymer to harvest a large fraction of photons from the solar spectrum, hence the high J_{SC} of 16.8 mA cm^{-2}. The inclusion of the selenophene units also allows a reasonable fill factor of 0.62 due to the reasonably high mobilities that can be achieved. However, the increased HOMO level leads to a narrower bandgap and a rather low V_{OC} of 0.69 V, which limits the PCE to 7.2%.

BDT has also been copolymerised with TPD to form push–pull copolymers. The **PBDTTPD** polymers reported by Beaujuge *et al.* made use of side-chain engineering to improve the performance of the polymers.[89] **PBDTTPD (2EH/C8)** and **PBDTTPD (2EH/C7)** make use of 2-ethylhexyloxy chains on the BDT unit and linear *n*-octyl and *n*-heptyl chains, respectively, on the TPD moiety. Both polymers, like others based on BDT and TPD,[90] exhibit an impressive V_{OC} >0.9 V due to their low-lying HOMOs; however, the moderate bandgap means that only a J_{SC} of around 12.5 mA cm^{-2} can be achieved in both polymers. **PBDTTPD (2EH/C7)** has one less methylene in its solubilising chain, thereby reducing the ratio of insulating material to conjugated polymer. This led to an improvement in the fill factor from 0.65 to 0.70, resulting in an improvement in the PCE from 7.6% to 8.6%.

Chou *et al.* reported a BDT-based copolymer, **PBDT-TFQ**, incorporating a fluorinated quinoxaline flanked by thiophenes as the acceptor moiety.[91] The fluorination of the quinoxaline unit stabilised the HOMO and LUMO levels, allowing a reasonable V_{OC} of 0.76 V to be achieved. The interaction of the fluorine atoms with the lone pair on the thiophenes acted to planarise the polymer backbone. As a result, a high J_{SC} (17.9 mA cm^{-2}) was reached and, despite a low fill factor, this polymer was able to achieve 7.9% PCE in devices.

Some work has also been carried out to produce polymers that utilise larger fused-ring donor moieties that are derivatives of BDT. Dithieno[2,3-*d*:2′,3′-*d*′]benzo[1,2-*b*:4,5-*b*′]dithiophene (DTBDT) has been incorporated into push–pull copolymers with the aim of improving the ability of the polymer to π–π stack by extending the conjugation of the donor unit. **PDT-S-T** and **PTDBD2** both make use of the DTBDT donor unit with the thieno[3,4-*b*]thiophene acceptor unit.[92,93] **PDT-S-T** is better able to π–π stack than polymers that use BDT and thieno[3,4-*b*]thiophene, improving the charge carrier mobility, and hence a reasonably high J_{SC} (16.6 mA cm^{-2}) can be obtained. The modest

V_{OC} of 0.76 V is a result of the fairly high-lying HOMO achieved when the thieno[3,4-*b*]thiophene unit is not fluorinated. The overall result was a PCE of 7.8%. **PTDBD2** makes use of a fluorinated thieno[3,4-*b*]thiophene, which acts to lower the HOMO and therefore improve the V_{OC} that can be achieved to 0.89 V. However, this polymer utilises linear alkyl solubilising chains rather than the alkylthiophenyl chains and therefore has a lower degree of conjugation. It therefore has a wider bandgap and lower J_{SC} of 13.0 mA cm^{-2}. This reduction in the ability of the polymer to harvest photons led to a poorer PCE (7.5%) despite the improvement in V_{OC}.

PDBT-T1, reported by Sun and co-workers, also makes use of DTBDT as the donor moiety; but instead uses the strongly electron-deficient 1,3-bis(thiophen-2-yl)-benzo-[1,2,-*c*:4,5-*c*′]dithiophene-4,8-dione (T1).[94] This polymer has a moderate bandgap of 1.85 eV and therefore the low-lying HOMO led to an impressive V_{OC} (0.92 V). Despite the relatively wide bandgap of this polymer compared with many other push–pull copolymers, **PDBT-T1** is still able to achieve a J_{SC} of 14.1 mA cm^{-2}. Sun and co-workers suggested that this is due to the relatively high charge carrier mobility achieved by the coplanarity in this polymer, which might also explain the exceptional fill factor of 0.75. Despite a modest J_{SC}, the high V_{OC} and fill factor allowed this polymer to achieve PCEs as high as 9.7%.

Takimiya *et al.* made use of a different BDT derivative: naphtho[1,2-*b*:5,6-*b*′] dithiophene (NDT). NDT was utilised as the donor with the NT naphthobis-thiadiazole acceptor moiety, flanked by thiophenes, to produce the polymer **PNNT-12HD**.[95] The reasonably high J_{SC} of 16.6 mA cm^{-2} was achieved due to the ability to produce 300 nm thick active layers, which still show optimum blend morphology. **PNNT-12HD** was also able to achieve a V_{OC} of 0.82 V due to a relatively stabilised HOMO and a fill factor of 0.64 was possible due to the favourable morphology found with the use of the hexyldecyl chains on the flanking thiophenes. Overall, this led to a PCE of 8.2%.

PBDTTS-FTAZ incorporated the 2D BDTT unit with a thiophene-flanked benzotriazole acceptor unit, producing a donor polymer capable of achieving 8.3% in devices.[96] Despite a relatively wide bandgap among donor polymers, a J_{SC} of 14.0 mA cm^{-2} was still possible and a good fill factor of 0.71was measured. The authors attributed this to the inclusion of the linear *n*-octylthio side-chains, which led to improved charge carrier mobility in the polymer due to a strong tendency to π–π stack.

Another interesting development in the design of BDT-based donor polymers was reported by Yang *et al.*, where the advantages of both 1D and a 2D symmetrical BDT units were combined to produce 1D–2D asymmetrical BDT donor units.[97] They reported three polymers, all of which made use of a benzodithiophene containing a phenyl (or alkylphenyl) solubilising chain on one side and an alkoxy side-chain on the other side. **PBDTDTffBT-H** and **PBDTDTffBT-R** contain the asymmetrical BDT donor unit copolymerised with an ffBT unit flanked on either side by thiophenes. In **PBDTTT-H**, the asymmetrical BDT was copolymerised with a fluorinated thieno[3, 4-*b*]thiophene. By including the phenyl-type solubilising chain on one, the conjugation could be extended and a deep HOMO achieved, allowing a reasonably high V_{OC} of >0.8 V to be accessible for all three polymers. However, by also including the 1D alkoxy side-chain, it was possible to improve the solubility of the polymers and tune the morphology in blends. This led to preferential ordering of the polymer and improved π–π stacking, all of which resulted in the attainment of reasonably high J_{SC} and fill factors. These polymers were all able to achieve PCEs >9% in devices, with **PBDTDTffBT-H** and **PBDTTT-H** achieving an impressive 9.4%.

3.3.5 Push–Pull Copolymers Based on Fluorene, Silafluorene, Carbazole and Other Bridged Biphenyl Derivatives

Despite being among the earliest donor units utilised in low bandgap push–pull copolymers, fluorene and other bridged biphenyl donor units are still able to yield some impressive performances in OPV devices. As discussed for the bridged bithiophene-containing polymers, variation of the bridging atom allows several opportunities to tune the optoelectronic and structural properties of the bridged biphenyl-based donor polymer. In addition to the group 14 bridging atoms most commonly utilised in bridged bithiophenes, nitrogen atoms have also been used to great effect in polymers that include bridged biphenyl units. The lone pair of the nitrogen atom donates into the conjugated system, thereby raising the HOMO. In addition, the sp^2 hybridisation of the bridging nitrogen atom allows the inclusion of only one solubilising alkyl chain; as such, poorer solubility is often observed with carbazole moieties unless a long, branched alkyl chain is utilised. The 9-alkylidene-9H-fluorene (AF) unit also makes use of a sp^2 hybridised bridging unit. However, in this case it allows the solubilising alkyl chains to sit planar to the conjugated backbone of the polymer, whereas the alkyl chains found in 9,9-dialkyl fluorene and *N*-alkyl-2,7-carbazole polymers are perpendicular to the backbone. This facilitates π–π stacking and increases the crystallinity in the AF-containing polymers. It must also be noted that the bridged biphenyl donor units are less donating than their bridged bithiophene analogues due to the stronger aromaticity of the phenyl rings. These polymers are typically less planar due to the increased steric clash of α-hydrogens of the phenyl rings with those on the acceptor unit of the copolymer. This results in wider bandgap polymers when bridged biphenyl units are used rather than bridged bithiophenes and their derivatives. From the structures, summarised in Figure 3.13 and Table 3.4, it can be seen that bridged biphenyl donor units have most successfully been incorporated into push–pull copolymers that include the BT acceptor moiety.

PSiFDBT consists of a 9,9-dioctylsilafluorene (SiF) donor unit copolymerised with dithienyl 2,1,3-benzothiadiazole (DTBT).[98] The inclusion of the thiophene units flanking the BT ensures that the polymer backbone is planar, increasing the conjugation, and therefore lowering the bandgap, while also improving the polymer's ability to π–π stack. Compared with its carbon-bridged analogue, the silafluorene-based polymer can obtain a higher V_{OC} (0.90 V) due to the lower HOMO and improved hole mobility resulting from the increased crystallinity of the polymer, as such, an improved efficiency of

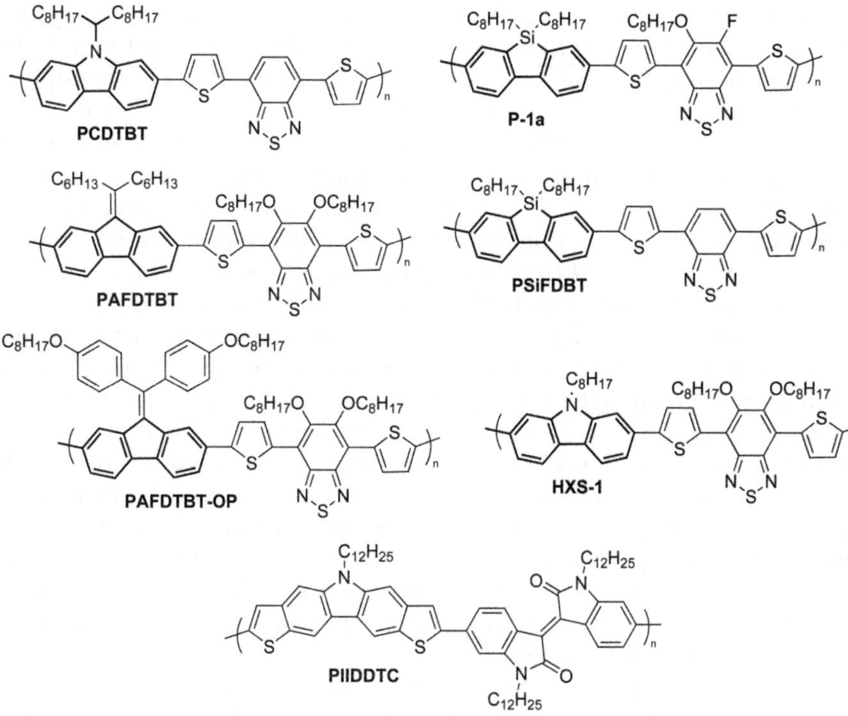

Figure 3.13 Bridged biphenyl-based push–pull donor copolymers.

Table 3.4 *J–V* characteristics and bandgap of bridged biphenyl-based push–pull copolymers.

Donor polymer	E_g (eV)	V_{OC} (V)	J_{SC} (mA cm^{-2})	Fill factor	PCE (%)	Acceptor	Ref.
PSiFDTBT	1.82	0.90	9.5	0.51	4.4	$PC_{60}BM$	98
PSiFDTAfBT	1.99	0.92	12.1	0.58	6.5	$PC_{70}BM$	99
PAFDTBT	1.84	0.89	9.9	0.70	6.2	$PC_{70}BM$	100
PAFDTBT-OP	2.00	0.90	12.7	0.57	6.5	$PC_{70}BM$	101
HXS-1	1.95	0.81	9.6	0.69	5.4	$PC_{70}BM$	102
PIIDDTC	1.61	0.78	15.2	0.69	8.2	$PC_{70}BM$	103

4.4% can be achieved in devices.[98] The low fill factor of 0.51 obtained in these devices, however, limits the efficiency that can be achieved. Such an analogue, **PSiFDTAfBT**, was developed by Bo *et al.*, where the BT unit was functionalised to include a fluorine and an alkoxy chain.[99] The inclusion of the fluorine atom yielded a slight improvement in V_{OC} to 0.92 V due to the lowering of the HOMO, but more significantly a greatly improved J_{SC} (12.1 mA cm^{-2}) and fill factor (0.58) were achieved by improving the solubility of the polymer without disrupting its packing. This resulted in an improved PCE of 6.5%.

Another attempt to improve upon **PSiFDTBT** was a copolymer reported by Zhang *et al.*, which instead made use of the AF donor unit for polymer

PAFDTBT.[100] The sp^2 hybridised alkylidene fluorene improved the crystallinity of the polymer, and therefore its hole mobility, which led to a slight improvement in J_{SC}. The most remarkable feature of **PAFDTBT** is the impressive fill factor of 0.70 that can be achieved, leading to a PCE of 6.2% in devices. The authors partially attribute the improved fill factor to the use of the additive DIO to improve the blend morphology, which also contributed to the increased crystallinity of the polymer, with an overall effect of promoting the formation of nanoscale donor and acceptor domains in the blend. An analogue of **PAFDTBT** with an arylidene fluorene donor unit has also been reported: **PAFDTBT-OP.**[101] The phenyl substituents of the arylidene sit in the plane of the conjugated backbone of the polymer and act to extend the lateral conjugation of the polymer chains. The authors suggested that this improved the inter-chain charge transport in the polymer, leading to an improvement in the J_{SC} to 12.7 mA cm^{-2} and in the PCE to 6.5%.

HXS-1, reported by Zhang *et al.*,[102] makes use of the carbazole moiety and, as has been the case with many of the successful bridged biphenyl-containing copolymers, is paired with the dithienyl benzothiadiazole unit. The noticeably lower V_{OC} of 0.81 V is due to the higher lying HOMO of **HXS-1**, caused by the donation of the nitrogen lone pair into the conjugated system in the carbazole unit. Despite this, a high fill factor (0.69) was obtained in devices, leading to a respectable 5.4%.

N-alkyldithiopheno[3,2-*b*:6,7-*b*]carbazole (DTC) was copolymerised with isoindigo (IID) by Wang *et al.* to yield a very rigid and planar donor polymer, **PIIDDTC.**[103] The high degree of planarity and the strong push–pull character exhibited by the polymer afforded a reasonably low bandgap of 1.61 eV. As a result, a high J_{SC} (15.2 mA cm^{-2}) can be achieved, much higher than normally seen in devices using bridged biphenyl-containing copolymers. It was also observed that a favourable blend morphology was able to form in devices using **PIIDDTC**; this contributed to the impressive J_{SC} achieved and also the fill factor of 0.69. However, the narrow bandgap of the acceptor led to a modest V_{OC} of 0.78 V; which to some extent limits this polymer from achieving efficiencies even higher than 8.2%. To improve the V_{OC}, the authors fluorinated the isoindigo unit and, although this did increase the V_{OC} to 0.88 V, the J_{SC} suffered as the blend morphology became worse. The loss of interpenetrating donor and acceptor domains on the order of the exciton diffusion length was suggested to be a result of the reduced solubility of the polymer, hence the $PC_{70}BM$ domains were frozen out more quickly and the interpenetrating networks could not form.[103]

3.3.6 Push–Pull Copolymers Based on Indacenodithiophene and Its Derivatives

Extended fused-ring units, such as indaceno[1,2-*b*:5,6-*b*′]dithiophene (IDT), have also attracted a lot of interest as donor units in push–pull copolymers. These moieties are very rigid and planar in nature, which leads to enhanced crystallinity and therefore improved charge carrier mobilities. IDT and its derivatives have relatively deep-lying HOMO levels, allowing an impressive V_{OC}, usually >0.85 V, to be achieved in devices. As for the bridged biaryl units, variation of the bridging atom can allow tuning of the structural and optoelectronic properties of the polymers. Variation of the S atoms in the thiophene moieties, located on the periphery of the donor unit, with other group 16 atoms allows control over the bandgap, as for the thiophene-based copolymers. As is the case with the bridged bithiophene and bridged biphenyl acceptors, Figure 3.14 shows that IDT and its derivatives have been most successfully copolymerised with BT acceptor units. Many of the highest performing IDT-based push–pull copolymers are summarised in Table 3.5.

α-PTPTBT is a push–pull copolymer that makes use of the IDT donor and BT acceptor moieties.[104] A relatively high V_{OC} (0.85 V) can be achieved due to the low-lying HOMO often associated with IDT. An impressive fill factor (0.67) can be attained as a favourable morphology can form when the blend is solvent-annealed, leading to a PCE of 6.4%. An attempt was made by Jen *et al.* to improve the V_{OC} using the difluorobenzothiadiazole acceptor unit

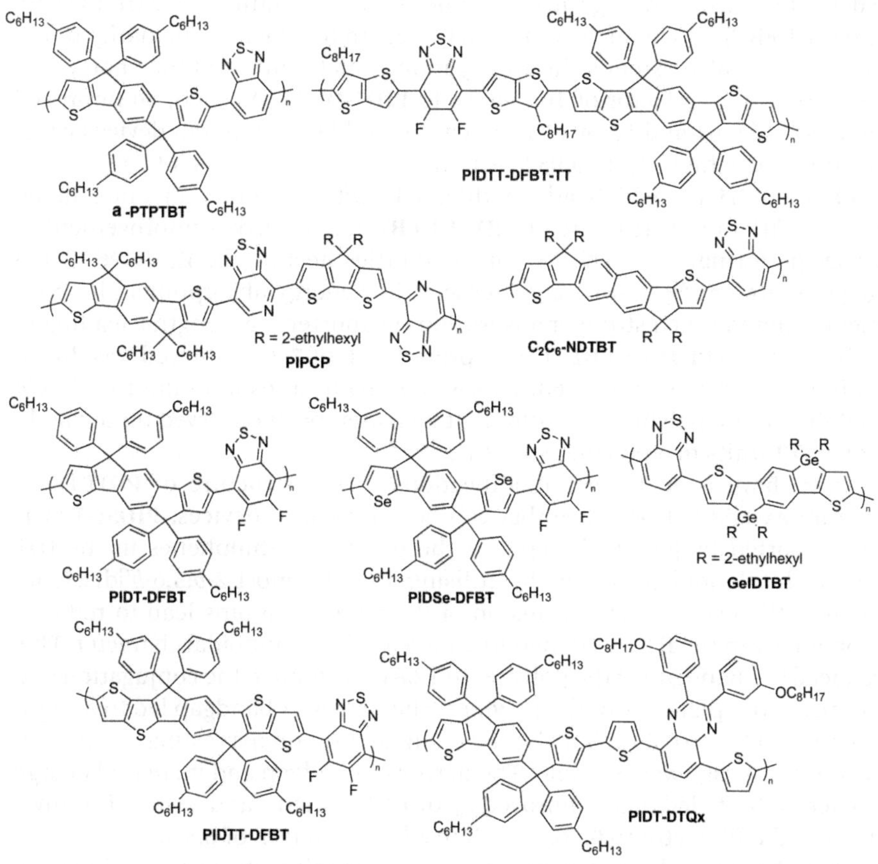

Figure 3.14 Indacenodithiophene-based push–pull donor copolymers.

Table 3.5 *J–V* characteristics and bandgap of indacenodithiophene-based push–pull copolymers.

Donor polymer	E_g (eV)	V_{OC} (V)	J_{SC} (mA cm^{-2})	Fill factor	PCE (%)	Acceptor	Ref.
α-PTPTBT	1.75	0.85	11.2	0.67	6.4	$PC_{70}BM$	104
PIDT-DFBT	1.78	0.97	11.2	0.55	6.0	$PC_{70}BM$	105
PIDTT-DFBT	1.78	0.95	12.2	0.61	7.1	$PC_{70}BM$	105
PIDTT-DFBT-TT	1.81	0.96	11.9	0.63	7.2	$PC_{70}BM$	106
PIDSe-DFBT	1.68	0.89	13.7	0.56	6.8	$PC_{70}BM$	109
GeIDTBT	1.74	0.85	14.0	0.55	6.5	$PC_{70}BM$	110
PIDTDTQx	1.74	0.87	12.3	0.70	7.5	$PC_{70}BM$	111
C_2C_6-NDTBT	1.81	0.92	15.1	0.54	7.5	$PC_{70}BM$	112

to further stabilise the HOMO. Although PIDT-DFBT could indeed achieve a high V_{OC} of 0.97 V, the overall PCE could not be improved (6.0%) due to a modest fill factor of 0.55.[105] The authors speculated that the poor fill factor is a result of the low charge mobility of PIDT-DFBT and they therefore designed a derivative incorporating the indacenodithieno[3,2-*b*]thiophene (IDTT) core (PIDTT-DFBT).[105] The inclusion of IDTT acts to improve the planarity of the polymer and also extends the conjugation, which improved the charge carrier mobility by a factor of 10 relative to **PIDT-DFBT**. As such, an improved fill factor of 0.61 and J_{SC} of 12.1 mA cm^{-2} could be achieved in devices while maintaining a high V_{OC} of 0.95 V, leading to an improved PCE of 7.0%.

PIDTT-DFBT-TT, which added thieno[3,2-*b*]thiophene (TT) units flanking the ffBT in the structure of **PIDTT-DFBT**, led to further improvement in device performance.[106] The TT units had little effect on the V_{OC} (0.96 V) and J_{SC} (11.9 mA cm^{-2}); despite a decrease in low-energy absorptions, the polymer exhibited much stronger absorptions at shorter wavelengths, leading to a similar J_{SC} to **PIDTT-DFBT**. The improved PCE of 7.2% achieved was due to an improved fill factor of 0.63. This was probably a result of improved hole mobility in the polymer, a feature that has often been observed on the addition of TT units to a polymer structure.[107,108]

There have been further modifications to the structure of **PIDT-DFBT** to improve the performance that can be achieved in devices. **PIDSe-DFBT**, first reported by Jen *et al.*, replaces the peripheral thiophenes in the IDT core with selenophenes, thereby utilising an indaceno[1,2-*b*:5,6-*b*′]diselenophene (IDSe) core.[109] The inclusion of the larger Se atoms lead to reduced aromaticity in the IDSe and therefore increased the quinoidal character. This reduced the bandgap of the polymer to 1.68 eV, extended the conjugation and improved the planarity of the polymer. The narrower bandgap led to a slight decrease in V_{OC} to 0.89 V, but also led to higher photon harvesting as a greater spectral coverage was possible. The narrower bandgap and increased charge carrier mobility led to an improved J_{SC} of 13.7 mA cm^{-2} and a small improvement in the fill factor to 0.56, overall resulting in a PCE of 6.8%.

There have also been attempts to use Ge bridging atoms in the indacenodithiophene core. Germaindaceno[1,2-*b*:5,6-*b*′]dithiophene (GeIDT) was copolymerised with BT by McCulloch *et al.* to produce **GeIDTBT**.[110] The effect of including Ge atoms is primarily to increase the degree of crystallinity in the polymer and a reasonably high J_{SC} of 14.0 mA cm^{-2} was achieved. Despite a V_{OC} of 0.85 V, this polymer was limited to 6.5% PCE by the low fill factor exhibited in devices. It is likely that the more crystalline nature of the Ge-containing polymer led to larger domains and therefore more recombination in the BHJ.

BT is clearly the most common acceptor moiety to be used with IDT and its derivatives; however, there are examples of high-performing donor polymers that utilise other acceptor units. **PIDTDTQx**, reported by Hou *et al.*, consists of IDT units copolymerised with dithiophenyl quinoxaline (DTQx).[111] Like many of the other IDT-based copolymers, **PIDTDTQx** has a reasonably narrow bandgap and low-lying HOMO, which allow a respectable J_{SC} (12.3 mA cm^{-2}) and V_{OC} (0.87 V) to be achieved. The reason that this is the highest performing IDT-based push–pull copolymer is that an impressive fill factor of 0.70 can be achieved. The authors attributed the high fill factor to the formation of an interpenetrating network of donor and acceptor domains on the length scale of exciton diffusion and balanced electron and hole mobilities in the blend.

McCulloch *et al.* reported a donor polymer that included a 4,4,10,10-tetraalkyl-4,10-dihydro-naphtha-[3″,2″:3,4; 7″,6″:3′,4′]-dicyclopenta-[2,1-*b*:2′,1′-*b*′]-dithiophene (NDT) donor unit and BT as the acceptor unit.[112] The inclusion of the NDT unit lowers the HOMO, which results in a V_{OC} of 0.92 V. The high J_{SC} (15.0 mA cm^{-2}) can be attributed to the reasonably high mobility that this polymer can achieve and the inverted architecture used in devices. Overall, this polymer was able to achieve a PCE of 7.5%, among the highest for a polymer from the IDT family.

3.4 Conclusions and Outlook

From the discussion in this chapter, it is clear that high-performance OPV devices are now possible as a result of the rapid progression of donor polymers over the last decade. The use of the push–pull effect has paved the way to developing low bandgap polymers, which make use of a greater fraction of the solar spectrum. Through careful design of the polymers, by considering the monomers and solubilising side-chains, it is possible to tune the energies of the FMOs of the polymer, the morphology achieved in blends and to improve the charge carrier mobilities in devices. Although some of these high-performance (>10%) polymers, such as **PffBT4T-2OD** and **PTB7-Th**,[61,80] are starting to become commercially available, there are still significant hurdles to overcome before the promise of OPV technologies can be realised. Donor polymers that can be produced from a relatively short and simple synthetic pathway will lead to low-cost donor materials becoming available. Improving the stability of donor polymers is another crucial goal in donor polymer design, both the oxidative stability of the organic semiconductors and the morphological stability of the active layer blends need to be addressed to produce devices with long-term stability. Unfortunately, this is not straightforward to achieve through the rational design of the donor polymers, but requires a deeper understanding of the degradation and diffusion processes that currently limit the stability of OPV devices.[113–115] The ability to solution-process the active layer from environmentally benign solvents would also be advantageous in the future. The highly planar and conjugated nature of most donor polymers renders them insoluble in most hydrocarbon solvents, instead requiring chlorinated solvents such as chloroform, chlorobenzene and ortho-dichlorobenzene to dissolve them. Halogenated solvents have a variety of issues associated with them, including their costly synthesis, disposal issues and their toxicity to both humans and the environment,[116,117] thus it is highly desirable to design polymers that do not require halogenated solvents to dissolve and process devices.

Most of the donor polymers that have been reported to date have been specifically designed to work well with fullerene acceptors. However, the limitations of fullerenes – such as their poor absorption in the visible region of the solar spectrum, their long-term morphological instability and the high synthetic costs associated with their synthesis[118–120] – have led to the development of a new class of organic semiconducting materials: non-fullerene acceptors. These non-fullerene acceptors are now able to achieve exceptional PCEs, matching and, in some cases, surpassing the performance of fullerene acceptors, despite being paired with donor polymers that have been designed with PCBM and its analogues.[22,121,122] Polymers that have been designed to work effectively with non-fullerene acceptors, both in terms of their optoelectronic and morphological characteristics, will play a crucial part in the advancement of the field of OPV devices.

References

1. J. J. Conti, P. D. Holtberg, J. R. Diefenderfer, S. A. Napolitano, A. M. Schaal, J. T. Turnure and L. D. Westfall, *Annual Energy Outlook 2015*, DOE/EIA, 2015, pp. 1–154.
2. F. C. Krebs, N. Espinosa, M. Hösel, R. R. Søndergaard and M. Jørgensen, *Adv. Mater.*, 2014, **26**, 29–39.
3. N. Espinosa, M. Hösel, D. Angmo and F. C. Krebs, *Energy Environ. Sci.*, 2012, **5**, 5117–5132.
4. B. A. Gregg and M. C. Hanna, *J. Appl. Phys.*, 2003, **93**, 3605–3614.
5. S. Karg, W. Reiss, V. Dyakonov and M. Schwoerer, *Synth. Met.*, 1993, **54**, 427–433.
6. C. W. Tang, *Appl. Phys. Lett.*, 1986, **48**, 183–185.
7. D. Baran, N. Li, A. C. Breton, A. Osvet, T. Ameri, M. Leclerc and C. J. Brabec, *J. Am. Chem. Soc.*, 2014, **136**, 10949–10955.
8. K. Vandewal, S. Albrecht, E. T. Hoke, K. R. Graham, J. Widmer, J. D. Douglas, M. Schubert, W. R. Mateker, J. T. Bloking, G. F. Burkhard, A. Sellinger, J. M. J. Fréchet, A. Amassain, M. K. Riede, M. D. McGehee, D. Neher and A. Salleo, *Nat. Mater.*, 2014, **13**, 63–68.
9. S. Gélinas, A. Rao, A. Kumar, S. L. Smith, A. W. Chin, J. Clark, T. S. van der Poll, G. C. Bazan and R. H. Friend, *Science*, 2014, **343**, 512–516.
10. S. M. Menke and R. J. Holmes, *Energy Environ. Sci.*, 2014, **7**, 499–512.
11. K. M. Pelzer and S. B. Darling, *Mol. Syst. Des. Eng.*, 2016, **1**, 10–24.

12. G. Yu, J. Gao, J. C. Hummelen, F. Wudl and A. J. Heeger, *Science*, 1995, **270**, 1789–1791.
13. K. Vandewal, S. Himmelberger and A. Salleo, *Macromolecules*, 2013, **46**, 6379–6387.
14. W. Ma, J. R. Tumbleston, M. Wang, E. Gann, F. Huang and H. Ade, *Adv. Energy Mater.*, 2013, **3**, 864–872.
15. M. P. de Jong, L. J. van Ijzendoorn and M. J. A. de Voigt, *Appl. Phys. Lett.*, 2000, **77**, 2255–2257.
16. F. C. Krebs, S. A. Gevorgyan and J. Alstrup, *J. Mater. Chem.*, 2009, **19**, 5442–5451.
17. S. K. Hau, H. L. Yip, N. S. Baek, J. Zou, K. O'Malley and A. K. Y. Jen, *Appl. Phys. Lett.*, 2008, **92**, 253301.
18. T. Ameri, P. Khoram, T. Heumuller, D. Baran, F. Machui, A. Troeger, V. Sgobba, D. M. Guldi, M. Halik, S. Rathgeber, U. Scherf and C. J. Brabec, *J. Mat. Chem. A*, 2014, **2**, 19461–19472.
19. T. Ameri, P. Khoram, J. Min and C. J. Brabec, *Adv. Mater.*, 2013, **25**, 4245–4266.
20. L. Lu, M. A. Kelly, W. You and L. Yu, *Nat. Photonics*, 2015, **9**, 491–500.
21. C. B. Nielsen, S. Holliday, H. Y. Chen, S. J. Cryer and I. McCulloch, *Acc. Chem. Res.*, 2015, **48**, 2803–2812.
22. S. Holliday, R. S. Ashraf, D. Baran, S. A. Yousaf, C. B. Nielsen, C. H. tan, S. D. Dimitrov, Z. Shang, N. Gasparini, M. Alamoudi, F. Laquai, C. J. Brabec, A. Salleo, J. R. Durrant and I. McCulloch, *Nat. Commun.*, 2015, **7**, 11585.
23. D. Baran, R. S. Ashraf, D. A. Hanifi, M. Abdelsamie, N. Gasparini, J. A. Röhr, S. Holliday, A. Wadsworth, S. Lockett, M. Neophytou, C. J. M. Emmott, J. Nelson, C. J. Brabec, A. Amassian, A. Salleo, T. Kirchartz, J. R. Durrant and I. McCulloch, *Nat. Mater.*, 2017, **16**, 363–369.
24. T. Ameri, G. Dennler, C. Lugenschmied and C. J. Brabec, *Energy Environ. Sci.*, 2009, **2**, 347–363.
25. F. Guo, N. Li, F. W. Fecher, N. Gasparini, C. O. R. Quiroz, C. Bronnbauer, Y. Hou, V. V. Radmilovic, V. R. Radmilovic, E. Spiecker, K. Forberich and C. J. Brabec, *Nat. Commun.*, 2015, **6**, 7730.
26. N. Li, D. Baran, K. Forberich, M. Turbiez, T. Ameri, F. C. Krebs and C. J. Brabec, *Adv. Energy Mater.*, 2013, **3**, 1597–1605.
27. J. H. Kim, J. B. Park, S. C. Yoon, I. H. Jung and D. H. Hwang, *J. Mat. Chem. C*, 2016, **4**, 2170–2177.
28. T. E. Kang, T. Kim, C. Wang, S. Yoo and B. J. Kim, *Chem. Mater.*, 2015, **27**, 2653–2658.
29. M. Jørgensen and F. C. Krebs, *Polym. Bull.*, 2003, **51**, 20–23.
30. J. Wolf, F. Cruciani, A. El Labban and P. M. Beaujuge, *Chem. Mater.*, 2015, **27**, 4184–4187.
31. G. Li, V. Shrotriya, J. Huang, Y. Yao, T. Moriarty, K. Emery and Y. Yang, *Nat. Mater.*, 2005, **4**, 864–868.
32. A. Moliton and J. M. Nunzi, *Polym. Int.*, 2006, **55**, 583–600.
33. H. Zhou, L. Yang, S. Liu and W. You, *Macromolecules*, 2010, **43**, 10390–10396.

34. R. S. Ashraf, I. Meager, M. Nikolka, M. Kirkus, M. Planells, B. C. Schroeder, S. Holliday, M. Hurhangee, C. B. Nielsen, H. Sirringhaus and I. McCulloch, *J. Am. Chem. Soc.*, 2015, **137**, 1314–1321.
35. I. Meager, R. S. Ashraf, S. Mollinger, B. C. Schroeder, H. Bronstein, D. Beatrup, M. S. Vezie, T. Kirchartz, A. Salleo, J. Nelson and I. McCulloch, *J. Am. Chem. Soc.*, 2013, **135**, 11537–11540.
36. I. Meager, R. S. Ashraf, S. Rossbauer, H. Bronstein, J. E. Donaghey, J. Marshall, B. C. Schroeder, M. Henney, T. D. Anthopoloulos and I. McCulloch, *Macromolecules*, 2013, **46**, 5961–5967.
37. J. C. Hummelen, B. W. Knight, F. LePeq, F. Wudl, J. Yao and C. L. Wilkins, *J. Org. Chem.*, 1995, **60**, 532–538.
38. M. Lenes, G. J. A. H. Wetzelaer, F. B. Kooistra, S. C. Veenstra, J. C. Hummelen and P. W. M. Blom, *Adv. Mater.*, 2008, **20**, 2116–2119.
39. Y. He, H. Y. Chen, J. Hou and Y. Li, *J. Am. Chem. Soc.*, 2010, **132**, 1377–1382.
40. G. Zhao, Y. He and Y. Li, *Adv. Mater.*, 2010, **22**, 4355–4358.
41. M. M. Wienk, J. M. Kroon, W. J. H. Verhees, J. Knol, J. C. Hummelen, P. A. van Hal and R. A. J. Janssen, *Angew. Chem., Int. Ed.*, 2003, **42**, 3371–3375.
42. D. Sun, D. Meng, Y. Cai, B. Fan, Y. Li, W. Jiang, L. Hou, Y. Sun and Z. Wang, *J. Am. Chem. Soc.*, 2015, **137**, 11156–11162.
43. H. Lin, S. Chen, Z. Li, J. Y. L. Lai, G. Yang, T. McAfee, K. Jiang, Y. Li, Y. Liu, H. Hu, J. Zhao, W. Ma, H. Ade and H. Yan, *Adv. Mater.*, 2015, **27**, 7299–7304.
44. D. Meng, D. Sun, C. Zhong, T. Liu, B. Fan, L. Huo, Y. Li, W. Jiang, H. Choi, T. Kim, J. Y. Kim, Y. Sun, Z. Wang and A. J. Heeger, *J. Am. Chem. Soc.*, 2016, **138**, 375–380.
45. W. Zhao, D. Qian, S. Zhang, S. Li, O. Inganänas, F. Gao and J. Hou, *Adv. Mater.*, 2016, **28**, 4734–4739.
46. V. D. Mihailetchi, P. W. M. Blom, J. C. Hummelen and T. C. Rispens, *J. Appl. Phys.*, 2003, **94**, 6849–6854.
47. E. Bundgaard and F. C. Krebs, *Sol. Energy Mater. Sol. Cells*, 2007, **91**, 954–985.
48. J. L. Brédas, D. Beljonne, V. Coropceanu and J. Cornil, *Chem. Rev.*, 2004, **104**, 4971–5004.
49. M. Al-Ibrahim, H. K. Roth, U. Zhokavets, G. Gobsch and S. Sensfuss, *Sol. Energy Mater. Sol. Cells*, 2005, **85**, 13–20.
50. S. Shoaee, S. Subramaniyan, H. Xin, C. Keiderling, P. S. Tuladhar, F. Jamieson, S. A. Jenekhe and J. R. Durrant, *Adv. Funct. Mater.*, 2013, **23**, 3286–3298.
51. J. M. Nunzi, *C. R. Phys.*, 2002, **3**, 523–542.
52. D. Bartesaghi, I. del Carmen Pérez, J. Kniepert, S. Roland, M. Turbiez, D. Neher and L. Jan Anton Koster, *Nat. Commun.*, 2015, **6**, 7083.
53. H. Yokoyama, E. J. Kramer, M. H. Rafailovich, J. Sokolov and S. A. Schwarz, *Macromolecules*, 1998, **31**, 8826.
54. K. H. Hendriks, G. H. L. Heintges, V. S. Gevaerts, M. M. Wienk and R. A. J. Janssen, *Angew. Chem., Int. Ed.*, 2013, **52**, 8341–8344.

55. J. C. Bijleveld, A. P. Zoombelt, S. G. J. Mathijssen, M. M. Wienk, M. Turbiez, D. M. de Leeuw and R. A. J. Janssen, *J. Am. Chem. Soc.*, 2009, **131**, 16616–16617.
56. J. C. Bijleveld, V. S. Gevaerts, D. Di Nuzzo, M. Turbiez, S. G. J. Mathijssen, D. M. de Leeuw, M. M. Wienk and R. A. J. Janssen, *Adv. Mater.*, 2010, **22**, E242–E246.
57. W. Li, K. H. Hendricks, A. Furlan, W. S. C. Roelofs, M. M. Wienk and R. A. J. Janssen, *J. Am. Chem. Soc.*, 2013, **135**, 18942–18948.
58. H. Choi, S. J. Ko, T. Kim, P. O. Morin, B. Walker, B. H. Lee, M. Leclerc, J. Y. Kim and A. J. Heeger, *Adv. Mater.*, 2015, **27**, 3318–3324.
59. T. Ma, K. Jiang, S. Chen, H. Hu, H. Lin, Z. Li, J. Zhao, Y. Liu, Y. M. Chang, C. C. Hsiao and H. Yan, *Adv. Energy Mater.*, 2015, **5**, 1501282.
60. X. Guo, N. Zhou, S. J. Lou, J. Smith, D. B. Tice, J. W. Hennek, R. P. Ortiz, J. T. L. Navarette, S. Li, J. Strzalka, L. X. Chen, R. P. H. Chang, A. Facchetti and T. J. Marks, *Nat. Photonics*, 2013, **7**, 825–833.
61. Y. Liu, J. Zhao, Z. Li, C. Mu, W. Ma, H. Hu, K. Jiang, H. Lin, H. Ade and H. Yan, *Nat. Commun.*, 2014, **5**, 5293.
62. J. Zhao, Y. Li, G. Yang, K. Jiang, H. Lin, H. Ade, W. Ma and H. Yan, *Nat. Energy*, 2016, **1**, 15027.
63. E. H. Jung and W. H. Jo, *Energy Environ. Sci.*, 2014, **7**, 650–654.
64. Y. Kim, H. R. Yeom, J. Y. Kim and C. Yang, *Energy Environ. Sci.*, 2013, **6**, 1909–1916.
65. J. Cao, Q. Liao, X. Du, J. Chen, Z. Xiao, Q. Zuo and L. Ding, *Energy Environ. Sci.*, 2013, **6**, 3224–3228.
66. Q. Fan, W. Su, X. Guo, B. Guo, W. Li, Y. Zhang, K. Wang, M. Zhang and Y. Li, *Adv. Energy Mater.*, 2016, **6**, 1600430.
67. S. Albrecht, S. Janietz, W. Schindler, J. Frisch, J. Kurpiers, J. Kneipert, S. Inal, P. Pingel, K. Fostiropoulos, N. Koch and D. Neher, *J. Am. Chem. Soc.*, 2012, **134**, 14932–14944.
68. J. Peet, J. Y. Kim, N. E. Coates, W. L. Ma, D. Moses, A. J. Heeger and G. C. Bazan, *Nat. Mater.*, 2007, **6**, 497–500.
69. H. Lv, X. Zhao, Z. Li, D. Yang, Z. Wang and X. Yang, *Polym. Chem.*, 2014, **5**, 6279–6286.
70. L. Dou, C. C. Chen, K. Yoshimura, K. Ohya, W. H. Chang, J. Gao, Y. Liu, E. Richard and Y. Yang, *Macromolecules*, 2013, **46**, 3384–3390.
71. Z. Li, S. W. Tsang, X. Du, L. Scoles, G. Robertson, Y. Zhang, F. Toll, Y. Tao, J. Lu and J. Ding, *Adv. Funct. Mater.*, 2011, **21**, 3331–3336.
72. T. Y. Chu, J. Lu, S. Beaupre, Y. Zhang, J. R. Pouliot, S. Wakim, J. Zhou, M. Leclerc, Z. Li, J. Ding and Y. Tao, *J. Am. Chem. Soc.*, 2011, **133**, 4250–4253.
73. C. E. Small, S. Chen, J. Subbiah, C. M. Amb, S. W. Tsang, T. H. Lai, J. R. Reynolds and F. So, *Nat. Photonics*, 2012, **6**, 115–120.
74. X. Guo, N. Zhou, S. J. Lou, J. W. Hennek, R. P. Ortiz, M. R. Butler, P. L. T. Boudreault, J. Strzalka, P. O. Morin, M. Leclerc, J. T. L. Navarette, M. A. Ratner, L. X. Chen, R. P. H. Chang, A. Facchetti and T. J. Marks, *J. Am. Chem. Soc.*, 2012, **134**, 18427–18439.

75. H. Y. Chen, J. Hou, S. Zhang, Y. Liang, G. Yang, Y. Yang, L. Yu, Y. Wu and G. Li, *Nat. Photonics*, 2009, **3**, 649–653.
76. Y. Liang, Z. Xu, J. Xia, S. T. Tsai, Y. Wu, G. Li, C. Ray and L. Yu, *Adv. Energy Mater.*, 2010, **22**, E135–E138.
77. Z. He, C. Zhong, S. Su, M. Xu, H. Wu and Y. Cao, *Nat. Photonics*, 2012, **6**, 591–595.
78. L. Huo, S. Zhang, X. Guo, F. Xu, Y. Li and J. Hou, *Angew. Chem.*, 2011, **123**, 9871–9876.
79. S. Zhang, L. Ye, W. Zhao, D. Liu, H. Yao and J. Hou, *Macromolecules*, 2014, **47**, 4653–4659.
80. Z. He, B. Xiao, F. Liu, Y. Yang, S. Xiao, C. Wang, T. P. Russell and Y. Cao, *Nat. Photonics*, 2015, **9**, 174–179.
81. C. Cui, W. Y. Wong and Y. Li, *Energy Environ. Sci.*, 2014, **7**, 2276–2284.
82. C. Cui, Z. He, Y. Wu, X. Cheng, H. Wu, Y. Li, Y. Cao and W. Y. Wong, *Energy Environ. Sci.*, 2016, **9**, 885–891.
83. H. Yao, W. Zhou, Z. Zheng, Y. Cui, J. Zhang, Z. Wei and J. Hou, *J. Mater. Chem. A*, 2016, **4**, 1708–1713.
84. K. Li, Z. Li, K. Feng, X. Xu, L. Wang and Q. Peng, *J. Am. Chem. Soc.*, 2013, **135**, 13549–13557.
85. H. Zhou, L. Yang, A. C. Stuart, S. C. Price, S. Liu and W. You, *Angew. Chem.*, 2011, **123**, 3051–3054.
86. N. Wang, Z. Chen, W. Wei and Z. Jiang, *J. Am. Chem. Soc.*, 2013, **135**, 17060–17068.
87. M. Zhang, Y. Gu, X. Guo, F. Liu, S. Zhang, L. Huo, T. P. Russell and J. Hou, *Adv. Mater.*, 2013, **25**, 4944–4949.
88. L. Dou, W. H. Chang, J. Gao, C. C. Chen, J. You and Y. Yang, *Adv. Mater.*, 2013, **25**, 825–831.
89. C. Cabanetos, A. El Labban, J. A. Bartlet, J. D. Douglas, W. R. Mateker, J. M. J. Frechet, M. D. McGhee and P. M. Beaujuge, *J. Am. Chem. Soc.*, 2013, **135**, 4656–4659.
90. C. Piliego, T. W. Holcombe, J. D. Douglas, C. H. Woo, P. M. Beaujuge and J. M. J. Frechet, *J. Am. Chem. Soc.*, 2010, **132**, 7595.
91. H. C. Chen, Y. H. Chen, C. C. Liu, Y. C. Chien, S. W. Chou and P. T. Chou, *Chem. Mater.*, 2012, **24**, 4766–4772.
92. Y. Wu, Z. Li, W. Ma, Y. Huang, L. Huo, X. Guo, M. Zhang, H. Ade and J. Hou, *Adv. Mater.*, 2013, **25**, 3449–3455.
93. H. J. Son, L. Lu, W. Chen, T. Xu, T. Zheng, B. Carsten, J. Strzalka, S. B. Darling, L. X. Chen and L. Yu, *Adv. Mater.*, 2013, **25**, 838–843.
94. L. Huo, T. Liu, X. Sun, Y. Cai, A. J. Heeger and Y. Sun, *Adv. Mater.*, 2015, **27**, 2938–2944.
95. I. Osaka, T. Kakara, N. Takemura, T. Koganezawa and K. Takimiya, *J. Am. Chem. Soc.*, 2013, **135**, 8834–8837.
96. Z. Genene, J. Wang, X. Meng, W. Ma, X. Xu, R. Yang, W. Mammo and E. Wang, *Adv. Electron. Mater.*, 2016, **2**, 1600084.
97. D. Liu, Q. Zhu, C. Gu, J. Wang, M. Qiu, W. Chen, X. Bao, M. Sun and R. Yang, *Adv. Mater.*, 2016, **28**, 8490–8498.

98. E. Wang, L. Wang, L. Lan, C. Luo, W. Zhuang, J. Peng and Y. Cao, *Appl. Phys. Lett.*, 2008, **92**, 33307.
99. G. Li, C. Kang, X. Gong, J. Zhang, C. Li, Y. Chen, H. Dong, W. Hu, F. Li and Z. Bo, *Macromolecules*, 2014, **47**, 4645–4652.
100. C. Du, C. Li, W. Li, X. Chen, Z. Bo, C. Veit, Z. Ma, U. Wuerfel, H. Zhu, W. Hu and F. Zhang, *Macromolecules*, 2011, **44**, 7617–7624.
101. Q. Liu, C. Li, E. Jin, Z. Lu, Y. Chen, F. Li and Z. Bo, *ACS Appl. Mater. Interfaces*, 2014, **6**, 1601–1607.
102. R. Qin, W. Li, C. Li, C. Du, C. Veit, H. F. Schleiermacher, M. Andersson, Z. Bo, Z. Liu, O. Inganas, U. Wuerfel and F. Zhang, *J. Am. Chem. Soc.*, 2009, **131**, 14612–14613.
103. Y. Deng, J. Liu, J. Wang, L. Liu, W. Li, H. Tian, X. Zhang, Z. Xie, Y. Geng and F. Wang, *Adv. Mater.*, 2014, **26**, 471–476.
104. Y. C. Chen, C. Y. Yu, Y. L. Fan, L. I. Hung, C. P. Chen and C. Ting, *Chem. Commun.*, 2010, **46**, 6503–6505.
105. Y. X. Xu, C. C. Chueh, H. L. Yip, F. Z. Ding, Y. X. Li, C. Z. Li, X. Li, W. C. Chen and A. K. Y. Jen, *Adv. Mater.*, 2012, **24**, 6356–6361.
106. J. J. Intemann, K. Yao, Y. X. Li, H. L. Yip, Y. X. Xu, P. W. Liang, C. C. Chueh, F. Z. Ding, X. Yang, X. Li, Y. Chen and A. K. Y. Jen, *Adv. Funct. Mater.*, 2014, **24**, 1465–1473.
107. Y. Huang, X. Guo, F. Liu, L. Huo, Y. Chen, T. P. Russell, C. C. Han, Y. Li and J. Hou, *Adv. Mater.*, 2012, **24**, 3383.
108. X. Guo, S. R. Puniredd, M. Baumgarten, W. Pisula and K. Müllen, *J. Am. Chem. Soc.*, 2012, **134**, 8404.
109. J. J. Intemann, K. Yao, H. L. Yip, Y. X. Xu, Y. X. Li, P. W. Liang, F. Z. Ding, X. Li and A. K. Y. Jen, *Chem. Mater.*, 2013, **25**, 3188–3195.
110. R. S. Ashraf, B. C. Schroeder, H. A. Bronstein, Z. Huang, S. Thomas, R. J. Kline, C. J. Brabec, P. Rannou, T. D. Anthopoulos, J. R. Durrant and I. McCulloch, *Adv. Mater.*, 2013, **25**, 2029–2034.
111. X. Guo, M. Zhang, J. Tan, S. Zhang, L. Huo, W. Hu, Y. Li and J. Hou, *Adv. Mater.*, 2012, **24**, 6536–6541.
112. A. C. Knall, R. S. Ashraf, M. Nikolka, C. B. Nielsen, B. Purushothaman, A. Sadnanala, M. Hurhangee, K. Broch, D. J. Harkin, J. Novak, M. Neophytou, P. Hayoz, H. Sirringhaus and I. McCulloch, *Adv. Funct. Mater.*, 2016, **26**, 6961–6969.
113. M. Jørgensen, K. Norrman and F. C. Krebs, *Sol. Energy Mater. Sol. Cells*, 2008, **92**, 686–714.
114. F. C. Krebs, *Stability and Degradation of Organic and Polymer Solar Cells*, Wiley, Chichester, 2012.
115. C. J. Brabec, S. Gowrisanker, J. J. M. Halls, D. Laird, S. Jia and S. P. Williams, *Adv. Mater.*, 2010, **22**, 3839–3856.
116. T. R. Andersen, T. T. Larsen-Olsen, B. Andreasen, A. P. L. Böttiger, J. E. Carlé, M. Helgesen, E. Bundgaard, K. Norrman, J. W. Andreasen, M. Jørgensen and F. C. Krebs, *ACS Nano*, 2011, **5**, 4188–4196.
117. P. G. Jessop, *Green Chem.*, 2011, **13**, 1391–1398.
118. Y. Lin and X. Zhan, *Mater. Horiz.*, 2014, **1**, 470–488.

119. S. Dupont, M. Oliver, F. C. Krebs and R. Dauskardt, *Sol. Energy Mater. Sol. Cells*, 2012, **97**, 171–175.
120. M. Mojica, J. Alonso and F. Mendez, *J. Phys. Org. Chem.*, 2013, **26**, 526–539.
121. Y. Lin, J. Wang, Z. G. Zhang, H. Bai, Y. Li, D. Zhu and X. Zhan, *Adv. Mater.*, 2015, **27**, 1170–1174.
122. W. Zhao, D. Qian, S. Zhang, S. Li, O. Inganas, F. Gao and J. Hou, *Adv. Mater.*, 2016, **28**, 4734–4739.

CHAPTER 4

p-Type Molecular Materials for Organic Solar Cells

ADAM C. YEATS, RUPERT G. D. TAYLOR AND PETER J. SKABARA[†]*

WestCHEM, University of Strathclyde, Department of Pure and Applied Chemistry, 295 Cathedral Street, Glasgow, G1 1XL, United Kingdom
*E-mail: peter.skabara@glasgow.ac.uk

4.1 Introduction

Since the discovery of the photoelectric effect by Becquerel in 1839,[1,2] interest in the generation of electricity from light has grown tremendously, with an exponential increase in the number of publications in the field. Although the most prominently reported organic materials are polymeric, there is an increasing focus on conjugated oligomers.[3,4] These two fields have grown rapidly and at the same rate for the past ten years, as highlighted in Figure 4.1. One of the largest areas of research is the development of small molecule donors for organic photovoltaic (OPV) devices because they show high potential for the construction of low-cost energy-generating devices.[5,6]

The interest in conjugated oligomers has been brought about due to the numerous advantages they hold over their polymeric counterparts. The main driving force is that the cost of synthesis for small molecules is far lower than that for polymers because lengthy purification steps such as Soxhlet

[†]New address: School of Chemistry, University of Glasgow, Glasgow, G12 8QQ, United Kingdom.

Nanoscience & Nanotechnology Series No. 45
Nanostructured Materials for Type III Photovoltaics
Edited by Peter Skabara and Mohammad Azad Malik

Published by the Royal Society of Chemistry, www.rsc.org

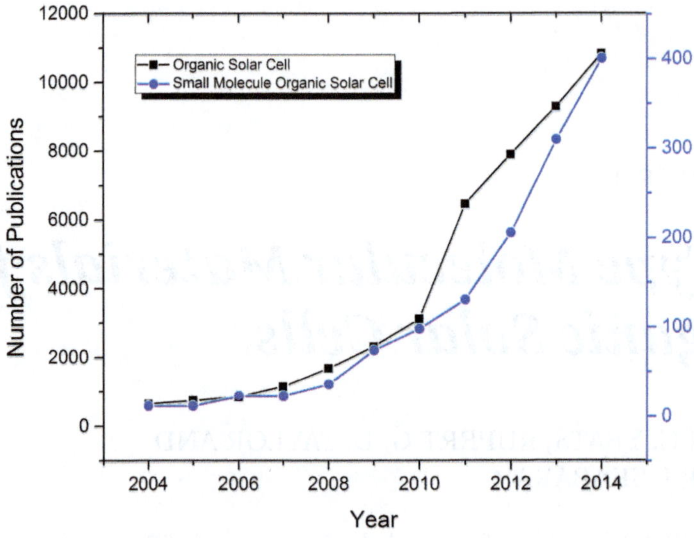

Figure 4.1 Number of hits from the search terms "organic solar cell" and "small molecule solar cell" from 2004–2014. Data collected from Web of Knowledge.

extraction can be avoided and lower temperatures can be used for some materials. In addition, oligomeric materials have well-defined, monodispersed structures, which can readily be reproduced and the purification of oligomers is typically less troublesome than that of polymers.

Although the champion power conversion efficiencies (PCEs) obtained using polymeric materials in the early years of the field out-performed those of small molecules by a large margin, this has been narrowing over the last few years, as demonstrated in Figure 4.2.

The first reported organic molecule fabricated into a donor–acceptor (D–A) OPV device was chlorophyll A by Tang *et al.*[7] in 1975. The device was prepared by the electrodeposition of chlorophyll onto a quartz disc coated by metals with a different work function. This first organic solar cell exhibited a PCE of around 0.001%. Building on this early work, metal phthalocyanines became the focus of study. The field has advanced immensely since this time and current small molecule devices are achieving PCEs in the region of 10%.

4.1.1 Operating Principles

The basic operating principles of an organic solar cell are outlined in Figure 4.3. The active layer material must first be reached by an incident photon of sufficient energy such that a bound electron–hole pair is formed – this is known as an exciton. Once formed, the exciton begins to migrate from the donor layer towards the D–A interface. At this stage, if the exciton does not reach an interface within a small enough distance (typically ~10 nm),[8–10] the coulombically bound electron and hole will recombine and decay, resulting

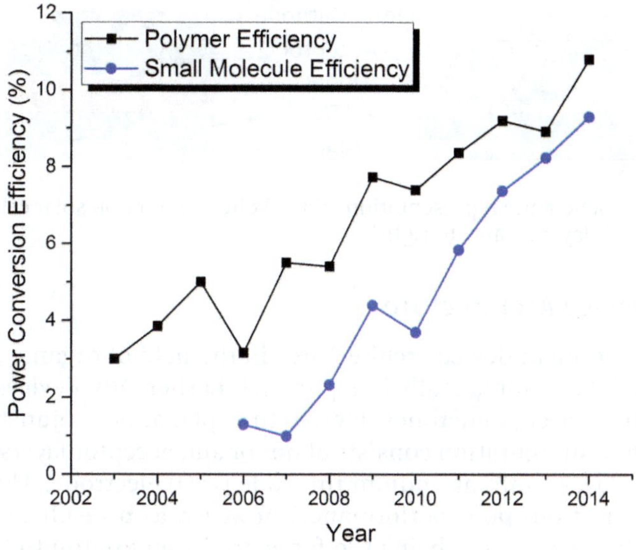

Figure 4.2 Highest reported PCEs of polymer and small molecule solar cells. Data from ref. 137–139.

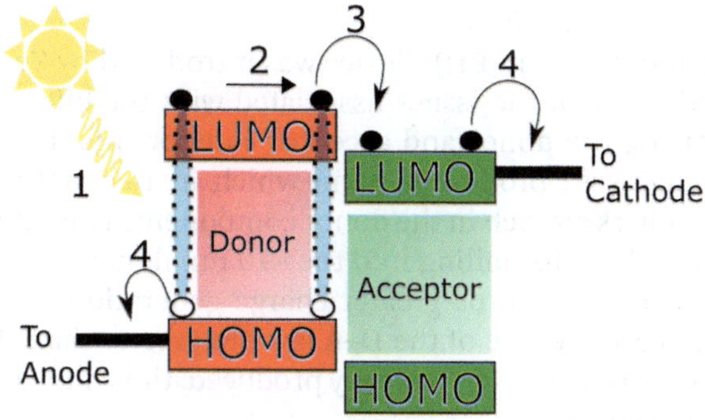

Figure 4.3 Operating principle of organic solar cells. 1, Absorption of a photon to create a coulombically bound electron–hole pair (exciton). The photoexcitation occurs in the donor material. 2, Diffusion of the electron–hole pair to the D–A interface. Any excitons that do not reach the interface do not contribute to the generation of current. 3, Separation of the electron–hole pair into free charge carriers. 4, Transport of the charge carriers to their respective external electrodes to generate charge within the circuit.

in no charge generation. If the exciton encounters an interface within a short enough distance, the bound electron–hole pair may then separate into a free electron and hole. The electron–hole pair then begin the transport of charge to their corresponding electrodes. If the free charges are successfully transported through the material and reach the electrodes, a charge is then produced within the circuit.

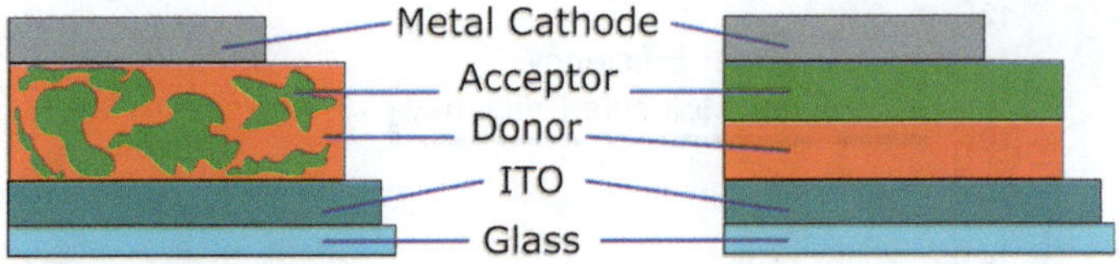

Figure 4.4 Schematic representation of a bulk heterojunction solar cell (left) and a bilayer solar cell (right).

4.1.2 Device Architecture

There are two main device architectures in the field of organic solar cells, represented diagrammatically in Figure 4.4. Earlier OPV devices were prepared in the bilayer (sometimes referred to as planar heterojunction) architecture. This configuration consists of donor and acceptor layers deposited as discrete layers onto an indium tin oxide (ITO) electrode. This configuration suffers from poor performance, however, as a result of the donor and acceptor layers often being too far apart for an exciton to travel to an interface. Reducing the overall thickness of the layers increases the probability of the exciton reaching an interface, but a thin layer will not absorb as much light. Accordingly, a balance must be struck in the thickness of the layers for good light absorption while still allowing for adequate charge drift so that current may flow.

The bulk heterojunction (BHJ) device was introduced by Yu *et al.*[11] in 1995 and addressed many of the issues associated with the bilayer device architecture. By mixing the donor and acceptor components in a single layer, a number of domains are produced – some which are rich in the acceptor component, while others are rich in the donor component. This allows far shorter exciton diffusion lengths, falling into the ~10 nm domain range required to avoid exciton decay, and thus greater charge generation. In addition, it is possible to increase the size of the D–A layer to greater than the 50–100 nm range in which bilayer cells are typically produced, thus increasing the potential photon harvesting.

4.2 Designing OPV Materials

The design of materials expected to be useful as active components in OPV devices has become more refined over the past decade, with numerous strategies being utilised and a deeper understanding of the processes involved. A number of factors are taken into account when novel materials are designed, such as the gap between the highest occupied molecular orbital (HOMO) and the lowest unoccupied molecular orbital (LUMO) of the material (E_g), the absorption of the solar spectrum, solubility and ease of synthesis.

The ideal material should show broad absorption over the visible to near-IR range of the solar spectrum to allow the maximum harvesting of photons. The bandgap of an ideal material should be in the region of 1–2 eV,[12] with an offset of ~0.3 eV[13] between the LUMO of the donor material and the LUMO

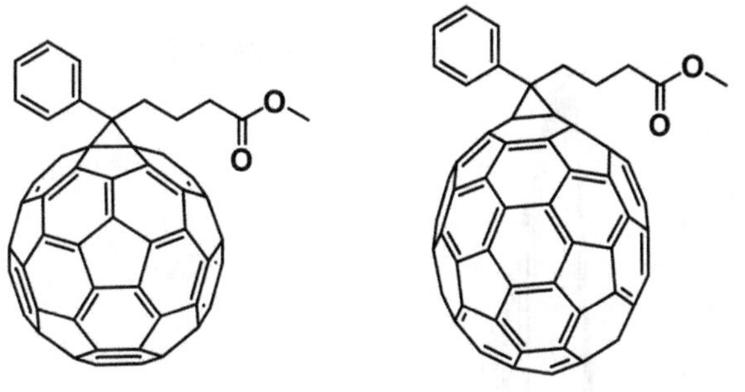

$PC_{60}BM$ $PC_{70}BM$

Figure 4.5 Chemical structures of the fullerene acceptors [6,6]-phenyl-C61-butyric acid methyl ester ($PC_{60}BM$) and [6,6]-phenyl-C71-butyric acid methyl ester ($PC_{70}BM$).

of the acceptor material (such as the commonly used $PC_{60}BM$ and $PC_{70}BM$; Figure 4.5).[14] The material should ideally be soluble in "green" solvents such as water or ethanol for lower costs and safer processing, and should ideally be stable in air so it may be used without the need of an encapsulant. When devices are being measured, the standard conditions are to illuminate under artificial sunlight, specifically the conditions set out by the American Society for Testing and Materials in the reference Air Mass 1.5 Global spectra (AM1.5G, Figure 4.6).[15] This spectrum accounts for both direct and diffuse radiation from the Sun with an integrated intensity of 1000 W m^{-2}.

4.2.1 Bandgap Engineering

One key principle of the design of organic semiconductors for use within photovoltaics is the ability to tune the bandgap of the material.[16–20] The bandgap is a crucial property of an organic solar cell material because it directly relates to the potential PCE. A bandgap of 1–2 eV[12] is ideal for solar cell applications, as demonstrated by Schockley and Queisser.[21]

The bandgap can be tuned by adjusting five components of a molecular system. This is outlined for a thiophene system in Figure 4.7. For a more thorough explanation of this concept, the reader is directed towards the excellent publications by Roncali, which explore the following terms and methods in greater depth.[22,23]

The factors relating to the bandgap, E_g, are as follows.

1. $E_{\delta r}$ – the energy associated with the bond length alternation. The bond length alternation refers to the difference in the average bond length of single and multiple bonds. There is no alternation in a system where the multiple C–C bonds are perfectly delocalised and therefore the material has no bandgap, effectively acting as a metal. Conjugated organic materials such as those used in OPV devices do not have perfectly delocalised electrons and therefore they do exhibit a bandgap.

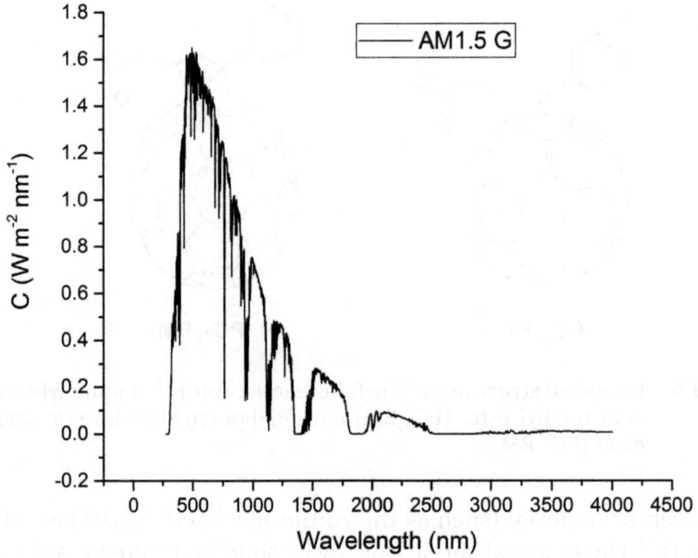

Figure 4.6 The American Society for Testing and Materials (ASTM) AM1.5G reference solar spectrum, which is used for standard OPV device measurements. Reproduced with permission from ref. 15, copyright ASTM International, 100 Barr Harbor Drive, West Conshohocken, PA 19428.

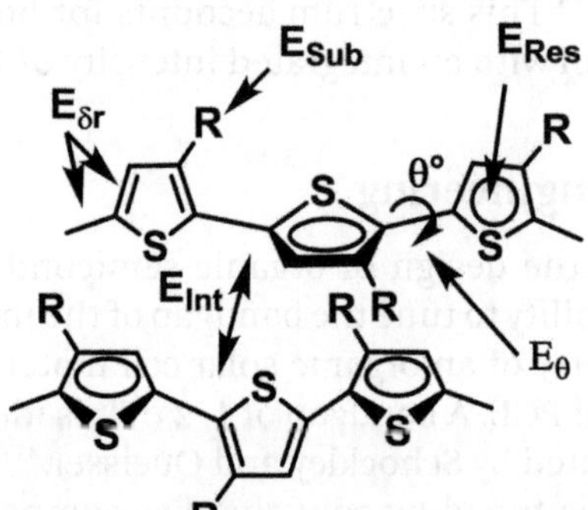

Figure 4.7 Factors influencing the energy bandgap for organic solar cell materials, shown using a terthiophene molecular system. Adapted with permission from ref. 23. Copyright (1997) American Chemical Society.

2. E_{Sub} – the energy associated with the resonance and/or inductive effects of substituents along the conjugated chain. Varying the substituents of the material is an incredibly versatile technique that may be used to alter the bandgap of a material. Variations in the substituents can have a marked effect on the material as a result of the potentially large changes they can impart to the electron density of the system. By introducing an electron-rich substituent to the conjugated chain,

the electron density is increased, thereby increasing the energy levels of the HOMO and LUMO to different degrees. Electron-deficient units pull electron density out of the system and therefore result in a reduction in the energy of the LUMO (and HOMO) levels. Some examples of electron-donating and electron-accepting fragments are shown in Figure 4.8.

3. E_θ – the energy associated with the dihedral angle between the units. In a hypothetical conjugated molecule in which the dihedral angles are all 0°, the effective conjugation length is maximised and the bandgap is minimised due to the greatly enhanced overlap of the π-orbitals. Conversely, a large dihedral angle leads to inferior π-orbital overlap between the sp^2 or sp units and a larger bandgap results. We can alter this property by forcing planarisation through fused ring systems (Figure 4.9) or by invoking non-covalent intramolecular interactions between the heteroatoms on adjacent units (Figure 4.10).[24–27]
4. E_{Res} – the energy associated with the resonance in each unit. Aromatic systems are more stable than their quinoidal counterparts and, as a result, there is an associated energy barrier that must be met to reach

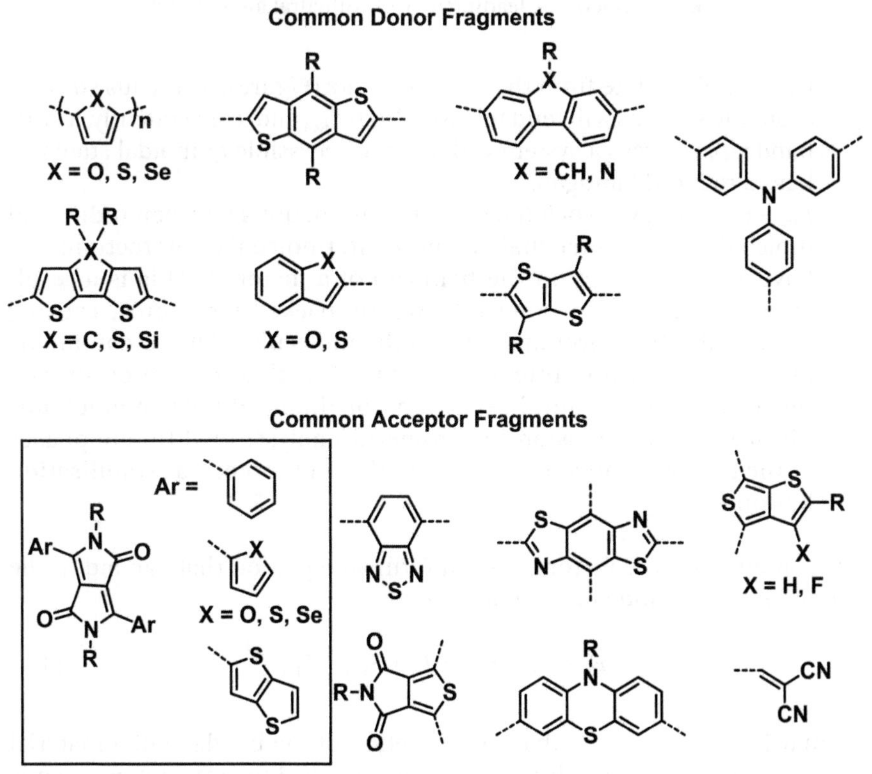

Figure 4.8 Common donor and acceptor fragments used in OPV devices to help to control the bandgap.

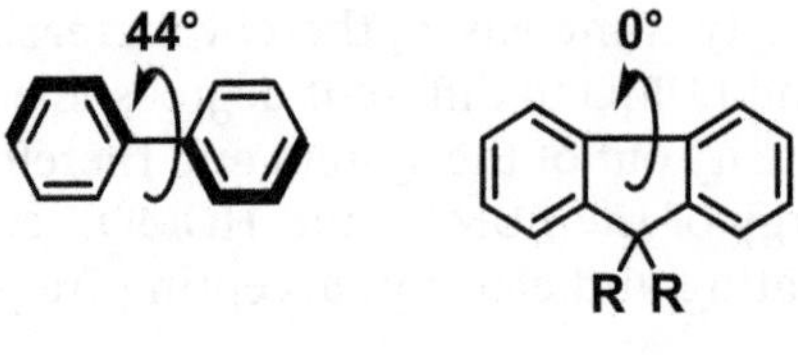

Figure 4.9 Planarisation through the fused ring systems. Biphenyl (left) has a dihedral angle of 44°, whereas fluorene (right) is planar. Data taken from ref. 140.

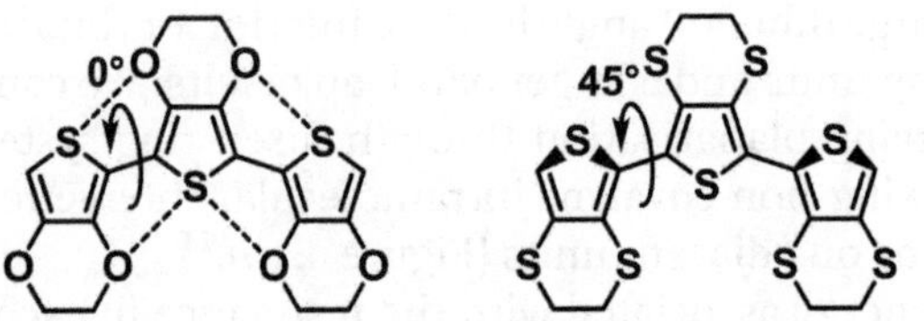

Figure 4.10 Planarisation using favourable non-covalent interactions. In this example, tris-EDOT (left) exhibits favourable S--O interactions, resulting in a dihedral angle of 0°,[141,142] whereas tris-EDTT (right) lacks these interactions, leading to a 45° dihedral angle.[141,142]

the quinoidal state from the aromatic state (Figure 4.11). Thus strongly aromatic systems will tend to have a high E_{Res} and, consequently, a large bandgap, whereas a system with a more accessible quinoidal state will have a reduced bandgap.

5. E_{Int} – the energy associated with the interactions between individual molecules or polymer chains. These intermolecular interactions can have a large impact on the bandgap of a material. This is a result of the molecules forming into tightly packed structures, causing increased orbital overlap between the molecules. This is commonly due to π–π stacking interactions, but it is also seen to occur as a result of heteroatomic interactions or dipole–dipole interactions. Absorption and emission spectrometric analyses on different phases of the material show the effect that the intermolecular stabilisation has on E_g.

By combining these factors, we can form an equation that can aid in the estimation of the bandgap of a material:

$$E_g = E_{\delta r} + E_{Sub} + E_{\theta} + E_{Res} + E_{Int} \tag{4.1}$$

It may intuitively appear that the reduction of the bandgap of a material to as close to 0 eV as possible would provide an ideal material for an organic solar cell; however, this is not the case, as demonstrated by Shockley and Queisser.[21]

Figure 4.11 The aromatic form of polythiophene (left) is lower in energy than the quinoidal form (right), but this is not necessarily the case for more complex polymers such as PITN.[143–145]

4.2.2 Power Conversion Efficiency

The ultimate measure of performance in the field of solar cells is the PCE, η, because this represents the amount of harvested sunlight that is converted into electrical energy. This is the data point that researchers wish to maximise through molecular design, device structure and any other methods at their disposal, and is the definitive measure of success for a material.

The PCE of a device can be calculated from eqn (4.2):

$$\eta = \frac{V_{OC} \times J_{SC} \times FF}{P_{in}} \tag{4.2}$$

V_{OC} is defined as the open-circuit voltage. This is the voltage produced by the cell when there is no current flowing (*i.e.* $J = 0$) under illumination. Work by Scharber[12] has shown that, to maximise the PCE, there is a key dependence on the offset of the HOMO of the donor material and the LUMO of the acceptor material that correlates with the V_{OC} of the overall system; a low V_{OC} is associated with poor morphology and/or low charge transport.[10] The V_{OC} may be estimated as:

$$V_{OC} = \frac{1}{e}\left(\left|\text{HOMO}_{\text{Donor}}\right| - \left|\text{LUMO}_{\text{Acceptor}}\right|\right) - 0.3\ \text{V} \tag{4.3}$$

where e is the elementary charge and 0.3 V is an empirical factor determined by Scharber *et al.*[12] in a study of 26 devices using conjugated materials as donors and $PC_{71}BM$ as the acceptor. This value corresponds to the typical loss found in a BHJ solar cell.[13]

J_{SC} is the short-circuit current density of the device, which is the current density at short circuit (*i.e.* $V = 0$) under illumination. The J_{SC} relates to the generation of free charge carriers, which are formed from photon absorption. Low values of J_{SC} may be due to short exciton diffusion lengths, poor charge transfer between the donor and acceptor, or charge traps. In the ideal situation (no losses at contacts) the J_{SC} is equal to:

$$J_{SC} = \frac{ne\mu E}{A} \tag{4.4}$$

where n is the charge carrier density, e is the elementary charge (1.602×10^{-19} C), μ is the carrier mobility of the device – a measure of the ease with which a charge carrier can move through the material in the presence of an applied electric field – E is the electric field and A is the area of the device.[10] The J_{SC} of a device can also be expressed in terms of the light absorbed by integrating the photon flux over the absorbed wavelengths of the solar spectrum within a given external quantum efficiency (EQE) limit:

$$J_{SC} = A_{Cell}^{-1} \int_{\lambda=0}^{\lambda=\lambda_g} \phi_p(\lambda)\eta_{EQE}(\lambda)d\lambda \tag{4.5}$$

where A_{Cell} is the area of the OPV cell, λ is the wavelength of the light, λ_g is the longest absorbed wavelength (corresponding to the optical bandgap), $\phi_p(\lambda)$ is the photon flux and η_{EQE} is the EQE limit for the given wavelength.[28]

The J_{SC} can be increased by an improvement in the device mobility, which is achieved by optimising the nanoscale morphology of the active layer:[29–31]

$$FF = \frac{P_{Max}}{J_{SC} \times V_{OC}} = \frac{J_{Max} \times V_{Max}}{J_{SC} \times V_{OC}} \tag{4.6}$$

The fill factor is defined as the percentage of power generated by the device relative to the maximum theoretically generated power. High-efficiency solar cells require the fill factor to reach as close to the maximum value of 1 as possible. Methods to increase the fill factor tend to involve altering the preparation of the device, such as thermal or solvent annealing. As a result of methodical, meticulous processing, there have been reports of increased PCEs compared with the same material processed without annealing.[32–34] Other techniques to control the morphology of the active layer – for example, the use of additives such as 1,8-diiodooctane (DIO) – have a marked effect on the fill factor. When DIO is used as an additive in the fabrication of a solar cell, it helps to solubilise the fullerene acceptor and also encourages the formation of smaller, better mixed regions of donor and acceptor domains. By determining the optimum amount of DIO to add (often in the range 1–3%), it is possible to strike a balance between the size of the donor and acceptor domains that is favourable for the generation and collection of charge carriers, as highlighted in Figure 4.12. A detailed analysis of the effect of the quantity of DIO used was published in 2015 by Magerle and co-workers.[35]

These values are extracted from a current–voltage (J–V) plot of the fabricated device under illumination, typically under the standard AM1.5G conditions. A typical J–V curve is shown schematically in Figure 4.13.

As eqn (4.2) shows, to achieve a high PCE the values of V_{OC}, J_{SC} and the fill factor must be maximised simultaneously. To maximise V_{OC}, the frontier orbitals (HOMO and LUMO) must lie at suitable energies so that they are well aligned with those of the acceptor. There must be an offset between the $LUMO_{Donor}$ and the $LUMO_{Acceptor}$ of at least 0.3 eV to act as a driving force for exciton dissociation – that is, the $LUMO_{Donor}$ must be at least 0.3 eV higher in energy than the $LUMO_{Acceptor}$.[12,13] A balance must be struck between the energy of the $HOMO_{Donor}$ and $LUMO_{Acceptor}$ while maintaining this >0.3 eV difference in the $LUMO_{Donor}$ and $LUMO_{Acceptor}$. This is outlined diagrammatically in Figure 4.14.

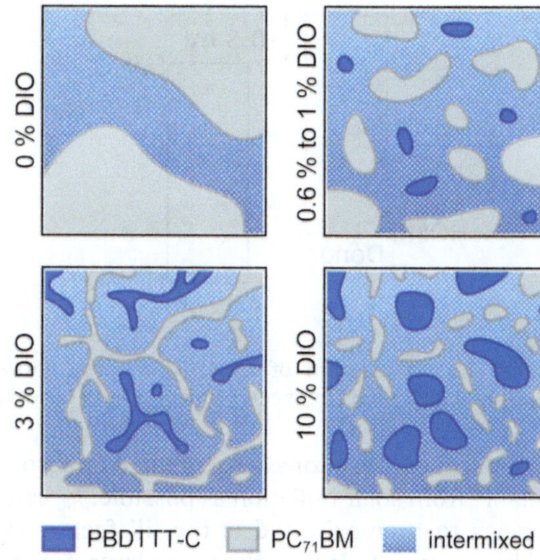

Figure 4.12 Schematic representation of the effects of various quantities of DIO. Reprinted by permission from Macmillan Publishers Ltd: *Scientific Reports* (ref. 35), copyright 2015.

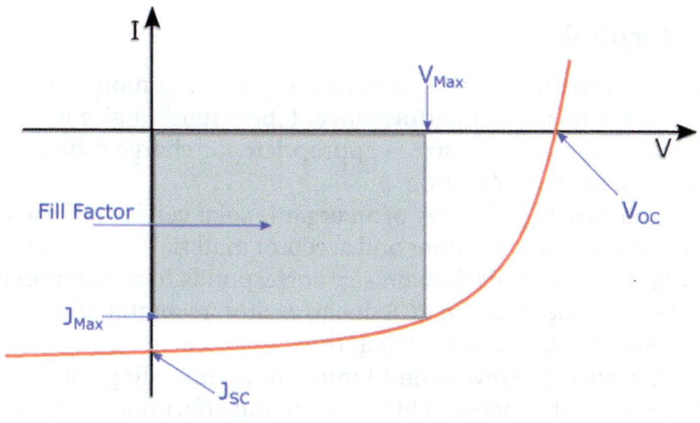

Figure 4.13 Representative *J–V* curve for an OPV device.

It is important to note that the described estimation is based on empirical evidence gathered by Scharber *et al.*[12] There are a number of other models and empirically derived relationships addressing the source of the 0.3 eV loss, but this is still a source of debate. A comprehensive review of the origin and mechanisms related to the V_{OC} has been published by Elumalai and Uddin.[36]

The J_{SC} of a device is the maximum current that can be extracted from a device after taking into consideration losses from poor morphology and losses due to series and shunt resistance. A broad absorbance must be realised across the spectrum to maximise this parameter; see eqn (4.5). By ensuring

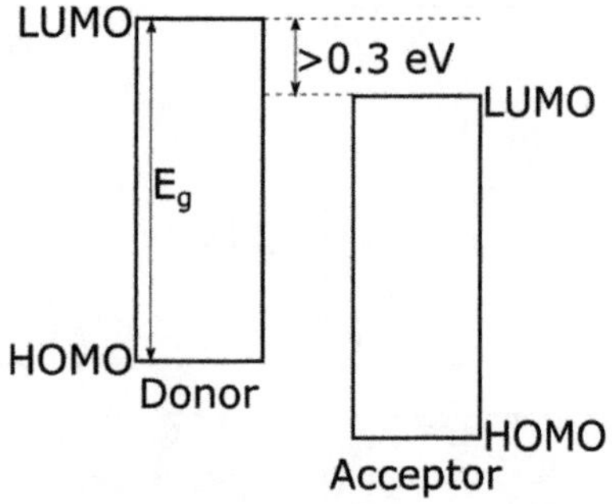

Figure 4.14 HOMO–LUMO diagram of an OPV material showing E_g and the required 0.3 eV offset between $LUMO_{Donor}$ and $LUMO_{Acceptor}$.

that the maximum number of excitons are generated by absorbing as many of the photons available from solar radiation as possible, J_{SC} can be maximised.

The final property to be maximised is the fill factor. The fill factor is strongly affected by the morphology of the active layer. A number of processing techniques are used to try to improve the morphology and thus increase the fill factor, such as solvent or thermal annealing, interface engineering or the use of solvent additives.[37,38]

4.2.3 Morphology

In addition to the materials used in OPV devices, the morphological structure plays a crucial part in performance. Fibres tend to give a good performance because the domain size is appropriate for charge diffusion without the occurrence of recombination.

The ideal surface morphology of an organic solar cell is one in which there are well-mixed domains of donor and acceptor material with a domain size of the order of 8–10 nm. This domain size corresponds to the exciton diffusion length – there will be losses in PCE due to exciton recombination in a system where the domain size is greater than this.

Techniques such as solvent and temperature annealing can have a large effect on device performance. This is due to the effect that annealing has on the material in terms of crystallinity, light harvesting and molecular ordering.

4.3 Thiophene-based Materials

Oligothiophenes play a vital part in the field of OPV devices and make up a huge amount of the reported donor materials. These small molecules have been extensively studied; for a more thorough background, the reader is directed to the comprehensive 2009 review article by Bäuerle and co-workers.[39]

One of the early pieces of work in the field of small molecule OPV materials was published by Roncali in 2006. In this work, the group outlined a series of silicon-centred oligothiophenes, achieving a PCE of 0.29% in a BHJ device with $PC_{60}BM$ as the acceptor with their best material **4.3.1b** (Figure 4.15).

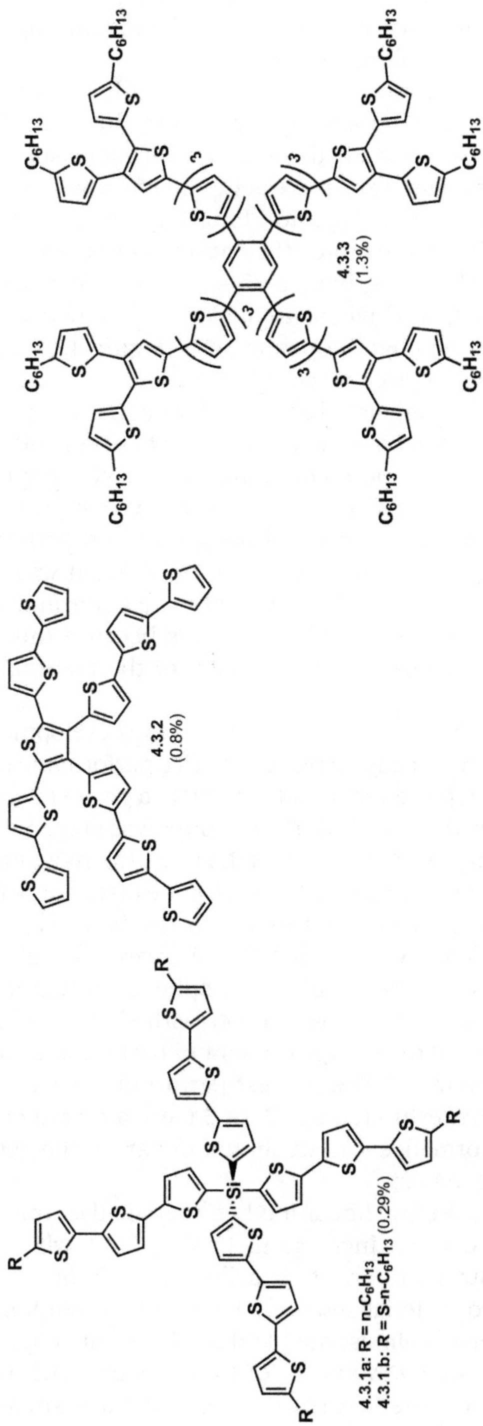

Figure 4.15 Early examples of oligothiophene materials for OPV devices.

This result was far behind the high-performance polymers of the time,[40,41] but was a leap forward for small molecules, demonstrating that they could indeed be used in photovoltaic devices.

The work on oligothiophenes continued rapidly, with Sun *et al.* reaching 0.8% PCE in 2006 using a thiophene centre instead of silicon (**4.3.2**; Figure 4.15).[42] The group studied the effects of increasing the number of thiophene units per chain, thus increasing the effective conjugation length. They found that a longer conjugation length typically leads to better performing materials due to the more efficient transfer of electrons from longer chain oligomers.[42] This slow, but notable, trend in increasing PCEs using small molecules continued when Kopidakis and co-workers prepared the analogous phenyl-cored oligothiophene **4.3.3** (Figure 4.15) and successfully surpassed the 1% PCE barrier, reaching 1.3%.[43]

The extension of the conjugated chain of thiophenes was pushed further by Bäuerle and co-workers, who created a series of dendritic oligothiophenes.[44] The group created dendrimers containing up to 90 thiophene units and found that the general trend of increased effective conjugation length lead to an improvement in performance, although the best performing oligomer was one containing 42 thiophene units (1.7% PCE). It was concluded that the PCE could be improved further by developing dendrimers that absorbed at lower energies than the described systems because this could produce better matching of the absorption spectrum of the material with the solar spectrum.

These early results helped to pave the way for the development of oligothiophene materials, with a steady increase in device performance being recorded since these initially reported materials. In 2010, a group of simple oligothiophenes with terminal dicyanovinyl (DCV) groups was prepared and tested (see Figure 4.17 for examples). Compound **4.3.4a** was the most impressive material, producing a promising PCE of 3.7%, with the fill factor being the limiting parameter.[45] This result compared well with the bare oligomer, α-sexithiophene (**4.3.5a**),[46] which saw a PCE of 2.38% increased to 5.2% when the DCV groups were appended to the peripheral thiophenes in compound **4.3.5b.**[47]

Replacement of the DCV groups of **4.3.4a** with electron-withdrawing alkyl cyanoacetate groups led to an improvement in performance with each of the three alkyl chain variants.[48] The highest performing device, **4.3.4b** bearing an octylcyanoacrylate, exhibited a PCE of 5.08%, a record efficiency at that time. The high performance was attributed in part to the good fill factor of 0.55 and J_{SC} of 10.74 mA cm^{-2}.

Incorporation of a 3-ethylrhodanine heterocyclic dye unit **4.3.4c** by Chen and co-workers[49] led to an increase in PCE to 6.1% while maintaining a simple molecular structure and the facile synthesis of **4.3.4a** and **4.3.4b.** The rhodanine-based material showed a high molar absorption coefficient of 9×10^4 L mol^{-1}, which is almost double that of the bare oligothiophene, and exhibited a good V_{OC} of 0.92 V and J_{SC} of 13.98 mA cm^{-2}, with a moderate fill factor of 0.47. This was the first reported example of a small molecule BHJ OPV device to surpass a PCE of 6%.

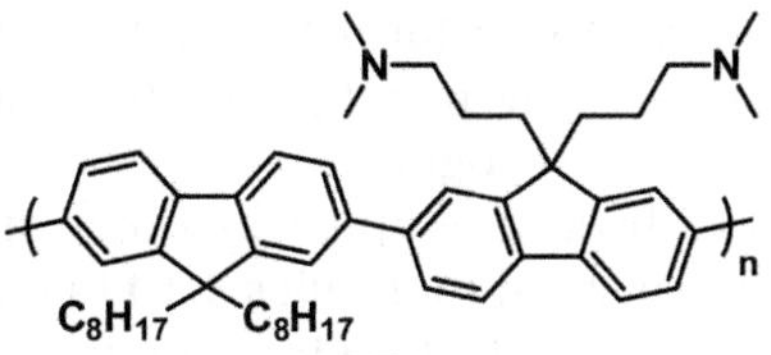

Figure 4.16 Structure of acceptor molecule PFN.

Conversion of the thione functionality within the rhodanine heterocycle to a DCV, while keeping the molecular skeleton unchanged, allowed Chen and co-workers to reach a PCE of 9.3%.[50] The excellent performance of **4.3.6** was attributed in part to the formation of a highly crystalline network of interpenetrating fibrils of the order of 10 nm, leading to very little loss through recombination of electron–hole pairs. In addition, the devices were prepared using the polymeric electron transport layer poly[(9,9-bis(3′-(*N*,*N*-dimethylamino)propyl)-2,7-fluorene)-*alt*-2,7-(9,9-dioctylfluorene)] (PFN; Figure 4.16),[51] which resulted in improved charge transport and collection.

Two more structurally similar groups of material were subsequently prepared, both of which featured a benzodithiophene–terthiophene–rhodanine framework, differing in the alkyl chain substituents. The first of these, **4.3.7a**, was prepared by Yang and co-workers[52] in 2013 and achieved a PCE of 7.2%, which they then increased to 8.1% using polydimethylsiloxane[53] as an additive to improve the film morphology, notably increasing the J_{SC} by almost 10% from 11.4 to 12.5 mA cm^{-2}. The second molecule, **4.3.8**, was prepared in 2015 by Jones *et al.*[54] and showed a very high fill factor of 0.77 and a PCE of 9.3%. These values were among the top reported at the time and suggested that the benzodithiophene–terthiophene–rhodanine framework had excellent potential for further improvement in achieving a high performance in devices that could be commercialised.

More recently, work by Wang and co-workers demonstrated the potential for dramatic improvements in fill factor by the incorporation of fluorine atoms.[55] In this work, a series of D–A–D–A–D materials was prepared with varying fluorine substitutions. Ultimately, a PCE of 8.1%, V_{OC} of 0.9 V and fill factor of 0.76 were achieved with compound **4.3.9c**, which compared favourably with the non-fluorinated **4.3.9a**, which showed a PCE of 2.5%. The excellent photovoltaic properties were attributed in part to the vapour annealing process, which led to an improvement in J_{SC} and therefore the PCE. Increasing the thiophene chain length also showed an improvement: adding one additional bridging thiophene unit per side resulted in a PCE improvement from 6.8% for **4.3.9b** to 8.1% for **4.3.9c**.

The same group investigated the effect of altering the bridging thiophene unit to a furan and selenophene. Under optimised conditions, the thiophene-bridged molecule **4.3.10a** produced the best device, with a PCE of 8.70%, followed by the selenophene (8.41%) material **4.3.10b**. The furan spacer **4.3.10c** was thought to perform less well (PCE 6.01%) due to the shorter effective conjugation length and poorer film morphology compared with the sulfur and selenium analogues.

Atom replacement has been attempted by other groups – for example, Bäuerle *et al.* produced two small molecules **4.3.11a** and **4.3.11b**, which differed in the end-group.[56] Molecule **4.3.11a** (containing terminal benzothiadiazole acceptor groups) achieved a 1.73% PCE in a bilayer configuration (*versus* 1.50% in a BHJ device due to a decrease in J_{SC} from 3.1 to 2.7 mA cm^{-2}). The thiadiazolopyridine-terminated **4.3.11b** showed a better performance, with a PCE of 3.15% and a good fill factor of 0.67 in a bilayer configuration. Compound **4.3.11b** also showed a poorer performance in the BHJ configuration, which was attributed to a different molecular organisation when blended with into films with $PC_{60}BM$.

Benzodithiophene (BDT) is another popular core for small molecule donor materials. One such example from Chen and co-workers is **4.3.12**, which is the BDT core analogue of compound **4.3.4b**. When prepared into a BHJ device with $PC_{60}BM$ as the acceptor, an optimised PCE of 5.44% was reached.[57] Increasing the blend ratio from 1:0.5 to 1:0.8 gave a fill factor of 0.64, which was the highest reported fill factor for solution-processed small molecule BHJ devices at the time. However, the PCE decreased to 4.87%, along with an associated decrease in J_{SC} from 9.77 to 8.25 mA cm^{-2}.

BDT also served as the core block in a report by Li and co-workers of four new thiophene-based oligomers (**4.3.13a–d**) with terminal indenedione groups.[58] By varying the number of bridging thiophene units and using either a 5-hexylthiophene or a 2-ethylhexyloxy central branching unit, PCEs between 4.15 and 6.75% were attained. These findings showed that the materials with bridging bithiophenes out-performed their monothiophene counterparts and that the alkoxy side-chains from the BDT core gave a lower photovoltaic performance than the corresponding thiophenes. Notably, the highest performance device (**4.3.13c**) had a high fill factor of 0.66, which was among the highest reported at the time, and **4.3.13a** had a high V_{OC} of 1.03 V, also among the highest reported at the time for solution-processed OPV devices.

Based on the encouraging results reported initially by Chen and co-workers[50] with material **4.3.6**, a series of oligomers was produced by the same group (**4.3.14a–f**). This series of materials was based around an oligothiophene with rhodanine-like end-capping groups and, at the time of publication, included the material that exhibited the highest reported PCE for a small molecule device: **4.3.14b** reached a PCE of 10.08%.[59] With the exception of the shortest chain analogue (**4.3.14a**), all the molecules performed well; compound **4.3.14a** formed poor films and a PCE of only 0.24% was attained. It was noted that molecules with an odd number of thiophene units benefited from an increase in J_{SC} and PCE compared with their even-numbered counterparts. This was attributed to the formation of a fibrillar network in the bulk of the material, as well as their axisymmetrical structure, which resulted in a large difference in dipole moment between the ground and excited states.

Table 4.1 Performance data relating to the thiophene-based materials discussed in Section 4.3.

Compound	PCE (%)	Fill factor	V_{OC} (V)	J_{SC} (mA cm^{-2})	E_g (eV)	Ref.
4.3.1b	0.29	0.24	0.85	1.13	–	146
4.3.2	0.8	0.26	0.85	3.65	2.6	42
4.3.3	1.3	0.40	0.94	3.35	2.10	43
4.3.4a	3.7	0.34	0.88	12.4	1.68	45
4.3.4b	5.08	0.55	0.86	10.74	1.74	48
4.3.4c	6.10	0.47	0.92	13.98	1.69	49
4.3.5a	2.38	0.45	0.58	9.2	2.2	46
4.3.5b	5.2	0.49	0.97	11.1	1.56	47
4.3.6	9.30	0.69	0.91	14.87	1.62	50
4.3.7a	8.1	0.66	0.93	12.5	1.72	147
4.3.7b	6.79	0.59	0.96	11.92	1.76	147
4.3.7c	8.02	0.72	0.92	12.09	1.76	147
4.3.7d	7.38	0.65	0.93	12.21	1.74	147
4.3.8	9.3	0.77	0.90	13.90	1.82	54
4.3.9a	2.5	0.38	0.90	7.2	1.77	55
4.3.9b	6.8	0.62	0.94	11.7	1.86	55
4.3.9c	8.1	0.76	0.90	11.9	1.76	55
4.3.10a	8.70	0.75	0.89	13.06	1.81	148
4.3.10b	8.41	0.72	0.87	13.40	1.77	148
4.3.10c	6.01	0.61	0.89	11.12	1.85	148
4.3.11a	1.73	0.57	0.98	3.1	2.00	56
4.3.11b	3.15	0.67	1.00	4.7	1.87	56
4.3.12	5.44	0.60	0.93	9.77	1.83	57
4.3.13a	5.67	0.55	1.03	10.07	1.61	58
4.3.13b	4.15	0.48	0.91	9.47	1.59	58
4.3.13c	6.75	0.66	0.92	11.05	1.60	58
4.3.13d	5.11	0.65	0.92	8.58	1.60	58
4.3.14a	0.24	0.38	0.90	0.70	1.88	59
4.3.14b	10.08	0.68	0.92	15.66	1.81	59
4.3.14c	6.33	0.58	0.92	11.45	1.60	59
4.3.14d	9.30	0.68	0.90	14.77	1.64	59
4.3.14e	6.50	0.68	0.86	10.80	1.57	59
4.3.14f	7.86	0.68	0.81	13.77	1.53	59

It is clear that oligothiophenes are vitally important to the progress of small molecule OPV devices and, with current state-of-the-art devices exceeding the 10% PCE mark for single-junction devices, they are a viable alternative to their polymeric counterparts. The performance data and associated chemical structures for the thiophene-based materials described here are shown in Table 4.1 and Figure 4.17, respectively.

4.4 Diketopyrrolopyrrole-based Materials

Another heavily explored class of compounds is those based on the diketopyrrolopyrrole (DPP) dye material. Such materials have been shown to produce OPV devices with good performance while retaining the ability to be synthetically modified to a large degree based on well-established chemistry.

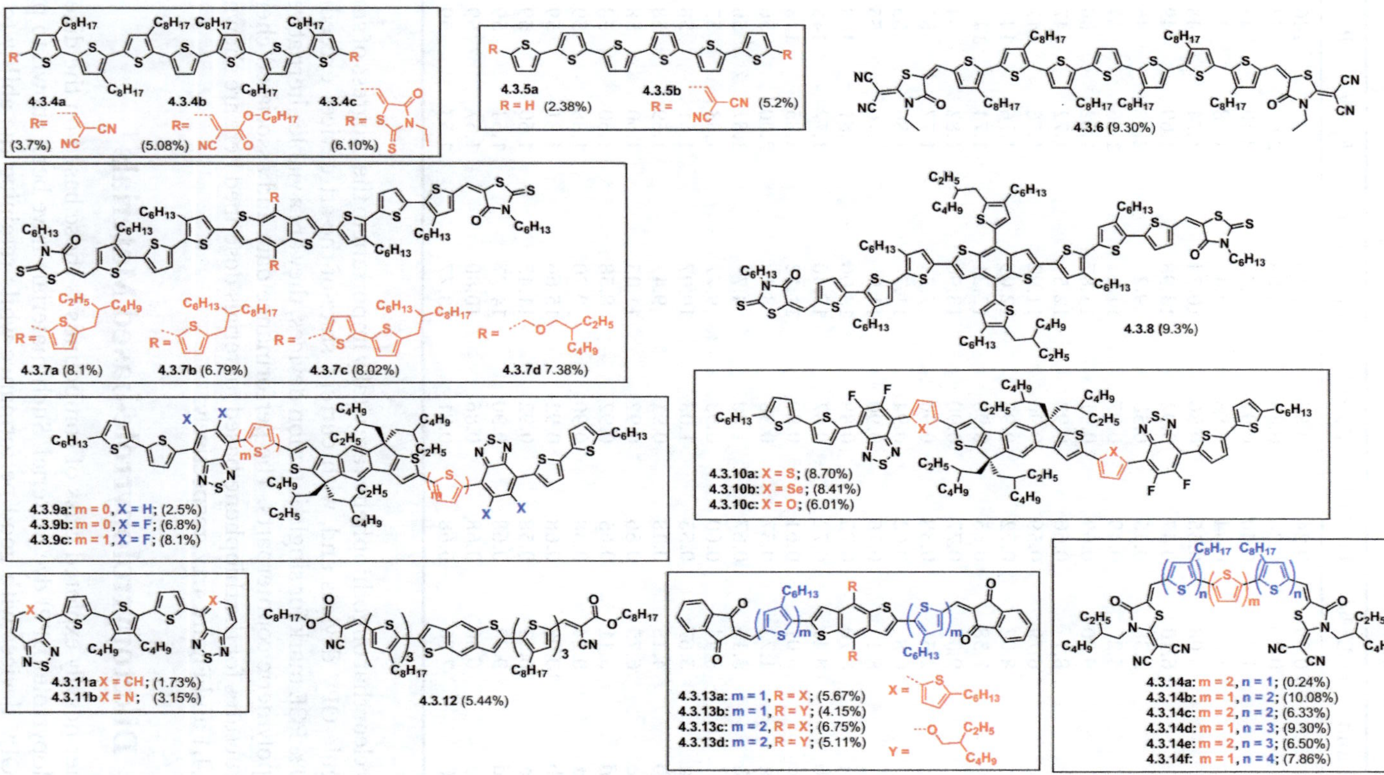

Figure 4.17 Chemical structures of thiophene materials discussed in Section 4.3.

This allows for parameters such as solubility, and therefore processability, to be tailored by the chemist to produce finely tuned materials for exploration.

Although the DPP class is well explored, these materials have typically shown slightly lower PCEs than thiophene-based materials. Nevertheless, efficiencies >6% are reported, showing that the scope for high-performance DPP-based materials is large. There are two main methods of functionalising DPPs: using them as a core (such as in a D–A–D type material) or where the DPP unit acts as an end-group (such as in an A–D–A system). Both approaches have yielded good results and both benefit from well-explored chemistry. In addition, because DPP has already found widespread use in industry as a dye material, it can be purchased in large quantities from numerous commercial suppliers.

A small selection of sample DPP materials is outlined in Table 4.2 and Figure 4.18. In 2015, the group of Huang and co-workers developed a relatively simple V-shaped D–A–D–A–D type material (**4.4.1**), which encompassed a central triphenylamine (TPA) unit with DPP moieties bridging to bithiophene end-groups.[60] A moderate PCE of 3.81% was obtained in a BHJ small molecule organic solar cell device when blended with $PC_{70}BM$ and a small amount of DIO additive, which improved the phase-separated morphology as well as aiding the formation of well-defined structures within the BHJ layer.

Also in 2015, a study by Wu and co-workers examined the effects of fluorine atom substitution on DPP-based oligomers.[61] Two derivatives were prepared that utilised DPP as the strong acceptor core, bridged with an indecenedithiophene unit to either a benzothiadiazole or difluorobenzothiadiazole unit, which acted as a weaker acceptor group, and peripheral bithiophenes capped with alkyl chains. The group achieved optimum performance for **4.4.2a** in a 1:3 blend with $PC_{70}BM$, which reached a PCE of 2.7% with a fill factor, V_{OC} and J_{SC} of 0.40, 0.80 V and 8.4 mA cm^{-2}, respectively. The fluorinated **4.4.2b** exceeded this performance with a PCE of 5.4%, a fill factor of 0.69, a V_{OC} of

Table 4.2 Performance data relating to DPP-based materials discussed in Section 4.4.

Compound	PCE (%)	Fill factor	V_{OC} (V)	J_{SC} (mA cm^{-2})	Eg (eV)	Ref.
4.4.1	3.81	0.63	0.64	9.48	1.45	60
4.4.2a	2.7	0.40	0.80	8.4	1.60	61
4.4.2b	5.4	0.69	0.74	10.5	1.61	61
4.4.3a	5.29	0.62	0.72	11.86	1.71	149
4.4.3b	3.65	0.51	0.70	10.21	1.57	62
4.4.3c	4.96	0.58	0.69	12.48	1.56	62
4.4.3d	6.00	0.64	0.69	13.60	1.55	62
4.4.4	4.30	0.51	0.69	12.24	1.57	65
4.4.5	4.4	0.48	0.92	10.0	1.8	66
4.4.6	4.02	0.54	0.72	10.4	1.69	70
4.4.7a	5.44	0.55	0.82	11.90	1.71	71
4.4.7b	1.45	0.32	0.97	4.09	1.98	71
4.4.8	6.88	0.61	0.80	14.12	1.50	72

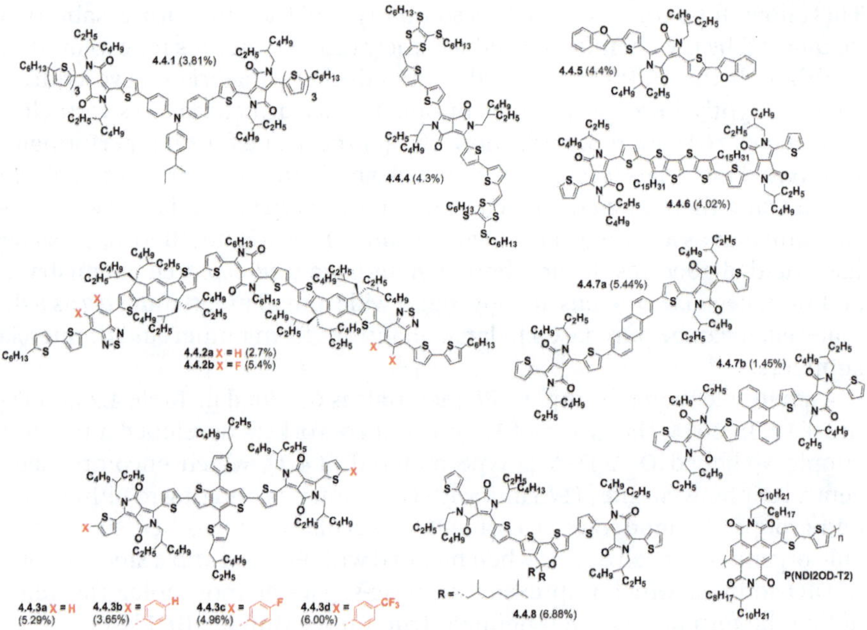

Figure 4.18 Chemical structures of DPP materials discussed in Section 4.4.

0.74 V and a J_{SC} of 10.5 mA cm^{-2}. The excellent improvement in photovoltaic performance was particularly impressive as this result was among the top performing DPP-based small molecule BHJ solar cells, despite no additive being used during processing.

The effect of substituting fluorine atoms was studied further by Hwang *et al.*, who prepared three new DPP-containing materials (**4.4.3a–c**), differing only by the terminal units.[62] The group studied the effect of conversion of the terminal group from a proton to a phenyl group, a *p*-fluorophenyl group and a *p*-trimethylfluoro phenyl group. In doing so, they produced materials with PCEs in the range 3.65–6.00% (using 1-chloronaphthalene as an additive);[63,64] note that these materials had an increasing fluorine content across the series, with the highest PCE generated using the molecule with the highest loading of fluorine. The enhancement in PCE was attributed to an improvement in the fill factor and J_{SC} resulting from morphological changes caused by the end-groups. The inclusion of the fluorine atoms aided the active layer, forming fibrous domains beneficial to charge transport. The group concluded by stating that they felt the inclusion of fluorine-containing end-groups may be one of the most efficient ways to improve OPV device performance, particularly for small molecule systems.

Singh and co-workers prepared **4.4.4**, a D–π–A–π–D material containing terminal dithiafulvalene donor groups bridged to a central DPP acceptor with thiophene units.[65] A maximum PCE of 4.30% was achieved when prepared as a 1:1.5 ratio with $PC_{70}BM$, exhibiting a fill factor of 0.51, a V_{OC} of 0.69 V and a J_{SC} of 12.24 mA cm^{-2}. The low bandgap of 1.57 eV, strong absorption in the near-IR and broad absorption aided the good preliminary performance of the material. The group aims to improve the photovoltaic performance further by optimising the processing conditions and altering the cathode material.

The well-known compound DPP(TBFu)$_2$ (**4.4.5**) was initially prepared in 2009 by Nguyen and co-workers, attaining a PCE of 4.4%, which, at the time, was the highest reported for solution-processed small molecule BHJ solar cells.[66] When annealed at 110 °C the group attained a J_{SC} of 10.0 mA cm^{-2}, a fill factor of 0.48 and a V_{OC} of 0.92 V in a device with PC$_{70}$BM as the acceptor in a 3 : 2 D : A ratio. Matsuo and co-workers improved on this performance by incorporating a 0.8 nm LiF interlayer to improve electron injection.[67,68] With PC$_{70}$BM as the acceptor, the group achieved a PCE of 4.69% with a fill factor of 0.5, a V_{OC} of 0.89 V and a J_{SC} of 10.74 mA cm^{-2}. On replacement of the acceptor with the 56 π-electron fullerene analogue methano-indene fullerene,[69] further enhancement of the photovoltaic properties was realised: a PCE of 5.08%, a fill factor of 0.52, a V_{OC} of 1.03 V and a J_{SC} of 9.52 mA cm^{-2} were the best results obtained when using a 3 : 2 mix of donor:acceptor.

A fused tetrathienoacene core flanked with DPP units (**4.4.6**) was prepared by Facchetti *et al.* in 2015.[70] Inverted devices were prepared with PC$_{70}$BM as the acceptor, with DIO being investigated as a solvent additive. At a DIO concentration of 1.0 vol%, a PCE of 4.02% was reached, with a fill factor, V_{OC} and J_{SC} of 0.54, 0.72 V and 10.4 mA cm^{-2}, respectively. The good photovoltaic performance was, in part, attributed to the well-positioned HOMO and LUMO levels (−5.12 and −3.17 eV, respectively) and the propensity for efficient π-stacking interactions between the planar tetrathienoacene core units. These features of the electron-rich core aided the formation of a favourable morphology for charge transport and allowed intermolecular interactions to occur throughout the bulk of the material. The crystallinity of the thin films formed by spin-coating was greatly improved by adding a small amount of DIO.

The related compound **4.4.7a** was prepared by Yang and co-workers later the same year.[71] This material differed from **4.4.6** in that the central core of the molecule was an anthracene entity rather than a fused thiophene system. The group attained a PCE of 5.44% – an improvement of >30% achieved by replacing only the core. The solution-processed device with PC$_{70}$BM as the acceptor and a 1% DIO additive achieved a V_{OC} of 0.82 V, a fill factor of 0.55 and a J_{SC} of 11.90 mA cm^{-2}. Interestingly, the isomeric material **4.4.7b** connected through the 9- and 10-positions of the anthracene core exhibited a significantly lower photovoltaic performance, achieving a maximum of only 1.45% PCE, a fill factor of 0.32, a V_{OC} of 0.97 V and a J_{SC} of 4.09 mA cm^{-2}. By measuring the EQE of each of the photovoltaic cells, the group found that **4.4.7a** showed a significantly improved EQE of 66.2% compared with isomer **4.4.7b** (EQE 30.8%), which explained the large difference in J_{SC} and PCE of the two materials.

Work by Jo *et al.* produced the novel compound **4.4.8**, which had a dithienopyran donor as the core flanked by two peripheral DPP units.[72] The optimised device showed an ideal morphology for charge transport due to a nanoscale interpenetrating network. This, along with broad and intense light absorption, led to a device performance among the top reported for DPP-containing small molecule organic solar cells, achieving a high PCE of 6.88% (fill factor 0.61, V_{OC} 0.80 V and a high J_{SC} of 14.12 mA cm^{-2}). The group also tested the performance of the material with the non-fullerene acceptor P(NDI2OD-T2)[73,74] (Figure 4.18), in which an encouraging PCE of 4.82% was attained, demonstrating that **4.4.8** is compatible with both fullerene and non-fullerene acceptors in the production of high-performance devices. The performance data and chemical structures for the DPP-based materials discussed in this section are outlined in Table 4.2 and Figure 4.18, respectively.

4.5 Oligoacene-based Materials

Oligoacenes have been investigated as potential organic semiconductors since the 1960s,[75,76] although it was not until much later that their applicability as organic donor materials for photovoltaic cells was considered. In 1998, Signerski *et al.* demonstrated a pentacene-based solar cell using a perylene diimide acceptor with a PCE <0.001%.[77]

In 2004, Yoo *et al.* reported a device based on a pentacene active layer **4.5.1** (Figure 4.19), which attained a PCE of 1.50% under AM1.5G illumination,[78] exhibiting a high EQE (58%) throughout the visible spectrum. The high EQE was attributed to the large exciton diffusion length shown by the polycrystalline pentacene, but the PCE was limited by the low V_{OC} of 0.36 V. When pentacene was exchanged for tetracene (**4.5.2**) in a bilayer cell, a PCE of 2.3% was reached, due to an increase in V_{OC} from 0.36 to 0.58 V.[79]

Additional functionalisation of the fused acene systems led to further enhancements of the photovoltaic properties – for instance, by appending two thiophene units onto the centre of the pentacene (**4.5.3**), Nuckolls and co-workers achieved a PCE of 1.4% in a bilayer device, increasing the V_{OC} to 0.63 V, but lowering the J_{SC} and fill factor to 4.20 mA cm^{-2} and 0.53, respectively.[80]

By incorporating trialkylsilylethynyl groups, Anthony *et al.* produced soluble pentacene derivatives known as TIPS-pentacenes.[81] A pentacene core with these solubilising groups attached to the central benzene ring (**4.5.4**), was prepared by Malliaras and co-workers and a device using C_{60} as an acceptor achieved a PCE of 0.52%.[82] By modifying the core to include two fused thiophene units at the periphery of the molecule, forming the anthradithiophene **4.5.5**, the PCE was increased to 1.0% with a good V_{OC} of 0.84 V.[83]

Work in 2008 by Valentini *et al.*, in which an anthracene core was solubilised by incorporating hexyloxy-substituted benzene rings (**4.5.6**), resulted in a device with a PCE of 1.12%.[84] This was improved by the group in 2010 through device optimisation to give a PCE of 1.27%.[85]

In the same year, Wong *et al.* reported three hexa-*peri*-hexabenzocoronene oligomers.[86] The large, heavily conjugated core unit had varying numbers of fluorene units appended at differing positions. The group achieved a poor PCE of 0.03% with the hexa-substituted system **4.5.7a**. The group concluded that this was a consequence of poor columnar stacking and poor intermolecular associations. By contrast, compounds **4.5.7b** and **4.5.7c** performed

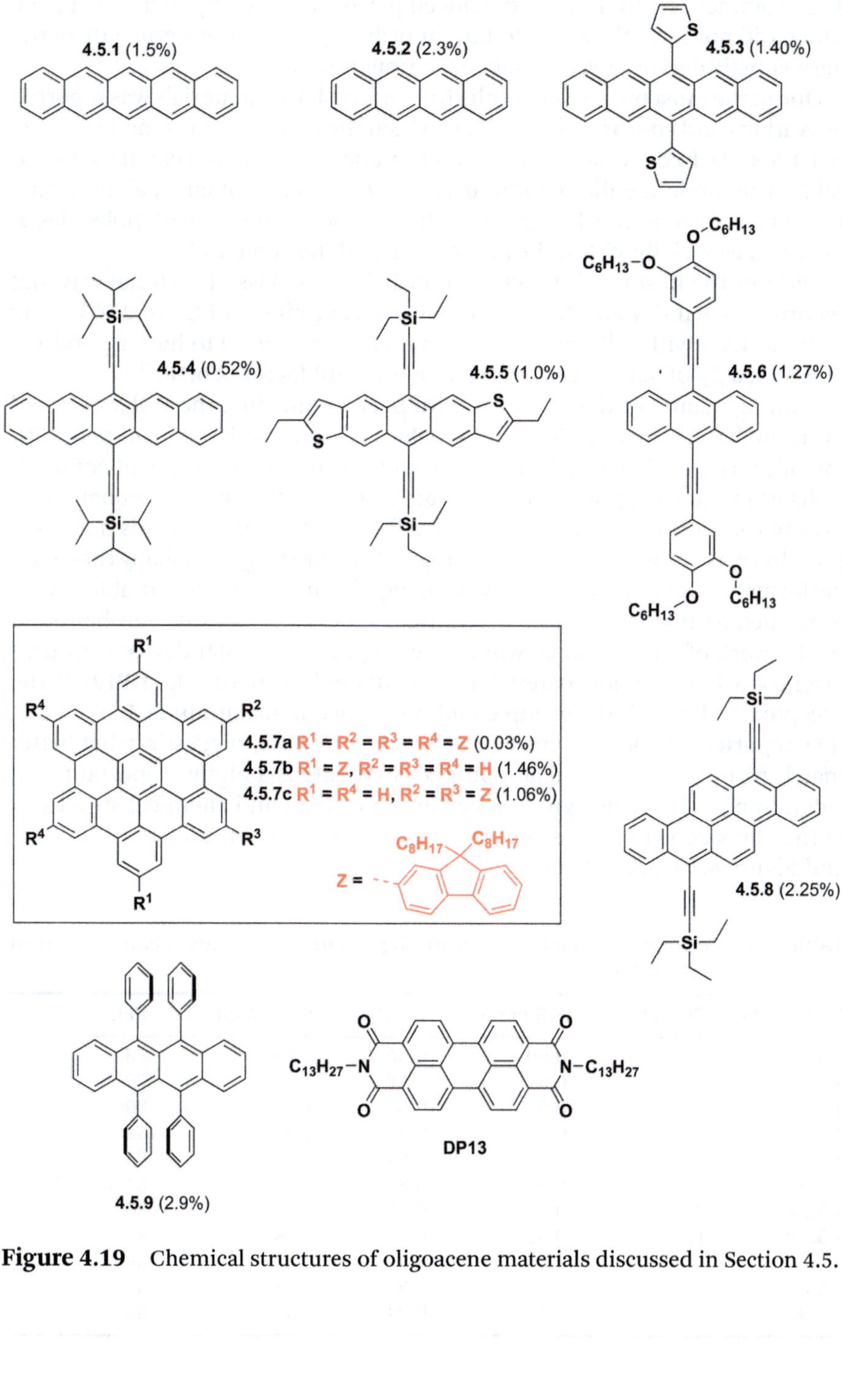

Figure 4.19 Chemical structures of oligoacene materials discussed in Section 4.5.

much better, achieving PCEs of 1.46 and 1.06%, respectively. Ultimately, the group associated the strong intramolecular self-assembly of **4.5.7b** and **4.5.7c** as a prominent factor in their enhanced performance compared with **4.5.7a**, which did not form the same columnar ordering due to the steric bulk of the hexa-substituted hexa-*peri*-hexabenzocoronene core.

One of the most promising results from acene-based materials was reported by Watkins and co-workers in 2009, in which the reported chrysene-like material **4.5.8** reached a PCE of 2.25% when prepared in a spin-coated BHJ solar cell.[87] The moderate fill factor of 0.41 and J_{SC} of 6.55 mA cm^{-2}, along with a V_{OC} of 0.83 V, led to this being one of the best performing small molecules at the time, especially within the acene class of donor materials.

Perhaps the best result in terms of PCE for this class of material was that reported by Pandey and Nunzi, who, in 2007, attained a PCE of 2.9% using rubrene **4.5.9** with fullerene C_{60}. The material was found to have a good V_{OC} of 0.94 V, a J_{SC} of 5.3 mA cm^{-2} and a moderate fill factor of 0.56.[88]

Although some of these results have been promising, the PCEs achieved by acene-based small molecules for solar cells are too low to compete with the other types of materials discussed in this chapter. In terms of commercialisation, this suggests that acenes are not likely to achieve adequate photovoltaic performance when used with a fullerene acceptor. That is not to say that this class of material is not worthy of further study; it is likely that good performance may be achieved by utilising alternative, more suitable acceptors, such as planar perylene derivatives. An example of this can be found by the work of Nunzi and co-workers, who prepared a BHJ device with pentacene **4.5.1** as a donor material and perylenedicarboxyimide (**DP13**) as the acceptor.[89] Although the group could only reach a maximum PCE of 0.52%, they reported a good J_{SC}, which exceeded 5.0 mA cm^{-2}, suggesting that better matching of the donor and acceptor components is a highly important consideration in device design. The performance data and chemical structures of the oligoacene materials covered in this section are presented in Table 4.3 and Figure 4.19, respectively.

Table 4.3 Performance data relating to oligoacene-based materials discussed in Section 4.5.

Compound	PCE (%)	Fill factor	V_{OC} (V)	J_{SC} (mA cm^{-2})	Ref.
4.5.1	1.50	0.50	0.36	8.20	78
4.5.2	2.30	0.57	0.58	7.00	79
4.5.3	1.4	0.53	0.63	4.20	80
4.5.4	0.52	0.52	0.47	1.9	82
4.5.5	1.0	0.40	0.84	2.96	83
4.5.6	1.27	0.45	0.89	3.10	84 and 85
4.5.7a	0.03	0.32	0.61	0.17	86
4.5.7b	1.46	0.61	0.90	2.68	86
4.5.7c	1.06	0.60	0.87	2.03	86
4.5.8	2.25	0.41	0.83	6.55	87
4.5.9	2.9	0.56	0.94	5.3	88

4.6 Triphenyl Amine-based Materials

Another extremely versatile series of materials is those based on the TPA structure, which was among the first reported classes of material for use in OPV devices.[90,91] The TPA unit is often used as a core in OPV materials as a result of its high chemical flexibility, which allows easy and extensive synthetic modifications. The uncommon star-shaped structure is also interesting because the inherent increased dimensionality from linear materials should lead to higher device performance as a result of improved charge separation.[92]

The two main structural classes of TPA materials are defined based on whether the TPA unit makes up the core of the molecule or whether it is used as a peripheral group. Both styles have been used to produce devices with good photovoltaic properties, with efficiencies for the top performing materials reaching ~6–7%.[93]

Wong and co-workers produced a TPA core with one arm substituted with a dithienosilole and terminated with a DCV group (**4.6.1**).[94] When assembled into a hybrid heterojunction device with $PC_{70}BM$ as the electron acceptor, a PCE of 3.82% was measured, along with a V_{OC} of 0.83V, a J_{SC} of 9.53 mA cm^{-2} and a fill factor of 0.48.

The group also attempted to replace the dithienosilole with a benzothiadiazole unit, maintaining the DCV capping group.[93] Through the use of careful structural modifications on the other arms of the TPA core, the group were able to finely tune the photovoltaic parameters of the fabricated devices. The protonated analogue (**4.6.2a**) showed a PCE of 5.6% with a J_{SC} of 11.45 mA cm^{-2}, a fill factor of 0.49 and a V_{OC} of 1.00 V, but the methylated structure (**4.6.2b**) produced a higher PCE of 6.8%, a J_{SC} of 13.48 mA cm^{-2}, a fill factor of 0.53 and a V_{OC} of 0.93 V. The higher V_{OC} from **4.6.2a** is due to the lower HOMO level of the material because the values for J_{SC} correlated well with the bandgap of the material. The larger PCE of **4.6.2b** was due to an increased J_{SC} and therefore improved fill factor compared with **4.6.2a**. These results highlight the importance that judicious molecular structure engineering has on device performance.

A TPA core containing an uncommon tricyanofuran group was prepared in 2011 by Prasad and co-workers (**4.6.3**).[95] The strongly electron-deficient tricyanofuran entity in combination with the TPA electron-donating core provided a tuneable material with a bandgap of 1.92 eV and well-aligned HOMO and LUMO levels for $PC_{70}BM$. A BHJ cell based on this material achieved a PCE of 2.44% (with a fill factor of 0.30, a V_{OC} of 0.93 V and a J_{SC} of 8.74 mA cm^{-2}). The low fill factor was in this case the limiting factor for the material, which suggested poor charge extraction.

Compound **4.6.4** was first reported by Roncali and co-workers.[96] In their initial report, the group reached a PCE of 1.02% in a bilayer device with $PC_{60}BM$ as the acceptor – a result which, at the time, was one of a very few examples of a small molecule capable of surpassing a 1% PCE barrier. This result was improved by adjusting the anodic buffer layer (ABL).[97] When a gold ABL was replaced by a molybdenum oxide layer, the PCE improved from 0.33% to 1.13%. This result was the first to demonstrate that molybdenum oxide is a more versatile material for use as an ABL than pure gold because molybdenum oxide is less affected by the HOMO of the donor. Additional studies on the buffer layer were examined by the same group in 2013, in which the use of a hybrid CuI/MoO_3 ABL in a device based on **4.6.4** as the donor and fullerene C_{60} as the acceptor was found to change the fill factor, J_{SC}, V_{OC} and, ultimately, PCE of the device.[98] By using a double ABL the group found a large increase in photovoltaic parameters, ultimately reaching a PCE of 2.5% with a fill factor of 0.55, a V_{OC} of 0.78 V and a J_{SC} of 6.6 mA cm^{-2}. By utilising the enhanced J_{SC} and fill factor due to the CuI layer, as well as the V_{OC} improvement from the MoO_3 layer, a balance was found that allowed a good enhancement in the hole collection efficiency and therefore device performance.

Roncali and co-workers continued to study TPA core donor materials and produced the two related materials **4.6.5a** and **4.6.5b** consisting of a TPA core with one arm substituted by a DCV-terminated thiophene (**4.6.5a**) or the analogous phenyl-fused **4.6.5b**.[99] Bilayer devices fabricated with $PC_{60}BM$ reached PCEs of 2.53 and 2.97% for **4.6.5a** and **4.6.5b**, respectively. By fusing a phenyl ring into the DCV–thiophene acceptor, the LUMO of **4.6.5b** was reduced by 0.43 eV while simultaneously red shifting the absorption associated with intramolecular charge transfer (from 501 to 610 nm) and halving the molar absorption coefficient from 33 900 M^{-1} cm^{-1} to 16 200 M^{-1} cm^{-1}.

A more recent study on TPA-based materials by Roncali and co-workers assessed the effect of rigidifying parts of the material through covalent linkages, as shown in compounds **4.6.6a** and **4.6.6b**.[100] By converting the terminal DCV group of the thiophene unit into a rigid DCV-substituted cyclic unit, the group saw an improvement in the PCE from 1.47% to 2.23%, with an associated increase in fill factor (from 0.32 to 0.55) and V_{OC} (from 0.70 to 0.82 V) and a decrease in J_{SC} (from 5.88 to 4.87 mA cm^{-2}) when fabricated into a bilayer cell with $PC_{60}BM$ as an acceptor. The reduction in J_{SC} was a result of the reduction in the light-harvesting ability shown by the fused material **4.6.6b**. In addition, **4.6.6b** showed a 0.1 eV increase in the LUMO level while the HOMO remained unchanged (−5.67 eV), leading to a HOMO–LUMO gap of 2.08 eV for **4.6.6b** compared with 1.86 eV for **4.6.6a**.

Interestingly, when Roncali and co-workers prepared the analogous material **4.6.7** with no bridging thienyl unit, a much lower PCE of 0.85% was obtained.[101] Although the performance was by no means state-of-the-art, importantly it demonstrated that a device based on a C_{60} acceptor, rather than the higher performing C_{70}, with no additives and a simple device configuration (a planar bilayer device), was capable of achieving near 1% efficiency with a donor material that is simple to make (three-step synthesis, 62% overall yield). The device data and chemical structures of the TPA-based materials covered in this section are outlined in Table 4.4 and Figure 4.20, respectively.

Table 4.4 Performance data relating to triphenylamine-based materials discussed in Section 4.6.

Compound	PCE (%)	Fill factor	V_{OC} (V)	J_{SC} (mA cm^{-2})	E_g (eV)	Ref.
4.6.1	3.82	0.48	0.83	9.53	–	94
4.6.2a	5.6	0.49	1.00	11.45	2.14	93
4.6.2b	6.8	0.53	0.93	13.48	2.08	93
4.6.3	2.44	0.30	0.93	8.74	1.92	95
4.6.4	2.5	0.55	0.78	6.6	–	96 and 98
4.6.5a	2.53	0.42	0.92	5.77	2.03	99
4.6.5b	2.97	0.52	0.97	5.32	1.59	99
4.6.6a	1.47	0.32	0.70	5.88	1.86	100
4.6.6b	2.23	0.50	0.82	4.87	2.08	100
4.6.7	0.85	0.37	0.67	3.04	–	101

4.7 Dyes and Other Miscellaneous Materials

4.7.1 Squaraine Dye Materials

The squaraine class of dyes has primarily been used within the dye-sensitised solar cell field; however, some work has been carried out with respect to their applicability as donor materials for small molecule BHJ devices.

The early work on squaraine materials for OPV applications began in 1976 when Merritt and Hovel demonstrated a 0.02% PCE with compound **4.7.1.1** (Figure 4.21) under AM0 illumination conditions.[102] The AM0 illumination conditions are non-standard and differ slightly from AM1.5G; the integrated power of the AM0 spectrum gives a power of 1366.1 W m^{-2} compared with the integrated power of 1000 W m^{-2} for AM1.5G. The AM0 spectrum is typically only used for space applications because it does not account for atmospheric absorption. Significant development of the squaraine class of compounds was achieved when Silvestri *et al.* surpassed the 1% PCE mark in 2008 with compound **4.7.1.2a**, which reached 1.24% efficiency, suffering predominantly from a poor fill factor and J_{SC}, suggesting poor charge transport.[103] The same work reported the analogous material **4.7.1.2b**, in which the branched 2-ethylhexyl solubilising alkyl group was replaced with a straight *n*-dodecyl group. The effects of this replacement were small decreases in the J_{SC}, V_{OC} and fill factor, which led to a lower PCE of 0.89%. At the time, however, these efficiencies were among the highest reported (~1.3%).[43] Work on these materials was continued by Bagnis *et al.* in 2010 when they prepared materials with hexyl and hexenyl groups on the pyrrolic nitrogen – materials **4.7.1.2c** and **4.7.1.2d**, respectively.[104] These new materials demonstrated an improvement over the earlier compounds, reaching PCEs of 1.47 and 2.05% in a BHJ configuration with $PC_{70}BM$ as the acceptor for **4.7.1.2c** and **4.7.1.2d**, respectively. The better performance of **4.7.1.2d** was attributed to the non-covalent intermolecular contact between the alkenyl and phenyl units of the molecule and represented the best performing squaraine-based material of the time.

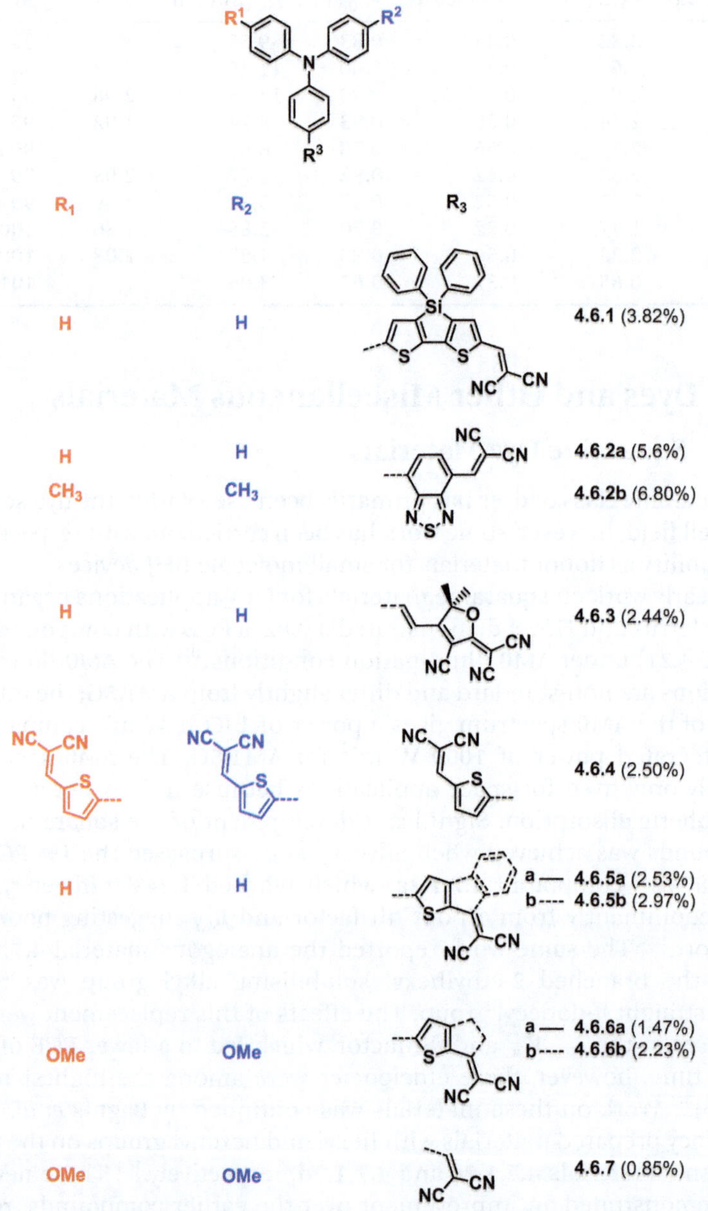

Figure 4.20 Chemical structures of triphenylamine materials discussed in Section 4.6.

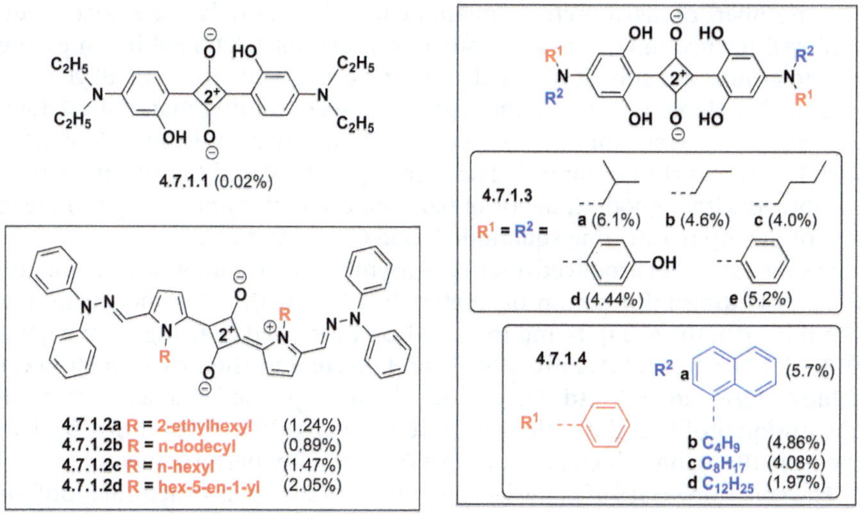

Figure 4.21 Chemical structures of squaraine materials discussed in Section 4.7.1.

The following year saw an increase in the reported PCE to 3.2% by Wang *et al.* in a bilayer configuration using the doubly sublimed squaraine material **4.7.1.3a.**[105] By attaching hydroxyl groups to the phenyl rings, the group could sublime the materials **4.7.1.3a–e** and avoid decomposition, which was attributed to the intramolecular hydrogen bonding between the hydroxyl groups and the CO groups of the four-membered ring.[106] The material **4.7.1.3a** was studied further and in 2012 work by Sasabe and co-workers achieved a PCE of 6.1% with a device using $PC_{70}BM$ as the acceptor after further optimisation.[107]

The series of squaraine-based OPV materials **4.7.1.3a–e** shows a range of performance, with **4.7.1.3a** having been fabricated into various devices that achieved PCEs in the range 2.7–6.1%.[105,107–112] Altering the alkyl chains on this material allowed simple tuning of the molecule, including its chemical, physical and photovoltaic properties. By using a shorter alkyl group in **4.7.1.3b**, Forrest and co-workers achieved a 4.6% efficiency (with a fill factor of 0.69, a V_{OC} of 0.68 V and a J_{SC} of 9.5 mA cm^{-2}), which was lower than the best achieved for the tetraphenyl-substituted derivative **4.7.1.3e** in the same paper with a PCE of 5.2% after annealing.[113]

Sasabe and co-workers prepared **4.7.1.3c** with longer *N*-alkyl chains in a 2014 report, along with other materials not covered here.[114] In this work, the group achieved a 4.0% PCE for material **4.7.1.3c** with a moderate fill factor of 0.49, a V_{OC} of 0.86 V and a J_{SC} of 9.4 mA cm^{-2} under AM1.5G conditions with $PC_{70}BM$ as the acceptor. Utilising a phenyl hydroxyl group in place of the alkyl chains (**4.7.1.3d**) led to a PCE of 4.44% with $PC_{70}BM$ as the acceptor.[115] The material exhibited a moderate fill factor of 0.53 with a good V_{OC} and J_{SC} of 0.84 V and 10.01 mA cm^{-2}, respectively.

A number of asymmetric squaraine-based materials have also been explored for use in OPV devices, such as materials **4.7.1.4a–d** in which the amino groups are substituted with a benzene ring and either an alkyl chain or a naphthalene unit. Using the naphthalene-containing material **4.7.1.4a**, the group of Forrest and co-workers achieved a very promising PCE of 5.7% with $PC_{60}BM$ as the acceptor.[116] The high V_{OC} (0.90 V) and J_{SC} (10.0 mA cm^{-2}), combined with a good fill factor of 0.64 achieved after annealing, produced one of the top performing squaraine-based OPV materials.

Sasabe *et al.* also produced a series of asymmetrical squaraine-based materials, providing information on materials **4.7.1.4b–d.**[117] On increasing the chain length, the group found that the PCE decreased, starting from a good PCE of 4.86% for **4.7.1.4b** to 4.08% for **4.7.1.4c** and then to a much lower value of 1.97% for **4.7.1.4d.** The shorter chain lengths led to a red shift in the absorption profile of the molecule, as well as a broadened absorption peak in the near-IR region, which is favourable for high OPV performance.

The squaraine class of materials provide an interesting structural motif for the synthetic chemist to work with, providing a large allowance for tuning the bandgap of the materials. Their high absorption coefficients and strong absorption in the visible to near-IR region, along with the ease of synthetic preparation and good stability, suggests that squaraines are a potentially promising material for small molecule OPV devices. With state-of-the-art squaraine materials reaching PCEs >5%[118] for a single-junction device, it is clear that much progress has been made. Often limited by a poor fill factor and J_{SC} values, material scientists should strive to maximise these to realise the full potential of this interesting class of materials. The photovoltaic performance data and associated chemical structures of the squaraine-based materials covered in this section are highlighted in Table 4.5 and Figure 4.21, respectively.

Table 4.5 Performance data relating to squaraine-based materials discussed in Section 4.7.1.

Compound	PCE (%)	Fill factor	V_{OC} (V)	J_{SC} (mA cm^{-2})	E_g (eV)	Ref.
4.7.1.1	0.02	0.20	–	–	–	102
4.7.1.2a	1.24	0.35	0.62	5.70	1.7	103
4.7.1.2b	0.89	0.32	0.59	4.72	1.7	103
4.7.1.2c	1.47	0.37	0.55	7.16	1.7	104
4.7.1.2d	2.05	0.37	0.57	9.32	1.7	104
4.7.1.3a	6.1	0.52	0.86	13.6	1.7	107
4.7.1.3b	4.6	0.69	0.68	9.5	1.5	113
4.7.1.3c	4.0	0.49	0.86	9.4	1.85	114
4.7.1.3d	4.44	0.53	0.84	10.01	1.4	115
4.7.1.3e	5.2	0.72	0.94	7.40	1.9	113
4.7.1.4a	5.7	0.64	0.90	10.0	1.6	116
4.7.1.4b	4.86	0.45	0.98	11.01	1.7	117
4.7.1.4c	4.08	0.41	0.99	9.96	1.7	117
4.7.1.4d	1.97	0.36	0.96	5.72	1.7	117

4.7.2 BODIPY-based Materials

The compound 4,4-difluoro-4-bora-3a,4a-diaza-*s*-indacene (BODIPY) is a fluorescent dye used in dye-sensitised solar cells and, less commonly, as a donor material in BHJ devices. Similar to many other classes of compound under heavy investigation as a potential photovoltaic compound, the BODIPY family provides a wide scope for structural modification and therefore molecular tuning. The high absorption coefficients, in some cases >125 000 M^{-1} cm^{-1}, and excellent stability are particular advantages of this family of compounds.

Roncali and co-workers prepared three deeply coloured BODIPY derivatives (**4.7.2.1a–c**; Figure 4.21) and incorporated these into solar cells, attaining PCEs of 1.17–2.17%,[119] results which were among the top reported PCEs at the time. By incorporating a bithiophene group into the BODIPY core (**4.7.2.1c**), the group hoped to improve the hole-transporting properties of the molecule while leaving the electronic properties effectively unchanged.[120] The observed increase in PCE and J_{SC} was evidence that the bithiophene unit did improve the hole-transporting properties; this was confirmed by preparing "hole only" devices and measuring the hole mobility.

Kraner *et al.* investigated the influence of side groups on the BODIPY core by preparing solar cells using three similar compounds.[121] The group found that molecules with a low level of substitution gave the highest charge carrier mobility and PCE. A PCE of 3.8% was achieved for the phenyl-substituted derivative **4.7.2.2a**, decreasing to 2.7 and 1.7% for molecules containing an additional methyl **4.7.2.2b** and methoxy **4.7.2.2c** side group, respectively.

Thompson and co-workers produced **4.7.2.3** – a BODIPY similar to that reported by Kraner, but with methyl substituents in the α-position instead of a phenyl group.[122] As predicted by Kraner's finding that molecules substituted with fewer groups produced higher PCEs, the group reached an efficiency of 4.5% with a moderate fill factor of 0.51, a V_{OC} of 0.81 V and a J_{SC} of 8.7 mA cm^{-2} in a bilayer device configuration.

Ziessel and co-workers produced a series of four planar BODIPY molecules **4.7.2.4a–d** with various alkyl-thiophene and ethylene glycol solubilising chains.[123] BHJ solar cells consisting of these donors and $PC_{60}BM$ as the acceptor gave PCEs up to a maximum of 4.7% for **4.7.2.4b** (fill factor 0.47, V_{OC} 0.70 V and J_{SC} 14.3 mA cm^{-2}). The group found that materials **4.7.2.4a** and **4.7.2.4b**, which contained the alkyl-thiophene terminal groups, outperformed the ethylene glycol solubilised materials **4.7.2.4c** and **4.7.2.4d**; this was attributed to poor charge separation and collection because the materials also had low hole mobilities in organic field-effect transistor devices. This conclusion was reinforced by the large decrease in performance when increasing the thickness of the active layer.

A series of four D–A–D molecules with a central BODIPY unit bridging to either a diphenyl-amino carbazole or diphenyl-amino fluorine donor unit has been reported by Lin *et al.*[124] The work showed that materials **4.7.2.5a** and **4.7.2.5b** with an alkyne entity performed better in general than **4.7.2.5a** and **4.7.2.5b**, which contained only a thiophene linker, when fabricated into an OPV device with $PC_{70}BM$ as the acceptor. Incorporation of the alkenyl unit was found to alleviate the large dihedral twist between the BODIPY core and the thiophene ring while acting as a spacer unit, increasing the planarity and conjugation length of the molecule and simultaneously red shifting the absorption, which improved device efficiency. Ultimately, the group achieved the highest BODIPY-based OPV device efficiency for the time with a PCE of 3.22% from compound **4.7.2.5a** and $PC_{70}BM$ as the acceptor (a fill factor of 0.40, a high V_{OC} of 0.99 V and a J_{SC} of 8.25 mA cm^{-2}) without the use of thermal or solvent annealing or any additives. The morphology of the films was studied by atomic force microscopy and was found to be very smooth, with **4.7.2.5a** and **4.7.2.5b** containing large domains of an interpenetrating network between the donor and $PC_{70}BM$ acceptor. As no annealing was performed and no additive was used, it is expected that an improved photovoltaic performance can be obtained from these same materials.

The current highest performance device involving BODIPY was prepared in 2015 by Misra and co-workers when this group produced a series of three carbazole-containing materials (**4.7.2.7a–c**).[125] All three materials exhibited good performance when fabricated into a device with $PC_{70}BM$ as the acceptor. The top performing material **4.7.2.7a** reached a good PCE of 5.05% with a fill factor of 0.55, a V_{OC} of 0.90 V and a J_{SC} of 10.20 mA cm^{-2}, out-performing **4.7.2.7c**, which showed a higher V_{OC} (0.96 V), but lower fill factor and J_{SC} (0.52 and 9.64 mA cm^{-2}, respectively).

The performance data and chemical structures of the BODIPY-based materials covered in this section are presented in Table 4.6 and Figure 4.22, respectively.

Table 4.6 Performance data relating to BODIPY-based materials discussed in Section 4.7.2.

Compound	PCE (%)	Fill factor	V_{OC} (V)	J_{SC} (mA cm^{-2})	E_g (eV)	Ref.
4.7.2.1a	1.17	0.34	0.80	4.43	1.95	119
4.7.2.1b	1.34	0.44	0.75	4.14	1.70	119
4.7.2.1c	2.17	0.38	0.74	6.93	1.70	120
4.7.2.2a	3.8	0.59	0.81	8.0	1.55	121
4.7.2.2b	2.7	0.55	0.71	7.0	1.55	121
4.7.2.2c	1.7	0.52	0.61	5.4	1.52	121
4.7.2.3	4.5	0.51	0.81	8.7	–	122
4.7.2.4a	1.4	0.31	0.76	5.84	1.60	123
4.7.2.4b	4.7	0.47	0.70	14.3	1.45	123
4.7.2.4c	0.9	0.30	0.56	5.1	1.42	123
4.7.2.4d	1.5	0.32	0.55	8.5	1.48	123
4.7.2.5a	3.22	0.40	0.99	8.25	1.99	124
4.7.2.5b	2.56	0.37	0.93	7.51	1.98	124
4.7.2.6a	1.51	0.32	0.91	5.22	2.06	124
4.7.2.6b	1.85	0.33	0.90	6.16	2.03	124
4.7.2.7a	5.05	0.55	0.90	10.20	1.72	125
4.7.2.7b	2.24	0.34	1.08	6.10	1.88	125
4.7.2.7c	4.80	0.52	0.96	9.64	1.85	125

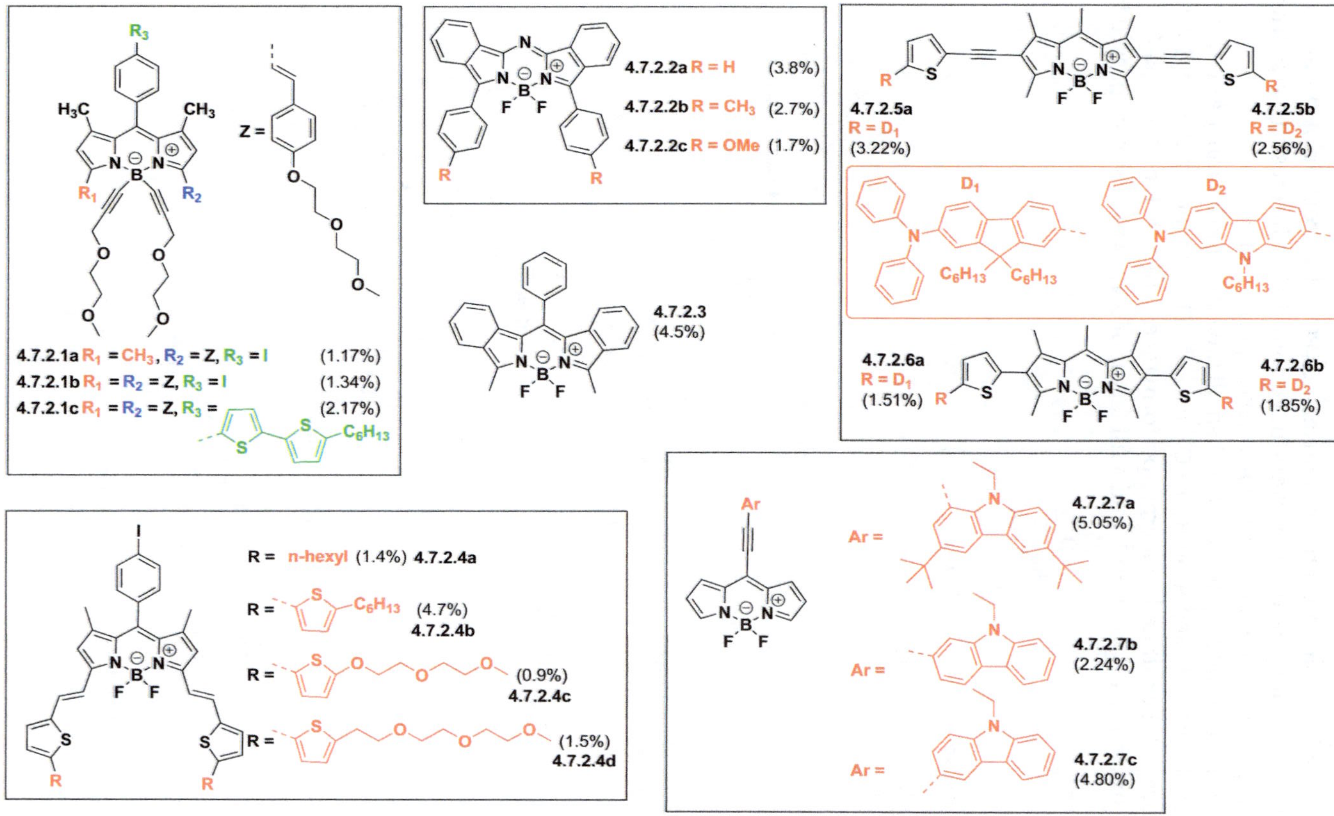

Figure 4.22 Chemical structures of BODIPY materials discussed in Section 4.7.2.

4.7.3 Miscellaneous Materials

A relatively unexplored class of compounds that has produced some promising results is based on the phenoxazine (POZ) heterocycle. A number of studies by Cheng and co-workers have found this interesting structure capable of attaining impressive OPV device performance.

In 2014 the group reported a D–π–A small molecule and a similar A–π–D–π–A molecule, both based around a POZ core.[126] The smaller molecule **4.7.3.1** (Figure 4.23) exhibited moderate performance with a PCE of 3.58%, a J_{SC} of 10.5 mA cm^{-2}, a V_{OC} of 0.85 V and a fill factor of 0.40, which was surpassed by

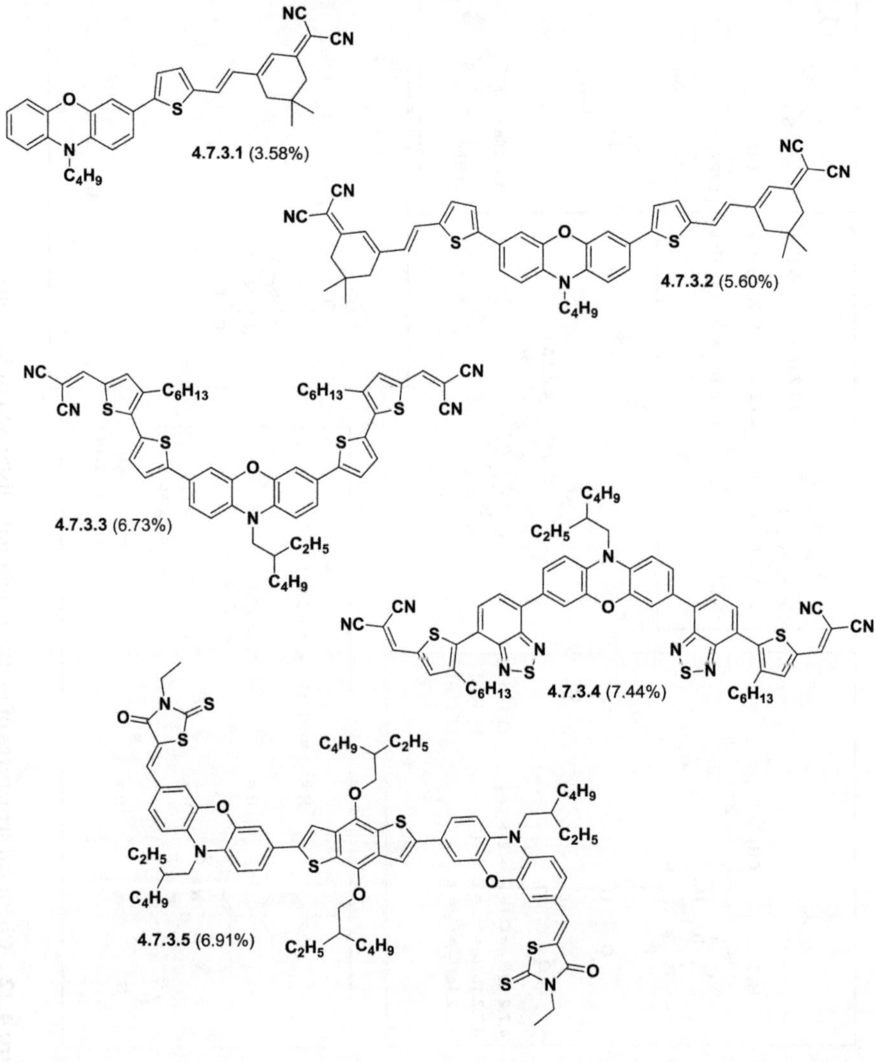

Figure 4.23 Chemical structures of phenoxazine materials discussed in Section 4.7.3.

4.7.3.2 showing a PCE of 5.60%, a J_{SC} of 13.1 mA cm^{-2}, a V_{OC} of 0.86 V and a fill factor of 0.50. The improvement in V_{OC} was attributed to the lower lying HOMO level of the larger molecule, whereas the improvements in J_{SC} were thought to be due to the improved light absorption.[126]

By incorporating additional thiophene units into the arms of the POZ core, the group achieved a PCE of 6.73% for small molecule **4.7.3.3**, with a J_{SC} of 10.5 mA cm^{-2}, a V_{OC} of 0.89 V and a fill factor of 0.72.[127] By including a benzothiadiazole moiety in place of one of the thiophene units, the group achieved a PCE of 7.44% with a J_{SC} of 11.9 mA cm^{-2}, a V_{OC} of 0.91 V and a fill factor of 0.69 for devices fabricated using compound **4.7.3.4.**[127] The improved performance was thought to be due to the electron-deficient benzothiadiazole groups improving hole mobility and conductivity, as well as broadening the absorption range of the material such that more of the solar spectrum was absorbed.

The same group utilised the planar benzodithiophene unit as a molecular core and appended POZ units as linkers to end-capping with 3-ethylrhodanine groups. This planar material, **4.7.3.5**, reached a PCE of 6.25%, which was improved to 6.91% by the addition of a 10 nm polyethylenimine 80% ethoxylated buffer layer.[128] The performance data of the POZ-based materials discussed in this section are shown in Table 4.7 and the chemical structures of the materials are shown in Figure 4.23.

A number of promising results have also been achieved from materials based around the dithienylsilole (DTS) core, which often produces a photovoltaic performance significantly higher than the non-silicon-containing analogue. One example of this is the work by Chen and co-workers published in 2015, in which they prepared both the silicon and carbon variants of A–D–A type materials with rhodanine end-groups.[129] The results showed that the silicon analogue (**4.7.3.6b**) produced a marked improvement compared with the carbon-centred **4.7.3.6a**, achieving a PCE of 8.02% – an astonishing improvement over the PCE of 0.75% of **4.7.3.6a** and one which was among the highest reported for DTS-based organic solar cells. The excellent improvement was attributed to **4.7.3.6b** exhibiting more favourable packing and light absorption than **4.7.3.6a**. Studies on the morphology of the films showed that **4.7.3.6b** formed 20–30 nm domains as well as adopting a face-on orientation, which is ideal for charge transport. These observations were consistent with the measured J_{SC} and fill factor values for both materials.

Table 4.7 Performance data relating to phenoxazine-based materials discussed in Section 4.7.3.

Compound	PCE (%)	Fill factor	V_{OC} (V)	J_{SC} (mA cm^{-2})	E_g (eV)	Ref.
4.7.3.1	3.58	0.40	0.85	10.5	1.72	126
4.7.3.2	5.60	0.50	0.86	13.1	1.65	126
4.7.3.3	6.73	0.72	0.89	10.5	1.67	127
4.7.3.4	7.44	0.69	0.91	11.9	1.59	127
4.7.3.5	6.91	0.62	0.90	12.5	1.84	128

In the same year, Brabec and co-workers assessed the effects of altering the terminal alkyl group on a DTS-based material for organic solar cells.[130] The work produced two materials – **4.7.3.7a** and **4.7.3.7b** – which differed only by the terminal alkyl chain. The group found that the longer hexyl chain in **4.7.3.7a** led to a lower PCE (1.25%) as a result of poor electron extraction, which reduced the J_{SC} and fill factor. The methyl terminal system allowed a high PCE of 5.30% to be achieved for **4.7.3.7b**, with a fill factor of 0.62, J_{SC} of 10.05 mA cm^{-2} and a V_{OC} of 0.85 V. The group reported that the large alkyl chains led to a poor morphology, which, in turn, led to a large amount of non-geminate recombination, explaining the lower PCE for **4.7.3.7a**.

Extension of the central DTS unit into a trimer with three different terminal units was studied by Yang and co-workers in 2016.[131] The authors presented three novel materials sharing a common trimeric core, with differing terminal acceptor units. The three electron-accepting units used in this rigid system were difluorobenzothiadiazole (**4.7.3.8a**), fluorobenzothiadiazole (**4.7.3.8b**) and 3-ethylrhodanine (**4.7.3.8c**). Compounds **4.7.3.8a** and **4.7.3.8b** produced PCEs <2%, due, in part, to an absorption profile that was less broad than that of **4.7.3.8c**. The high PCE of 7.56% achieved by **4.7.3.8c** was credited not only to the very broad absorption, but also to the enhanced planarity of the molecule as a result of the more rigid backbone. This added rigidity led to more efficient charge transport because the molecules adopted a preferential molecular packing motif.

Perhaps one of the most thoroughly explored DTS-based OPV devices is compound **4.7.3.9b**, which has been studied by a number of groups.[132–135] The analogous material containing the pyridyl sub-unit (**4.7.3.9a**) achieved a record efficiency at the time publication of 6.7%.[133] The high J_{SC} (14.4 mA cm^{-2}) and fill factor (0.59) combined with the moderate V_{OC} of 0.78 V resulted in a high PCE of 6.7% when fabricated into a BHJ device with $PC_{70}BM$ and a small amount (0.25% *v*/*v*) of DIO additive.

The values for **4.7.3.9b** have been among the highest reported since 2012, when a PCE of 6.7% was achieved by Bazan and co-workers[132] in a device with $PC_{70}BM$. This represented one of the top performing devices at that time and attracted a much attention. In 2013 the work was continued when they optimised the thickness of the calcium layer between the active layer and the cathode. A thickness of 20 nm was found to be the optimum and the group reported a PCE of 8.01% with a very high fill factor of 0.73, a J_{SC} of 14.2 mA cm^{-2} and a V_{OC} of 0.78 V, another top performing device of the time.[133]

In 2013 the same group studied the use of an alternative cathode layer in the device.[134] Using barium as the cathode layer for a device with **4.7.3.9b** as the donor material, the group reached some of the highest reported photovoltaic parameters, with a fill factor of 0.75, a V_{OC} of 0.78 V, a J_{SC} of 15.47 mA cm^{-2} and an exceptional PCE of 9.02%. This astounding enhancement in performance was attributed to the reduction in series resistance and an increase in shunt resistance when using barium as the cathode layer, with barium being an excellent hole-blocking material. The performance data of

Table 4.8 Performance data for DTS-based materials discussed in Section 4.7.3.

Compound	PCE (%)	Fill factor	V_{OC} (V)	J_{SC} (mA cm^{-2})	E_g (eV)	Ref.
4.7.3.6a	0.75	0.30	0.85	2.74	1.71	129
4.7.3.6b	8.02	0.69	0.82	13.67	1.66	129
4.7.3.7a	1.25	0.32	0.91	4.37	–	130
4.7.3.7b	5.30	0.62	0.85	10.05	–	130
4.7.3.8a	1.96	0.41	0.82	5.57	1.76	131
4.7.3.8b	1.29	0.33	0.83	4.16	1.75	131
4.7.3.8c	7.56	0.68	0.88	12.39	1.67	131
4.7.3.9a	6.7	0.68	0.81	12.8	1.8	132, 133
4.7.3.9b	9.02	0.75	0.78	15.47	1.78	132–135

the DTS-based materials discussed in this section are shown in Table 4.8 and the chemical structures of the materials are shown in Figure 4.24.

4.8 Future Prospects and Outlook

This chapter has provided a brief overview of the current state of small molecular systems for p-type photovoltaic devices. Although by no means comprehensive, it has covered a number of molecular classes and current state-of-the-art devices. There are many strategies for improving these devices to even higher efficiencies, with techniques such as molecular tuning, the replacement of fullerene-type acceptors, the fabrication of multiple-junction devices and varying the device structure or utilising alternative anode and cathode materials.

At present, the top performing material is an oligothiophene (**4.3.14b**), with a PCE >10%,[59] a value comparable with many high-performance polymeric materials. Many other classes of material also show much promise. DTS-based materials have reached PCEs of ~9%,[132,134,135] both DPP-based materials and TPA-based materials have achieved up to 7% PCE[72,93] and dye-based materials have shown PCEs of up to ~6%.[116,125] Although the oligoacene class of materials does not tend to demonstrate high PCEs when prepared with a fullerene acceptor (2–3%),[79,87] this is not to say that a good performance cannot be extracted from them. By using different acceptors it may be possible to reach higher photovoltaic parameters. A number of materials remain relatively under-explored, such as the POZ-based materials, which have already demonstrated PCEs >7%.[127] This highlights the enormous versatility of small molecule organic solar cells and the wide range of properties that can be tuned through careful molecular engineering.

With the 10% PCE mark successfully passed,[59] materials scientists around the world should now focus on reaching 15%,[136] an efficiency certainly within reach, especially with technologies such as tandem devices and hybrid systems and with targeted scientific progress. A continued effort to improve these small molecule systems is crucial to ensuring their success and eventual commercialisation, a realisation that will lead to a cheap renewable source of energy.

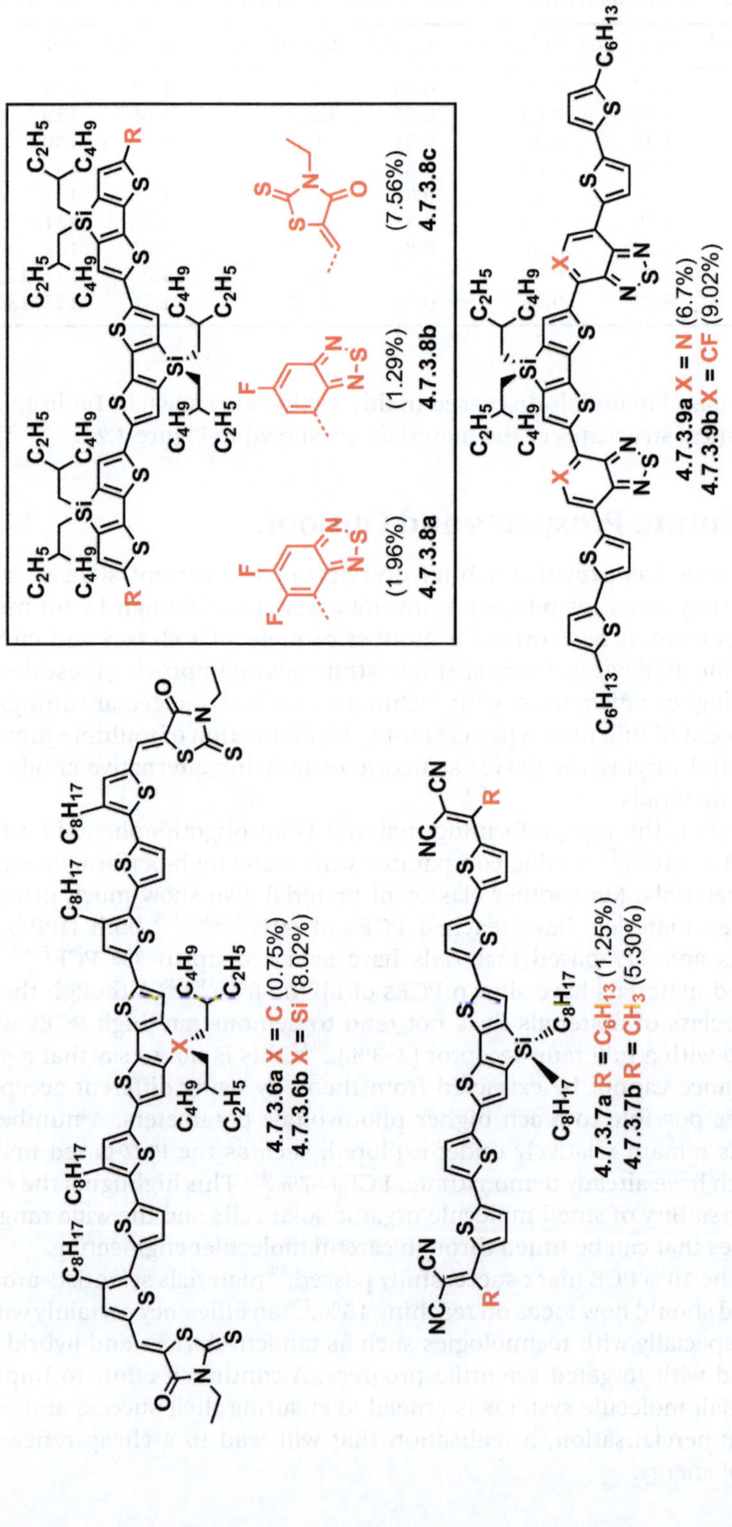

Figure 4.24 Chemical structures of DTS-based materials discussed in Section 4.7.3.

References

1. E. A. Becquerel, *C. R. Acad. Sci.*, 1839, **9**, 145–149.
2. R. Williams, *J. Chem. Phys.*, 1960, **32**, 1505.
3. J. Roncali, P. Leriche and P. Blanchard, *Adv. Mater.*, 2014, **26**, 3821–3838.
4. A. Mishra and P. Bäuerle, *Angew. Chem., Int. Ed.*, 2012, **51**, 2020–2067.
5. S. B. Darling and F. You, *RSC Adv.*, 2013, **3**, 17633.
6. N. Espinosa, M. Hösel, D. Angmo and F. C. Krebs, *Energy Environ. Sci.*, 2012, **5**, 5117–5132.
7. C. W. Tang, *Appl. Phys. Lett.*, 1986, **48**, 183.
8. *Organic Solar Cells: Fundamentals, Devices, and Upscaling*, ed. B. P. Rand and H. Richter, Pan Stanford, 1st edn, 2014.
9. W. Tress, *Organic Solar Cells: Theory, Experiment, and Device Simulation*, Springer International Publishing, Cham, 2014, vol. 208.
10. S. Günes, H. Neugebauer and N. S. Sariciftci, *Chem. Rev.*, 2007, **107**, 1324–1338.
11. G. Yu, J. Gao, J. C. Hummelen, F. Wudl and A. J. Heeger, *Science*, 1995, **270**, 1789–1791.
12. M. C. Scharber, D. Muehlbacher, M. Koppe, P. Denk, C. Waldauf, A. J. Heeger and C. J. Brabec, *Adv. Mater.*, 2006, **18**, 789–794.
13. M. C. Scharber and N. S. Sariciftci, *Prog. Polym. Sci.*, 2013, **38**, 1929–1940.
14. Q. Qiao, *Organic Solar Cells: Materials, Devices, Interfaces, and Modeling*, CRC Press, 1st edn, 2015.
15. ASTM International, *ASTM G 173-03. Standard Tables for Reference Solar Spectral Irradiances: Direct Normal and Hemispherical on 37° Tilted Surface*, West Conshohoken, PA, 2012.
16. G. Chen, H. Sasabe, T. Igarashi, Z. Hong and J. Kido, *J. Mater. Chem. A*, 2015, **3**, 14517–14534.
17. P.-L. T. Boudreault, A. Najari and M. Leclerc, *Chem. Mater.*, 2011, **23**, 456–469.
18. C. R. Belton, A. L. Kanibolotsky, J. Kirkpatrick, C. Orofino, S. E. T. Elmasly, P. N. Stavrinou, P. J. Skabara and D. D. C. Bradley, *Adv. Funct. Mater.*, 2013, **23**, 2792–2804.
19. J. E. Coughlin, Z. B. Henson, G. C. Welch and G. C. Bazan, *Acc. Chem. Res.*, 2014, **47**, 257–270.
20. J. Chen and Y. Cao, *Acc. Chem. Res.*, 2009, **42**, 1709–1718.
21. W. Shockley and H. J. Queisser, *J. Appl. Phys.*, 1961, **32**, 510.
22. J. Roncali, *Macromol. Rapid Commun.*, 2007, **28**, 1761–1775.
23. J. Roncali, *Chem. Rev.*, 1997, **97**, 173–206.
24. Y. H. Wijsboom, A. Patra, S. S. Zade, Y. Sheynin, M. Li, L. J. W. Shimon and M. Bendikov, *Angew. Chem., Int. Ed.*, 2009, **48**, 5443–5447.
25. Y. H. Wijsboom, Y. Sheynin, A. Patra, N. Zamoshchik, R. Vardimon, G. Leitus and M. Bendikov, *J. Mater. Chem.*, 2011, **21**, 1368–1372.
26. H. J. Spencer, P. J. Skabara, M. Giles, I. McCulloch, S. J. Coles and M. B. Hursthouse, *J. Mater. Chem.*, 2005, **15**, 4783.

27. G. Conboy, H. J. Spencer, E. Angioni, A. L. Kanibolotsky, N. J. Findlay, S. J. Coles, C. Wilson, M. B. Pitak, C. Risko, V. Coropceanu, J.-L. Brédas and P. J. Skabara, *Mater. Horiz.*, 2016, **3**, 333–339.
28. J. D. Servaites, M. A. Ratner and T. J. Marks, *Appl. Phys. Lett.*, 2009, **95**, 2009–2011.
29. H. Hoppe, T. Glatzel, M. Niggemann, W. Schwinger, F. Schaeffler, A. Hinsch, M. C. Lux-Steiner and N. S. Sariciftci, *Thin Solid Films*, 2006, **511–512**, 587–592.
30. T. Martens, J. D'Haen, T. Munters, Z. Beelen, L. Goris, J. Manca, M. D'Olieslaeger, D. Vanderzande, L. De Schepper and R. Andriessen, *Synth. Met.*, 2003, **138**, 243–247.
31. J. K. J. van Duren, X. Yang, J. Loos, C. W. T. Bulle-Lieuwma, A. B. Sieval, J. C. Hummelen and R. A. J. Janssen, *Adv. Funct. Mater.*, 2004, **14**, 425–434.
32. Y. Zhao, Z. Xie, Y. Qu, Y. Geng and L. Wang, *Synth. Met.*, 2008, **158**, 908–911.
33. G. Li, V. Shrotriya, Y. Yao and Y. Yang, *J. Appl. Phys.*, 2005, **98**, 43704.
34. G. D. Sharma, P. Suresh, S. S. Sharma, Y. K. Vijay and J. A. Mikroyannidis, *ACS Appl. Mater. Interfaces*, 2010, **2**, 504–510.
35. A. Zusan, B. Gieseking, M. Zerson, V. Dyakonov, R. Magerle and C. Deibel, *Sci. Rep.*, 2015, **5**, 8286.
36. N. K. Elumalai and A. Uddin, *Energy Environ. Sci.*, 2016, **9**, 391–410.
37. C. J. Brabec, M. Heeney, I. McCulloch and J. Nelson, *Chem. Soc. Rev.*, 2011, **40**, 1185–1199.
38. S. J. Lou, J. M. Szarko, T. Xu, L. Yu, T. J. Marks and L. X. Chen, *J. Am. Chem. Soc.*, 2011, **133**, 20661–20663.
39. A. Mishra, C.-Q. Ma and P. Baüerle, *Chem. Rev.*, 2009, **109**, 1141–1276.
40. W. Ma, C. Yang, X. Gong, K. Lee and A. J. Heeger, *Adv. Funct. Mater.*, 2005, **15**, 1617–1622.
41. M. A. Green, K. Emery, D. L. King, Y. Hishikawa and W. Warta, *Prog. Photovoltaics: Res. Appl.*, 2007, **15**, 35–40.
42. X. Sun, Y. Zhou, W. Wu, Y. Liu, W. Tian, G. Yu, W. Qiu, S. Chen and D. Zhu, *J. Phys. Chem. B*, 2006, **110**, 7702–7707.
43. N. Kopidakis, W. J. Mitchell, J. van de Lagemaat, D. S. Ginley, G. Rumbles, S. E. Shaheen and W. L. Rance, *Appl. Phys. Lett.*, 2006, **89**, 103524.
44. C.-Q. Ma, M. Fonrodona, M. C. Schikora, M. M. Wienk, R. A. J. Janssen and P. Bäuerle, *Adv. Funct. Mater.*, 2008, **18**, 3323–3331.
45. B. Yin, L. Yang, Y. Liu, Y. Chen, Q. Qi, F. Zhang and S. Yin, *Appl. Phys. Lett.*, 2010, **97**, 23303.
46. J. Sakai, T. Taima, T. Yamanari and K. Saito, *Sol. Energy Mater. Sol. Cells*, 2009, **93**, 1149–1153.
47. R. Fitzner, E. Reinold, A. Mishra, E. Mena-Osteritz, H. Ziehlke, C. Körner, K. Leo, M. Riede, M. Weil, O. Tsaryova, A. Weiß, C. Uhrich, M. Pfeiffer and P. Bäuerle, *Adv. Funct. Mater.*, 2011, **21**, 897–910.
48. Y. Liu, X. Wan, F. Wang, J. Zhou, G. Long, J. Tian, J. You, Y. Yang and Y. Chen, *Adv. Energy Mater.*, 2011, **1**, 771–775.
49. Z. Li, G. He, X. Wan, Y. Liu, J. Zhou, G. Long, Y. Zuo, M. Zhang and Y. Chen, *Adv. Energy Mater.*, 2012, **2**, 74–77.

50. Q. Zhang, B. Kan, F. Liu, G. Long, X. Wan, X. Chen, Y. Zuo, W. Ni, H. Zhang, M. Li, Z. Hu, F. Huang, Y. Cao, Z. Liang, M. Zhang, T. P. Russell and Y. Chen, *Nat. Photonics*, 2014, **9**, 35–41.
51. Z. He, C. Zhong, X. Huang, W.-Y. Wong, H. Wu, L. Chen, S. Su and Y. Cao, *Adv. Mater.*, 2011, **23**, 4636–4643.
52. Y. Liu, C.-C. Chen, Z. Hong, J. Gao, Y. (Michael) Yang, H. Zhou, L. Dou, G. Li and Y. Yang, *Sci. Rep.*, 2013, **3**, 3356.
53. K. R. Graham, J. Mei, R. Stalder, J. W. Shim, H. Cheun, F. Steffy, F. So, B. Kippelen and J. R. Reynolds, *ACS Appl. Mater. Interfaces*, 2011, **3**, 1210–1215.
54. K. Sun, Z. Xiao, S. Lu, W. Zajaczkowski, W. Pisula, E. Hanssen, J. M. White, R. M. Williamson, J. Subbiah, J. Ouyang, A. B. Holmes, W. W. H. Wong and D. J. Jones, *Nat. Commun.*, 2015, **6**, 6013.
55. J.-L. Wang, Q.-R. Yin, J.-S. Miao, Z. Wu, Z.-F. Chang, Y. Cao, R.-B. Zhang, J.-Y. Wang, H.-B. Wu and Y. Cao, *Adv. Funct. Mater.*, 2015, **25**, 3514–3523.
56. S. Steinberger, A. Mishra, E. Reinold, J. Levichkov, C. Uhrich, M. Pfeiffer and P. Bäuerle, *Chem. Commun.*, 2011, **47**, 1982.
57. Y. Liu, X. Wan, F. Wang, J. Zhou, G. Long, J. Tian and Y. Chen, *Adv. Mater.*, 2011, **23**, 5387–5391.
58. S. Shen, P. Jiang, C. He, J. Zhang, P. Shen, Y. Zhang, Y. Yi, Z. Zhang, Z. Li and Y. Li, *Chem. Mater.*, 2013, **25**, 2274–2281.
59. B. Kan, M. Li, Q. Zhang, F. Liu, X. Wan, Y. Wang, W. Ni, G. Long, X. Yang, H. Feng, Y. Zuo, M. Zhang, F. Huang, Y. Cao, T. P. Russell and Y. Chen, *J. Am. Chem. Soc.*, 2015, **137**, 3886–3893.
60. S. Wang, J. Yang, Z. Zhang, Y. Hu, X. Cao, H. Li, Y. Tao, Y. Li and W. Huang, *RSC Adv.*, 2015, **5**, 68192–68199.
61. Q.-R. Yin, J.-S. Miao, Z. Wu, Z.-F. Chang, J.-L. Wang, H.-B. Wu and Y. Cao, *J. Mater. Chem. A*, 2015, **3**, 11575–11586.
62. J.-H. Kim, J. B. Park, H. Yang, I. H. Jung, S. C. Yoon, D. Kim and D.-H. Hwang, *ACS Appl. Mater. Interfaces*, 2015, **7**, 23866–23875.
63. C. Cabanetos, A. El Labban, J. A. Bartelt, J. D. Douglas, W. R. Mateker, J. M. J. Fréchet, M. D. McGehee and P. M. Beaujuge, *J. Am. Chem. Soc.*, 2013, **135**, 4656–4659.
64. J. Warnan, C. Cabanetos, R. Bude, A. El Labban, L. Li and P. M. Beaujuge, *Chem. Mater.*, 2014, **26**, 2829–2835.
65. K. Narayanaswamy, A. Venkateswararao, V. Gupta, S. Chand and S. P. Singh, *Chem. Commun.*, 2016, **52**, 210–213.
66. B. Walker, A. B. Tamayo, X.-D. Dang, P. Zalar, J. H. Seo, A. Garcia, M. Tantiwiwat and T.-Q. Nguyen, *Adv. Funct. Mater.*, 2009, **19**, 3063–3069.
67. J. W. Ryan and Y. Matsuo, *Sci. Rep.*, 2015, **5**, 8319.
68. L. S. Hung, C. W. Tang and M. G. Mason, *Appl. Phys. Lett.*, 1997, **70**, 152.
69. Y. Matsuo, J. Kawai, H. Inada, T. Nakagawa, H. Ota, S. Otsubo and E. Nakamura, *Adv. Mater.*, 2013, **25**, 6266–6269.
70. N. Zhou, S. Vegiraju, X. Yu, E. F. Manley, M. R. Butler, M. J. Leonardi, P. Guo, W. Zhao, Y. Hu, K. Prabakaran, R. P. H. Chang, M. A. Ratner, L. X. Chen, A. Facchetti, M.-C. Chen and T. J. Marks, *J. Mater. Chem. C*, 2015, **3**, 8932–8941.

71. X. Duan, M. Xiao, J. Chen, X. Wang, W. Peng, L. Duan, H. Tan, G. Lei, R. Yang and W. Zhu, *ACS Appl. Mater. Interfaces*, 2015, **7**, 18292–18299.
72. J. W. Jung, T. P. Russell and W. H. Jo, *Chem. Mater.*, 2015, **27**, 4865–4870.
73. Z. Chen, Y. Zheng, H. Yan and A. Facchetti, *J. Am. Chem. Soc.*, 2009, **131**, 8–9.
74. Z. Li, J. D. A. Lin, H. Phan, A. Sharenko, C. M. Proctor, P. Zalar, Z. Chen, A. Facchetti and T.-Q. Nguyen, *Adv. Funct. Mater.*, 2014, **24**, 6989–6998.
75. N. Geacintov, *J. Chem. Phys.*, 1966, **45**, 2639.
76. W. Moore and M. Silver, *J. Chem. Phys.*, 1960, **33**, 1671.
77. R. Signerski, G. Jarosz and J. Godlewski, *Synth. Met.*, 1998, **94**, 135–137.
78. S. Yoo, B. Domercq and B. Kippelen, *Appl. Phys. Lett.*, 2004, **85**, 5427.
79. C.-W. Chu, Y. Shao, V. Shrotriya and Y. Yang, *Appl. Phys. Lett.*, 2005, **86**, 243506.
80. A. A. Gorodetsky, M. Cox, N. J. Tremblay, I. Kymissis and C. Nuckolls, *Chem. Mater.*, 2009, **21**, 4090–4092.
81. J. E. Anthony, J. S. Brooks, D. L. Eaton and S. R. Parkin, *J. Am. Chem. Soc.*, 2001, **123**, 9482–9483.
82. M. T. Lloyd, A. C. Mayer, A. S. Tayi, A. M. Bowen, T. G. Kasen, D. J. Herman, D. A. Mourey, J. E. Anthony and G. G. Malliaras, *Org. Electron.*, 2006, **7**, 243–248.
83. M. T. Lloyd, A. C. Mayer, S. Subramanian, D. A. Mourey, D. J. Herman, A. V. Bapat, J. E. Anthony and G. G. Malliaras, *J. Am. Chem. Soc.*, 2007, **129**, 9144–9149.
84. L. Valentini, D. Bagnis, A. Marrocchi, M. Seri, A. Taticchi and J. M. Kenny, *Chem. Mater.*, 2008, **20**, 32–34.
85. F. Silvestri, A. Marrocchi, M. Seri, C. Kim, T. J. Marks, A. Facchetti and A. Taticchi, *J. Am. Chem. Soc.*, 2010, **132**, 6108–6123.
86. W. W. H. Wong, T. B. Singh, D. Vak, W. Pisula, C. Yan, X. Feng, E. L. Williams, K. L. Chan, Q. Mao, D. J. Jones, C.-Q. Ma, K. Müllen, P. Bäuerle and A. B. Holmes, *Adv. Funct. Mater.*, 2010, **20**, 927–938.
87. K. N. Winzenberg, P. Kemppinen, G. Fanchini, M. Bown, G. E. Collis, C. M. Forsyth, K. Hegedus, T. B. Singh and S. E. Watkins, *Chem. Mater.*, 2009, **21**, 5701–5703.
88. A. K. Pandey and J.-M. Nunzi, *Adv. Mater.*, 2007, **19**, 3613–3617.
89. A. K. Pandey, K. N. N. Unni and J.-M. Nunzi, *Thin Solid Films*, 2006, **511–512**, 529–532.
90. A. Cravino, S. Roquet, O. Alévêque, P. Leriche, P. Frère and J. Roncali, *Chem. Mater.*, 2006, **18**, 2584–2590.
91. Y. Shirota, *J. Mater. Chem.*, 2005, **15**, 75–93.
92. B. A. Gregg, *J. Phys. Chem. Lett.*, 2011, **2**, 3013–3015.
93. Y.-H. Chen, L.-Y. Lin, C.-W. Lu, F. Lin, Z.-Y. Huang, H.-W. Lin, P.-H. Wang, Y.-H. Liu, K.-T. Wong, J. Wen, D. J. Miller and S. B. Darling, *J. Am. Chem. Soc.*, 2012, **134**, 13616–13623.
94. H.-W. Lin, L.-Y. Lin, Y.-H. Chen, C.-W. Chen, Y.-T. Lin, S.-W. Chiu and K.-T. Wong, *Chem. Commun.*, 2011, **47**, 7872.

95. M. J. Cho, J. Seo, H. S. Oh, H. Jee, W. J. Kim, K. H. Kim, M. H. Hoang, D. H. Choi and P. N. Prasad, *Sol. Energy Mater. Sol. Cells*, 2012, **98**, 71–77.
96. S. Roquet, A. Cravino, P. Leriche, O. Alévêque, P. Frère and J. Roncali, *J. Am. Chem. Soc.*, 2006, **128**, 3459–3466.
97. J. C. Bernède, L. Cattin, S. O. Djobo, M. Morsli, S. R. B. Kanth, S. Patil, P. Leriche, J. Roncali, A. Godoy, F. R. Diaz and M. A. del Valle, *Phys. Status Solidi*, 2011, **208**, 1989–1994.
98. J. C. Bernède, L. Cattin, M. Makha, V. Jeux, P. Leriche, J. Roncali, V. Froger, M. Morsli and M. Addou, *Sol. Energy Mater. Sol. Cells*, 2013, **110**, 107–114.
99. A. Leliège, C.-H. Le Régent, M. Allain, P. Blanchard and J. Roncali, *Chem. Commun.*, 2012, **48**, 8907.
100. D. Demeter, V. Jeux, P. Leriche, P. Blanchard, Y. Olivier, J. Cornil, R. Po and J. Roncali, *Adv. Funct. Mater.*, 2013, **23**, 4854–4861.
101. V. Jeux, D. Demeter, P. Leriche and J. Roncali, *RSC Adv.*, 2013, **3**, 5811.
102. V. Y. Merritt and H. J. Hovel, *Appl. Phys. Lett.*, 1976, **29**, 414.
103. F. Silvestri, M. D. Irwin, L. Beverina, A. Facchetti, G. A. Pagani and T. J. Marks, *J. Am. Chem. Soc.*, 2008, **130**, 17640–17641.
104. D. Bagnis, L. Beverina, H. Huang, F. Silvestri, Y. Yao, H. Yan, G. a. Pagani, T. J. Marks and A. Facchetti, *J. Am. Chem. Soc.*, 2010, **132**, 4074–4075.
105. S. Wang, E. I. Mayo, M. D. Perez, L. Griffe, G. Wei, P. I. Djurovich, S. R. Forrest and M. E. Thompson, *Appl. Phys. Lett.*, 2009, **94**, 233304.
106. M. Tian, M. Furuki, I. Iwasa, Y. Sato, L. S. Pu and S. Tatsuura, *J. Phys. Chem. B*, 2002, **106**, 4370–4376.
107. G. Chen, H. Sasabe, Z. Wang, X.-F. Wang, Z. Hong, Y. Yang and J. Kido, *Adv. Mater.*, 2012, **24**, 2768–2773.
108. G. Wei, S. Wang, K. Renshaw, M. E. Thompson and S. R. Forrest, *ACS Nano*, 2010, **4**, 1927–1934.
109. G. Chen, H. Sasabe, Z. Wang, X. Wang, Z. Hong, J. Kido and Y. Yang, *Phys. Chem. Chem. Phys.*, 2012, **14**, 14661.
110. G. Wei, S. Wang, K. Sun, M. E. Thompson and S. R. Forrest, *Adv. Energy Mater.*, 2011, **1**, 184–187.
111. G. Wei, R. R. Lunt, K. Sun, S. Wang, M. E. Thompson and S. R. Forrest, *Nano Lett.*, 2010, **10**, 3555–3559.
112. G. Chen, D. Yokoyama, H. Sasabe, Z. Hong, Y. Yang and J. Kido, *Appl. Phys. Lett.*, 2012, **101**, 83904.
113. G. Wei, X. Xiao, S. Wang, K. Sun, K. J. Bergemann, M. E. Thompson and S. R. Forrest, *ACS Nano*, 2012, **6**, 972–978.
114. G. Chen, H. Sasabe, Y. Sasaki, H. Katagiri, X.-F. Wang, T. Sano, Z. Hong, Y. Yang and J. Kido, *Chem. Mater.*, 2014, **26**, 1356–1364.
115. S. Karak, P. J. Homnick, A. M. Della Pelle, Y. Bae, V. V. Duzhko, F. Liu, T. P. Russell, P. M. Lahti and S. Thayumanavan, *ACS Appl. Mater. Interfaces*, 2014, **6**, 11376–11384.
116. G. Wei, X. Xiao, S. Wang, J. D. Zimmerman, K. Sun, V. V. Diev, M. E. Thompson and S. R. Forrest, *Nano Lett.*, 2011, **11**, 4261–4264.

117. H. Sasabe, T. Igrashi, Y. Sasaki, G. Chen, Z. Hong and J. Kido, *RSC Adv.*, 2014, **4**, 42804–42807.
118. J. D. Zimmerman, B. E. Lassiter, X. Xiao, K. Sun, A. Dolocan, R. Gearba, D. A. Vanden Bout, K. J. Stevenson, P. Wickramasinghe, M. E. Thompson and S. R. Forrest, *ACS Nano*, 2013, 7, 9268–9275.
119. T. Rousseau, A. Cravino, T. Bura, G. Ulrich, R. Ziessel and J. Roncali, *Chem. Commun.*, 2009, 1673.
120. T. Rousseau, A. Cravino, E. Ripaud, P. Leriche, S. Rihn, A. De Nicola, R. Ziessel and J. Roncali, *Chem. Commun.*, 2010, **46**, 5082.
121. S. Kraner, J. Widmer, J. Benduhn, E. Hieckmann, T. Jägeler-Hoheisel, S. Ullbrich, D. Schütze, K. Sebastian Radke, G. Cuniberti, F. Ortmann, M. Lorenz-Rothe, R. Meerheim, D. Spoltore, K. Vandewal, C. Koerner and K. Leo, *Phys. Status Solidi*, 2015, **212**, 2747–2753.
122. J. J. Chen, S. M. Conron, P. Erwin, M. Dimitriou, K. McAlahney and M. E. Thompson, *ACS Appl. Mater. Interfaces*, 2015, **7**, 662–669.
123. T. Bura, N. Leclerc, S. Fall, P. Lévêque, T. Heiser, P. Retailleau, S. Rihn, A. Mirloup and R. Ziessel, *J. Am. Chem. Soc.*, 2012, **134**, 17404–17407.
124. H.-Y. Lin, W.-C. Huang, Y.-C. Chen, H.-H. Chou, C.-Y. Hsu, J. T. Lin and H.-W. Lin, *Chem. Commun.*, 2012, **48**, 8913.
125. T. Jadhav, R. Misra, S. Biswas and G. D. Sharma, *Phys. Chem. Chem. Phys.*, 2015, **17**, 26580–26588.
126. M. Cheng, X. Yang, C. Chen, Q. Tan and L. Sun, *J. Mater. Chem. A*, 2014, **2**, 10465.
127. M. Cheng, C. Chen, X. Yang, J. Huang, F. Zhang, B. Xu and L. Sun, *Chem. Mater.*, 2015, **27**, 1808–1814.
128. M. Cheng, B. Xu, C. Chen, X. Yang, F. Zhang, Q. Tan, Y. Hua, L. Kloo and L. Sun, *Adv. Energy Mater.*, 2015, **5**, 1401720.
129. W. Ni, M. Li, F. Liu, X. Wan, H. Feng, B. Kan, Q. Zhang, H. Zhang and Y. Chen, *Chem. Mater.*, 2015, **27**, 6077–6084.
130. J. Min, Y. N. Luponosov, N. Gasparini, M. Richter, A. V. Bakirov, M. A. Shcherbina, S. N. Chvalun, L. Grodd, S. Grigorian, T. Ameri, S. A. Ponomarenko and C. J. Brabec, *Adv. Energy Mater.*, 2015, **5**, 1500386.
131. W. Li, W. Deng, K. Wu, G. Xie, C. Yang, H. Wu and Y. Cao, *J. Mater. Chem. C*, 2016, **4**, 1972–1978.
132. T. S. Van Der Poll, J. A. Love, T. Q. Nguyen and G. C. Bazan, *Adv. Mater.*, 2012, **24**, 3646–3649.
133. Y. Sun, G. C. Welch, W. L. Leong, C. J. Takacs, G. C. Bazan and A. J. Heeger, *Nat. Mater.*, 2011, **11**, 44–48.
134. V. Gupta, A. K. K. Kyaw, D. H. Wang, S. Chand, G. C. Bazan and A. J. Heeger, *Sci. Rep.*, 2013, **3**, 1965.
135. A. K. K. Kyaw, D. H. Wang, V. Gupta, W. L. Leong, L. Ke, G. C. Bazan and A. J. Heeger, *ACS Nano*, 2013, **7**, 4569–4577.
136. G. Dennler, M. C. Scharber, T. Ameri, P. Denk, K. Forberich, C. Waldauf and C. J. Brabec, *Adv. Mater.*, 2008, **20**, 579–583.
137. Y. Chen, X. Wan and G. Long, *Acc. Chem. Res.*, 2013, **46**, 2645–2655.

138. B. Kan, Q. Zhang, M. Li, X. Wan, W. Ni, G. Long, Y. Wang, X. Yang, H. Feng and Y. Chen, *J. Am. Chem. Soc.*, 2014, **136**, 15529–15532.
139. Y. Liu, J. Zhao, Z. Li, C. Mu, W. Ma, H. Hu, K. Jiang, H. Lin, H. Ade and H. Yan, *Nat. Commun.*, 2014, **5**, 5293.
140. F. Grein, *J. Mol. Struct.: THEOCHEM*, 2003, **624**, 23–28.
141. H. Pang, P. J. Skabara, S. Gordeyev, J. J. W. McDouall, S. J. Coles and M. B. Hursthouse, *Chem. Mater.*, 2007, **19**, 301–307.
142. A. Cravino, H. Neugebauer, A. Petr, P. J. Skabara, H. J. Spencer, J. J. W. McDouall, L. Dunsch and N. S. Sariciftci, *J. Phys. Chem. B*, 2006, **110**, 2662–2667.
143. F. Wudl, M. Kobayashi and A. J. Heeger, *J. Org. Chem.*, 1984, **49**, 3382–3384.
144. J. L. Brédas, A. J. Heeger and F. Wudl, *J. Chem. Phys.*, 1986, **85**, 4673–4678.
145. J. L. Bredas, B. Themans, J. M. Andre, A. J. Heeger and F. Wudl, *Synth. Met.*, 1985, **11**, 343–352.
146. S. Roquet, R. de Bettignies, P. Leriche, A. Cravino and J. Roncali, *J. Mater. Chem.*, 2006, **16**, 3040.
147. J. Zhou, Y. Zuo, X. Wan, G. Long, Q. Zhang, W. Ni, Y. Liu, Z. Li, G. He, C. Li, B. Kan, M. Li and Y. Chen, *J. Am. Chem. Soc.*, 2013, **135**, 8484–8487.
148. J.-L. Wang, F. Xiao, J. Yan, Z. Wu, K.-K. Liu, Z.-F. Chang, R.-B. Zhang, H. Chen, H.-B. Wu and Y. Cao, *Adv. Funct. Mater.*, 2016, **26**, 1803–1812.
149. J. Huang, C. Zhan, X. Zhang, Y. Zhao, Z. Lu, H. Jia, B. Jiang, J. Ye, S. Zhang, A. Tang, Y. Liu, Q. Pei and J. Yao, *ACS Appl. Mater. Interfaces*, 2013, **5**, 2033–2039.

CHAPTER 5

Fullerenes and New Acceptors for Organic Solar Cells

JIAYU WANG AND XIAOWEI ZHAN*

Peking University, Department of Materials Science and Engineering, College of Engineering, Beijing, 100871, P. R. China
*E-mail: xwzhan@pku.edu.cn

5.1 Introduction

Organic solar cells (OSCs) have attracted considerable attention in recent years and are regarded as a promising alternative for the conversion of solar energy to electricity. They present a number of advantages over silicon-based solar cells, such as low cost, light weight, flexibility and the ability to fabricate cells with a large area.[1–4] The heterojunction, which is the interface between two layers or regions of donor and acceptor material, is the basis of typical OSCs. The intrinsic properties of the donors and acceptors are therefore one of the essential factors affecting device performance. The four main roles of the acceptors in OSCs are as follows.

1. Light absorption. Although donors absorb photons and generate excitons, acceptors can also absorb photons to produce excitons and contribute to the photocurrent. Complementary absorption between donors and acceptors can therefore be an effective method of increasing light harvesting.

Nanoscience & Nanotechnology Series No. 45
Nanostructured Materials for Type III Photovoltaics
Edited by Peter Skabara and Mohammad Azad Malik

Published by the Royal Society of Chemistry, www.rsc.org

2. Exciton dissociation. The excitons generated under illumination diffuse to the donor–acceptor interface and dissociate into holes and electrons. The driving force for the dissociation of excitons comes from the difference in energy between the highest occupied molecular orbital (HOMO) and the lowest unoccupied molecular orbital (LUMO) of the donor and acceptor. The HOMO and LUMO offsets should be large enough to ensure the efficient dissociation of excitons. Too large an offset causes a large energy loss during the dissociation of excitons, which leads to lower open circuit voltage (V_{OC}) and lower power conversion efficiency (PCE). An empirical energy level difference of 0.3–0.5 eV is favorable for both the efficient dissociation of excitons and a suitable V_{OC}.
3. Electron transport. After the dissociation of excitons, the electrons are transported to the cathode through the acceptor channels. As a result of the limited lifetime (nanoseconds to microseconds) of the charge carriers, the mobility of the acceptor should be high enough to ensure that the electron can be transported to and extracted by the cathode within its lifetime, otherwise recombination between the hole and electron will occur. A high mobility facilitates the enhancement of the short circuit current density (J_{SC}) and fill factor and promotes the PCE. The presence of π–π stacking leads to a higher mobility due to the strong electronic coupling and excellent orbital overlap between the π-systems on adjacent molecules, which is a general design concept for high-mobility materials.
4. Morphology control. Morphology control of the active layer is a crucial issue in bulk heterojunction (BHJ) OSCs. As a result of their dissimilarity, donor and acceptor materials tend to aggregate with homogeneous molecules and separate from heterogeneous molecules, resulting in spontaneous phase separation. The phase separation provides donor–acceptor interfaces for exciton dissociation and channels for hole/electron transport. Small-scale phase separation offers larger area donor–acceptor interfaces at a distance within the exciton diffusion length, which promotes exciton dissociation, but restricts charge transport. Large-scale phase separation and crystalline domains promote the formation of continuous channels for charge transport, but lower the efficiency of exciton dissociation. Thus the optimized morphology of the active layer is an interpenetrating donor–acceptor network with phase separation of around 10–20 nm in length scale, which balances both exciton dissociation and charge transport.[5]

Research on electron acceptors began with heterojunction OSCs. Tang fabricated the first bilayer heterojunction solar cell in 1986, in which a perylene diimide (PDI) derivative *trans*-3,4,9,10-perylene tetracarboxylic bis-benzimidazole acceptor was stacked on the top of a copper phthalocyanine (Figure 5.1) donor.[6] In 1995, Hummelen, Wudl *et al.* first synthesized one

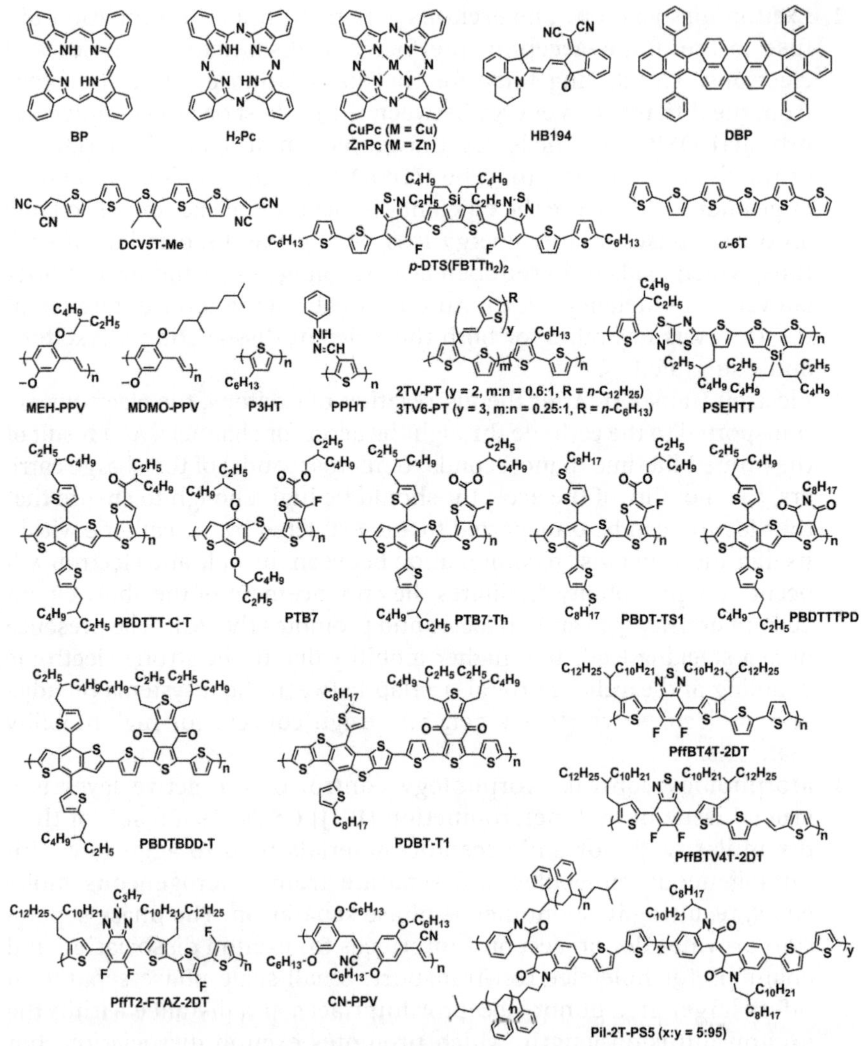

Figure 5.1 Chemical structures of donor materials.

of the most successful fullerene acceptors: phenyl-C_{61}-butyric acid methyl ester ($PC_{61}BM$; Figure 5.2).[7] Heeger *et al.*,[8] who developed the concept of the BHJ with Wudl and coworkers, blended poly[2-methoxy-5-(2′-ethylhexyloxy-1,4-phenylenevinylene)] (MEH-PPV; Figure 5.1) with $PC_{61}BM$ to fabricate OSCs, achieving a PCE of 2.9%. Friend *et al.*[9] blended a MEH-PPV donor with a polymer acceptor poly(2,5-hexyloxy-1,4-phenylene cyanovinylene) (CN-PPV; Figure 5.1) to fabricate all-polymer solar cells.

The successful application of $PC_{61}BM$ has attracted considerable attention. $PC_{61}BM$ and other fullerene derivatives, especially the C_{70} homologue phenyl-C_{71}-butyric acid methyl ester ($PC_{71}BM$; Figure 5.2), gradually became

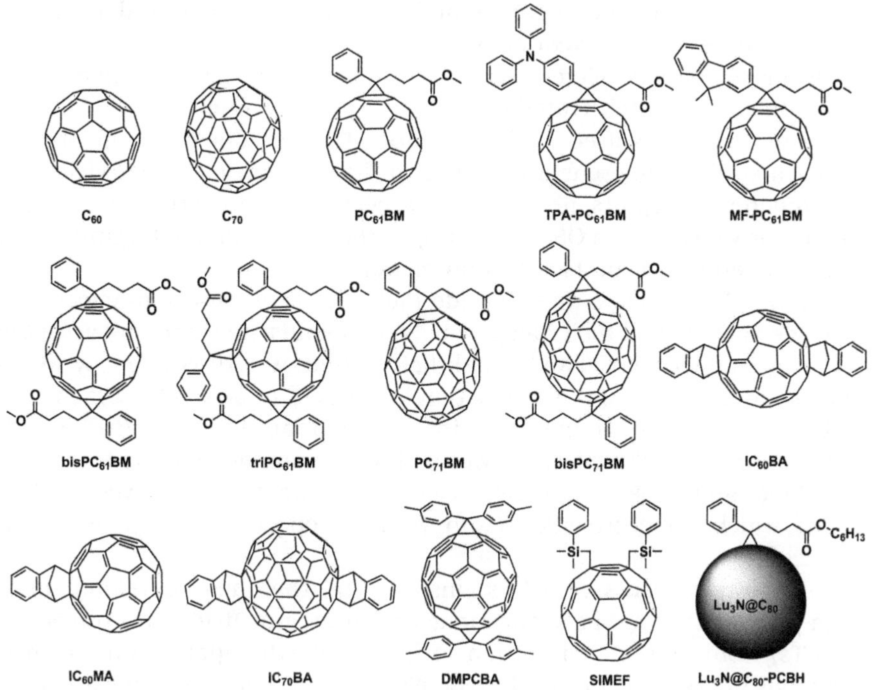

Figure 5.2 Chemical structures of fullerene acceptors.

the dominant acceptors in OSCs. The reasons why fullerene derivatives exhibit such excellent acceptor properties can be summarized as follows: (1) they have a large electron affinity and strong tendency to accept electrons from donor materials; (2) they have high electron mobility, even in composite systems; (3) they exhibit isotropic charge transport; (4) they undergo reversible electrochemical reduction; and (5) they have the ability to form favorable nanoscale morphological networks with donor materials. The best PCEs of fullerene-based OSCs achieved so far are: 8.3% for vacuum-deposited single-junction OSCs (C_{60} as acceptor);[10] 11.1% for vacuum-deposited tandem OSCs (C_{60} and C_{70} as acceptors);[11] 10.95% for solution-processed single-junction OSCs ($PC_{71}BM$ as acceptor);[12] and 11.83% for solution-processed tandem OSCs ($PC_{71}BM$ and indene-C_{60} bisadduct ($IC_{60}BA$; Figure 5.2) as acceptors).[13]

Although fullerene derivatives have some advantages, their insufficiencies cannot be ignored and need to be overcome: (1) a narrow and weak absorption in the visible region, which limits their contribution to the photocurrent; (2) limited bandgap variability; (3) difficulty in tuning the absorption and bandgap by chemical modification; (4) ease of migration and a high tendency to aggregate in thin films, which decreases the morphological stability and long-term device stability; and (5) their high cost, especially for C_{70} derivatives.[14] Therefore there are still incentives to develop new electron acceptors that will not only retain the favorable properties of fullerenes, but also overcome their shortcomings.

The history of non-fullerene acceptors can be traced back to the first heterojunction solar cell fabricated by Tang in 1986, in which a perylene-based small molecule was used as the acceptor. However, due to the great progress achieved by fullerene acceptors and the very low efficiency yielded by non-fullerene acceptors, non-fullerene acceptors did not attract much attention in early research on OSCs, resulting in their development lagging behind that of fullerene acceptors and donor materials.

Boron subphthalocyanine chloride (SubPc), boron subnaphthalocyanine chloride (SubNc) and their halogenated compounds are successful acceptors for vacuum-deposited OSCs. Subphthalocyanine was discovered by Meller and Ossko in 1972,[15] and halogenation has proved to be an effective method of transforming p-type materials into n-type materials.[16] In 2014, Cnops *et al.* fabricated a three-layer device with α-sexithiophene (α-6T; Figure 5.1) as a donor and SubPc and SubNc as acceptors. The device achieved a PCE of 8.40%, which is a record efficiency for vacuum-deposited single-junction OSC devices.[17]

For solution-processed OSCs, inspired by the rapid development of high-performance n-type semiconductors for organic field-effect transistors (OFETs), some researchers began to investigate the applications of these materials in OSCs. Before 2011, BHJ OSCs based on non-fullerene acceptors generally exhibited PCEs $<1.5\%$; from 2011 to 2014, many different systems—such as PDI, naphthalene diimide (NDI) and indacenodithiophene (IDT)—were investigated and the PCEs increased to 6%. More exciting results were published in 2015 and PCEs have now exceeded 8%.[18,19]

This significantly enhanced performance indicates a bright future for non-fullerene acceptors. It can be inferred that the PCEs of non-fullerene-based OSCs will soon exceed 10% and be comparable with those of fullerenes, which will provide us with an alternative to solve existing challenges and realize the large-scale manufacture and practical utility of OSCs.

5.2 Fullerenes and Derivatives

Fullerenes are composed of carbon allotropes and have various structures, such as C_{28}, C_{32}, C_{60}, C_{70}, C_{76}, C_{78}, C_{84}, C_{240} and C_{540}. Fullerenes and their derivatives have been widely used as acceptors in OSCs due to their high electron affinity, high electron mobility and isotropic charge transport. The parent fullerenes are mainly used in vacuum-deposited OSCs as a result of their insolubility in common organic solvents, whereas their soluble derivatives are mainly used in solution-processed OSCs. Figure 5.1 illustrates all the chemical structures of the donor materials discussed in this chapter. The chemical structures of the fullerene acceptors are shown in Figure 5.2 and the performance data of these fullerene acceptors in vacuum-deposited and solution-processed OSCs are listed in Tables 5.1 and 5.2, respectively.

Table 5.1 Performance of fullerene acceptors in vacuum-deposited OSCs under an illumination of 100 mW cm^{-2}.

Donor–acceptor	V_{OC} (V)	J_{SC} (mA cm^{-2})	Fill factor	PCE (%)	Ref.
ZnPc–C_{60}	0.56	7.52	0.543	2.27	22
ZnPc–C_{70}	0.56	9.88	0.522	2.87	22
HB194–C_{60}	0.77	12.1	0.45	4.5	23
HB194–C_{70}	0.50	6.8	0.33	1.4	23
DBP–C_{70}[a]	0.91	13.8	0.55	6.4	24
DBP–C_{70}[b]	0.93	13.2	0.66	8.1	24
DCV5T-Me–C_{60}	0.96	13.20	0.658	8.3	10

[a]Without buffer layer.
[b]With buffer layer.

Table 5.2 Performance of fullerene acceptors in solution-processed OSCs under an illumination of 100 mW cm^{-2}.

Donor–acceptor	V_{OC} (V)	J_{SC} (mA cm^{-2})	Fill factor	PCE (%)	Ref.
P3HT–$PC_{61}BM$	0.64	10.6	0.55	3.7	26
P3HT–$PC_{71}BM$	0.61	12.2	0.55	4.5	26
P3HT–$PC_{85}BM$	0.34	2.11	0.44	0.25	28
P3HT–$PC_{61}BM$	0.58	–	–	3.8	29
P3HT–bis$PC_{61}BM$	0.724	9.14	0.68	4.5	29
P3HT–$PC_{61}BM$	0.61	8.94	0.60	2.4	31
P3HT–bis$PC_{61}BM$	0.73	7.30	0.63	2.4	31
P3HT–tri$PC_{61}BM$	0.81	0.99	0.37	0.21	31
P3HT–bis$PC_{71}BM$	0.75	7.03	0.62	2.3	31
P3HT–$PC_{61}BM$	0.63	10.4	0.64	4.2	32
P3HT–TPA-$PC_{61}BM$	0.65	9.9	0.62	4.0	32
P3HT–MF-$PC_{61}BM$	0.65	9.8	0.59	3.8	32
P3HT–$PC_{61}BM$	0.58	10.8	0.62	3.88	33
P3HT–$IC_{60}MA$	0.63	9.66	0.64	3.89	33
P3HT–$IC_{60}BA$	0.84	9.67	0.67	5.44	33
	0.83	15.16	0.62	7.86	35
P3HT–$PC_{71}BM$	0.58	10.04	0.68	3.96	34
P3HT–$IC_{71}BA$	0.84	9.73	0.69	5.64	34
	0.87	11.35	0.75	7.40	36
P3HT–DMPCBA	0.87	9.05	0.655	5.2	37
BP–SIMEF	0.75	10.5	0.65	5.2	38
P3HT–Lu_3N@C_{80}-PCBH	0.81	8.64	0.61	4.2	41

5.2.1 C_{60} and C_{70}

C_{60} was discovered by Kroto *et al.* in 1985.[20] The symmetrical spherical geometry of C_{60} endows it with an excellent ability to transport electrons in all directions, but leads to weak absorption in the visible region due to symmetry-forbidden transitions. Compared with C_{60}, C_{70} has a lower symmetry and thus exhibits stronger visible absorption owing to a relaxation of symmetry–forbidden transitions. The mobilities of C_{60} and C_{70} measured by OFETs are 10^{-2}–10^{-1} and 10^{-3}–10^{-2} cm^2 V^{-1} s^{-1}, respectively.[21]

Leo and coworkers reported that the PCEs of zinc phthalocyanine (ZnPc):C_{70}-based devices (2.87%) was 25% higher than that of ZnPc:C_{60}-based devices (2.27%).[22] The enhancement in the PCE may be attributed to a higher photocurrent (from 7.52 to 9.88 mA cm^{-2}) resulting from the stronger long-wavelength absorption of C_{70} than C_{60}. Würthner and coworkers fabricated vacuum-deposited OSCs with a merocyanine dye (HB194) as the donor and C_{60} or C_{70} as the acceptor. The C_{60}-based devices exhibited PCEs of 4.5%, whereas the C_{70}-based devices exhibited lower PCEs of 1.4% due to the lower mobility of C_{70}.[23]

The best results of single-junction vacuum-deposited OSCs based on fullerene acceptors have been reported by Forrest and coworkers and Leo and coworkers. Forrest and coworkers fabricated a planar-mixed heterojunction OSC with tetraphenyldibenzoperiflanthene (DBP) as a donor and C_{70} as an acceptor, which exhibited a PCE of 6.4%. When stacking a buffer on the active layer, the PCE significantly increased to 8.1%.[24] Leo and coworkers fabricated a BHJ OSC with dicyanovinyl-dimethyl-pentathiophene (DCV5T-Me) as a donor and C_{60} as an acceptor; the highest PCE of 8.3% was achieved by precisely optimizing the deposition rate and substrate temperature, resulting in the best morphology and optical tuning result.[10]

5.2.2 Aryl-fullerene-butyric Acid Methyl Ester

5.2.2.1 $PC_{61}BM$, $PC_{71}BM$ and $PC_{85}BM$

$PC_{61}BM$ was firstly synthesized and characterized by Hummelen *et al.* in 1995.[7] Heeger *et al.* later fabricated solution-processed BHJ OSCs with an MEH-PPV donor and $PC_{61}BM$ acceptor.[8] A deficiency of $PC_{61}BM$ as a photovoltaic material is its weak absorption in the visible region due to symmetry–forbidden transitions. To improve the visible absorption of $PC_{61}BM$, Hummelen *et al.* synthesized the corresponding C_{70} derivative ($PC_{71}BM$) in 2003.[25] $PC_{71}BM$ exhibits stronger absorption from 400 to 700 nm and the PCEs of $PC_{71}BM$-based OSCs are generally >10% higher than those of $PC_{61}BM$-based devices as a result of the enhancement of the photocurrent contributed by $PC_{71}BM$.[26] The energy levels of $PC_{61}BM$ and $PC_{71}BM$ are similar: the HOMO energy levels are −5.93 and −5.87 eV, respectively, and the LUMO energy levels are both −3.91 eV.[27] So far, most solution-processed OSCs with PCEs >10% use $PC_{61}BM$ or $PC_{71}BM$ as acceptors.

In addition to $PC_{61}BM$ and $PC_{71}BM$, other fullerene analogs have also been investigated. Hummelen *et al.* synthesized and investigated the photovoltaic properties of phenyl-C_{85}-butyric acid methyl ester ($PC_{85}BM$).[28] $PC_{85}BM$ exhibited absorption in the near-infrared (>800 nm) and a LUMO energy level 0.35 eV lower than $PC_{61}BM$. The poly[2-methoxy-5-(3,7-dimethyloctyloxy)-1,4-phenylenevinylene] (MDMO-PPV):$PC_{85}BM$-based OSCs exhibited a PCE of 0.25%. The undesirable performance was mainly attributed to a low V_{OC}

and poor morphology caused by the decreased solubility of $PC_{85}BM$ as a result of the larger size of the C_{84} cage.

5.2.2.2 *$PC_{61}BM$ and $PC_{71}BM$ Multi-adduct*

In the synthesis of $PC_{61}BM$ (or $PC_{71}BM$), the product is a mixture of $PC_{61}BM$ (or $PC_{71}BM$) and its multi-adduct, such as $PC_{61}BM$-bis-adduct (bis$PC_{61}BM$), $PC_{61}BM$-tris-adduct (tri$PC_{61}BM$), $PC_{61}BM$-tetra-adduct and higher adducts. These multi-adducts can also be purified and applied as acceptors in OSCs.

In 2008, Blom and coworkers introduced bis$PC_{61}BM$ as a new acceptor and investigated its photovoltaic properties.[29] The LUMO level energy of bis$PC_{61}BM$ was 0.1 eV higher than that of $PC_{61}BM$, which increased the V_{OC}.[30] The OSCs based on poly(3-hexylthiophene) (P3HT) and bis$PC_{61}BM$ exhibited PCEs of 4.5% (3.8% for $PC_{61}BM$) with a V_{OC} of 0.724 V (0.58 V for $PC_{61}BM$). Blom and coworkers further investigated the photovoltaic properties of tri$PC_{61}BM$ and $PC_{71}BM$-bisadduct (bis$PC_{71}BM$).[31] P3HT : tri$PC_{61}BM$-based OSCs exhibited a V_{OC} of 0.81 V, a J_{SC} of 0.99 mA cm^{-2}, a fill factor of 0.37 and a PCE of 0.21%; P3HT : bis$PC_{71}BM$-based OSCs exhibited a V_{OC} of 0.75 V, a J_{SC} of 7.03 mA cm^{-2}, a fill factor of 0.62 and a PCE of 2.3%. tri$PC_{61}BM$-based devices exhibited a higher V_{OC} as a result of their higher LUMO energy levels. The presence of a high multitude of isomers in the trisadduct caused disorder in the fullerene phase and low mobility, leading to a low J_{SC} and fill factor.

5.2.2.3 *Aryl-substituted PCBM*

Replacing the phenyl unit with other aryl units is an effective method of tuning the energy levels as well as the aggregation states of the acceptors. Jen and coworkers replaced the phenyl group with triphenylamine (TPA) and 9,9-dimethylfluorene (MF) to give two amorphous acceptors, TPA-$PC_{61}BM$ and MF-$PC_{61}BM$.[32] The TPA-$PC_{61}BM$- and MF-$PC_{61}BM$-based OSCs exhibited higher V_{OC} because the electron-donating TPA and MF up-shifted the LUMO energy levels of the acceptors. The PCEs of TPA-$PC_{61}BM$- and MF-$PC_{61}BM$-based OSCs were 4.0 and 3.8%, which were comparable with the controlled $PC_{61}BM$-based devices (4.2%). The PCEs of TPA-$PC_{61}BM$- and MF-$PC_{61}BM$-based OSCs exhibited no significant degradation after 10 h at 150 °C because the bulky TPA and MF moieties restricted the crystallization of the acceptors, which suppressed the crystallization-driven phase separation between the donor and acceptor during thermal annealing and improved the stability of the OSCs.

5.2.3 Indene-fullerene Adduct

A relatively low V_{OC} limits the efficiency of P3HT : fullerene-based OSCs. To increase the V_{OC}, Li and coworkers developed a series of indene-fullerene adducts, including indene-C_{60}-monoadduct ($IC_{60}MA$), indene-C_{60}-bisadduct

($IC_{60}BA$),[33] indene-C_{70}-monoadduct ($IC_{70}MA$) and indene-C_{70}-bisadduct ($IC_{70}BA$).[34] $IC_{60}BA$ and $IC_{70}BA$ are regarded as the most effective acceptors for P3HT-based OSCs.

$IC_{60}MA$ and $IC_{70}MA$ are the byproducts in the synthesis of $IC_{60}BA$ and $IC_{70}BA$. $IC_{60}MA$ can be applied as an acceptor, but $IC_{70}MA$ cannot as a result of its poor solubility. The LUMO energy levels of $IC_{60}MA$ (−3.86 eV), $IC_{60}BA$ (−3.74 eV), $IC_{70}MA$ (−3.85 eV) and $IC_{70}BA$ (−3.72 eV) are higher than those of $PC_{61}BM$ and $PC_{71}BM$ (−3.91 eV). The P3HT : $IC_{60}BA$-based devices exhibited a ~45% higher V_{OC} (0.84 V) than that of P3HT : $PC_{61}BM$-based devices (0.58 V), whereas the J_{SC} and fill factor did not differ much, leading to enhanced PCEs of 3.88–5.44%.[33] Tang and coworkers further improved the performance of P3HT : $IC_{60}BA$-based devices by forming a biomimetic moth eye nanostructure in a quasi-periodic gradient shape active layer through soft nanoimprint lithography; the optimized devices exhibited PCEs of 7.86%.[35] Similar to P3HT : $IC_{60}BA$-, P3HT : $IC_{70}BA$-based devices showed a better performance (5.64%) than P3HT : $PC_{71}BM$-based devices (3.96%).[34] By additive and pre-thermal annealing processing, the PCEs of P3HT : $IC_{70}BA$-based devices were improved to 7.40%.[36]

5.2.4 Other Fullerene Derivatives

Hsu and coworkers synthesized a di(4-methylphenyl)methano-C_{60} bisadduct (DMPCBA).[37] The LUMO energy level of DMPCBA (−3.85 eV) was 0.1 eV higher than that of $PC_{61}BM$. The BHJ OSCs based on P3HT : DMPCBA exhibited a PCE of 5.2%. The phenyl planes were parallel to the fullerene surface, which protected the C_{60} core from severe intermolecular aggregation, rendering it morphologically amorphous. The DMPCBA-based devices exhibited good thermal stability, with 80% of the original PCE retained after heating at 160 °C for 20 h.

A Si-containing fullerene derivative, bis(dimethylphenylsilylmethyl)[60] fullerene (SIMEF), was synthesized by Nakamura and coworkers in 2009.[38] SIMEF exhibited good thermal stability and its LUMO was up-shifted by 0.1 eV relative to that of $PC_{61}BM$. They fabricated a three-layered device with a p–i–n geometry through a solution process, where p was a layer of tetrabenzoporphyin (BP), i was an interdigitated BHJ layer of BP and SIMEF and n was a layer of SIMEF. The BP : SIMEF-based devices exhibited a PCE of 5.2% (2.0% for the controlled BP : $PC_{61}BM$-based devices fabricated by the same process).

Endohedral fullerenes exhibit much higher LUMO energy levels than their corresponding empty-cage fullerenes, which has been proved both theoretically[39] and experimentally.[40] Drees and coworkers synthesized a series of soluble PCBM-like $Lu_3N@C80$ derivatives and the LUMO energy level of $Lu_3N@C_{80}$–PCBH was 0.28 eV higher thah that of $PC_{61}BM$.[41] The devices based on P3HT : $Lu_3N@C_{80}$–PCBH showed a PCE of 4.2% (3.4% for controlled P3HT : $PC_{61}BM$ devices).

5.3 New Acceptors

Although fullerene acceptors have achieved great success, there are still incentives to develop new acceptors that will not only retain the favorable properties of fullerenes, but also overcome their shortcomings, such as weak absorption in the visible region, a narrow spectral breadth, a limited band gap variability, poor morphological stability and high costs.

5.3.1 Subphthalocyanine Derivatives

SubPc was serendipitously synthesized by Meller and Ossko in 1972 as they were trying to obtain boron-phthalocyanine.[15] SubPcs exhibit excellent thermal and chemical stability, strong absorption comparable with metal-phthalocyanine complexes, and flexibility in their optical and electronic properties through chemical modification. SubPcs are thus widely used as both donors and acceptors in vacuum-deposited OSCs.[42] The chemical structures and performance data of SubPc-based acceptors are shown in Figure 5.3 and Table 5.3, respectively.

In 2011, Jones and coworkers reported a bilayer heterojunction OSC based on the **a1** donor (Figure 5.3) and a hexachloro-SubPc acceptor (**a2**).[43] The LUMO energy level (−3.7 eV) and HOMO energy level (−5.8 eV) of **a2** were 0.3 eV lower than those of **a1**. The bilayer device exhibited a PCE of 2.68%.

Verreet *et al.* reported a fluorinated SubPc dimer (**a3**, an 1 : 1 *syn*/*anti* mixture) with strong and complementary absorption to that of SubPc.[44] The LUMO and HOMO energy levels of **a3** were −3.95 eV and −5.7 eV, respectively.

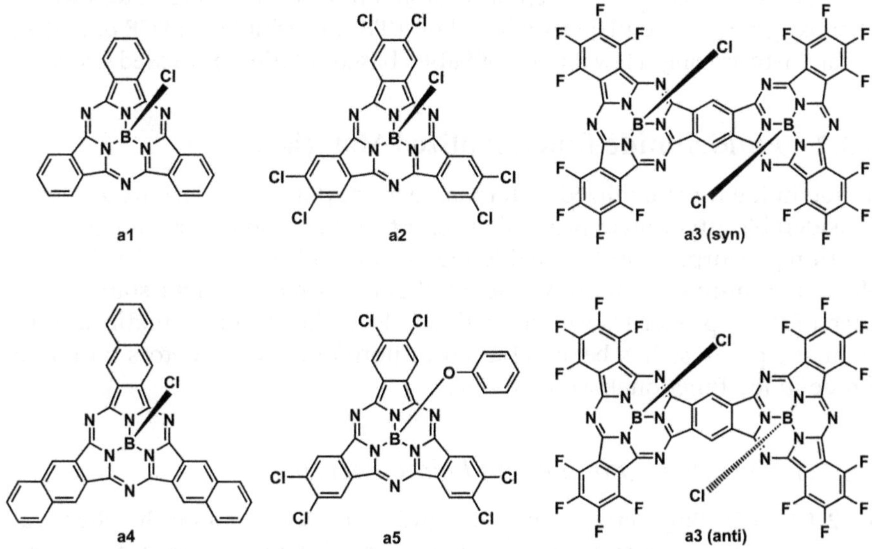

Figure 5.3 Chemical structures of subphthalocyanine-based acceptors.

Table 5.3 Performance of subphthalocyanine-based acceptors in OSCs under an illumination of 100 mW cm^{-2}.

Donor–acceptor	V_{OC} (V)	J_{SC} (mA cm^{-2})	Fill factor	PCE (%)	Ref.
a1:**a2**	1.31	3.53	0.58	2.68	43
a1:**a3**	0.95	7.8	0.54	4.0	44
α-6T–**a1**	1.09	7.46	0.579	4.69	17
α-6T–**a4**	0.94	12.04	0.539	6.02	17
α-6T–**a4**–**a1**	0.96	14.55	0.610	8.40	17
PTB7:**a5**	0.935	7.8	0.48	3.51	45

The devices with a C_{60} electron transport layer stacking on an **a1**/**a3** active layer exhibited the best PCE of 4.0%.

In 2014, Cnops *et al.* reported that SubPc and SubNc could act as acceptors with α-6T as a donor.[17] Relative to the controlled devices based on α-6T/C_{60} (V_{OC} 0.42 V, J_{SC} 4.56 mA cm^{-2}, fill factor 0.546 and PCE 1.03%), the bilayer OSCs based on α-6T/**a1** and α-6T/**a4** exhibited significantly improved V_{OC} and J_{SC}, resulting in much higher PCEs of 4.69 and 6.02%, respectively. Because of the small absorption overlap between **a1** and **a4**, a simple three-layer architecture of α-6T/**a1**/**a4**, where α-6T was a donor and **a1** and **a4** were two acceptors, could cover the entire visible region. The three-layer devices exhibited a PCE of 8.40%, which is a record efficiency for vacuum-deposited single-junction OSC devices.

In most cases, SubPcs are used in vacuum-deposited OSCs, although some SubPc derivatives have excellent solubility, which can be used in solution-processed OSCs. Samuel and coworkers synthesized a halogenated SubPc-based acceptor (**a5**) by replacing the chlorine atom on the central boron atom with a phenoxy group.[45] The BHJ OSCs based on PTB7:**a5** exhibited a PCE of 3.51%. So far, there are only a few reports of SubPc-based solution-processed OSCs.

5.3.2 Amide/imide-functionalized Materials

Amide/imide-functionalized materials, such as rylene diimide, have attracted considerable attention as promising acceptors due to some advantages, such as strong absorption in the visible region, variability of energy levels, high electron affinity and mobility, and excellent thermal and light stability.[46,47] Each of these properties can be readily tailored by chemical modifications. Dramatic progress has been achieved in non-fullerene acceptors based on amide/imide-functionalized materials.

5.3.2.1 Perylene Diimide and Related Acceptors

5.3.2.1.1 PDI Polymers. Figure 5.4 and Table 5.4 illustrate the chemical structures and performance data, respectively, of PDI polymer acceptors. Studies on rylene diimide-based polymers began with the research of Zhan *et al.* in 2007, using the PDI and dithienothiophene (DTT) alternating copolymer

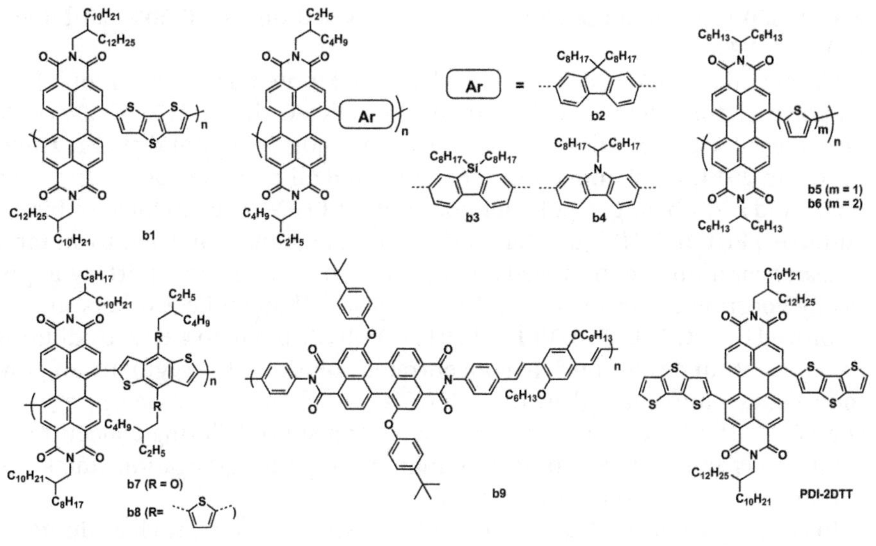

Figure 5.4 Chemical structures of PDI polymer acceptors.

Table 5.4 Performance of PDI polymer acceptors in OSCs under an illumination of 100 mW cm^{-2}, except for **b9**.

Donor	Acceptor	V_{OC} (V)	J_{SC} (mA cm^{-2})	Fill factor	PCE (%)	Ref.
b1	2TV-PT	0.63	4.2	0.39	1.03	48
	PBDTTT-C-T	0.752	8.55	0.515	3.45	50
	PBDTTT-C-T[a]	0.765	9.77	0.609	4.63	51
b2	3TV6-PT	0.76	1.77	0.43	0.58	52
b3	3TV6-PT	0.72	2.80	0.40	0.81	52
b4	3TV6-PT	0.70	6.35	0.50	2.23	52
b5	PiI-2T-PS5	1.04	8.77	0.46	4.21	53
r-b6	P3HT	0.52	7.65	0.55	2.17	54
i-b6	P3HT	0.49	6.25	0.51	1.55	54
b7	PTB7-Th	0.72	10.14	0.376	2.75	55
b8	PTB7-Th	0.80	11.51	0.511	4.71	55
b9	PPHT	0.60	2.98[b]	0.39	2.32	56

[a]Diluting concentrated solution.
[b]Under illumination of 30 mW cm^{-2}.

PPDIDTT (**b1**) as the electron acceptor.[48] The LUMO and HOMO energy levels of **b1** were −3.9 and −5.9 eV, respectively. The thin film of **b1** showed strong absorption throughout the visible and extending into the near-infrared region, with a low optical band gap of 1.7 eV arising from a strong intramolecular charge transfer of the donor–acceptor arrangement of the main chain. The electron mobility of **b1** was as high as 0.01 $cm^2\ V^{-1}\ s^{-1}$ measured by OFETs. The blended film of a bis(thienylenevinylene)-substituted polythiophene 2TV-PT donor and **b1** acceptor exhibited very broad absorption from

250 to 850 nm; the all-polymer solar cells based on 2TV-PT:**b1** exhibited a PCE of 1.03%.

Shuai and coworkers conducted a dynamic Monte Carlo simulation of the all-polymer solar cells based on **b1** and predicted that a PCE of 5% could be achieved with optimized phase separation morphology and charge transport.[49] In 2014, Zhan and coworkers fabricated all-polymer solar cells based on **b1** and a low bandgap polymer donor PBDTTT-C-T. Treatment with binary additives, PDI-2DTT (Figure 5.4) and 1,8-diiodooctane (DIO), led to suitable phase separation and improved and balanced the charge transport, synergistically boosting the PCEs from 1.18 to 3.45%.[50] Zhan and coworkers further improved the PCE of a PBDTTT-C-T : **b1** : PDI-2DTT blend to 4.63% by diluting the concentrated solution.[51] In concentrated solutions, the segments of polymers contact and entangle with each other, which is a favorable aggregation state for both phase separation and charge transport. Diluting concentrated solutions is an effective approach to maintaining the aggregation states and dynamics during solvent evaporation.

In addition to the DTT unit, some other fused-ring units, such as fluorene, dibenzosilole (DBS) and carbazole, can be used as building blocks to synthesize PDI copolymer acceptors. Hashimoto and coworkers synthesized a series of copolymers based on PDI and fluorene (**b2**), DBS (**b3**) and carbazole (**b4**).[52] When blended with a tris(thienylenevinylene)-substituted polythiophene 3TV6-PT, the all-polymer solar cells based on **b2**, **b3** and **b4** exhibited PCEs of 0.58, 0.81 and 2.23%, respectively.

Copolymers of PDI and oligothiophene were also used as acceptors in OSCs. Bao and coworkers applied a copolymer of PDI and thiophene (**b5**) as the acceptor and a polystyrene-side-chain-containing copolymer of isoindigo and bithiophene (PiI-2T-PS5) as the donor to fabricate all-polymer solar cells, which exhibited a PCE of 4.21%.[53] Generally, the bromination of PDI affords a mixture of 1,6- and 1,7-regioisomers of PDI dibromides. As the two isomers are difficult to isolate, most PDI polymers are synthesized using the mixture as the monomer. Zhao and coworkers isolated and purified the 1,7-dibromo PDI isomer *via* repeated column chromatography and then synthesized a regioregular 1,7-PDI-bithiophene copolymer **r-*b6***.[54] Compared with its irregular counterpart **i-b6**, **r-b6** exhibited higher electron mobility (5×10^{-4} *versus* 3×10^{-4} $cm^2 V^{-1} s^{-1}$) and a higher PCE (2.17 *versus* 1.55%).

Li and coworkers synthesized two PDI and benzodithiophene (BDT) alternating copolymers **b7** and **b8**.[55] **b7** was a one-dimensional conjugated polymer, whereas **b8** was two-dimensional. The two-dimensional conjugated polymer exhibited a 10 nm red-shifted absorption, similar LUMO energy level and higher electron mobility relative to the one-dimensional conjugated polymer. When blending with PTB7-Th, the PCE of the **b8**-based device was 4.71%, higher than that of **b7** (2.75%).

Most of the PDI polymers are polymerized *via* the bay positions (1,6,7,12 positions) of the perylene core; however, PDI can also be polymerized *via* the imide position. Mikroyannidis *et al.* synthesized an alternating PDI–phenylenevinylene copolymer (**b9**), which, when blended with a poly(3-phenylhydrazone thiophene) (PPHT) donor, gave devices that exhibited a PCE of 2.32% under white light illumination calibrated to an AM 1.5 intensity of 30 mW cm^{-2}.[56]

5.3.2.1.2 PDI and Related Small Molecules. PDI small molecules are the earliest and one of the most common non-fullerene acceptors used in OSCs. The chemical structures and performance data of PDI and related small molecule acceptors are shown in Figure 5.5 and Table 5.5, respectively. In 1986, Tang fabricated a bilayer heterojunction OSC with a copper phthalocyanine donor and *trans*-3,4,9,10-perylene tetracarboxylic bisbenzimidazole (*trans*-PTCBI; Figure 5.5) acceptor, which yielded a PCE of 1%.[6] Introducing solubilizing groups on the imide nitrogen atoms can make the PDI small molecules suitable for solution processing. 3-Pentyl-substituted PDI (**c1**) was a popular non-fullerene acceptor in the early research of solution-processed OSCs.[57] Devices based on P3HT/**c1** exhibited a PCE of 0.19%.[58] The poor performance was mainly caused by the strong tendency of **c1** to form large crystalline domains (micrometer scale) in the blended films, which acted as electron traps and led to a decreased photocurrent.

Restricting the crystallinity without adversely affecting the charge transport properties is proposed as an important design principle for PDI acceptors. Laquai and coworkers introduced four alkyl chains on the non-bay

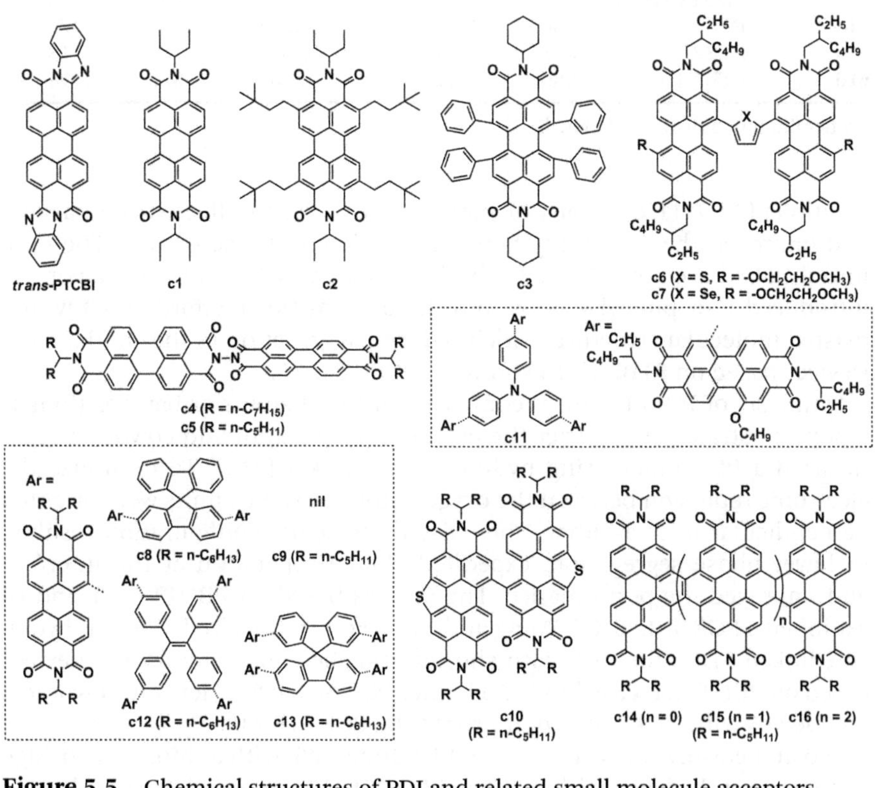

Figure 5.5 Chemical structures of PDI and related small molecule acceptors.

Table 5.5 Performance of PDI and related small molecule acceptors in OSCs under an illumination of 100 mW cm^{-2}, except for **c1**.

Acceptor	Donor	V_{OC} (V)	J_{SC} (mA cm^{-2})	Fill factor	PCE (%)	Ref.
c1	P3HT	0.40	0.12[a]	0.39	0.19	58
c2	P3HT	0.75	1.74	0.38	0.5	59
c3	PTB7-Th	0.87	10.1	0.464	4.1	60
c4	PBDTTT-C-T	0.77	9.0	0.46	3.20	62
c5	PBDT-TS1	0.82	12.5	0.53	5.40	63
	PTB7-Th	0.79	13.12	0.60	6.41	64
c6	PBDTTT-C-T	0.85	8.86	0.541	4.03	65
	PBDTTT-C-T	0.84	12.83	0.564	6.08	66
c7	PBDTTT-C-T	0.79	10.60	0.479	4.01	67
c8	P3HT	0.61	5.92	0.65	2.35	68
	PffBT4T-2DT	0.99	11.0	0.58	6.3	69
c9	PBDTTT-C-T	0.73	10.58	0.468	3.63	70
	PBDTBDD-T	0.87	8.26	0.611	4.39	71
	PTB7	0.78	10.51	0.58	5.00	64
	PTB7-Th	0.79	12.86	0.54	5.56	64
	PTB7-Th	0.80	11.98	0.59	5.90	72
c10	PDBT-T1	0.90	11.98	0.661	7.16	73
c11	PBDTTT-C-T	0.88	11.92	0.336	3.32	74
c12	PTB7-Th	0.91	11.7	0.52	5.53	75
c13	PffBT4T-2DT	0.93	11.03	0.511	5.27	76
	PffBTV4T-2DT	0.90	12.02	0.542	5.98	76
c14	PTB7-Th	0.803	13.3	0.566	6.05	77
c15	PTB7-Th	0.81	14.5	0.67	7.9	18
c16	PTB7-Th	0.80	15.2	0.68	8.3	18

[a]Under illumination of 10 mW cm^{-2}.

positions (2,5,8,11 positions of perylene) (**c2**), which affected the packing and improved the miscibility between the donor and acceptor.[59] The optimized OSCs based on P3HT:**c2** exhibited a PCE of 0.5%. Sun and coworkers introduced four phenyl groups into the bay positions, affording **c3** with a twisted molecular structure, which reduced aggregation in the solid state.[60] Devices based on PTB7-Th:**c3** showed a PCE of 4.1%.

Synthesis of PDI oligomers connected at the N atoms or bay positions is a popular strategy to restrict the crystallinity. Narayan and coworkers synthesized a PDI dimer using hydrazine as a linker (**c4**).[61] To minimize the electronic repulsion between the oxygen atoms, two PDI units were oriented perpendicular to each other. The twisted molecular configuration resulted in lower phase segregation, expedited charge separation at the interface and improved the performance. The devices based on PBDTTT-C-T and **c4** exhibited a PCE of 3.20%.[62] Wang and coworkers synthesized a similar hydrazine-linked PDI dimer (**c5**) with shorter branched alkyl groups; the devices based on PBDT-TS1 : **c5** exhibited a higher PCE of 5.40%.[63] Jen and coworkers blended **c5** with PTB7-Th and improved the PCE to 6.41%.[64]

Yao and coworkers synthesized a PDI dimer (**c6**) with a thiophene bridge connected at the bay position.[65] The PBDTTT-C-T : **c6**-based devices showed a PCE of 0.77% without any post-treatment. After optimizing the film morphology by adding 5% DIO, the PCE was improved to 4.03%. To further improve the device performance, they conducted solvent vapor annealing for the blended film. Solvent vapor annealing reconstructed the film morphology, which was helpful for the phase separation and aggregation of the donor and acceptor, resulting in higher hole and electron mobilities. The optimized devices exhibited a PCE of 6.08%.[66] After replacing the thiophene bridge with selenophene (**c7**), the devices based on PBDTTT-C-T:**c7** exhibited a PCE of 4.01%.[67]

Zhao and coworkers developed a series of PDI dimers with different arylene linkers at the bay position.[68] These PDI dimers showed similar absorption spectra in solution and thin film, indicating minimal intermolecular aggregation. Among these dimers, **c8** with a spirobifluorene linker exhibited non-planar molecular geometries and the OSCs based on P3HT:**c8** showed a PCE of 2.35%. Yan and coworkers fabricated BHJ OSCs based on **c8** and PffBT4T-2DT, which exhibited a higher PCE of 6.3%.[69]

Wang and coworkers synthesized a bay-linked PDI dimer **c9** and the PBDTTT-C-T:**c9**-based devices exhibited a PCE of 3.63%.[70] Blending **c9** with PBDTBDD-T, PTB7 and PTB7-Th improved the PCE to 4.39,[71] 5.00 and 5.56%,[64] respectively. By using a $PC_{61}BM$ self-assembled monolayer on zinc oxide to enhance electron extraction, the PCE of PTB7-Th:**c9**-based devices could be improved to 5.90%.[72]

Wang and coworkers modified **c9** by inserting two thiophene units at the bay positions, affording **c10**. The molecule **c10** has a more twisted molecular configuration, with two PDI planes almost perpendicular to each other (a dihedral angle of 80°) and a higher LUMO energy than **c9**. **c10** exhibited strong absorption from 400 to 550 nm; when blending with a polymer donor PDBT-T1 with complementary absorption from 500 to 700 nm, the devices exhibited a PCE of 7.16%.[73]

Apart from linear PDI dimers, three-dimensional/quasi-three-dimensional PDI oligomers—such as those with a star-/propeller-like, tetrahedral or helical configuration—have attracted much attention and exhibited a promising performance. Zhan and coworkers synthesized a novel star-shaped PDI acceptor (**c11**) with a triphenylamine core.[74] Compound **c11** had a three-dimensional molecular structure and exhibited weak intermolecular interaction and molecular aggregation. OSCs based on PBDTTT-C-T:**c11** exhibited a PCE of 3.22%.

Yan and coworkers synthesized a four-wing propeller-shaped acceptor with a tetraphenylethylene (TPE) core and four PDIs (**c12**).[75] As a result of the twisted nature of TPE, **c12** exhibited weak molecular aggregation. The three-dimensional structure of **c12** facilitated three-dimensional charge transport as well as relatively high mobility (1×10^{-3} cm^2 V^{-1} s^{-1}). The OSCs based on PTB7-Th:**c12** exhibited a PCE of 5.53%.

Cho and coworkers reported a three-dimensional small molecule acceptor based on a spirobifluorene core and four PDIs (**c13**).[76] The molecule **c13** showed good solubility and a three-dimensional interlocking geometry.

Two polymers, PffBT4T-2DT and PffBTV4T-2DT, were applied as donors as a result of their matched energy levels and complementary absorption with **c13**. OSCs based on PffBT4T-2DT:**c13** and PffBTV4T-2DT:**c13** exhibited PCEs of 5.27 and 5.98%, respectively.

Nuckolls and coworkers synthesized a series of PDI helices in the form of a dimer (**c14**),[77] a trimer (**c15**) and a tetramer (**c16**).[18] PTB7-Th was chosen as the donor due to its complementary absorption with these helical acceptors. Femtosecond transient absorption spectroscopy revealed both electron and hole transfer processes at the donor–acceptor interfaces, which indicated that charge carriers were generated from the photogenerated excitons in both the donor and acceptor phases and contributed to photocurrent. These helical conjugated molecules could lead to ultrafast charge separation, just like fullerenes. The PCE of a PTB7-Th:**c14**-based device was 6.05%, whereas PTB7-Th:**c15**- and PTB7-Th:**c16**-based devices exhibited a higher J_{SC} and fill factor, resulting in significantly enhanced PCEs of 7.9 and 8.3%, respectively.

5.3.2.2 Naphthalene Diimide Acceptors

Compared with its large-sized analog PDI, NDI is readily brominated to yield a pure 1,5-dibromo isomer, enabling the synthesis of a regioregular polymeric backbone. As a result of the higher regioregularity, NDI copolymers have enhanced electron delocalization and π-stacking, resulting in better electron transport relative to their PDI counterparts. The chemical structures and performance data of NDI acceptors are shown in Figure 5.6 and Table 5.6, respectively.

In 2009, Facchetti and coworkers synthesized an NDI-bithiophene copolymer P(NDI2OD-T2) (**d1**), the so called Polyera ActivInk N2200, which exhibited an electron mobility as high as 0.85 $cm^2\ V^{-1}\ s^{-1}$ under ambient conditions.[78] In 2011, Loi and coworkers were the first to use **d1** as an acceptor in OSCs.[79] The device based on a P3HT:**d1** blended film exhibited high fill factor (~0.70), suggesting highly balanced hole and electron mobilities in the polymer blend. However, the device exhibited a very low J_{SC}, leading to a very low PCE (<0.16%). Neher and coworkers demonstrated that suppressing the aggregation of NDI copolymers at the early stage of film formation could improve the intermixing of the donor and acceptor, leading to an enhanced device performance.[80] By choosing different donors and optimizing the processing methods, the performance of **d1**-based OSCs was significantly improved to ~6%.[81–86]

Jen and coworkers improved the performance of **d1** by introducing two fluorine atoms on the bithiophene unit, affording **d2**.[87] The polymer **d2** exhibited a higher electron mobility ($5.3 \times 10^{-3}\ cm^2\ V^{-1}\ s^{-1}$) than **d1** ($0.7 \times 10^{-3}\ cm^2\ V^{-1}\ s^{-1}$) and the PCE of PTB7-Th:**d2**-based devices (6.29%) was higher than that of PTB7-Th:**d1**-based devices (5.28%). Replacing the 2-octyldodecyl side-chains in **d2** with 2-decyltetradecyl afforded polymer **d3**, which had an increased crystallinity and electron-transporting capability and improved the PCE to 6.71%.

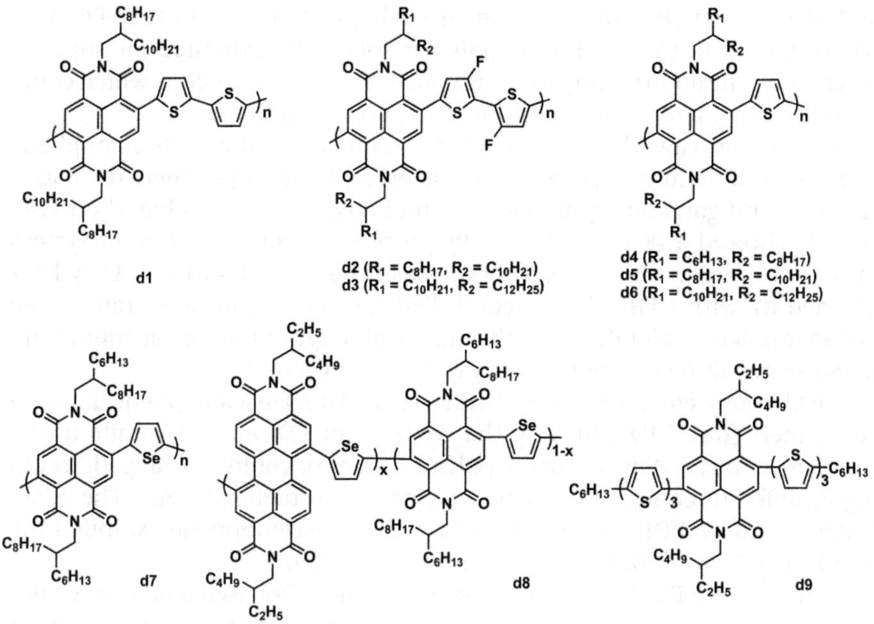

Figure 5.6 Chemical structures of NDI acceptors.

Table 5.6 Performance of NDI acceptors in OSCs under an illumination of 100 mW cm^{-2}.

Acceptor	Donor	V_{OC} (V)	J_{SC} (mA cm^{-2})	Fill factor	PCE (%)	Ref.
d1	P3HT	0.56	3.77	0.65	1.4	80
	PBDTBDD-T	0.87	11.7	0.575	5.8	86
d1	PTB7-Th	0.79	11.92	0.56	5.28	87
d2	PTB7-Th	0.81	12.32	0.63	6.29	87
d3	PTB7-Th	0.81	13.53	0.62	6.71	87
d4	PTB7-Th	0.79	13.46	0.56	5.96	88
	PBDTTTPD	1.06	11.22	0.56	6.64	89
d5	PTB7-Th	0.80	11.97	0.53	5.05	88
d6	PTB7-Th	0.81	7.81	0.52	3.25	88
d7	PSEHTT	0.76	7.78	0.55	3.26	90
	PSEHTT	0.76	10.47	0.60	4.81	91
	PTB7-Th	0.81	18.80	0.51	7.73	92
d8	PBDTTT-C-T	0.79	18.55	0.45	6.29	93
d9	P3HT	0.82	3.43	0.53	1.5	94
	P3HT	0.81	2.95	0.48	1.15	95

Kim and coworkers synthesized a series of NDI and thiophene alternating copolymers with different side alkyl chains on NDI (**d4–d6**).[88] Different alkyl chains significantly affected the film morphology. The devices based on **d4**, **d5** and **d6** with PTB7-Th as a donor exhibited PCEs of 5.96, 5.05 and 3.25%, respectively. The all-polymer solar cells based on another donor PBDTTTPD and **d4** exhibited a PCE of 6.64%, which was higher than PBDTTTPD : $PC_{61}BM$-based devices (6.12%).[89] The all-polymer solar cells exhibited far superior mechanical durability compared with the fullerene-based OSCs, which could be a promising alternative in flexible and portable devices.

Jenekhe and coworkers synthesized a semi-crystalline NDI-selenophene copolymer **d7**. Compared with an NDI-thiophene copolymer, **d7** had a smaller band gap and higher electron mobility. The devices based on PSE-HTT:**d7** showed a PCE of 3.26%.[90] By processing from a co-solvent system to optimize the morphology, the PCE was improved to 4.81%.[91] They later blended **d7** with PTB7-Th and controlled the self-organization rate of the polymer/polymer blended film through a film aging process at room temperature, resulting in a significantly enhanced PCE of 7.73%.[92]

Jenekhe and coworkers also synthesized PDI/NDI-selenophene random copolymers (**d8**).[93] By adjusting the contents of PDI and NDI units in the copolymer, the crystallinity of the polymer acceptor could be changed, resulting in different compatibility and morphology in blended films. The copolymer with 30% PDI-selenophene and 70% NDI-selenophene exhibited the best PCE of 6.29% when blending with PBDTTT-C-T.

Relative to the PDI-based small molecules, the NDI-based small molecules were less successful as electron acceptors in solution-processed OSCs due to their high planarity and strong intermolecular interactions, which led to large crystalline domains in the blended films. Jenekhe and coworkers reported a series of small molecules with NDI as the core and oligothiophene as the arms for BHJ OSCs.[94] Compound **d9** with terthiophene arms exhibited a bicontinuous nanoscale morphology when blended with P3HT and the optimized P3HT : **d9**-based device showed a PCE of 1.5%. These NDI-based small molecules could form highly crystalline nanowires in the thin films by self-assembly; the all-nanowire BHJ OSCs based on P3HT : **d9** showed a PCE of 1.15%.[95]

5.3.2.3 Other Amide/imide-functionalized Acceptors

Figure 5.7 and Table 5.7, respectively, show the chemical structures and performance data of other amide/imide-functionalized acceptors. Diketopyrrolopyrrole (DPP) is a common amide functionalized building block in organic semiconductors. Janssen and coworkers developed a series of DPP-based polymer acceptors.[96] When blending with P3HT, the DPP and fluorene copolymer **e1** gave a PCE of 0.37%. The low PCE resulted from the low electron mobility (5×10^{-6} $cm^2 V^{-1} s^{-1}$) of the acceptor and the inefficient generation of long-lived free charge carriers. Zhan and coworkers synthesized a star-shaped DPP trimer **e2** with triphenylamine as the core, which was the first example of a three-dimensional non-fullerene acceptor for BHJ OSCs.[97] Devices based on P3HT : **e2** showed a PCE of 1.20% with a very high V_{OC} of 1.18 V. Zhan and coworkers also reported a DPP dimer (**e3**) with dibenzosilole as a linker.[98] The devices based on P3HT : **e3** showed a PCE of 2.05%. Wan and coworkers synthesized a series of X-shaped DPP oligomers with spirobifluorene as the core (**e4**) and the devices based on P3HT : **e4** showed a PCE of 3.63%.[99]

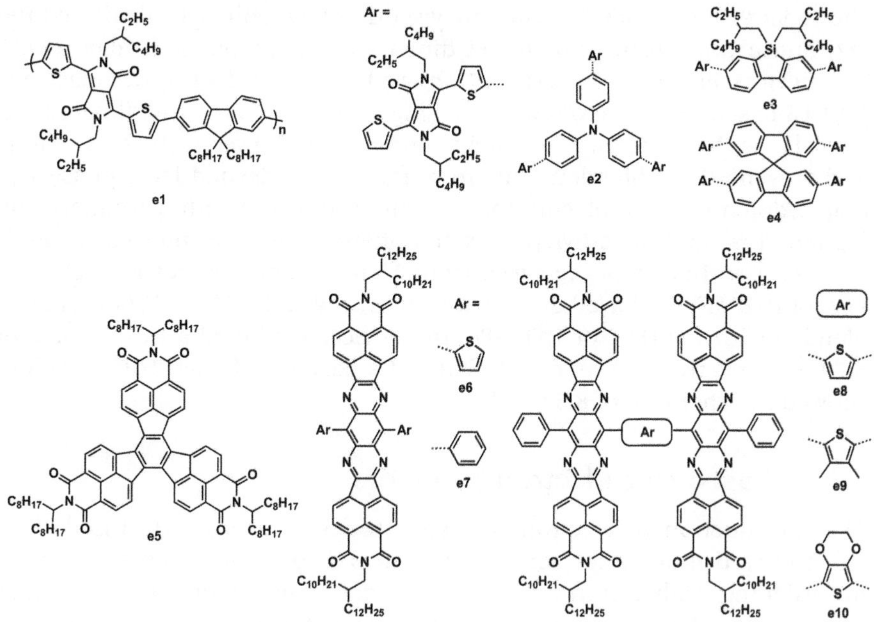

Figure 5.7 Chemical structures of other amide/imide-functionalized acceptors.

Table 5.7 Performance of other amide/imide-functionalized acceptors in OSCs under an illumination of 100 mW cm^{-2}.

Acceptor	Donor	V_{OC} (V)	J_{SC} (mA cm^{-2})	Fill factor	PCE (%)	Ref.
e1	P3HT	0.90	1.63	0.25	0.37	96
e2	P3HT	1.18	2.68	0.379	1.2	97
e3	P3HT	0.97	4.91	0.43	2.05	98
e4	P3HT	1.10	6.96	0.475	3.63	99
e5	P3HT	0.58	4.87	0.57	1.60	100
e6	PSEHTT	0.79	5.14	0.44	1.80	101
e7	PSEHTT	0.94	3.16	0.49	1.44	102
e8	PSEHTT	0.86	10.14	0.58	5.04	102
e9	PSEHTT	0.92	12.56	0.55	6.37	103
e10	PSEHTT	0.93	13.82	0.63	8.10	19
	PTB7-Th	0.95	13.99	0.51	6.70	19
	PSEHTT–PTB7-Th	0.91	15.67	0.60	8.52	19

The amide/imide-functionalized aromatic compounds usually contain one or two electron-deficient amide/imide groups. Multiple amide/imide substituents may increase the electron affinity of the parent material, thereby facilitating electron injection and charge transport. Wudl and coworkers developed a series of acceptors based on decacyclene triimides with three electron-deficient imide groups. Devices based on P3HT:**e5** exhibited a PCE of 1.60%.[100]

Jenekhe and coworkers developed a series of acceptors (**e6–10**) based on tetraazabenzodifluoranthene diimide (BFI). OSCs based on PSEHTT:**e6** showed a PCE of 1.80%.[101] Later, they synthesized a phenyl-substituted BFI (**e7**) and a phenyl-substituted BFI dimer with a thiophene bridge (**e8**).[102] The devices based on PSEHTT:**e7** showed a PCE of 1.44%, whereas PSEHTT:**e8**-based devices showed a much higher PCE of 5.04%, attributed to its non-planar three-dimensional architecture, which lowered the crystallinity and prevented the blended films from forming undesired large phase-separated domains. By replacing the bridging thiophene with 3,4-dimethylthiophene (**e9**) and 3,4-ethylenedioxythiophene (**e10**), the interplanar angle between two BFI units increased from 33 to 62 and 76°, respectively. The PCE of PSEHTT:**e9**-based devices was improved to 6.37%.[103] When **e10** was blended with PSEHTT or PTB7-Th, the devices exhibited a PCE of 8.10 or 6.70%, respectively. Ternary blended OSCs based on PSEHTT/PTB7-Th/**e10** showed a higher PCE of 8.52%.[19]

5.3.3 Fused-ring Electron Acceptors

The introduction of electron-deficient groups into aromatic fused rings has attracted much attention and has proved to be an efficient approach to designing high-performance acceptor materials.[5] Aromatic fused rings provide extended conjugation and effective intermolecular π–π overlaps, enhancing intermolecular charge hopping and charge transport. Tailoring fused rings using electron-deficient groups does not only lower the LUMO and HOMO energy levels of semiconductors, but also forms push–pull electronic structures and intramolecular charge transfer, which are beneficial to enhancing absorption and intermolecular alignment. The chemical structures and performance data of fused-ring electron acceptors are shown in Figure 5.8 and Table 5.8, respectively.

Lim and coworkers synthesized a fluorene-based small molecule flanked with 3-ethylrhodanine (**f1**) *via* a thiophene bridge.[104] The molecule **f1** exhibited strong and broad absorption in the visible region; the LUMO and HOMO energy levels of **f1** were −3.53 and −5.58 eV, respectively, matching with those of P3HT. The devices based on P3HT:**f1** exhibited a PCE of 3.08%.

McCulloch and coworkers modified the molecular structure of **f1** by replacing the electron-donating thiophene unit with an electron-accepting benzothiadiazole (BT) unit affording **f2**, resulting in lower energy levels (LUMO/HOMO −3.57/−5.70 eV).[105] The molecule **f2** showed a non-planar structure with a large dihedral angle of 35° between the fluorene core and the BT unit, which reduced the tendency to crystallize and restricted the formation of large crystalline domains in the BHJ blend. The PCE of P3HT:**f2**-based devices was 4.11%, beyond those of the P3HT:$PC_{61}BM$ control devices (3.53%).

Zhan and coworkers synthesized a BT and 3-ethylrhodanine flanked small molecule acceptor with indacenodithiophene (IDT) as the core (**f3**).[106] The LUMO and HOMO energy levels of **f3** were −3.69 and −5.52 eV, respectively. Different from **f2**, the backbone of **f3** exhibited a planar geometry, which was beneficial to charge transport. The four hexylphenyl groups exhibited a dihedral angle of ~115° to the backbone plane, facilitating the formation of favorable interpenetrating networks without severe self-aggregation when

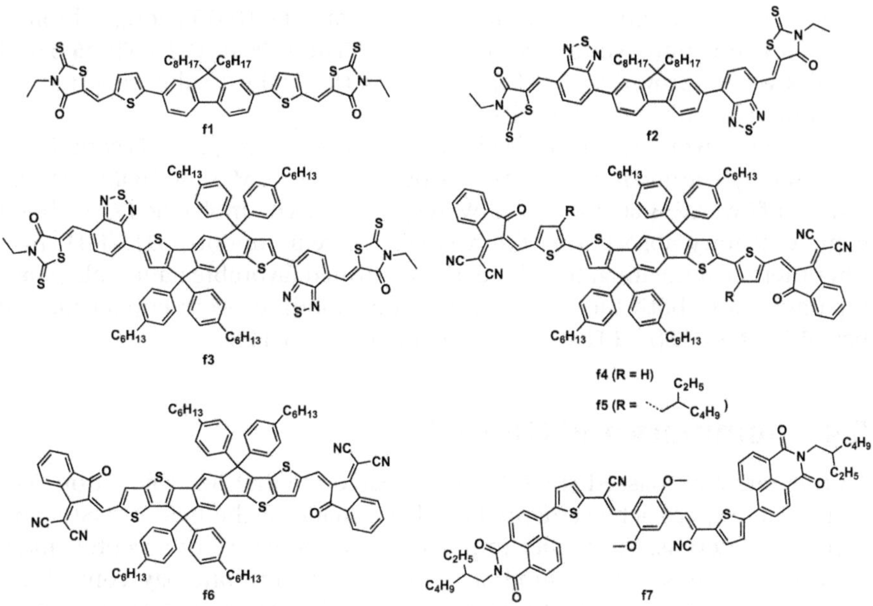

Figure 5.8 Chemical structures of fused-ring electron acceptors.

Table 5.8 Performance of fused-ring electron acceptors in OSCs under an illumination of 100 mW cm^{-2}.

Acceptor	Donor	V_{OC} (V)	J_{SC} (mA cm^{-2})	Fill factor	PCE (%)	Ref.
f1	P3HT	1.03	5.70	0.52	3.08	104
f2	P3HT	0.82	7.95	0.63	4.11	105
f3	P3HT	0.84	8.91	0.681	5.12	106
f4	PBDTTT-C-T	0.90	8.33	0.523	3.93	110
f5	PTB7-Th	0.97	13.55	0.48	6.31	111
	PffT2-FTAZ-2DT	0.998	12.2	0.59	7.3	112
f6	PTB7-Th	0.81	14.21	0.591	6.80	113
f7	*p*-DTS(FBTTh$_2$)$_2$	0.85	9.68	0.66	5.44	114

mixing with P3HT. The PCE of P3HT : **f3**-based devices was 5.12%, which was much higher than that of the P3HT : PC$_{61}$BM control devices (3.71%) and the highest value reported for non-fullerene P3HT-based solar cells.

Zhan and coworkers synthesized a series of fused-ring electron acceptors based on IDT and a strong electron-withdrawing group, 1,1-dicyanomethylene-3-indanone.[107–109] Devices based on PBDTTT-C-T : **f4** exhibited a PCE of 3.93%.[110] The incorporation of an alkyl group on the thiophene unit in **f5** improved the solution processability and crystallinity. OSCs based on PTB7-Th : **f5**, using a PDI derivative as a cathode interlayer showed a promising PCE of 6.31%.[111] Yan and coworkers improved the PCE of **f5** to 7.3% by choosing another donor PffT2-FTAZ-2DT, which exhibited complementary absorption with **f5**.[112] The molecule **f6**, which can be regarded as the modified product of **f4** by fusing the two thiophene rings into the IDT core to form a larger size indacenodithieno[3,2-*b*]thiophene (IDTT) core, exhibited an even better photovoltaic performance.[113] The PCE of PTB7-Th : **f6**-based devices was improved to 6.80%, which was the highest value reported for non-fullerene polymer solar cells.

Park and coworkers synthesized a small molecule acceptor (**f7**) containing cyanomethylene and naphthalimide groups.[114] The LUMO and HOMO energy levels of **f7** were −3.66 and −5.75 eV, respectively. The molecule **f7** exhibited complementary absorption with a small molecule donor *p*-DTS$(FBTTh_2)_2$. After thermal annealing, the blended film formed favorable nanoscale phase separation and both the hole and electron mobilities were enhanced. The best PCE of *p*-DTS$(FBTTh_2)_2$: **f7**-based devices was 5.44%.

5.4 Summary and Outlook

Thirty years have passed since Tang fabricated the first heterojunction OSC in 1986 and great achievements have been made in the field of OSCs and acceptor materials. To further improve the performance of acceptor materials, molecular structure modification and film morphology control are two major methods from the chemical and physical perspectives, respectively. Future research on electron acceptors should focus on the following aspects:[115] (1) the synthesis of new acceptors with superior properties, such as high electron affinity and mobility, strong absorption complementary to the donors and tunable morphology; (2) improvement of the chemical, light, thermal, morphological and mechanical stability; (3) theoretical studies on device physics, especially for the electronic process of the new acceptors, which still lacks in-depth research; (4) new approaches to optimize the morphology beyond the existing method (annealing, solvent additive) and morphological studies; and (5) issues related to practical applications, such as the manufacture of large area devices and modules and cost reduction.

There is no doubt that fullerenes and their derivatives are the most successful acceptors so far. However, new acceptors have already shown a promising performance comparable with that of fullerenes. The best PCE reported for non-fullerene acceptor based single-junction OSCs is 8.7%,[116] whereas the best PCE reported for fullerene acceptor based single-junction OSCs is 11.7%.[117] In the past, the dramatic development of OSCs has been mainly ascribed to the development of donor materials. If acceptors can also make a significant contribution, then more striking achievements are to be expected.

References

1. A. J. Heeger, *Chem. Soc. Rev.*, 2010, **39**, 2354.
2. Y. Lin, Y. Li and X. Zhan, *Chem. Soc. Rev.*, 2012, **41**, 4245.
3. L. Lu, T. Zheng, Q. Wu, A. M. Schneider, D. Zhao and L. Yu, *Chem. Rev.*, 2015, **115**, 12666.
4. F. C. Krebs, *Sol. Energy Mater. Sol. Cells*, 2009, **93**, 394.

5. Y. Lin and X. Zhan, *Adv. Energy Mater.*, 2015, **5**, 1501063.
6. C. W. Tang, *Appl. Phys. Lett.*, 1986, **48**, 183.
7. J. C. Hummelen, B. W. Knight, F. LePeq, F. Wudl, J. Yao and C. L. Wilkins, *J. Org. Chem.*, 1995, **60**, 532.
8. G. Yu, J. Gao, J. C. Hummelen, F. Wudl and A. J. Heeger, *Science*, 1995, **270**, 1789.
9. J. J. M. Halls, C. A. Walsh, N. C. Greenham, E. A. Marseglia, R. H. Friend, S. C. Moratti and A. B. Holmes, *Nature*, 1995, **376**, 498.
10. R. Meerheim, C. Körner and K. Leo, *Appl. Phys. Lett.*, 2014, **105**, 063306.
11. X. Che, X. Xiao, J. D. Zimmerman, D. Fan and S. R. Forrest, *Adv. Energy Mater.*, 2014, **4**, 1400568.
12. J. Huang, J. H. Carpenter, C.-Z. Li, J.-S. Yu, H. Ade and A. K. Y. Jen, *Adv. Mater.*, 2016, **28**, 967.
13. A. R. b. M. Yusoff, D. Kim, H. P. Kim, F. K. Shneider, W. J. da Silva and J. Jang, *Energy Environ. Sci.*, 2015, **8**, 303.
14. Y. Lin and X. Zhan, *Mater. Horiz.*, 2014, **1**, 470.
15. A. Meller and A. Ossko, *Monatsh. Chem.*, 1972, **103**, 150.
16. J. E. Anthony, A. Facchetti, M. Heeney, S. R. Marder and X. Zhan, *Adv. Mater.*, 2010, **22**, 3876.
17. K. Cnops, B. P. Rand, D. Cheyns, B. Verreet, M. A. Empl and P. Heremans, *Nat. Commun.*, 2014, **5**, 3406.
18. Y. Zhong, M. T. Trinh, R. Chen, G. E. Purdum, P. P. Khlyabich, M. Sezen, S. Oh, H. Zhu, B. Fowler, B. Zhang, W. Wang, C.-Y. Nam, M. Y. Sfeir, C. T. Black, M. L. Steigerwald, Y.-L. Loo, F. Ng, X. Y. Zhu and C. Nuckolls, *Nat. Commun.*, 2015, **6**, 8242.
19. Y.-J. Hwang, H. Li, B. A. E. Courtright, S. Subramaniyan and S. A. Jenekhe, *Adv. Mater.*, 2016, **28**, 124.
20. H. W. Kroto, J. R. Heath, S. C. Obrien, R. F. Curl and R. E. Smalley, *Nature*, 1985, **318**, 162.
21. J. N. Haddock, X. Zhang, B. Domercq and B. Kippelen, *Org. Electron.*, 2005, **6**, 182.
22. S. Pfuetzner, J. Meiss, A. Petrich, M. Riede and K. Leo, *Appl. Phys. Lett.*, 2009, **94**, 223307.
23. V. Steinmann, N. M. Kronenberg, M. R. Lenze, S. M. Graf, D. Hertel, K. Meerholz, H. Bürckstümmer, E. V. Tulyakova and F. Würthner, *Adv. Energy Mater.*, 2011, **1**, 888.
24. X. Xiao, K. J. Bergemann, J. D. Zimmerman, K. Lee and S. R. Forrest, *Adv. Energy Mater.*, 2014, **4**, 1301557.
25. M. M. Wienk, J. M. Kroon, W. J. H. Verhees, J. Knol, J. C. Hummelen, P. A. van Hal and R. A. J. Janssen, *Angew. Chem., Int. Ed.*, 2003, **42**, 3371.
26. P. A. Troshin, H. Hoppe, J. Renz, M. Egginger, J. Y. Mayorova, A. E. Goryachev, A. S. Peregudov, R. N. Lyubovskaya, G. Gobsch, N. S. Sariciftci and V. F. Razumov, *Adv. Funct. Mater.*, 2009, **19**, 779.
27. Y. He and Y. Li, *Phys. Chem. Chem. Phys.*, 2011, **13**, 1970.
28. F. B. Kooistra, V. D. Mihailetchi, L. M. Popescu, D. Kronholm, P. W. M. Blom and J. C. Hummelen, *Chem. Mater.*, 2006, **18**, 3068.

29. M. Lenes, G.-J. A. H. Wetzelaer, F. B. Kooistra, S. C. Veenstra, J. C. Hummelen and P. W. M. Blom, *Adv. Mater.*, 2008, **20**, 2116.
30. V. D. Mihailetchi, J. K. J. van Duren, P. W. M. Blom, J. C. Hummelen, R. A. J. Janssen, J. M. Kroon, M. T. Rispens, W. J. H. Verhees and M. M. Wienk, *Adv. Funct. Mater.*, 2003, **13**, 43.
31. M. Lenes, S. W. Shelton, A. B. Sieval, D. F. Kronholm, J. C. Hummelen and P. W. M. Blom, *Adv. Funct. Mater.*, 2009, **19**, 3002.
32. Y. Zhang, H.-L. Yip, O. Acton, S. K. Hau, F. Huang and A. K. Y. Jen, *Chem. Mater.*, 2009, **21**, 2598.
33. Y. He, H.-Y. Chen, J. Hou and Y. Li, *J. Am. Chem. Soc.*, 2010, **132**, 1377.
34. Y. He, G. Zhao, B. Peng and Y. Li, *Adv. Funct. Mater.*, 2010, **20**, 3383.
35. J.-D. Chen, L. Zhou, Q.-D. Ou, Y.-Q. Li, S. Shen, S.-T. Lee and J.-X. Tang, *Adv. Energy Mater.*, 2014, **4**, 1301777.
36. X. Guo, C. Cui, M. Zhang, L. Huo, Y. Huang, J. Hou and Y. Li, *Energy Environ. Sci.*, 2012, **5**, 7943.
37. Y.-J. Cheng, M.-H. Liao, C.-Y. Chang, W.-S. Kao, C.-E. Wu and C.-S. Hsu, *Chem. Mater.*, 2011, **23**, 4056.
38. Y. Matsuo, Y. Sato, T. Niinomi, I. Soga, H. Tanaka and E. Nakamura, *J. Am. Chem. Soc.*, 2009, **131**, 16048.
39. J. M. Campanera, C. Bo, M. M. Olmstead, A. L. Balch and J. M. Poblet, *J. Phys. Chem. A*, 2002, **106**, 12356.
40. C. M. Cardona, B. Elliott and L. Echegoyen, *J. Am. Chem. Soc.*, 2006, **128**, 6480.
41. R. B. Ross, C. M. Cardona, D. M. Guldi, S. G. Sankaranarayanan, M. O. Reese, N. Kopidakis, J. Peet, B. Walker, G. C. Bazan, E. Van Keuren, B. C. Holloway and M. Drees, *Nat. Mater.*, 2009, **8**, 208.
42. C. G. Claessens, D. González-Rodríguez, M. S. Rodríguez-Morgade, A. Medina and T. Torres, *Chem. Rev.*, 2014, **114**, 2192.
43. P. Sullivan, A. Duraud, I. Hancox, N. Beaumont, G. Mirri, J. H. R. Tucker, R. A. Hatton, M. Shipman and T. S. Jones, *Adv. Energy Mater.*, 2011, **1**, 352.
44. B. Verreet, B. P. Rand, D. Cheyns, A. Hadipour, T. Aernouts, P. Heremans, A. Medina, C. G. Claessens and T. Torres, *Adv. Energy Mater.*, 2011, **1**, 565.
45. B. Ebenhoch, N. B. A. Prasetya, V. M. Rotello, G. Cooke and I. D. W. Samuel, *J. Mater. Chem. A*, 2015, **3**, 7345.
46. X. Zhan, A. Facchetti, S. Barlow, T. J. Marks, M. A. Ratner, M. R. Wasielewski and S. R. Marder, *Adv. Mater.*, 2011, **23**, 268.
47. X. Zhao and X. Zhan, *Chem. Soc. Rev.*, 2011, **40**, 3728.
48. X. Zhan, Z. a. Tan, B. Domercq, Z. An, X. Zhang, S. Barlow, Y. Li, D. Zhu, B. Kippelen and S. R. Marder, *J. Am. Chem. Soc.*, 2007, **129**, 7246.
49. L. Meng, Y. Shang, Q. Li, Y. Li, X. Zhan, Z. Shuai, R. G. E. Kimber and A. B. Walker, *J. Phys. Chem. B*, 2009, **114**, 36.
50. P. Cheng, L. Ye, X. Zhao, J. Hou, Y. Li and X. Zhan, *Energy Environ. Sci.*, 2014, 7, 1351.
51. P. Cheng, W. Ma, C. Yan, Y. Li and X. Zhan, *Energy Environ. Sci.*, 2015, **8**, 2357.

52. E. Zhou, J. Cong, Q. Wei, K. Tajima, C. Yang and K. Hashimoto, *Angew. Chem., Int. Ed.*, 2011, **50**, 2799.
53. Y. Zhou, T. Kurosawa, W. Ma, Y. Guo, L. Fang, K. Vandewal, Y. Diao, C. Wang, Q. Yan, J. Reinspach, J. Mei, A. L. Appleton, G. I. Koleilat, Y. Gao, S. C. B. Mannsfeld, A. Salleo, H. Ade, D. Zhao and Z. Bao, *Adv. Mater.*, 2014, **26**, 3767.
54. Y. Zhou, Q. Yan, Y.-Q. Zheng, J.-Y. Wang, D. Zhao and J. Pei, *J. Mater. Chem. A*, 2013, **1**, 6609.
55. Y. Zhang, Q. Wan, X. Guo, W. Li, B. Guo, M. Zhang and Y. Li, *J. Mater. Chem. A*, 2015, **3**, 18442.
56. J. A. Mikroyannidis, M. M. Stylianakis, G. D. Sharma, P. Balraju and M. S. Roy, *J. Phys. Chem. C*, 2009, **113**, 7904.
57. L. Schmidt-Mende, A. Fechtenkötter, K. Müllen, E. Moons, R. H. Friend and J. D. MacKenzie, *Science*, 2001, **293**, 1119.
58. J. Li, F. Dierschke, J. Wu, A. C. Grimsdale and K. Mullen, *J. Mater. Chem.*, 2006, **16**, 96.
59. V. Kamm, G. Battagliarin, I. A. Howard, W. Pisula, A. Mavrinskiy, C. Li, K. Müllen and F. Laquai, *Adv. Energy Mater.*, 2011, **1**, 297.
60. Y. Cai, L. Huo, X. Sun, D. Wei, M. Tang and Y. Sun, *Adv. Energy Mater.*, 2015, **5**, 1500032.
61. S. Rajaram, R. Shivanna, S. K. Kandappa and K. S. Narayan, *J. Phys. Chem. Lett.*, 2012, **3**, 2405.
62. R. Shivanna, S. Shoaee, S. Dimitrov, S. K. Kandappa, S. Rajaram, J. R. Durrant and K. S. Narayan, *Energy Environ. Sci.*, 2014, **7**, 435.
63. L. Ye, K. Sun, W. Jiang, S. Zhang, W. Zhao, H. Yao, Z. Wang and J. Hou, *ACS Appl. Mater. Interfaces*, 2015, **7**, 9274.
64. C.-H. Wu, C.-C. Chueh, Y.-Y. Xi, H.-L. Zhong, G.-P. Gao, Z.-H. Wang, L. D. Pozzo, T.-C. Wen and A. K. Y. Jen, *Adv. Funct. Mater.*, 2015, **25**, 5326.
65. X. Zhang, Z. Lu, L. Ye, C. Zhan, J. Hou, S. Zhang, B. Jiang, Y. Zhao, J. Huang, S. Zhang, Y. Liu, Q. Shi, Y. Liu and J. Yao, *Adv. Mater.*, 2013, **25**, 5791.
66. X. Zhang, C. Zhan and J. Yao, *Chem. Mater.*, 2014, **27**, 166.
67. X. Zhang, J. Yao and C. Zhan, *Chem. Commun.*, 2014, **51**, 1058.
68. Q. Yan, Y. Zhou, Y.-Q. Zheng, J. Pei and D. Zhao, *Chem. Sci.*, 2013, **4**, 4389.
69. J. Zhao, Y. Li, H. Lin, Y. Liu, K. Jiang, C. Mu, T. Ma, J. Y. Lin Lai, H. Hu, D. Yu and H. Yan, *Energy Environ. Sci.*, 2015, **8**, 520.
70. W. Jiang, L. Ye, X. Li, C. Xiao, F. Tan, W. Zhao, J. Hou and Z. Wang, *Chem. Commun.*, 2014, **50**, 1024.
71. L. Ye, W. Jiang, W. Zhao, S. Zhang, D. Qian, Z. Wang and J. Hou, *Small*, 2014, **10**, 4658.
72. Y. Zang, C.-Z. Li, C.-C. Chueh, S. T. Williams, W. Jiang, Z.-H. Wang, J.-S. Yu and A. K. Y. Jen, *Adv. Mater.*, 2014, **26**, 5708.
73. D. Sun, D. Meng, Y. Cai, B. Fan, Y. Li, W. Jiang, L. Huo, Y. Sun and Z. Wang, *J. Am. Chem. Soc.*, 2015, **137**, 11156.
74. Y. Lin, Y. Wang, J. Wang, J. Hou, Y. Li, D. Zhu and X. Zhan, *Adv. Mater.*, 2014, **26**, 5137.

75. Y. Liu, C. Mu, K. Jiang, J. Zhao, Y. Li, L. Zhang, Z. Li, J. Y. L. Lai, H. Hu, T. Ma, R. Hu, D. Yu, X. Huang, B. Z. Tang and H. Yan, *Adv. Mater.*, 2015, **27**, 1015.
76. J. Lee, R. Singh, D. H. Sin, H. G. Kim, K. C. Song and K. Cho, *Adv. Mater.*, 2016, **28**, 69.
77. Y. Zhong, M. T. Trinh, R. Chen, W. Wang, P. P. Khlyabich, B. Kumar, Q. Xu, C.-Y. Nam, M. Y. Sfeir, C. Black, M. L. Steigerwald, Y.-L. Loo, S. Xiao, F. Ng, X. Y. Zhu and C. Nuckolls, *J. Am. Chem. Soc.*, 2014, **136**, 15215.
78. H. Yan, Z. Chen, Y. Zheng, C. Newman, J. R. Quinn, F. Dotz, M. Kastler and A. Facchetti, *Nature*, 2009, **457**, 679.
79. S. Fabiano, Z. Chen, S. Vahedi, A. Facchetti, B. Pignataro and M. A. Loi, *J. Mater. Chem.*, 2011, **21**, 5891.
80. M. Schubert, D. Dolfen, J. Frisch, S. Roland, R. Steyrleuthner, B. Stiller, Z. Chen, U. Scherf, N. Koch, A. Facchetti and D. Neher, *Adv. Energy Mater.*, 2012, **2**, 369.
81. N. Zhou, H. Lin, S. J. Lou, X. Yu, P. Guo, E. F. Manley, S. Loser, P. Hartnett, H. Huang, M. R. Wasielewski, L. X. Chen, R. P. H. Chang, A. Facchetti and T. J. Marks, *Adv. Energy Mater.*, 2014, **4**, 1300785.
82. D. Mori, H. Benten, I. Okada, H. Ohkita and S. Ito, *Adv. Energy Mater.*, 2014, **4**, 1301006.
83. J. W. Jung, T. P. Russell and W. H. Jo, *Chem. Mater.*, 2015, **27**, 4865.
84. C. Mu, P. Liu, W. Ma, K. Jiang, J. Zhao, K. Zhang, Z. Chen, Z. Wei, Y. Yi, J. Wang, S. Yang, F. Huang, A. Facchetti, H. Ade and H. Yan, *Adv. Mater.*, 2014, **26**, 7224.
85. D. Mori, H. Benten, I. Okada, H. Ohkita and S. Ito, *Energy Environ. Sci.*, 2014, **7**, 2939.
86. L. Ye, X. Jiao, M. Zhou, S. Zhang, H. Yao, W. Zhao, A. Xia, H. Ade and J. Hou, *Adv. Mater.*, 2015, **27**, 6046.
87. J. W. Jung, J. W. Jo, C.-C. Chueh, F. Liu, W. H. Jo, T. P. Russell and A. K. Y. Jen, *Adv. Mater.*, 2015, **27**, 3310.
88. C. Lee, H. Kang, W. Lee, T. Kim, K.-H. Kim, H. Y. Woo, C. Wang and B. J. Kim, *Adv. Mater.*, 2015, **27**, 2466.
89. T. Kim, J.-H. Kim, T. E. Kang, C. Lee, H. Kang, M. Shin, C. Wang, B. Ma, U. Jeong, T.-S. Kim and B. J. Kim, *Nat. Commun.*, 2015, **6**, 8547.
90. T. Earmme, Y.-J. Hwang, N. M. Murari, S. Subramaniyan and S. A. Jenekhe, *J. Am. Chem. Soc.*, 2013, **135**, 14960.
91. T. Earmme, Y.-J. Hwang, S. Subramaniyan and S. A. Jenekhe, *Adv. Mater.*, 2014, **26**, 6080.
92. Y.-J. Hwang, B. A. E. Courtright, A. S. Ferreira, S. H. Tolbert and S. A. Jenekhe, *Adv. Mater.*, 2015, **27**, 4578.
93. Y.-J. Hwang, T. Earmme, B. A. E. Courtright, F. N. Eberle and S. A. Jenekhe, *J. Am. Chem. Soc.*, 2015, **137**, 4424.
94. E. Ahmed, G. Ren, F. S. Kim, E. C. Hollenbeck and S. A. Jenekhe, *Chem. Mater.*, 2011, **23**, 4563.
95. G. Ren, E. Ahmed and S. A. Jenekhe, *J. Mater. Chem.*, 2012, **22**, 24373.
96. M.-F. Falzon, A. P. Zoombelt, M. M. Wienk and R. A. J. Janssen, *Phys. Chem. Chem. Phys.*, 2011, **13**, 8931.

97. Y. Lin, P. Cheng, Y. Li and X. Zhan, *Chem. Commun.*, 2012, **48**, 4773.
98. Y. Lin, Y. Li and X. Zhan, *Adv. Energy Mater.*, 2013, **3**, 724.
99. X.-F. Wu, W.-F. Fu, Z. Xu, M. Shi, F. Liu, H.-Z. Chen, J.-H. Wan and T. P. Russell, *Adv. Funct. Mater.*, 2015, **25**, 5954.
100. T. V. Pho, F. M. Toma, M. L. Chabinyc and F. Wudl, *Angew. Chem., Int. Ed.*, 2013, **52**, 1446.
101. H. Li, F. S. Kim, G. Ren, E. C. Hollenbeck, S. Subramaniyan and S. A. Jenekhe, *Angew. Chem., Int. Ed.*, 2013, **52**, 5513.
102. H. Li, T. Earmme, G. Ren, A. Saeki, S. Yoshikawa, N. M. Murari, S. Subramaniyan, M. J. Crane, S. Seki and S. A. Jenekhe, *J. Am. Chem. Soc.*, 2014, **136**, 14589.
103. H. Li, Y.-J. Hwang, B. A. E. Courtright, F. N. Eberle, S. Subramaniyan and S. A. Jenekhe, *Adv. Mater.*, 2015, **27**, 3266.
104. Y. Kim, C. E. Song, S.-J. Moon and E. Lim, *Chem. Commun.*, 2014, **50**, 8235.
105. S. Holliday, R. S. Ashraf, C. B. Nielsen, M. Kirkus, J. A. Röhr, C.-H. Tan, E. Collado-Fregoso, A.-C. Knall, J. R. Durrant, J. Nelson and I. McCulloch, *J. Am. Chem. Soc.*, 2015, **137**, 898.
106. Y. Wu, H. Bai, Z. Wang, P. Cheng, S. Zhu, Y. Wang, W. Ma and X. Zhan, *Energy Environ. Sci.*, 2015, **8**, 3215.
107. H. Bai, P. Cheng, Y. Wang, L. Ma, Y. Li, D. Zhu and X. Zhan, *J. Mater. Chem. A*, 2014, **2**, 778.
108. Y. Lin, J. Wang, S. Dai, Y. Li, D. Zhu and X. Zhan, *Adv. Energy Mater.*, 2014, **4**, 1400420.
109. H. Bai, Y. Wu, Y. Wang, Y. Wu, R. Li, P. Cheng, M. Zhang, J. Wang, W. Ma and X. Zhan, *J. Mater. Chem. A*, 2015, **3**, 20758.
110. H. Bai, Y. Wang, P. Cheng, J. Wang, Y. Wu, J. Hou and X. Zhan, *J. Mater. Chem. A*, 2015, **3**, 1910.
111. Y. Lin, Z.-G. Zhang, H. Bai, J. Wang, Y. Yao, Y. Li, D. Zhu and X. Zhan, *Energy Environ. Sci.*, 2015, **8**, 610.
112. H. Lin, S. Chen, Z. Li, J. Y. L. Lai, G. Yang, T. McAfee, K. Jiang, Y. Li, Y. Liu, H. Hu, J. Zhao, W. Ma, H. Ade and H. Yan, *Adv. Mater.*, 2015, **27**, 7299.
113. Y. Lin, J. Wang, Z.-G. Zhang, H. Bai, Y. Li, D. Zhu and X. Zhan, *Adv. Mater.*, 2015, **27**, 1170.
114. O. K. Kwon, J.-H. Park, D. W. Kim, S. K. Park and S. Y. Park, *Adv. Mater.*, 2015, **27**, 1951.
115. Y. Lin and X. Zhan, *Acc. Chem. Res.*, 2016, **49**, 175.
116. Y. Lin, Q. He, F. Zhao, L. Huo, J. Mai, X. Lu, C. Su, T. Li, J. Wang, J. Zhu, Y. Sun, C. Wang and X. Zhan, *J. Am. Chem. Soc.*, 2016, **138**, 2973.
117. J. Zhao, Y. Li, G. Yang, K. Jiang, H. Lin, H. Ade, W. Ma and H. Yan, *Nat. Energy*, 2016, **1**, 15027.

CHAPTER 6

Structure/Property/Processing Relationships for Organic Solar Cells

MATTHEW DYSON[a], RENEE KROON[b], ALEXANDER B. SIEVAL[c], MARIANO CAMPOY-QUILES[d], CHRISTIAN MÜLLER[b] AND NATALIE STINGELIN*[e,f]

[a]Imperial College London, Department of Physics and Centre for Plastic Electronics, Exhibition Road, London, SW7 2AZ, UK; [b]Chalmers University of Technology, Department of Chemistry and Chemical Engineering, 41296 Göteborg, Sweden; [c]Solenne BV, Zernikepark 6, 9747AN Groningen, The Netherlands; [d]Institute of Materials Science of Barcelona, ICMAB-CSIC, Campus UAB, 08193 Bellaterra, Spain; [e]Imperial College London, Department of Materials and Centre for Plastic Electronics, Exhibition Road, London, SW7 2AZ, UK; [f]Georgia Institute of Technology, School of Materials Science and Engineering and School of Chemical and Biomolecular Engineering, 311 Ferst Drive, Atlanta, GA 30332, USA
*E-mail: natalie.stingelin@mse.gatech.edu

6.1 Introduction

Organic photovoltaic (OPV) cells[1–5] generally consist of a blend of an electron donor with an electron acceptor (small molecule or macromolecule), which usually form a thin film architecture with a complex solid state microstructure and phase morphology: a so-called bulk heterojunction (BHJ) (Figure 6.1).

Nanoscience & Nanotechnology Series No. 45
Nanostructured Materials for Type III Photovoltaics
Edited by Peter Skabara and Mohammad Azad Malik

Published by the Royal Society of Chemistry, www.rsc.org

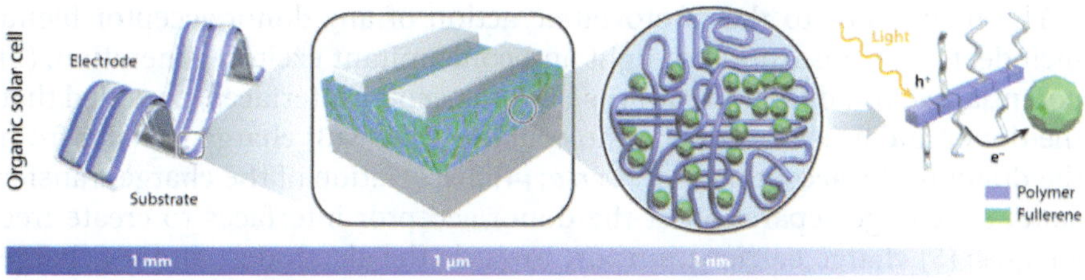

Figure 6.1 OPVs based on a donor:acceptor blend need to be assembled over all length scales: the performance of discrete devices or integrated structures such as modules (left) depend on features on the nanometre to micrometre scale. Most OPVs are based on BHJs (centre left), where the donor and acceptor form complex phase morphologies of mixed and relatively phase-pure phases (centre right). In addition, structural features on the molecular level, including the molecular arrangement (right) can have drastic effects on solar cell function and the resulting performance (adapted from ref. 34 with permission from *Annual Reviews*).

BHJs depend on a number of factors, including the composition, the specific donor, acceptor and processing additives used, the processing history, the substrate and substrate treatment.[6–10] This complexity has made efforts to obtain a full picture of the overall microstructure and to establish valid structure/processing/property interrelationships an intricate task. Nonetheless, materials systems and processes have now been advanced that allow the fabrication of working single-junction devices with efficiencies close to 12%.[11,12] It is clear, however, that enhancing our understanding of the interplay between desirable chemical features, microstructural characteristics and the resulting optoelectronic processes is essential to create a complete knowledge base that allows development – from the outset – of strategies for the design and processing of a diverse range of materials into efficient photovoltaic devices and to enable future increases in efficiency. This chapter attempts to review some key concepts regarding materials design and processing and to provide an overview of the crucial features that dictate the interplay between structure, processing, material properties and the device characteristics of OPV blends, with the main focus on polymer-based systems.

6.1.1 Microscopic Considerations in Organic Photovoltaics

We start our discussion with an overview of the general device mechanism of organic cells and how this relates to their (micro-)structural features. Principally, organic photovoltaic devices are excitonic photovoltaic devices, meaning that photoinduced electron/hole pairs (excitons) are created that have large binding energies (in part due to the relatively low dielectric constants of organic materials, leading to weak Coulomb screening). As a consequence of the latter, efficient exciton dissociation is usually only obtained when there is intimate mixing of two materials with very different electronegativities – a donor and an acceptor – when a BHJ is formed (Figure 6.1).[1–5]

The main steps to the photovoltaic action of any donor:acceptor blend include: (1) the absorption of light and concomitant exciton generation; (2) potential exciton diffusion towards donor/acceptor interfaces, provided that the donor and acceptor are not molecularly mixed; (3) charge transfer from the donor to the acceptor or *vice versa*; (4) dissociation of the charge transfer state, *i.e.* charge separation at the donor/acceptor interfaces to create free charges; (5) charge carrier transport towards the electrodes; and (6) charge extraction at the electrodes[1,5] (Figure 6.2).

Despite the complexity of the structure and processes, charge separation in donor:acceptor blends is often efficient, with dissociation occurring much faster than recombination, although the origin of why such relatively high efficiencies can be obtained is still unclear and has led to controversies over the years.[13] What is unambiguous is that various mechanisms, most probably often acting synergistically, contribute to OPVs performing well. Phenomena to consider include: the formation of interfacial dipoles; high, local mobilities of the created charge carriers that limit recombination and assist charge extraction;[14] cascading energy levels from the interface and/or intermixed phases towards more phase-pure regions;[15] and extra kinetic energy effects of hot electrons that promote the photovoltaic process.[13,16–18]

Structural features from the molecular to the micrometre range play an important part in many of these phenomena. For instance, interfacial dipoles may form in a situation where donor and acceptor molecules are not only sufficiently close to each other, but also positioned at a given relative orientation.[19] Variations in the local interface orientation may statistically average out this contribution. Locally, organic molecules may feature higher charge carrier mobilities, as time-resolved microwave conductivity measurements on single polymer chains indicate,[14] with the exact value depending on the chemical structure and conformation. In addition, charge transfer into an ordered (perhaps crystalline) acceptor-rich domain may further contribute to high local charge transport.[15] Exciton dissociation may also depend on

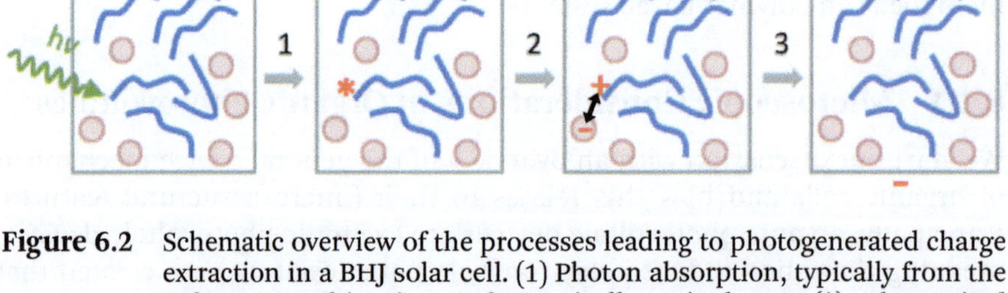

Figure 6.2 Schematic overview of the processes leading to photogenerated charge extraction in a BHJ solar cell. (1) Photon absorption, typically from the polymer, resulting in an electronically excited state (*). The excited state, if required, then diffuses to interfaces and dissociates leading to (2) the formation of unbound positive and negative charge carriers *via* an intermediate charge transfer state represented by the black arrow. (3) The charge carriers then travel to their respective electrodes, where they are collected.

local structural features; for instance, in scenarios where molecularly disordered domains border gradually more ordered regions, *e.g.* in three-phase systems where intermixed and more phase-pure domains are present, this may lead to an energy level cascade that naturally results in the separation of electron/hole pairs beyond their coulombic attraction.[20] Other pathways involve hot electrons; the rate at which they lose kinetic energy relates to the energetic penalty of each hop, which depends on the energy landscape of the two components and hence on the local microstructure and/or degree of molecular order.[13,16–18]

Longer scale charge transport, also crucial in the efficient operation of solar cell devices by assisting charge extraction and limiting charge recombination, is greatly influenced by structural features, from the molecular level to larger scale packing and phase morphology of the final architecture. Generally, higher charge carrier mobilities are found in molecularly ordered and well-interconnected domains, although short-range order, often not detectable in X-ray diffraction experiments, may be sufficient.[21] Macroscopic charge transport relevant for charge extraction can, however, be limited by interchain hopping phenomena and grain boundaries. The provision of an interconnected network minimises these effects. A compromise might be found between large domains, which improve transport, and small domains, which ensure a high charge generation yield. The domain size can be manipulated, to a certain extent, by the molecular weight of the donor (in case a polymer is used).[22–24] In addition, it is assumed that the charge transport pathways for each type of carrier are ideally based on relatively phase-pure regions to limit bimolecular recombination.[12]

Other structural considerations also have to be taken into account. For instance, in many cases organic molecules adopt a preferential orientation with respect to the substrate plane, resulting in a molecular texture and hence optical anisotropy. This is often particularly pronounced when working with macromolecules and is important to consider for solar cell devices because, as the proportion of backbones orientated parallel to the substrate increases, the absorption of light at normal incidence often also increases (provided that the angle subtended by the dipole moment and the chain backbone is small, which is often the case).[25]

Chain conformation (in the case of a macromolecular donor) also plays a part. For example, it has been demonstrated that materials with a higher number of segments exhibiting planar conformations can lead to a 50% increase in the strength of the absorption coefficient.[26] An additional factor is the degree of crystallinity, which affects both the spectral dependence and magnitude of absorption (typically red shifting and enhancing with increasing crystallinity). These dependencies become more complex when producing donor:acceptor blends because the conformation and packing arrangement differ from the neat materials. In the case of polymer:fullerene blends, for instance, the addition of the acceptor to the donor can lead to vitrification,[27] often strongly reducing the degree of texture, both of which reduce light absorption compared with the neat polymer.

Moreover, charge transfer efficiencies between donor and acceptor molecules can be affected by blending and their respective molecular orientation.[28] When planar acceptors (such as small rigid conjugated molecules) are used, the degree of overlap between the electronic wavefunctions of both types of molecules (which determines the charge transfer efficiency) strongly depends on the relative orientation of both molecules, in terms of both relative rotations and translations. This is less pronounced for solar cell blends comprising fullerenes as the acceptor, where charge transfer is often considered to be high.

6.1.2 Macroscopic Considerations in OPV Devices

The photocurrent generated by an OPV device will be proportional to the product of the efficiencies of the microscopic processes outlined in the preceding section, *i.e.* the efficiencies for light absorption, charge transfer, charge separation and charge collection. Therefore the photocurrent itself will depend on the chemical structure of the donor and acceptor (because it defines the light absorption and packing characteristics) and phase behaviour (the number, connectivity and purity of phases). Macroscopically, there is an optimum active layer thickness that enables good optical absorption while maintaining relatively low levels of the bimolecular recombination that appears due to the characteristic low charge carrier mobilities of these systems. The fill factor will depend on the transport properties and therefore also on the microsctructure. The fill factor also depends on the leakage current. Although this is often dealt with at the device architecture level when designing charge-blocking layers, the spontaneous vertical segregation of components in the BHJ can also lead to a reduced dark current provided that the right component ends up next to the right electrode (a donor-rich sub-layer at the cathode and an acceptor-rich sub-layer at the anode). The open circuit voltage will be defined by the component's energy level, provided that electrodes with appropriate workfunctions are chosen. The microstructure, however, can slightly modify these energy levels and thus small variations can appear. For instance, the highest occupied molecular orbital (HOMO) of amorphous regiorandom poly(3-hexylthiophene-2,5-diyl) (P3HT) is shifted 100 meV with respect to that of semicrystalline regioregular P3HT.

Achieving the microstructure that optimises device performance is therefore a challenging task that depends on the chemical nature of the donor and acceptor, as well as the kinetic and thermodynamic behaviour of the system. This also includes the solution from which the films are cast and the fabrication protocol of the layer itself. As OPV devices are intended for large area applications, the control of the structure should equally be extended over large areas. Moreover, long lifetimes are required to ensure a high energy return on investment for this technology. Although a general discussion on OPV device stability goes beyond the scope of this chapter, we will address the issue of structure-related stability. This refers to the evolution over time of the optimum structure generated for an optimised solar cell just after fabrication, evolution that is typically accelerated with the mild temperature soaking originated by solar heating and device operation.

In short, OPV technologies could have a tremendous impact on our society. To further advance the rapid development of this versatile photovoltaic platform, it is of great importance to gain further fundamental insights into the relevant structure/processing/performance interrelations, and which strategies and processes we can use to control from the outset the structural – and hence, electronic – features of the relevant device components. We review here the most important understanding that has been gained on the crucial structural features of OPV systems, focusing on polymer donor–acceptor blends, and how these features affect the final device performance.

6.2 Chemical Design Criteria for Polymers Targeted for OPV Applications

We first discuss structure/property correlations in macromolecular matter in general to help us understand the function of polymeric components in OPVs.[29,30] We start with some definitions used to describe this interesting, but complex, class of materials, before giving examples of how their optoelectronic properties can be fine-tuned, processing parameters manipulated and thermal transitions engineered, and how these factors affect structure formation in both OPV blends and their neat components.

Important features at the molecular level that can affect the overall properties of polymeric matter include the constitution of their chains, *e.g.* the make-up and number of repeating units; their configuration, which describes how the repeating units are coupled; their chain conformation due to bond rotation between backbone moieties;[31] and the length of individual chains (*i.e.* their molecular weight)[30] and the chain length distribution (summarised by the dispersity *Đ*).[29,32,33] Although the constitution, configuration and chain length/chain length distribution of polymers can be influenced by the synthetic pathway selected, their conformation – while also dictated chemically – can often evolve differently depending on the specific thermodynamic and kinetic conditions they are exposed to and hence this is highly dependent on the processing conditions.[27,29,34] Generally, all these features will affect how macromolecules assemble into supramolecular structures. These supramolecular structures may involve only a number of chains or chain segments, but can also consist of the majority of the macromolecules in the system, leading to short-range ordered aggregates, large numbers of well-ordered crystalline moieties, liquid crystalline domains or anisotropic melts.[29]

6.2.1 Manipulation of Optoelectronic Properties Based on Chemical Structure

The structural features that polymers adopt can have drastic effects on their properties, including their optoelectronic properties in the case of the π-conjugated species used in OPVs. Polymers that are π-conjugated feature backbones with alternating single and double bonds.[33] As a result of the covalent coupling of conjugated monomers into long polymer chains, their π and π* orbitals split, leading to two energy bands with molecular orbitals delocalised along each conjugated segment. The π-orbitals evolve into a valence band, bordered by the HOMO, while a conduction band emerges from the π* orbitals, bordered by the lowest unoccupied molecular orbital (LUMO).[35,36] The energy difference between the HOMO and LUMO is called the bandgap, E_g[33] (Figure 6.3).[37] The σ bonds preserve the structural integrity of the polymer chain, thus allowing the π electrons to freely undergo optical and electronic interactions.

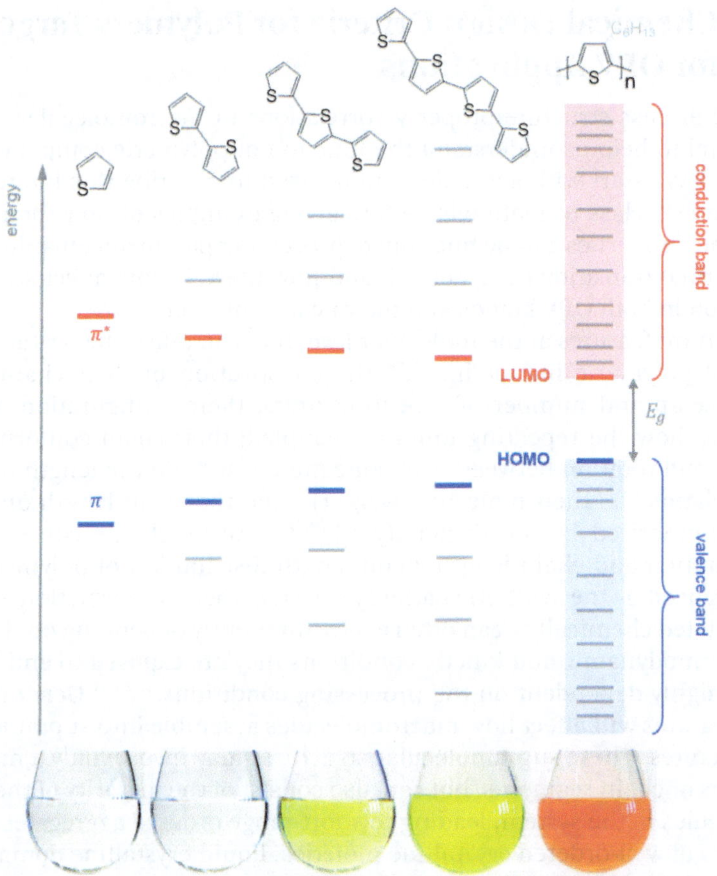

Figure 6.3 Schematic diagram of π and π* orbital splitting in thiophenes on coupling into extended structures, leading to a conduction band and a valence band. The bandgap decreases with increasing chain length (as long as sufficient orbital overlap is retained), leading to changes in the optoelectronic properties – for example, the optical absorption is represented by the colour progression observed for extended thiophene structure solutions (reproduced from ref. 37 with permission from The Royal Society of Chemistry).

For solar cells, molecular design can be used to influence specific device properties. Photocurrent generation, reflected in the short circuit current, J_{SC}, is believed to benefit from a sufficiently large $LUMO_{donor}$–$LUMO_{acceptor}$ offset (Figure 6.4), but also complementary absorption of the donor with respect to the acceptor and/or a broad absorption spectrum of one component (usually the donor polymer) that can be facilitated through a low bandgap. A large photovoltage, reflected in the open circuit voltage, V_{OC}, is believed to be obtained through a large $HOMO_{donor}$–$LUMO_{acceptor}$ offset.

To date, the vast majority of conjugated polymers that have been developed for use in solar cells are p-type donor materials, although acceptor polymers have also been advanced. One prominent example of the latter category is the naphthalene diimide based copolymer P(NDI2OD-T2). The bandgap and the HOMO and LUMO levels of these solar cell materials can be molecularly engineered. One widely used approach is based on alternating donor–acceptor copolymer architectures (Figure 6.5), where electron-rich and electron-poor monomers hybridise when coupled to form an intermediate electronic structure.[38] Other tools used to tune energy levels include: the introduction of electron-donating or electron-accepting side groups; the use of fused monomers to improve the planarity of the backbone and π stacking; and the incorporation of heteroatoms to tune energy levels and to induce or enhance interchain interactions.[39]

With all these design tools, a myriad of exotic conjugated polymer designs can be developed and synthesised that display the desired energy levels and bandgaps. However, a balance needs to be found to control the physicochemical properties that can affect processing (most prominently solubility), provide options for the targeted manipulation of microstructural features and hence device performance, and will assist in increasing the stability of the solar cell.

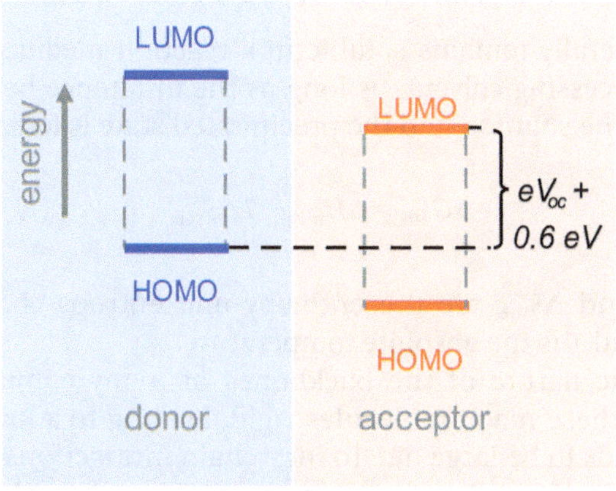

Figure 6.4 Simplistic diagram showing how the relative positions of the HOMO and LUMO levels affect specific OPV characteristics.

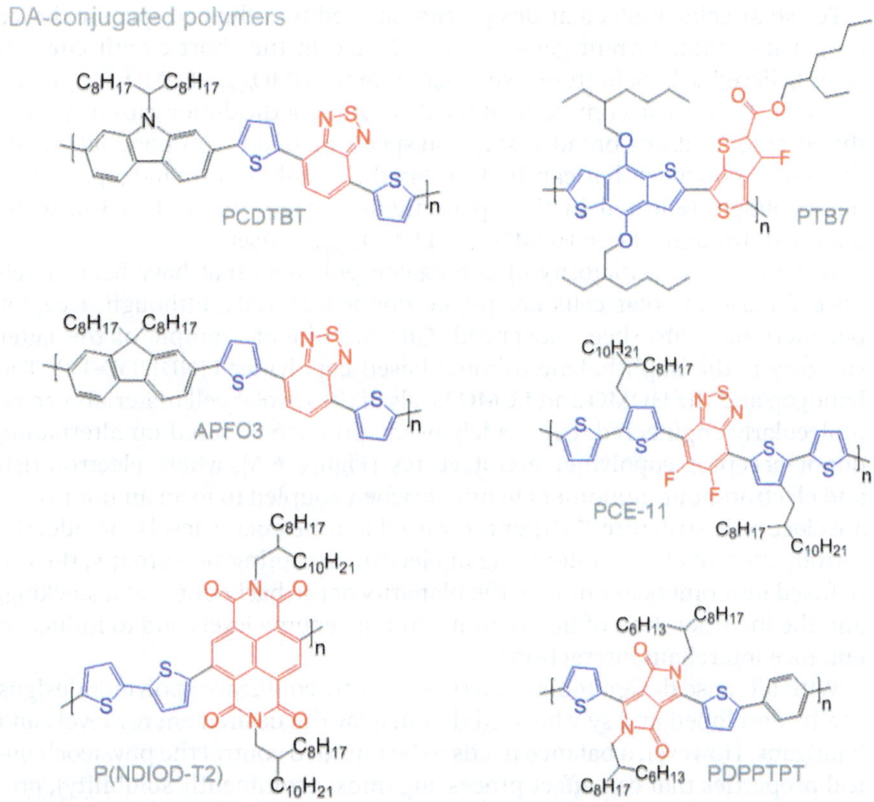

Figure 6.5 Examples of molecular donor (blue) and acceptor (red) combinations used to engineer small bandgap polymers with a range of HOMO and LUMO levels.

6.2.2 Influence of Chemical Structure on Solution Behaviour

A polymer generally remains soluble in a reaction medium, or can be dissolved in a processing solvent, as long as the difference between the Gibbs free energy of the solution and the precipitated state is less than zero:

$$\Delta G_{\text{mix}} = \Delta H_{\text{mix}} - T\Delta S_{\text{mix}} < 0 \tag{6.1}$$

where ΔH_{mix} and ΔS_{mix} are the enthalpy and entropy change on mixing, respectively, and T is the absolute temperature.

The aromatic nature of the backbones of many conjugated polymers often renders these macromolecules rigid, leading to a low ΔS_{mix}. In addition, ΔH_{mix} tends to be large due to interchain interactions in the form of π stacking resulting from the often-pursued regular, planar backbone design. The incorporation of fused and polar conjugated monomers, such as diketopyrrolopyrrole or naphthalene diimide, increases ΔH_{mix} even further

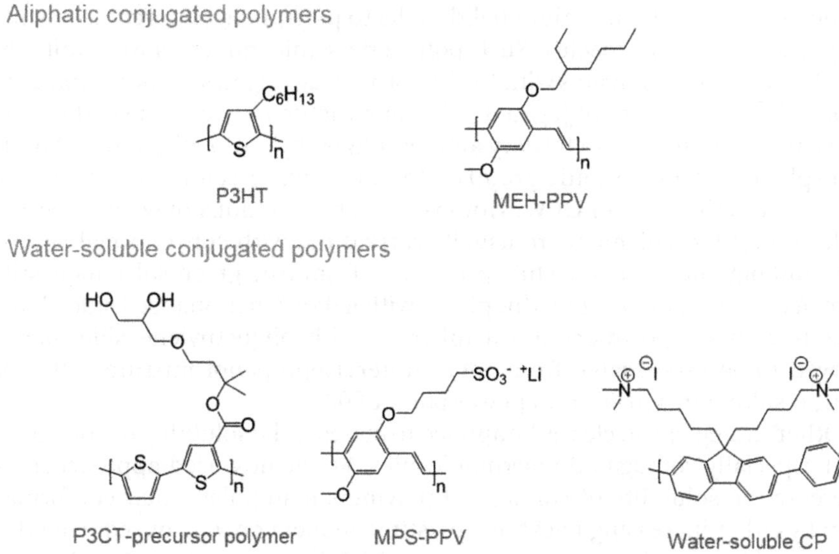

Figure 6.6 Side-chain engineering can control the solubility of polymers in organic solvents with straight or branched alkyl side-chains (top) or render them soluble in water or 'green' solvents (bottom) through side-chain functionalisation with polar or ionic groups.

(Figure 6.6).[40,41] The combination of a low ΔS_{mix} and high ΔH_{mix} tends to lead to $\Delta G_{mix} \geq 0$, which makes dissolution unfavourable. As a result, early conjugated polymers such as poly(*para*-phenylene vinylene) (PPV) and polythiophene were intractable after synthesis. Over the years, strategies have been developed that allow the manipulation of ΔS_{mix} and ΔH_{mix} from a structural point of view.

The most widely used approach to produce soluble conjugated polymers is *via* decoration of their aromatic backbone with (often branched) alkyl side-chains. Their flexible nature increases ΔS_{mix} and their ability to inhibit π stacking of the backbone by steric hindrance lowers ΔH_{mix}, together enhancing the solubility of the resulting polymers. Early examples of soluble conjugated polymers include methoxy-ethylhexyloxy-PPV (MEH-PPV) and poly(3-alkyl thiophene) (P3AT).[42,43] These were usually soluble in halogenated and/or high-boiling aromatic solvents, such as chloroform, chlorobenzenes and xylenes. A plethora of materials has since been synthesised, many with high charge carrier mobilities, high solar cell efficiencies and other interesting features that can be exploited in bioelectronics applications, thermoelectrics and beyond.

A transition into other applications, in addition to enabling 'green' processing, can be realised by the use of polar side-chains. Such materials have attracted attention as they can facilitate dissolution in environmentally more benign solvents, such as ethanol or water. Chemical modifications are mainy targeted at attaching charged functional groups to the side-chains – for example, sulfonic acid salts introduced into a PPV derivative (MPS-PPV) or quaternary alkylammonium halide salts to poly(phenyl-fluorene) – resulting in water-soluble species. Such polymeric semiconductors were initially used to form bioconjugates with DNA or protein aggregates for detection purposes.[44,45] Many other conjugated polymer structures have been synthesised since then – among them tertiary amines, oligoethylene oxide, and carboxyl, phosphate and zwitterionic groups – to obtain water-soluble polymers for use in the active layer of OPVs. However, most evaluations of water-processable conjugated polymers are usually carried out with conventional acceptors and organic solvents, while processing from true green solvents is still rare. A combination of a polythiophene with a diol-functionalised side-chain (P3CT-precursor polymer) and a fullerene with oligoethylene side-chains proved to be processable from a 1 : 1 water : isopropanol mixture with 5% THF, resulting in a maximum power point of 0.7%.[46]

Other strategies can clearly be applied to increase the solubility of polymers, and especially conjugated macromolecules. Frequently used approaches to increase the solubility of conjugated polymers manipulate their conformation by either increasing backbone twisting, reducing ΔH_{mix}, or reducing the rotational barriers between monomers, which increases ΔS_{mix}. Copolymerisation (including random copolymerisation) or the control of regioregularity can also be applied. The control of chain length, *i.e.* molecular weight, which can assist to a certain extent in changing the solution behaviour, is another pathway, although it is a chemical design approach more often exploited to control the microstructure of the resulting blends.

6.2.3 Effects of Solvent(s)

When processing conjugated polymers into devices such as OPVs, not only the solute(s), but also the solvent system plays a crucial part. In general, the quality of a solvent determines the conformation of a dissolved polymer chain. In a good solvent, strong interactions with solvent molecules swell the polymer chain, whereas a bad solvent leads to a collapsed coil.[47] An intermediate case is represented by so-called theta solvents, where chain segments follow a random walk. The size of the macromolecular coil, *i.e.* the radius of gyration (R_g) of the polymer molecules, scales with their molecular weight M according to:

$$R_g \propto M^{\nu} \tag{6.2}$$

where $\nu = 3/5$, 1/2 or 1/3 in the case of a good, theta or bad solvent, respectively.

The solubility of any polymer, including π-conjugated species, can typically be improved through the choice of a good solvent, by reducing the solute concentration and/or, in most cases, an increase in temperature. In a good solvent ($\nu = 3/5$), the polymer chains extend the most, which can lead to entanglements in the case of sufficiently high molecular weights.[30] The associated strong increase in solution viscosity because of these entanglements has a pronounced impact on the formation of the solid state structure, both during melt processing and solution-casting of thin films.[30] By contrast, in a bad solvent ($\nu = 1/3$), the polymer chains collapse, which results in the formation of pre-aggregates in solution and a concomitant reduction in solution viscosity.

These effects must be considered when interpreting solidification processes during solution-casting (or processing from the melt). A number of studies on the processing of OPVs have explored the impact of chain conformation in the processing solvent on the solid state structure of conjugated polymers and polymer:fullerene blends. To give a few examples: Wolfer *et al.* explored the solution viscosity of PCDTBT, in particular the influence of the solution thermal history (the solution cooling rate after preparation) on the amount of entanglements in PCDTBT and PCDTBT:$PC_{71}BM$ solutions.[48] Rapid cooling of PCDTBT:$PC_{71}BM$ solutions resulted in suppressed disentanglement and thus higher solution viscosities after passing a critical temperature between 80 and 100 °C. This effect was more pronounced for systems consisting of PCDTBT of higher molecular weights and when solutions of higher polymer concentrations surpassing the semi-dilute concentration regime were used. Indeed, OPVs based on such rapidly cooled higher viscosity solutions displayed higher short circuit current densities. A follow-up paper by the same group attributed this result to slower $PC_{71}BM$ diffusion within the fast-cooled, more viscous PCDTBT:$PC_{71}BM$ solutions, resulting in more $PC_{71}BM$ being 'trapped' in the polymer-rich phase and thus increasing the polymer/fullerene interfacial area.[49] Another example is given by Hu *et al.*, who explored sonication as a tool to disentangle P3HT chains in solution.[50] The effect of disentanglement on charge carrier mobility in organic field-effect transistors was found to be more pronounced in higher molecular weight P3HT and only when theta solvents such as toluene and chloroform were used. This observation was explained by the fact that a greater amount of entanglements can be 'loosened' in materials of high molecular weight, hence allowing more free movement of polymer chains compared with systems containing more entanglements, leading to the formation of more ordered aggregates in solution.

6.2.4 Controlling Structure Formation in the Solid State Through Chemical Design

Both intrinsic and extrinsic factors determine the solid state structure of a polymer. First, the nature of its chemical structure – predominantly given by chain regularity, molecular weight, chain length distribution, type of side-chain substitution and tacticity[29] – affects its ability to order. Second, the thermodynamic and kinetic conditions used during solidification can influence the crystalline domain size, the degree of mixing and the overall phase morphology. In materials targeted for OPVs, chemical design considerations must take into account both the primary optoelectronic properties and the effect on the final thin film microstructure, as this will, in turn, influence the energy levels, charge transport properties and many other optoelectronic features.

6.2.4.1 Regioregularity and Coupling Defects

Polymers with a regular chain configuration usually have a stronger ability to crystallise,[29] with regioregularity predominantly determined by the polymerisation method. The most prominent example is the synthesis of P3HT. P3HT can be chemically synthesised *via* oxidative polymerisation with $FeCl_3$, which results in regiorandom (RRa) P3HT. As a result of the randomness of the reaction, which can either occur at the head or tail, four triads can be formed (Figure 6.7). This random coupling can be circumvented by using either the McCullough (Kumada coupling) or Rieke (Negishi coupling) method (Figure 6.8a), which are both able to generate near defect-free, regioregular (RR) P3HT.[51] Because of this different coupling, RRa-P3HT and RR-P3HT display markedly different properties. The former is a material of low molecular order with poor electronic transport and a broad, featureless absorption spectrum, whereas the latter is generally of a semicrystalline nature featuring good charge transport properties and an absorption with relatively well-defined characteristic 0–0, 0–1 and 0–2 vibronic transitions[52,53] (Figure 6.7b).

Another example of chemically induced defects can occur during Stille and Suzuki cross-coupling, which are the polymerisation methods most often used to produce alternating copolymers. Both methods utilise a combination of a dibromo monomer and either distannyl (Stille) or diboronic ester (Suzuki) functionalised monomers (Figure 6.8b).

A number of studies have explored the impact of homocoupling defects on the photovoltaic performance of polymer:fullerene blends. In the case of DPP based polymers, a (deliberately) suboptimum catalyst to ligand ratio was correlated with a higher defect density.[54] The observed decrease in photovoltaic performance was explained by the defects' more localised LUMO levels acting as charge traps. For the donor polymer, PTB7, it was observed

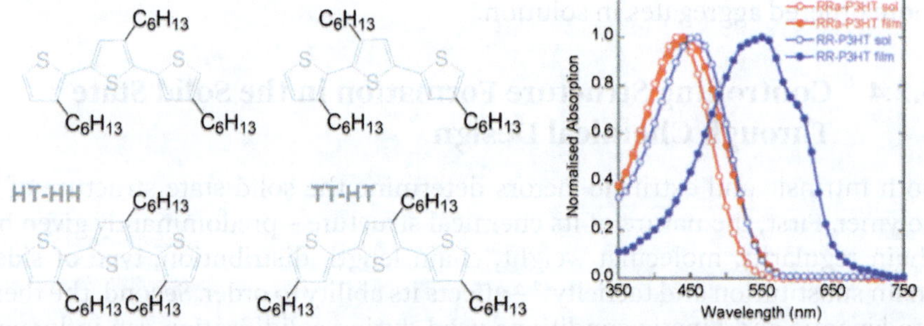

Figure 6.7 (a) Influence of regioregularity and coupling defects on (b) the optical absorption of solutions and films of RRa and RR-P3HT. Adapted with permission from ref. 53. Copyright 2011 American Chemical Society.

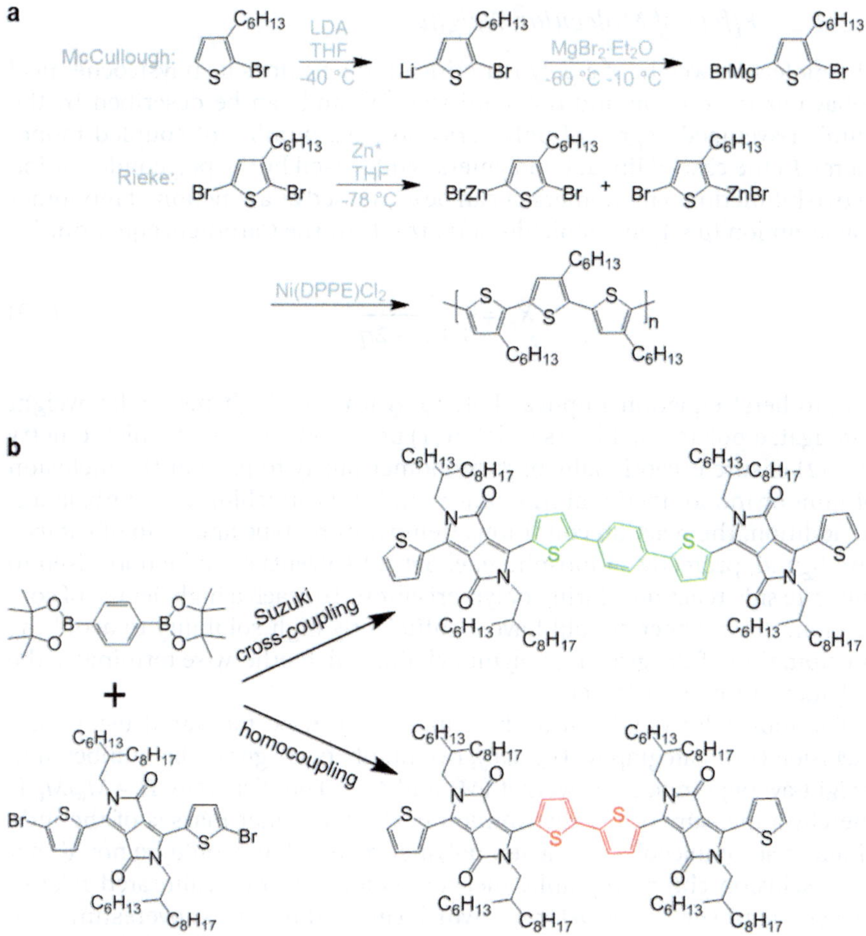

Figure 6.8 (a) Nearly defect-free P3HT can be produced either through the McCullough method (Kumada cross-coupling) or the Rieke method (Negishi cross-coupling). (b) Example of a homocoupling defect (red moiety) that potentially occurs during Suzuki polymerisation instead of the desired coupling product (green moiety) of a DPP based conjugated polymer.

that homocoupling defects resulted in a lower molecular weight and broader dispersity (*i.e.* chain length distribution), with a concomitant decrease in the solar cell's maximum power point.[55,56] Lombeck *et al.* were able to vary the number of homocoupling defects for PCDTBT from 0 to 8 mol% and found that solar cells based on defect-free polymers displayed the highest device performance.[57] Details about how molecular weight and dispersity affect the solid state microstructure and blend phase morphology are reviewed in the following sections.

6.2.4.2 Effect of Molecular Weight

The molecular weight of a polymer critically determines its physicochemical behaviour in solution and the solid state[29,30] and can be described by the number-averaged degree of polymerisation ($\bar{X}_n$, number of coupled monomers). In the case of linear copolymers synthesised by the polycondensation of two bifunctional monomers, it can be expressed as a function of monomer consumption ($p \leq 1$) and stoichiometry ($r \leq 1$) by the Carothers equation:[31]

$$\bar{X}_n = \frac{1+r}{1+r-2rp} \tag{6.3}$$

Carothers' equation implies that, to synthesise high molecular weight conjugated polymers, it is essential to (1) utilise an equimolar stoichiometry and (2) ensure exceptionally high monomer purity to prevent the inclusion of non- or monofunctional monomers and thus stoichiometric imbalance. In addition, the reaction conditions (temperature, type and ratio of catalyst and ligand, protective atmosphere, choice of solvent) must be optimised to mitigate side reactions during polymerisation. To reach a high degree of conversion, the polymer should have a sufficiently high solubility to avoid the precipitation of the growing polymer chain, which otherwise terminates the polymerisation prematurely.

The molecular weight of a given material can be measured using size-exclusion chromatography. Typically, two numbers are given, the number- and weight-average molecular weight (M_n and M_w). The dispersity $Đ = M_w/M_n$ is thereby a measure of the distribution of the molecular masses of the individual macromolecules in a given polymer material. It should be noted that size-exclusion chromatography measurements are often calibrated relative to a standard such as polystyrene, which can lead to a gross overestimate of the molecular weight. For instance, the molecular weight of P3HT is typically overestimated by a factor of two.[58]

Many physicochemical properties tend to be strongly affected by the molecular weight when short-chain oligomers are used; this effect becomes less molecular weight dependent for long-chain polymers. One example is the melting temperature T_m.[30,59] For any polymer, the melting temperature initially increases with molecular weight. This is because T_m is directly correlated with the thickness of the crystalline moieties within a polymeric (or oligomeric) material, which, in the regime where the individual macromolecules are not sufficiently long, form chain-extended crystals. In the case of P3HT, T_m reaches about 230 °C when processed from the melt for materials featuring M_n ~25 kg mol^{-1}.[30] Once the molecules become longer, chain folding sets in, leading to thinner, interconnected crystals embedded within an amorphous matrix, which limit the lamellar thickness. The reduction in thickness of these crystals leads directly to a reduction in T_m.

The optoelectronic properties of conjugated polymers follow a comparable behaviour. For instance, analogous to the melting temperature, the charge carrier mobilities (measured in field-effect transistors)[30] as well as charge generation (deduced from photoconductivity measurements)[60] initially increase with molecular weight. Once a critical molecular weight is reached (in neat P3HT at $M_n > 25$ kg mol^{-1}), they plateau. Solar cells based on P3HT:PCBM (phenyl-C_{61}-butyric acid methyl ester) display a similar behaviour, with the optimum molecular weight of P3HT for device fabrication reported to be around M_n ~20 to 30 kg mol^{-1}.[22–24] The degree of preferential orientation of polymer chains parallel to the substrate, which defines the optical anisotropy and the concomitant absorption for normal incidence light, also increases with molecular weight, as has been shown for systems such as MEH-PPV, F8BT and DPP-TT-T and IDTBT.[26]

A pronounced molecular weight dependence of the photovoltaic performance of polymer:fullerene blends has also been observed for several donor–acceptor copolymers, including APFO3, PTB7 and DPP-TT-T, with higher molecular weight materials typically resulting in superior results.[23,24,61,62] This further highlights the importance of controlling and optimising the molecular weight of the polymeric components used in OPVs.

6.2.4.3 *Effect of Glass Transition Temperature and Tendency for Vitrification*

A material that has been rapidly quenched below its glass transition temperature T_g, where segmental relaxation is suppressed, usually features a high degree of disorder. Blending can further assist such a vitrification, for both kinetic and thermodynamic reasons.[27,63,64] The former is partly determined by the T_g of the individual components (in OPV devices, the donor and the acceptor), while their mutual miscibility, strongly influenced by their chemical nature and leading to thermodynamic mixing, dictates the second phenomena.

Re-heating above T_g allows order to evolve on an experimental timescale. In donor:acceptor blends that feature a three-phase morphology comprised of an acceptor-rich, a donor-rich and an intermixed phase,[27] this usually constitutes the T_g of the intermixed phase.[63] For instance, in the case of spin-cast P3HT:PCBM blends that form largely disordered films directly after casting, annealing above the T_g of the intermixed phase leads to crystallisation – often of both components.[65,66] The glass transition temperature of the neat components, as well as the intermixed phases created by them, is therefore an important parameter in the selection of annealing protocols. They also have a strong effect on the thermal stability of the final device.[67]

In neat components, a number of structural parameters affect the glass transition temperature. The empirical Flory–Fox equation[62,64] can be used for polymers; it describes the increase in T_g with molecular weight, which increases sharply for low molecular weight materials such as oligomers, but approaches a value independent of chain length for higher molecular weight polymers:

$$T_g(M_n) = T_g(M_\infty) - \frac{K}{M_n} \tag{6.4}$$

where $T_g(M_\infty)$ is the glass transition temperature of a polymer with infinite molecular weight (infinite chain length) and K is a polymer-specific constant.[62,64] Other important parameters that affect T_g are the backbone stiffness and the nature of the side-chains (length, linear *versus* branched). For instance, P3ATs display a relatively low T_g because their backbones consist of thiophene repeating units with comparatively low energetic barriers to rotation. The T_g can thereby be further manipulated by the selection of side-chain length. Increasing their length from butyl to dodecyl reduces the T_g of P3ATs from ~40 to −15 °C.[64] By contrast, donor polymers with more rigid backbones display significantly higher glass transition temperatures, often in excess of 100 °C.[64]

6.2.4.4 Molecular Order and Preferred Orientation (Texture)

The various nuances of molecular ordering strongly influence many important optoelectronic properties of conjugated polymers, both neat and in blends. As such, much effort has been devoted to both control their molecular arrangement and packing, and to develop methodologies to obtain insights into the various microstructural features.

The chemical nature of the material has a profound effect on its tendency to order. For instance, in polymers, the chain rigidity – which can be controlled by the introduction of aromatic moieties – plays an important part.[32,68] Flexible polymers, such as RR-P3HT, are semicrystalline and can feature relatively extended crystalline domains that can be characterised *via* X-ray diffraction and/or thermal analysis because they usually feature well-distinguishable melting and crystallisation temperatures, T_m and T_c, on heating and cooling, respectively. More rigid polymers often do not feature such long-range order.[32] Rather, they can be characterised by a high degree of local order due to the strong π stacking of the chains.[21] The melting of rigid polymers typically occurs at higher temperatures than for more flexible chain materials because of the associated low ΔS_m and high ΔH_m (as alluded to earlier),[32] which leads to a high T_m according to:

$$T_m = \frac{\Delta H_m}{\Delta S_m} \tag{6.5}$$

Backbone rigidity can also lead to a liquid crystalline character,[32] shown, for example, by the conjugated polymers PBTTT and APFO3,[62,69] often resulting in specific molecular orientations and, as a consequence, in many scenarios, to long-range ordered domains.

Another crucial aspect that affects order is the molecular weight.[30,32] As already alluded to, sufficiently long chains in higher molecular weight materials can interconnect ordered regions, including crystalline moieties, as depicted in Figure 6.9. These so-called tie-chains can be important as they provide charge transport 'highways' between adjacent crystallites.[21,70] As a consequence, the number and thickness of these ordered domains have a profound impact on charge transport,[30,71,72] charge generation[60,73,74] and on exciton dissociation.[75–77]

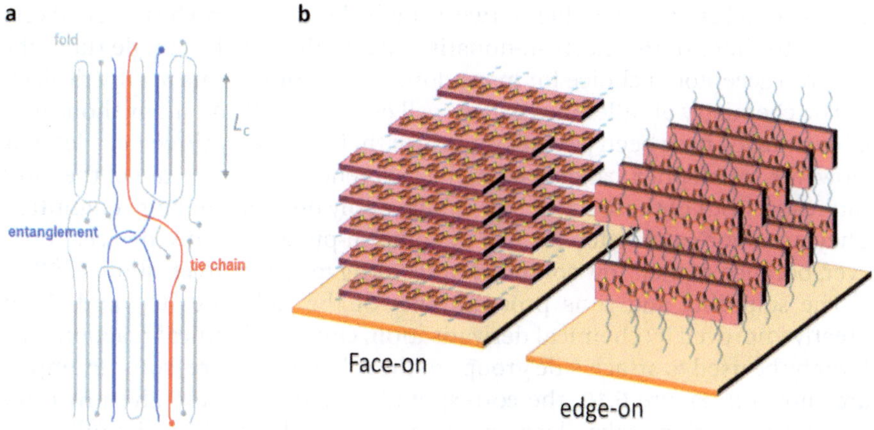

Figure 6.9 (a) Semicrystalline polymers are made of crystalline moieties interconnected with amorphous regions. Molecules that link individual crystal lamellae, such as the one highlighted in red, are called 'tie chain' molecules. (b) Semicrystalline polymers often display a preferred orientation (*i.e.* texture) of the crystalline lamellae with respect to the substrate, *i.e.* face-on or edge-on. Reproduced from ref. 37 with permission from The Royal Society of Chemistry.

The latter has been related to the electronically heterogenous nature of such systems that can lead to a kind of energy cascade, leading to the distribution of charge–pair distances and hence differences in the exciton lifetime. In addition to molecular weight, other factors play pivotal parts. The orientation of the ordered regions relative to the substrate and within the bulk material (*i.e.* their texture; Figure 6.9b) tends, for instance, to correlate with device performance, with preferential out-of-plane π stacking (face-on) typically leading to a higher photocurrent in vertical devices (such as OPVs) and edge-on stacking improving transport in lateral devices, such as organic field-effect transistors.[78–80]

Not only optoelectronic features are affected by the solid state microstructure and phase morphologies of neat and blend systems. A semicrystalline nature, as often found in polymers of relatively high linearity/regularity, such as P3HT of sufficiently high molecular weight,[30,32] provides the materials with a certain mechanical toughness – at least when used above the glass transition temperature. Such toughness is a prerequisite if truly flexible applications are targeted.[81] Environmental stability, thermal phase behaviour and other characteristics, such as mechanical ductility, are also affected.[82]

6.2.5 Chemical Design of Fullerene Acceptors to Manipulate Device Performance

Although many of the tools discussed so far can be applied in a similar manner to both fullerene and non-fullerene acceptors, fullerene-based systems will be discussed here separately. Fullerenes are close-caged carbon spheres and are good electron conductors with energy levels that can be matched very well with those of the electron-donating materials. This has made them the electron acceptor of choice for many donor:acceptor systems in OPV devices.

The chemistry of fullerenes is now well established[83] and many thousands of derivatives have been synthesised since their discovery three decades ago. However, for all practical purposes, the fullerenes C_{60} and C_{70} are of the most interest.[84] Both of these fullerenes can be easily obtained in large quantities. They are poor candidates for use in solution-processed OPV devices, however, because of their low solubility and strong tendency to crystallise.[85,86]

The solubility and thus processability of the fullerenes C_{60} and C_{70} is greatly improved by chemical derivatisation. Organic chemical reactions can thereby be used to attach side groups to the fullerene sphere (some examples are shown in Figure 6.10, the corresponding solubilities are given in Table 6.1). Such reactions take place on one of the double bonds of the fullerene. In C_{60}, all 30 double bonds are equal, which means that in most reactions a single reaction product is formed. By contrast, reactions on C_{70} lead to the formation of a mixture of structural isomers because the 35 double bonds in C_{70} are not all equal due to the lower symmetry of the fullerene sphere.

Although other synthetic modifications are sometimes used, for OPV applications the fullerenes are usually modified by cycloaddition reactions. These are easy to perform and systematic structural variations of the adducts can be realised by changing the reagent, even on a large scale. Accordingly, materials can be synthesised that have properties that are well suited and easily tuned for application in fullerene-based OPV systems.

Three different types of cycloaddition reactions are commonly used to modify fullerenes. The first is the addition of a substituted cyclopropane ring, for which various synthetic routes exist. A well-known example of this type of derivative is PCBM (Figure 6.10), one of the omnipresent molecules in polymer-based OPV systems. The second class are Diels–Alder type adducts, which lead to compounds such as ICBA (Figure 6.10). The third class is the so-called fulleropyrrolidines or Prato-adducts, widely used in covalently linked donor–acceptor systems.[87]

The tunability of the solubility of fullerene is important, as shown by a large study of 27 different fullerenes blends with P3HT. Troshin *et al.* found that one of the most important factors determining the efficiency of the corresponding solar cell was the relative solubility of the two compounds in a given solvent system, as shown schematically in Figure 6.11.[7] They concluded that, to obtain efficient solar cells, the solubilities of the donor and acceptor molecules in a given solvent should be both sufficiently high and similar (around 50 mg mL^{-1}). In addition, high solubilities were reported to enable the fabrication of sufficiently thick, homogenous films, while also limiting the degree of phase separation, which is known to be unfavourable for charge generation.

Other physical properties can be manipulated by derivatisation, such as the molecules' dipole moment, the overall absorption, and the HOMO and LUMO levels. The latter observation is important because, for many fullerene

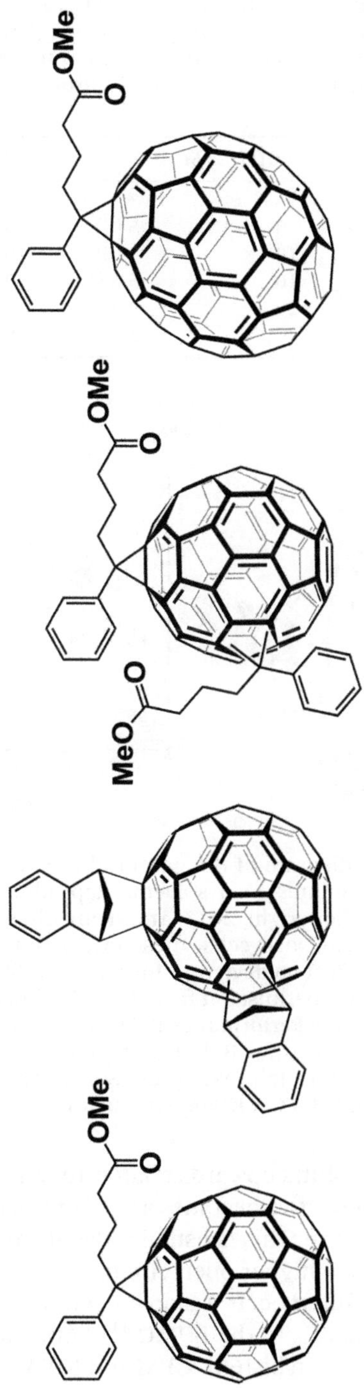

Figure 6.10 Chemical structures of commonly used fullerene derivatives, including, from left to right, [60]PCBM (abbreviated to PCBM throughout), bis[60]PCBM, ICBA and [70]PCBM (major isomer).

Table 6.1 Practical dissolution guidelines for C_{60}, C_{70}, [60]PCBM, [60]ThCBM and [70]PCBM (in mg mL^{-1}). Values for C_{60} and C_{70} are taken from previously reported work or were determined at Solenne (Groningen, the Netherlands). The values for fullerene derivatives were determined at Solenne by calibrated high-performance liquid chromatography after stirring with excess solid for three days.

Solvent	[60]PCBM	[60]ThCBM	C_{60}	[70]PCBM	C_{70}
Toluene	10	5	3	20	1.4
***p*-Xylene**	5	5	5 (S)	10	4
***o*-Xylene**	15	10	7 (S)	30	11
Chlorobenzene	25	10	5.5 (S)	40	–
Chloroform	25	20	0.2	30	–
ODCB	35	20	24	70	36

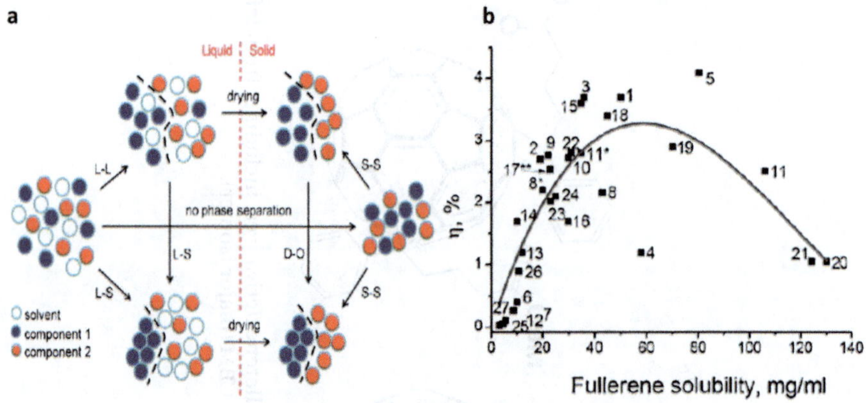

Figure 6.11 (a) Schematic diagram of the potential interactions that can occur between two solutes in a solvent. Reproduced with permission from ref. 92. Copyright 2013 American Chemical Society. (b) In polymer:fullerene solar cells, these interactions in ternary systems have a strong effect on device performance, as found by Troshin *et al.* Reproduced with permission from P. A. Troshin *et al.*, Material Solubility-Photovoltaic Performance Relationship in the Design of Novel Fullerene Derivatives for Bulk Heterojunction Solar Cells, *Advanced Functional Materials*, John Wiley and Sons, Copyright © 2009 WILEY-VCH Verlag GmbH & Co. KGaA, Weinheim.

derivatives, the LUMO level shifts upwards relative to that of C_{60} (or C_{70}). This means that the fullerene derivative becomes a weaker electron acceptor compared with the parent fullerene. This is usually advantageous in OPV devices because the device will give a higher open circuit voltage than a device containing the unsubstituted fullerene. It is also known that different modifications can lead to different shifts in the LUMO.[88–90] Attaching more than one group to the fullerene, such as in bis[60]PCBM and ICBA, can shift the LUMO level further upwards;[88–90] however, doing so often reduces charge transport, usually because these derivatives have a lower ability to crystallise.[88–90]

6.3 Physicochemical Tools to Manipulate OPV Blends

The process of solidification of an OPV blend is usually a highly non-equilibrium transformation from a solution into the solid state. Although the chemical design influences many factors of the solidification process, it is also strongly affected by thermodynamic and kinetic phenomena dictated by the selected processing conditions and solidification environment. Here we discuss the current understanding of how to manipulate the relevant phase transitions and solidification mechanisms physiochemically (see also ref. 91) and how kinetic factors can be controlled to induce the desired optoelectronic response.

6.3.1 Solubility in Solution, Evaporation Rate and Drying Conditions

The interaction of the donor and acceptor molecules with the third component, the solvent, as well as the removal of the solvent during evaporation (shown schematically in Figure 6.11a) are the two key parameters that determine how close the resulting system will become to thermodynamic equilibrium.[27,92,93]

We first consider the solidification process of an OPV blend from solution. Before the solvent starts evaporating, the solubility of each component within the solvent will determine the nature of the aggregates that form at a given operational temperature and concentration. The aggregate size will also be dictated by these conditions, although kinetic factors such as the rate of cooling and solvent evaporation will also have an effect. More specifically, according to the IUPAC Gold Book, the solubility of a solute is 'the analytical composition of a saturated solution, expressed in terms of the proportion of a designated solute in a designated solvent'.[33] If controlled, early aggregation in solution can allow control of the extent of phase separation and the size of the resulting phases in the solid state. This is, for instance, the case for P3BT, which forms nanofibres (or 'whiskers') in comparatively poor solvents, leading to a relatively controlled phase morphology in blends with PCBM,[94] or a more recent example that exploited the pre-aggregation of the donor polymer PCE11 to produce some of the most efficient devices to date.[12]

From a materials design point of view, the solubility of each component will depend on the specific side-chains selected (including the length of the side-chains and their nature, *e.g.* linear or branched) and is often reduced with increasing molecular weight in the case of polymers, as described in detail in the previous section. For a given solute/solvent system, the solubility can be described in terms of the Hansen solubility parameters.[95] Each molecule is defined by three parameters: the energy associated to dispersion (δ_d), as well as polar and hydrogen bond forces (δ_p and δ_h respectively) between molecules. According to this theory, two substances will be more soluble within each other the closer they lie within the Hansen parameter space (*i.e.* 'like dissolves like').

Experimentally, the dispersion forces in organic semiconductors tend to be large ($\delta_d \approx 18 - 20$) MPa compared with the other two kinds of contributions ($\delta_p \approx \delta_h \approx 3$–9) MPa. Interestingly, this parameterisation of solubility can be used to theoretically find non-halogenated and green solvents.[96,97] The solubility of a given molecular species can be enhanced through change in temperature or entropic/enthalpic means (*e.g.* the solubility of mixtures of fullerenes in a given solvent can be greater than the solubility of each separate component).[98]

As the solvent evaporates, the solution concentration increases until it reaches saturation (*i.e.* the solubility limit). Beyond this point, the solute starts to aggregate (in some cases crystallising). As the donor and acceptor in OPV blends are likely to have slightly different solubilities, one of them may start solidifying at an earlier stage, leading to a rougher phase separation. Hence control of not only the solidification sequence, but also the solidification time, is important. This has been clearly demonstrated for semiconducting:insulating blends, but is also applicable to the solidification of OPV donor:acceptor systems.

If the miscibility of the donor and acceptor is large, a thoroughly mixed, often vitrified, solid phase is created,[27,63] especially when drying occurs rapidly. This can result in systems that feature efficient charge transfer, but which also often undergo prominent recombination if mixing is too pronounced due to the hindrance of the formation of relatively phase-pure regions that can act as charge extraction pathways.[12,15,20]

Because of the importance of the final phase morphology of OPV blends, *i.e.* the relative distribution and size of the phases present,[99–102] a large body of work has been devoted to controlling the drying kinetics of donor:acceptor systems to induce an optimum degree of phase separation.[65,103] Various pathways have been investigated, including post-deposition treatments[103–105] and/or the use of processing additives. For instance, in a vitrified donor-acceptor blend (vitrified, for example, because of a high mutual miscibility of the two components in the solvent system and/or rapid solvent removal that quenched the system below the relevant glass transition temperatures),[27,64] higher temperature or solvent annealing processes can provide the required molecular mobility of at least one component. This will allow these molecules to diffuse and aggregate, leading to a certain degree of phase separation. In this process, a knowledge of the glass transition temperatures of the individual components and that of the intermixed phase is important[63,64] and will assist in the selection of the most appropriate annealing temperatures. It is important to note that the components generally will phase separate over time, which can be disastrous for long-term stability of the resulting devices.

6.3.2 Use of Co-solvents and Solvent Additives

The use of co-solvents and solvent additives also provides a mechanism for controlling the progress of solidification. They often change the solubility of one, if not both, components, or the mutual miscibility between the donor and acceptor.[106–109] Other types of additive, such as nucleating agents (affecting the nucleation process) and/or organogelators have also been successfully applied.[74,80,110–112] Whatever their action, co-solvents and solvent additives are used because they generally assist in controlling the extent of phase separation. They either assist in inducing a smaller phase separation when the tendency of the donor and acceptor to segregate is too large, or *vice versa*, provoking some degree of phase separation when the two materials have a strong tendency to vitrify and/or mix (Figure 6.12).

Although formulating OPV 'inks' allows some flexibility with respect to the amount of additive used in polymer-based blends,[106] the additive content in small molecule OPVs often has to be selected judiciously because small deviations, often by 0.5 wt% or less, can cause the device performance to deteriorate drastically, as illustrated by the example of T1:PCBM:diiodooctane (DIO) solar cells.[107,113] Here, the use of an additional additive (a nucleation agent) in combination with the solvent additive DIO can be highly useful, rendering the resulting microstructure (and hence device performance) less sensitive to the processing conditions and deviations in blend formulation.[113]

Additives can also help to avoid the liquid–liquid phase separation that sometimes occurs during the drying of these solutions.[92] Liquid–liquid phase separation can often be identified from the micron-sized spherical domains that form in solution; these are undesirable in achieving good

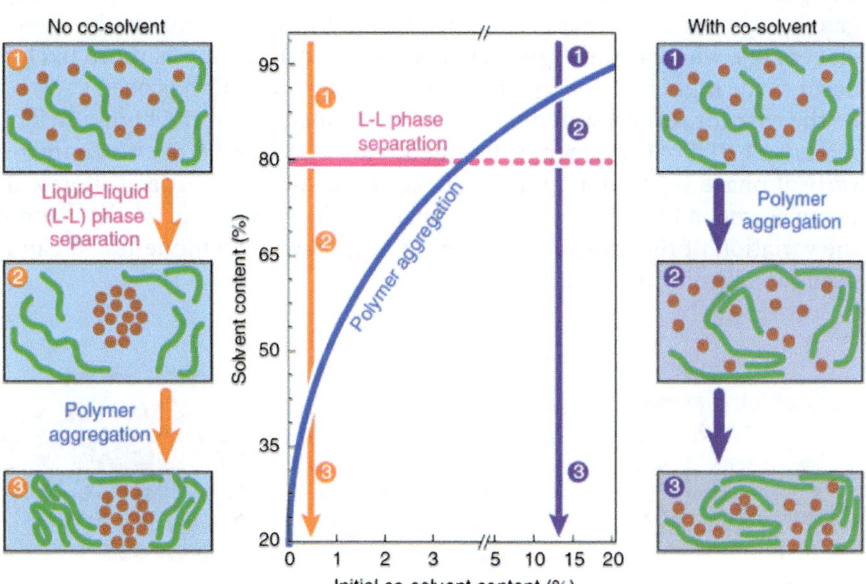

Figure 6.12 Illustration of how additives may affect the solidification of donor:acceptor blends. Reprinted by permission from Macmillan Publishers Ltd: *Nature Communications*, ref. 108, copyright 2015.

charge generation and charge extraction. This is the case, for instance, for DPP5T:[70]PCBM solutions in chloroform, which show liquid–liquid phase separation on the order of hundreds of nanometres. The use of ~2–4% of DIO or the addition of 3–20% of the co-solvent oDCB strongly supresses this undesirable phenomenon.[108] When using co-solvents, however, care must be taken to avoid undesirable Marangoni flows, which could result in inhomogeneous film thicknesses and lateral variations in concentration, although the latter can, to a certain degree, be avoided by the use of nucleating agents[74] as in the T1:PCBM:DIO system.[113]

6.3.3 Surface Energy Effects

Surface energy features can result in additional drying effects. For example, it has been shown that the surface energy of the substrate can induce vertical phase separation.[114–116] In specific work by Campoy *et al.*,[114] the composition near the substrate for blend films deposited on top of a hydrophilic substrate, *i.e.* poly(3,4-ethylenedioxythiophene) polystyrene sulfonate (PEDOT:PSS) or quartz, varied from polymer-rich, slightly fullerene-rich to largely fullerene-rich depending on the surface energy of the substrate.

The surface energy of the substrate can also induce specific nucleation phenomena at the interfaces (Figure 6.13).[117] For a typical semicrystalline conjugated polymer, the lowest free energy crystal faces are the alkyl-dominated (100) faces, while the π–π packing direction (010) is characterised by a higher energy.[29,32] This means that more pronounced edge-on packing of conjugated polymers can be expected at the air interface to minimise the interfacial energy, whereas the higher energy face-on orientation is more likely to occur at the higher free energy PEDOT:PSS interface. This has been shown to be the case for several systems, such as P3HT blends and the high-efficiency material PBTffT4T-2OD when blended with PC70BM.[118] In the context of vertical phase separation and the effect of the substrate surface energy, the use of solvent additives,[119] the inclusion of nucleation agents[74,80,112,113] and/or the variation of the molecular weight of at least one component[11,120] can be used to minimise these effects.

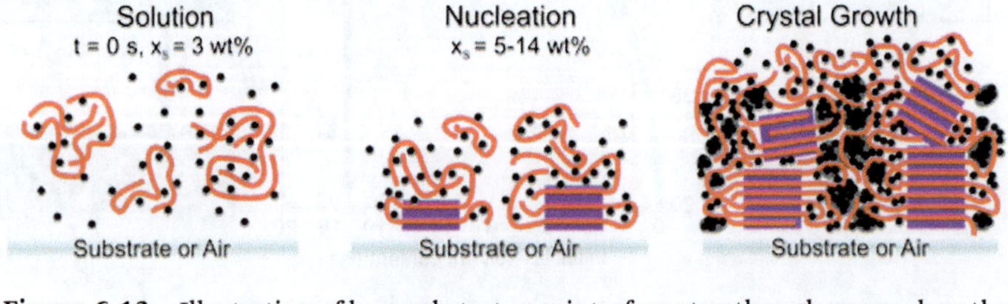

Figure 6.13 Illustration of how substrates or interfaces to other phases such as the air environment can affect film formation. Reprinted with permission from ref. 117. Copyright 2011 American Chemical Society.

6.3.4 Wetting and Film Formation Phenomena

Achieving a homogeneous thickness and composition in thin film structures is a technological challenge in which the interfaces, the solidification process and solution wetting play important parts.[119] The surface energy of the substrate, which can have a strong effect on the vertical distribution of the donor and acceptor components in OPV devices, also plays a crucial part in wetting. The spreading of a solution on top of an existing substrate/layer is dominated by surface energy interactions, often characterised by the wettability of the solution.

Wetting is defined by the IUPAC Gold Book as the 'process by which an interface between a solid and a gas is replaced by an interface between the same solid and a liquid'.[33] The wettability of an ink on a given substrate can be predicted from the surface tension of the ink and the surface energy wetting envelope of the target substrate. The wettability of a solid surface by a liquid can be deduced from the shape and size of the wetting envelope according to the Owens–Wendt–Kälble method. The area covered by the wetting envelope is determined from the dispersive and polar components of the surface tension, which can be experimentally measured using, for example, sessile and hanging drop methods. The boundaries of the resulting area define the wettability region in terms of the surface tension that a liquid must overcome to completely wet the solid surface.[121–123] The polar and dispersive components of the surface tension of the solution can be measured by combining the pendant droplet method and the contact angle on several substrates using the sessile droplet method. These are required for the definition of the wetting envelope within the Fowkes approach. Accordingly, the effect of surface treatments, such as plasma treatments and other cleaning protocols, on the wettability characteristics of substrates can be assessed. In the case of an improvement, the wetting envelope is found to broaden. Analogously, we can study whether co-solvents or processing additives can affect wetting of a given ink formulation.

Other phenomena can also play an important part. During film drying, the so-called coffee stain effect and Marangoni flows are important phenomena that need to be considered in the formation of homogenous films, especially over larger areas. The coffee ring effect describes a situation in which most of the solute material is deposited at the rim of the drop or wet film, favouring the appearance of defects in the concave central area. Physically, it is the consequence of unbalanced solvent evaporation during the drying process and of a pinned contact line, generating an internal solvent flow within the drop from its central region towards the interfacial edges with the substrate. Such an effect may be minimised and a few strategies have been reported. Kajiya *et al.* increased the vapour pressure around the drying feature by exposing the droplets to the solvent vapour for a controlled time; in this way, they rendered the ring-like profiles of the droplets fully flat.[124] Oh *et al.* introduced a minor fraction of co-solvent with a higher boiling point and lower surface tension than the majority solvent of the ink formulation with analogous success.[125] A similar approach has been used by Tseng *et al.* when inkjet printing TIPS-pentacene using a dichlorobenzene:dodecane solvent mixture (3 : 1 volume ratio). This led to uniform films in organic thin fim transistors.[126] Another strategy relied on the use of nucleating agents, where the addition of a minute amount of additive allowed the realisation of inkjet-printed TIPS-pentacene devices on very high surface energy substrates.[74]

Marangoni[33] forces appear when there is a gradient in surface tension, such as in a drying wet film cast from two solvents with different boiling points. The use of low boiling point solvents results in Marangoni forces, which favour spreading and act against aggregation during drying, thus compensating for coffee stain effects. By contrast, surfactants are able to disrupt the Marangoni forces driven by surface tension when present at low concentrations (from as little as 0.1 vol%).

Another highly undesirable effect that needs to be prevented in OPV thin film deposition is when the ink re-dissolves or swells part of the underlying layer/material, leading to strong intermixing or even the complete removal of the underlying layer. Therefore it is of utmost importance to use orthogonal solvents, *i.e.* solvents that do not affect the underlying layer; however, this is not always a simple task because it can lead to low wettability (and adhesion of the resulting film). Thus a careful balance is required for a practical solution that assures the maximum quality at the interfaces. Chemical modification of the first-produced layer can assist the deposition of the second layer. It is important to note, however, that even after the successful fabrication of multilayers, the interdiffusion of matter between the layers can lead to less well-defined interfaces and, in some cases, degradation. It is known, for instance, that the acidic nature of the widely used conductive polymer PEDOT:PSS can degrade both the underlying indium tin oxide and the active semiconducting layer. Indium atoms may also diffuse into the PEDOT:PSS as well as the organic semiconductor.

A multitude of other phenomena can occur, although these are often highly dependent on the deposition method. Examples include bulging, satellite droplet formation and overfilling in the case of inkjet printing, or viscous fingering (Saffman–Taylor instability)[127,128] for roll-to-roll transfer techniques (gravure, offset or flexographic printing).

6.4 Formation of Solid State Structure in OPV Blends

The performance of an OPV blend is determined by the precise distribution of its individual components. This distribution decides the size of the domains formed, their purity and the potential preferred orientation of more ordered regions. Techniques that enable characterisation of the often complex resulting phase morphologies have received considerable attention over the last decade, leading to important information on which factors dictate the formation of the solid state structure and how this affects charge generation, recombination and extraction.

6.4.1 Solid State Miscibility and the Presence of Mixed Phases

Early work on BHJ blends promoted the picture of an interpenetrating network of a pure donor and a pure acceptor phase.[129] It is now widely established that the majority of donor:acceptor blends feature a mixed phase, *i.e.* a number of distinct phases can coexist in OPV blend films.[130] 'Phases', according to the IUPAC Gold Book, describe 'an entity of a material system which is uniform in chemical composition and physical state'.[33] Regions composed of a certain phase are referred to as phase domains, with their make-up and connectivity strongly dependent on both their kinetic and thermodynamic properties. For the latter, we have to consider whether the individual components: (1) are completely miscible, leading to a one-phase system, with the IUPAC Gold Book defining miscibility as the 'capability of a mixture to form a single phase over certain ranges of temperature, pressure, and composition';[33] (2) are only partially miscible, leading to a three-phase morphology where intermixed regions coexist with relatively phase-pure phases of the individual components; or (3) are completely immiscible, leading, in binary systems, to a two-phase morphology[91] (see Figure 6.14). It is therefore important to consider that fast solidification rates and/or high compatibilisation of the components in solution can quench a system into a fully, or at least partly, mixed state, even when they have a low miscibility in the solid state.

Consequently, the majority of donor:acceptor blends within OPVs feature, in addition to relatively phase-pure domains, a mixed phase that can arise due to two principal reasons: (1) the blend components are at least partially miscible; and/or (2) the blend becomes vitrified because of rapid solidification and/or because the initially homogeneous distribution of the blend components in the solution state is 'frozen in' below T_g, which results in a molecular glass.[27,131] It is important to distinguish between these two effects because if the mixed phase forms because of high miscibility between the two components, it is thermodynamically stable over a certain range of compositions and temperatures, whereas a kinetically trapped nanostructure can coarsen over time, especially when heated above T_g.

For polymer:fullerene blends the presence of a mixed phase was first deduced from spectroscopic measurements, which indicated ultra-fast charge transfer from the acceptor to the donor in a time frame <200 fs.[132,133] These data could not be explained by the presence of two relatively pure phases, which would require excitons to diffuse to an interface between relatively phase-pure domains formed by the two components, leading to delayed exciton dissociation. It took a few years until direct evidence was reported of the presence of relatively high miscibility between specific donors and acceptors, which can lead to intermixed phases. Treat *et al.* found, for instance,

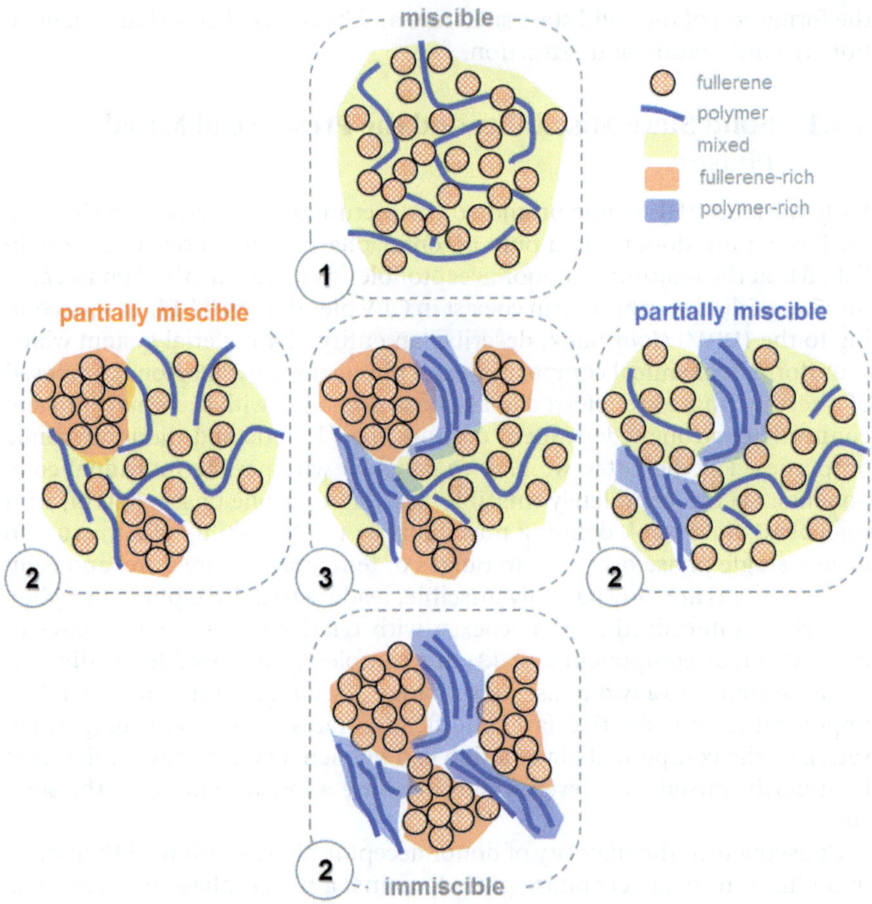

Figure 6.14 Schematic diagrams showing representations of the different phase morphologies possible within two-component systems such as donor:acceptor blends. In the case of full miscibility, a one-phase morphology is obtained, whereas in a scenario where the two components are not at all miscible, two phases are obtained. Interim morphologies will occur in partially miscible systems, from two-phase morphologies where an intermixed phase coexists with the relatively phase-pure domains of one of the components, to three-phase morphologies where both components form relatively phase-pure regions in addition to an intermixed phase.

that deuterated PCBM can laterally diffuse into an initially pristine layer of P3HT; the diffusion coefficient D_{PCBM} of PCBM increases by a factor of 50 when increasing the temperature from 50 to 110 °C.[134,135] Fischer *et al.* investigated the influence of cross-linking on the uptake of C_{60} by a polyfluorene derivative and found that D_{PCBM} decreases by three orders of magnitude with increasing cross-linking density.[136] Another interesting approach was to monitor the local composition around micrometre-sized fullerene crystals, which tend to grow in polymer:fullerene thin films on thermal annealing. The diffusion of PCBM to the growth front leads to a gradient that could be described by Fick's law of diffusion. The concentration of fullerene remaining in the polymer phase close to the crystal surface was identified as the amount of PCBM that is miscible at the annealing temperature, *i.e.* 19% PCBM at 140 °C in a blend with P3HT.[104] Other evidence was revealed in a study combining transient absorption, UV–visible and photoluminescence spectrometry with dynamic secondary ion mass spectrometry and high-angle annular dark field scanning transmission electron micrsocopy, which provided information about how diffusion, phase separation and, often, the interplay between crystallisation and vitrification influence important photophysical processes. These include polaron formation and geminate recombination and were shown to be tuneable by the processing routes used and the characteristics of the materials used (*e.g.* molecular weight, side-chain substitution).[130]

6.4.2 Use of Phase Diagrams

One useful method of gaining insights into the multifaceted structure of OPV blends is the use of temperature/composition phase diagrams. They often allow a picture to be acquired of the microstructure that forms directly from solution, what phase transformations can occur in the solid state and what phases may form. Information on potential kinetic processes can also be obtained.[91]

One relatively convenient technique to establish temperature/composition phase diagrams for organic matter is *via* differential scanning calorimetry, including flash differential scanning calorimetry measurements that allow the characterisation of thin films.[137] The phase diagrams of several polymer:fullerene blends have been constructed in this manner, most prominently for P3HT:PCBM (Figure 6.15).[15,131,138,139]

Both regioregular P3HT and PCBM feature a distinct melting endotherm with a peak around 230 and 280 °C, respectively, which assists in establishing the phase diagram. Following the development of the melting temperature with composition gives an insight into whether the components are miscible. In the case of P3HT:PCBM, blending with the second component leads to a considerable reduction in melting temperature for both species. This is a typical behaviour for a eutectic system, *i.e.* a system that displays full miscibility in the liquid state and at least partial immiscibility of the two components in the solid state.[29,91] A composition/temperature combination therefore exists, the so-called eutectic point, where 100% of the liquid will undergo a transition into two solid phases. As this process is diffusion-limited, eutectic decomposition leads to finely phase-separated structures.

The eutectic composition (c_e) and eutectic temperature (T_e) for the P3HT:PCBM system are at ~65 wt% P3HT and ~205 °C, respectively.[131] The optimum photovoltaic performance is typically observed close to the

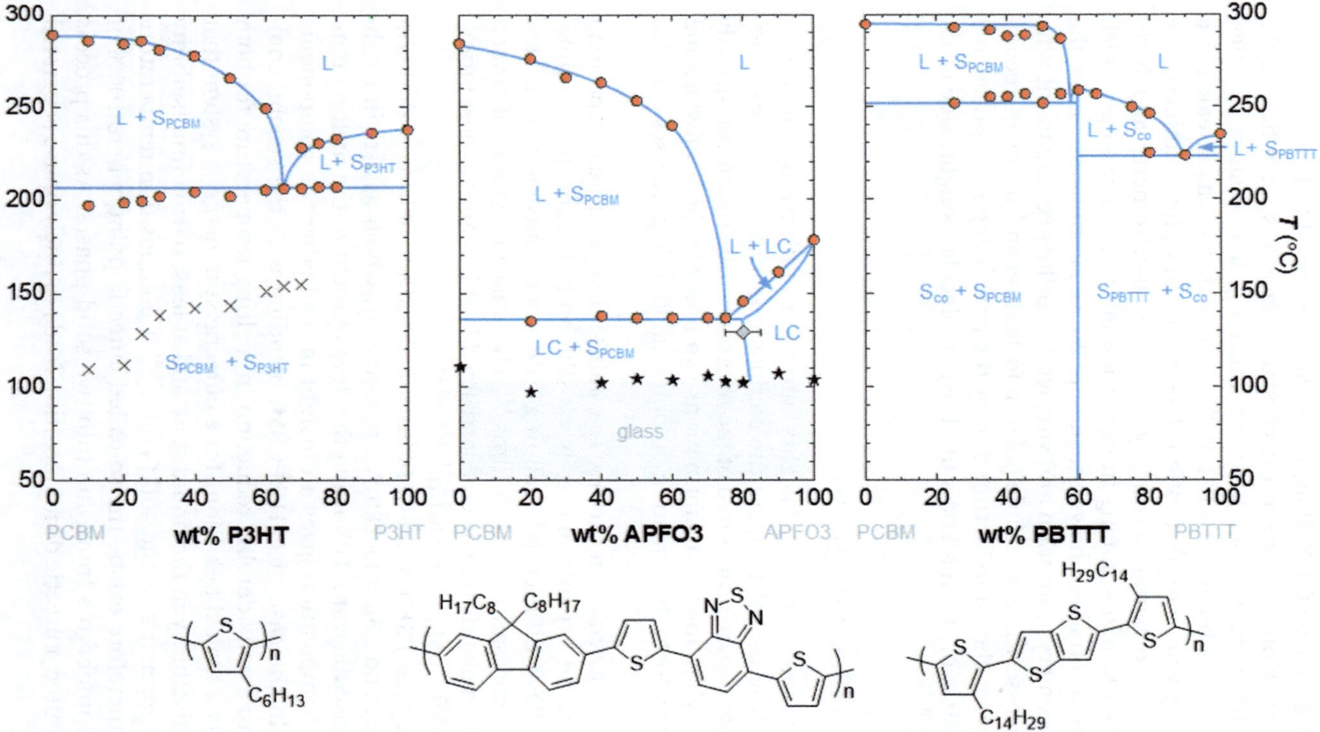

Figure 6.15 Phase diagrams of three different donor polymer systems (P3HT, APFO3 and PBTTT) with PCBM (left, adapted with permission from C. Müller *et al.*, Binary Organic Photovoltaic Blends: A Simple Rationale for Optimum Compositions, *Advanced Materials*,[131] John Wiley and Sons, Copyright © 2008 WILEY-VCH Verlag GmbH & Co. KGaA, Weinheim; centre, adapted from ref. 67 with permission from The Royal Society of Chemistry; and right, adapted from ref. 15 with permission from The Royal Society of Chemistry).

eutectic composition, *i.e.* at 50–60 wt% P3HT. A slight excess of the acceptor appears to be beneficial because of both its positive impact on electron transport and in assisting charge generation, as indicated by work on the related PBTTT:PCBM system.[20]

Because of the way phase diagrams are established, it is important to note that no direct information on the presence of potential amorphous phases – such as amorphous intermixed phases – is obtained. This is because the phase diagrams are established from endotherms and exotherms measured by differential scanning calorimetry and these are related to the melting and crystallisation processes of the crystalline moieties in the system.

Other characterisation methods may also be used. One effective option is variable-temperature ellipsometry, which allows direct monitoring of the thermal evolution of thin films,[67,140–142] thus providing information on potential thin film/substrate interactions that change the phase behaviour and/or confinement effects. Information on phase transitions, including the melting temperature, glass transition temperature and eutectic temperature can also be obtained and can be compared with differential scanning calorimetry data if these are available.

To illustrate the usefulness of variable-temperature ellipsometry, we discuss here the example of APFO3:PCBM thin films. For this system, a distinct T_g ~98 °C was found, with a T_e ~ 138 °C and a eutectic point at c_e ~ 75 wt% APFO3 using variable-temperature ellipsometry.[67] A mixed phase consisting of an APFO3 matrix that dissolved up to 20 wt% PCBM was inferred from photo- and electroluminescence spectrometry, with strong quenching of the polymer emission indicating intimate mixing of the donor and acceptor. PCBM-rich blends with a composition $c < c_e$ featured phase-separated, fullerene-rich domains embedded in the mixed phase as for P3HT:PCBM; a high fullerene content of 80 wt% PCBM was required to maximise the photovoltaic performance (as expected from the phase diagram). On thermal annealing, however, the growth of micrometre-sized fullerene crystals was observed.[15,20,67,100,102,143,144]

Phase diagrams can also predict more complex phase behaviour, including compound formation. For instance, donor:acceptor combinations have been reported that form co-crystals. This means that, at certain compositions, both blend components are incorporated into a single ordered phase. One prominent example of this scenario are blends consisting of the polythiophene derivative PBTTT, which allows the intercalation of PCBM between its side-chains.[145] Jamieson *et al.* studied the temperature/composition phase behaviour of this system and found that two eutectic compositions occur in this binary system (Figure 6.15): one eutectic is formed by PBTTT and the PBTTT:PCBM co-crystal and the other eutectic by the co-crystal and the fullerene.[15] The first eutectic point occurs at a temperature T_{e1} ~ 250 °C and a composition c_{e1} ~ 60 wt%, whereas the second eutectic point was identified to be at T_{e2} ~ 220 °C and c_{e2} ~ 90 wt% PBTTT. We can deduce from the phase diagram that at temperatures <220 °C, the PBTTT and PCBM phases do not coexist. Either PBTTT phases are present beside co-crystal domains, or co-crystal regions occur together with PCBM. It is in the latter regime that efficient charge separation is observed.

These examples illustrate that non-equilibrium temperature/composition diagrams (deduced for specific processing conditions) enable the establishment of protocols to dictate the solidification of OPV blends. These allow tailoring of the final phase morphology and enable manipulation of the overall solid state microstructure from the outset. More specifically, phase diagrams assist in providing guidelines about materials design (*e.g.* how the thermal properties need to be manipulated to realise a specific eutectic composition), processing (*e.g.* what annealing temperatures to use) and device stability.[146] This has rendered them a powerful tool in identifying promising material combinations and blend compositions, in addition to giving insights into the origin of the vitrification of specific blends and why specific binaries lead to inferior device performance than expected from their optoelectronic characteristics.

6.4.3 Thermal Stability of OPV Blends

In many cases, the highest photovoltaic performance for a given OPV blend is realised from a processing protocol that freezes the system into a metastable state. Vitrification of P3HT:fullerene blends are good examples of this, as are more recent systems based on PCE11, where the low miscibility of the donor and acceptor leads to a rapid burn-in of device performance.[146] Specific processing methodologies can further aggravate this phenomenon. Spin-coating often extracts the solvent at a too high a rate, whereas high-throughput coating techniques usually require the rapid removal of the solvent through heating, leading to fast evaporation. This means that intermixed microstructures are obtained, despite the two components being of low mutual miscibility in the solid state. As a consequence, the donor and acceptor will eventually phase-separate, as the exemplary electron micrographs of the thiophene–quinoxaline copolymer TQ1 blended with PCBM indicate (Figure 6.16). This process will proceed faster at higher temperatures. Because industrial standards demand that solar cells can withstand cycling between −40 and +85 °C to be certified as thermally stable, this metastability can be highly undesirable, as the PCE11:PCBM example prominently illustrates.

Various strategies exist to at least slow down the degradation process – especially if it is based on a phase separation process over time. Polymers with a high glass transition temperature of at least 100 °C may be used.[64,67,147–150] In such systems, the molecular dynamics and mobility of the materials at room temperature will be low, limiting phase separation. A compromise must be found with this approach because, usually, a high T_g is a consequence of stiffer backbones, which lowers the material's entropy and hence tendency for mixing. As T_g depends on the molecular weight, it might be possible to increase T_g without varying the backbone structure by increasing the molecular weight, although this also leads to a lower miscibility.

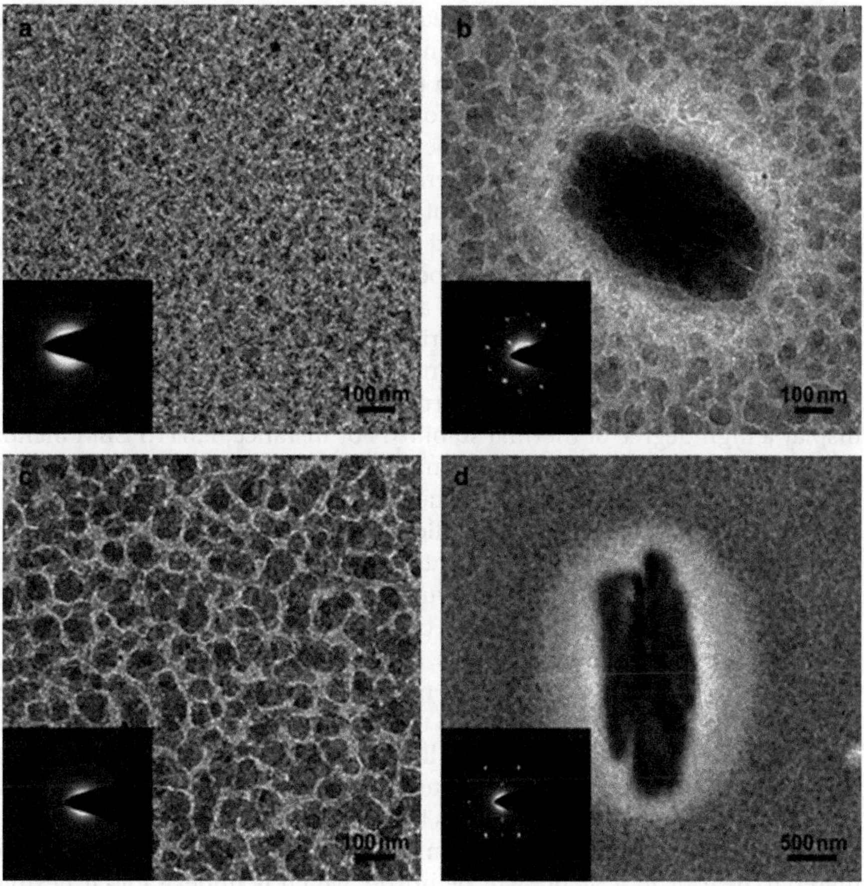

Figure 6.16 Transmission electron micrographs and electron diffraction patterns (insets) of a 1 : 1 TQ1 : PCBM film that displays a $T_g^{(\text{intermixed phase})} \approx$ 110 °C: (a) as-cast; (b, c) annealed at 130 °C; and (d) annealed at 150 °C. Reproduced from ref. 149 with permission from The Royal Society of Chemistry.

The glass transition temperature of the acceptor can also be varied. Fullerene derivatives, which generally have T_g >100 °C and show at least partial miscibility with many donor polymers, are strong vitrifiers for the latter.[27,63,64] As a consequence, binaries with fullerene derivatives as the acceptor often lead to intermixed systems, often with the formation of only a small fraction of phase-pure domains. The amount of intermixing (and phase-pure domains) is thereby dictated by thermodynamic and kinetic factors, as has been shown in a comparison of binary blends of P3HT with PCBM, bis-PCBM and ICBA.[27] It is often not the T_g of the pure component that dictates the final behaviour of such blend systems over time, but the T_g of the intermixed phase, which is in between the T_g of the pure components.[63,64,151] This is because molecular diffusion must occur in this specific phase for phase separation to be initiated. In addition, it is crucial to realise, especially in the context of device stability, that the T_g does not represent a sharp transition, but rather a temperature range over which relaxation processes start to occur on an experimentally accessible timescale. Therefore, given enough time, donor:acceptor blend microstructures can evolve, even below T_g.[150] Thus for long-term stability, a T_g significantly above the operating temperature should be targeted, as demonstrated for a ternary blend of PTB7:C_{60}:C_{70} with $T_g > 200$ °C.[152] However, as alluded to earlier, this should be realised without drastically changing the compatibility of the two components, which is an intricate task.

Mechanistically, the coarsening of a donor:acceptor blend structure over time can predominantly have two origins: spinodal (or binodal) demixing and crystallisation. Whereas the former is a continuous process that is difficult to avoid, the latter results in structures which, if carefully chosen, can display a high degree of thermal stability. For instance, P3HT:PCBM blends are relatively resilient to further thermal stress once the blend has been thermally annealed, which results in crystallisation of the donor polymer.[103] Control over the crystallisation of the fullerene acceptor is more challenging to achieve due to a very low nucleation rate,[149] leading to the detrimental growth of micrometre-sized crystals that disrupt the active layer of the solar cell (Figure 6.16). Routes that limit fullerene crystallisation include glass formation through the use of fullerene mixtures, and the addition of nucleating agents, such as C_{60} or commercially available nucleating agents developed for isotactic polypropylene.[74,80,111,149,152,153] The use of the latter nucleating agents can also result in a finer distribution of fullerene crystallites, assist in controlling vertical phase separation and/or the texture of crystalline moieties, or accelerate the crystallisation process, thus limiting vitrification.[74,80,111,112]

T_g also strongly affects the mechanical properties. Below T_g, any organic material, neat or in a blend, will be brittle, which is undesirable if flexible applications are targeted. Blends are more ductile above T_g and thus display improved bonding to adjacent layers, such as electrodes in a photovoltaic device.[81,154] Thus a compromise must be found between thermal stability and ductility, which are more readily obtained below and above T_g, respectively.

6.5 Conclusions

The microstructure and phase morphology of OPV blends can be highly complex, with many aspects that have to be taken into account, both at the molecular level (such as the molecular conformation) and at the micrometre scale (such as the volume fraction and composition of the phases that form, the phase distribution and the domain size). All these structural features define the electronic landscape of the final architecture, but can also have significant effects on other important characteristics, including their mechanical properties.

We have discussed the relevant strategies and approaches that can be used to manipulate molecular ordering, *e.g.* through chemical design and/or a knowledge of the relevant phase transitions and the solidification process in general. Our overall goal was to provide some insights into the structural features of single- and multicomponent systems that are relevant to the function of photovoltaic devices from a materials science point of view for use by a broad range of researchers at the interface of chemistry, materials science, physics and device engineering and to enable future progress in the rapidly maturing field of OPVs.

References

1. J. Nelson, *Mater. Today*, 2011, **14**, 462–470.
2. M. D. McGehee and M. A. Topinka, *Nat. Mater.*, 2006, **5**, 675–676.
3. J. Nelson, *Curr. Opin. Solid State Mater. Sci.*, 2002, **6**, 87–95.
4. R. H. Friend and J. J. M. Halls, in *Clean Electricity from Photovoltaics (Series on Photoconversion of Solar Energy, Volume 1)*, ed. M. D. Archer and R. D. Hill, Imperial College Press, London, UK, 2001, pp. 377–445.
5. H. Hoppe and N. S. Sariciftci, *J. Mater. Res.*, 2004, **19**, 1925–1945.
6. M. A. Faist, S. Shoaee, S. Tuladhar, G. F. A. Dibb, S. Foster, W. Gong, T. Kirchartz, D. D. C. Bradley, J. R. Durrant and J. Nelson, *Adv. Energy Mater.*, 2013, **3**, 744–752.
7. P. A. Troshin, H. Hoppe, J. Renz, M. Egginger, J. Y. Mayorova, A. E. Goryachev, A. S. Peregudov, R. N. Lyubovskaya, G. Gobsch, N. S. Sariciftci and V. F. Razumov, *Adv. Funct. Mater.*, 2009, **19**, 779–788.
8. Z. Fei, R. S. Ashraf, Z. Huang, J. Smith, R. J. Kline, P. D'Angelo, T. D. Anthopoulos, J. R. Durrant, I. McCulloch and M. Heeney, *Chem. Commun.*, 2012, **48**, 2955–2957.
9. C. M. Amb, S. Chen, K. R. Graham, J. Subbiah, C. E. Small, F. So and J. R. Reynolds, *J. Am. Chem. Soc.*, 2011, **133**, 10062–10065.
10. A. A. Y. Guilbert, L. X. Reynolds, A. Bruno, A. MacLachlan, S. P. King, M. A. Faist, E. Pires, J. Emyr Macdonald, N. Stingelin, S. A. Haque and J. Nelson, *ACS Nano*, 2012, **6**, 3868–3875.
11. W. Ma, G. Yang, K. Jiang, J. H. Carpenter, Y. Wu, X. Meng, T. McAfee, J. Zhao, C. Zhu, C. Wang, H. Ade and H. Yan, *Adv. Energy Mater.*, 2015, **5**, 1501400.
12. Y. Liu, J. Zhao, Z. Li, C. Mu, W. Ma, H. Hu, K. Jiang, H. Lin, H. Ade and H. Yan, *Nat. Commun.*, 2014, **5**, 5293.
13. C. Silva, *Nat. Mater.*, 2013, **12**, 5–6.
14. T. M. Burke and M. D. McGehee, *Adv. Mater.*, 2014, **26**, 1923–1928.
15. F. C. Jamieson, E. B. Domingo, T. McCarthy-Ward, M. Heeney, N. Stingelin and J. R. Durrant, *Chem. Sci.*, 2012, **3**, 485–492.
16. A. A. Bakulin, A. Rao, V. G. Pavelyev, P. H. M. van Loosdrecht, M. S. Pshenichnikov, D. Niedzialek, J. Cornil, D. Beljonne and R. H. Friend, *Science*, 2012, **335**, 1340–1344.
17. J. Lee, K. Vandewal, S. R. Yost, M. E. Bahlke, L. Goris, M. A. Baldo, J. V. Manca and T. Van Voorhis, *J. Am. Chem. Soc.*, 2010, **132**, 11878–11880.
18. G. Grancini, M. Maiuri, D. Fazzi, A. Petrozza, H.-J. Egelhaaf, D. Brida, G. Cerullo, G. Lanzani, M. Binda, L. Criante and S. Perissinotto, *Nat. Mater.*, 2013, **12**, 29–33.

19. K. R. Graham, C. Cabanetos, J. P. Jahnke, M. N. Idso, A. El Labban, G. O. N. Ndjawa, T. Heumueller, K. Vandewal, A. Salleo, B. F. Chmelka, A. Amassian, P. M. Beaujuge and M. D. McGehee, *J. Am. Chem. Soc.*, 2014, **136**, 9608–9618.
20. E. Buchaca-Domingo, A. J. Ferguson, F. C. Jamieson, T. McCarthy-Ward, S. Shoaee, J. R. Tumbleston, O. G. Reid, L. Yu, M.-B. Madec, M. Pfannmöller, F. Hermerschmidt, R. R. Schröder, S. E. Watkins, N. Kopidakis, G. Portale, A. Amassian, M. Heeney, H. Ade, G. Rumbles, J. R. Durrant and N. Stingelin, *Mater. Horiz.*, 2014, **1**, 270–279.
21. R. Noriega, J. Rivnay, K. Vandewal, F. P. V. Koch, N. Stingelin, P. Smith, M. F. Toney and A. Salleo, *Nat. Mater.*, 2013, **12**, 1038–1044.
22. F. Liu, D. Chen, C. Wang, K. Luo, W. Gu, A. L. Briseno, J. W. P. Hsu and T. P. Russell, *ACS Appl. Mater. Interfaces*, 2014, **6**, 19876–19887.
23. A. M. Ballantyne, L. Chen, J. Dane, T. Hammant, F. M. Braun, M. Heeney, W. Duffy, I. McCulloch, D. D. C. Bradley and J. Nelson, *Adv. Funct. Mater.*, 2008, **18**, 2373–2380.
24. C. Liu, K. Wang, X. Hu, Y. Yang, C. Hsu, W. Zhang, S. Xiao, X. Gong and Y. Cao, *ACS Appl. Mater. Interfaces*, 2013, **5**, 12163–12167.
25. B. P. Rand, D. Cheyns, K. Vasseur, N. C. Giebink, S. Mothy, Y. Yi, V. Coropceanu, D. Beljonne, J. Cornil, J. L. Brédas and J. Genoe, *Adv. Funct. Mater.*, 2012, **22**, 2987–2995.
26. M. S. Vezie, S. Few, I. Meager, G. Pieridou, B. Dörling, R. Shahid, A. R. Goñi, H. Bronstein, I. Mcculloch, S. C. Hayes and J. Nelson, *Nat. Mater.*, 2016, **15**, 746–753.
27. P. Westacott, N. D. Treat, J. Martin, J. H. Bannock, J. C. de Mello, M. Chabinyc, A. B. Sieval, J. J. Michels and N. Stingelin, *J. Mater. Chem. A*, 2017, **5**, 2689–2700.
28. J. R. Tumbleston, B. A. Collins, L. Yang, A. C. Stuart, E. Gann, W. Ma, W. You and H. Ade, *Nat. Photonics*, 2014, **8**, 385–391.
29. B. Wunderlich, *Macromolecular Physics, Vol. 1: Crystal Structure, Morphology, Defects*, Academic Press, New York, USA, 1973.
30. F. P. V. Koch, J. Rivnay, S. Foster, C. Müller, J. M. Downing, E. Buchaca-Domingo, P. Westacott, L. Yu, M. Yuan, M. Baklar, Z. Fei, C. Luscombe, M. A. McLachlan, M. Heeney, G. Rumbles, C. Silva, A. Salleo, J. Nelson, P. Smith and N. Stingelin, *Prog. Polym. Sci.*, 2013, **38**, 1978–1989.
31. W. H. Carothers, *Trans. Faraday Soc.*, 1936, **32**, 39–49.
32. A. A. Virkar, S. Mannsfeld, Z. Bao and N. Stingelin, *Adv. Mater.*, 2010, **22**, 3857–3875.
33. A. Wilkinson and A. D. McNaught, *IUPAC. Compendium of Chemical Terminology (The 'Gold Book')*, XML on-line corrected version, http://goldbook.iupac.org (2006-) created by M. Nic, J. Jirat, B. Kosata, updates compiled by A. Jenkins, Last update: 2014-02-24, version: 2.3.3, Blackwell Scientific Publications, Oxford, UK, 2nd edn, 1997.
34. N. D. Treat, P. Westacott and N. Stingelin, *Annu. Rev. Mater. Res.*, 2015, **45**, 459–490.

35. W. A. Little, *Phys. Rev.*, 1964, **134**, 1416–1424.
36. M. Pope and C. E. Swenberg, *Electronic Processes in Organic Crystals and Polymers*, Oxford University Press, Oxford, UK, 2nd edn, 1999.
37. R. Kroon, D. A. Mengistie, D. Kiefer, J. Hynynen, J. D. Ryan, L. Yu and C. Müller, *Chem. Soc. Rev.*, 2016, **45**, 6147–6164.
38. H. A. M. van Mullekom, J. A. J. M. Vekemans, E. E. Havinga and E. W. Meijer, *Mater. Sci. Eng.*, 2001, **32**, 1–40.
39. C. Liu, K. Wang, X. Gong and A. J. Heeger, *Chem. Soc. Rev.*, 2016, **45**, 4825–4846.
40. M. M. Wienk, M. Turbiez, J. Gilot and R. A. J. Janssen, *Adv. Mater.*, 2008, **20**, 2556–2560.
41. J. Hou, H.-Y. Chen, S. Zhang, R. I. Chen, Y. Yang, Y. Wu and G. Li, *J. Am. Chem. Soc.*, 2009, **131**, 15586–15587.
42. F. Wudl, P. M. Allemand, G. Srdamov and D. McBranch, *Materials for Nonlinear Optics*, 1991, ACS Sympos, pp. 683–688.
43. S. Hotta, S. Rughooputh, A. Heeger and F. Wudl, *Macromolecules*, 1987, **20**, 212–215.
44. B. S. Gaylord, A. J. Heeger and G. C. Bazan, *Proc. Natl. Acad. Sci. U. S. A.*, 2002, **99**, 10954–10957.
45. L. Chen, D. W. McBranch, H.-L. Wang, R. Helgeson, F. Wudl and D. G. Whitten, *Proc. Natl. Acad. Sci. U. S. A.*, 1999, **96**, 12287–12292.
46. R. Søndergaard, M. Helgesen, M. Jørgensen and F. C. Krebs, *Adv. Energy Mater.*, 2011, **1**, 68–71.
47. P. J. Flory, *Principles of Polymer Chemistry*, Cornell University Press, Ithaca, NY, 1953.
48. P. Wolfer, A. Armin, A. Pivrikas, M. Velusamy, P. L. Burn and P. Meredith, *J. Mater. Chem. C*, 2014, **2**, 71.
49. A. J. Clulow, E. P. Gilbert, P. Wolfer, P. L. Burn and I. R. Gentle, *Macromolecules*, 2015, **48**, 8331–8336.
50. H. Hu, K. Zhao, N. Fernandes, P. Boufflet, J. H. Bannock, L. Yu, J. C. de Mello, N. Stingelin, M. Heeney, E. P. Giannelis and A. Amassian, *J. Mater. Chem. C*, 2015, **3**, 7394–7404.
51. R. D. McCullough and R. D. Lowe, *J. Chem. Soc., Chem. Commun.*, 1992, 70–72.
52. X. M. Jiang, R. Osterbacka, O. Korovyanko, C. P. An, B. Horovitz, R. A. J. Janssen and Z. V. Vardeny, *Adv. Funct. Mater.*, 2002, **12**, 587.
53. W. C. Tsoi, S. J. Spencer, L. Yang, A. M. Ballantyne, P. G. Nicholson, A. Turnbull, A. G. Shard, C. E. Murphy, D. D. C. Bradley, J. Nelson and J. S. Kim, *Macromolecules*, 2011, **44**, 2944–2952.
54. K. H. Hendriks, W. Li, G. H. L. Heintges, G. W. P. Van Pruissen, M. M. Wienk and R. A. J. Janssen, *J. Am. Chem. Soc.*, 2014, **136**, 11128–11133.
55. T. Vangerven, P. Verstappen, J. Drijkoningen, W. Dierckx, S. Himmelberger, A. Salleo, D. Vanderzande, W. Maes and J. V. Manca, *Chem. Mater.*, 2015, **27**, 3726–3732.
56. L. Lu, T. Zheng, T. Xu, D. Zhao and L. Yu, *Chem. Mater.*, 2015, **27**, 537–543.

57. F. Lombeck, H. Komber, D. Fazzi, D. Nava, J. Kuhlmann, D. Stegerer, K. Strassel, J. Brandt, A. D. de Zerio Mendaza, C. Müller, W. Thiel, M. Caironi, R. Friend and M. Sommer, *Adv. Energy Mater.*, 2016, **6**, 1–9.
58. F. P. V Koch, P. Smith and M. Heeney, *J. Am. Chem. Soc.*, 2013, **135**, 13695–13698.
59. F. P. V. Koch, M. Heeney and P. Smith, *J. Am. Chem. Soc.*, 2013, **135**, 13699–13709.
60. O. G. Reid, J. A. N. Malik, G. Latini, S. Dayal, N. Kopidakis, C. Silva, N. Stingelin and G. Rumbles, *J. Polym. Sci., Part B: Polym. Phys.*, 2012, **50**, 27–37.
61. D. J. D. Moet, M. Lenes, J. D. Kotlarski, S. C. Veenstra, J. Sweelssen, M. M. Koetse, B. de Boer and P. W. M. Blom, *Org. Electron.*, 2009, **10**, 1275–1281.
62. C. Müller, E. Wang, L. M. Andersson, K. Tvingstedt, Y. Zhou, M. R. Andersson and O. Inganäs, *Adv. Funct. Mater.*, 2010, **20**, 2124–2131.
63. D. Leman, M. A. Kelly, S. Ness, S. Engmann, A. Herzing, C. Snyder, H. W. Ro, R. J. Kline, D. M. DeLongchamp and L. J. Richter, *Macromolecules*, 2015, **48**, 383–392.
64. C. Müller, *Chem. Mater.*, 2015, **27**, 2740–2754.
65. G. Li, V. Shrotriya, J. Huang, Y. Yao, T. Moriarty, K. Emery and Y. Yang, *Nat. Mater.*, 2005, **4**, 864–868.
66. J. G. Labram, E. B. Domingo, N. Stingelin, D. D. C. Bradley and T. D. Anthopoulos, *Adv. Funct. Mater.*, 2011, **21**, 356–363.
67. C. Müller, J. Bergqvist, K. Vandewal, K. Tvingstedt, A. S. Anselmo, R. Magnusson, M. I. Alonso, E. Moons, H. Arwin, M. Campoy-Quiles and O. Inganäs, *J. Mater. Chem.*, 2011, **21**, 10676.
68. I. Botiz and N. Stingelin, *Materials*, 2014, 7, 2273–2300.
69. I. McCulloch, M. Heeney, C. Bailey, K. Genevicius, I. Macdonald, M. Shkunov, D. Sparrowe, S. Tierney, R. Wagner, W. Zhang, M. L. Chabinyc, R. J. Kline, M. D. McGehee and M. F. Toney, *Nat. Mater.*, 2006, **5**, 328–333.
70. S. A. Mollinger, B. A. Krajina, R. Noriega, A. Salleo and A. J. Spakowitz, *ACS Macro Lett.*, 2015, **4**, 708–712.
71. M. A. Baklar, F. Koch, A. Kumar, E. B. Domingo, M. Campoy-Quiles, K. Feldman, L. Yu, P. Wobkenberg, J. Ball, R. M. Wilson, I. McCulloch, T. Kreouzis, M. Heeney, T. Anthopoulos, P. Smith and N. Stingelin, *Adv. Mater.*, 2010, **22**, 3942–3947.
72. C. Müller, N. D. Zhigadlo, A. Kumar, M. A. Baklar, J. Karpinski, P. Smith, T. Kreouzis and N. Stingelin, *Macromolecules*, 2011, **44**, 1221–1225.
73. H. S. Marsh, O. G. Reid, G. Barnes, M. Heeney, N. Stingelin and G. Rumbles, *J. Polym. Sci., Part B: Polym. Phys.*, 2014, **52**, 700–707.
74. N. D. Treat, J. A. N. Malik, O. Reid, L. Yu, C. G. Shuttle, G. Rumbles, C. J. Hawker, M. L. Chabinyc, P. Smith and N. Stingelin, *Nat. Mater.*, 2013, **12**, 628–633.
75. F. Paquin, H. Yamagata, N. J. N. J. Hestand, M. Sakowicz, N. Bérubé, M. Côté, L. X. Reynolds, S. A. Haque, N. Stingelin, F. C. Spano and C. Silva, *Phys. Rev. B*, 2013, **88**, 155202.

76. F. Paquin, G. Latini, M. Sakowicz, P. L. Karsenti, L. Wang, D. Beljonne, N. Stingelin and C. Silva, *Phys. Rev. Lett.*, 2011, **106**, 197401.
77. F. Paquin, J. Rivnay, A. Salleo, N. Stingelin and C. Silva, *J. Mater. Chem. C*, 2015, **3**, 10715–10722.
78. E. D. Gomez, K. P. Barteau, H. Wang, M. F. Toney and Y.-L. Loo, *Chem. Commun.*, 2011, **47**, 436–438.
79. I. Osaka, M. Saito, T. Koganezawa and K. Takimiya, *Adv. Mater.*, 2014, **26**, 331–338.
80. J. A. Nekuda Malik, N. D. Treat, M. Abdelsamie, L. Yu, R. Li, D.-M. Smilgies, A. Amassian, C. J. Hawker, M. L. Chabinyc and N. Stingelin, *Org. Photonics Photovoltaics*, 2014, **2**, 59–69.
81. C. Bruner, F. Novoa, S. Dupont and R. Dauskardt, *ACS Appl. Mater. Interfaces*, 2014, **6**, 21474–21483.
82. S. Savagatrup, D. Rodriquez, A. D. Printz, A. B. Sieval, J. C. Hummelen and D. J. Lipomi, *Chem. Mater.*, 2015, **27**, 3902–3911.
83. A. Hirsch and M. Brettreich, *Fullerenes: Chemistry and Reactions*, Wiley-VCH Verlag GmbH, Wienheim, Germany, 2005.
84. R. Taylor, *Lecture Notes on Fullerene Chemistry: A Handbook for Chemists*, Imperial College Press, London, UK, 1999.
85. M. T. Beck and G. Mándi, *Fullerene Sci. Technol.*, 1997, **5**, 291–310.
86. N. Sivaraman, R. Dhamodaran, I. Kaliappan, T. G. Srinivasan, P. R. P. V. Rao and C. K. C. Mathews, *Fullerene Sci. Technol.*, 1994, **2**, 233–246.
87. N. Martín, L. Sánchez, B. Illescas and I. Pérez, *Chem. Rev.*, 1998, **98**, 2527–2548.
88. A. B. Sieval and J. C. Hummelen, in *Organic Photovoltaics*, ed. C. Brabec, U. Scherf and V. Dyakonov, Wiley-VCH Verlag GmbH, Weinheim, Germany, 2014.
89. M. Lenes, S. W. Shelton, A. B. Sieval, D. F. Kronholm, J. C. Hummelen and P. W. M. Blom, *Adv. Funct. Mater.*, 2009, **19**, 3002–3007.
90. A. B. Sieval, N. D. Treat, D. Rozema, J. C. Hummelen and N. Stingelin, *Chem. Commun.*, 2015, **51**, 8126–8129.
91. N. Stingelin, *Polym. Int.*, 2012, **61**, 866–873.
92. S. Kouijzer, J. J. Michels, M. van den Berg, V. S. Gevaerts, M. Turbiez, M. M. Wienk, R. A. J. Janssen, M. Van Den Berg and S. Gevaerts, *J. Am. Chem. Soc.*, 2013, **135**, 12057–12067.
93. S. Nilsson, A. Bernasik, A. Budkowski and E. Moons, *Macromolecules*, 2007, **40**, 8291–8301.
94. K. J. Ihn, J. Moulton and P. Smith, *J. Polym. Sci., Part B: Polym. Phys.*, 1993, **31**, 735–742.
95. C. Hansen, *The Three Dimensional Solubility Parameter and Solvent Diffusion Coefficient and Their Importance in Surface Coating Formulation*, Danish Technical Press, Copenhagen, Denmark, 1967.
96. F. Machui, S. Langner, X. Zhu, S. Abbott and C. J. Brabec, *Sol. Energy Mater. Sol. Cells*, 2012, **100**, 138–146.
97. I. Burgués-Ceballos, F. Machui, J. Min, T. Ameri, M. M. Voigt, Y. N. Luponosov, S. a. Ponomarenko, P. D. Lacharmoise, M. Campoy-Quiles and C. J. Brabec, *Adv. Funct. Mater.*, 2014, **24**, 1449–1457.

98. A. D. De Zerio Mendaza, A. Melianas, S. Rossbauer, O. Bäcke, L. Nordstierna, P. Erhart, E. Olsson, T. D. Anthopoulos, O. Inganäs and C. Müller, *Adv. Mater.*, 2015, **27**, 7325–7331.
99. B. A. Collins, Z. Li, J. R. Tumbleston, E. Gann, C. R. McNeill and H. Ade, *Adv. Energy Mater.*, 2013, **3**, 65–74.
100. M. Causa', J. De Jonghe-Risse, M. Scarongella, J. C. Brauer, E. Buchaca-Domingo, J.-E. Moser, N. Stingelin and N. Banerji, *Nat. Commun.*, 2016, **7**, 12556.
101. F. Dou, E. Buchaca-Domingo, M. Sakowicz, E. Rezasoltani, T. McCarthy-Ward, M. Heeney, X. Zhang, N. Stingelin and C. Silva, *J. Mater. Chem. C*, 2015, **3**, 3722–3729.
102. M. Scarongella, J. De Jonghe-Risse, E. Buchaca-Domingo, M. Causa, Z. Fei, M. Heeney, J. E. Moser, N. Stingelin and N. Banerji, *J. Am. Chem. Soc.*, 2015, **137**, 2908–2918.
103. W. Ma, C. Yang, X. Gong, K. Lee and A. J. Heeger, *Adv. Funct. Mater.*, 2005, **15**, 1617–1622.
104. B. Watts, W. J. Belcher, L. Thomsen, H. Ade and P. C. Dastoor, *Macromolecules*, 2009, **42**, 8392–8397.
105. N. D. Treat, C. G. Shuttle, M. F. Toney, C. J. Hawker and M. L. Chabinyc, *J. Mater. Chem.*, 2011, **21**, 15224–15231.
106. N. Shin, L. J. Richter, A. A. Herzing, R. J. Kline and D. M. DeLongchamp, *Adv. Energy Mater.*, 2013, **3**, 938–948.
107. L. A. Perez, K. W. Chou, J. A. Love, T. S. Van Der Poll, D. M. Smilgies, T. Q. Nguyen, E. J. Kramer, A. Amassian and G. C. Bazan, *Adv. Mater.*, 2013, **25**, 6380–6384.
108. J. J. van Franeker, M. Turbiez, W. Li, M. M. Wienk and R. A. J. Janssen, *Nat. Commun.*, 2015, **6**, 6229.
109. Y. Sun, G. C. Welch, W. L. Leong, C. J. Takacs, G. C. Bazan and A. J. Heeger, *Nat. Mater.*, 2011, **11**, 44–48.
110. S. Masi, A. Rizzo, R. Munir, A. Listorti, A. Giuri, C. Esposito Corcione, N. D. Treat, G. Gigli, A. Amassian, N. Stingelin and S. Colella, *Adv. Energy Mater.*, 2017, 1602600.
111. C. Lindqvist, J. Bergqvist, C. C. Feng, S. Gustafsson, O. Bäcke, N. D. Treat, C. Bounioux, P. Henriksson, R. Kroon, E. Wang, A. Sanz-Velasco, P. M. Kristiansen, N. Stingelin, E. Olsson, O. Inganäs, M. R. Andersson and C. Müller, *Adv. Energy Mater.*, 2014, **4**, 1301437.
112. A. Sharenko, N. D. Treat, J. A. Love, M. F. Toney, N. Stingelin and T.-Q. Nguyen, *J. Mater. Chem. A*, 2014, **2**, 15717–15721.
113. N. D. Treat, O. G. Reid, S. Fearn, G. Rumbles, C. J. Hawker, M. L. Chabinyc and N. Stingelin, 2017, (Under Review).
114. M. Campoy-Quiles, T. Ferenczi, T. Agostinelli, P. G. Etchegoin, Y. Kim, T. D. Anthopoulos, P. N. Stavrinou, D. D. C. Bradley and J. Nelson, *Nat. Mater.*, 2008, **7**, 158–164.
115. A. J. Pearson, T. Wang and D. G. Lidzey, *Rep. Prog. Phys.*, 2013, **76**, 22501.

116. D. S. Germack, C. K. Chan, R. Joseph Kline, D. A. Fischer, D. J. Gundlach, M. F. Toney, L. J. Richter and D. M. Delongchamp, *Macromolecules*, 2010, **43**, 3828–3836.
117. B. Schmidt-Hansberg, M. Sanyal, M. F. G. Klein, M. Pfaff, N. Schnabel, S. Jaiser, A. Vorobiev, E. Müller, A. Colsmann, P. Scharfer, D. Gerthsen, U. Lemmer, E. Barrena and W. Schabel, *ACS Nano*, 2011, **5**, 8579–8590.
118. H. W. Ro, J. M. Downing, S. Engmann, A. A. Herzing, D. M. DeLongchamp, L. J. Richter, S. Mukherjee, H. Ade, M. Abdelsamie, L. K. Jagadamma, A. Amassian, Y. Liu and H. Yan, *Energy Environ. Sci.*, 2016, **9**, 2835–2846.
119. A. Guerrero, B. Dörling, T. Ripolles-Sanchis, M. Aghamohammadi, E. Barrena, M. Campoy-Quiles and G. Garcia-Belmonte, *ACS Nano*, 2013, **7**, 4637–4646.
120. J. W. Kingsley, P. P. Marchisio, H. Yi, A. Iraqi, C. J. Kinane, S. Langridge, R. L. Thompson, A. J. Cadby, A. J. Pearson, D. G. Lidzey, R. A. L. Jones and A. J. Parnell, *Sci. Rep.*, 2014, **4**, 5286.
121. D. Y. Kwok and A. W. Neumann, *Adv. Colloid Interface Sci.*, 1999, **81**, 167–249.
122. D. Y. Chung, J. Huang, D. D. C. Bradley and A. J. Campbell, *Org. Electron.*, 2010, **11**, 1088–1095.
123. M. M. Voigt, R. C. I. MacKenzie, C. P. Yau, P. Atienzar, J. Dane, P. E. Keivanidis, D. D. C. Bradley and J. Nelson, *Sol. Energy Mater. Sol. Cells*, 2011, **95**, 731–734.
124. T. Kajiya, W. Kobayashi, T. Okuzono and M. Doi, *Langmuir*, 2010, **26**, 10429–10432.
125. Y. Oh, J. Kim, Y. J. Yoon, H. Kim, H. G. Yoon, S. N. Lee and J. Kim, *Curr. Appl. Phys.*, 2011, **11**, 359–363.
126. H.-Y. Tseng, B. Purushothaman, J. Anthony and V. Subramanian, *Org. Electron.*, 2011, **12**, 1120–1125.
127. P. G. Saffman and G. Taylor, *Proc. R. Soc. A*, 1958, **245**, 312–329.
128. P. Tabeling, G. Zocchi and A. Libchaber, *J. Fluid Mech.*, 1987, **177**, 67–82.
129. G. Yu, J. Gao, J. C. Hummelen, F. Wudl and A. J. Heeger, *Science*, 1995, **270**, 1789–1791.
130. P. Westacott, J. R. Tumbleston, S. Shoaee, S. Fearn, J. H. Bannock, J. B. Gilchrist, S. Heutz, J. DeMello, M. Heeney, H. Ade, J. Durrant, D. S. McPhail and N. Stingelin, *Energy Environ. Sci.*, 2013, **6**, 2756–2764.
131. C. Müller, T. A. M. Ferenczi, M. Campoy-Quiles, J. M. Frost, D. D. C. Bradley, P. Smith, N. Stingelin-Stutzmann and J. Nelson, *Adv. Mater.*, 2008, **20**, 3510–3515.
132. S. De, T. Pascher, M. Maiti, K. G. Jespersen, F. Zhang, O. Ingans, A. Yartsev, V. Sundström and C. F. Blends, *J. Am. Chem. Soc.*, 2007, **129**, 8466–8472.
133. J. Piris, T. E. Dykstra, A. A. Bakulin, P. H. M. Van Loosdrecht, W. Knulst, M. T. Trinh, J. M. Schins and L. D. A. Siebbeles, *J. Phys. Chem. C*, 2009, **113**, 14500–14506.

134. N. Treat, T. Mates and C. Hawker, *Macromolecules*, 2013, **46**, 1002–1007.
135. N. D. Treat, M. A. Brady, G. Smith, M. F. Toney, E. J. Kramer, C. J. Hawker and M. L. Chabinyc, *Adv. Energy Mater.*, 2011, **1**, 82–89.
136. F. Fischer, T. Hahn, H. Bässler, I. Bauer, P. Strohriegl and A. Köhler, *Adv. Funct. Mater.*, 2014, **24**, 6172–6177.
137. M. Abdelsamie, N. D. Treat, K. Zhao, C. McDowell, M. A. Burgers, R. Li, D. M. Smilgies, N. Stingelin, G. C. Bazan and A. Amassian, *Adv. Mater.*, 2015, **27**, 7285–7292.
138. J. Zhao, A. Swinnen, G. Van Assche, J. Manca, D. Vanderzande and B. Van Mele, *J. Phys. Chem. B*, 2009, **113**, 1587–1591.
139. P. E. Hopkinson, P. A. Staniec, A. J. Pearson, A. D. F. Dunbar, T. Wang, A. J. Ryan, R. A. L. Jones, D. G. Lidzey and A. M. Donald, *Macromolecules*, 2011, **44**, 2908–2917.
140. M. Campoy-Quiles, M. Sims, P. G. Etchegoin and D. D. C. Bradley, *Macromolecules*, 2006, **39**, 7673–7680.
141. M. Campoy-Quiles, M. I. Alonso, D. D. C. Bradley and L. J. Richter, *Adv. Funct. Mater.*, 2014, **24**, 2116–2134.
142. X. Wang, T. Ishwara, W. Gong, M. Campoy-Quiles, J. Nelson and D. D. C. Bradley, *Adv. Funct. Mater.*, 2012, **22**, 1454–1460.
143. L. M. Andersson, F. Zhang and O. Inganäs, *Appl. Phys. Lett.*, 2007, **91**, 2005–2008.
144. M. Scarongella, A. A. Paraecattil, E. Buchaca-Domingo, J. D. Douglas, S. Beaupré, T. McCarthy-Ward, M. Heeney, J.-E. Moser, M. Leclerc, J. M. J. Fréchet, N. Stingelin and N. Banerji, *J. Mater. Chem. A*, 2014, **2**, 6218–6230.
145. A. C. Mayer, M. F. Toney, S. R. Scully, J. Rivnay, C. J. Brabec, M. Scharber, M. Koppe, M. Heeney, I. McCulloch and M. D. McGehee, *Adv. Funct. Mater.*, 2009, **19**, 1173–1179.
146. N. Li, J. D. Perea, T. Kassar, M. Richter, T. Heumueller, G. J. Matt, Y. Hou, N. S. Güldal, H. Chen, S. Chen, S. Langner, M. Berlinghof, T. Unruh and C. J. Brabec, *Nat. Commun.*, 2017, **8**, 14541.
147. S. Bertho, I. Haeldermans, A. Swinnen, W. Moons, T. Martens, L. Lutsen, D. Vanderzande, J. Manca, A. Senes and A. Bonfiglio, *Sol. Energy Mater. Sol. Cells*, 2007, **91**, 385–389.
148. S. Bertho, G. Janssen, T. J. Cleij, B. Conings, W. Moons, A. Gadisa, J. D'Haen, E. Goovaerts, L. Lutsen, J. Manca and D. Vanderzande, *Sol. Energy Mater. Sol. Cells*, 2008, **92**, 753–760.
149. C. Lindqvist, A. Sanz-Velasco, E. Wang, O. Bäcke, S. Gustafsson, E. Olsson, M. R. Andersson and C. Müller, *J. Mater. Chem. A*, 2013, **1**, 7174.
150. J. Bergqvist, C. Lindqvist, O. Bäcke, Z. Ma, Z. Tang, W. Tress, S. Gustafsson, E. Wang, E. Olsson, M. R. Andersson, O. Inganäs and C. Müller, *J. Mater. Chem. A*, 2014, **2**, 6146–6152.
151. F. A. A. Nugroho, A. Diaz de Zerio Mendaza, C. Lindqvist, T. J. Antosiewicz, C. Müller and C. Langhammer, *Anal. Chem.*, 2017, **89**, 2575–2582.
152. A. Diaz de Zerio Mendaza, A. Melianas, F. A. A. Nugroho, O. Bäcke, E. Olsson, C. Langhammer, O. Inganäs and C. Müller, *J. Mater. Chem. A*, 2017, **5**, 4156–4162.

153. C. Lindqvist, J. Bergqvist, O. Bäcke, S. Gustafsson, E. Wang, E. Olsson, O. Inganäs, M. R. Andersson and C. Müller, *Appl. Phys. Lett.*, 2014, **104**, 153301.
154. N. R. Tummala, C. Bruner, C. Risko, J. L. Brédas and R. H. Dauskardt, *ACS Appl. Mater. Interfaces*, 2015, 7, 9957–9964.

CHAPTER 7

Charge Generation and Recombination in Organic Solar Cells

J. C. BRAUER[a], M. CAUSA'[b] AND N. BANERJI*[b]

[a]University of Fribourg, Department of Chemistry, Chemin du Musée 9, 1700 Fribourg, Switzerland; [b]University of Bern, Department of Chemistry and Biochemistry, Freiestrasse 3, 3012 Bern, Switzerland
*E-mail: natalie.banerji@dcb.unibe.ch

7.1 Introduction

Organic semiconductors consist of carbon-based, π-conjugated molecules or polymers, which can be solution-processed or vacuum-deposited to obtain thin films for optoelectronic applications. Because of the low dielectric constant in organic solids, they are excitonic in nature, meaning that the electrons and holes are strongly bound in the excited state, unlike inorganic materials, where free charges can be directly generated by light absorption. In the 1980s, the first organic photovoltaic (OPV) device with an interface between an electron donor layer and an electron acceptor layer was reported.[1] The junction allows the bound excitons to dissociate. Another breakthrough in the field was the demonstration of the bulk heterojunction (BHJ) in 1995.[2,3] This involves blending the donor and acceptor materials to form a nanoscale interpenetrating network, allowing both a large interfacial area for exciton dissociation and a percolating network for charge transport to the electrodes.

Nanoscience & Nanotechnology Series No. 45
Nanostructured Materials for Type III Photovoltaics
Edited by Peter Skabara and Mohammad Azad Malik

Published by the Royal Society of Chemistry, www.rsc.org

The efficiency of OPV devices now exceeds 11%.[4] The main advantage of this technology is the low-cost manufacturing of solution-processed devices, which are large-area, flexible and lightweight. The most commonly used materials in organic solar cells are π-conjugated polymer electron donors, blended with electron-accepting fullerene derivatives such as [6,6]-phenyl-C_{61}-butyric-acid-methyl-ester (PCBM). These materials are dissolved in common organic solvents and then cast or printed as highly absorbing thin films of ~100 nm thickness. We focus here on such solution-cast polymer : fullerene solar cells, although highly promising small molecule solar cells (vacuum- or solution-processed),[5,6] as well as non-fullerene acceptors, have also been reported.[7]

When light is converted to electricity in polymer : fullerene OPV devices, the generation mechanism of the charges, which competes with various recombination losses, plays a predominant part in determining the power conversion efficiency. Figure 7.1 shows an over-simplified picture of a BHJ blend. The following processes are assumed to occur when photoexcitation takes place: (1) light is absorbed by the polymer; (2) the exciton diffuses to a fullerene interface; (3) the exciton dissociates to an interfacial electron–hole pair (often referred to as a polaron pair, charge transfer state or charge transfer exciton); (4) the charge pair spatially dissociates to separate charges (also called free charges or mobile carriers); and (5) the free charges are transported to the electrodes.[8]

In reality, the detailed mechanism of charge generation is much more complex and controversial. Some of the charge generation steps might compete with relaxation processes (and occur from hot states), delocalisation phenomena might be involved and the arrangement of the donor and acceptor, which has proved much more complex than the simple two-phase picture in Figure 7.1, strongly affects charge generation, recombination and transport.

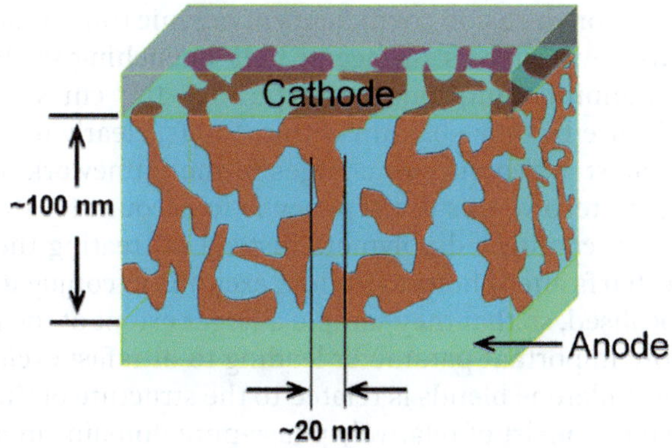

Figure 7.1 Schematic representation of an OPV device in which the BHJ active layer is sandwiched between two electrodes. The polymer (donor) phase is shown in blue and the fullerene (acceptor) phase in brown; they form a nanoscale interpenetrating network. Reprinted with permission from ref. 179. Copyright (2007) American Chemical Society.

We review here the most recent insights and hypotheses that have been proposed to explain efficient separate charge generation in polymer : fullerene solar cells and provide an overview of how charges can be lost *via* geminate or non-geminate recombination processes.

7.2 From Light Absorption to Exciton Dissociation

We start by describing the steps that lead from light absorption, *via* possibly exciton diffusion, to the generation of the interfacial electron–hole pair. According to Figure 7.1, the rate of electron–hole pair formation depends on the rate of exciton diffusion and on the intrinsic rate of exciton dissociation by electron transfer from the photoexcited polymer to the fullerene. A transient absorption experiment with very high time resolution (<10 fs) from 2001 showed that charges appear within 45 fs[9] and it was for a long time assumed that exciton quenching in polymer : fullerene blends is always ultrafast. This is consistent with the intrinsic timescale of electron transfer, which has been shown to be between 40 and 240 fs in an investigation where different fullerene derivatives were used to vary the driving force for exciton dissociation, given roughly by the lowest occupied molecular orbital (LUMO) level offset between the polymer and the fullerene (Figure 7.2).[10] The rate of electron transfer was found to increase with the driving force (the Marcus inverted region was not observed) and to depend on the electronic coupling between the donor and the acceptor. The minimum driving force for efficient electron transfer is generally believed to be ~0.3 eV, although lower values have been reported.[11]

The ultrafast exciton quenching in polymer : fullerene blends is not consistent with exciton diffusion. In the traditional view of the BHJ shown in Figure 7.1, the polymer domain size is ~10–20 nm, which means that to dissociate, excitons would need to diffuse several nanometres before reaching a fullerene interface. Exciton diffusion coefficients in organic thin films can be measured in various ways (such as surface or bulk quenching studies, or using exciton–exciton annihilation)[12] and are typically ~10^{-3} $cm^2 s^{-1}$. This means that excitons diffuse by only ~0.1 nm within 100 fs, clearly not sufficient to explain the ultrafast appearance of charges in the framework of Figure 7.1. One hypothesis to resolve this discrepancy is to account for the delocalisation of excitons in conjugated polymers instead of treating them as point-like. Up to 100–200 fs after photoexcitation, excitons in conjugated polymers are highly delocalised, so that they sample a larger extent of the BHJ.[13] Possibly an even more important parameter leading to ultrafast exciton dissociation in polymer : fullerene blends is related to the structure of the BHJ, which has been shown to consist of relatively phase-pure domains in addition to a molecularly mixed polymer : fullerene phase, where excitons can dissociate without any need for diffusion.[14] Exciton quenching is delayed if the excitons are not generated in the intermixed phase and need to diffuse through neat domains. Charges can also be generated in OPV blends following excitation of the fullerene, by hole transfer to the polymer.

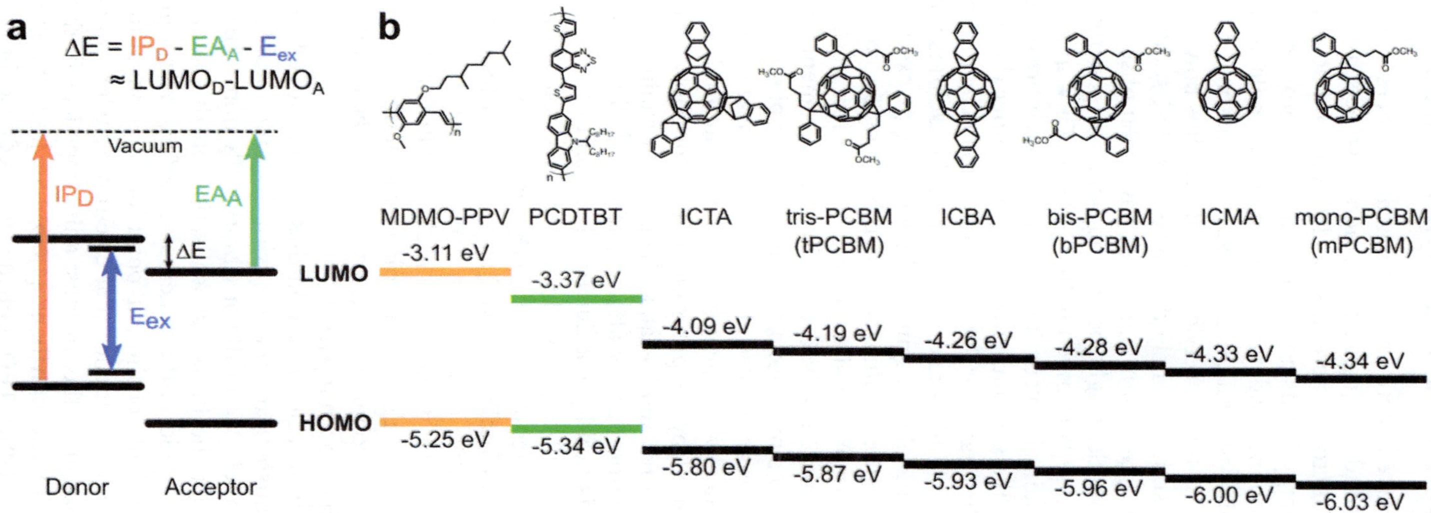

Figure 7.2 (a) Schematic diagram of the HOMO and LUMO levels of the donor and acceptor in an organic solar cell. The energetic driving force for exciton dissociation (ΔE) is defined as the difference between the ionisation potential of the donor (IP_D), the electron affinity of the acceptor (EA_A) and the energy of the photogenerated exciton (E_{EX}). It is roughly equal to the LUMO energy offset of the donor and acceptor. (b) Molecular structure of typical conjugated polymers and fullerene derivatives with their respective HOMO and LUMO energy levels. Reprinted with permission from ref. 10. Copyright (2016) American Chemical Society.

7.2.1 Light Absorption and Exciton Delocalisation

The nature of the excited state in conjugated polymers has long been debated.[13,15] On one hand, conjugated polymers have been treated as one-dimensional semiconductors, where mobile electrons and holes are directly photogenerated in energy bands and rapidly localise into a bound exciton state.[16–18] On the other hand, evidence exists for a more molecular approach, where the primary photoexcitation is already an exciton.[19–21] For highly ordered systems, such as polydiacetylene quantum wires polymerised in an ordered monomer crystal, extremely delocalised excited states with quantum coherence reaching tens of micrometres have been observed.[22] In this case, an energy-band model might be appropriate. It is generally accepted that, for solution-processed organic thin films used in OPV applications, the conjugated polymer chains break up into much smaller, more molecular-like chromophores of variable size, either because of disorder that disrupts the π-conjugation along the chain[23] or due to coupling to nuclear modes.[24,25]

Exciton diffusion in thin films takes place when the excitons hop incoherently between the chromophore segments by Förster-type excitation energy transfer (intra-chain or inter-chain). Multistep hopping to sites with lower energy typically leads to a progressive red shift of the fluorescence spectrum, extensively studied by time-resolved emission measurements, often combined with Monte Carlo simulations.[26–32] Shortly after photoexcitation (up to 100–200 fs), the excitons are not yet localised on a single segment because they are either electronically not relaxed (found in a manifold of higher lying delocalised states formed by coupling of the chromophoric units) and/or they are structurally not relaxed (because the polymer chains are still in the ground state conformation directly after excitation).[23] This initial delocalisation of the excitons is rapidly lost due to relaxation to the more localised lowest energy state and self-trapping of the electronic states that couple to nuclear modes during structural relaxation (Figure 7.3). The dynamic localisation has been experimentally observed as an ultrafast loss of anisotropy during polarisation-sensitive emission and transient absorption measurements,[33–40] as well as by the decay of the peak shift in three-pulse photon echo experiments.[41–44] It has also been predicted theoretically.[23–25,35,45] Before this ultrafast localisation, coherent processes such as the quantum-assisted transport of excitons have been reported.[46]

Given the ultrafast timescale of exciton dissociation in OPV blends, it is likely that electron transfer can compete with dynamic localisation of the exciton and occur from electronically or vibrationally non-relaxed states. It has been suggested that the initial delocalisation on this short timescale allows the exciton to couple to the fullerene interface over a greater spatial extent without need of incoherent point-like exciton diffusion.[33,47–49] There is solid experimental evidence that electron transfer can occur from hot states.[40,49–52] For example, the ultrafast anisotropy loss in the transient absorption signatures of the poly(benzo[1,2-*b*:4,5-*b*′]dithiophene-*alt*-thieno[3,4-*c*]pyrrole-4,6-dione) (PBDTTPD) polymer was reduced in the presence of PCBM, showing that electron transfer to the fullerene was faster than

Figure 7.3 Evolution of the polymer excited state in a single chain: Excitation into a manifold of higher lying electronic states leads to initial delocalisation (left), the exciton localises due to electronic and structural relaxation (middle) and then hops incoherently between chromophoric subunits (right). Reprinted with permission from ref. 23. Copyright (2011) American Chemical Society.

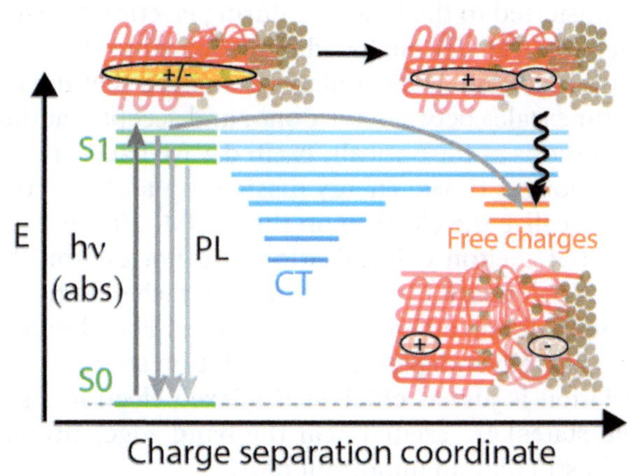

Figure 7.4 Dissociation of a hot, delocalised polymer exciton at the fullerene interface, leading to a delocalised charge transfer state and ultrafast generation of separate charges. Reprinted with permission from ref. 49. Copyright (2013) American Chemical Society.

the relaxation leading to the anisotropy decay.[40] A broadband time-resolved emission investigation (based on an optical Kerr shutter) comparing the population and relaxation dynamics in neat regioregular poly(3-hexylthiophene) (P3HT) films and P3HT : PCBM blends established that exciton dissociation occurs to a large extent from hot delocalised states (Figure 7.4).[49] Applying a Perrin-type quenching model to the exciton–exciton annihilation dynamics observed for the neat film at high excitation intensities, these researchers also concluded that the excitons initially occupy a large volume (delocalisation up to 10 nm along the polymer chain) and travel very quickly.

Coherent processes can take place on the ultrafast timescale of electron transfer. The role of vibrational quantum coherence and non-equilibrium nuclear configurations during electron transfer in suspended P3HT : PCBM nanoparticles was investigated using two-dimensional electronic spectroscopy.[50] It was found that part of the electron transfer is much faster than vibrational relaxation (>300 fs) and that vibrational coherence in the P3HT exciton is directly transferred to the P3HT hole. This was seen as strong oscillations in the dynamics of spectral signatures representing the excitons and the holes due to coherence in the C=C stretching mode. The retention of vibrational coherence shows that electron transfer must occur from a non-thermalised state (because thermal relaxation is an incoherent process and coherence would be lost afterwards) and that the hole exhibits the same or less delocalisation than the exciton (because higher delocalisation would diminish the amplitude of the oscillations).

In a related study using transient absorption spectroscopy with a very high time resolution (<15 fs), supported by time-dependent density functional theory simulation, coherent oscillations in the vibrational modes of PCBM were observed after the photoexcitation of P3HT, showing that the vibrational coherence is transferred to the fullerene during electron transfer. This points to coherent electron transfer mediated by strong vibronic coupling between the polymer and fullerene. The simulations showed that the charge is oscillating at short timescales between the donor and acceptor at the frequency of the coupled vibrations as the optically excited wave-packet delocalises across the interface. At longer times, energy dissipation and structural relaxation are expected to localise the electron on the PCBM. Although there is strong evidence for initial exciton delocalisation and hot exciton dissociation in polymer : fullerene blends, it is currently not clear whether this is really a necessary condition to obtain a high yield of charges. The photocurrent yield in OPV devices is generally independent of excitation energy,[53,54] implying that exciton dissociation is just as efficient if relaxed electronic states (and not hot delocalised states) are excited near the band edge, although structural relaxation might still play an important part.

7.2.2 Role of Phase Morphology

The observation of charges in polymer : fullerene blends on the ultrafast timescale is also related to another parameter. The structure of the BHJ (*i.e.* the arrangement of the donor and acceptor from the molecular to micrometre length scale) has a primordial role in the charge generation mechanism of OPV devices and is also largely responsible for the ultrafast timescale typically observed for exciton dissociation. Many empirical ways to enhance the efficiency of organic solar cells by influencing the BHJ structure during solution processing have been developed, such as casting solvent optimisation, the use of additives or thermal/solvent annealing.[55–57] With recent advances in the structural characterisation of organic thin films[58,59] it has become clear that the simple two-phase picture shown in Figure 7.1 is not correct. Not only can the polymer chains and fullerene molecules arrange into either amorphous or crystalline neat domains of variable size, but an additional intermixed polymer : fullerene phase has been identified.[60] The three-phase morphology, consisting of coexisting intermixed polymer : fullerene regions with neat polymer and fullerene domains, often leads to the highest OPV efficiency.[61–64]

Several femtosecond transient absorption studies of different polymer: fullerene blends have revealed that exciton dissociation is ultrafast if mainly the intermixed phase is present, but shows delayed components if excitons are generated in neat polymer domains further away from a fullerene interface. For example, the exciton dynamics were probed in blends of the poly[2,7-(9,9-dioctyl-fluorene)-*alt*-5,5-(4′,7′-di-2-thienyl-2′,1′,3′-benzothiadiazole)] (APFO) polymer with different concentrations of PCBM. At low fullerene loading, where neat polymer domains are expected to exist, the quenching of the exciton emission was not complete and exciton quenching dynamics of a few picoseconds were observed, which were related to exciton diffusion by one or more hops prior to their dissociation.[65] In another study, the transient absorption data of P3HT : PCBM films were compared with and without thermal annealing, which leads to phase separation and the formation of more ordered P3HT domains.[66] Although exciton quenching was faster than the time resolution for the non-annealed sample, where the polymer and fullerene are intermixed, it was slower (~3 ps time constant) in the annealed case because some excitons were generated further from PCBM.

Poly[2,5-bis(3-hexadecyl-thiophen-2-yl)thieno[3,2-*b*]thiophene] (pBTTT) was used as a model system to precisely correlate the exciton dissociation dynamics in polymer : fullerene blends to the arrangement of the donor and the acceptor into neat and intermixed phases.[67–69] This polymer is known to have a highly ordered microstructure and excellent mobility in field-effect transistors, although it is generally a less well performing solar cell material.[70,71] Nevertheless, fullerene molecules intercalate between the side-chains of the ordered pBTTT domains, forming a well-defined intermixed co-crystal phase.[62,64] The phase morphology of the pBTTT : PCBM blends can then be precisely characterised and manipulated, either by varying the polymer : fullerene weight ratio or by using fatty acid methyl ester processing additives of different alkyl chain lengths (Me 7, Me 12 and Me 14).[62,64,72]

Four phase morphologies (Figure 7.5a) were investigated by femtosecond transient absorption spectroscopy: fully intermixed (1 : 1 pBTTT : PCBM weight ratio); intermixed co-crystal regions and neat PCBM clusters (1 : 4 weight ratio); partially phase-separated (1 : 1 weight ratio processed with Me 7); and predominantly phase-separated (1 : 1 weight ratio processed with Me 12 or Me 14).[67–69] For the fully intercalated 1 : 1 pBTTT : PCBM blend, >99% of the polymer emission was quenched due to efficient electron transfer to PCBM in the intermixed co-crystal phase[62] and ultrafast spectroscopy showed that this occurred with a ~100 fs time constant (Figure 7.5b).[68] The presence of neat pBTTT domains in the phase-separated samples processed with the additives decreased the quenching efficiency of the pBTTT excitons (the polymer emission was only ~75% quenched in the predominantly phase-separated blend).[72] The transient absorption measurements revealed multiphasic quenching (with time constants of 0.6, 9 and 125 ps; Figure 7.5b) related to delayed exciton dissociation after exciton diffusion though the neat pBTTT domains.[67] The quenching was slower in the more phase-separated sample treated with Me 12 than in the sample treated with Me 7.

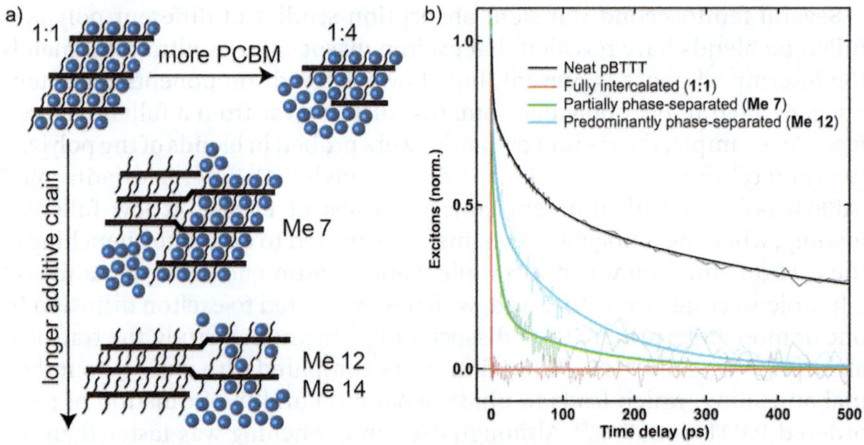

Figure 7.5 (a) Schematic representation of the different phase morphologies obtained by manipulating the processing conditions during the preparation of pBTTT : PCBM thin films. In the 1 : 1 blend, only the intermixed co-crystal phase is present; in the 1 : 4 blend there is coexistence of the co-crystal phase and neat PCBM clusters. The introduction of a short chain additive Me7 to the 1 : 1 blend yields a three-phase morphology due to partial phase separation (neat polymer and fullerene domains as well as an intermixed phase), whereas long alkyl chain additives (Me12 and Me14) lead to a predominantly phase-separated blend. (b) Time-evolution of the exciton population, reflecting the exciton quenching dynamics by electron transfer in the different phase morphologies. Reprinted by permission from Macmillan Publishers Ltd: *Nature Communications* (ref. 67), copyright 2016.

About 10–20% of the excitons generated within the neat polymer domains were lost to the ground state by exciton recombination before reaching a quenching site. This represents a significant loss mechanism in phase-separated pBTTT : PCBM blends and might contribute to the low solar cell efficiency of this polymer.

In other more efficient systems, exciton recombination is negligible because even the delayed exciton dissociation typically occurs within a few picoseconds, which is not consistent with point-like exciton diffusion.[40,66] For example, in an optimised PBDTTPD : PCBM blend, which is known to have a three-phase morphology (with neat polymer and fullerene domains as well as intermixed regions),[61] 25% of the excitons dissociate promptly in the intermixed phase, while 75% of the electron transfer is delayed, but still occurs with a time constant of 1 ps.[40] This was related to good exciton delocalisation even after the initial relaxation because structural self-trapping is limited by a planar backbone. Based on exciton–exciton annihilation studies, the PBDTTPD domain size was estimated to be 8 nm and the size of the exciton to be 2 nm, meaning that excitons generated in the centre of a polymer region could reach the interface with a single hop of about 1 ps.

7.2.3 Hole Transfer

Electron transfer is not the only pathway for charge generation in polymer:fullerene blends. Alternatively, the fullerene can be excited, which is followed by hole transfer to the polymer (Figure 7.6). Just as with electron transfer, this leads to the formation of an electron–hole pair, although the driving force is this time given by the highest occupied molecular orbital (HOMO) level offset between the acceptor and the donor. Hole transfer is just as efficient as electron transfer and the intrinsic time constant for the process is as short as 30 fs.[73] Such prompt hole transfer can be observed in intermixed regions of polymer blends with PCBM when the fullerene is selectively excited during transient absorption measurements.[40,68] When neat fullerene clusters are present, hole transfer is delayed due to relatively slow diffusion of the fullerene excitons to a polymer interface.[40,67,68,74] This is typically seen as a gradual rise (within tens of picoseconds) of polymer ground state bleaching and charge absorption in transient absorption experiments, although the fullerene is initially excited. For example, when selectively exciting the PCBM in the 1:4 pBTTT:PCBM blend consisting of the intermixed phase and neat PCBM clusters (Figure 7.5a), both prompt and delayed appearance of the charges are observed. As PCBM is present both in the co-crystal regions and the neat fullerene clusters, a proportion of the fullerene excitons need to diffuse over various distances through the neat PCBM domains prior to their dissociation by hole transfer to the polymer, leading to strongly multiphasic quenching dynamics with time constants ranging from sub-picoseconds to hundreds of picoseconds. It is likely that some of the fullerene excitons never reach a quenching site during their lifetime, especially if the fullerene domains are larger than the 5 nm exciton diffusion length of PCBM.[75] In the highly efficient PBDTTPD:PCBM blend, exciting the fullerene leads to slightly faster delayed hole transfer (about 75 ps average time constant) and there is no loss due to PCBM exciton recombination, showing that the

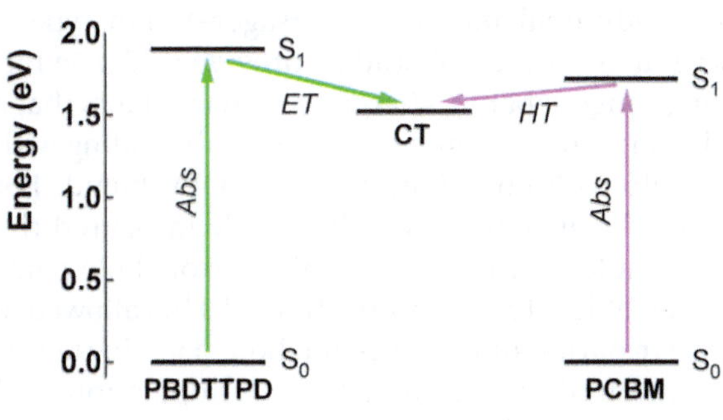

Figure 7.6 State diagram showing the ground and excited state of the polymer donor (PBDTTPD) and fullerene acceptor (PCBM) components of the blend, as well as the charge transfer state, generated either by electron transfer (ET) or hole transfer (HT). Reprinted with permission from ref. 40. Copyright (2014) American Chemical Society.

PCBM domains are small enough for complete quenching.[40] Especially when the acceptor in the BHJ absorbs in the visible range (for example with more soluble C_{70} derivatives), a significant fraction of photocurrent generation in the corresponding devices proceeds *via* hole transfer.[76] There is evidence suggesting that excitons in PCBM clusters spontaneously split to charges at room temperature, especially in the presence of an applied field and above an excitation energy threshold.[77,78] This process can contribute to the photocurrent, so it is likely that not only PCBM excitons, but also PCBM holes, diffuse towards the polymer interface when delayed hole transfer is observed.

7.3 From Bound to Separate Charges

The dissociation of the exciton between the donor and the acceptor component, as described in the previous section, leads to the formation of an electron–hole pair across the donor–acceptor interface. This is, in principle, still bound by the Coulomb interaction. Given the relatively low dielectric constant of organic thin films (ε_r = 3–4), the dissociation of the electron–hole pair to separate charges is an uphill energetic process. Assuming a point-charge model and an initial electron–hole separation of 1–2 nm, a Coulomb binding energy of 100–500 meV can be estimated, which is clearly more than the room temperature thermal energy (~25 meV).[79] Nevertheless, essentially all electron–hole pairs end up being successfully extracted from efficient solar cell blends. We will now discuss different mechanisms that have been proposed to explain the separation of electron–hole pairs in OPV systems and examine the closely related question of the timescale for separate charge formation. Reports of sub-picosecond generation of separate charges contrast with studies implying slower, Onsager–Braun type dissociation of electron–hole pairs.[80]

7.3.1 Evidence for Ultrafast Long-range Charge Separation

Challenging the traditional picture of Onsager–Braun type electron–hole separation,[79] several experimental studies provide evidence for the ultrafast generation of long-range separate charges, *i.e.* charge pairs that are separated by a sufficient distance to overcome their Coulomb binding within hundreds of femtoseconds after photoexcitation of the OPV blend. For example, a nanosecond-resolved transient absorption study measured the recombination rate (by monomolecular tunnelling) of electron–hole pairs, the motion of which was frozen at low temperature (10 K).[81] This allowed an estimation of the distance distribution of the electron–hole pairs in their initial conformation, which was found to be bimodal with a population of closely bound charges and a population of long-range (3–4 nm) separate charges (Figure 7.7a). The yield of the two populations was extracted by fitting the recombination dynamics with a double Gaussian model, where the ~3 ns lifetime (associated with the peak of the fast Gaussian rate distribution) corresponds to bound states, while the second slow Gaussian component, with a lifetime

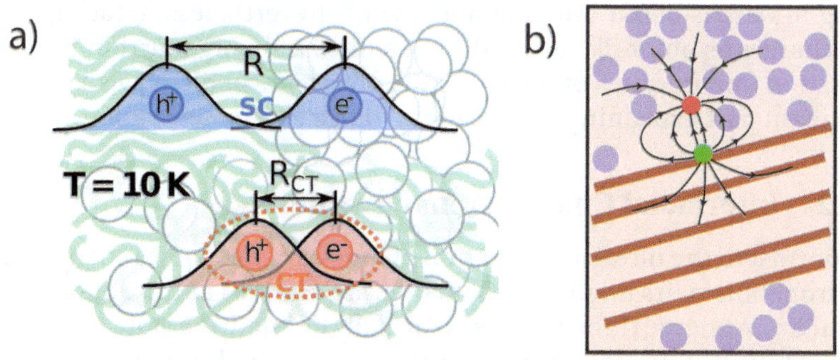

Figure 7.7 (a) Bimodal distance distribution of charge pairs frozen in their initial conformation at low temperature (10 K), as deduced from the analysis of the monomolecular recombination rate. Either separate charges (SC) or bound charge transfer (CT) states are instantaneously generated. Reprinted with permission from ref. 81. Copyright (2014) American Chemical Society. (b) Sketch of a photoinduced dipole-like electric field generated around an electron–hole pair at the donor–acceptor interface (polymer chains are red lines, fullerene molecules are purple spheres). Adapted with permission from ref. 68. Copyright (2014) American Chemical Society.

of ~1 μs, accounts for separate charges. The measured rate distribution was then transformed into a distance distribution by considering the appropriate tunnelling attenuation factor ($\beta \approx 0.3$ Å), leading to the conclusion that the energetic barrier between bound and separate charges is overcome *via* long-range charge separation on the ultrafast thermalisation timescale.

Similarly, a femtosecond-resolved stimulated Raman investigation of a polymer:fullerene blend evidenced spectral signatures of positive charges (holes) within 300 fs, which were identified by comparison with the spectra of doped polymer films.[82] The absence of significant structural relaxation on the 50 ps timescale, which would be expected during charge pair dissociation, indicated that the holes were already free from any Coulomb attraction at 300 fs. In a femtosecond transient absorption experiment, the appearance of an electroabsorption signature was exploited to estimate the electron–hole separation of photogenerated charge pairs with high time resolution.[83] The observed electroabsorption signal results from the electric field produced by photogenerated charges at the donor–acceptor interface (Figure 7.7b) and was shown to be a sensitive probe for the distance between the electron and hole. The charges were found to have reached a distance of 4 nm within the first 100 fs to account for the amplitude of the electroabsorption signature observed in certain blends. At this distance, the Coulomb binding energy is below the thermal energy and the charges are separated. Theoretically, long-range electron transfer from an exciton localised at a certain distance from the donor–acceptor interface could explain the prompt appearance of separate charges. This effect, which would bypass any interfacial states by allowing the exciton to dissociate before reaching the interface, was modelled using a Marcus-based hopping approach.[84] Nevertheless, treating excitons and charges as point-like is usually not sufficient in organic thin films. Delocalisation (in hot states or related to the phase morphology) also plays a predominant part explaining ultrafast long-range charge separation.

7.3.2 Interfacial Charge Transfer States

In contrast to the ultrafast formation of long-range separate charges, bound electron–hole pairs might populate interfacial states at the donor–acceptor boundary. A charge transfer state is formally defined as a state that is related to the ground state *via* a charge transfer transition, thus a transition involving partial charge transfer from an electron donor to an electron acceptor.[85] As a result of electronic interactions between the polymer and fullerene in OPV blends, such charge transfer states exist at the donor–acceptor interface. Given their coupling to the ground state, they show characteristic absorption and emission behaviour, although the oscillator strength of the transitions is typically very low. The existence of charge transfer states in polymer : fullerene blends has nevertheless been evidenced using highly sensitive absorption techniques (photothermal deflection spectroscopy (PDS)), emission measurements and electroluminescence spectroscopy.[86,87] It has been shown that the energy of the lowest-lying charge transfer state in polymer : fullerene blends is related to the open circuit voltage (V_{OC}) of the corresponding solar cell devices.[85] Charge transfer states also play a predominant part in the recombination processes in OPV materials. For example, geminate charge recombination (gCR) of electron–hole pairs that stem from the same exciton dissociation event typically occurs *via* a charge transfer state.[11] It has also been suggested that the bimolecular encounter of separate charges can lead to the formation of a charge pair in a charge transfer state, which then recombines non-radiatively or radiatively (leading to delayed fluorescence) to the ground state or undergoes intersystem crossing to the triplet state.[86,88–90]

A question that has raised intense debate is the involvement of charge transfer states in the spatial separation of electron–hole pairs. According to one view, the charge transfer state should be considered as an intermediate state populated by the electron–hole pair after exciton dissociation, which then, in turn, subsequently dissociates to separate charges.[87] According to the other view, the charge transfer state is bypassed (by various possible mechanisms) and separate charges are directly generated on an ultrafast timescale (in agreement with experimental evidence). In this case, the charge transfer state is considered as a trap and any electron–hole pairs ending up in this state geminately recombine without a chance of spatial separation.[80] The first view was supported by early transient absorption measurements, which claimed distinct spectral signatures for charges in the charge transfer state *versus* separate charges and reported conversion between the two within picoseconds.[91,92] However, careful analysis of more recent transient absorption data points to a negligible difference between the spectral signatures of the two species, making it difficult to directly use transient absorption spectroscopy to evaluate the role of the charge transfer state.[40,68,80]

Insights were nevertheless gained by applying two opposite kinetic models to the experimental time profiles obtained by transient absorption measurements at different excitation densities (Figure 7.8).[80] In the first model, dissociation and gCR were mediated by the relaxed charge transfer state, which separated or recombined on a relatively long timescale of ~5 ns (Onsager–Braun mechanism). The second model proposed that the majority of excitons directly dissociate to separate charges within hundreds of femtoseconds from non-relaxed charge transfer states, whereas only a small fraction of electron–hole pairs populate the relaxed charge transfer state and these exclusively recombine on the nanosecond timescale. Therefore the rate equations describing this second mechanism considered the number of bound and separate charges as an initial condition determined within the instrument response of the experiment.

Annealed and non-annealed P3HT : PCBM films were compared. The observed kinetics for both systems were consistent with the second mechanism of ultrafast separate charge generation. However, unlike for the less efficient non-annealed blend, the fraction of populated relaxed charge transfer state and the gCR were strongly suppressed in the annealed blend with

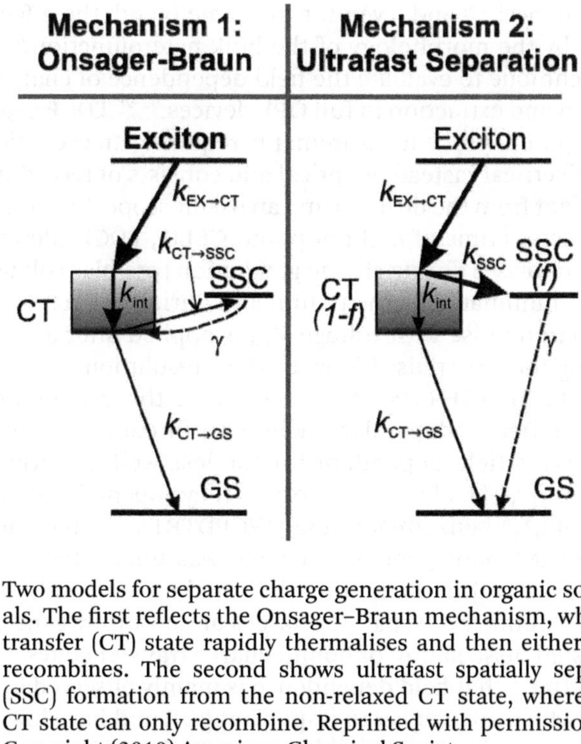

Figure 7.8 Two models for separate charge generation in organic solar cell materials. The first reflects the Onsager–Braun mechanism, where the charge transfer (CT) state rapidly thermalises and then either dissociates or recombines. The second shows ultrafast spatially separated charge (SSC) formation from the non-relaxed CT state, whereas the relaxed CT state can only recombine. Reprinted with permission from ref. 80. Copyright (2010) American Chemical Society.

high OPV efficiency, which had 85% yield of separate charges. The main finding from this analysis was therefore that two distinct populations of separate charges and charges in the charge transfer state are generated within <100 fs when the excitons dissociate, whereby only the former can be extracted as photocurrent and the latter undergo gCR. The branching ratio between these populations is determined by the phase morphology of the blend.

Another way to assess the involvement of charge transfer states in the spatial separation of electron–hole pairs is to consider how the yield of separate charges (or the extent of gCR) depends on the voltage bias of the solar cell devices. With increasing reverse bias, the presence of an electric field within the active layer is expected to assist the dissociation of the bound charge transfer state according to the Onsager–Braun formalism. Numerous studies to investigate this field dependence under solar cell operating conditions and in reverse bias have been conducted by optical techniques (*e.g.* femtosecond- or nanosecond-resolved transient absorption)[93,94] or by photocurrent methods (*e.g.* time-delayed collection-field (TDCF) or charge extraction by linearly increasing voltage (CELIV)).[95–98] Thereby contrasting observations ranging from field-independence to strong field dependence were made. This originates from the difficulty in separating the field dependence of electron–hole pair dissociation from charge extraction and non-geminate recombination.[11,95]

Different polymer:fullerene blends typically show a very different extent of bias-dependence[97,98] and, even for the same blend, the effect of the field is influenced by the morphology of the bulk heterojunction.[67,95–97] TDCF is an elegant technique to evaluate the field dependence of charge generation, recombination and extraction in full OPV devices.[95–98] TDCF experiments are equivalent to pump–probe measurements requiring an excitation pulse, but the probe is electrical instead of optical and consists of recording the photocurrent transient from the device using an oscilloscope. In contrast to related techniques, such as time-of-flight or photo-CELIV, TDCF allows the application of different biases during charge generation (pre-bias voltage, V_{pre}, when the sample is illuminated by the pump) and during extraction, when a rectangular collection pulse with voltage V_{coll} is applied after a controlled time delay following the laser pulse (down to 10 ns resolution).

TDCF results for P3HT:PCBM showed that the generation of extractable charges is field-independent, whereas, in contrast, the charge separation process is field-dependent for the less well performing blend of PCBM with poly[2,6-(4,4-bis(2-ethylhexyl)-4H-cyclopenta[2,1-*b*; 3,4-*b*″]dithiophene)-*alt*-4,7-(2,1,3-benzothiadiazole)] (PCPDTBT).[97,99] In the former blend, the total extracted photogenerated charge was unaffected by V_{pre} ranging from −0.4 to 0.4 V within the first few hundred nanoseconds, implying that a considerable number of separate charges are promptly generated without the field-assisted dissociation of bound charge transfer states. Overall, there seems to be a trend that field dependence is mainly observed in “poor” solar cell systems (which are subject to important gCR and have low efficiency).[11] Here, even the built-in electric field present at short circuit conditions can contribute to dissociate the charge transfer state.[79] However, in "good" polymer:fullerene blends (which show little gCR), the generation of separate charges is often independent of the field. The electric field is not needed in these efficient systems because the separate charges are probably generated by an intrinsic mechanism other than field-dependent Onsager–Braun type dissociation of the charge transfer state.

7.3.3 Involvement of Hot States

One hypothesis that has been put forward to explain the ultrafast appearance of separate charges by a field-independent mechanism is that the charge transfer state might be bypassed *via* short-lived hot states (Figure 7.9a).[100–102] The (electronically and/or vibrationally) hot states have excess energy and/or have not relaxed to localised polarons, so their enhanced delocalisation favours the separation of the electron and hole. Convincing evidence was

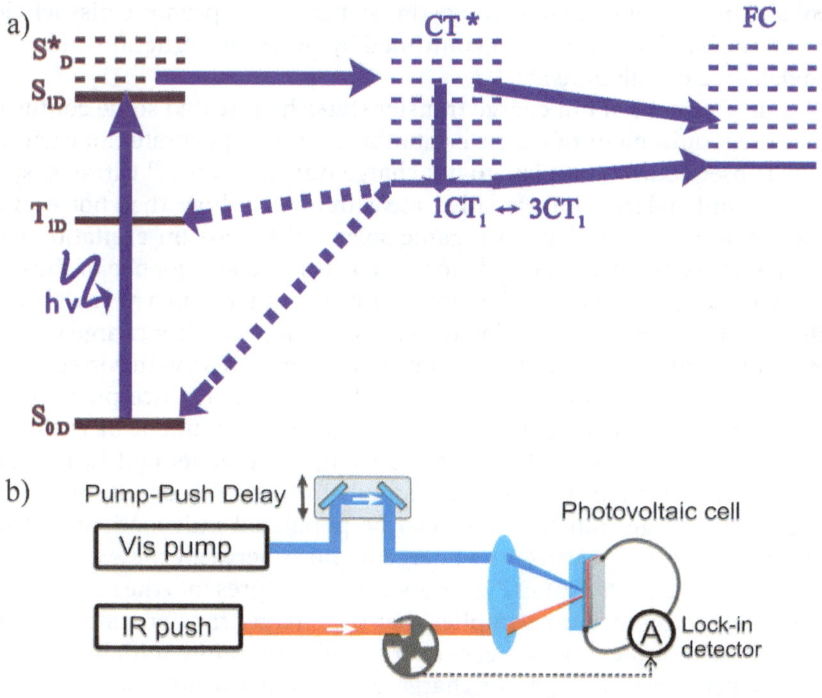

Figure 7.9 (a) Energy state diagram of an OPV system describing free charge (FC) generation *via* hot charge transfer (CT) states (CT*) or *via* the relaxed CT state (CT_1). The latter process competes with recombination mechanisms involving singlet or triplet states. Adapted from ref. 79 with permission of the Royal Society of Chemistry. (b) Schematic representation of the experimental set-up used for pump-push-photocurrent spectroscopy. Reprinted with permission from ref. 100. Copyright (2013) American Chemical Society.

obtained using the pump–push–photocurrent technique (Figure 7.9b).[102] In this experiment, full OPV devices were excited by a femtosecond pump pulse, followed by an IR push pulse, which re-populated the hot charge transfer states. The increase in photocurrent due to the push pulse was recorded as a function of the pump–push delay. A significant increase in photocurrent was observed when the electron–hole pairs were given a second chance to separate *via* the re-populated hot charge transfer states, especially in poorly performing polymer : fullerene blends.

Anisotropy measurements and atomistic modelling showed that the push pulses promoted the charges back to delocalised states similar to those formed just after exciton dissociation and which play a major part in allowing the electron–hole pairs to decouple and move apart from the heterojunction. Evidence of short-lived hot photocarriers was also obtained using terahertz spectroscopy, which showed a direct link between excess excitation energy and charge carrier mobility.[103] The hot charges showed high terahertz conductivity, which decayed on a picosecond timescale. The importance of delocalisation (to reduce the Coulomb barrier for electron–hole pair separation) and of vibrationally hot charge transfer states (to promote dissociation beyond the barrier) was further confirmed by electronic structure and quantum dynamic calculations.[104]

The involvement of hot charge transfer states has created some confusion about the requirement of excess excitation energy to generate hot excitons, which subsequently dissociate to hot charge transfer states.[51] Ultrafast spectroscopy and polarisation-sensitive measurements show that hot exciton dissociation can take place in organic solar cell blends on excitation with energies above the bandgap.[40,51] However, this is a consequence of the exciton dissociation being faster than its relaxation, rather than a necessary condition for the efficient generation of separate charges *via* hot charge transfer states. Spectrally flat internal quantum efficiencies (IQEs), with correct determination of electro-optical cavity effects such as interference phenomena, show that the photocurrent yield is essentially independent of excitation energy.[53,54] This suggests that separate charges are generated just as efficiently if the precursor excitons are initially formed with or without excess energy. Hot charge transfer states can be populated independently of hot exciton dissociation because the energetic offset between the exciton state and the charge transfer state (Figure 7.9a), which represents the driving force for the exciton dissociation, implies that the charge transfer state is always populated with more or less excess energy and is therefore initially hot.

The dependence of separate charge generation *via* hot charge transfer states on this driving force was further explored using pump–push photocurrent spectroscopy.[100,101] In one case, the charge transfer state dynamics was investigated for both electron transfer and hole transfer, which have different driving forces roughly according to the LUMO–LUMO offset (for electron transfer) and to the HOMO–HOMO offset (for hole transfer) between the donor and the acceptor.[100] Although the lifetime of the charge transfer state generated *via* the hole transfer and electron transfer processes was identical, showing its similar nature, the yield of separate charges (insensitive to the IR push pulse) compared with the yield of bound charges in the relaxed charge transfer state (which can be dissociated by the push pulses to increase the photocurrent) was strongly related to the driving energies of the two charge transfer pathways. It was demonstrated that the formation of spatially separated charges increased when the energy difference driving the exciton dissociation (HOMO–HOMO or LUMO–LUMO) increased because this excess energy increases the probability of populating hot charge transfer states.

In the second study, a low bandgap polymer with almost zero driving force for electron transfer with PCBM was used, which led to a dependence of separate charge generation on the excitation wavelength.[101] The agreement between IQE, transient absorption and pump–push–photocurrent measurements further confirmed the model wherein the efficiency of separate charge generation was not determined by quenching of the excitons, which is an energetically independent step, but depends on the amount of excess energy in the initially formed hot charge transfer states. Both investigations support the idea that hot charge transfer states are able to overcome their Coulomb attraction, whereas relaxed interfacial charge transfer states are unable to dissociate efficiently and act as traps.

The proposed mechanism, that separate charges are generated from short-lived hot charge transfer states, has nevertheless been severely challenged by an investigation combining highly sensitive photothermal deflection and IQE measurements.[105] It was shown that separate charges, leading to a photocurrent, are efficiently generated from the lowest-lying relaxed charge transfer state (CT_1) without the need to go through hot states. The charge transfer state can be directly excited from the ground state below the band edge of the polymer and the corresponding absorption spectrum in different polymer : fullerene blends was recorded using PDS. However, given the vertical nature of electronic transitions and the horizontal offset between the potential energy curves of the ground state and the charge transfer state, absorption into the charge transfer state always occurs with excess energy (Figure 7.10a). To directly excite the relaxed CT_1 state implies starting the transition from the hot, higher energy level of the ground state, which is not populated at a room temperature Boltzmann distribution. Therefore the direct absorption of the CT_1 state was calculated from its emission, as measured by electroluminescence spectroscopy. The external quantum efficiency was then sensitively recorded from the low-energy to high-energy region and converted to the IQE using the absorption spectra obtained from electroluminescence and PDS measurements. This yielded the IQE for direct CT_1 excitation, charge transfer excitation with excess energy as well as for polymer and fullerene excitation. The IQE was found to be spectrally flat, proving that even CT_1 dissociated with the same yield and produced a photocurrent (Figure 7.10b). Similar results were obtained for high-performance and low-performance OPV blends. This led to two main conclusions. First, the results did not reconcile with a scenario in which the CT_1 state acts as a trap (which would lead to a decrease in the corresponding IQE value). Second, the relaxed

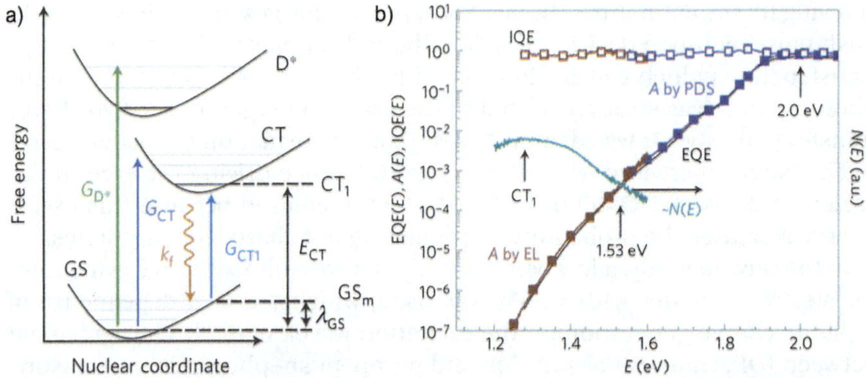

Figure 7.10 (a) Potential energy curves of the ground state (GS), polymer excited state (D*) and charge transfer (CT) state with vertical absorption transitions. (b) External quantum efficiency (EQE) spectrum (black line), electroluminescence emission spectrum (green line), low-energy absorption spectrum calculated from the electroluminescence spectrum (red line with closed squares) and high-energy absorption spectrum measured by PDS (blue lines with closed squares), as well as the flat IQE spectrum obtained as EQE/absorption (for PBDTTPD : PCBM). Reprinted by permission from Macmillan Publishers Ltd: *Nature Materials* (ref. 87), copyright (2013).

charge transfer state and hot charge transfer states generate separate charges with equal efficiency. Given the contradictory evidence for and against the importance of hot charge transfer states, this question is thus unresolved.

7.3.4 Impact of Phase Morphology and Interface Structure

The structure of the OPV blend, from the molecular level arrangement of the donor and acceptor to the micrometre-scale ordering of relatively phase-pure (neat) regions and intermixed domains, plays a predominant part in determining the charge generation efficiency. We will start by discussing the role of the phase morphology, *i.e.* the coexistence of neat donor and acceptor regions with an intermixed phase, on the spatial separation of electron–hole pairs. As a result of the proximity of the donor and acceptor, excitons generated in the intermixed phase do not have to diffuse, so the exciton dissociation is ultrafast and exciton recombination is prevented.[14,68] However, the spatial separation of electron–hole pairs is unfavourable when the donor and acceptor are intimately mixed. Therefore charge pairs in the intermixed phase tend to undergo gCR *via* interfacial charge transfer states. The extent of gCR can be significantly reduced if neat fullerene and/or polymer domains are present as well as the intermixed regions. The neat regions favour the generation of separate charges. The beneficial impact of having both neat and intermixed phases on the spatial separation of electron–hole pairs in polymer : fullerene blends explains why the best OPV efficiency is usually obtained in three-phase morphologies combining intermixed regions with neat polymer and fullerene domains.[14,61–64,106,107]

This has been extensively investigated for the pBTTT:PCBM model system, where molecular level mixing of the polymer and fullerene leads to a well-defined co-crystal phase, the coexistence of which with neat phases can be manipulated (Figure 7.5a).[62,64,72] A femtosecond transient absorption study revealed that ~90% of charge pairs undergo gCR if only the co-crystal phase is present. This decreases to 50% when the blend is phase-separated to a three-phase morphology (intermixed, neat pBTTT and neat PCBM) using processing additives and to 30% in the blend containing a large excess of PCBM, which coexists in fullerene clusters with the intermixed phase.[67–69] The trend correlates inversely with the solar cell efficiency of the corresponding photovoltaic devices.[72,108] The extent of gCR also scales with the amplitude of direct charge transfer state absorption measured by PDS,[96,109] confirming that the recombination occurs *via* interfacial charge transfer states. In agreement with the femtosecond study, transient absorption measurements on the longer (microsecond) timescale indicated a higher yield of long-lived free charges in the pBTTT:PCBM samples with neat PCBM domains and in the phase-separated blends compared with the fully intercalated blend.[62,72] It was also shown that the nature of the long-lived pBTTT holes is more localised in the intermixed phase and more delocalised in the neat pBTTT domains.[110]

A similar dependence of the extent of gCR on the phase morphology was observed in other polymer:fullerene blends, where the coexistence of fullerene clusters (and possibly neat polymer domains) with the intermixed phase were found to assist separate charge generation.[10,74,80,81] The phase morphology also affects the field dependence of photocurrent generation.[67,96,97] TDCF investigations conducted for different polymer:fullerene blends arrived at the concurring conclusion that the field dependence of separate charge generation is much weaker for optimised phase morphologies containing neat domains than for non-optimised blends undergoing important gCR.[96,97] This suggests that optimising the phase morphology leads to an efficient, field-independent mechanism to dissociate electron–hole pairs and to prevent gCR. Different mechanisms to explain the beneficial impact of having coexisting neat and intermixed regions have been proposed, such as improved transport in the neat regions,[106,108] electron delocalisation into fullerene clusters[10,83,111] or energy level cascades driving the charges from the intermixed to the neat regions.[62,106,112–114]

The experimentally observed ultrafast, field-independent, morphology influenced generation of separate charges has been explained by delocalisation of the electron wavefunction into neat fullerene aggregates.[10,81,83,96,111] This intermolecular delocalisation enhances the electron–hole separation and thus assists the spatial separation of the charge pairs in optimised blends containing neat fullerene clusters. Blends of two polymers, poly-[2-methoxy-5-(3′,7′-dimethyloctyloxy)-1,4-phenylenevinylene] (MDMO-PPV) and poly[*N*-9′-heptadecanyl-2,7-carbazole-*alt*-5,5-(4′,7′-di-2-thienyl-2′,1′,3′-benzothiadiazole)] (PCDTBT) (Figure 7.2), with different substituted fullerene derivatives in various polymer : fullerene ratios allowed the role of fullerene aggregation and the electron transfer driving force to be addressed simultaneously (impacting the population of hot charge transfer states).[10] Both transient absorption spectroscopy with <20 fs resolution and pump–push–photocurrent spectroscopy were used. Given the high miscibility of both polymers with the fullerenes, only the intermixed phase was present at low fullerene loading, whereas neat fullerene domains coexisted with the intermixed phase at high fullerene loading. In both cases, the diffusion of polymer excitons prior to exciton dissociation by electron transfer was not necessary, so the intrinsic electron transfer rate could be deduced. This rate of exciton dissociation by electron transfer (in the 40–240 fs range) was found to depend on the driving force (the LUMO level offset of the donor and acceptor) as well as on the presence of fullerene aggregates, which probably improve the electronic coupling between the polymer and fullerene. However, the yield of separate charges had a minor dependence on the driving force, but a major dependence on the fullerene aggregate size and packing, pointing to the primordial importance of delocalisation.

In another study of a small molecule : fullerene (C_{60}) system with different fullerene concentrations, the static and optical dielectric constant (determined by capacitance and ellipsometry measurements) was found to increase with fullerene loading.[115] This was related to the properties of the charge transfer state (energy, lifetime, polarisability). Electroabsorption experiments revealed that fullerene crystallites >4 nm (at high fullerene concentration) resulted in charge transfer states with strong polarisability, increased delocalisation and occupying a greater volume. Charge separation was suggested to occur *via* those delocalised charge transfer states.

The delocalisation of the electron wavefunction over several fullerenes and their role in the dissociation of photogenerated electron–hole pairs has been extensively addressed by theory.[83,111,116] Electronic structure calculations of PCBM clusters of varying size showed that a high density of low-lying delocalised charged states exists when the cluster size is increased.[116] When those states are aligned with the exciton state of the polymer, this can give rise to resonant electron transfer directly into the delocalised fullerene states, allowing ultrafast separate charge generation by a non-Marcus type mechanism and bypassing more localised charge transfer states that lead to gCR. The study was complemented by pump–push–photocurrent spectroscopy of MDMO-PPV : PCBM blends and it was suggested that the role of the push pulse was to re-populate the delocalised PCBM states, thus increasing the photocurrent yield. The delocalisation over several fullerenes was essentially independent of rotational and translational disorder, making spherical fullerenes particularly robust electron acceptors for OPV applications.

Although this investigation considered the delocalised fullerene states in the framework of electronically "hot" charge transfer dissociation, we note that the delocalised states might also be thermally accessible or accessible at ultrafast times before structural localisation to form polarons.[79,83,117] In principle, the resonant transfer of the electron to the short-lived delocalised acceptor states can take place from either a localised donor exciton (incoherent transition) or from the delocalised superposition of initial states (fully coherent transition).[83] The latter was investigated by a quantum mechanical lattice model simulating the electronic dynamics of photoexcitation in polymer : fullerene heterojunctions, including the effects of polymer π-stacking, energetic disorder and vibrational relaxation.[111] The initial delocalised exciton on the polymer donor, treated as a coherent superposition of energy eigenstates, evolved according to the time-dependent Schrödinger equation. It was found that noise fluctuations in the lattice vibrations that couple to the exciton induce couplings between the exciton and resonant delocalised charge transfer states across the heterojunction. This allowed for long-range tunnelling of the electron away from the initial excitation, directly into photocurrent producing states, within 35 fs.

Although much attention has been drawn to the part played by intermolecular electron delocalisation in fullerene clusters, there is experimental evidence that a similar beneficial effect for spatial charge separation is obtained by hole delocalisation in neat polymer domains. Based on recombination rates in frozen electron–hole pairs at low temperature, blends of P3HT and PCDTBT with minimal PCBM content (5%, not allowing any fullerene aggregation) were found to efficiently produce long-range separate charges, possibly *via* the coupling of delocalised excitons to delocalised charges in the polymer.[81] We note that this study indicates that hole transfer from photoexcited PCBM also leads to ultrafast separate charge generation in the presence of neat fullerene domains, even if this mechanism has been much less studied. More evidence for the importance of neat polymer domains was obtained by combining time-resolved microwave conductivity, time-resolved photoluminescence and femtosecond transient absorption spectroscopy.[118] Donor–acceptor blends containing either only polymer aggregates with a dilute non-aggregated acceptor (non-fullerene SiPc), or containing only PCBM aggregates with a non-aggregated polymer (regiorandom P3HT) were investigated. It was found that long-lived separate charges were only generated if intermolecular aggregates were present, but it did not matter whether the aggregates formed in the polymer or fullerene phase. It still needs to be established whether the delocalisation in the polymer has to be truly intermolecular, or whether intramolecular delocalisation enhanced by better chain ordering in the aggregates is sufficient.

This discussion has focused on the importance of having neat fullerene or polymer domains to allow delocalisation for ultrafast electron–hole pair separation. The intermixed phase is, in principle, not necessary in this mechanism. Another proposed mechanism to explain why the spatial separation of electron–hole pairs is specifically enhanced if both neat and intermixed phases coexist concerns the involvement of an energy level cascade between the different phases. The energy level cascade was first proposed for the pBTTT : PCBM system, where a blend consisting of only the intermixed co-crystal phase (with important gCR) was compared with a blend consisting of intermixed and neat fullerene regions (with less gCR and higher solar cell efficiency).[62] It was shown by cyclic voltammetry that PCBM aggregation in the neat fullerene regions leads to a 100–200 meV increase in its electron affinity compared with non-aggregated fullerenes in the intermixed regions. This energetic offset (cascade) favours the migration of electrons from the intermixed to the neat regions and thus represents a driving force for the spatial separation of electron–hole pairs that are efficiently generated by exciton dissociation in the co-crystal regions.

The concept of having an energetic landscape due to the coexistence of different phases in polymer:fullerene blends has been extended to hole migration in the polymer.[112] Using a combination of cyclic voltammetry, steady-state absorption and ultraviolet photoelectron spectroscopy, a variety of different polymers (including P3HT and pBTTT) blended with two C_{70}-fullerene derivatives were investigated. It was found that different degrees of disorder between aggregated and amorphous neat polymer domains result in an energetic offset (ΔE_{DI}) due to bandgap widening of up to 150 meV, whereas intermolecular interactions between the polymer and fullerene can shift the energy levels of the polymer (and possibly fullerene) by >350 meV (ΔE_{IM}) (Figure 7.11).

These intermolecular effects might include charge transfer and electrostatic interactions. Overall, the energetic offsets drive the holes away from (amorphous) intermixed regions to electron-poor aggregated polymer domains, thus favouring spatial charge separation and preventing recombination. The emerging picture suggests that the energy cascade to separate electron–hole pairs can be optimised by having a phase morphology where

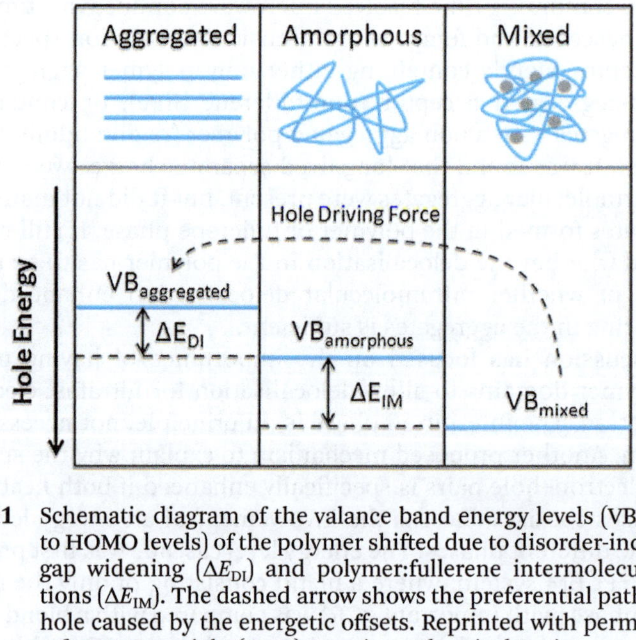

Figure 7.11 Schematic diagram of the valance band energy levels (VB, equivalent to HOMO levels) of the polymer shifted due to disorder-induced bandgap widening (ΔE_{DI}) and polymer:fullerene intermolecular interactions (ΔE_{IM}). The dashed arrow shows the preferential pathway for the hole caused by the energetic offsets. Reprinted with permission from ref. 112. Copyright (2014) American Chemical Society.

neat polymer and fullerene regions are separated by a narrow intermixed phase at the interface. However, a recent study on model donor–acceptor bilayer systems with controlled interfacial "cascade" layers suggested that,[119] while being beneficial for electron–hole pair separation, an interfacial intermixed layer can favour gCR and repel excitons from the interface due to unfavourable energetics for Förster energy transfer towards the intermixed region.[114] Alternatively, an interlayer of a neat material distinct from the bulk donor or acceptor (another polymer or functionalised fullerene derivative) could be used to obtain an optimised energy level cascade. It was shown that even a monolayer might be sufficient to efficiently drive the separation of electron–hole pairs.[113]

Both the concept of optimising the energetic landscape at the interface and the concept of optimising the aggregation of the fullerene and polymer domains for charge delocalisation imply that the donor–acceptor interface must be engineered at the molecular level. The molecular arrangement at the interface also impacts closely related properties that might contribute to spatial charge separation, such as interfacial electric fields, disorder or entropy effects.[79,119] The distance, orientation and packing of the donor and acceptor determines their electronic coupling (orbital overlap), which has a significant influence on the performance of OPV devices because charge separation/recombination, delocalisation and intermolecular interactions that shift the energy levels are all affected by this coupling.[120–122] Microscopic electrostatic effects occur near the interface due to a mismatch in dielectric constants of the donor and acceptor, molecular packing and static or induced dipoles around the photogenerated charges.[119,123] The related electric fields at the interface affect the energy levels (generating cascades) and provide screening that reduces the Coulomb binding of the electron–hole pair, driving the charges away from the interface. The nanoscale disorder at the interface also shapes the energetic landscape and, together with delocalisation, this makes the transfer of charges into neat domains (with a higher density of states) entropically favourable.[119,124–126]

7.3.5 Effect of Transport Properties on Electron–Hole Separation

Irrespective of the mechanism of free charge separation, the dissociation of the electron–hole pair implies that the charges must have a pathway for separation, *i.e.* that they can get away from each other *via* a hopping mechanism or delocalisation. Transport therefore plays a primordial part in allowing the spatial separation of the electron–hole pair. If there are no percolating electron or hole transport pathways, this can severely hamper the OPV efficiency, as has been observed for PBDTTPD : PCBM blends containing a low fullerene concentration in the intermixed regions.[61] According to the Onsager–Braun formalism, there is kinetic competition between the dissociation of the electron–hole pair as the charges hop apart, and of gCR when the charges are in close proximity.[80] The spatial charge separation is therefore expected to depend on the drift of the carriers, thus on parameters such as their mobility, the electric field and the temperature. When used to reproduce experimental data, this model requires the lifetime of the charge transfer state to be excessively long (of the order of microseconds), although the measured lifetime is typically only a few nanoseconds. Therefore various extensions and adaptations have been proposed.[79] Although the Onsager–Braun model stipulates that the dissociation rate depends on the sum of the electron and hole mobilities (and is thus dominated by the faster carrier), a recent study suggests that it is actually the slower carrier which determines the efficiency of charge separation.[127] A careful analysis of the separate charge yield (*via* IQE and transient absorption measurements), together with the charge mobilities (obtained by resistance-dependent photovoltage and intensity-dependent photocurrent experiments) was used. It was concluded that it is entropically favourable if both charges (not just the faster one) move away from the interface simultaneously.

A difficulty in evaluating the effect of charge mobility on the separation of electron–hole pairs arises from the fact that mobility in OPV blends is not a unique value, but changes over distance and time.[128–130] This dispersive charge transport implies that the mobility on the nanometre length scale, which is the scale that matters for electron–hole pair separation, can be orders of magnitude higher than the mobility measured in, for example, transistors over micrometre distances. On the very short length scale, the charges can travel intra-molecularly (*e.g.* within a single polymer chain), but with increasing distance they will first have to hop between molecules and will then encounter an increasing number of grain boundaries and traps, which gradually slows them down. There might be relaxation of the charges within the disordered density of states of the available hopping sites, which leads to a decrease in mobility with time.[128]

Contactless spectroscopic techniques, such as terahertz (THz) or microwave conductivity experiments, allow direct access to the local, short-range carrier mobility.[130–132] The drift of the charges within the oscillating electric field of the THz or microwave radiation (period 1–100 ps) is measured, which corresponds to the nanometre length scale. The local THz mobility of regiorandom and regioregular P3HT films was 0.1 and 0.4 $cm^2\ V^{-1}\ s^{-1}$, respectively, compared with values of about 2×10^{-5} and 0.06 $cm^2\ V^{-1}\ s^{-1}$ reported for field-effect transistors.[130] The regioregularity strongly affects intermolecular hopping because the polymer packing is influenced by the side-chain arrangement. This explains why the long-range transistor mobility is extremely low in the amorphous, regiorandom sample. The difference in local mobility compared with the regioregular film is much less pronounced for the THz measurements because the predominantly intra-chain local transport is only slightly affected by differences in backbone planarity due to more disorder in the regiorandom film. The short-range THz mobility is generally found to be independent of the time delay after charge photogeneration (up to 1 ns) and of temperature.[130,133] This shows that any vibrational or electronic relaxation of the photogenerated charges, or relaxation within the disordered density of states of available hopping sites, does not affect the short-range mobility within the polymer chains.

In a kinetic Monte Carlo investigation simulating the dissociation of electron–hole pairs in OPV blends, the high local THz mobility was used to account for the hopping apart of the charges according to an Onsager–Braun-related charge transfer state dissociation mechanism.[106] The dissociation efficiency was determined by the probability of a splitting event (single hop) being successful, multiplied by the average number of splitting events before competing recombination, given by the local charge mobility multiplied by the lifetime of the charge transfer state, which was corrected to represent only the gCR rate and not the dissociation. The probability that a hop leads to successful splitting was determined by the energetic landscape, *i.e.* by the presence of energetic offsets due to the coexistence of neat and intermixed regions. It was found that using high local mobilities and corrected charge transfer state lifetimes (and possibly energy level cascades if the recombination rate was too fast), efficient spatial separation of the electron–hole pairs could be explained without any need for the involvement of delocalisation or hot states.

A direct way of visualising the separation of the electron–hole pair in full OPV devices under reverse bias consists of electromodulated differential absorption (EDA) and time-resolved electric-field-induced second harmonic generation (TREFISH) spectroscopy.[128,134–138] Both these femtosecond pump-probe approaches rely on the decrease in the electric field in the active layer of a complete device when the photogenerated charges drift towards the electrodes and gradually screen the internal electric field (the sum of the externally applied reverse field and the intrinsic built-in field). Thus the average distance of the electron–hole separation as well as charge mobility can be extracted from the electric field dynamics. EDA and TREFISH differ only in the way of probing the decrease in the electric field. In EDA (Figure 7.12a), the electroabsorption (Stark effect) of the active layer is probed, which typically has a quadratic field dependence. In TREFISH, the reduction of the field-induced second harmonic generation is exploited. The dispersive, slowing down transport of the charges as they sample greater distances with time can be obtained with high time resolution and the electron–hole separation distance in the direction of the electrodes is followed from hundreds of femtosecond to nanosecond timescales.

The electron–hole separation in P3HT:PCBM devices has been investigated using TREFISH spectroscopy.[136] It was shown that charge pairs reach an electron–hole separation of about 3 nm within 10 ps. By calculating the diffusion distance from the charge mobility using the Einstein relation and modelling the carrier motion using Monte Carlo simulations, it was concluded that fast charge diffusion in the energetically disordered medium drives this initial separation by increasing the entropy of the system. At later times (hundreds of picoseconds), the Coulomb interaction of the electron–hole pairs is effectively overcome and the carrier separation is still dominated by diffusion, although drift in the internal electric field contributes

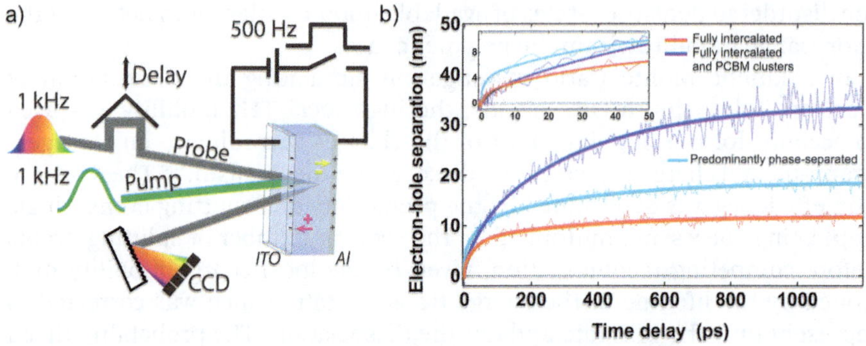

Figure 7.12 (a) Schematic diagram of the experimental set-up for EDA measurements. (b) Time-resolved electron–hole separation in pBTTT:PCBM blends with different phase morphologies (fully intercalated 1:1 blend, 1:4 blend consisting of co-crystal regions and PCBM clusters, phase-separated blend treated with Me 12; see Figure 7.5). Adapted by permission from Macmillan Publishers Ltd: *Nature Communications* (ref. 67), Copyright 2016.

more significantly. Overall, the experimental results, together with Monte Carlo simulations, showed that fast carrier motion, such as along the polymer chain, rather than an initial large electron–hole pair separation was the main reason for high dissociation efficiency. The same experimental approach was also applied to study the charge carrier motion in poly[2,7-(9,9-dioctylfluorene)-*alt*-5,5-(4,7′-di-2-thienyl-2′,1′,3′-benzothiadiazole)] (APFO3) blends with PCBM.[135] Different APFO3:PCBM ratios were considered to separately study the electron and hole motion dynamics in different phase morphologies. High initial mobilities (comparable with THz mobilities) were observed at short times, which then gradually decreased due to dispersive charge motion. It was found that the electron mobility depended strongly on the blend stoichiometry. The 1:4 blend (with 80% PCBM) showed electron extraction in <100 ps at 4 V applied voltage, whereas in the 2:1 sample (with 33% PCBM), the electron extraction took tens to hundreds of nanoseconds and required a higher applied field to be efficient. By contrast, hole extraction changed only from several tens of nanoseconds to >100 ns from 67 to 20% of polymer content, making the hole mobility less crucial in solar cell efficiency. These observations led to the conclusion that high OPV performance, as for the APFO3:PCBM 1:4 blend, is due to the fast electron motion in fullerene clusters, preventing gCR even at low applied fields.

The motion of electrons and holes was also directly followed in pBTTT:PCBM blends with different phase morphologies (co-crystal phase, neat donor and acceptor phase, obtained by manipulating the microstructure with additives and with excess PCBM content; Figure 7.5) using EDA spectroscopy.[67] EDA spectroscopy is sensitive only to those electron–hole pairs that dissociate to separate charges and are transported to the electrodes (in contrast with transient absorption measurements, where both bound and separate charges are seen). Therefore the voltage drop obtained by the EDA technique scaled with the yield of separate charges and was highest in the optimised 1:4 blend consisting of intermixed phase and neat fullerene clusters. When extractable charges were generated in the intermixed co-crystal phase, the electrons reached the neat fullerene domains (present in the 1:4 sample with 80% PCBM) after 10 ps, whereas the holes arrived in the neat pBTTT domains (present in the phase-separated samples processed with additives) in <200 fs, mainly due to the high hole mobility along the backbone of polymer chains spanning from the intermixed to the neat regions (Figure 7.12b). In the fully intermixed 1:1 blend (without any neat regions), an electron–hole separation of only 10 nm was achieved within 1 ns, whereas this increased to 20–30 nm after 1 ns in the presence of neat pBTTT and/or PCBM domains.

In the same study, the transport of electrons and holes between the different phases of the blends was also confirmed by the evolution of the electroabsorption signature extracted from femtosecond transient absorption measurements.[67] It was demonstrated that the observed electroabsorption signal was essentially associated with the presence of holes in the co-crystal phase, whereas holes in neat pBTTT regions and electrons on PCBM showed a negligible electroabsorption contribution in the transient absorption spectra. The evolution of the electroabsorption signal was therefore a convenient probe for estimating in which phase the charges were formed and how they moved. The results (obtained without bias) agreed with the EDA measurements.

Another conclusion from the EDA experiment was that, under conditions that favoured a higher yield of separate charges (optimised phase morphology, high reverse bias), an increased voltage drop in the devices was observed from the earliest measurable times after photoexcitation. This implies that the branching between electron–hole pairs that will undergo gCR and those that will be able to separate was determined instantaneously, at the moment of their generation by exciton dissociation. Nearby neat phase domains could shift the branching ratio towards more separating charge generation. Nevertheless, even if the ability of charge pairs to dissociate was determined on the sub-100 fs timescale, the spatial dissociation of the charges to reach the ~5 nm electron–hole separation needed to overcome their Coulomb attraction was relatively slow and could take 5–30 ps depending on the phase morphology (Figure 7.12b). Overall, the investigation reconciled the previously reported contradictory hypotheses about the dissociation rate of electron–hole pairs. It showed that it is instantaneously determined (by the molecular arrangement and local environment of the charge pairs) whether the electron–hole pairs are able to dissociate. However, the actual spatial dissociation of the charges if fairly slow. The existence of charge transfer states with different abilities to dissociate, depending on the local environment, has also been reported in a photo- and electroluminescence study of various polymer:fullerene blends.[139] The general idea is consistent with a theoretical investigation revealing the presence of different localised and delocalised charge transfer states that account for ultrafast branching between the dissociating and recombining charges.[117]

7.4 Non-geminate Recombination of Separate Charges

We have discussed how separate charges are generated in polymer : fullerene OPV blends and will now overview the loss mechanisms that can lead to the recombination of the separate charges. In the ideal case, all the absorbed photons dissociate to separate charges, are transported through percolated donor and acceptor domains, and are collected at the respective electrodes, which would translate into a unity internal quantum yield. In reality, there are several loss or recombination mechanisms that will lower the IQE and affect the open circuit voltage (V_{OC})[140] and fill factor of the devices.[141,142] The terminology distinguishes between geminate and non-geminate recombination. The former describes the recombination of an electron and a hole originating from a single photon absorption event, such as exciton recombination or the gCR of bound electron–hole pairs. After the charge pairs have split into separate charges, they have to make it to their respective electrodes. During this time, the charges can undergo non-geminate recombination with opposite free or trapped charges originating from a different photon absorption event or from charge carrier injection through the electrodes. Consequently, the efficiency of OPV devices and the carrier dynamics depend on the competition between the charge extraction and recombination rates, as demonstrated in a TDCF study in P3HT : PCBM blends.[99] From these definitions of geminate and non-geminate recombination, it is evident that they can be distinguished based on their kinetic order with respect to the excitation density (number of absorbed photons per unit volume). Geminate recombination is a monomolecular process that occurs with dynamics independent of the excitation density, whereas non-geminate recombination might depend on the intensity of light excitation. This has been elegantly demonstrated in a transient absorption study that investigated charge generation and recombination in P3HT : PCBM blends for different excitation fluences and a kinetic model taking into account the monomolecular recombination of charge pairs and the bimolecular recombination of separate charges.[80]

The following sections focus on different non-geminate recombination mechanisms, which can be distinguished by their respective dependence on the charge carrier density.[11,143] For example, monomolecular trap-assisted recombination shows a linear dependence, Langevin free carrier recombination a quadratic dependence and Auger-type recombination a cubic dependence on the excitation density. Various transient and steady-state techniques working on functional devices, including transient photovoltage,[144–146] photo-CELIV,[147,148] time-of-flight,[149] TDCF[99] and steady-state current–voltage measurements, in addition to optical techniques such as pump–probe spectroscopy with or without bias, are available to study the recombination processes.

It is evident that non-geminate recombination and therefore charge carrier collection efficiency at the electrodes strongly depend on the applied bias. Once an electron–hole pair is dissociated, the separate charges drift towards the electrodes due to the internal electric field in the device. Increasing the applied forward bias from zero to V_{OC} decreases the internal electric field as the driving force for charge extraction so that non-geminate recombination losses increase. At V_{OC}, all free charge carriers are lost due to non-geminate recombination and no charge is collected at the electrodes. At an intermediate bias, the field dependence is reflected by the fill factor. If the fill factor is > 0.65, an applied forward bias has little effect on the collection efficiency at the maximum power point.[11] Using steady-state and time-resolved techniques (TDCF), the relation between the fill factor and a single parameter θ, describing the ratio of the recombination and extraction rate, was demonstrated.[142] A broad range of materials and different processing conditions were investigated. In this way, material parameters such as the mobility, recombination rate and internal field were varied. Non-geminate bimolecular recombination also has an influence on V_{OC}.[150,151] In a metal–insulator–metal model of OPV device data, V_{OC} was shown to scale with the difference between the acceptor LUMO and the donor HOMO level according to $V_{OC} = e^{-1}\,(E_{LUMO} - E_{HOMO}) + \Delta V$. ΔV is usually an empirical term obtained from testing many materials, but it can be expressed as a function depending on the balance between the generation current and the non-geminate recombination current according to $\Delta V = -n_{id}k_bTe^{-1}\ln(J_{BI}/J_{SC})$, with n_{id}, $k_{b,}$ J_{BI} and J_{SC} being the diode ideality factor, the Boltzmann constant, the non-geminate recombination current and the short circuit current, respectively.

7.4.1 Bimolecular Langevin Recombination

In optimised solar cells with efficient electron–hole pair dissociation, the bimolecular recombination of separate charges is the most common loss mechanism. Here a free electron recombines non-geminately with a free hole. Such a recombination of two mobile carriers can be seen as a two-step process. First, the two opposite charges have to find each other, then they actually recombine. In materials with low electron and hole mobilities, the recombination rate is diffusion-limited. The faster the charges are moving, the faster they will encounter each other. The rate of such recombination in a three-dimensional system can be described by the Langevin expression:[152]

$$R = \frac{q}{\varepsilon}\left(\mu_n + \mu_p\right)\left(np - n_{int}p_{int}\right) \tag{7.1}$$

Here, q is the elementary charge, ε is the dielectric constant, $\mu_{p/n}$ is the mobility of the hole and electron, n and p are their respective densities, and n_{int} and p_{int} are the intrinsic carrier concentrations. As n_{int} and p_{int} are usually small in OPV systems, they can be neglected to simplify the expression. Langevin recombination has been proposed and experimentally observed in different OPV systems.[153] Eqn (7.1) suggests that the recombination should decrease with decreasing mobility, but decreasing mobility also slows down charge extraction and consequently increases the carrier concentration. Therefore a decreasing bimolecular recombination yield is generally observed with increasing mobility.

The Langevin expression is based on two assumptions. First, the Coulomb capture radius has to be larger than the mean free path of the charges. This assumption is often met in organic materials due to their low mobility. Second, charge carrier transport is supposed to be homogeneous within the active layer. This assumption is clearly not met in complex BHJs. The hole and electron densities are correlated and the phase separation of the donor and acceptor leads to a spatial charge distribution and percolative transport of the holes and electrons in the donor and acceptor materials, respectively. Therefore observed bimolecular recombination rates are often lower than predicted by the Langevin model. This deviation is accounted for by a Langevin reduction factor. The reduced rate is generally explained by the phase separation in OPV devices and the fact that bimolecular recombination occurs at the donor–acceptor interface. The electron is confined in the acceptor phase while the hole resides in the donor phase. This spatial separation decreases the probability of an electron–hole encounter.

The effect of phase segregation in P3HT : PCBM films was investigated by transient voltage measurements and Monte Carlo simulations. A reduction in the Langevin recombination rate by two to three orders of magnitude was found in annealed phase-segregated samples, whereas in non-annealed films without phase separation the Langevin and measured recombination rates were of the same order of magnitude.[154] A first approach to account for phase separation was to replace the sum of electron and hole mobility by a spatial average.[155] Later, the fill factor for systems with different electron and hole mobilities was modelled and it was shown that the recombination rate only depends on the mobility of the slower carrier.[156] The faster charge carrier approaches and becomes stuck at the donor–acceptor interface until the slower charge carrier arrives to undergo recombination. The bimolecular recombination rate was also found to be directly proportional to the carrier concentration, mobility anisotropy, dimensionality and spatial confinement of the charges.[157] The influence of the interface in the BHJ was further considered by introducing the concept of image charges known from metal–semiconductor interfaces.[158,159]

The effect of molecular weight on the bimolecular recombination in PCDTBT : fullerene systems was studied using the high-intensity resistance-dependent photovoltage technique.[146] The results showed that recombination for low molecular weight PCDTBT follows the Langevin model, whereas recombination is strongly reduced in the devices prepared from high molecular weight PCDTBT. Such a reduction in the bimolecular recombination rate compared with the three-dimensional Langevin model also occurs if recombination is taking place in two-dimensional structures. Mobilities in P3HT can differ by >100 times along and across the semicrystalline lamellar structures formed by the polymer chains in the solid state, rendering charge transport essentially two-dimensional.[160] In this case, the bimolecular recombination coefficient is reduced according to $\beta_{2D} = 3/4\pi^{1/2}l^{3/2}n^{1/2}\beta_{3D}$ (where l is the lamellar spacing) and the recombination order increases from 2 (bimolecular) to 2.5.[80,161]

7.4.2 Recombination *via* Triplet States

Organic materials are particularly prone to spin-related phenomena because spin–orbit coupling is weak in the absence of heavy atoms. Therefore spin selection rules are rigorously followed and influence the OPV device properties. As discussed earlier, bimolecular recombination of separate charges in polymer : fullerene blends occurs mainly at the interface of the donor and acceptor. As the electron and hole are uncorrelated, both singlet (^{1}CT) and triplet (^{3}CT) charge transfer states are formed on the encounter of the charges in a 1 : 3 ratio as dictated by spin statistics. The ^{1}CT state can recombine directly to the ground state radiatively or non-radiatively, or re-dissociate to form separate charges again. The direct transition from the ^{3}CT state to the ground state is spin-forbidden, but if the triplet state of the polymer is lower than the ^{3}CT state, then it represents an efficient recombination channel by triplet charge annihilation (Figure 7.9a). From either ^{1}CT or ^{3}CT, there is a kinetic competition to either recombine or to form separate charges again.

The role of triplet-mediated recombination in the polymer : fullerene active layer was investigated using transient absorption spectroscopy to follow the interconversion of charges to triplet states.[90] BHJs composed of different donor and acceptor materials, varying in energy level alignment and charge transfer state delocalisation, were investigated. It was possible to demonstrate that recombination is governed by spin and delocalisation in addition to energetics. Triplet state formation during non-geminate recombination in PCPDTBT : fullerene blends with different processing conditions and morphologies was also studied by transient absorption spectroscopy and two-dimensional nuclear magnetic resonance sprectrometry.[162] The yield of triplet state formation was found to be influenced by the blend microstructure and probably charge transfer state delocalisation. A dielectric detection scheme allowed probing of the spin-selective recombination of charge transfer states by their dielectric polarisability under magnetic resonance.[163] The high magnetic fields of the microwave radiation suppressed the hyperfine splitting between electron and nuclear spins, leading to a Rabi splitting in the population response. A quantitative model to describe the contribution of singlet and triplet rates was developed.

7.4.3 Trap-based Recombination

Although the average distance between opposite charges is low and bimolecular recombination is the dominant recombination mechanism at high charge carrier densities, this picture changes at low charge densities and recombination becomes trap-mediated.[164] In this case, the reduction of the Langevin recombination factor by more than one order of magnitude cannot be accounted for by taking into consideration only mobility mismatch, energetic disorder and domain size, so that recombination *via* trap states has been invoked.[157] Bimolecular recombination sets in at charge densities where all the localised trap states are filled.[165] In the trap-assisted recombination mechanism, a charge is first immobilised in a trap before a second opposite charge recombines with the former. Therefore the rate of the recombination depends on the density of trap states and the capture rate of the trapped charge. The trap-assisted recombination kinetics can be dependent or independent of the charge carrier density (it is often monomolecular).[80]

A model for the recombination of trapped charges with opposite free charges in inorganic semiconductors was proposed by Shockley and Read and is known as the Shockley–Read–Hall mechanism.[166,167] The Shockley–Read–Hall mechanism has been incorporated into a model of Langevin-type recombination for organic thin films.[156,168] A strong electron acceptor (acting as a trap) was added into MDMO-PPV : PCBM blends and the light intensity dependence of V_{OC} was investigated. Assuming Langevin recombination to be the only non-geminated recombination pathway, V_{OC} depended linearly on the absorbed photon flux with a slope of kT/q. The introduction or presence of trap states increased the slope, which could be modelled by incorporating the Shockley–Read–Hall term.

7.4.4 Auger and Higher Order Recombination

Auger recombination might occur in OPV systems at very high charge carrier densities.[169] The Auger recombination mechanism requires the participation of three particles. An electron and a hole recombine, transferring the energy to a third carrier, which is excited to a higher energetic state. This process is well established for inorganic materials[170] and the new emerging class of hybrid inorganic–organic perovskites,[171] but not yet for organic materials.[172] Higher order (>2) recombination kinetics have been observed in OPV systems,[80,145,173,174] but they are attributed to a rate constant dependent on the charge carrier density rather than to a real trimolecular process. The stronger dependence on charge carrier density might be due to charge carrier dependent mobility, trap states or restricted carrier diffusion due to the morphology. The Langevin rate constant depends on the mobility, which, in turn, depends on the charge carrier density.[175,176]

Transport in amorphous organic materials is commonly described by the variable-range hopping concept, *i.e.* a carrier either hops over a small distance with a high activation energy or over a larger distance with a lower activation energy. In amorphous semiconductors, the states participating in transport are not at a single energy level, but usually present a Gaussian distribution of energy states. At low charge densities, lower lying states are required to participate in the transport, whereas higher energy states are populated at increasing charge carrier densities, so that the activation energy for transport decreases and the mobility increases.[177] In addition to mobility effects, the spatial carrier concentration and contact effects can lead to an increased recombination order. Modelling the thickness dependence of the recombination order in OPV devices showed massively increased reaction orders for devices with $d < 70$ nm.[178] The restriction of charge transport to two dimensions is also a reason for the observation of a higher kinetic order.

7.5 Conclusions

In organic solar cells consisting of polymer : fullerene blends, charge generation and recombination processes play the primary part in determining how efficiently light can be converted to photocurrent. We have discussed here how light absorption in conjugated polymers leads to an initially hot and delocalised state, which relaxes within 100–200 fs to a bound exciton. This exciton dissociates at a donor–acceptor junction to an electron–hole pair. The ultrafast rate of exciton quenching is typically not consistent with exciton diffusion towards a fullerene interface in 10–20 nm phase-segregated bulk heterojunctions. It can, however, be reconciled with the splitting of delocalised excitons (possibly before their initial relaxation), or by the presence of an intermixed donor–acceptor phase, where exciton diffusion is not necessary.

Once the electron–hole pair is formed, it needs to spatially dissociate to separate charges against the Coulomb interaction. Mechanisms to account for this include the involvement of hot charge transfer states (although even relaxed charge transfer states have been shown to dissociate), delocalisation into aggregated fullerene or polymer domains (leading to ultrafast long-range charge separation), energy level cascades due to the coexistence of different phases with a rich energetic landscape in the blend morphology, high local charge mobilities, or entropic effects. It is probable that several of these mechanisms occur depending on the system.

The emerging picture suggests that the branching of electron–hole pairs into a fraction that can dissociate and a fraction that will geminately recombine has already been determined at the moment of exciton dissociation. Once separate charges are generated, they have to be collected at the device electrodes, which competes with non-geminate recombination during their transport to the electrodes. The most common mechanisms include bimolecular recombination according to a Langevin-type mechanism, possibly involving triplet states, and trap-assisted recombination. Overall, the discussion here demonstrates that optimising charge generation while eliminating recombination for the highest OPV efficiency implies carefully balancing the material, processing and device properties for optimum energy level alignments and structure of the BHJ.

Acknowledgements

We thank all the collaborators with whom we have had the privilege to investigate organic solar cells over recent years, in particular Natalie Stingelin (Georgia Institute of Technology), Martin Heeney (Imperial College London) and Mario Leclerc (Laval Université, Quebec). We acknowledge funding from the Swiss National Science Foundation (PP00P2_150536) and the University of Fribourg.

References

1. C. W. Tang, *Appl. Phys. Lett.*, 1986, **48**(2), 183.
2. G. Yu, J. Gao, J. C. Hummelen, F. Wudl and A. J. Heeger, *Science*, 1995, **270**(5243), 1789.
3. J. J. M. Halls, C. A. Walsh, N. C. Greenham, E. A. Marseglia, R. H. Friend, S. C. Moratti and A. B. Holmes, *Nature*, 1995, **376**(6540), 498.
4. M. A. Green, K. Emery, Y. Hishikawa, W. Warta and E. D. Dunlop, *Prog. Photovolt: Res. Appl.*, 2015, **23**(7), 805.
5. Q. Zhang, B. Kan, F. Liu, G. Long, X. Wan, X. Chen, Y. Zuo, W. Ni, H. Zhang, M. Li, Z. Hu, F. Huang, Y. Cao, Z. Liang, M. Zhang, T. P. Russell and Y. Chen, *Nat. Photonics*, 2015, **9**(1), 35.
6. R. Meerheim, C. Körner, B. Oesen and K. Leo, *Appl. Phys. Lett.*, 2016, **108**(10), 103302.
7. C. B. Nielsen, S. Holliday, H.-Y. Chen, S. J. Cryer and I. McCulloch, *Acc. Chem. Res.*, 2015, **48**(11), 2803.
8. B. C. Thompson and J. M. J. Fréchet, *Angew. Chem., Int. Ed.*, 2008, **47**(1), 58.
9. C. J. Brabec, G. Zerza, G. Cerullo, S. De Silvestri, S. Luzzati, J. C. Hummelen and S. Sariciftci, *Chem. Phys. Lett.*, 2001, **340**(3–4), 232.
10. A. C. Jakowetz, M. L. Böhm, J. Zhang, A. Sadhanala, S. Huettner, A. A. Bakulin, A. Rao and R. H. Friend, *J. Am. Chem. Soc.*, 2016, **138**(36), 11672.
11. C. M. Proctor, M. Kuik and T.-Q. Nguyen, *Prog. Polym. Sci.*, 2013, **38**(12), 1941.
12. J. D. A. Lin, O. V. Mikhnenko, J. Chen, Z. Masri, A. Ruseckas, A. Mikhailovsky, R. P. Raab, J. Liu, P. W. M. Blom, M. A. Loi, C. J. García-Cervera, I. D. W. Samuel and T.-Q. Nguyen, *Mater. Horiz.*, 2014, **1**(2), 280.
13. N. Banerji, *J. Mater. Chem. C*, 2013, **1**(18), 3052.
14. P. Westacott, J. R. Tumbleston, S. Shoaee, S. Fearn, J. H. Bannock, J. B. Gilchrist, S. Heutz, J. deMello, M. Heeney, H. Ade, J. Durrant, D. S. McPhail and N. Stingelin, *Energy Environ. Sci.*, 2013, **6**(9), 2756.
15. N. S. Sariciftci, *Primary Photoexcitations in Conjugated Polymers: Molecular Exciton versus Semiconductor Band Model*, World Scientific, Singapore, 1997.
16. A. J. Heeger, N. S. Sariciftci and E. B. Namdas, *Semiconducting and Metallic Polymers*, Oxford University Press, Oxford, UK, 2010.
17. A. J. Heeger, *Chem. Soc. Rev.*, 2010, **39**(7), 2354.
18. A. J. Heeger, S. Kivelson, J. R. Schrieffer and W. P. Su, *Rev. Mod. Phys.*, 1988, **60**(3), 781.
19. H. Bässler, *Macromol. Symp.*, 1996, **104**(1), 269.
20. H. Bässler, V. I. Arkhipov, E. V. Emelianova, A. Gerhard, A. Hayer, C. Im and J. Rissler, *Synth. Met.*, 2003, **135–136**, 377.
21. I. G. Scheblykin, A. Yartsev, T. Pullerits, V. Gulbinas and V. Sundström, *J. Phys. Chem. B*, 2007, **111**(23), 6303.
22. F. Dubin, R. Melet, T. Barisien, R. Grousson, L. Legrand, M. Schott and V. Voliotis, *Nat. Phys.*, 2006, **2**(1), 32.

23. I. Hwang and G. D. Scholes, *Chem. Mater.*, 2011, **23**(3), 610.
24. O. R. Tozer and W. Barford, *J. Phys. Chem. A*, 2012, **116**(42), 10310.
25. W. Barford, I. Boczarow and T. Wharram, *J. Phys. Chem. A*, 2011, **115**(33), 9111.
26. J. Pina, J. S. de Melo, H. D. Burrows, T. W. Bunnagel, D. Dolfen, C. J. Kudla and U. Scherf, *J. Phys. Chem. B*, 2009, **113**(49), 15928.
27. J. Sperling, F. Milota and H. F. Kauffmann, *Opt. Spectrosc.*, 2005, **98**(5), 729.
28. J. Sperling, F. Milota, A. Tortschanoff, C. Warmuth, B. Mollay, H. Bässler and H. F. Kauffmann, *J. Chem. Phys.*, 2002, **117**(23), 10877.
29. S. C. J. Meskers, J. Hübner, M. Oestreich and H. Bässler, *J. Phys. Chem. B*, 2001, **105**(38), 9139.
30. R. Kersting, B. Mollay, M. Rusch, J. Wenisch, G. Leising and H. F. Kauffmann, *J. Chem. Phys.*, 1997, **106**(7), 2850.
31. R. Kersting, U. Lemmer, R. F. Mahrt, K. Leo, H. Kurz, H. Bässler and E. O. Göbel, *Phys. Rev. Lett.*, 1993, **70**(24), 3820.
32. S. Heun, R. F. Mahrt, A. Greiner, U. Lemmer, H. Bassler, D. A. Halliday, D. D. C. Bradley, P. L. Burn and A. B. Holmes, *J. Phys.: Condens. Matter*, 1993, **5**(2), 247.
33. N. Banerji, S. Cowan, E. Vauthey and A. J. Heeger, *J. Phys. Chem. C*, 2011, **115**(19), 9726.
34. J. M. Guo, H. Ohkita, H. Benten and S. Ito, *J. Am. Chem. Soc.*, 2009, **131**(46), 16869.
35. T. E. Dykstra, E. Hennebicq, D. Beljonne, J. Gierschner, G. Claudio, E. R. Bittner, J. Knoester and G. D. Scholes, *J. Phys. Chem. B*, 2009, **113**(3), 656.
36. M. H. Chang, M. J. Frampton, H. L. Anderson and L. M. Herz, *Phys. Rev. Lett.*, 2007, **98**(2), 027402.
37. N. P. Wells, B. W. Boudouris, M. A. Hillmyer and D. A. Blank, *J. Phys. Chem. C*, 2007, **111**(42), 15404.
38. A. Ruseckas, P. Wood, I. D. W. Samuel, G. R. Webster, W. J. Mitchell, P. L. Burn and V. Sundström, *Phys. Rev. B*, 2005, **72**(11), 115214.
39. M. M. L. Grage, Y. Zaushitsyn, A. Yartsev, M. Chachisvilis, V. Sundström and T. Pullerits, *Phys. Rev. B*, 2003, **67**(20), 205207.
40. A. A. Paraecattil and N. Banerji, *J. Am. Chem. Soc.*, 2014, **136**(4), 1472.
41. X. Yang, T. E. Dykstra and G. D. Scholes, *Phys. Rev. B*, 2005, **71**(4), 045203.
42. T. E. Dykstra, V. Kovalevskij, X. Yang and G. D. Scholes, *Chem. Phys.*, 2005, **318**(1–2), 21.
43. J. Sperling, A. Nemeth, P. Baum, F. Sanda, E. Riedle, H. F. Kauffmann, S. Mukamel and F. Milota, *Chem. Phys.*, 2008, **349**(1–3), 244.
44. N. P. Wells and D. A. Blank, *Phys. Rev. Lett.*, 2008, **100**(8), 086403.
45. W. J. D. Beenken and T. Pullerits, *J. Phys. Chem. B*, 2004, **108**(20), 6164.
46. E. Collini and G. D. Scholes, *J. Phys. Chem. A*, 2009, **113**(16), 4223.
47. S. R. Cowan, N. Banerji, W. L. Leong and A. J. Heeger, *Adv. Funct. Mater.*, 2012, **22**(6), 1116.
48. L. G. Kaake, J. J. Jasieniak, R. C. Bakus, G. C. Welch, D. Moses, G. C. Bazan and A. J. Heeger, *J. Am. Chem. Soc.*, 2012, **134**(48), 19828.

49. K. Chen, A. J. Barker, M. E. Reish, K. C. Gordon and J. M. Hodgkiss, *J. Am. Chem. Soc.*, 2013, **135**(49), 18502.
50. Y. Song, S. N. Clafton, R. D. Pensack, T. W. Kee and G. D. Scholes, *Nat. Commun.*, 2014, **5**, 4933.
51. G. Grancini, M. Maiuri, D. Fazzi, A. Petrozza, H. J. Egelhaaf, D. Brida, G. Cerullo and G. Lanzani, *Nat. Mater.*, 2013, **12**(1), 29.
52. A. E. Jailaubekov, A. P. Willard, J. R. Tritsch, W.-L. Chan, N. Sai, R. Gearba, L. G. Kaake, K. J. Williams, K. Leung, P. J. Rossky and X.-Y. Zhu, *Nat. Mater.*, 2013, **12**(1), 66.
53. A. Armin, M. Velusamy, P. Wolfer, Y. Zhang, P. L. Burn, P. Meredith and A. Pivrikas, *ACS Photonics*, 2014, **1**(3), 173.
54. M. Scharber, *Nat. Mater.*, 2013, **12**(7), 594.
55. J. Peet, J. Y. Kim, N. E. Coates, W. L. Ma, D. Moses, A. J. Heeger and G. C. Bazan, *Nat. Mater.*, 2007, **6**(7), 497.
56. W. Ma, C. Yang, X. Gong, K. Lee and A. J. Heeger, *Adv. Funct. Mater.*, 2005, **15**(10), 1617.
57. S. E. Shaheen, C. J. Brabec, N. S. Sariciftci, F. Padinger, T. Fromherz and J. C. Hummelen, *Appl. Phys. Lett.*, 2001, **78**(6), 841.
58. Y. Huang, E. J. Kramer, A. J. Heeger and G. C. Bazan, *Chem. Rev.*, 2014, **114**(14), 7006.
59. F. Liu, Y. Gu, X. Shen, S. Ferdous, H.-W. Wang and T. P. Russell, *Prog. Polym. Sci.*, 2013, **38**(12), 1990.
60. P. Westacott, J. R. Tumbleston, S. Shoaee, S. Fearn, J. H. Bannock, J. B. Gilchrist, S. Heutz, M. J. de, M. Heeney, H. Ade, J. Durrant, D. S. McPhail and N. Stingelin, *Energy Environ. Sci.*, 2013, **6**, 2756.
61. J. A. Bartelt, Z. M. Beiley, E. T. Hoke, W. R. Mateker, J. D. Douglas, B. A. Collins, J. R. Tumbleston, K. R. Graham, A. Amassian, H. Ade, J. M. J. Fréchet, M. F. Toney and M. D. McGehee, *Adv. Energy Mater.*, 2013, **3**, 364.
62. F. C. Jamieson, E. B. Domingo, T. McCarthy-Ward, M. Heeney, N. Stingelin and J. R. Durrant, *Chem. Sci.*, 2012, **3**(2), 485.
63. B. A. Collins, J. R. Tumbleston and H. Ade, *J. Phys. Chem. Lett.*, 2011, **2**(24), 3135.
64. A. C. Mayer, M. F. Toney, S. R. Scully, J. Rivnay, C. J. Brabec, M. Scharber, M. Koppe, M. Heeney, I. McCulloch and M. D. McGehee, *Adv. Funct. Mater.*, 2009, **19**(8), 1173.
65. S. De, T. Kesti, M. Maiti, F. Zhang, O. Inganäs, A. Yartsev, T. Pascher and V. Sundström, *Chem. Phys. Lett.*, 2008, **350**(1–3), 14.
66. R. A. Marsh, J. M. Hodgkiss, S. Albert-Seifried and R. H. Friend, *Nano Lett.*, 2010, **10**(3), 923.
67. M. Causa', J. De Jonghe-Risse, M. Scarongella, J. C. Brauer, E. Buchaca-Domingo, J.-E. Moser, N. Stingelin and N. Banerji, *Nat. Commun.*, 2016, **7**, 12556.
68. M. Scarongella, J. De Jonghe-Risse, E. Buchaca-Domingo, M. Causa', Z. Fei, M. Heeney, J.-E. Moser, N. Stingelin and N. Banerji, *J. Am. Chem. Soc.*, 2015, **137**(8), 2908.

69. M. Scarongella, A. A. Paraecattil, E. Buchaca-Domingo, J. D. Douglas, S. Beaupré, T. McCarthy-Ward, M. Heeney, J.-E. Moser, M. Leclerc, J. M. J. Fréchet, N. Stingelin and N. Banerji, *J. Mater. Chem. A*, 2014, **2**(17), 6218.
70. B. H. Hamadani, D. J. Gundlach, I. McCulloch and M. Heeney, *Appl. Phys. Lett.*, 2007, **91**(24), 243512.
71. I. Mcculloch, M. Heeney, C. Bailey, K. Genevicius, I. McDonald, M. Shkunov, D. Sparrowe, S. Tierney, R. Wagner, W. M. Zhang, M. L. Chabinyc, R. J. Kline, M. D. Mcgehee and M. F. Toney, *Nat. Mater.*, 2006, **5**(4), 328.
72. E. Buchaca-Domingo, A. J. Ferguson, F. C. Jamieson, T. McCarthy-Ward, S. Shoaee, J. R. Tumbleston, O. G. Reid, L. Yu, M. B. Madec, M. Pfannmoller, F. Hermerschmidt, R. R. Schröder, S. E. Watkins, N. Kopidakis, G. Portale, A. Amassian, M. Heeney, H. Ade, G. Rumbles, J. R. Durrant and N. Stingelin, *Mater. Horiz.*, 2014, **1**(2), 270.
73. A. A. Bakulin, J. C. Hummelen, M. S. Pshenichnikov and P. H. M. van Loosdrecht, *Adv. Funct. Mater.*, 2010, **20**(10), 1653.
74. F. Etzold, I. A. Howard, N. Forler, D. M. Cho, M. Meister, H. Mangold, J. Shu, M. R. Hansen, K. Muellen and F. Laquai, *J. Am. Chem. Soc.*, 2012, **134**(25), 10569.
75. D. R. Kozub, K. Vakhshouri, S. V. Kesava, C. Wang, A. Hexemer and E. D. Gomez, *Chem. Commun.*, 2012, **48**, 5859.
76. Y. Zhang, A. K. Pandey, K. Tandy, G. K. Dutta, P. L. Burn, P. Meredith, E. B. Namdas and S. Patil, *Appl. Phys. Lett.*, 2013, **102**(22), 223302.
77. G. F. Burkhard, E. T. Hoke, Z. M. Beiley and M. D. McGehee, *J. Phys. Chem. C*, 2012, **116**(50), 26674.
78. S. Cook, H. Ohkita, Y. Kim, J. J. Benson-Smith, D. D. C. Bradley and J. R. Durrant, *Chem. Phys. Lett.*, 2007, **445**(4–6), 276.
79. S. Few, J. M. Frost and J. Nelson, *Phys. Chem. Chem. Phys.*, 2015, **17**(4), 2311.
80. I. A. Howard, R. Mauer, M. Meister and F. Laquai, *J. Am. Chem. Soc.*, 2010, **132**(42), 14866.
81. A. J. Barker, K. Chen and J. M. Hodgkiss, *J. Am. Chem. Soc.*, 2014, **136**(34), 12018.
82. F. Provencher, N. Berube, A. W. Parker, G. M. Greetham, M. Towrie, C. Hellmann, M. Cote, N. Stingelin, C. Silva and S. C. Hayes, *Nat. Commun.*, 2014, **5**, 4288.
83. S. Gélinas, A. Rao, A. Kumar, S. L. Smith, A. W. Chin, J. Clark, T. S. van der Poll, G. C. Bazan and R. H. Friend, *Science*, 2014, **343**(6170), 512.
84. D. Caruso and A. Troisi, *Proc. Natl. Acad. Sci. U. S. A.*, 2012, **109**(34), 13498.
85. K. Vandewal, A. Gadisa, W. D. Oosterbaan, S. Bertho, F. Banishoeib, I. Van Severen, L. Lutsen, T. J. Cleij, D. Vanderzande and J. V. Manca, *Adv. Funct. Mater.*, 2008, **18**(14), 2064.
86. C. Piliego and M. A. Loi, *J. Mater. Chem.*, 2012, **22**(10), 4141.
87. K. Vandewal, S. Albrecht, E. T. Hoke, K. R. Graham, J. Widmer, J. D. Douglas, M. Schubert, W. R. Mateker, J. T. Bloking, G. F. Burkhard, A. Sellinger, J. M. J. Fréchet, A. Amassian, M. K. Riede, M. D. McGehee, D. Neher and A. Salleo, *Nat. Mater.*, 2013, **13**, 63.

88. T. M. Burke, S. Sweetnam, K. Vandewal and M. D. McGehee, *Adv. Energy Mater.*, 2015, **5**(11), 1500123.
89. P. E. Keivanidis, V. Kamm, C. Dyer-Smith, W. Zhang, F. Laquai, I. McCulloch, D. D. C. Bradley and J. Nelson, *Adv. Mater.*, 2010, **22**(45), 5183.
90. A. Rao, P. C. Y. Chow, S. Gelinas, C. W. Schlenker, C.-Z. Li, H.-L. Yip, A. K. Y. Jen, D. S. Ginger and R. H. Friend, *Nature*, 2013, **500**(7463), 435.
91. I.-W. Hwang, C. Soci, D. Moses, Z. Zhu, D. Waller, R. Gaudiana, C. J. Brabec and A. J. Heeger, *Adv. Mater.*, 2007, **19**(17), 2307.
92. S. De, T. Pascher, M. Maiti, K. G. Jespersen, T. Kesti, F. Zhang, O. Inganäs, A. Yartsev and V. Sundström, *J. Am. Chem. Soc.*, 2007, **129**(27), 8466.
93. F. C. Jamieson, T. Agostinelli, H. Azimi, J. Nelson and J. R. Durrant, *J. Phys. Chem. Lett.*, 2010, **1**(23), 3306.
94. R. A. Marsh, J. M. Hodgkiss and R. H. Friend, *Adv. Mater.*, 2010, **22**, 3672.
95. A. A. Bakulin, Y. Xia, H. J. Bakker, O. Inganäs and F. Gao, *J. Phys. Chem. C*, 2016, **120**(8), 4219.
96. A. Zusan, K. Vandewal, B. Allendorf, N. H. Hansen, J. Pflaum, A. Salleo, V. Dyakonov and C. Deibel, *Adv. Energy Mater.*, 2014, **4**(17), 1400922.
97. S. Albrecht, W. Schindler, J. Kurpiers, J. Kniepert, J. C. Blakesley, I. Dumsch, S. Allard, K. Fostiropoulos, U. Scherf and D. Neher, *J. Phys. Chem. Lett.*, 2012, **3**(5), 640.
98. J. Kniepert, M. Schubert, J. C. Blakesley and D. Neher, *J. Phys. Chem. Lett.*, 2011, **2**(7), 700.
99. J. Kniepert, M. Schubert, J. C. Blakesley and D. Neher, *J. Phys. Chem. Lett.*, 2011, **2**(7), 700.
100. A. A. Bakulin, S. D. Dimitrov, A. Rao, P. C. Y. Chow, C. B. Nielsen, B. C. Schroeder, I. McCulloch, H. J. Bakker, J. R. Durrant and R. H. Friend, *J. Phys. Chem. Lett.*, 2013, **4**(1), 209.
101. S. D. Dimitrov, A. A. Bakulin, C. B. Nielsen, B. C. Schroeder, J. Du, H. Bronstein, I. McCulloch, R. H. Friend and J. R. Durrant, *J. Am. Chem. Soc.*, 2012, **134**(44), 18189.
102. A. A. Bakulin, A. Rao, V. G. Pavelyev, P. H. M. van Loosdrecht, M. S. Pshenichnikov, D. Niedzialek, J. Cornil, D. Beljonne and R. H. Friend, *Science*, 2012, **335**(6074), 1340.
103. P. A. Lane, P. D. Cunningham, J. S. Melinger, O. Esenturk and E. J. Heilweil, *Nat. Commun.*, 2015, **6**, 7558.
104. H. Tamura and I. Burghardt, *J. Am. Chem. Soc.*, 2013, **135**(44), 16364.
105. K. Vandewal, S. Albrecht, E. T. Hoke, K. R. Graham, J. Widmer, J. D. Douglas, M. Schubert, W. R. Mateker, J. T. Bloking, G. F. Burkhard, A. Sellinger, J. M. J. Fréchet, A. Amassian, M. K. Riede, M. D. McGehee, D. Neher and A. Salleo, *Nat. Mater.*, 2014, **13**(1), 63.
106. T. M. Burke and M. D. McGehee, *Adv. Mater.*, 2014, **26**(12), 1923.
107. P. M. Beaujuge and J. M. J. Fréchet, *J. Am. Chem. Soc.*, 2011, **133**(50), 20009.
108. I.-W. Hwang, J. Y. Kim, S. Cho, J. Yuen, N. Coates, K. Lee, M. Heeney, I. McCulloch, D. Moses and A. J. Heeger, *J. Phys. Chem. C*, 2008, **112**(21), 7853.

109. E. Buchaca-Domingo, K. Vandewal, Z. Fei, S. E. Watkins, F. H. Scholes, J. H. Bannock, J. C. de Mello, L. J. Richter, D. M. DeLongchamp, A. Amassian, M. Heeney, A. Salleo and N. Stingelin, *J. Am. Chem. Soc.*, 2015, **137**(16), 5256.
110. F. Dou, E. Buchaca-Domingo, M. Sakowicz, E. Rezasoltani, T. McCarthy-Ward, M. Heeney, X. Zhang, N. Stingelin and C. Silva, *J. Mater. Chem. C*, 2015, **3**(15), 3722.
111. E. R. Bittner and C. Silva, *Nat. Commun.*, 2014, **5**, 3119.
112. S. Sweetnam, K. R. Graham, G. O. Ngongang Ndjawa, T. Heumuller, J. A. Bartelt, T. M. Burke, W. Li, W. You, A. Amassian and M. D. McGehee, *J. Am. Chem. Soc.*, 2014, **136**(40), 14078.
113. S. Izawa, K. Nakano, K. Suzuki, K. Hashimoto and K. Tajima, *Adv. Mater.*, 2015, **27**(19), 3025.
114. K. Nakano, K. Suzuki, Y. Chen and K. Tajima, *Sci. Rep.*, 2016, **6**, 29529.
115. B. Bernardo, D. Cheyns, B. Verreet, R. D. Schaller, B. P. Rand and N. C. Giebink, *Nat. Commun.*, 2014, **5**, 3245.
116. B. M. Savoie, A. Rao, A. A. Bakulin, S. Gelinas, B. Movaghar, R. H. Friend, T. J. Marks and M. A. Ratner, *J. Am. Chem. Soc.*, 2014, **136**(7), 2876.
117. G. D'Avino, L. Muccioli, Y. Olivier and D. Beljonne, *J. Phys. Chem. Lett.*, 2016, **7**(3), 536.
118. H. M. Feier, O. G. Reid, N. A. Pace, J. Park, J. J. Bergkamp, A. Sellinger, D. Gust and G. Rumbles, *Adv. Energy Mater.*, 2016, **6**(6), 1502176.
119. F. Castet, G. D'Avino, L. Muccioli, J. Cornil and D. Beljonne, *Phys. Chem. Chem. Phys.*, 2014, **16**(38), 20279.
120. J. R. Tumbleston, B. A. Collins, L. Yang, A. C. Stuart, E. Gann, W. Ma, W. You and H. Ade, *Nat. Photonics*, 2014, **8**(5), 385.
121. Y. T. Fu, C. Risko and J. L. Bredas, *Adv. Mater.*, 2013, **25**(6), 878.
122. K. R. Graham, C. Cabanetos, J. P. Jahnke, M. N. Idso, A. El Labban, G. O. Ngongang Ndjawa, T. Heumueller, K. Vandewal, A. Salleo, B. F. Chmelka, A. Amassian, P. M. Beaujuge and M. D. McGehee, *J. Am. Chem. Soc.*, 2014, **136**(27), 9608.
123. S. R. Yost and T. Van Voorhis, *J. Phys. Chem. C*, 2013, **117**(11), 5617.
124. N. R. Monahan, K. W. Williams, B. Kumar, C. Nuckolls and X.-Y. Zhu, *Phys. Rev. Lett.*, 2015, **114**(24), 247003.
125. T. M. Clarke and J. R. Durrant, *Chem. Rev.*, 2010, **110**(11), 6736.
126. S. Ono and K. Ohno, *Phys. Rev. B*, 2016, **94**(7), 075305.
127. M. Stolterfoht, A. Armin, S. Shoaee, I. Kassal, P. Burn and P. Meredith, *Nat. Commun.*, 2016, **7**, 11944.
128. A. Melianas, V. Pranculis, A. Devižis, V. Gulbinas, O. Inganäs and M. Kemerink, *Adv. Funct. Mater.*, 2014, **24**(28), 4507.
129. R. Noriega, J. Rivnay, K. Vandewal, F. P. V. Koch, N. Stingelin, P. Smith, M. F. Toney and A. Salleo, *Nat. Mater.*, 2013, **12**(11), 1038.
130. M.-J. Sher, J. A. Bartelt, T. M. Burke, A. Salleo, M. D. McGehee and A. M. Lindenberg, *Adv. Electron. Mater.*, 2016, **2**(3), 1500351.
131. P. Pingel, A. Zen, R. D. Abellon, F. C. Grozema, L. D. A. Siebbeles and D. Neher, *Adv. Funct. Mater.*, 2010, **20**(14), 2286.

132. T. J. Savenije, A. J. Ferguson, N. Kopidakis and G. Rumbles, *J. Phys. Chem. C*, 2013, **117**(46), 24085.
133. C. S. Ponseca, A. Yartsev, E. Wang, M. R. Andersson, D. Vithanage and V. Sundström, *J. Am. Chem. Soc.*, 2012, **134**(29), 11836.
134. A. Devizis, J. De Jonghe-Risse, R. Hany, F. Nuesch, S. Jenatsch, V. Gulbinas and J. E. Moser, *J. Am. Chem. Soc.*, 2015, **137**(25), 8192.
135. V. Pranculis, Y. Infahsaeng, Z. Tang, A. Devizis, D. A. Vithanage, C. S. Ponseca Jr, O. Inganas, A. P. Yartsev, V. Gulbinas and V. Sundstrom, *J. Am. Chem. Soc.*, 2014, **136**(32), 11331.
136. D. A. Vithanage, A. Devizis, V. Abramavicius, Y. Infahsaeng, D. Abramavicius, R. C. I. MacKenzie, P. E. Keivanidis, A. Yartsev, D. Hertel, J. Nelson, V. Sundström and V. Gulbinas, *Nat. Commun.*, 2013, **4**, 2334.
137. J. Cabanillas-Gonzalez, T. Virgili, A. Gambetta, G. Lanzani, T. D. Anthopoulos and D. M. de Leeuw, *Phys. Rev. Lett.*, 2006, **96**(10), 106601.
138. V. Gulbinas, R. Kananavicius, L. Valkunas and H. Bässler, *Phys. Rev. B*, 2002, **66**(23), 233203.
139. K. Tvingstedt, K. Vandewal, F. Zhang and O. Inganäs, *J. Phys. Chem. C*, 2010, **114**(49), 21824.
140. A. Maurano, R. Hamilton, C. G. Shuttle, A. M. Ballantyne, J. Nelson, B. O'Regan, W. M. Zhang, I. McCulloch, H. Azimi, M. Morana, C. J. Brabec and J. R. Durrant, *Adv. Mater.*, 2010, **22**(44), 4987.
141. R. Mauer, I. A. Howard and F. Laquai, *J. Phys. Chem. Lett.*, 2010, **1**(24), 3500.
142. D. Bartesaghi, I. D. Perez, J. Kniepert, S. Roland, M. Turbiez, D. Neher and L. J. A. Koster, *Nat. Commun.*, 2015, **6**, 19794.
143. G. Lakhwani, A. Rao and R. H. Friend, *Annu. Rev. Phys. Chem.*, 2014, **65**, 557.
144. C. G. Shuttle, B. O'Regan, A. M. Ballantyne, J. Nelson, D. D. C. Bradley and J. R. Durrant, *Phys. Rev. B*, 2008, **78**(11), 113201.
145. C. G. Shuttle, B. O'Regan, A. M. Ballantyne, J. Nelson, D. D. C. Bradley, J. de Mello and J. R. Durrant, *Appl. Phys. Lett.*, 2008, **92**(9), 093311.
146. B. Philippa, M. Stolterfoht, R. D. White, M. Velusamy, P. L. Burn, P. Meredith and A. Pivrikas, *J. Chem. Phys.*, 2014, **141**(5), 054903.
147. G. Juska, K. Arlauskas, M. Viliunas and J. Kocka, *Phys. Rev. Lett.*, 2000, **84**(21), 4946.
148. A. J. Mozer, N. S. Sariciftci, L. Lutsen, D. Vanderzande, R. Osterbacka, M. Westerling and G. Juska, *Appl. Phys. Lett.*, 2005, **86**(11), 112104.
149. W. E. Spear, *Adv. Phys.*, 1974, **23**(3), 523.
150. D. Credgington and J. R. Durrant, *J. Phys. Chem. Lett.*, 2012, **3**(11), 1465.
151. D. Credgington, R. Hamilton, P. Atienzar, J. Nelson and J. R. Durrant, *Adv. Funct. Mater.*, 2011, **21**(14), 2744.
152. P. Langevin, *Ann. Chim. Phys.*, 1903, **28**, 433.
153. U. Albrecht and H. Bassler, *Phys. Status Solidi B*, 1995, **191**(2), 455.
154. R. Hamilton, C. G. Shuttle, B. O'Regan, T. C. Hammant, J. Nelson and J. R. Durrant, *J. Phys. Chem. Lett.*, 2010, **1**(9), 1432.
155. C. L. Braun, *J. Chem. Phys.*, 1984, **80**, 4157.

156. L. J. A. Koster, V. D. Mihailetchi and P. W. M. Blom, *Appl. Phys. Lett.*, 2006, **88**(5), 052104.
157. C. Groves and N. C. Greenham, *Phys. Rev. B*, 2008, **78**(15), 155205.
158. J. C. Scott and G. G. Malliaras, *Chem. Phys. Lett.*, 1999, **299**(2), 115.
159. J. Szmytkowski, *Chem. Phys. Lett.*, 2009, **470**(1–3), 123.
160. H. Sirringhaus, P. J. Brown, R. H. Friend, M. M. Nielsen, K. Bechgaard, B. M. W. Langeveld-Voss, A. J. H. Spiering, R. A. J. Janssen, E. W. Meijer, P. Herwig and D. M. de Leeuw, *Nature*, 1999, **401**(6754), 685.
161. G. Juska, K. Genevicius, N. Nekrasas, G. Sliauzys and R. Osterbacka, *Appl. Phys. Lett.*, 2009, **95**(1), 013303.
162. F. Etzold, I. A. Howard, N. Forler, A. Melnyk, D. Andrienko, M. R. Hansen and F. Laquai, *Energy Environ. Sci.*, 2015, **8**(5), 1511.
163. S. L. Bayliss, N. C. Greenham, R. H. Friend, H. Bouchiat and A. D. Chepelianskii, *Nat. Commun.*, 2015, **6**, 8534.
164. M. M. Mandoc, W. Veurman, L. J. A. Koster, B. de Boer and P. W. M. Blom, *Adv. Funct. Mater.*, 2007, **17**(13), 2167.
165. I. Montanari, A. F. Nogueira, J. Nelson, J. R. Durrant, C. Winder, M. A. Loi, N. S. Sariciftci and C. Brabec, *Appl. Phys. Lett.*, 2002, **81**(16), 3001.
166. W. Shockley and W. T. Read, *Phys. Rev.*, 1952, **87**(5), 835.
167. R. N. Hall, *Phys. Rev.*, 1952, **87**(2), 387.
168. M. M. Mandoc, F. B. Kooistra, J. C. Hummelen, B. de Boer and P. W. M. Blom, *Appl. Phys. Lett.*, 2007, **91**(26), 263505.
169. V. I. Klimov, A. A. Mikhailovsky, D. W. McBranch, C. A. Leatherdale and M. G. Bawendi, *Science*, 2000, **287**(5455), 1011.
170. S. Sze and M. K. Lee, *Semiconductor Devices: Physics and Technology*, Wiley & Sons Inc., New York, 3rd edn, 2012.
171. M. B. Johnston and L. M. Herz, *Acc. Chem. Res.*, 2016, **49**(1), 146.
172. C. Deibel and V. Dyakonov, *Rep. Prog. Phys.*, 2010, **73**(9), 096401.
173. C. Deibel, A. Baumann and V. Dyakonov, *Appl. Phys. Lett.*, 2008, **93**(16), 163303.
174. A. Pivrikas, N. S. Sariciftci, G. Juska and R. Österbacka, *Prog. Photovolt: Res. Appl.*, 2007, **15**(8), 677.
175. C. Tanase, P. W. M. Blom and D. M. de Leeuw, *Phys. Rev. B*, 2004, **70**(19), 193202.
176. C. Tanase, E. J. Meijer, P. W. M. Blom and D. M. de Leeuw, *Phys. Rev. Lett.*, 2003, **91**(21), 216601.
177. M. C. J. M. Vissenberg and M. Matters, *Phys. Rev. B*, 1998, **57**(20), 12964.
178. T. Kirchartz and J. Nelson, *Phys. Rev. B*, 2012, **86**(16), 165201.
179. X. Yang and J. Loos, *Macromolecules*, 2007, **40**(5), 1353.

CHAPTER 8

Dye-sensitised Solar Cells

Y. HU†, A. IVATURI‡ AND N. ROBERTSON*

University of Edinburgh, EaStCHEM School of Chemistry, Joseph Black Building, Edinburgh, EH9 3FJ, UK
*E-mail: neil.robertson@ed.ac.uk

8.1 Introduction

Dye-sensitised solar cells (DSSC) with practical levels of power conversion efficiency first emerged in the early 1990s through the pioneering work of Grätzel and O'Regan using high surface area mesoporous metal oxides combined with Ru dyes, stimulating an ever-increasing growth in the academic study of these devices. DSSCs were initially recognised as potential low cost alternatives to the incumbent solar technologies, bearing in mind that the cost of silicon solar cells at that time was still prohibitive for most markets. However, as studies progressed, DSSCs were also highlighted as potential flexible, lightweight, semi-transparent and attractively coloured solar cells, offering help in opening up photovoltaic devices to new markets and new niche application areas. Ongoing studies addressed some of the key challenge areas in the field, including improvements to stability, reductions in cost, manufacture and scale-up, and a better understanding of the underpinning thermodynamic and kinetic factors. A complex interplay has unfolded between new materials development

†Current address: Wuhan National Laboratory for Optoelectronics (WNLO), Huazhong University of Science and Technology (HUST), 1037 Luoyu Road, Wuhan 430074, P.R. China.
‡Current address: University of Strathclyde, Department of Pure and Applied Chemistry, Thomas Graham Building, 295 Cathedral Street, Glasgow, G1 1XL, UK.

Nanoscience & Nanotechnology Series No. 45
Nanostructured Materials for Type III Photovoltaics
Edited by Peter Skabara and Mohammad Azad Malik

Published by the Royal Society of Chemistry, www.rsc.org

and detailed optical, optoelectronic, structural and performance investigations to assess and understand outcomes. Progress in the field is often expressed through historical improvements to efficiency, but it is important to also recognise the important advances made in these other essential aspects of developing practical, working photovoltaic products. This chapter aims to give an overview and flavour of these advances by systematically summarising recent progress in each aspect of the cell structure in turn and by considering the scale-up and commercialisation activities and the current state of the technology.

8.2 Structure and Mechanism of DSSCs

The basic DSSC has five main components: (1) a conductive transparent anode, typically fluorine-doped tin dioxide (SnO_2:F or FTO) coated glass; (2) a nanocrystalline semiconducting oxide with a highly porous structure and high surface area, typically titanium dioxide (TiO_2); (3) a sensitising dye such as a Ru bipyridyl complex; (4) an organic electrolyte containing a redox mediator, usually a mixture of iodide/triiodide (I^-/I_3^-) in a low viscosity organic solvent such as acetonitrile; and (5) another FTO-coated glass with a Pt catalyst coating as the counter electrode.

The original implementation of the DSSC contained a liquid electrolyte; however, solid state DSSCs have also been developed in which the liquid electrolyte is replaced by solid state organic or p-type conducting polymer hole transporting materials (HTMs).

Figure 8.1 shows the overall assembled structure of a DSSC and Figure 8.2 summarises the working processes I–VIII. The details of these processes are as follows.

(I) Absorption of a photon by the dye (D) leads to the formation of its electronically excited state D*:

$$D + h\nu \rightarrow D^* \tag{8.1}$$

(II) The molecule in the excited state undergoes oxidative quenching, injecting electrons into the conduction band (CB) of TiO_2 and leading to the oxidation of the sensitiser:

$$D^* \rightarrow D^+ + e^-CB \tag{8.2}$$

(III) The injected electrons travel through the mesoporous network of particles to reach the back-collector electrode and then propagate around the external circuit to the Pt-coated cathode.

(IV) The electrons reaching the counter electrode through the external circuit reduce the oxidised iodide (I_3^-) back to iodide (I^-). The Pt catalyst enables this reaction to proceed with minimal over-potential:

$$I_3^- + 2e^- \rightarrow 3I^- \tag{8.3}$$

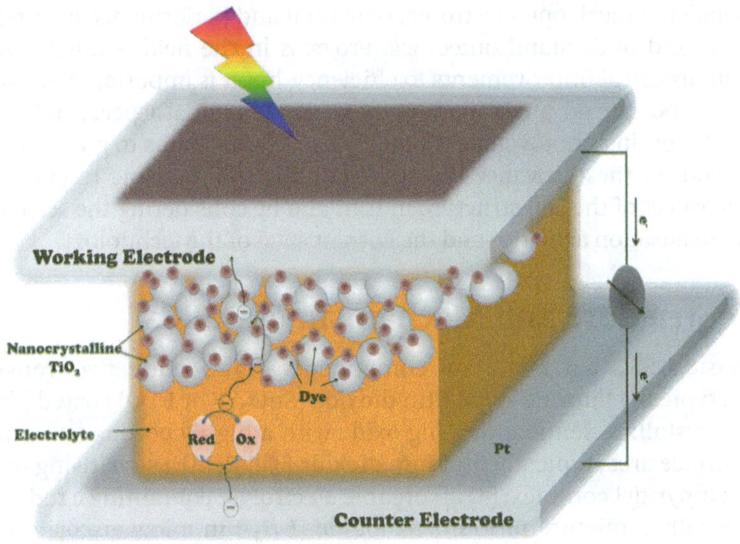

Figure 8.1 Schematic structure of a DSSC.

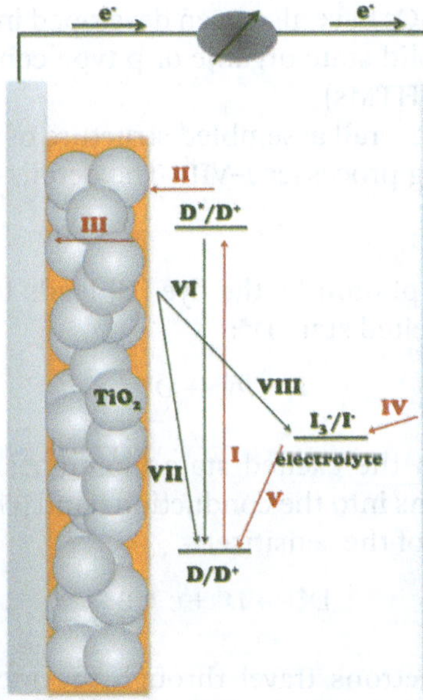

Figure 8.2 Summary of working processes in a DSSC.

(V) The oxidised dye is regenerated rapidly to the neutral state by the electrolyte:

$$2D^+ + 3I^- \rightarrow 2D + I_3^- \tag{8.4}$$

(VI) The dye in the excited state can decay back to the ground state when there is competition between the electron injection into the metal oxide and the radiative and non-radiative decay of the excited state dye:

$$D^* \rightarrow D + h\nu/\text{heat} \tag{8.5}$$

(VII) The oxidised dye is recombined with an injected electron in the conduction band of TiO_2:

$$D^+ + e^-CB \rightarrow D \tag{8.6}$$

(VIII) The electron in the conduction band recombines with the oxidised redox species in the electrolyte:

$$I_3^- + 2e^-CB \rightarrow 3I^- \tag{8.7}$$

For a high efficiency cell, it is essential that forward processes of light absorption, electron injection, dye regeneration and charge transport I–V (red arrows) should be kinetically favoured over the competing loss pathways of excited state decay to ground and electron recombination processes VI–VIII (green arrows).[1,2]

8.3 Substrates for DSSCs

Different substrates can be used in DSSCs, including conducting glass, metal foil and flexible polymer film. Among these, glass substrates are most widely used in academic research because they confer good protection against the penetration of oxygen or water. The glass used has a surface coating of a transparent conductive oxide (TCO). TCOs are used in a wide range of applications, such as optoelectronic devices, light-emitting diodes and transparent contacts for solar cells due to their low sheet resistance and high transparency to solar radiation in the visible–IR region. TCOs are generally n-type wide bandgap semiconductors consisting of a relatively high concentration of free electrons in the conduction band. The high optical transmittance is provided by the wide bandgap and the electrical conductivity is increased by the free electrons. The most commonly used TCOs in DSSCs are indium-doped tin oxide (In:SnO_2, ITO) and FTO. ITO is the gold standard of TCOs due to its high transmittance (~80–90%) and high conductivity.[3] However, when ITO is exposed to temperatures >300 °C, which are required for sintering of the mesoporous oxide, there will be a decrease in the number of oxygen vacancies, leading to a decrease in the number of electric carriers and a dramatic drop in conductivity. FTO is preferable when making DSSCs, not only because of its good stability at high temperatures, but also because of its competitive cost. The transmittance of FTO can be as high as 80–85% depending on the thickness and the resistivity can be as low as 2×10^{-4} ohms per square. Other alternative materials, such as aluminium-doped zinc oxide are also used.

Rigid TCO glass substrates have some drawbacks in terms of frangibility, weight and rigidity, which limit the integration of DSSCs into portable electronic devices and roll-to-roll mass production. These substrates also represent about 20–30% of the total material costs in a DSSC.[4] Thus alternative, low cost, lightweight and flexible substrates have been investigated, such as plastic films and thin metal foils. Polyethylene terephthalate and polyethylene naphthalate polymers coated with ITO are the most common choices of substrate.[5] Different materials, such as carbon nanotubes,[6] graphene[7] and conductive polymers,[8] have also been used to replace FTO. A limitation of plastic substrates is the requirement for temperatures <150 °C to prevent the thermal degradation of the plastic. Thin metal foils such as aluminium, copper, iron, nickel, zinc, titanium and stainless steel have been studied because they can withstand higher temperatures.[9] However, the stability of cells based on these substrates remain a problem, mainly due to the corrosive electrolyte.[10]

8.4 Nanocrystalline Semiconductors

Semiconductor oxides used in DSSCs serve as the carrier for the monolayers of dye and the path for electrons diffusing to the conducting substrate. Many wide bandgap oxide semiconductors, including TiO_2, ZnO, SnO_2 and Nb_2O_5, have been examined as potential candidates for DSSCs. TiO_2 is the most popular choice due to its chemical stability, ready availability, non-toxicity, low cost and biocompatibility. TiO_2 has a wide 3.2 eV bandgap and can be found in nature as anatase, rutile or brookite. Anatase is the most common structure used in high-performance solar cells.

Various techniques are used to synthesise nanocrystalline TiO_2 particles, including vapour–liquid–solid, sol–gel and solution-based processes, electrodeposition and polymer-assisted growth. Of these, the sol–gel procedure is the most commonly used.[11]

Three traditional deposition methods – spin-coating, doctor blading and screen printing – are used to prepare the nanocrystalline TiO_2 film on the conducting glass before sintering at high temperatures. Four layers are usually needed for a good TiO_2 electrode: (1) a dense TiO_2 blocking layer coating on the whole FTO glass to prevent contact between the redox mediator in the electrolyte and the FTO; (2) a nanocrystalline sensitiser absorption layer for dye absorption and electron transport; (3) a light-scattering layer consisting of larger particles to give effective light scattering; and (4) another thin coating layer over the whole assembly. However, traditional methods are no longer suitable for DSSCs based on organic flexible substrates because the substrates are not durable at high temperatures. New methods have thus been studied and reported, such as low-temperature annealing[12] at <150 °C and lift-off techniques[13] to transfer pre-sintered porous layers from the conducting glass substrate to a flexible substrate while maintaining the original electrical properties.

The typical film thickness is 2–15 μm using nanosized particles of 10–30 nm. A large surface area allows the film to absorb more dye, but also increases the recombination in process VIII (Figure 8.2). Research to coat the TiO_2 film with a thin overcoat of another metal oxide with a higher conduction band edge has been investigated in an attempt to solve this problem by increasing the physical separation of injected electrons from the oxidised dye.[14,15]

8.5 Electrolytes

The electrolyte plays a very important part in DSSCs. It is responsible for regenerating the oxidised dye and transporting the electrons. According to the different states of matter, electrolytes can be divided into three types: liquid electrolytes; quasi-solid state electrolytes; and solid electrolytes.

8.5.1 Liquid Electrolytes

Liquid electrolytes are the classic electrolytes used in DSSCs and record efficiencies are always achieved using liquid electrolyte DSSCs. A liquid electrolyte usually contains four elements: electron mediators such as I^-/I_3^-; cations working as the counter ions for the electron mediators; additives such as *tert*-butylpyridine; and organic solvents to dissolve these components. Different redox couples have been studied. The most common choice is iodide/triiodide dissolved in a mixture of acetonitrile and valeronitrile. However, the triiodide absorbs a non-negligible portion of the visible light and thus is competitive with the dye. Also, the low redox potential of this couple sets a limit on the open circuit voltage (V_{OC}) and its corrosive nature also causes problems. Efforts have been made to find new electron mediators and polypyridyl cobalt(II/III) complexes as well as the $(SeCN)_3^-/SeCN^-$ couple have been found to be the most promising replacements. An efficiency of 12.3% (uncertified) was achieved by incorporating a Co(II/III) *tris*(bipyridyl)-based redox electrolyte in conjunction with a custom synthesised donor–π–bridge–acceptor zinc porphyrin dye **YD-2-o-C8** and organic dye **Y123** as co-sensitiser (Figure 8.3).[16] Despite the high efficiency, co-sensitising makes the fabrication and optimisation of these devices technically challenging. The idea of developing a single sensitiser with a panchromatic light response in a DSSC led to the new record (uncertified) efficiency at that time of 13% at full Sun illumination using a bulky donor–π–bridge–acceptor zinc porphyrin dye **SM315** without the requirement of a co-sensitiser (Figure 8.3).

8.5.2 Ionic Liquid and Quasi-solid State Electrolytes

A room temperature ionic liquid is a salt in the liquid state at room temperature. These can have excellent thermal and chemical stability, negligible vapour pressure, high ionic conductivity and a wide electrochemical window.

Most of the ionic liquids applied in DSSCs are iodide salts, especially alkylimidazolium-based materials. Their special properties allow them to be both the source of iodide and the solvent. The limitation of ionic liquids is their viscosity, which slows down the transport of I_3^- in the electrolyte and

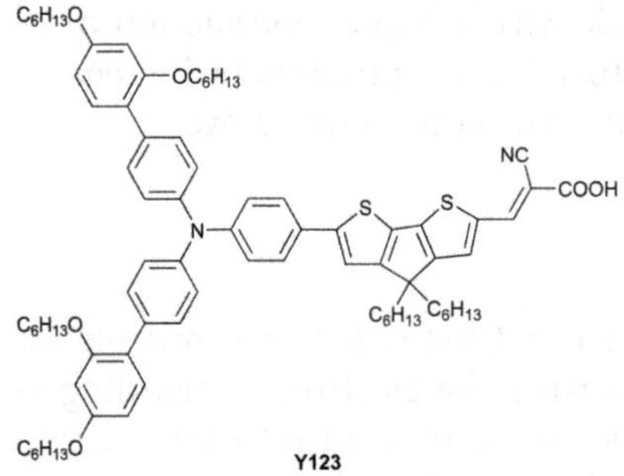

Y123

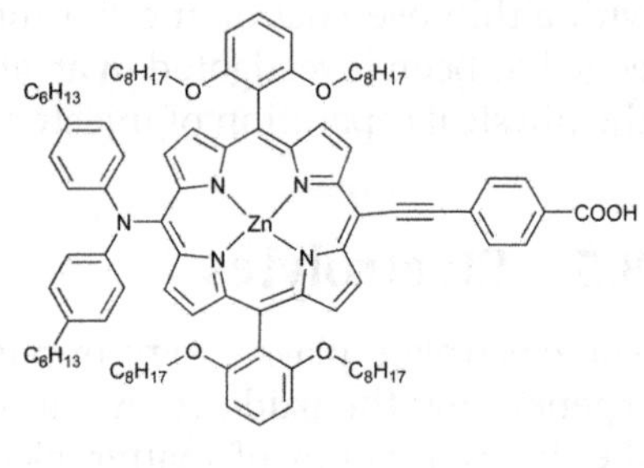

YD-2-o-C8

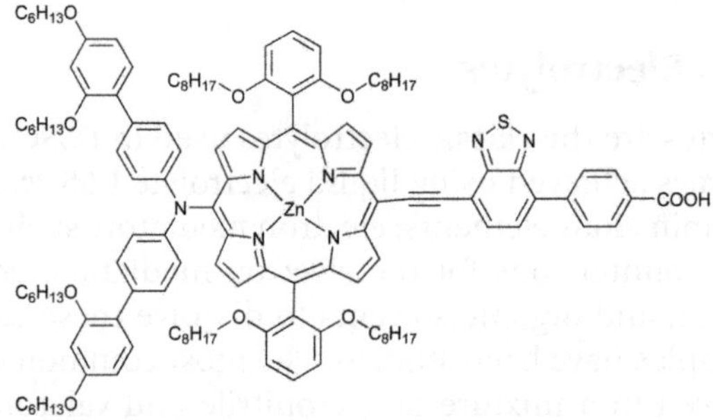

SM315

Figure 8.3 Molecular structures of **YD-2-o-C8, Y123** and **SM315.**

decreases the current density and solar efficiency. It has been shown that the viscosity of the ionic liquid increases with increasing alkyl chain length due to van der Waals forces, whereas the conductivity decreases with increasing viscosity.[17] Low viscosity ionic liquids were developed to improve ion mobility[18] and different ionic liquids were used together as a mixture to adjust the viscosity and conductivity. A high efficiency of 7.6% was achieved by Kuang *et al.* in 2004 using **K77** as the dye and the binary ionic liquids (Figure 8.4) 1-propyl-3-methylimidazolium iodide (**PMII**) and 1-ethyl-3-methylimidazolium tetracyanoborate (**EMIB(CN)$_4$**) with I^-/I_3^- as the electrolyte.[19] Two additives are also often used in the electrolyte, *N*-butylbenzoimidazole (NBB) and guanidinium thiocyanate (GNCS). An impressive efficiency of 9.1% with high stability was achieved using DMII/EMII/EMIB(CN)$_4$/I$_2$/NBB/GNCS (molar ratios 12 : 12 : 16 : 1.67 : 3.33 : 0.67) as the electrolyte and **C103** as the dye (Figure 8.4).[20] An alternative redox couple has also been studied. By using a low viscosity ionic liquid **EMISeCN** with an iodine-free $SeCN^-/(SeCN)_3^-$ electrolyte, an efficiency of 7.5% was reached.[21]

8.5.3 Solid Electrolytes

To increase the long-term performance and durability of DSSCs and to fully solve the inevitable problems of any liquid electrolyte – such as leakage, corrosion, volatility and photoreactivity – purely solid state DSSCs need to be

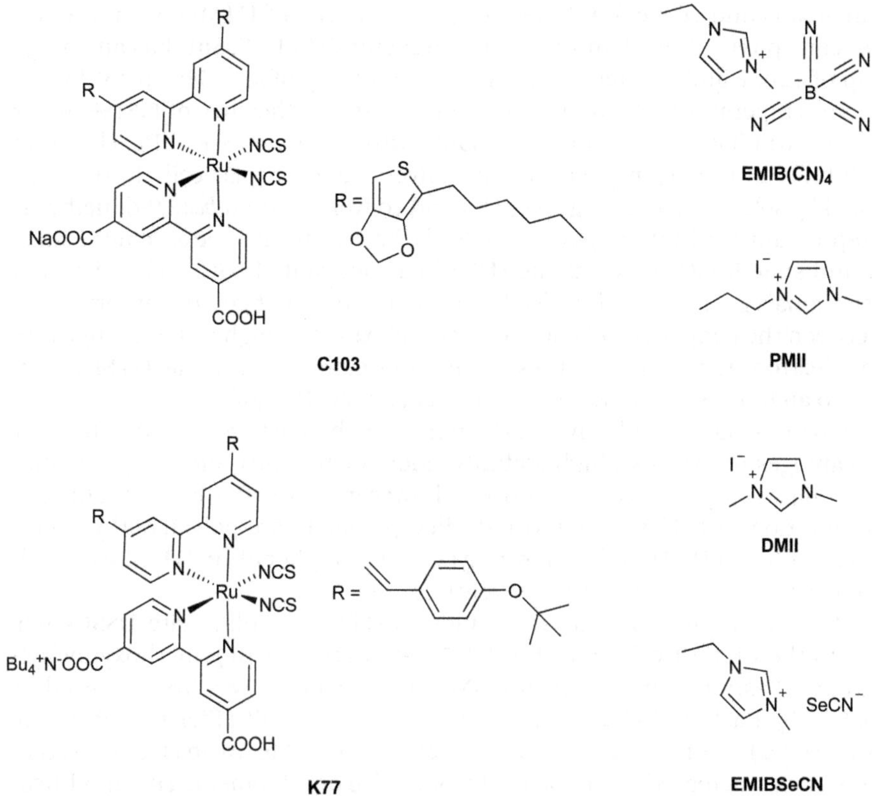

Figure 8.4 Molecular structures of **C103** and **K77** and the ionic liquids used with these dyes.

developed. Efforts are currently focused on using inorganic, conducting polymers or organic HTMs.

The HTMs used for DSSCs should meet a number of requirements: (1) the redox potential of the HTM should be less positive than that of the dye so it can accept holes from the oxidised dye before electron recombination can take place; (2) the HTM needs to have good hole mobility; (3) the HTM must have good permeation into the pores of the nanoparticle film; (4) there must be a way to deposit the HTM without dissolving or degrading the dye on TiO_2; and (5) the HTM should not absorb light in the visible spectrum, where the dye absorbs light.[22,23]

Copper(I) iodide (CuI),[24–26] copper(I) thiocyanate (CuSCN)[27] and NiO[28] are representative inorganic HTMs used in DSSCs. Kumara and co-workers used 1-ethyl-3-methylimidazolium thiocyanate as a CuI crystal growth inhibitor and used fine CuI crystallites to fill the pores. The cells could be optimised to 3.75% (AM1.5). The highest conversion efficiency using CuI as the HTM was 7.4% reported by Sakamoto *et al.*[29] Compared with CuI, CuSCN has even better stability, but results in low efficiency. NiO also gave low efficiency cells. This was mainly due to the low hole mobilities of these materials. In 2012,

Kanatzidis and co-workers[30] developed a new type of HTM: a solution-processable p-type direct bandgap semiconductor $CsSnI_3$. $CsSnI_3$ has an energy gap of 1.3 eV and a remarkably high hole mobility of $\mu_h = 585\ cm^2\ V^{-1}\ s^{-1}$ at room temperature. The optimum cell exhibits an efficiency of 8.51% under AM 1.5 irradiation when doping $CsSnI_3$ with 5% F and 5% SnF_2. The good solubility of $CsSnI_3$ in polar organic solvents allows solar cells to be fabricated by solution processing. Hagfeldt and co-workers used *bis*(2,9-dimethyl-1,10-phenanthroline) copper(I/II) ($Cu^{(I/II)}(dmp)_2$) as the redox couple in a liquid state DSSC as well as the HTM in a solid state DSSC.[31] The distorted tetragonal geometry minimised the structural change in conformation between the copper(I) and copper(II) complexes. The high self-exchange rate for electron transfer makes these complexes suitable for use as HTMs, leading to an impressive power conversion efficiency (PCE)of 8.2%.

Polymers have also been investigated over the past decades due to their advantages of low cost, high stability, good adaptability and the possibility of using an easy spin-coating process; however, they suffer from insufficient penetration into TiO_2. The most studied polymers are poly(3,4-ethylenedioxythiophene) (**PEDOT**)[32,33] and poly(3-hexylthiophene) (**P3HT**),[34] with efficiencies of 6.8 and 7.4%, respectively (Figure 8.5).

The use of small organic molecules as HTMs in solid state DSSCs has led to the highest efficiency for 2,2′,7,7′-tetrakis(*N*,*N*-di-*p*-methoxyphenylamine)9,9′-spirobifluorene (**spiro-OMeTAD**).[27] Since it was first reported in DSSCs by Bach and Grätzel in 1998, giving an overall efficiency of 0.74% based on a Ru complex dye, the efficiency of spiro-OMeTAD-based solid state DSSCs has increased dramatically through dye development, TiO_2 modification and the use of additives. It has now reached 5.0% using the Ru complex dye **C106** and 7.2% using the metal-free organic dye **Y123** (Figure 8.6).

Spiro-OMeTAD has been used as a standard HTM in both solid state DSSCs and in perovskite solar cells. However, it still needs to be investigated further and improved. For example, the slow charge transport in spiro-OMeTAD limits device performance.[35] One solution to this problem is to add strong oxidants as p-dopants to increase the conductivity of spiro-OMeTAD.[36,37] Other alternatives would be to design new materials to improve hole mobility.[38]

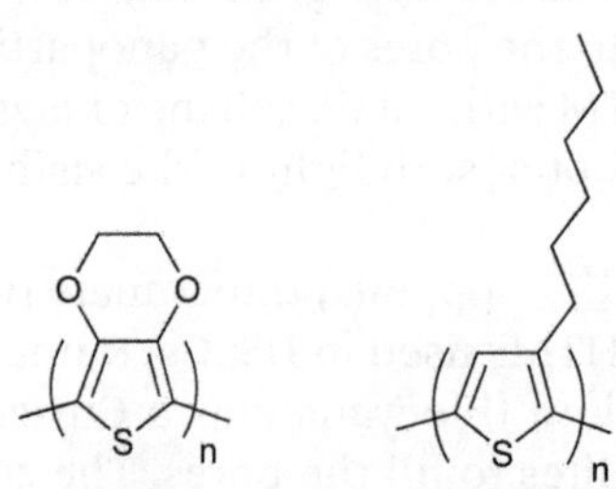

Figure 8.5 Molecular structures of representative polymers used as hole transporting materials.

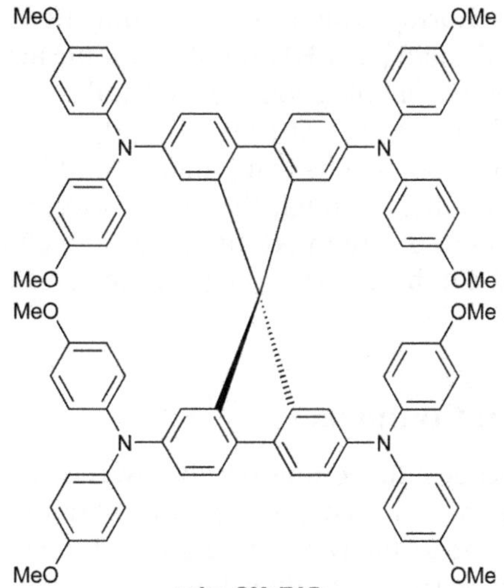

spiro-OMeTAD

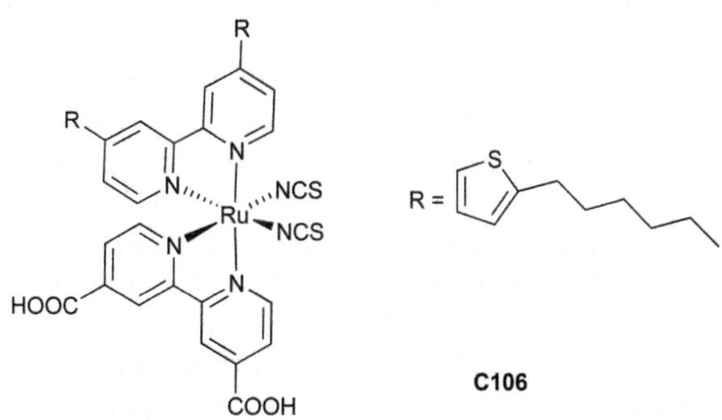

C106

Figure 8.6 Molecular structures of **spiro-OMeTAD** and the dyes used with it to reach high efficiency.

The oxidation potential of these HTMs can be precisely controlled and tuned by introducing different electron-donating functional groups.[39]

8.6 Sensitising Dyes

One of the most important components in DSSCs is the sensitising dye. A number of features are needed for the dye to enable a high PCE in the DSSC: (1) the dye should have a panchromatic absorption envelope in the visible and near-IR region with a high molar extinction coefficient – ideally, all photons below a wavelength of 920 nm should be harvested and converted into an electric current; (2) the ground state and excited state redox potentials should be positioned for both injection and regeneration; (3) the dye excitation should be a charge transfer transition with the electron density moving towards the anchoring group on the TiO_2 surface to facilitate electron injection – several anchoring groups are commonly used, such as carboxylic acid (–COOH), sulfonic acid (–SO_3H) and phosphonic acid (–PO_3H_2); and (4) the dyes need to have reasonably good thermal and chemical stabilities.[40,41]

There are two main categories of sensitising dyes: metal complexes and metal-free organic dyes. Solar cells using metal complexes lead to high efficiencies and good stability, although the purification is more complicated. Organic dyes are easier to purify and design, but their efficiency and stability are still debated and the best performing examples often require complex multi-step synthesis.

8.6.1 Ruthenium Complexes

Over the past two decades, Ru complexes endowed with thiocyanate ligands have led the performance in DSSCs over hundreds of dyes. The reasons why Ru is chosen as the metal for DSSC dyes are as follows: (1) its octahedral geometry allows ligands to be introduced in a controllable way; (2) the metal can exist in different oxidation states with good stability; and (3) the photophysical, photochemical and electrochemical properties of these complexes can be predictably tuned.

The first DSSC to show an efficiency >10% used the complex $Ru(dcbpy)_2(NCS)_2$ (dcbpy = 4,4′-dicarboxy-2,2′-bipyridine), better known as **N3** (Figure 8.7). It is now often used as a standard for the comparative assessment of new dyes. In **N3**, the highest occupied molecular orbital (HOMO) is localised primarily on the metal–thiocyanate fragment, whereas the lowest unoccupied molecular orbital (LUMO) is positioned on the dcbpy anchoring ligands. This allows efficient charge injection to TiO_2 after photoexcitation. This complex was discovered to have a broad visible absorption spectrum with maximum absorption at 380 and 518 nm, depending on intra-ligand absorption and singlet metal-to-ligand charge transfer, respectively. In addition, this complex has a sufficiently long excited state lifetime and strong adsorption on TiO_2. Hundreds of dyes have been designed and studied based on variations of **N3**.

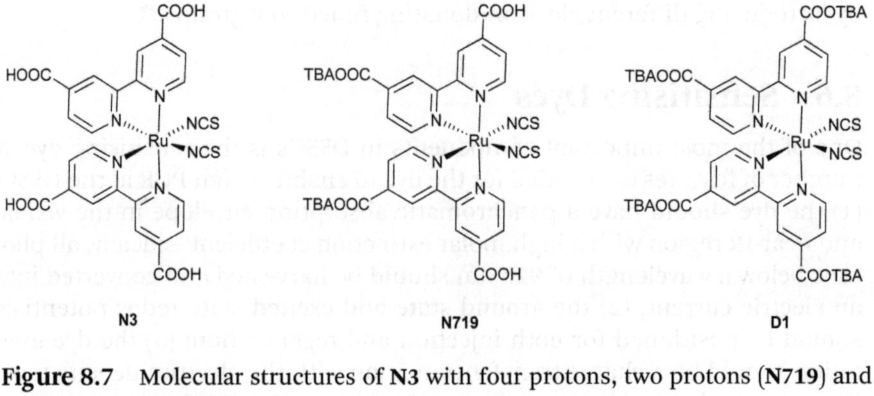

Figure 8.7 Molecular structures of **N3** with four protons, two protons (**N719**) and no proton.

Protons carried by the dye will have an effect on the performance of DSSCs because the conduction band energy of TiO_2 depends on the pH.[42,43] The effect was systematically studied by comparing the **N3** dye with no proton (**D1**), two protons (**N719**) and four protons (**N3**).[40,44,45] It was found that the fully protonated sensitiser gave a higher photocurrent, but lower V_{OC} because of the downward shift of the conduction band edge induced by TiO_2 surface protonation. However, the sensitiser that carried no proton showed a high V_{OC} compared with the fully protonated sensitiser at the sacrifice of a lower short circuit photocurrent. Thus an optimum degree of protonation of the sensitiser to obtain both a good short circuit photocurrent and V_{OC} is the doubly protonated **N3** analogue, also known as the dye **N719**.[45] In **N719**, tetrabutylammonium was used as the cation (Figure 8.7). Other cations, such as sodium or potassium ions, can also be used and a single conclusion cannot be drawn about which cation is the best because the effect of cation substitution depends on the electrolyte composition.

Two different bpy ligands can be used in Ru complexes. One can be the dcbpy ligand to attach the dye to the TiO_2 and the other can be used to tune the properties of the dye. For example, the length of alkyl chains on the bpy ligand has an effect, mainly on the stability of the solar cells. The sensitiser undergoes water-induced desorption from the TiO_2 surface over time. By introducing long alkyl chains in the ligand, the hydrophobic properties increase and thus increase the thermal stability of the solar cells. A key example is **Z907** (Figure 8.8).[46] The long chains of the sensitiser can also retard the charge recombination process (VIII) by forming an aliphatic network to prevent the oxidised redox couple I_3^- in the electrolyte from reaching the surface of TiO_2. The dye regeneration process (V) is suppressed at the same time, resulting in a reduced PCE.

The expansion of π-conjugation of the polypyridyl ligand will enhance the molar extinction coefficient of the sensitiser. By endowing it further with electron-donating alkoxy groups, the LUMO energy level can be fine-tuned and the directionality in the excited state can be guaranteed. Good examples for this

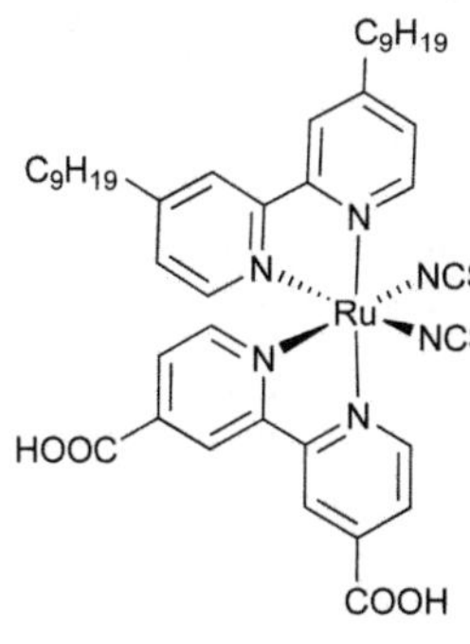

Z907

Figure 8.8 Molecular structure of sensitiser with long alkyl chains (**Z907**).

theory are dyes with conjugated aromatic rings, including, but not limited to, the methoxystyryl group (**Z910**[47] and **N945H**),[48] thiophene (**C101**[49] and **C106**),[50] 3,4-ethylenedioxythiophene (**C103**),[1] selenophene (**C105**)[51] and triarylamine (**D2** and **D3**).[52] Other dyes use a tridentate ligand to make **N749**[53] (also called the black dye) and a tetradentate ligand **N886**[54] has also been tried. These dyes give an obvious panchromatic absorption spectrum (up to 900 nm) (Figure 8.9).

Thiocyanate (NCS^-) ligands have also been considered. These have been used as the electron-donating group for most efficient Ru dyes. However,

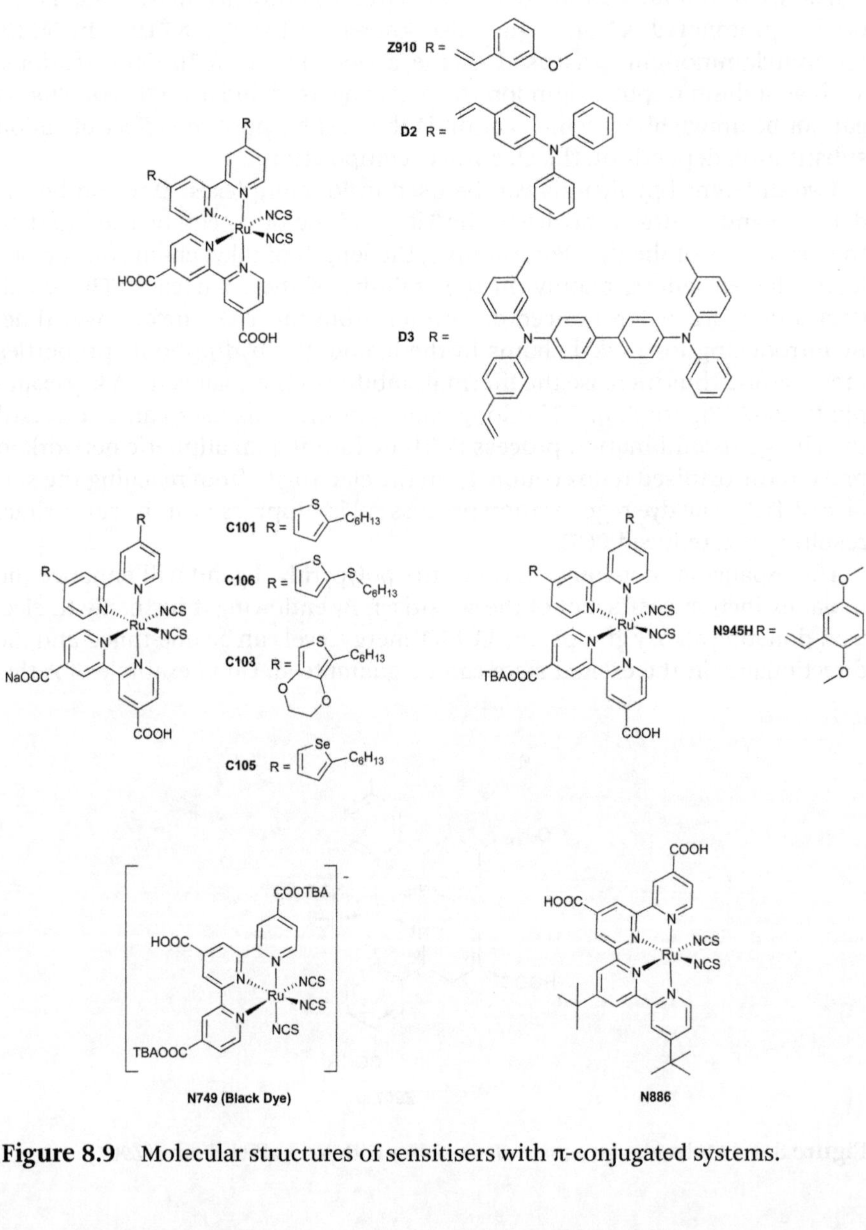

Figure 8.9 Molecular structures of sensitisers with π-conjugated systems.

from the prospect of chemical stability, NCS^- seems to be the most labile part of the complex. Over time, the dyes will lose the monodentate, labile NCS^- ligands. Efforts have been made to overcome this problem by replacing NCS^- with other ligands. In 2011, the efficiency of this kind of DSSCs reached 10.7% using dye **TF-3**[55] containing a 4,4′,4″-tricarboxy-2,2′:6,2″-terpyridine ligand and a styrene-substituted pyridylpyrazolate auxochrome. Many other examples of thiocyanate-free Ru dyes have also been reported (Figure 8.10).[56]

8.6.2 Porphyrin and Phthalocyanine Dyes

Phthalocyanine is an intensely blue-green coloured aromatic macrocyclic compound. In contrast to Ru complexes, porphyrin and phthalocyanine dyes exhibit an intense Soret band in the 400–450 nm region and a Q-band in the 500–700 nm region with good chemical, photochemical and thermal stability. Among all the porphyrin and phthalocyanine dyes, the diamagnetic metalloporphyrins containing zinc with carboxylic binding groups have shown the best performance. Nazeeruddin and co-workers used the dye **D4** (Figure 8.11) to achieve a high incident photon to converted electron efficiency of 75% and an overall efficiency of 4.8%.[57] One problem in this kind of dye is that the HOMO and LUMO levels are both localised in the macrocycle, so charge separation is not good and this makes it difficult for the electrons to inject into TiO_2. To separate the HOMO and LUMO levels, strong electron donors and acceptors with anchoring groups are introduced. Dye **YD-2** (Figure 8.11) is designed based on this idea and its derivative **YD-2-o-C8** (Figure 8.3) is one of the dyes used in top-performing cells.

8.6.3 Metal-free Organic Dyes

Metal-free organic dyes typically show higher molar extinction coefficients than Ru polypyridyl complexes. The structures are very diverse and can be easily designed. In addition, without using a rare metal, the cost of this kind of dye may also be lower. The metal-free organic dye is most commonly

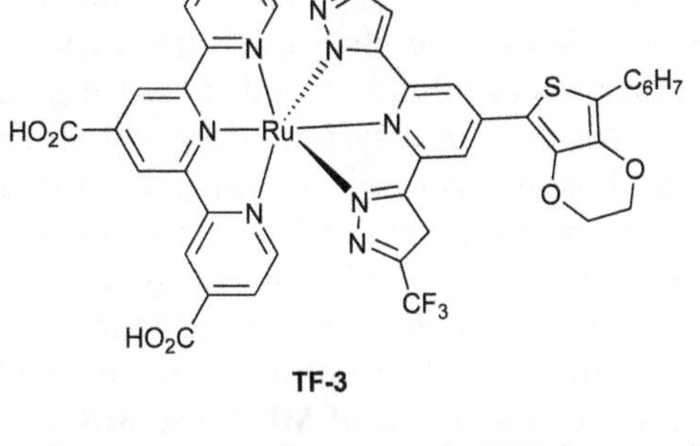

TF-3

Figure 8.10 Molecular structures of an example NCS-free sensitiser.

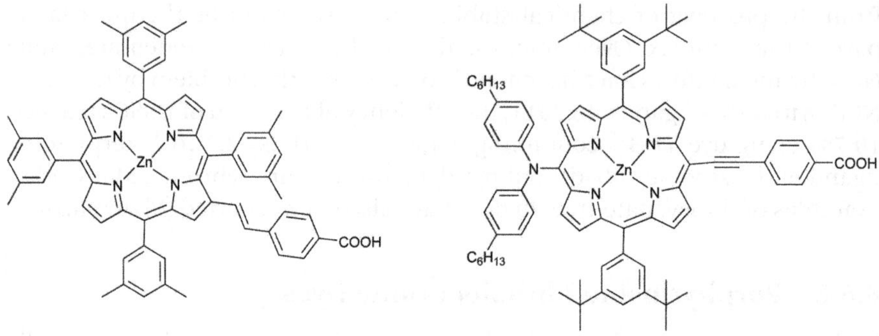

Figure 8.11 Molecular structures of asymmetrical D–A-type Zn porphyrin dyes.

designed based on a donor–π spacer–acceptor (D–π–A) architecture. In this case, the donor and the π spacer part will contain the HOMO of the dye, whereas the LUMO is distributed on the acceptor part and the anchoring group. On excitation, the electrons move from the donor to the acceptor through the π bridge. Cyanoacrylic acid is a commonly used acceptor that decreases the gap between the HOMO and LUMO levels, leading to a red shift in the lowest energy absorption maximum.[58] Various organic dyes have been investigated, including anthraquinones, boradiazaindacenes, carbazoles, coumarins, *N*,*N*-diakylanilines, hemicyanines, heteroanthracenes, indolines, merocyanines, oligothiophenes, tetrahydroquinolines, triarylamines, squaraines, perylenes, polymers and natural dyes.[59]

The problems that organic dyes have are dye aggregation due to strong π–π stacking, a lack of absorption in the red region of the visible spectrum and their low stability. Molecular aggregation has a negative influence on the solar cell efficiency because it may lead to quenching of the intermolecular excited state. Also, the aggregated molecules are not directly attached to the TiO_2 and are therefore not efficient in charge transfer.[60] Although dye aggregation causes red shifts and broadening of the absorption spectra, it results in lowering of the J_{SC} in the solar cell.

To prevent aggregation, sterically hindered substituents, such as long alkyl chains and bulky aromatic units, have been introduced into the dye molecules to disturb the π–π stacking. The alkyl substitution of dyes can block the oxidised species in the electrolyte from the injected electrons, thus preventing charge recombination and increasing the V_{OC} of the cell. Hara and co-workers carried out a systematic study on the how thiophene alkyl chain length and positions in the π-conjugated bridge of organic dyes affect the photovoltaic performance and the kinetics of electron injection and charge recombination.[61] The series of MK dyes synthesised has a carbazole derivative as an electron donor and a cyanoacrylic acid moiety as an electron acceptor. The donor and acceptor are connected by *n*-hexyl-substituted oligothiophenes. Comparison of **MK-1** and **MK-3**, or **MK-2** and **MK-4** or **MK-5** and **MK-7**, it was observed that fewer hexyl chains in the dye led to a blue shift accompanied by a broadening of the absorption spectrum, suggesting an influence of the number and position of hexyl chains on the intermolecular interaction of the dyes. The relatively sharp absorption peaks on the TiO_2 surface of **MK-1**, **MK-2** and **MK-8** also suggested a decrease in π–π stacking when more hexyl chains were present. In addition, the V_{OC} of dyes **MK-1** and **MK-2** were 80 mV higher than those of **MK-3**, **MK-4**, **MK-5** and **MK-7** due to their longer electron lifetimes as a result of the alkyl chains (Figure 8.12).

Studies have also been carried out on the electron donor groups. A bulky donor group is usually preferred to reduce electron recombination between the redox electrolyte and the TiO_2 surface and to stabilise the oxidised dye. The influence of the donor size in a D–π–A dye was studied by Yang *et al.*[62] A higher voltage was observed when they introduced a bulky indoline moiety as the donor into the reported **IQ4** dye (Figure 8.13).[63] The *bis*(2,4-dihexyloxy) benzene substitution on the quinoxaline of **YA421** and **YA422** reduced electron recombination between the oxidised species in the electrolyte and the injected electrons in TiO_2, especially when using Co(II/III) electrolyte.

Some groups have also introduced another electron-donating group to help inhibit the I_3^- ions approaching the TiO_2 surface, thus forming a D–D–π–A architecture. This will increase the distance between the dye cation centre and the TiO_2 surface and retard charge recombination.[64–67]

Other solutions to solve the aggregation problem include adding a co-adsorbent, chenodeoxylcholic acid (CDCA) into the dye solution. However, as a co-adsorbent, CDCA competes with the dyes adsorbed on the TiO_2, resulting in a reduction in injected electrons. It is important to tune the amount of CDCA to effectively reduce dye aggregation.

Most D–π–A structured dyes were first designed for DSSCs based on liquid redox electrolytes. For those cells, the electron-donating group is necessary due to the reasons discussed earlier. However, for solid state DSSCs, where the liquid electrolyte is replaced by mild oxidant HTMs, the donor group may be unnecessary. Abate *et al.*[68] investigated a donor-free dye **oligo-3HT** and applied it in a solid state DSSC (Figure 8.14). Given the same basic molecular structure, **oligo-3HT** was compared with **MK-2**. A significantly higher V_{OC} was achieved, whereas J_{SC} remained nearly the same. It was discovered that under operating conditions, the **oligo-3HT** dye has a significant fraction of dye in the oxidised form and the higher V_{OC} was attributed to this. This result suggested a new strategy to obtain efficient solid state solar cells.

8.6.4 Co-sensitisers

To further improve the spectral response of the solar cells, multiple sensitisers with different absorption spectra can be used to extend the light absorption spectrum. A co-sensitiser should meet the following requirements: (1) it needs to have strong light absorption in the desired region; (2) the co-sensitiser should not cause competition with the other dye when anchoring with TiO_2; and (3) no aggregation should result among the different dyes.[69]

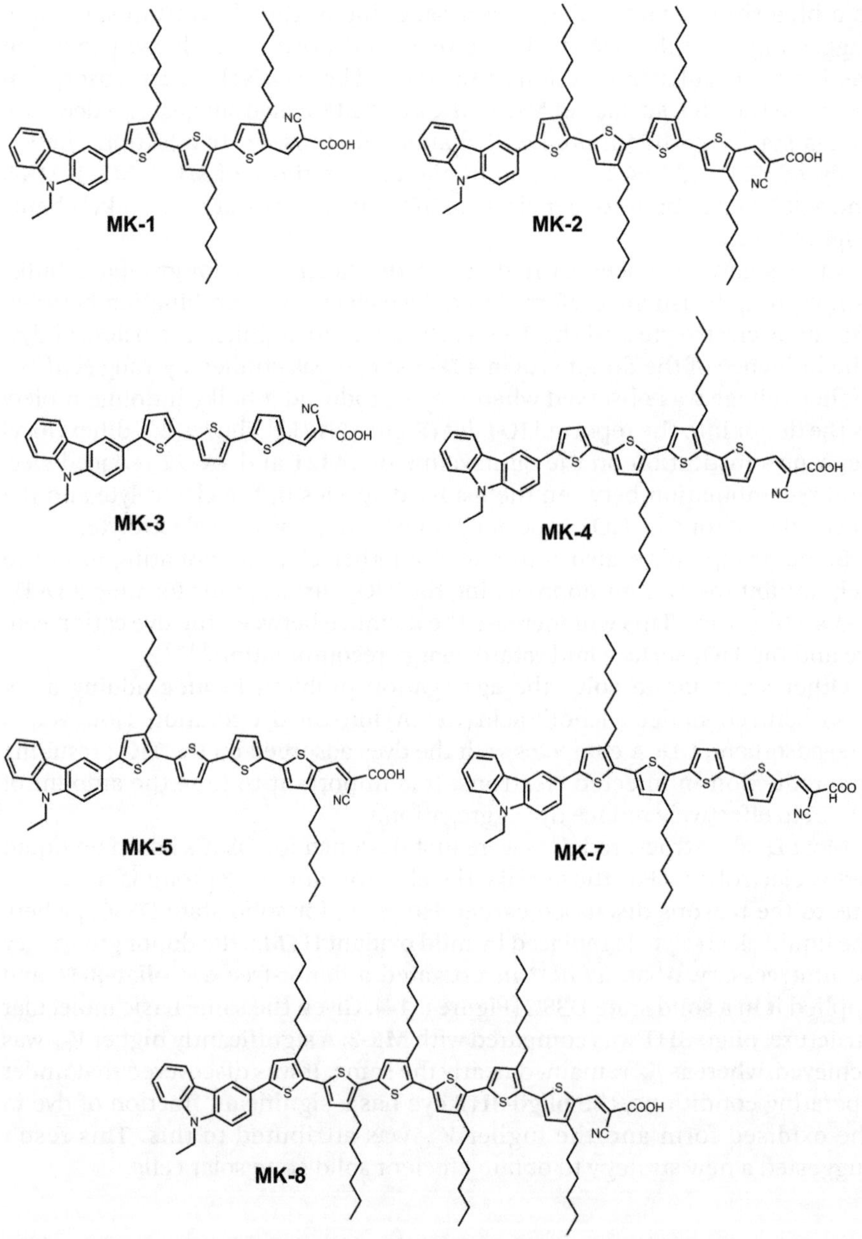

Figure 8.12 Molecular structures of organic dye **MK** series.

8.7 Counter Electrodes

The counter electrode in DSSCs basically functions as a catalyst for the regeneration of the redox couple and as a collector of electrons from the external circuit. Pt-coated FTO is widely used as the counter electrode in DSSCs with

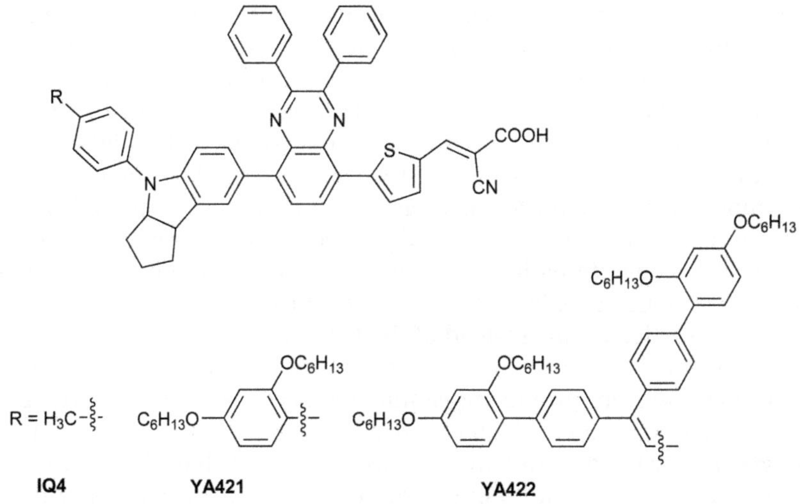

Figure 8.13 Molecular structures of quinoxaline-based dyes.

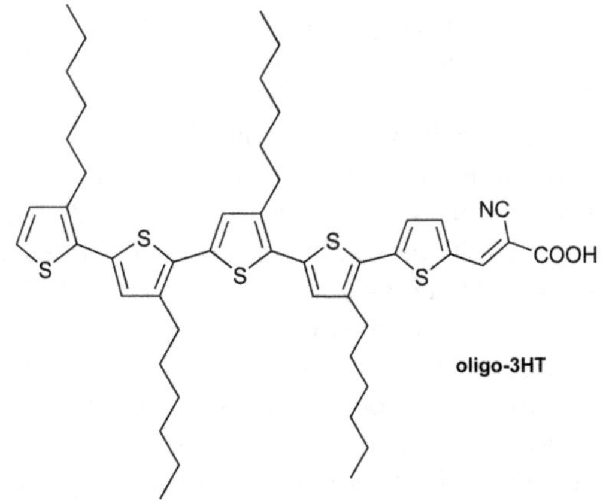

Figure 8.14 Molecular structures of the oligomeric dye **oligo-3HT**.

a liquid electrolyte, mainly because of its excellent catalytic activity towards reducing I_3^-. The Pt layer is commonly deposited at room temperature *via* sputtering, which requires a bulk Pt disc as the target material. Alternative methods, which require low quantities of Pt (~10–100 mg cm^{-2}), but high temperatures (~400–500 °C) post-treatment include the electrochemical reduction of hexachloroplatinate or the thermal decomposition of chloroplatinic acid, resulting in Pt nanoparticle films.[70–73] Periodically arranged three-dimensional nanostructures with high surface areas have emerged as promising materials compared with planar Pt films.[74–76] For example, Pt nanocups with a controlled diameter (300–600 nm) and pitch size (400–800 nm) showed a PCE of 9.75%, whereas the planar Pt showed only a 7.87% PCE.[77] It is interesting to note that the atomic arrangement on the exposed crystal facets of the Pt nanostructures is also important in its catalytic activity.[78] Density functional theory results on the catalytic reaction processes of triiodide reduction over Pt (100), (111) and (411) facets indicate the order of activity as Pt (111) > Pt (411) > Pt (100). This was in agreement with experiments, in which a Pt (111) counter electrode showed the highest PCE of 6.91%, followed by Pt (411) and Pt (100).[79] Although Pt is the counter electrode most often used in DSSCs, it has major limitations, especially for large area and long-term usage – specifically its cost and the fact that it corrodes in electrolytes containing iodide to generate PtI_4.[80–82] Interestingly, the PCE reported in DSSCs using Pt as the counter electrode is as high as 12.3% using the Co(II/III) redox couple instead of the I^-/I_3^- couple.[16]

Of the various materials explored to replace Pt as the counter electrode, carbon materials are promising candidates because of their merits of low cost, high catalytic activity, high electrical conductivity, high thermal stability and good corrosion resistance. However, DSSCs based on graphite,[83] carbon nanotubes,[84] amorphous carbon[85] and hard carbon spherules[83] as catalyst materials have, to date, shown lower PCEs than those achieved with Pt. However, materials based on graphene, such as graphene nanoplatelets,[86] nitrogen-doped graphene nanoplatelets,[87] reduced graphene oxide[88–91] and hybrid structures of reduced graphene oxide–graphene nanoplatelets–carbon nanotubes[92,93] exploited as counter electrodes exhibited PCEs close to or exceeding those based on Pt. In fact, the record (uncertified) PCE of 14.3% reported for DSSCs was achieved by exploiting graphene nanoplatelets in combination with Au on FTO as the counter electrode.[94]

Other inorganic materials, such as transition metal dichalcogenides, metal oxides, metal carbides, nitrides, sulphides and selenides, as well as hybrid systems, have also been tested as counter electrode in DSSCs, although with lower PCEs than the state-of-the-art DSSCs.[95]

In addition to inorganic materials, organic conducting polymers have also been explored as counter electrodes in DSSCs. These include PEDOT–PSS,[96] polypyrrole[97] and polyaniline,[98] which showed PCEs close to or exceeding those based on Pt. Organic polymers are promising transparent flexible counter electrodes that could replace Pt, but the durability of such polymeric counter electrodes has still to be explored.

8.8 Scaling-up of DSSCs

It is important to note that nearly all of the fundamental optimisation studies on DSSCs in laboratories around the world are carried out on small area cells with an active area ≤1 cm^2. However, for technological exploitation it is important to scale up to sub-modules, large modules and panels, according to the needs of industry and the applications concerned. However, when the active area of a DSSC is increased, there is an invariable loss in efficiency as reported for other photovoltaic technologies. For example, the best certified PCEs reported to date for DSSCs are 11.9 ± 0.4% on a single cell with active area of 1 cm^2 and 8.8 ± 0.3% on a module with an area of ~400 cm^2.[99] This is mainly due to the loss of some photogenerated charge carriers *via* recombination and/or trap processes. On scaling up DSSCs, the conductivity of the TCO substrates significantly influences the internal resistance of the solar cells and induces considerable reduction of the fill factor and hence the PCE. To mitigate these scaling-up issues, attempts have been made to optimise the material preparation, the mode of deposition of the photoelectrode, the counter electrode design and the ways of interconnecting small area cells in the several-module architectures developed.[11]

8.8.1 Device Architectures of DSSC Modules

The DSSC module designs explored so far mostly consist of large rectangular cells (area ≥3 cm^2) interconnected *via* two major types of connections (Figure 8.15): (1) series connections, which provide a high output voltage, including monolithic designs, W-type designs and Z-type designs; and (2) parallel connections, which provide a high photocurrent, such as parallel grid connections. Other connections, such as ball–grid connections (based on ball–grid array connections in electronic circuits), have also been reported.[11,100]

Some attempts have also been made to use a combination of both series and parallel connections, but these cells have very small active areas. The best performing dually interconnected device resulted in a PCE ~6% (V_{OC} ~ 1.4 V, J_{SC} = 287 mA and fill factor ~56%) in a device area of 90 cm^2.[101]

8.8.1.1 Series Connections

8.8.1.1.1 Monolithic Designs. In the monolithic configuration (also called the Kay cell, named after the inventor Andrea Kay, who reported the first DSSC monolithic module in 1996),[102] the semiconductor oxide layer (*e.g.* TiO_2), a porous insulating spacer layer (*e.g.* ZrO_2) and a porous conductive

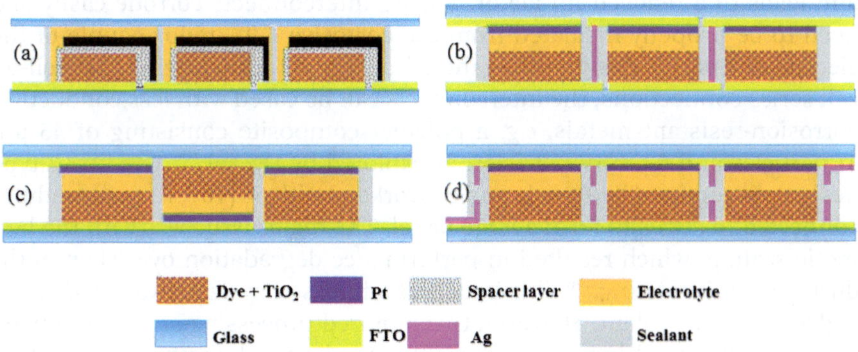

Figure 8.15 Schematic cross-section of the most common DSSC module architectures using the most widely used materials: (a) monolithic architecture; (b) Z-type design; (c) W-type design; and (d) parallel connection.

layer (*e.g.* graphite or carbon black) are all deposited on the same conductive substrate. The adjacent cells are separated from each other *via* laser scribing on the TCO. The counter electrode (a porous carbon layer) of the first cell is connected to the working electrode of the adjacent cell (and *vice versa*) and the device is sealed using a non-conducting substrate and filled with the electrolyte through a hole. The major advantages of this design are: (1) the estimated 20–30% cost reduction in materials because it replaces the expensive Pt with a carbon-based counter electrode and the design requires only one conducting glass; and (2) the high active to total area ratio.[103] This design is also highly compatible with roll-to-roll processing for flexible DSSC modules.[104] The major drawbacks of this design are a much lower J_{SC}, high opacity and poor sealing of the device. The first monolithic DSSC module consisted of 10 μm TiO_2, a 10 μm rutile spacer and 60 μm of carbon black (six cells of 4.7×0.7 cm^2 with a total area of 21 cm^2) and yielded a PCE of 5.3% (V_{OC} = 3.90 V, J_{SC} = 1.3 mA cm^{-2} and fill factor ~61%). ZrO_2 was later explored to replace the opaque rutile spacer, but the transparency was lower than for a conventional Pt electrode.[104] Improved transparency was later reported by the company Aisin Seiki, replacing the opaque rutile spacer by an SiO_2 layer and the carbon counter electrode by a highly transparent paste of In_2O_3:Sn and Pt nanoparticles. The semi-transparent monolithic DSSC modules (with 12 cells, total area ~91 cm^2) yielded a PCE <3%, limited mainly by the J_{SC} (~1.7 mA cm^{-2}) despite the high V_{OC} of ~8 V.[105]

8.8.1.1.2 Z-Type Designs. The Z-type configuration is a sandwich-type design requiring two TCO substrates. The adjacent cells are isolated by laser scribing on the TCO and the individual cells are interconnected internally *via* conducting current collective fingers (usually Ag) so that the working electrode of the first cell on one TCO is connected to the counter electrode of the neighbouring cell, which is on the opposite TCO, and *vice versa*. The major disadvantage of this configuration is the relatively low active area due to additional vertical metallic connections, which also adds to the series resistance and leads to a reduced fill factor. The Ag interconnects corrode easily and need to be properly insulated from the corrosive I^-/I_3^- redox couple of the electrolyte. As this design offers the advantage of particularly short cell-to-cell series connections, the interconnects can be cased with less conductive, corrosion-resistant metals, *e.g.* a polymer composite consisting of 45 μm Ti particles and 5 μm W particles, as explored by Dyesol.[106] The first Z-type design was reported by Toyoda and co-workers, with 64 (10×10 cm^2) modules connected in series to form a large panel. The major issue was with the hermetic sealing, which resulted in performance degradation over six months during stability testing.[106] This issue was addressed by Sastrawan *et al.* using a glass frit as a sealant. However, this required processing at high temperatures >500 °C, resulting in the development of a novel pumping method for dye anchoring after module sealing.[107] The DSSC module resulted in a PCE of ~3.5% (V_{OC} ~20 V, J_{SC} ~168 mA and fill factor ~53%) with an active area of 680 cm^2. By restricting the width of each cell to <1 cm, a DSSC module with significantly enhanced PCE ~ 6.6% (V_{OC} ~ 8 V, J_{SC} ~ 1.23 mA cm^{-2} and fill factor ~67%) has been reported by Jun *et al.*[108] Similarly, a PCE of ~7% on an aperture area of 45 cm^2 has been reported by Giordano *et al.*, with an optimised photoelectrode geometry using a back-reflector.[109]

8.8.1.1.3 W-Type Designs. The W-type configuration is also a sandwich-type design that requires two TCO substrates. The adjacent cells are isolated by laser scribing on the TCO. The individual cells are not interconnected internally *via* conducting current collective fingers, but are interconnected *via* the TCO layer such that the working electrode of the first cell on one TCO is connected to the counter electrode of the neighbouring cell on the same TCO. As a result, there are two types of cell configuration – front-illuminated (TiO_2 side) and back-illuminated (Pt side) cells – on the same module face, leading to current mismatch problems between the adjacent cells. This mismatch could be eliminated by properly reducing the width of the front-illuminated cells with respect to that of the back-illuminated cells.[110] The W-type design has the advantage of a higher active area than the Z-type design due to the absence of additional interconnections, which also results in a higher device fill factor. The highest certified DSSC modules with W-type interconnections have been reported by Sharp Co. (PCE ~ 8.2% with respect to the active area of 25.45 cm^2).[11]

8.8.1.2 Parallel Connections

The parallel grid configuration is also a sandwich-type design that requires two TCO substrates. However, the adjacent cells are not isolated by laser scribing on the TCO and the individual cells are not interconnected. In parallel connections, charge is collected not only from the bottom of the electrodes in contact with the FTO, but also from the sides using metal grids (Ag, Ni, Cu and Au),[111–115] resulting in enhanced module performance. Parallel-connected DSSC modules (27 modules with a total area of ~100 cm^2) with a PCE ~4.3% have been reported by Späth *et al.*[116] Using Ag in the current collecting grids resulted in a PCE ~7.4%.[117] The parallel-type design gives a higher performance in DSSC modules than a series connection with the same electrode width.

All these design considerations are suitable for any TCO glass/polymer/metal combinations as long as one of the substrates is transparent.

8.9 Printing/Coating Techniques

DSSCs and modules can be fabricated by adopting all the main printing techniques on both rigid and flexible substrates. Transparency, colour, shape, layout and other DSSC features can be easily varied by changing the printing parameters and the paste/ink formulations used in the printing process. The printing and coating techniques used in the fabrication of DSSCs and modules include: screen printing, doctor blading, slot-die coating, pad printing, inkjet printing and roll-to-roll methods (Gravure printing, flexographic printing and rotary screen printing). A review article discussing each of these techniques in detail has been published by Paolo *et al.*[118] The most common methods are the screen printing techniques, which basically involve a screen with a pre-designed pattern to transfer printable materials (in the form of paste/ink) onto the substrates by squeegee pressure through the mesh. The pattern is realised by filling the screen with an emulsion impervious to the ink/paste in the area to which the printable materials must not transfer. The ink/paste formulation for screen printing requires a high viscosity, high boiling point solvent and binder to avoid percolation under the mesh and early hardening of the material to help pore formation inside the film network.[119] The precision of the desired pattern, thickness and quality of the printed material can be optimised *via* intrinsic parameters, *e.g.* the speed, pressure and angle of the squeegee, the number of squeegee runs, the distance between the screen and the substrate (snap-off) and its velocity. The thickness of the screen printed material can be adjusted with a precision of ±0.5 μm.[120] This technique is useful for large area DSSCs – for example, DSSC panels of 70 cm × 200 cm have been manufactured at Fraunhoder ISE using a screen printing technique to produce prototype modules of 6000 cm^2 or 900 cm^2.[121,122] A widespread physical technique used in DSSC fabrication is spray-coating, which is used to coat blocking layers and for counter electrodes. This technique has been explored to deposit graphene for fabricating graphene-based, large area (>90 cm^2 with an active area of ~43.2 cm^2) DSSC modules in Z-connection with a PCE of 3.5%, with a cost reduction of four orders of magnitude compared with Pt-based modules.[123]

8.10 Manufacturing of DSSC Modules

Although laboratory procedures for DSSC preparation have been developed to yield relatively reproducible results, most procedures are manually based and most of the processing times used in the laboratories are too long and therefore not directly transferable to industrial manufacturing environments. For industrial environments, all the manufacturing steps are designed for moderate to high line speeds (2 to >20 m min^{-1}) to obtain the maximum production yield, the lowest labour content and optimum capital investment. Over the last two decades, intensive research in the DSSC field has led to commercial developments of DSSC modules with target applications for power generation, building-integrated photovoltaics or for use in portable electronic devices. These are triggered as a result of the transparency of the DSSCs and their excellent performance under diffuse and indoor low light conditions, which remain less affected by the angle of light incidence. Global companies developing DSSCs and modules include G24 Power (UK), Solaronix (Switzerland), Dyesol (Australia), Dyepower (Italy), Samsung SDI (Korea), Fujikura (Japan), 3G Solar (Israel), CSIRO (Australia), Taiyo Yuden Co. Ltd (Japan), SONY Technology Centre (Japan), Toyota/Aisen Seiki (Japan), Pecccell Technologies, Inc. (Japan), Eneos Co. Ltd (Japan), NGK Spark Plug Co. Ltd

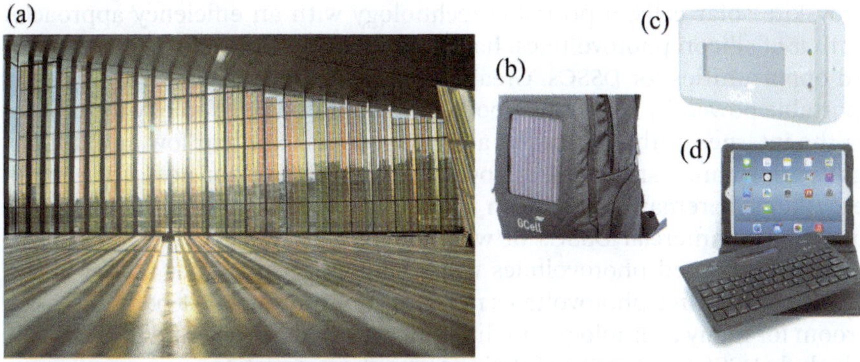

Figure 8.16 (a) DSSC modules with active area ~200 m^2 from the company Solaronix, installed as a semi-transparent façade at the SwissTech Convention Centre at the EPFL campus; (b) the Solar backpack; (c) the Solar iBeacon; and (d) the Solar keyboard from G24 Power.

(Japan), Panasonic Denko Co. Ltd (Japan), J. Touch Taiwan (Taiwan), Gunze Ltd (Japan), Yingkou OPV Tech New Enery Co. Ltd (China), Ricoh (Japan), Acrosol (Korea), Exeger (Sweden) and Dye Tec Solar (USA/global). The details of the kind of DSSC module architectures fabricated by these companies, along with their key achievements, are given in the a book on DSSCs,[11] with the latest updates reported in a review article.[103] Some applications of DSSCs for building-integrated photovoltaics and indoor electronics are shown in Figure 8.16a–d. Figure 8.16a shows DSSC modules with an active area ~200 m^2 from Solaronix, installed as a semi-transparent façade at the SwissTech Convention Centre at the EPFL campus in Lausanne in April 2014. These panels, in five different colours, are estimated to generate ~2000 kW h of annual solar electricity. Figure 8.16b–d are indoor applications of DSSC sub-modules from G24: the Solar backpack, Solar iBeacon and Solar keyboard.

8.11 Conclusions

Mesoscopic DSSCs have undergone a number of changes since their initial implementation 25 years ago. Some of the most important achievements and advances include: flexible, transparent electrode TiO_2 systems to enable roll-to-roll production; metal-free sensitisers with vibrant colours, broad spectral coverage and/or high absorption coefficients; hydrophobic dyes to improve device stability; gel, ionic liquid and solid state electrolytes or HTMs to improve stability; and non-Pt counter electrodes to improve cost and stability. This progress has led to the commercialisation of DSSCs in several niche areas. It is important to recognise that the general photovoltaics context has also radically altered since the first DSSCs were developed. The cost of silicon photovoltaics has plummeted, such that grid parity with incumbent silicon technology is now widespread and expected almost everywhere within several years. Also, over the last five years, the explosion of interest in perovskite solar cells, a printable technology with an efficiency approaching that of silicon photovoltaics, has also led to a rethink about the purpose and opportunities for DSSCs. Greater emphasis is now placed on DSSCs as a portable power for consumer goods or as an indoor technology for powering the internet of things, taking advantage of its excellent low and diffuse light performance and concerns over the personal/indoor use of Pb-based perovskites. It remains to be seen whether these will prove to be the main outlet for commercial DSSCs or whether mass power generation through building-integrated photovoltaics will also play a part. What seems likely, however, is that in a photovoltaics market expanding at ~40% per year, there is room for many technologies in different application areas and this is likely to include DSSCs as a result of their unique characteristics.

References

1. K. C. D. Robson, P. G. Bomben and C. P. Berlinguette, *Dalton Trans.*, 2012, **41**, 7814–7829.
2. A. Listorti, B. O'Regan and J. R. Durrant, *Chem. Mater.*, 2011, **23**, 3381–3399.
3. S. C. Chen, The properties of transparent ductivite oxides ITO deposited on flexible substrate, National Cheng Kung University, Tainan, 2003.
4. T. M. Brown, F. De Rossi, F. Di Giacomo, G. Mincuzzi, V. Zardetto, A. Reale and A. Di Carlo, *J. Mater. Chem. A*, 2014, **2**, 10788–10817.
5. W. A. MacDonald, M. K. Looney, D. MacKerron, R. Eveson, R. Adam, K. Hashimoto and K. Rakos, *J. Soc. Inf. Disp.*, 2007, **15**, 1075–1083.
6. K. Aitola, A. Kaskela, J. Halme, V. Ruiz, A. G. Nasibulin, E. I. Kauppinen and P. D. Lunda, *J. Electrochem. Soc.*, 2010, **157**, B1831–B1837.
7. S. W. Lee, D. H. Kang, K. S. Kim, J. H. Chung, J. K. Jo, J. Y. Lee, H. J. Cho, S. H. Lee, W. S. Ham, K. S. Lee and T. H. Kim, *J. Endourol.*, 2012, **26**, A67.
8. K. S. Lee, H. K. Lee, D. H. Wang, N. G. Park, J. Y. Lee, O. O. Park and J. H. Park, *Chem. Commun.*, 2010, **46**, 4505–4507.
9. K. Miettunen, X. L. Ruan, T. Saukkonen, J. Halme, M. Toivola, G. S. Huang and P. Lund, *J. Electrochem. Soc.*, 2010, **157**, B814–B819.
10. T. Watson, G. Reynolds, D. Wragg, G. Williams and D. Worsley, *Int. J. Photoenergy*, 2013, **2013**, 247–252.
11. *Dye-sensitized Solar Cells*, ed. K. Kalyanasundaram, EPFL Press, 2010.
12. C. Longo, A. F. Nogueira, M. A. De Paoli and H. Cachet, *J. Phys. Chem. B*, 2002, **106**, 5925–5930.
13. M. Durr, A. Schmid, M. Obermaier, S. Rosselli, A. Yasuda and G. Nelles, *Nat. Mater.*, 2005, **4**, 607–611.
14. M. L. Wang, Y. Wang and J. B. Li, *Chem. Commun.*, 2011, **47**, 11246–11248.
15. E. Palomares, J. N. Clifford, S. A. Haque, T. Lutz and J. R. Durrant, *J. Am. Chem. Soc.*, 2003, **125**, 475–482.
16. A. Yella, H. W. Lee, H. N. Tsao, C. Y. Yi, A. K. Chandiran, M. K. Nazeeruddin, E. W. G. Diau, C. Y. Yeh, S. M. Zakeeruddin and M. Gratzel, *Science*, 2011, **334**, 629–634.

17. D. B. Kuang, P. Wang, S. Ito, S. M. Zakeeruddin and M. Gratzel, *J. Am. Chem. Soc.*, 2006, **128**, 7732–7733.
18. Z. F. Fei, D. B. Kuang, D. B. Zhao, C. Klein, W. H. Ang, S. M. Zakeeruddin, M. Gratzel and P. J. Dyson, *Inorg. Chem.*, 2006, **45**, 10407–10409.
19. D. B. Kuang, C. Klein, Z. P. Zhang, S. Ito, J. E. Moser, S. M. Zakeeruddin and M. Gratzel, *Small*, 2007, **3**, 2094–2102.
20. D. Shi, N. Pootrakulchote, R. Z. Li, J. Guo, Y. Wang, S. M. Zakeeruddin, M. Gratzel and P. Wang, *J. Phys. Chem. C*, 2008, **112**, 17046–17050.
21. P. Wang, S. M. Zakeeruddin, J. E. Moser, R. Humphry-Baker and M. Gratzel, *J. Am. Chem. Soc.*, 2004, **126**, 7164–7165.
22. J. Garcia-Canadas, F. Fabregat-Santiago, H. J. Bolink, E. Palomares, G. Garcia-Belmonte and J. Bisquert, *Synth. Met.*, 2006, **156**, 944–948.
23. K. Tennakone, G. R. Kumara and A. R. Kumarasinghe, *Semicond. Sci. Technol.*, 1995, **10**, 1689–1693.
24. K. Tennakone, G. R. R. A. Kumara, I. R. M. Kottegoda and V. P. S. Perera, *Chem. Commun.*, 1999, 15–16.
25. T. Taguchi, X. T. Zhang, I. Sutanto, K. Tokuhiro, T. N. Rao, H. Watanabe, T. Nakamori, M. Uragami and A. Fujishima, *Chem. Commun.*, 2003, 2480–2481.
26. A. Konno, T. Kitagawa, H. Kida, G. R. A. Kumara and K. Tennakone, *Curr. Appl. Phys.*, 2005, **5**, 149–151.
27. U. Bach, D. Lupo, P. Comte, J. E. Moser, F. Weissortel, J. Salbeck, H. Spreitzer and M. Gratzel, *Nature*, 1998, **395**, 583–585.
28. J. Bandara and H. Weerasinghe, *Sol. Energy Mater. Sol. Cells*, 2005, **85**, 385–390.
29. H. Sakamoto, S. Igarashi, M. Uchida, K. Niume and M. Nagai, *Org. Electron.*, 2012, **13**, 514–518.
30. I. Chung, B. Lee, J. Q. He, R. P. H. Chang and M. G. Kanatzidis, *Nature*, 2012, **485**, 486–U494.
31. M. Freitag, Q. Daniel, M. Pazoki, K. Sveinbjornsson, J. Zhang, L. Sun, A. Hagfeldt and G. Boschloo, *Energy Environ. Sci.*, 2015, **8**, 2634–2637.
32. X. Z. Liu, W. Zhang, S. Uchida, L. P. Cai, B. Liu and S. Ramakrishna, *Adv. Mater.*, 2010, **22**, E150–E155.
33. J. Kim, J. K. Koh, B. Kim, S. H. Ahn, H. Ahn, D. Y. Ryu, J. H. Kim and E. Kim, *Adv. Funct. Mater.*, 2011, **21**, 4633–4639.
34. X. Guo, C. H. Cui, M. J. Zhang, L. J. Huo, Y. Huang, J. H. Hou and Y. F. Li, *Energy Environ. Sci.*, 2012, **5**, 7943–7949.
35. F. Fabregat-Santiago, J. Bisquert, L. Cevey, P. Chen, M. K. Wang, S. M. Zakeeruddin and M. Gratzel, *J. Am. Chem. Soc.*, 2009, **131**, 558–562.
36. J. Burschka, A. Dualeh, F. Kessler, E. Baranoff, N. L. Cevey-Ha, C. Y. Yi, M. K. Nazeeruddin and M. Gratzel, *J. Am. Chem. Soc.*, 2011, **133**, 18042–18045.
37. A. Abate, T. Leijtens, S. Pathak, J. Teuscher, R. Avolio, M. E. Errico, J. Kirkpatrik, J. M. Ball, P. Docampo, I. McPherson and H. J. Snaith, *Phys. Chem. Chem. Phys.*, 2013, **15**, 2572–2579.

38. A. Abate, A. Petrozza, G. Cavallo, G. Lanzani, F. Matteucci, D. W. Bruce, N. Houbenov, P. Metrangolo and G. Resnati, *J. Mater. Chem. A*, 2013, **1**, 6572–6578.
39. M. Planells, A. Abate, D. J. Hollman, S. D. Stranks, V. Bharti, J. Gaur, D. Mohanty, S. Chand, H. J. Snaith and N. Robertson, *J. Mater. Chem. A*, 2013, **1**, 6949–6960.
40. M. K. Nazeeruddin, R. Humphry-Baker, P. Liska and M. Gratzel, *J. Phys. Chem. B*, 2003, **107**, 8981–8987.
41. N. Robertson, *Angew. Chem.*, 2006, **45**, 2338–2345.
42. S. G. Yan and J. T. Hupp, *J. Phys. Chem.*, 1996, **100**, 6867–6870.
43. Y. Tachibana, S. A. Haque, I. P. Mercer, J. E. Moser, D. R. Klug and J. R. Durrant, *J. Phys. Chem. B*, 2001, **105**, 7424–7431.
44. G. Wolfbauer, A. M. Bond, G. B. Deacon, D. R. MacFarlane and L. Spiccia, *J. Am. Chem. Soc.*, 2000, **122**, 130–142.
45. M. K. Nazeeruddin, S. M. Zakeeruddin, R. Humphry-Baker, M. Jirousek, P. Liska, N. Vlachopoulos, V. Shklover, C.-H. Fischer and M. Graetzel, *Inorg. Chem.*, 1999, **38**, 6298–6305.
46. P. Wang, S. M. Zakeeruddin, I. Exnar and M. Gratzel, *Chem. Commun.*, 2002, 2972–2973.
47. P. Wang, S. M. Zakeeruddin, J. E. Moser, R. Humphry-Baker, P. Comte, V. Aranyos, A. Hagfeldt, M. K. Nazeeruddin and M. Gratzel, *J. Am. Chem. Soc.*, 2004, **125**, 7164–7165.
48. M. K. Nazeeruddin, Q. Wang, L. Cevey, V. Aranyos, P. Liska, E. Figgemeier, C. Klein, N. Hirata, S. Koops, S. A. Haque, J. R. Durrant, A. Hagfeldt, A. B. P. Lever and M. Gratzel, *Inorg. Chem.*, 2006, **45**, 787–797.
49. F. Gao, Y. Wang, D. Shi, J. Zhang, M. K. Wang, X. Y. Jing, R. Humphry-Baker, P. Wang, S. M. Zakeeruddin and M. Gratzel, *J. Am. Chem. Soc.*, 2008, **130**, 10720–10728.
50. F. F. Gao, Y. Wang, J. Zhang, D. Shi, M. K. Wang, R. Humphry-Baker, P. Wang, S. M. Zakeeruddin and M. Gratzel, *Chem. Commun.*, 2008, 2635–2637.
51. F. F. Gao, Y. M. Cheng, Q. J. Yu, S. Liu, D. Shi, Y. H. Li and P. Wang, *Inorg. Chem.*, 2009, **48**, 2664–2669.
52. S. Handa, H. Wietasch, M. Thelakkat, J. R. Durrant and S. A. Haque, *Chem. Commun.*, 2007, 1725–1727.
53. M. K. Nazeeruddin, P. Pechy, T. Renouard, S. M. Zakeeruddin, R. Humphry-Baker, P. Comte, P. Liska, L. Cevey, E. Costa, V. Shklover, L. Spiccia, G. B. Deacon, C. A. Bignozzi and M. Gratzel, *J. Am. Chem. Soc.*, 2001, **123**, 1613–1624.
54. C. Barolo, M. K. Nazeeruddin, S. Fantacci, D. Di Censo, P. Comte, P. Liska, G. Viscardi, P. Quagliotto, F. De Angelis, S. Ito and M. Gratzel, *Inorg. Chem.*, 2006, **45**, 4642–4653.
55. C. C. Chou, K. L. Wu, Y. Chi, W. P. Hu, S. J. Yu, G. H. Lee, C. L. Lin and P. T. Chou, *Angew. Chem., Int. Ed.*, 2011, **50**, 2054–2058.
56. K. L. Wu, Y. Hu, C. T. Chao, Y. W. Yang, T. Y. Hsiao, N. Robertson and Y. Chi, *J. Mater. Chem. A*, 2014, **2**, 19556–19565.

57. M. K. Nazeeruddin, R. Humphry-Baker, D. L. Officer, W. M. Campbell, A. K. Burrell and M. Gratzel, *Langmuir*, 2004, **20**, 6514–6517.
58. Z. S. Wang, Y. Cui, K. Hara, Y. Dan-Oh, C. Kasada and A. Shinpo, *Adv. Mater.*, 2007, **19**, 1138–1141.
59. A. Hagfeldt, G. Boschloo, L. C. Sun, L. Kloo and H. Pettersson, *Chem. Rev.*, 2010, **110**, 6595–6663.
60. A. C. Khazraji, S. Hotchandani, S. Das and P. V. Kamat, *J. Phys. Chem. B*, 1999, **103**, 4693–4700.
61. Z. S. Wang, N. Koumura, Y. Cui, M. Takahashi, H. Sekiguchi, A. Mori, T. Kubo, A. Furube and K. Hara, *Chem. Mater.*, 2008, **20**, 3993–4003.
62. J. B. Yang, P. Ganesan, J. Teuscher, T. Moehl, Y. J. Kim, C. Y. Yi, P. Comte, K. Pei, T. W. Holcombe, M. K. Nazeeruddin, J. L. Hua, S. M. Zakeeruddin, H. Tian and M. Gratzel, *J. Am. Chem. Soc.*, 2014, **136**, 5722–5730.
63. K. Pei, Y. Z. Wu, A. Islam, Q. Zhang, L. Y. Han, H. Tian and W. H. Zhu, *ACS Appl. Mater. Interfaces*, 2013, **5**, 4986–4995.
64. Z. J. Ning and H. Tian, *Chem. Commun.*, 2009, 5483–5495.
65. Z. J. Ning, Y. Fu and H. Tian, *Energy Environ. Sci.*, 2010, **3**, 1170–1181.
66. Z. J. Ning, Q. Zhang, W. J. Wu, H. C. Pei, B. Liu and H. Tian, *J. Org. Chem.*, 2008, **73**, 3791–3797.
67. Z. J. Ning, Q. Zhang, H. C. Pei, J. F. Luan, C. G. Lu, Y. P. Cui and H. Tian, *J. Phys. Chem. C*, 2009, **113**, 10307–10313.
68. A. Abate, M. Planells, D. J. Hollman, S. D. Stranks, A. Petrozza, A. R. S. Kandada, Y. Vaynzof, S. K. Pathak, N. Robertson and H. J. Snaith, *Adv. Energy Mater.*, 2014, **4**, 1400116.
69. S. F. Zhang, X. D. Yang, Y. H. Numata and L. Y. Han, *Energy Environ. Sci.*, 2013, **6**, 1443–1464.
70. X. Fang, T. Ma, G. Guan, M. Akiyama, T. Kida and E. Abe, *J. Electroanal. Chem.*, 2004, **570**, 257–263.
71. A. Hauch and A. Georg, *Electrochim. Acta*, 2001, **46**, 3457–3466.
72. S. Hao, J. Wu, J. Lin and Y. Huang, *Compos. Interfaces*, 2006, **13**, 899–909.
73. N. Papageorgiou, *Coord. Chem. Rev.*, 2004, **248**, 1421–1446.
74. T.-L. Hsieh, H.-W. Chen, C.-W. Kung, C.-C. Wang, R. Vittal and K.-C. Ho, *J. Mater. Chem.*, 2012, **22**, 5550–5559.
75. H. Zhang, W. Zhou, Y. Du, P. Yang and C. Wang, *Electrochem. Commun.*, 2010, **12**, 882–885.
76. V.-D. Dao, S.-H. Kim, H.-S. Choi, J.-H. Kim, H.-O. Park and J.-K. Lee, *J. Phys. Chem. C*, 2011, **115**, 25529–25534.
77. H. Jeong, Y. Pak, Y. Hwang, H. Song, K. H. Lee, H. C. Ko and G. Y. Jung, *Small*, 2012, **8**, 3757–3761.
78. R. Narayanan and M. A. El-Sayed, *Nano Lett.*, 2004, **4**, 1343–1348.
79. B. Zhang, D. Wang, Y. Hou, S. Yang, X. H. Yang, J. H. Zhong, J. Liu, H. F. Wang, P. Hu, H. J. Zhao and H. G. Yang, *Sci. Rep.*, 2013, **3**, 1836.
80. T. N. Murakami and M. Grätzel, *Inorg. Chim. Acta*, 2008, **361**, 572–580.
81. E. Olsen, G. Hagen and S. Eric Lindquist, *Sol. Energy Mater. Sol. Cells*, 2000, **63**, 267–273.

82. F. Cai, J. Liang, Z. Tao, J. Chen and R. Xu, *J. Power Sources*, 2008, **177**, 631–636.
83. Z. Huang, X. Liu, K. Li, D. Li, Y. Luo, H. Li, W. Song, L. Chen and Q. Meng, *Electrochem. Commun.*, 2007, **9**, 596–598.
84. G. Calogero, F. Bonaccorso, O. M. Marago, P. G. Gucciardi and G. Di Marco, *Dalton Trans.*, 2010, **39**, 2903–2909.
85. K. Imoto, K. Takahashi, T. Yamaguchi, T. Komura, J.-i. Nakamura and K. Murata, *Sol. Energy Mater. Sol. Cells*, 2003, **79**, 459–469.
86. S. Mathew, A. Yella, P. Gao, R. Humphry-Baker, F. E. CurchodBasile, N. Ashari-Astani, I. Tavernelli, U. Rothlisberger, K. NazeeruddinMd and M. Grätzel, *Nat. Chem.*, 2014, **6**, 242–247.
87. M. J. Ju, J. C. Kim, H.-J. Choi, I. T. Choi, S. G. Kim, K. Lim, J. Ko, J.-J. Lee, I.-Y. Jeon, J.-B. Baek and H. K. Kim, *ACS Nano*, 2013, **7**, 5243–5250.
88. W. Hong, Y. Xu, G. Lu, C. Li and G. Shi, *Electrochem. Commun.*, 2008, **10**, 1555–1558.
89. J. D. Roy-Mayhew, D. J. Bozym, C. Punckt and I. A. Aksay, *ACS Nano*, 2010, **4**, 6203–6211.
90. L. Kavan, J.-H. Yum and M. Graetzel, *ACS Appl. Mater. Interfaces*, 2012, **4**, 6999–7006.
91. E. Kymakis, C. Petridis, T. D. Anthopoulos and E. Stratakis, *IEEE J. Sel. Top. Quantum Electron.*, 2014, **20**, 106–115.
92. H. Choi, H. Kim, S. Hwang, W. Choi and M. Jeon, *Sol. Energy Mater. Sol. Cells*, 2011, **95**, 323–325.
93. J. Velten, A. J. Mozer, D. Li, D. Officer, G. Wallace, R. Baughman and A. Zakhidov, *Nanotechnology*, 2012, **23**, 085201.
94. K. Kakiage, Y. Aoyama, T. Yano, K. Oya, J.-i. Fujisawa and M. Hanaya, *Chem. Commun.*, 2015, **51**, 15894–15897.
95. M. Wu and T. Ma, *J. Phys. Chem. C*, 2014, **118**, 16727–16742.
96. J. M. Pringle, V. Armel and D. R. MacFarlane, *Chem. Commun.*, 2010, **46**, 5367–5369.
97. S. S. Jeon, C. Kim, J. Ko and S. S. Im, *J. Mater. Chem.*, 2011, **21**, 8146–8151.
98. Q. Tai, B. Chen, F. Guo, S. Xu, H. Hu, B. Sebo and X.-Z. Zhao, *ACS Nano*, 2011, **5**, 3795–3799.
99. M. A. Green, K. Emery, Y. Hishikawa, W. Warta and E. D. Dunlop, *Prog. Photovoltaics*, 2016, **24**, 905–913.
100. J. Takashima, T. Fujii and K. Furusaki, *Meet. Abstr.*, 2008, **478**, MA2008–02.
101. Y. Liu, H. Wang, H. Shen and W. Chen, *Appl. Energy*, 2010, **87**, 436–441.
102. A. Kay and M. Grätzel, *Sol. Energy Mater. Sol. Cells*, 1996, **44**, 99–117.
103. A. Fakharuddin, R. Jose, T. M. Brown, F. Fabregat-Santiago and J. Bisquert, *Energy Environ. Sci.*, 2014, **7**, 3952–3981.
104. L. Wang, X. Fang and Z. Zhang, *Renewable Sustainable Energy Rev.*, 2010, **14**, 3178–3184.
105. Y. Takeda, N. Kato, K. Higuchi, A. Takeichi, T. Motohiro, S. Fukumoto, T. Sano and T. Toyoda, *Sol. Energy Mater. Sol. Cells*, 2009, **93**, 808–811.
106. I. L. Skryabin, M. J. Sundman, G. E. Tulloch, G. Phani and S. M. Tulloch, Google Patents, 2005.

107. R. Sastrawan, J. Beier, U. Belledin, S. Hemming, A. Hinsch, R. Kern, C. Vetter, F. M. Petrat, A. Prodi-Schwab, P. Lechner and W. Hoffmann, *Sol. Energy Mater. Sol. Cells*, 2006, **90**, 1680–1691.
108. Y. Jun, J.-H. Son, D. Sohn and M. G. Kang, *J. Photochem. Photobiol., A*, 2008, **200**, 314–317.
109. F. Giordano, A. Guidobaldi, E. Petrolati, L. Vesce, R. Riccitelli, A. Reale, T. M. Brown and A. Di Carlo, *Prog. Photovoltaics*, 2013, **21**, 1653–1658.
110. F. Giordano, E. Petrolati, T. M. Brown, A. Reale and A. D. Carlo, *IEEE Trans. Electron Devices*, 2011, **58**, 2759–2764.
111. P. M. Sommeling, M. Späth, H. J. P. Smit, N. J. Bakker and J. M. Kroon, *J. Photochem. Photobiol., A*, 2004, **164**, 137–144.
112. S. Dai, J. Weng, Y. Sui, C. Shi, Y. Huang, S. Chen, X. Pan, X. Fang, L. Hu, F. Kong and K. Wang, *Sol. Energy Mater. Sol. Cells*, 2004, **84**, 125–133.
113. K. Okada, H. Matsui, T. Kawashima, T. Ezure and N. Tanabe, *J. Photochem. Photobiol., A*, 2004, **164**, 193–198.
114. J. M. Kroon, N. J. Bakker, H. J. P. Smit, P. Liska, K. R. Thampi, P. Wang, S. M. Zakeeruddin, M. Grätzel, A. Hinsch, S. Hore, U. Würfel, R. Sastrawan, J. R. Durrant, E. Palomares, H. Pettersson, T. Gruszecki, J. Walter, K. Skupien and G. E. Tulloch, *Prog. Photovoltaics*, 2007, **15**, 1–18.
115. T.-C. Wei, J.-L. Lan, C.-C. Wan, W.-C. Hsu and Y.-H. Chang, *Prog. Photovoltaics*, 2013, **21**, 1625–1633.
116. M. Späth, P. M. Sommeling, J. A. M. van Roosmalen, H. J. P. Smit, N. P. G. van der Burg, D. R. Mahieu, N. J. Bakker and J. M. Kroon, *Prog. Photovoltaics*, 2003, **11**, 207–220.
117. G. R. A. Kumara, S. Kawasaki, P. V. V. Jayaweera, E. V. A. Premalal and S. Kaneko, *Thin Solid Films*, 2012, **520**, 4119–4121.
118. M. Paolo, V. Luigi and C. Aldo Di, *Semicond. Sci. Technol.*, 2015, **30**, 104003.
119. L. Vesce and R. Riccitelli, *Prog. Photovoltaics*, 2012, **20**, 960–966.
120. S. Ito, P. Chen, P. Comte, M. K. Nazeeruddin, P. Liska, P. Péchy and M. Grätzel, *Prog. Photovoltaics*, 2007, **15**, 603–612.
121. A. Hinsch, S. Behrens, M. Berginc, H. Bönnemann, H. Brandt, A. Drewitz, F. Einsele, D. Faßler, D. Gerhard, H. Gores, R. Haag, T. Herzig, S. Himmler, G. Khelashvili, D. Koch, G. Nazmutdinova, U. Opara-Krasovec, P. Putyra, U. Rau, R. Sastrawan, T. Schauer, C. Schreiner, S. Sensfuss, C. Siegers, K. Skupien, P. Wachter, J. Walter, P. Wasserscheid, U. Würfel and M. Zistler, *Prog. Photovoltaics*, 2008, **16**, 489–501.
122. A. Hinsch, W. Veurman, H. Brandt, R. Loayza Aguirre, K. Bialecka and K. Flarup Jensen, *Prog. Photovoltaics*, 2012, **20**, 698–710.
123. S. Casaluci, M. Gemmi, V. Pellegrini, A. Di Carlo and F. Bonaccorso, *Nanoscale*, 2016, **8**, 5368–5378.

CHAPTER 9

Hybrid Solar Cells

SINCLAIR RYLEY RATNASINGHAM* AND RUSSELL BINIONS

School of Engineering and Materials Science, Queen Mary University of London, Mile End Road, London, E1 4NS, United Kingdom
*E-mail: s.r.ratnasingham@qmul.ac.uk

9.1 Hybrid Solar Cells

9.1.1 What Is a Hybrid Solar Cell?

A new development in solar cell technology is the hybrid solar cell, which is made of both inorganic and organic materials. This allows the combination of properties unique to each material type, such as the lower cost and processability of organic materials with the increased stability and performance of inorganic materials.[1] In most hybrid solar cells, the organic semiconducting material acts as the absorber and the inorganic component is responsible for charge transport. This forms a heterojunction, which facilitates the separation of charges. Several different architectures have been implemented to form this heterojunction, such as polymer–inorganic hybrid cells and dye-sensitised solar cells (DSSCs). In this chapter, the focus will be on DSSCs composed of inorganic and organic constituents and the perovskite-based solar cells that evolved from them.

Nanoscience & Nanotechnology Series No. 45
Nanostructured Materials for Type III Photovoltaics
Edited by Peter Skabara and Mohammad Azad Malik

Published by the Royal Society of Chemistry, www.rsc.org

9.1.2 DSSCs

The DSSCs are one of the more promising solar cell technologies and are also one of the earliest examples of hybrid solar cells, combining semiconducting metal oxides and organic dyes. They are photoelectron chemical cells, combining the photoexcitation of a semiconductor and a chemical reaction. Work by O'Regan and Grätzel in 1991[2] showed that this was a viable alternative to Si and GaAs photovoltaic devices. There have been several developments in this field over the years, with research focused on many areas, such as the dyes, the semiconducting material and the stability of the cells. This has led to some limited commercial availability and, by 2009, companies such as G24i had released cells to the market.[3] DSSC research has led to even higher performing solar cell technologies, such as organohalide perovskite cells. The highest reported DSSC efficiency in academia is 13%.[4]

9.1.2.1 Mechanisms of DSSCs

A traditional DSSC consists of three main components sandwiched between two electrodes: a semiconducting charge-transporting layer, an organic sensitiser and a redox electrolyte. The semiconducting layer is commonly made from metal oxides, such as ZnO and TiO_2, which are deposited with a certain nanoarchitecture, the most common of which is a mesoporous structure. The idea behind this layer and its structure is to provide a large surface area for the organic sensitiser to adsorb onto, forming a heterojunction over a larger area. The sensitiser is usually composed of an organic dye; however, many different materials have been used, such as quantum dots and organometal halide materials. The main role of the sensitiser is to absorb incoming light.

Under illumination, the dye is electronically excited and forms excitons. This process is like those found within organic solar cells. However, in this case the excitons are pulled apart at a heterojunction formed between the semiconductor and the sensitiser, with an electron or a hole being transferred into the semiconductor. Depending on the type of cell, an electron or hole is transferred. In an n-type DSSC the sensitiser transfers an electron and in a p-type cell a hole is transferred. The charge is then transferred from the semiconductor to an external circuit *via* an electrode material. The other process within a DSSC involves the redox electrolyte. Depending on the type of cell, the redox electrolyte will either donate or accept an electron, relaxing the dye back to its ground state. The counter electrode then replenishes the redox electrolyte.[6] Figure 9.1 shows an example of an n-type DSSC.

The efficiency of a DSSC depends heavily on the relative energy levels and kinetics of charge transfer between the interfaces of the semiconductor, the sensitiser and the electrolyte. The relative speed of charge carrier injection from the dye to the semiconductor must be faster than the decay of the dye to the relaxed state. The restoration of the sensitiser to its original state *via* the redox electrolyte must be faster than both the back-injection of the charge from the semiconducting to the sensitiser and the injection of charge into

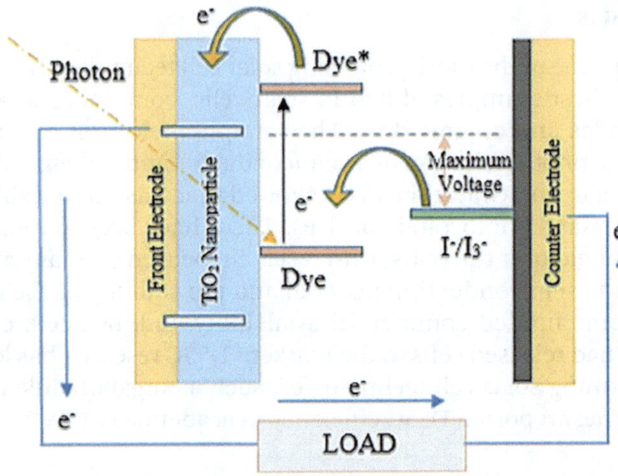

Figure 9.1 Schematic diagram of an n-type DSSC.[5] Reprinted from *Renewable and Sustainable Energy Reviews*, 52, V. Sugathan, E. John and K. Sudhakar, Recent improvements in dye sensitized solar cells: A review, 54–64, Copyright 2015, with permission from Elsevier.

the electrolyte. The reaction at the counter electrode must proceed fast enough to replenish the redox electrolyte, otherwise this may become a limiting step. Therefore, to create an efficient DSSC, we must consider the materials used, the morphology, the interfaces, the charge carrier kinetics and the chemical reactions of the electrolyte.[7]

9.1.2.2 The Beginning of DSSCs

O'Regan and Grätzel pioneered the development of the DSSC in the early 1990s. In their paper published in 1991, they reported the successful creation of the first DSSC. The maximum efficiencies of the resulting cells were reported to be between 7.1 and 7.9%, which was an eight-fold increase in efficiency compared with previous photoelectrochemical cells.[7] The large increase in efficiency was attributed to the unique construction of the cell and the use of organic and inorganic components was also one of the first examples of a viable hybrid solar cell.

The cell was constructed by coating a conductive glass substrate with a layer of nanoscale mesoporous TiO_2, forming the inorganic photoanode of the cell. This layer was designed to be mesoporous to increase its surface roughness and therefore its surface area. A layer of dye consisting of a trimeric ruthenium complex was then deposited on top of this mesoporous layer. A liquid iodide–triiodide redox electrolyte was then soaked into the cell and a conducting glass counter electrode was placed on top.[2]

The mechanism of this cell was based around utilising the heterojunction formed between the dye and the TiO_2. Under illumination, the dye absorbed light and became electronically excited. The excited dye then injected an electron into the conduction band of the TiO_2 and this was carried out of the cell *via* the electrode. As a result of the extraction of the electron from the dye, this type of cell is known as an n-type cell. The large surface area of the mesoporous TiO_2 dramatically increased the surface area available for light harvesting and the number of dye molecules that could be adsorbed onto the surface. This allowed more dye to be electronically excited and a larger interfacial area for charge to be transferred to the TiO_2. The role of the electrolyte was to reduce the oxidised dye. This function was also improved by the mesoporous structure of the TiO_2, which allowed the infiltration of the electrolyte into the structure and therefore ensured contact with the dye. The use of a mesoporous structure was one of the major innovations credited with the large step-up in efficiency.[2]

This work spurred on a whole new research area, culminating in the development of many variants, such as the solid state DSSC, which uses polymers and organic small molecules as hole transporting materials. Another recent development that evolved from DSSCs is the field of organometal halides.[8]

9.1.3 Materials and Architecture

A DSSC is constructed from several different components. Research has primarily focused on the semiconducting material, with less research on the sensitiser and far less on the electrolyte. Figure 9.2 shows the evolution of

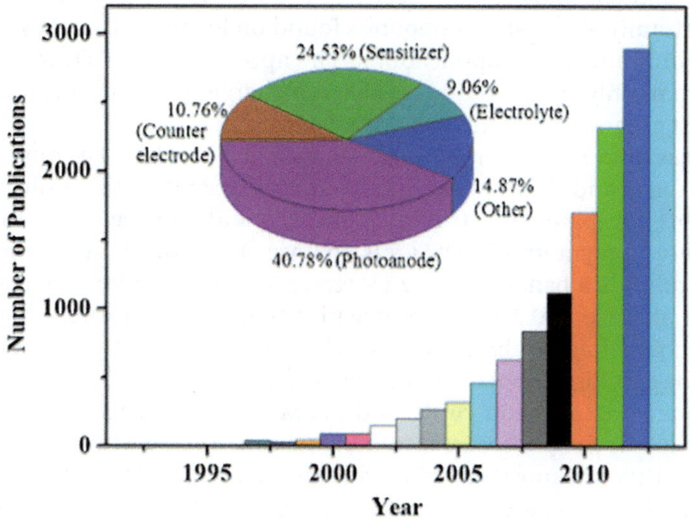

Figure 9.2 Evolution of the number of publications related to specific areas of DSSC research.[9] Reprinted from *Materials Today*, 18, M. Ye, X. Wen, M. Wang, J. Iocozzia, N. Zhang and C. Lin, Recent advances in dye-sensitized solar cells: from photoanodes, sensitizers and electrolytes to counter electrodes, 155–162, Copyright 2014, with permission from Elsevier.

publications related to DSSC research.[9] The following section discusses some of the most common materials and how they are used in DSSCs.

9.1.3.1 *Inorganic Layer*

The inorganic layer within a DSSC is primarily there to create a heterojunction and facilitate charge transport; however, it can also act as a blocking layer and as an architectural layer. The layer also acts as a high surface area support layer for the sensitiser, meaning that in most cases it is nanostructured to increase the surface area.

To create the heterojunction with the dye, the semiconductor must have appropriate band edge positions. For example, in an n-type system the conduction band of the semiconductor must allow the injection of electrons from the excited dye. Another requisite for the material is the ability to be nanostructured. High surface areas are key to the function of a DSSC. Therefore the material must be able to be deposited in layers with a high surface area, such as a mesoporous structure. Some of the more commonly used materials are TiO_2 and ZnO, which will be the focus of this chapter.

9.1.3.2 *TiO_2*

TiO_2 was the first semiconducting material used in DSSCs and was utilised in the original DSSC by Grätzel and O'Reagan.[2] TiO_2 is a wide bandgap (3.0–3.2 eV) semiconductor with a wide range of properties, such as photovoltaic activity, photosensitivity and electrical conductivity.[10] It is also one of the most naturally abundant compounds found on Earth. This makes it suitable for use in photovoltaic applications. The application of TiO_2 within DSSCs has been mainly concentrated on electron charge transport, meaning that it is almost exclusively for n-type DSSCs.

TiO_2 has several polymorphs: rutile, anatase and brookite. The most common natural form, and the most thermodynamically stable, is the rutile structure. Figure 9.3 compares the structures of rutile and anatase.[12] The most commonly used polymorph in DSSCs is anatase. TiO_2 consisting of the anatase polymorph has a bandgap of 3.2 eV *versus* 3.0 eV for rutile. The conduction band of anatase is 0.2 eV more negative than that of rutile, which allows more photovoltage to be preserved when using a similar redox electrolyte and leads to a higher open circuit voltage (V_{OC}) in a DSSC.[10]

Comparisons between rutile and anatase TiO_2 in DSSCs by Park *et al.* showed 30% lower photocurrents in DSSCs constructed with the rutile phase. However, this was mostly attributed to the structure of the films. The rutile layer was shown to have less surface area per volume, allowing less dye to be absorbed and less interparticle connectivity, both of which impede charge carrier generation and transport.[13] Other studies have shown slower electron diffusion in rutile than in anatase.[14] More recent research has demonstrated that a mesoporous layer containing a mixture of rutile and anatase phases leads to an increase in the efficiency of DSSCs. It is suggested that this effect is caused by interfacial electron transfer between the rutile and anatase

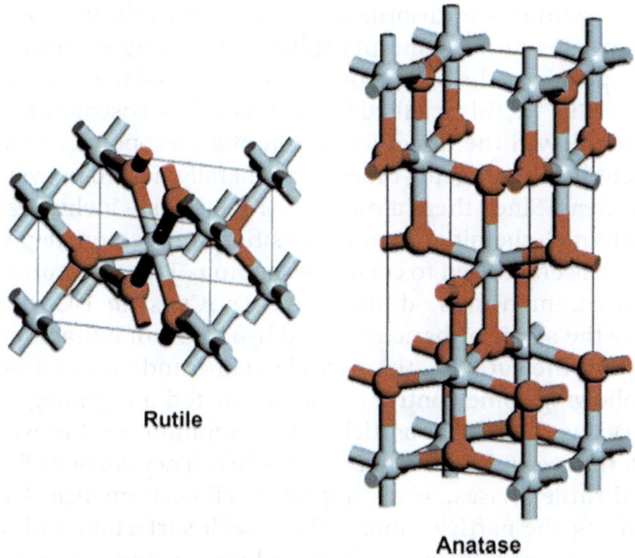

Figure 9.3 Structure of rutile (left) and anatase (right). Titanium atoms are grey and oxygen atoms are red.[11] Reprinted with permission from T. L. Thompson and J. T. Yates, *Chem. Rev.*, 2006, **106**(10), 4428–4453. Copyright 2006 American Chemical Society.

phases.[15] A good example of the mixed-phase effect is Degussa P25. This is a mixed-phase (anatase and rutile) nanopowder commonly used to create mesoporous layers in many high-performance cells.

Another way of improving the performance of the TiO_2 layer is *via* doping. One caveat with TiO_2 is that, as a large bandgap semiconductor, the band edge positions are not accessible to some dyes. It has been shown that doping with Sm leads to a lowered conduction band, which improved the driving force for electron injection.[16] Doping TiO_2 with copper has been shown to yield up to a 10% improvement in V_{OC} in DSSCs.[17] Nb doping has been shown to increase both the current density and V_{OC}.[18] A variety of other dopants have been trialled with promising results, such as Zn, Yt, Nb, N and Al.[19–23]

There are many examples of TiO_2 layers in DSSCs. In terms of the polymorph, there have been many examples utilising rutile, anatase and rutile–anatase mixtures. The morphologies used in DSSCs also vary, ranging from mesoporous layers to nanotubes. The highest recorded efficiency of a DSSC at the time of writing is 13%; this DSSC utilised a mesoporous rutile–anatase TiO_2 layer.[4]

9.1.3.3 *Mesoporous TiO_2 Nanostructures*

The most common TiO_2 morphology used in DSSCs is the mesoporous nanostructure. This is a structure consisting of pores between 2 and 50 nm in size that can be either ordered or disordered. Antonelli and Ying were first to report the synthesis of disordered mesoporous TiO_2 in 1995. They used a modified sol–gel process and phosphate templating chemistry, with titanium isopropoxide as the precursor and an alkylphosphate surfactant. After calcination at 500 °C, this resulted in a hexagonal mesostructure.[24] Although this method allowed the direct formation of a mesoporous structure from the sol, it required long ageing times and generally yielded uneven structures over larger areas. Since then, a number of methods – including hydrothermal, solvothermal, chemical vapour deposition and electron deposition techniques – have been utilised to create mesoporous TiO_2 structures.[6]

The most common method utilised for DSSCs is *via* TiO_2 nanoparticles produced by the acidic or basic catalysed hydrolysis of a titanium precursor. Depending on the catalyst, the particle shape and size can be altered,[25] therefore allowing some control of the disordered morphology in the final structure. Degussa P25 nanoparticles are a common, readily available commercial alternative and are often used in DSSCs. They consist of a mixture of anatase and rutile phases, which improves cell performance. The next step involves mixing the particles into a slurry with surfactant additives, which control the porosity of the layer. A balance between porosity and percolation pathways must be achieved because very high porosities lead to fewer interconnections between the particles and decrease charge collection.

A current state-of-the-art DSSC does not only utilise a standard mesoporous TiO_2 layer, but also includes several modifications, such as a compact TiO_2 blocking layer (~50 nm) and over-layers on top of the structure. The compact layer is designed to stop the redox electrolyte from contacting the fluorine-doped tin oxide (FTO) electrode and can be deposited in a number of ways, the most common of which is *via* sol–gel spin-coating. Over-layers are tasked to scatter more light, either by utilising larger particles (400 nm) to create a rougher surface or by using a porous over-layer.[6] DSSCs utilising a mesoporous TiO_2 layer currently hold the record for efficiency.

9.1.3.4 TiO_2 Nanowires

Nanowires are another type of structure that have garnered a lot of attention in research. Theoretically, one-dimensional nanostructures have a number of advantages. In a mesoporous architecture the electrolyte is permeated throughout the whole layer, resulting in localised areas of potential gradient or, in other words, localised heterojunctions. This means that once a charge is separated at those areas, it must diffuse to the electrode. The numerous grain boundaries of the mesoporous architecture contain trap states that may further impede the mobility of the charge. In a nanotube or nanowire architecture, there is a more direct and continuous pathway for charge transport once it has been separated at the interface. These structures also exhibit excellent light-scattering properties.[26]

There are a number of methods of synthesising nanowire arrays. One of the more novel synthesis approaches involves direct growth on FTO. Feng *et al.* reported the growth of ordered nanowire arrays using hydrothermal growth on FTO substrates. In this method, a precursor, such as $TiCl_4$, was mixed with a solution of HCl and heated to 180–220 °C for 24 h. The resulting nanowires were between 2 and 3 μm long and were composed of rutile. These arrays were then sensitised using the N179 dye and the resulting cell had an efficiency of ~5%.[27] Wu *et al.* later showed the hydrothermal growth of anatase nanorods on FTO. The power efficiency of a cell using N179 dye was 7.34%. The increase can be attributed to the lower rates of charge carrier recombination in anatase phase TiO_2.[28]

Disordered nanowires have also been shown to be a possible route because they can form films of high porosity. One simple approach to creating such structures is *via* electrospinning. Fujihara *et al.* spray-coated the FTO substrates with ground electrospun TiO_2 fibres dispersed in a methanol–ethanol solution. The coated substrates were then treated with $TiCl_4$ to improve the interconnectivity and heated to 500 °C. The resulting DSSCs constructed from these substrates had an efficiency of 5.8%. Although this method did not yield a high efficiency, it is a potentially scalable production route for TiO_2 anodes.[29]

9.1.3.5 TiO_2 Nanotubes

Nanotubes are an alternative one-dimensional structure that can be readily synthesised from TiO_2 and have been used as photoanodes in DSSCs. Compared with nanowires, nanotubes have a higher surface area and therefore potentially higher dye adsorption. The most common method used for fabrication is the anodic oxidation of titanium foil. This is followed by annealing in air (~450 °C) to turn the arrays into anatase.[26]

Macak *et al.* demonstrated a cell utilising such a nanotube array, which had an efficiency of 4.9%. The slightly lower efficiency can be partly explained by the way this cell was illuminated. As the arrays were grown on titanium foil, no light can pass through the front of the cell. Therefore this cell has to be illuminated from the opposite side.[30] This is a major drawback because a large amount of light is lost as it has to travel through the electrolyte before hitting the sensitised layer.

A potential solution to this problem was trialled by first growing the nanotubes on titanium foil and then removing the nanotube film *via* acid treatment and then re-depositing it on an FTO substrate. A $TiCl_4$ treatment was applied to the substrate to better adhere the nanotubes to the surface. This method allows light to pass through the front of the cell (the FTO layer). This method showed an improved efficiency of 7%. This synthetic procedure is complex, however, and may not be well suited to large-scale production.[31]

9.1.3.6 Other TiO_2 Nanostructures

There are a number of other TiO_2 nanostructures that have been explored in DSSCs. Shao *et al.* produced DSSCs with branched forest-like TiO_2 layers using an acid-assisted hydrothermal method. These cells had an efficiency of 7%. The layers were first synthesised on titanium foil, then removed and placed onto FTO substrates. This produced a structure resembling a forest, which consisted of crystalline nanorods with crystalline branches of a mixture of anatase and rutile phases. The branches filled the spaces in between the rods and therefore increased the surface area available for dye adsorption and light harvesting. The incident photon-to-electron conversion efficiency and electrical impedance spectroscopy measurements showed a lower charge carrier recombination than that of nanorods synthesised under similar conditions. This was attributed to more direct pathways for electron transport through the branched arrays. Absorption measurements showed an increase in the branched array, which is explained by the increase in surface area. This absorption increase led to an increased short circuit current (J_{SC}).[32]

In a comprehensive study by Mali *et al.*, a number of different nanostructures were produced *via* surfactant-free hydrothermal synthesis. By varying the reaction temperature between 100 and 190 °C, the morphology changed from nanoparticles to nanorods to three-dimensional nanospheres, and then to hollow three-dimensional nanostructures that resembled urchins. The urchin-like morphology exhibited the highest J_{SC} and highest incident photon-to-electron conversion efficiency, which was credited to greater light scattering and the larger surface area of these structures.[33]

9.1.3.7 ZnO

ZnO also has a long history of use in DSSC cells and was one of the first semiconductors to be used in such cells. The most common form of ZnO, and that used in DSSCs, is the wurtzite structure. Wurtzite ZnO has a similar bandgap (3.37 eV[34]) and band edge position to anatase TiO_2. It also has a one-magnitude advantage in bulk electron mobility.[35] Similar to TiO_2, it also has a history of synthesis in a number of nanostructures, such as nanowires, nanotubes, mesoporous and branched nanostructures.[26]

One major disadvantage of ZnO is that it has poor chemical stability, dissolving under both acidic and basic conditions. This has been shown to lead to some instability due to dyes reacting with and dissolving the ZnO layer. For instance, it has been shown that ruthenium-based dyes can form complexes with Zn^{2+}. Therefore the sensitiser and its anchoring groups must be considered carefully. Sensitisers with lower acidity and no complexing agent are more suitable for use with ZnO, which means that sensitiser options are more limited than in TiO_2.[35] ZnO-based DSSCs are less researched than their TiO_2 counterparts and currently lag behind in efficiency, with the most efficient published result being 7.5%.[36]

Like TiO_2, doping of ZnO has been an avenue of exploration. Although it has not been as extensively researched, a number of dopants have been trialled. Examples of dopants that have been shown to improve the electrical or absorption properties of ZnO include F, I, Ga, Pr and N.[37–39]

9.1.3.8 Mesoporous ZnO

Mesoporous ZnO can be prepared in a number of ways, such electrodeposition, doctor blading from nanoparticles and spray pyrolysis.[35] One of the more facile approaches is *via* the doctor blading method. Saito and Fujihara created a 6.6% efficiency DSSC using commercially available 20 nm ZnO powder. The powder was mixed in a slurry by dispersing it in a mixture of ethanol, acetic acid and water. This slurry was then doctor bladed onto an FTO substrate and sintered to form the unordered mesporous layer.[40]

Yoshida *et al.* demonstrated a novel approach to create ordered mesoporous ZnO. In this method, ZnO was cathodically electrodeposited from $ZnCl_2$ solution containing a quantity of dye. It was found that the dye molecules acted as a templating agent, which resulted in the ZnO layer having an ordered morphology. A number of other dye molecules were re-adsorbed into this structure to study their effects and the dye D149 had the highest performance of 5.6%. These layers were directly electrodeposited onto flexible substrates, showing a viable route to flexible DSSCs.[41]

The highest performing unmodified ZnO-based DSSC to date also uses a mesoporous layer. In this cell, the photoanode layer is composed of two ZnO layers, a compact buffer layer and a mesoporous top layer, with both layers deposited using spray pyrolysis. The buffer layer was produced by spraying a methanol–water solution of zinc acetate dihydrate onto a heated substrate. The mesoporous layer used a similar zinc acetate solution mixed with commercial ZnO nanoparticles. Cells were constructed with and without buffer layers and it was shown that the performance tripled with the inclusion of this layer. The buffer layer increased the electron lifetime, indicating that this layer stops recombination at the transparent conductive oxide interface. This cell is notable in that the spray pyrolysis method is a highly scalable production method.[36]

9.1.3.9 ZnO Nanowires

Nanowires of ZnO, in theory, have the same advantages as TiO_2 nanowires over mesoporous structures. The direct charge transport pathways allow improved electron transport and a more continuous electric field, which improves charge separation. Light scattering may also improve the amount of absorption.

One of the early examples of ZnO nanowires in a DSSC was the work of Law *et al.* The nanowires were directly synthesised on FTO *via* chemical bath deposition. The first step was the deposition of ZnO nanoparticles onto the FTO substrate. The substrates were then immersed in an aqueous solution of zinc acetate dehydrate and heated to 92 °C. The resulting cell had a relatively poor performance of 1.5%, largely attributed to the low fill factor (<0.6) of the cell.[42]

The highest performing cell using ZnO nanowires to date was produced by Barpuzary *et al.* The nanowires were produced *via* a seed layer method, which involved creating a layer of ZnO nanoparticles *via* spin-coating the substrate with a solution of zinc acetate dehydrate and then heating the substrate. The nanowires were grown on this substrate by immersing them in a zinc nitrate solution at elevated temperatures for 20 h. Although the resulting cell had a low fill factor, it achieved an efficiency >5%. This was partially attributed to good anchoring of the carbazole derivative SK1 dye to the ZnO, allowing efficient electron transfer.[43]

9.1.3.10 ZnO Nanotubes

ZnO nanotubes are another morphology commonly studied for use as photoanodes in DSSCs. They are promising because they combine the direct electron pathways of nanowires with a higher surface area for light absorption and dye adsorption. The highest efficiency reported to date is 6.0% using a ZnO nanotube array decorated with Au nanoparticles and 4.7% without the Au nanoparticles. The rods were electrodeposited directly on indium tin oxide (ITO) glass from aqueous solutions of $ZnCl_2$. The increase in efficiency can be attributed to the plasmonic effects of the Au nanoparticles.[44]

9.1.3.11 Other ZnO Nanostructures

There are a number of other ZnO morphologies that have been trialled in DSSCs, including nanoflakes, nanoplates, terapods, nanofibres and a variety of other hierarchical structures. We focus here on some of the higher performing examples.

An interesting nanourchin-like morphology was created by Zheng *et al.* The nanourchins were grown hydrothermally on a ZnO layer, immersed in a solution of zinc nitrate and hexamethylenetetramine and heated at 91 °C for 3 h. This gave a hierarchical structure of a forest of urchin-like growths on larger nanoparticles. This resulted in a photoanode with a large surface area for dye adsorption, large amounts of light scattering and good electron transport properties. This cell had a high efficiency, for ZnO-based photanodes, of 6.4%.[45]

Other morphologies showing promise are nanoplatelets and nanosheets. In 2011, Lin *et al.* created a DSSC using TiO_2 decorated ZnO nanosheets, which had an efficiency of >7%. Undecorated nanosheets had an efficiency of 6.06%. The sheets were synthesised by growing layered hydroxide zinc carbonate ($Zn_4CO_3(OH)_6 \cdot H_2O$), deposited *via* chemical bath deposition, on FTO substrates and then pyrolysing it at 300 °C to form ZnO sheets. The good efficiency was attributed to the high surface area, and therefore dye loading, and the high electron diffusion coefficient. The improved performance of the decorated films was credited to the increased surface area, which allowed more dye to be adsorbed.[46]

9.1.3.12 Other Inorganic Semiconductors

Although TiO_2 and ZnO are the most commonly used semiconductors in DSSCs, there are a number of other materials currently being investigated. SnO_2 is another stable semiconductor that has been explored. It is a stable oxide with a wide bandgap of 3.6 eV and has a 0.5 eV lower conduction band edge than anatase. This is an advantage because a lower conduction band edge may allow more dye to inject electrons, such as perylene synthesisers with their lower-lying lowest unoccupied molecular orbital (LUMO) levels.[6] However, this reduces the V_{OC} of the cell. SnO_2 has been shown to have a better electron conductivity and better UV stability than TiO_2. However, to date, the highest efficiency of a DSSC utilising SnO_2 is only 3.2%.[47] This poor performance may be due to faster recombination as well as less than ideal dye absorption with carboxyl acid groups. If these problems can be overcome, then SnO_2 could be another viable photoanode material.[48]

Other materials include zinc stanate (Zn_2SnO_4), which is another wide bandgap semiconductor (3.6 eV). Tan *et al.* demonstrated a cell consisting of a mesoporous layer of Zn_2SnO_4 that had an efficiency of 3.8%.[49] Nb_2O_5 is a another wide bandgap semiconductor (3.4 eV). It has a slightly higher conduction band edge than both TiO_2 and ZnO, which, in theory, increases a cell's V_{OC}. However, this may limit the amount of dye that may be able to inject electrons into this higher conduction band.[50]

Core–shell structures are a recent development in DSSC technology and are designed to supress interfacial charge recombination between the semiconductor and the electrolyte. They consist of a semiconducting core material made up of a certain nanomorphology – such as nanoparticles, nanowires or nanotubes – with an outer shell consisting of a different semiconducting or insulating material. The idea behind this is that the coating layer forms an energy barrier to recombination. An example using TiO_2 nanoparticles showed beneficial effects when coating them with Nb_2O_5, resulting in an increased V_{OC} and J_{SC}. It has been suggested that this occurs because the band bends as a result of the lower conduction band edge of the Nb_2O_5. This stops the back-transfer of electrons and lowers the amount of recombination. Other structures, such as Al_2O_3, ZnO, SnO_2 and $SrTiO_2$, have also been explored and have yielded varying amounts of success. It has been suggested that the observed effects may be more complex than just forming an energy barrier and that formation of a dipole or even the increased adsorption of dyes may influence the results.[26]

9.1.3.13 Sensitisers

The role of the sensitiser in a DSSC is one of the most important elements. For a sensitiser to perform well, it must meet a number of criteria. First, it must be able to absorb a large amount of the incoming solar radiation, mainly in the visible spectrum. It has to adsorb chemically onto the chosen semiconductor to ensure good charge transfer. It has to have appropriate highest occupied molecular orbital (HOMO) and LUMO levels to inject charge into the semiconductor and to be regenerated by the electrolyte. To ensure longevity of the cell, it has to have photo, thermal and chemical stability. Numerous dyes have been investigated and developed for DSSCs, most of which can be categorised into five main families: ruthenium polypyridyl, metal-free organic dyes, porphyrin dyes, quantum dots and perovskites.[6,9] Perovskite sensitisers are discussed in a separate section of this chapter.

9.1.3.14 Ruthenium Polypyridyl

Ruthenium polypyridyl dyes are the most investigated and applied in the field of DSSCs. They meet most of the criteria for a dye because they have suitable HOMO and LUMO levels for TiO_2 and have a broad absorption spectrum from the UV to the near-IR. Ruthenium polypyridyl dyes consist of a central ruthenium metal ion with ancillary pyridyl ligands. The strong light absorption of these dyes can be attributed to the metal to ligand charge transfer process. The ancillary ligands can be substituted to alter the photophysical and electronic properties of the dye. Anchoring groups, such as carboxyl or cyanoacrylic acid groups, are used to link the dye to the semiconductor surface and allow good charge transfer into the conduction band.[6]

One of the earliest, high-performance and popular dyes used in research is N3 and its salt derivative N719. These dyes consist of two bipyridyl groups with carboxyl acid groups and two thiocyanate groups forming a complex with the ruthenium ion. The full names of N3 and N719 are *cis-bis*(isothiocyanato) *bis*(2,2′-bipyridyl-4,4′-dicarboxylato) ruthenium(II) and di-tetrabutylammonium *cis-bis*(isothiocyanato)*bis*(2,2′-bipyridyl-4,4′-dicarboxylato) ruthenium(II), respectively. Although this group of dyes gives a good performance, with the best-performing device to date having an efficiency of 11.7%, they suffer from stability issues.[6,26,51]

These issues are primarily due to the thiocyanate, with studies showing its susceptibility to replacement with competing ligands from electrolyte additives such as 4-*tert*-butylpyridine. There have been a number of efforts to improve stability, such as using ancillary ligands or replacing the thiocyanate ligand altogether.[9] Wang *et al.* synthesised a thiocyanate-free ruthenium complex. The highest performing device, with an efficiency of 10.2% using a TiO_2 anode, was composed of a 2,2′-bipyridyl-4,4′-dicarboxylato ruthenium(II) complex with two pyridyl azolate ancillaries functionalised with a thiophene group. Previous studies by this group have shown that these types of molecule are stable under prolonged exposure to light at elevated temperatures. Although there has been progress on stability, the performance of N3 and N719 has not been matched to date.[52] The high cost of the dyes, due to the high cost of ruthenium, makes these dyes less favourable for large-scale applications.

9.1.3.15 Metal-free Organic Dyes

A promising alternative to metal complex dyes are metal-free organic dyes. These dyes have the promise of being more cost-effective, very tuneable and have been shown to have higher molar extinction coefficients. The focus of most of the research seems to be on donor–bridge–acceptor molecules. These molecules consist of an electron-rich component, such as a triarylamine, carbozoles or indulines, and an electron-withdrawing component, such as cyanoacrylic acid, rhodoamines or pyrydines. These are spaced out by π-conjugated parts, such as thiophenes and polyenes. The bridging molecule allows intramolecular charge separation, with the HOMO preferentially residing on the donor and the LUMO on the acceptor. This means that charge transfers from the donor unit to the acceptor unit, which then injects the charge into the semiconductor. By tuning the donor and acceptor parts to the redox potential of the electrolyte and semiconductor band edges, respectively, it is possible to design dyes preferentially suited to specific systems. It has been shown that modifications to the bridging component alter the HOMO and LUMO levels of the acceptor/donor and the absorption properties of the dye.[6,9]

There are two top-performing organic-sensitised DSSCs to date, both of which show an efficiency of 10.3%. The first was created by Zeng *et al.* in 2010. They used the sensitiser C219, which consists of an alkoxy-substituted triphenylamine electron donor and a cyanoacrylic acid electron acceptor. The bridging parts consisted of electron-rich 3,4-ethylenedioxythiophene, which lifts the HOMO level of the donor, and a dihexyl-substituted dithienosilole, which lowers the LUMO level. These units were chosen so that the HOMO and LUMO levels correctly lined up to the conduction band of the TiO_2 anode and the acetonitrile electrolyte.[53]

The more recent study of Yella *et al.* used a JF419 donor molecule attached to a 4H-cyclopenta[2,1-*b*:3,4-*b*]dithiophene bridge with a cyanoacrylic acid acceptor. In this study, the focus was to not only use the superior electron-donating properties of JF419, but also the steric hindrance caused by such a large molecule. This insulated the TiO_2 surface from the electrolyte, reducing back-recombination and increasing the electron lifetime.[54]

Organic dyes are a promising and more economical alternative to ruthenium dyes; however, a lot of development is still required. Longevity and stability in organic dyes are relatively understudied.

9.1.3.16 Porphyrin Dyes

Porphyrin dyes are a group of heterocyclic compounds composed of pyrrole subunits, which are often complexed with transition metals. This group also includes a wide variety of natural compounds, such as chlorophyll and haeme. The highest performing sensitiser to date has an efficiency of 13%. These dyes show good absorption from the UV to the near-IR region and have tunable redox potentials. They have good chemical and photostability and suitable HOMO and LUMO levels.[9]

One of the highest performing porphyrin is YD-2 and its derivatives. YD-2 consists of a Zn complex porphyrin ring that acts as a bridging unit in a donor–acceptor molecule. The donor is a diarylamino and ethynylbenzoic acid acts as an acceptor. Yella *et al.*[55] produced a DSSC using YD-2 in 2011. The cell used a double layer of TiO_2 as the photoanode, which was sensitised with YD-2. This was then coated with another light-reflecting layer of TiO_2 nanoparticles to improve light scattering. The next development of this family of dyes came in the form of YD-2-o-C8. This variant replaces the alkyl chains with alkoxyl chains, which reduces back-recombination and pushed the efficiency to 11.9%. This efficiency was increased further with a cell co-sensitised with YD-2-o-C8 and Y123 (an organic dye), which showed efficiencies >12%.[55]

The highest performing porphyrin-based DSSC, and current record holder in the field of DSSCs, used a very similar structure to the YD-2 Zn complex porphyrins. This dye, SM315, had one major alteration in its structure, with the acceptor unit being a benzothiadiazole group. This sensitiser was shown to be very broadly absorbing, leading to high photocurrents. This improvement resulted in a cell with an efficiency >13%.[56]

9.1.3.17 Electrolytes

The role of the electrolyte is to replenish the dye with charge carriers after they have been injected into the conduction band of the semiconductor. An electrolyte has a number of criteria that it must meet. It has to be chemically stable, not react with the semiconducting material and has to have good charge regeneration kinetics.[9]

9.1.3.18 Liquid Electrolytes

The *de facto* electrolyte for most DSSC systems is the iodide–triiodide (I_3^-/I^-) electrolyte. This is the same liquid electrolyte system that was used in the original DSSC by Grätzel and O'Reagan.[2] In their cell the iodide electrolyte was created using tetrapropylammonium iodide and iodine dissolved in ethylene carbonate and acetonitrile. The performance of cells using this electrolyte is fairly high, with an 11% efficiency.[51] This cell used lithium iodide and iodine as the electrolyte, with the presence of Li having been shown to be beneficial to charge carrier separation. This family of electrolytes has been shown to yield great results, although problems still exist. This type of electrolyte absorbs in the blue region of visible light (430 nm) and has been shown to cause corrosion of the counter electrode in DSSCs. To address these shortcomings, a number of other dyes have been investigated. Examples include: other halogen systems, such as Br_3^-/Br^-; transition metal systems, such as $Co^{II/III}$ and $Ni^{III/IV}$; and pseudo-halogen systems, such as $SCN^-/(SCN)_3^-$.[6,9] The highest performing DSSC to date (13%) used a $Co^{II/III}$-based electrolyte in the form of a bypyridyl complex (*tris*(2,2′-bipyridyl)cobalt$^{II/III}$ ($[Co(bpy)_3]^{2+/3+}$)).[4]

9.1.3.19 Solid State Electrolytes

One of the newest developments in electrolytes has been the advent of solid electrolytes. These take the form of a layer of hole transporting material that transports the hole to the opposite electrode, forming a solid state DSSC. The charge transport occurs *via* a hopping mechanism from molecule to molecule. These layers, in essence, serve the same function as the electrolyte. Solid electrolytes have a major advantage in that they do not suffer from the volatile solvent evaporation problems that liquid electrolytes have.

The disadvantages of solid electrolytes is their inability to completely fill the pores of the semiconducting layer.[6,9]

Common solid state electrolytes can be split into two categories: organic and inorganic p-type materials. The highest performing inorganic material is $CsSCI_3$, which has shown efficiencies >10% when paired with an N719 sensitised mesoporous TiO_2 layer.[57] The highest performing organic hole transporting material is spiro-OMeTAD, which, together with a perovskite-sensitised TiO_2, layer achieved an efficiency of 15%.[58]

9.2 Organometal Halide Perovskites

Organometal halide perovskites are a recent development and have attracted a lot of attention in the field of photovoltaics. The beginning of this field was the use of perovskite sensitisers in DSSCs. However, they are now considered as a separate research entity and show potential for achieving high efficiency at a low material cost. Over a span of only six years, from 2006 to 2015, the cell efficiency increased to 21%.[60] This makes it one of the fastest improving photovoltaic technologies currently under development, although much work still remains to be done in optimising both the material and the synthesis.

9.2.1 Structure and Properties

The crystal structure of $CaTiO_3$ was discovered by Gustav Rose in 1839 and was named perovskite in honour of the Russian mineralogist L. A. Perovski.[61] The structure of perovskites follows the general formula ABX_3. The ideal cubic unit cell is composed of five atoms, with the A cation occupying the cubo-octahedral site and the B cation occupying the octahedral site. The X anions occupy the face-centred positions (Figure 9.4).[62]

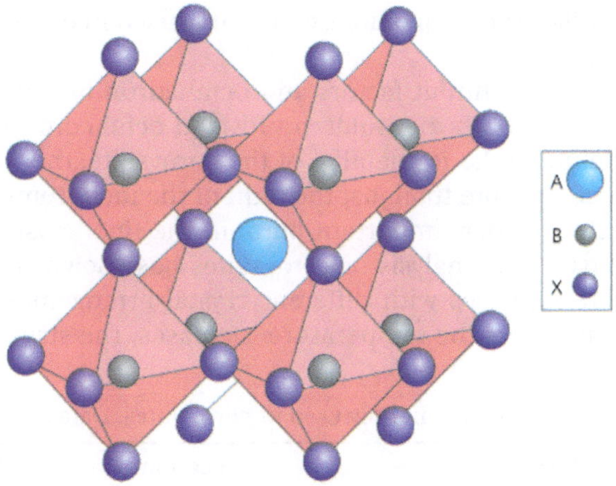

Figure 9.4 Schematic diagram of perovskite crystal structure. Reprinted by permission from Macmillan Publishers Ltd: *Nature Photonics* (ref. 59), Copyright 2014.

To form this crystal structure, the ionic radii of A, B and X need to meet certain values established by a geometric tolerance factor known as the Goldschmidt[63] tolerance factor (t). The tolerance factor is calculated as:

$$t = (r_A + r_X) / \left[\sqrt{2}(r_B + r_X)\right] \tag{9.1}$$

where r_A = radius of ion A, r_B = radius of ion B and r_X = radius of ion X.[64] If the tolerance factor is between 0.85 and 1.11, then a cubic structure will form, which is ideal for electronic properties.[63] Deviation of the factor from 1 may cause slight distortions in the crystal. Values of $t < 1$ result in rotation of the octahedra, which reduces the symmetry and affects the electronic properties. However, if $t > 1$, then there is a higher degree of symmetry and a stable hexagonal structure is created. The size of the A ion can also affect the dimensionality of the perovskite. If the ion is very large, then the three-dimensional network is broken and two- or one-dimensional structures are formed. Therefore if $t \ll 1$, the three-dimensional structure will be destroyed.[62]

Another value to consider in perovskites is the octahedron factor, μ, developed by Li *et al.*[63] This factor describes the ratio between the radii of the B cation and the X anion. It is calculated as:[63]

$$\mu = r_B / r_X \tag{9.2}$$

If $\mu < 0.442$, then the perovskite will not be formed because the octahedron will be too unstable.[63] In halide perovskites, the B site is usually occupied by large divalent metal cations, such as Pb^{2+}, Sn^{2+} or Ge^{2+}.[62,65] Pb^{2+} has an ionic radius of 1.19 Å. The X site is occupied by halide ions (Br^-, Cl^- or I^-; $r_{Br} \approx 1.60$ Å, $r_{Cl} \approx 1.58$ Å, $r_I \approx 1.64$ Å). The organic component occupies the A site and the most commonly used compounds are methyl ammonium cations ($CH_3NH_3^+$) (effective radius 1.8 Å) and formamidinium cations ($HC(NH_2)_2^+$) (radius 1.9–2.2 Å). Table 9.1 lists the radii of ions commonly used in organometal halide perovskites.

The resulting Goldschmidt factors and octahedron factors are shown in Figure 9.5. Although both the Goldschmidt and octahedron factors can be used as a guideline to the formability of the perovskite, the resulting structures are affected by more than just the radii of the ions. Temperature, pressure and defects also play important parts in the final crystal structure of a perovskite. Structural analysis of perovskites has shown that they generally form different phases with different crystal arrangements depending on the temperature. As the temperature increases, the structure becomes

Table 9.1 Radii (Å) of ion commonly used in perovskites. Data taken from ref. 62.

A cations			B cations			X anions			
EA	Formamidinium	Methyl ammonium	Cs	Pb	Sn	I	Br	Cl	F
2.3	1.9–2.2	1.8	1.67	1.19	1.1	2.2	1.96	1.81	1.33

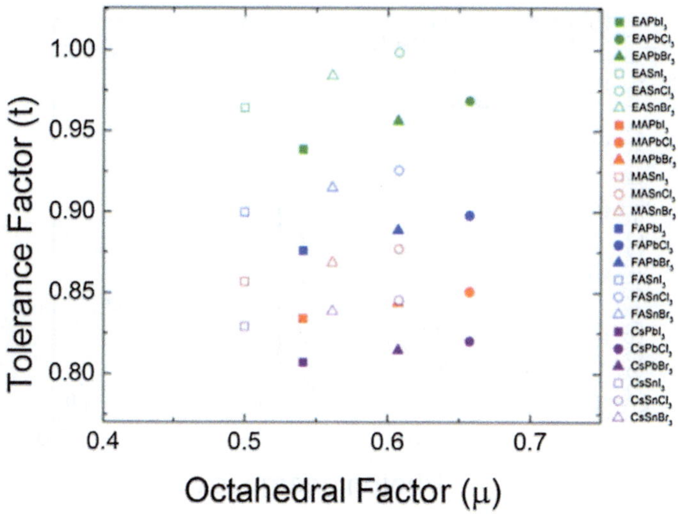

Figure 9.5 Calculated values for tolerance factor and octahedral factor. Reprinted from *Nano Today*, 10, Q. Chen, N. De Marco, Y. (M.) Yang, T.-B. Song, C.-C. Chen, H. Zhao, Z. Hong, H. Zhou and Y. Yang, Recent improvements in dye sensitized solar cells: A review, 54–64., Copyright 2015, with permission from Elsevier.

more ordered, moving from orthogonal to tetragonal and then to cubic. These phase transformations are classified as high temperature phase α, intermediate temperature phase β and low temperature phase γ.[66–68] As an example, the most commonly and initially studied halide perovskite, $CH_3NH_3PbI_3$, forms a tetragonal structure at room temperature, which is corroborated by the fact that the Goldschmidt factor is outside the ideal range for cubic formation (0.83).

9.2.1.1 *Optical Absorption Properties*

A property that makes halide perovskites a promising material for photovoltaic applications is their high absorption coefficient, which allows even thin absorbing layers to sufficiently absorb sufficient light. Perovskites absorb a broad band of light, which means that light absorbing layers as thin as 500 nm can be used.[69,70]

Thin layers also have economic benefits if the technology is scaled up because less material is required; Si requires far thicker layers to absorb sufficient light. The absorption of light in perovskites has a characteristic steep onset due to the direct bandgap of the material. Perovskites are compared with other films in Figure 9.6.

There is a steep onset of absorption in $CH_3NH_3PbI_3$ at around 800 nm (IR region), with a sharp shoulder forming at around 1.55 eV corresponding to its bandgap. Compared with competing solar cell technologies, $CH_3NH_3PbI_3$

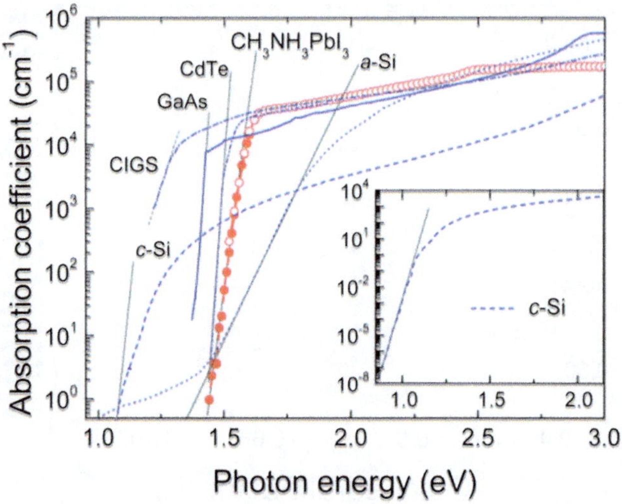

Figure 9.6 Effective absorption coefficient of $CH_3NH_3PbI_3$ thin film compared with other typical photovoltaic materials at room temperature.[71] Reprinted with permission from S. De Wolf, J. Holovsky, S.-J. Moon *et al.*, *J. Phys. Chem. Lett.*, 2014, 5(6), 1035–1039. Copyright 2014 American Chemical Society.

shows a stronger absorption than GaAs, CIGS, CdTe and Si.[71] Theoretical work by Yin *et al.*[72] shed some light on the high absorption in perovskite *versus* other direct bandgap materials. They theorised that the improved absorption was due to the increased density of states in the bottom of the conduction band of perovskite, created by the B site cation. This means that transitions from the valence band are more likely. This increased density of states was attributed to the B site p-orbitals of Pb.

The absorption of perovskites is also dependent on their composition. Changes in the metal B cation change the onset of absorption. For instance, although $CH_3NH_3PbI_3$ and $CH_3NH_3SnI_3$ show a similar steep onset, the position of the onset of absorption is different. The onset of absorption in $CH_3NH_3SnI_3$ is shifted into the near-IR up to 1000 nm. A mixed-metal perovskite, $CH_3NH_3Sn_xPb_{1-x}I_3$, showed a shift of the onset of absorption towards 1300 nm.[73]

Replacement of the organic component (A site cation) results in small shifts. Replacing the methyl ammonium cation with the formamidinium cation to give $HC(NH_2)_2PbI_3$ showed a shift in the onset of absorption towards 850 nm.[74,75] In addition, Cs-based perovskites, such as $CsPbBr_3$, have shown a blue-shifted onset of absorption to 540 nm.[76] A study by Grätzel *et al.* showed that a triple cation system containing methyl ammonium cation, a formamidinium cation and Cs had an onset of absorption at 775 nm.[77]

Changes to the halide (*X* site) also show variations in the onset of absorption, although the effect is dependent on the particular composition.

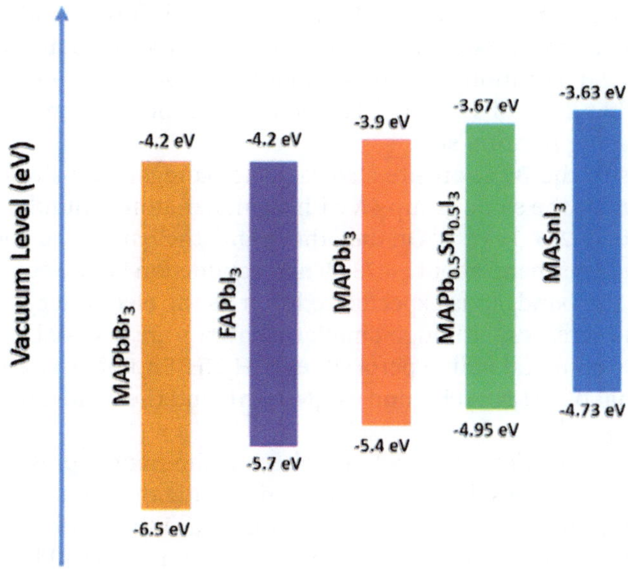

Figure 9.7 Band edge positions of various organometal halide perovskites.[62] Reprinted from *Nano Today*, Vol 10, Q. Chen, N. De Marco, Y. (M.) Yang, T.-B. Song, C.-C. Chen, H. Zhao, Z. Hong, H. Zhou and Y. Yang, Recent improvements in dye sensitized solar cells: A review, 54–64, Copyright 2015, with permission from Elsevier.

For $CH_3NH_3PbI_3$, the addition of Cl^- ions to form the mixed halide $CH_3NH_3PbI_{3-x}Cl_x$ resulted in very little difference in the absorption spectra.[78] Conversely, forming a mixed halide with Br ($CH_3NH_3PbI_{3-x}Br_x$) shifted the absorption point closer to 700 nm.[79] In $HC(NH_2)_2PbI_3$, the addition of Br, forming the mixed halide $HC(NH_2)_2PbI_{3-x}Br_x$, shifted the onset of absorption to 550 nm.[80] In $CsPbBr_xI_{1-x}$, the addition of Br shifted the absorption from 750 to 540 nm.[76]

9.2.1.2 *Bandgap*

The bandgaps of organometal halide perovskites are another reason why they are suitable as photovoltaic absorbers, with values corresponding to the visible light spectrum. The size of the bandgap has been shown to vary and depends heavily on the composition. Figure 9.7 shows the relative bandgaps and band edge positions of perovskites.

$CH_3NH_3PbI_3$ has a reported bandgap of 1.51–1.55 eV, which places it just above the ideal bandgap value, determined by the Shockley–Queisser equation, of 1.34 eV.[81] The valence band lies at −5.4 eV and the conduction band at −3.9 eV.[62]

The substitution of the methyl ammonium cation with the larger formamidinium cation, forming $HC(NH_2)_2PbI_3$, has been shown to reduce the bandgap to a more favourable 1.43–1.48 eV and shift the band edges downwards (valence band about −5.7 eV and conduction band about −4.2 eV).[62,74,75] Conversely, substitution with the smaller Cs cation in the A site leads to an increased bandgap of 1.73 eV.[82] This shows a trend of decreasing bandgap with increasing A cation size.[62]

Changes in the B cation site also vary the bandgap. Substituting Pb with Sn, an atom of the same group 4A with a smaller atomic number, lowers the bandgap to ~1.2–1.3 eV.[62,83] On the other hand, the band edge positions shift upwards (valence band about −4.73 eV and conduction band about −3.63 eV).[62] Generally, the bandgap is expected to lower as the electronegativity of the B site cation increases, although smaller B cations are less stable and rapidly oxidise. Mixed-metal halide perovskites ($CH_3NH_3Sn_xPb_{1-x}I_3$) have a bandgap between that of $CH_3NH_3PbI_3$ and $CH_3NH_3SnI_3$ and the bandgap can be tuned by changing x.[73,84]

Halide substitution also had an effect on the bandgap. Br substitution in a Pb halide perovskite produces a wide bandgap (about 1.7–2.3 eV).[85,86] It is possible to tune the bandgap in a mixed halide system of I and Br ($PbI_{3-x}Br_x$).[87] Similar tuning is also possible in a $HC(NH_2)_2PbI_{3-x}Br_x$ system and $CsPbBr_xI_{1-x}$.[76,88]

9.2.1.3 Charge Carriers in Halide Perovskites

The latest results for halide perovskites indicate that free charges are created on excitation *via* visible light. Using ultrafast spectroscopy techniques to analyse $CH_3NH_3PbI_3$, it has been shown that excitons disassociate to form free charge carriers within 2 ps. The resulting charge carriers have been shown to have high mobility and long carrier diffusion lengths.[89,90] A potential reason for this fast charge separation may be spontaneous polarisation due to ferroelectric domains, which create an internal electric field, pulling the charge carriers apart and forming transport pathways. However, more evidence is required. This effect also could explain the large carrier diffusion lengths and good carrier mobility in perovskites.[91]

The charge carrier mobility has been reported to range from ~10 $cm^2\ V^{-1}\ s^{-1}$ in $CH_3NH_3PbI_3$[62] and very long charge carrier lifetimes of the order of tens of microseconds have also been observed. A similar range of values for charge carrier mobility has been reported in $HC(NH_2)_2PbI_3$, $HC(NH_2)_2PbBr_3$ and mixed halide $HC(NH_2)_2PbBr_3$ systems. The charge carrier mobility was reported to be between 14 and 27 $cm^2\ V^{-1}\ s^{-1}$.[92] Sn-based perovskites have shown lower charge carrier mobility; this may, in part, be due to the instability of the Sn^{2+} ions.[93] There is a wide range of reported hole mobilities, probably due to differences in measurement techniques and differences in the device structure. However, most measurements are of the same order of magnitude.

Transient photoluminescence measurements on $CH_3NH_3PbI_3$ have shown that the diffusion length is ~100 nm. $CH_3NH_3PbI_{3-x}Cl_x$ has been shown to have diffusion lengths >1 μm.[94] Other halide perovskite systems show similar diffusion lengths. $CH_3NH_3PbBr_3$, for example, has been reported to have diffusion lengths from 100 to 360 nm.[95]

The hole transport seems to have higher diffusion lengths and mobility than the electrons. Edri *et al.*[96] investigated electron diffusion lengths in $CH_3NH_3PbI_3$ and $CH_3NH_3PbI_{3-x}Cl_x$ using a current induced by an electron beam. Results from the investigation indicated that the hole diffusion length may be >100 nm and that the electron diffusion length is the limiting factor. In the mixed halide $CH_3NH_3PbI_{3-x}Cl_x$ it was shown that the diffusion lengths exceeded 1 μm and, in contrast with $CH_3NH_3PbI_3$, both the electron and hole diffusion lengths were of similar values. The variance may have been caused by the grain structure differences between the perovskites, with $CH_3NH_3PbI_{3-x}Cl_x$ having larger and more orientated grains.[96]

This agrees with the results from a study by Chen *et al.*[97] comparing $CH_3NH_3PbI_{3-x}Cl_x$ and $CH_3NH_3PbI_3$, with the morphology differences between films minimised. The results indicated that Cl does not have an effect on the electron diffusion length of the material; however, the addition of Cl created a better interface with the electron transporting layer and therefore led to a better cell performance. This shows that halide substitution does not have a direct effect on the charge carrier diffusion lengths and that morphological effects dominate. Investigations of monocrystalline $CH_3NH_3PbI_3$ have shown large diffusion lengths of 175 μm.[98] However, the measurements did not consider crystal surface charge trapping, potentially leading to an overestimation. This result still corroborates the earlier work, however, demonstrating that morphological features such as defects and grain boundaries dominate diffusion lengths.

The large diffusion lengths are, in part, also due to the favourable recombination properties found in perovskites. An investigation conducted by Johnston *et al.*[99] in perovskite devices showed that unimolecular trap-assisted recombination is one of the largest recombination pathways. The trap states that contribute to this recombination are largely related to the morphology of the material, impurities and the interfacial properties between the functional layers of the cell. For instance, it is thought that mid-band trap levels in TiO_2, which acts as an electron transport layer, creates recombination centres.[62,100] From the data it can also be seen that Sn-based perovskites show a particularly high rate of monomolecular recombination. This is probably due to the increased number of holes, acting as traps for electrons, created by the reduction of Sn^{2+} ions to Sn^{4+}. Both bimolecular radiative recombination and auger recombination are very small in comparison and show similar values regardless of the composition of the material.

9.2.1.4 Hysteresis

Halide perovskite devices show an interesting property when subjected to current–voltage measurements. Depending on the scan direction, the current value obtained will be different, forming a hysteresis (Figure 9.8). This hysteresis has been shown to be affected by previous applications of bias and

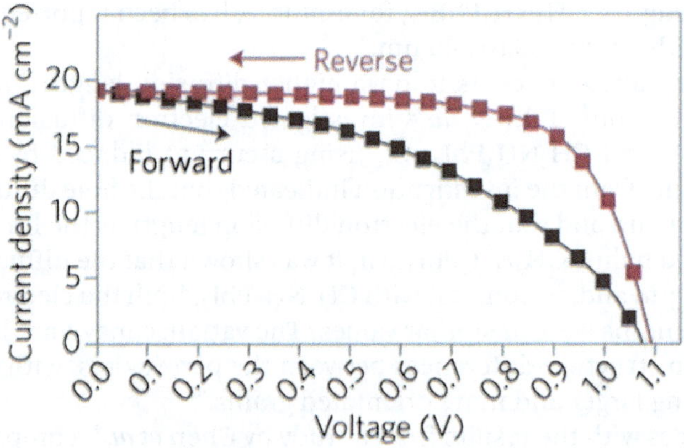

Figure 9.8 Example of J-V curve affected by hysteresis.[101] Reprinted by permission from Macmillan Publishers Ltd: *Nature Materials* (ref. 101), copyright 2014.

therefore the current results can vary depending on the test history of the cell. This can lead to some reported values of current being unrepresentative. The causes of this hysteresis are hotly debated, with three major issues being suggested.[62,83]

One possible cause is linked to the morphology of perovskite cells. Shao *et al.*[102] hypothesised that grain boundaries and surface defects create localised trap states, from which charge carriers diffuse and can lower cell performance. In their study, surfaces and grain boundaries were passivated using PCBM, which led to increased efficiency and the elimination of the hysteresis effect. It has also been shown that an increase in grain size leads to a lower hysteresis. This corroborates defect-related hysteresis because larger grain sizes mean that the material has fewer grain boundaries and therefore lower trap states. Increasing the thickness of the mesoporous TiO_2 layer and reducing the crystal size of the perovskite have also reduced the hysteresis effect. This suggests that the origin of the hysteresis effect may be the capacitive properties of perovskite.[103] These findings suggest that a larger grain size, interface passivation and an optimised transport layer thickness are important in reducing hysteresis. It has also been shown that the device architecture plays a part in hysteresis. Inverted designs do not suffer from this problem,[104] suggesting that the problem may be linked to TiO_2.

Ferroelectric properties may be another contributor to the hysteresis effect. The application of a bias may cause the displacement of A cations within the structure, creating ferroelectric domains within the material. These domains could affect the electron and hole mobility, leading to the hysteresis effect. It is thought that the orientation of A cations plays a part in the creation of domains.[62] Gottesman *et al.*[105] proposed that the A cation will align itself with the electric field under the application of a bias, inducing a distortion in the organic framework and therefore causing hysteresis effects.

Ion migration is another potential contributor to the ferroelectric effect and most of the recent investigations into hysteresis point to this being one of the leading causes. This effect occurs when mobile ions within the material migrate under illumination when an electric field is applied. Under the influence of the electric field, the ions migrate to the charge transport layers. Here they are thought to form space-charge zones, which act as localised p–n junctions.[106,107] This leads to higher currents under a forward bias and lower currents in reverse bias. This movement of ions could not only potentially provide an explanation of the hysteresis effect, but could also explain the S-shaped J–V curve observed in some results.[62]

As a result of hysteresis effects, it is important to measure cells in a manner that can more accurately represent their true properties. Measurements from single scans, even if hysteresis is not apparent, are not wholly reliable because they may inflate results. To obtain more accurate readings, it is advisable to measure the maximum power output of the cell for a lengthy period until the readings have stabilised. The highest stabilised efficiency reported to date is 18% after 250 h, which was achieved by a triple cation organometal halide.[77]

9.2.2 History of Organometal Halides

9.2.2.1 Halide Perovskites in DSSCs

The first use of halide perovskites in the field of photovoltaics was in a DSSC solar cell produced by Miyasaka *et al.* in 2009.[108] Two variations of halide perovskite, $CH_3NH_3PbBr_3$ and $CH_3NH_3PbI_3$, were used as sensitising layers on top of a TiO_2 layer.

The perovskites were deposited by coating the substrates, which already contained a mesoporous TiO_2 layer, with solutions containing stoichiometric amounts of CH_3NH_3X and PbX. This formed the cells' photoelectrode, with Pt-coated FTO acting as the cathode. The cells were then filled with a liquid electrolyte depending on which halide perovskite was used. The cell based on $CH_3NH_3PbI_3$ used LiI and I_2 dissolved in methoxyacetonitrile. The cell based on $CH_3NH_3PbBr_3$ used LiBr and Br_2 dissolved in acetonitrile. The cell functioned in a similar manner to a typical DSSC; however, the electrons were absorbed by the perovskite layer and then injected into the TiO_2 (Figure 9.9). The TiO_2 then transported the electron out of the cell to be harvested. The electron was replenished into the perovskite by the electrolyte.

The resulting cells had similar efficiencies, with the Br-based cell achieving 3.13% and the I cell achieving 3.83%. A high V_{OC} of 0.96 V was realised in the Br-based cell. One caveat reported by Miyasaka *et al.* was the rapid degradation that the cell underwent when exposed to light and air, a problem which is still being investigated.[108]

An evolution of the DSSC-based design was demonstrated by Park's research group in 2011.[109] In this DSSC design, $CH_3NH_3PbI_3$ was deposited in the form of nanocrystal quantum dots with diameters of 2–3 nm. The quantum dots

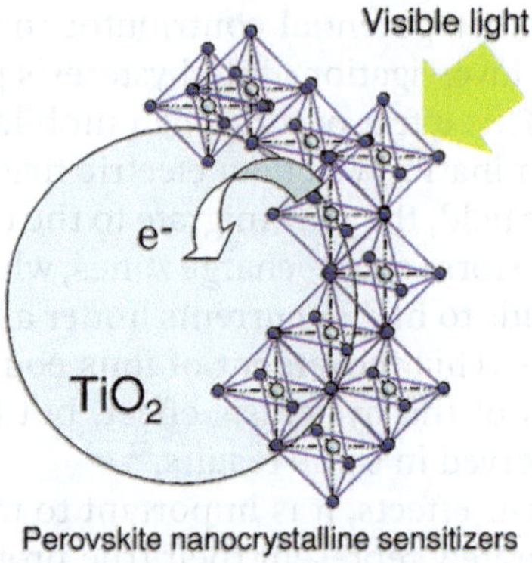

Figure 9.9 Schematic diagram of electron injection into TiO_2 from perovskites.[108] Reprinted with permission from A. Kojima, K. Teshima, Y. Shirai *et al.*, *J. Am. Chem. Soc.*, 2009, **131**(17), 6050–6051. Copyright (2009) American Chemical Society.

were synthesised by dissolving equimolar amounts of CH_3NH_3I and PbI_2 powders in γ-butyrolactone. The solutions were then filtered and deposited on a TiO_2 surface and left for one minute to aid penetration into the mesoporous TiO_2 before being dried at increased temperatures. The counter electrode used in the cell was FTO glass with a layer of Pt.

The efficiency of this cell was 6.2% with a V_{OC} of 0.629 and a fill factor of 0.617. The cell was further improved *via* the introduction of a $Pb(NO_3)_2$ blocking layer, which reduced recombination. This increased the efficiency to 6.5% and the V_{OC} to 0.706. However, the degradation due to the electrolyte that had plagued earlier perovskite solar cells was still present.

9.2.2.2 *Arrival of Solid State Cells*

A big breakthrough in terms of stability and efficiency was the development of solid state DSSC perovskite devices. The research groups of Park and Snaith both demonstrated perovskite-based cells in which the electrolyte was replaced by the hole transporting polymer 2,2′,7,7′-tetrakis-(*N*,*N*-di-*p*-methoxyphenylamine)9,9′-spirobifluorene (spiro-OMetTAD).

The perovskite used in a study by the Park group[110] was $CH_3NH_3PbI_3$ and was synthesised in the same way as their previous study using CH_3NH_3I and PbI_2 in γ-butyrolactone. This was then deposited onto a mesoporous layer of TiO_2. The layer of spiro-OMetTAD was deposited by dissolving the polymer in chlorobenzene and spin-coating the layer at 4000 rpm. This cell had an efficiency of 9.7% with a fill factor of 0.62 and a V_{OC} of 0.888 V, exceeding the efficiency achieved by the DSSC counterparts. The greatest improvement due to the abandonment of liquid electrolytes was increased stability. The stability of this cell was analysed and >500 h of illumination led to only a small decrease in J_{SC}.

The cell produced by Snaith's group[111] had a similar structure, although the perovskite was of a mixed halide form ($CH_3NH_3PbI_2Cl$). This was prepared using a dimethylformamide (DMF) solution of CH_3NH_3I and PbCl and spin-coated onto an FTO substrate with a layer of compact TiO_2 followed by a layer of mesoporous TiO_2. The resulting cell functioned as expected, but further investigation into the charge carrier properties, by replacement of the mesoporous TiO_2 layer with a mesoporous layer of Al_2O_3, led to the discovery of ambipolar charge carrier transport in halide perovskites. The cell also produced a better V_{OC}, indicating that less recombination occurred, and therefore achieved higher efficiencies than its DSSC counterpart. The final efficiency reported was 10.9%.[111]

This formed the basis of a new type of cell called the meso-superstructured solar cell (MSSC). A MSSC uses a layer of insulating mesoporous material – for example, Al_2O_3 – on which the perovskite is deposited. The mesoporous material acts like a scaffold for the perovskite layer, but cannot accept any electrons. In 2013, Snaith's group produced a mixed halide ($CH_3NH_3PbI_2Cl$) MSSC cell with an efficiency of 12.3%.[112]

9.2.2.3 Planar Perovskites

The discovery of ambipolar transport also opened up the possibility of removing the mesoporous structure completely and using a film of perovskite in a planar heterojunction structure. A number of cells were created and demonstrated efficiencies >10%.

One of the early heterojunction cells was produced by Snaith's group[112] in a study exploring the effect of the thickness of the mesoporous scaffold. In this study, a cell was made by omitting the mesoporous layer completely. The device fabrication was similar to that in previous MSSC studies, but the DMF solution of CH_3NH_3I and $PbCl_2$ was directly spin-coated onto a compact layer of TiO_2 before heating. A layer of spiro-OMeTAD was then deposited on top *via* spin-coating with a chlorobenzene solution containing the polymer. This formed a planar p–i–n junction across the cell. The resulting efficiency was 4.9% and, although not as efficient as the cells produced with scaffolds, it showed that the cell could function with only a perovskite absorbing layer.

Another variation of thin film perovskites cells utilised a heterojunction and was pioneered by Etgar *et al.* in 2012.[113] To create this junction, the p-type spiro-OMeTAD hole transporting layer was abandoned and a heterojunction was formed across the layers of TiO_2 and $CH_3NH_3PbI_3$. In this architecture, the perovskite was acting as both an absorber and a hole conductor. The cell was manufactured using $CH_3NH_3PbI_3$ nanoparticles produced from a solution of CH_3NH_3I and PbI_2 in γ-butyrolactone. The nanoparticles were deposited by spin-coating onto a layer of TiO_2 nanosheets. Au was evaporated on top of this a layer of as a back-contact. The efficiency of this cell was 5.5%, with a J_{SC} of 15.1 mA cm^{-2} and a V_{OC} of 0.631 V.[113] Both the p–i–n and heterojunction structures show great potential to create low-cost, high-efficiency solar cells due to their simpler structure.

9.2.2.4 Current State-of-the-art Mixed Perovskites

The highest reported efficiency at the time of writing this review is 21.1%, produced by Grätzel *et al.*[60] in December 2015. The noteworthy aspect of this study was the composition of the material. Unlike the previous mixed halide cells produced by Snaith's group,[111] these cells were composed of a mixed halide perovskite with three different cations. The composition explored was Cs_x(methyl ammonium$_{0.17}$formamidinium$_{0.83}$)$_{(100-x)}$Pb$(I_{0.83}Br_{0.17})_3$, with the amount of Cs varying between 0 and 15%. The cell with 5% Cs showed the highest efficiency of 21.1%. The cell was tested continuously for 250 h to give a stabilised efficiency of 18%. This is particularly noteworthy as the cell showed both high efficiency and good stability.

The material was synthesised by creating a solution of formamidinium iodide (1 M), PbI_2 (1.1 M), methyl ammonium bromide (0.2 M) and $PbBr_2$ dissolved in a DMF- dimethyl sulfoxide (DMSO) mixture (4 : 1 by volume), to which a solution of CsI (1.5 M in DMSO) was then added. This solution was spin-coated on a substrate with a layer of mesporous TiO_2, which itself was on a compact layer of TiO_2. The solution was spin-coated twice at 100 rpm and then 6000 rpm, with 100 μL of chlorobenzene added at the end of the second spin step. The films where then annealed for 1 h at 100 °C in a nitrogen-filled glove-box. The hole transporting layer was formed of spiro-OMeTAD and a final Au electrode layer was thermally evaporated on top. This formed the p–i–n junction. The maximum efficiency (21.1%) was achieved with a V_{OC} of 1.15 V, a J_{SC} of 23.5 mA cm^{-2} and a fill factor of 0.785.

9.2.3 Stability Issues

The stability of organohalide perovskites are a major disadvantage and hamper their potential for real-world commercialisation. Most perovskites currently utilised for photovoltaic applications are inherently unstable and degrade at ambient conditions, which reduces their performance over short periods of time. It has been reported that perovskites are affected in the presence of humidity, air and light.[62,83]

Early DSSC-based perovskite cells suffered rapid degradation due to the perovskite sensitising layers dissolving into the redox electrolyte. This was largely alleviated by transitioning to using spiro-OMeTAD and creating solid state cells. The stability of organometal halide cells is an area with much ongoing research and great advances have been made in understanding the degradation mechanisms. However, most of the studies have focused on the lead iodide organometal halide ($CH_3NH_3PbI_3$). The causes of degradation in organometal halides have primarily been shown to be *via* exposure to oxygen, light, moisture and increased temperatures.

9.2.3.1 Oxygen-induced Degradation

One unfortunate feature of the organometal halide materials is their sensitivity to air. This is particularly problematic because it means that real-world applications will always require a form of encapsulation. Research has indicated that the primary cause of the instability in air is due to the presence of oxygen.[114] Work by Haque with $CH_3NH_3PbI_3$ has shown that the interaction between photoexcited electrons and atmospheric oxygen leads to the formation of superoxide (O_2^-).[115] This superoxide then breaks down the $CH_3NH_3PbI_3$, resulting in the formation of methylamine (CH_3NH_2), lead iodide (PbI_2), iodine (I_2) and water (H_2O). It has also been shown that the diffusion of molecular oxygen within the TiO_2 electron transport layers can be instigated *via* exposure to UV radiation. This again leads to the formation of superoxides, which could lead to the degradation of the organometal halide layer.

9.2.3.2 *Moisture-induced Degradation*

Moisture has been shown to lead to the semi-irreversible breakdown of the organometal halide. Degradation due to ambient conditions is noticeable within a few hours and has been reported in humidity as low as 55%.[116] The ingress of moisture into $CH_3NH_3PbI_3$ has been also shown to occur rapidly, within seconds, at humidities as low as 10%.[117] The degradation pathways are yet to be fully understood, although recent studies have shown that at the time of initial moisture ingress a monohydrate ($CH_3NH_3PbI_3{\cdot}H_2O$) is formed at the grain boundaries within the $CH_3NH_3PbI_3$. It was shown that this was completely reversible by drying the cell in an inert atmosphere and that only excess moisture leading to complete conversion to the monohydrate phase would lead to irreversible degradation.[118] Exposure to excess moisture over long periods of time, however, leads to the formation of a dihydrate $(CH_3NH_3)_4PbI_6{\cdot}2H_2O$ and this process is not reversible. Hailegnaw *et al.* studied the effect of water exposure at different levels of pH on $CH_3NH_3PbI_3$.[119] Although the study focused on the environmental impacts of the cells, it showed that exposure led to complete degradation. Eqn (3) shows the general decomposition reaction of perovskites:[62]

$$BX_2(s) + AX(aq) \leftrightarrow ABX_3 \tag{9.3}$$

9.2.3.3 *Light-induced Degradation*

In terms of light exposure, UV illumination has been shown to be the most detrimental to the stability of organometal halides. Illumination with UV irradiation can facilitate oxygen degradation in cells using TiO_2. It has also been theorised that UV illumination of the M–X bond could create free halogen radicals, which break the perovskites down into dihalogens.

9.2.3.4 *Thermal Decomposition*

The reduction of device performance in organometal halide cells starts at a temperature of 140 °C, with some tests at 150 °C showing degradation in as little as one hour. Grätzel *et al.* reported that decomposition in $CH_3NH_3PbI_3$ occurs due to the loss of CH_3NH_3 and HI species.[120] The degradation has a visually observable effect, with films of $CH_3NH_3PbI_3$ turning from brown to yellow.[83]

This process is a result of the degradation of $CH_3NH_3PbI_3$ to PbI_2.[83] The thermal conductivity properties of $CH_3NH_3PbI_3$ have also been shown to contribute towards degradation because the low thermal conductivity leads to more mechanical stress and affects the lifetime of the material.[121] The thermal degradation pathways have not yet been extensively explored for all organometal halide systems, although this is an important area of study because the cells will have to withstand elevated temperatures over a long period of time in commercial use.

9.2.3.5 *Chemical Stability*

Perovskites are also susceptible to chemical exposure from solvents and other materials used during the creation of the cell. It has been reported that the TPB and acetonitrile used as additives in hole transporting layers have corrosive effects on perovskite films.[122,123] TPB has been shown to directly attack $CH_3NH_3PbI_3$ by dissolving the PbI_2. Acetonitrile, which is a commonly used solvent for Li-bis(trifluoromethanesulfonyl) imide, has also been shown to directly corrode the perovskite. Zhu *et al.* have reported that exposure to ammonia can causes major changes within a perovskite film.[122]

9.2.3.5.1 Compositional Influences on Stability. Pb-based perovskites mainly suffer from stability issues due to their high sensitivity to humidity. Sn-based perovskites show even more extreme stability issues. This is, in part, due to the instability of the Sn^{2+} cation, with exposure to oxidative environments reducing it to Sn^{4+}.[62,75] Total decomposition has been shown to occur within 1 day under ambient conditions.[124] However, the addition of Pb to form mixed-metal systems improves this stability issue. A cations also affect the stability, but, depending on which metal cation the system uses, the effects are opposing. In Pb-based perovskites, formamidinium demonstrates better thermal stability than methyl ammonium.[125] However, the moisture sensitivity is comparable between the two. This trend is seemingly reversed in Sn-based materials, with formamidinium showing a greater inclination to break down.[75,84] The addiction of Cs to formamidinium-based perovskites has been shown to improve their sensitivity to both moisture and increases in temperature[126] and triple cation perovskites have shown long-term (250 h) stability with high efficiency.[77]

9.2.3.5.2 Improving Stability. From the proposed mechanisms, it can be seen that degradation processes require moisture, oxygen and light. Removing some of these exposures is the key to stability. The stability of perovskites has been shown to greatly increase in an inert atmosphere, with long-term stability being reported under irradiation (1000 h).[83] However, this was carried out without a mesoporous TiO_2 layer. The inclusion of this layer leads to degradation even in inert atmospheres due to UV-induced mechanisms. It has also been shown that the use of a bilayer of mesoporous TiO_2 and ZnO_2 results in increased stability.[128] From these results it can be concluded that a sealed cell with modifications or alternative materials to the TiO_2 layer could be a feasible solution. Although temperature-related stability is still an issue, this could be addressed *via* compositional changes. Formamidinium-based perovskites have shown better thermal resistance and could potentially be a solution. In terms of the degradation caused by additives or solvents in hole transporting layers, the removal or replacement of solvents, or changing the hole transporting layer to a semiconducting oxide, could alleviate this issue. More research on the mechanisms of degradation and a focus on degradation under real-world conditions could provide clearer indicators of the requirements for increased perovskite stability.

9.2.4 Device Architectures

Perovskite-based solar cells evolved from their use as a component in DSSCs to forming their own architecture, referred to as a meso-structured solar cell. This architecture has now been superseded by planar heterojunction structures, entirely eliminating mesoporous scaffolds.

9.2.4.1 Mesoporous Scaffold Architectures

The first device architecture to fully utilise a perovskite-based absorber was the mesoporous scaffold architecture. The structure of these cells is similar to that of the DSSCs from which they evolved. The substrate is typically formed of glass coated with a transparent oxide, such as FTO or ITO. This is then coated with a dense layer of TiO_2 (>100 nm), which acts as the electron transporting layer. This layer is covered with a layer of mesoporous TiO_2 scaffold (~350 nm), which is the derivation of the cell's name. The perovskite layer (~350–500 nm) is deposited on top of this scaffold and partially infiltrates the scaffold layer. The next layer is typically 2,2′,7,7′-tetrakis[*N*,*N*-di(4-methoxyphenyl)amino]9,9′-spirobifluorene (spiro-OMeTAD) (~200 nm), which is the hole transporting layer. The final layer is a metal contact (~100 nm), although carbon black has been shown to be an effective electrode.[62] The resulting cell cross-section is shown in Figure 9.10. In this cell, the free electrons are generated in the perovskite sensitiser and are collected and injected into the mesoporous layer. The holes are transported through the hole transporting layer. The electron and hole transporting layers can also act as barriers to any charge carriers that they are not designed for.

9.2.4.2 Mesoporous Modifications

Although the perovskite heavily dictates their performance, in this device the TiO_2 scaffold has a significant effect on the resulting measurements. A variety of different forms of scaffold have been trialled, including nanowires, nanorods, nanocones, nanofibres, nanotubes and nanohelics.[62] Each morphology has been shown to have a different effect on light scattering and charge transport and therefore the J_{SC}. The injection/extraction and mobility

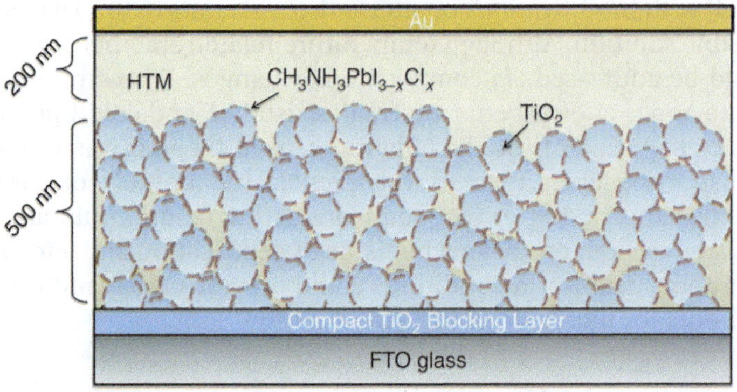

Figure 9.10 Cross-section of a mesoporous scaffold cell.[127] Reprinted by permission from Macmillan Publishers Ltd: *Nature Communications* (ref. 127), copyright 2013.

of carriers is also affected by the change in morphology and therefore fill factor and V_{OC}. Modification of the composition of the layer has also been extensively researched. TiO_2–C nanocomposites have shown improved electron transport properties and have led to higher efficiencies (~15%).[129] The carbon is in the form of reduced graphene and is blended with TiO_2 nanocrystals. TiO_2 doped with metals (Sn, Mg) has also been shown to improve charge extraction and boost efficiencies. Sn-doped TiO_2 has been indicated to increase the bandgap from 3.0 to 3.04 eV, which improves light harvesting.[130] However, Sn is both toxic and costly and is therefore not a desirable component. Mg is a good alternative and doping has been shown to increase V_{OC} and the efficiency.[131] It was thought that the elevation of the conduction band edge to a more favourable position contributed to this effect; however, it has been hypothesised that the formation of Mg oxide/hydroxide layers may passivate the trap states that exist in the TiO_2 interface and may have a larger effect on V_{OC}.[132] Trap state passivation may also occur with Sn-doped TiO_2 and further research into this area using different metals may lead to more efficient cells.

Alternative materials to TiO_2 have also been trialled. ZnO has been extensively explored because it has similar energy levels.[133] The valence band of ZnO sits just below that of most perovskites, allowing electron injection. The most commonly explored scaffolds of ZnO take the form of nanorods and have been shown to be effective in $CH_3NH_3PbI_3$.[134] In a similar manner to TiO_2, metal dopants have been shown to be effective and have boosted efficiencies. Al was used as a dopant for ZnO and increased V_{OC} in a similar fashion to metal doping in TiO_2.[135] ZnO can be doped with a number of transition metals (*e.g.* Mn, Co, Fe or Sn),[136] although no other system has been utilised in perovskite cells to date.

Another intriguing alternative is the use of perovskite oxides. In 2014, Bera *et al.* discovered that using $SrTiO_3$ as the mesoporous layer led to a 25%

higher V_{OC} with $CH_3NH_3PbI_{3-x}Cl_x$.[137] However, the large size of the $SrTiO_3$ particles reduced the surface area and affected the photocurrent. More recently, a different perovskite oxide has been explored. $BaSnO_3$ led to increased performance and formed a better interface with $CH_3NH_3PbI_3$.[138] However, even though perovskite oxides have a higher performance, the cost of these materials and the inclusion of expensive metals may prevent them from being competitive with TiO_2. Future work in this area should focus on perovskite oxides formed from common, non-toxic and cheap materials.

An inverted mesoporous structure is another form that has been investigated. In these cells, the mesoporous layer is formed of p-type material and electron collection is handled by an electron transporting layer. Studies have been conducted using NiO as a mesoporous scaffold. The electron transporting layer consists of either PCBM or a carbon electrode. This structure has been shown to have increased electron lifetimes and a good energy conversion efficiency (11.4%).[139]

It has also been shown that a cell can function without a hole transporting layer. Etgar *et al.*[113] demonstrated that a cell with nanoplatelets of TiO_2 with a sensitising layer of $CH_3NH_3PbI_3$ can be utilised with no hole transporting layer; instead, the holes are transported directly to the Au contact. This is a potential avenue to cheaper cells as hole transporting layers tend to be made from expensive polymers.

9.2.4.3 MSSCs

The meso-superstructured architecture was the next evolutionary step for perovskite cells and led to the discovery of the ambipolar charge characteristics of perovskite. In this cell structure, the n- or p-type mesoporous layer is replaced by an insulating mesoporous layer of Al_2O_3. In this configuration, the perovskite layer acts as both the electron and hole transporter, with the scaffold acting as the support for the layer. This structure has been shown to produce good efficiencies and to avoid UV instabilities due to the lack of TiO_2. Other insulators, such as SiO_2, have also been used as an effective scaffold.[140] This architecture has since been superseded by planar heterojunctions.

9.2.4.4 Planar Heterojunctions

A planar configuration has been developed as a result of the ambipolar nature of perovskites. In this architecture, the perovskite is sandwiched between an electron transporting layer and a hole transporting layer, creating a thin film solar cell. The layers form a p–i–n junction across the perovskite and charge separation occurs *via* the electric field that is subsequently formed across the absorber layer. Depending on which order the layers are deposited, the resulting structure will be p–i–n or n–i–p; note that both still form p–i–n junctions.[62] The n–i–p configuration is more common and is formed by depositing the electron transporting layer onto a transparent conducting oxide glass substrate. This is followed by the perovskite and then the electron transporting layer. The final layer consists of a metal back-contact. The light in this device enters from the n side. The inverse configuration, p–i–n, is formed by swapping the positions of the hole transporting and electron transporting layers. In this case, light enters from the p side. These cells have shown efficiencies surpassing those of both mesostructured and MSSCs.[62]

This architecture held the record for efficiency in 2015, with a thin film perovskite having a certified efficiency of 21%.[60] In addition to the high efficiency, another advantage of this architecture is its simplicity; by omitting the mesoporous structure, only a thin film of perovskite needs to be laid down. This means that the deposition techniques currently used in existing thin film technologies can be applied. Another advantage of this architecture is the ability to deposit on flexible substrates, allowing the possibility of flexible perovskite-based solar cells. Research into this area is still in its infancy, although Kim *et al.*[141] demonstrated a durable flexible cell with 12% efficiency. In this study, $CH_3NH_3PbI_{3-x}Cl_x$ was used in conjunction with a polyethylene naphthalate substrate. The resulting cell underwent 1000 cycles of moderate bending without losing efficiency.

A notable feature is that the omission of the electron transporting layer leads to a similar performance in these devices. This is similar to previous reports of a mesoscopic cell by Etgar *et al.*, where it was shown that a cell could still function in the absence of an electron transporting layer.[143] The omission of these layers is favourable because it leads to cheaper and simpler cells. It is still to be determined whether both an electron transporting layer and a hole transporting layer are required for very high efficiency cells and whether the interface and other defects limit the performance of cells.

9.2.4.5 Hole Transporting Materials

The most commonly used material is spiro-OMeTAD, which is a small organic molecule hole conducting material (Figure 9.11). Spiro-OMeTAD was first pioneered as an electrolyte replacement in solid state DSSCs. In its pristine condition, it suffers from low hole mobility and must be doped to achieve higher levels. Dopants such as Li and Co salts have been used successfully, greatly improving the hole mobility.[144] However, these dopants have been shown to negatively affect the stability of some perovskite systems.

Spiro-OMeTAD is typically spun onto the surface and forms a layer of amorphous glass. One of the reasons why this material is suitable as a hole conductor in perovskites is the location of its HOMO level. For a hole conductor to transport holes from the absorber, its valence band/HOMO level has to be more positive than that of the absorber. In the case of perovskites, the valence band is as positive as −4.73 eV in Sn-based perovskites and as low as −6.5 eV in $CH_3NH_3BrI_3$.[62] Spiro-OMeTAD has a HOMO level of −5.0 eV.[146]

One of the major disadvantages of this material is its high cost due to the complexity of processing. This has prompted a search for alternative materials. Other potential small molecules include the hole transporting material M1 designed by Cheng *et al.*[145] (Figure 9.12). This is an acceptor–donor–acceptor

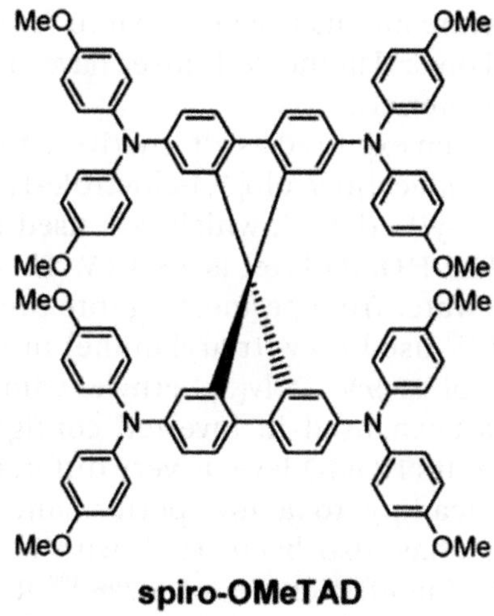

Figure 9.11 Chemical structure of spiro-OMeTAD.[142] Reproduced from ref. 142 with permission from the PCCP Owner Societies.

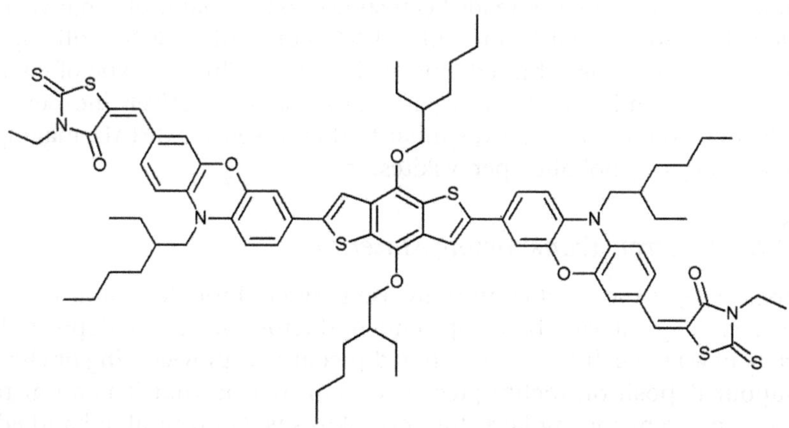

Figure 9.12 The small molecule hole transporting material M1.

phenoxazine-based small molecule with a similar hole mobility to spiro-OMeTAD (2.71×10^{-4} cm^2 V^{-1} s^{-1}). The HOMO level is at −5.29 eV, making it suitable for Pb-based perovskite. Donor–acceptor small molecules have also been explored and show higher hole mobilities than spiro-OMeTAD. DOR3T-TBDT is reported to have mobilities of 0.26 cm^2 V^{-1} s^{-1}, with a band edge position of −5.1 eV. This material is especially notable because it does not require doping to achieve mobility.[145] Other similar spiro-type molecules have also been synthesised; one such example is (4,4′,4″,4‴)-(2H,2′H,4H,4′H-3,3′-spiro-bi[thieno[3,4-b][1,4]dioxepine]-6,6′,8,8′-tetrayl)tetrakis(*N*,*N*-*bis*(4-methoxyphenyl)aniline) developed by Ganesan *et al.*[147] This molecule forms a quasi-spiro-like structure and has been shown to form a crystalline solid; small molecules based on arylamine derivatives have also been shown to be promising alternative materials.

Polymers have also been explored as alternatives. A recent development is poly[2,5-bis(2-decyldodecyl)pyrrolo[3,4-*c*]pyrrole-1,4(2H,5H)-dione-(E)-1,2-di(2,2′-bithiophen-5-yl) ethene], which was used in a perovskite cell by Ganesan *et al.*[148] The HOMO level is −5.4 eV. This polymer is hydrophobic and prevented water from permeating into the humidity-sensitive perovskite. Jeon *et al.*[149] used poly-(triarylamine) in perovskite cell. The cell had an efficiency of 17.9%. Poly(3,4-ethylenedioxythiophene) : polystyrene sulfonate has been used in inverted configurations due to its transparency; however, its HOMO level is very different from the valence level of perovskites, leading to a low performance. Poly(3-hexylthiophene-2,5-diyl) (P3HT) has also been used, with some earlier work in mesoporous cells showing efficiencies of 6.7%.[150] Recent developments have shown a perovskite cell with a graphdiyne-modified P3HT layer. The results showed improved solar cell performance, with a resulting efficiency >14%.[151]

Copper thiocyanate (CuSCN) is a highly transparent material and is easy to solution-process; it has already been shown to be a viable alternative hole transporting layer. A cell formed of $CH_3NH_3PbI_3$ and CuSCN, utilising an inverted structure, has achieved efficiencies >15%. This is a type of coordination polymer and is highly transparent, robust and relatively inexpensive. The HOMO level lies at −5.3 eV, similar to that of spiro-OMeTAD, making it ideal for many organohalide perovskites.[152,153]

9.2.4.6 Electron Transporting Materials

Compact TiO_2 is the most commonly used material for electron transporting layers. TiO_2 is a wide bandgap semiconductor and can be deposited in a variety of ways, such as spin-coating, dip-coating, spray-coating or chemical vapour deposition techniques. The main reason that it is an attractive electron transporting layer for perovskites is its favourable band edge positions. The conduction band is located at −4.0 eV, which allows good electron injection. The valence band sits at −7.0 eV, creating a barrier for holes.[154]

ZnO is a suitable substitute for TiO_2. It has similar band edge positions, but has the notable advantage of not inducing UV degradation in perovskites. Unlike TiO_2, it shows less sensitivity to the thickness of the layer and efficiencies of 15% have been reported.[155]

A further development is the inverse opal-like TiO_2 electron transport layer. The layer is produced by spin-coating two TiO_2 precursor solutions, one dissolved in water and the other in polystyrene. Once sintered, it forms a unique porous structure that allows more light to reach the perovskite layer.[156] Constant improvements to both the electron transporting and hole transporting layers is the key to improving efficiencies, although cost reduction is equally important.

9.3 Processing Perovskite Films

Organohalide perovskites are usually made from two precursors. The first component is an organic halide cation (OX^+, where X = Cl, Br or I and O = formamidinium or methyl ammonium). The second is an inorganic metal halide species (MX_2, where M = Pb or Sn and X= Cl, Br or I). These precursors can be mixed to form the resulting perovskite. These precursors can be deposited and mixed in a number of different ways.[62]

9.3.1 Techniques Based on Solution Processing

Solution processing was one of the original methods used to synthesise perovskites. This method has a number of advantages, including low cost and ease of production. It can be divided into two subcategories depending on how the precursors are mixed and deposited.

One-step processing involves first mixing the precursors and forming the perovskite solution. This is achieved by mixing an organic halide cation with a metal halide species in a polar solvent. The solvents used in this process are DMF, γ-butyrolactone or DMSO.[62] This forms the coating solution for a variety of deposition methods, such as spin-coating. After deposition, the film is annealed at elevated temperatures to form the perovskite. The morphology of the film can be altered by changing the precursor concentration, stoichiometry, drying temperatures and drying time.[62] Higher organic fractions in the precursor mixture produce smoother films with larger grains. Solvent engineering techniques developed by Jeon *et al.*[157] have further improved the performance of perovskite cells. Solvent engineering involves dissolving the precursors in a mixture of γ-butyrolactone, DMSO and toluene. This leads to the controlled nucleation of perovskite crystals, allowing large grains to form.

In two-step deposition the metal halide component is deposited first, again dissolved in a polar solvent. This is followed by the deposition of an organic halide cation dissolved in isopropanol, which is usually achieved *via* spin-coating. This leads to rapid crystallisation of the perovskite and the film is formed within a few seconds. This method has been shown to work for both thin film and mesoporous perovskites.[62] However, although solution processing is easy, it is not very scalable. The inability to create uniform films over large areas is the main drawback of this technology.

Chen *et al.*[158] demonstrated the successful deposition of vapour phase organic halide cations onto the metal halide component. This is a hybrid technique between vapour deposition and solution processing. The resulting grains were very large and this method is promising because it is suitable for larger scale production.

9.3.2 Lead Acetate Perovskites

Lead acetate has been shown to be an alternative to using lead iodide in the production of $CH_3NH_3PbI_3$.[159,160] Snaith's group produced a cell using spiro-OMeTAD and TiO_2 as the hole transporting and electron transporting layers, respectively, using a lead acetate precursor instead of lead iodide. The layers were very smooth and pinhole free and the resulting device had a peak efficiency of 14.0%, an improvement on a lead iodide film in the same study. The synthesis was very quick and required only a single deposition step. The film was formed *via* spin-coating a solution of lead acetate and methyl ammonium iodide in a 3:1 molar ratio. The films were then annealed for a very short time at 100 °C.[161]

These findings were corroborated by Qing *et al.*,[160] who created a similar cell with an efficiency of 12.71%. They also noted that the crystallisation of the film and the resulting structure depended very little on the substrate and was reproducible. Using this synthesis method may provide a promising route to the large-scale production of films with consistent qualities.

9.3.3 Vapour Deposition Techniques

Although solution-based processing is both a common and simple method, it is not very scalable. This is due to the difficulty in spreading a solution over a large area and to form a uniform film. In industry, vapour deposition is commonly used to create thin films over large areas. Perovskites films have also been synthesised using this method and it is a promising route to large-scale synthesis.

One of the early vapour deposition techniques was co-evaporation. In 2013, Liu *et al.*[162] demonstrated the formation of a thin film of $CH_3NH_3PbI_{3-x}Cl_x$ using a dual-source thermal evaporation system. The system functioned by placing a substrate with a compact TiO_2 layer in a vacuum chamber. Two crucibles containing 500 mg of CH_3NH_3I and 100 mg of $PbCl_2$ were within this chamber. The crucibles were heated and the contents evaporated and were deposited onto the substrate. The films were fully crystalised by annealing before a subsequent hole transporting layer was spin-coated on top. The resulting film was very smooth with crystals at a 100 nm scale. The disadvantage of this method is the requirement for a high vacuum system.

Sequential vapour deposition deposits the precursors separately. This method was pioneered by Chen *et al.*[163] and is essentially similar to two-step solution processing. $PbCl_2$ was sublimed onto a substrate under a vacuum. This was followed by a second step in which CH_3NH_3I was sublimed. The layer was annealed, resulting in an even perovskite film. However, unlike the co-evaporation method, the grains were on the micrometre scale. This system also requires a high vacuum.

One of the more recent vapour deposition techniques is one-step[164] chemical vapour deposition. The advantage of this system is that it does not require a high vacuum. This system uses a sealed reaction chamber with two different temperature zones (low and high temperature). The precursors are placed within this chamber in the high temperature zone and the substrate is placed in the lower temperature zone. The precursors are evaporated in the high temperature zone. Argon gas is passed through the chamber, which transports the evaporated precursors into the low temperature area where they are deposited onto the substrate. The substrates are then annealed to form the final film. Grain sizes were recorded on the micrometre scale and had lifetimes in the range of nanoseconds.

Aerosol-assisted vapour deposition is another system that does not require a high vacuum. The precursor is fed into the reaction chamber *via* a liquid-gas aerosol containing small droplets of the precursor. This is then passed through a sealed chamber with an inert carrier gas. This is a low-cost technique and is highly scalable, although the morphology of the resulting films is less than ideal. The films showed had poor coverage and a low density. Further developments of this technique have led to larger grain sizes (1–2 μm) *via* the use of a two-step process.[165] However, as yet, no device has been made using this technique and only perovskites have been deposited. Aerosol-assisted vapour deposition may prove to be a low-cost and scalable method for commercial perovskite production if it can be optimised.

9.4 Conclusions

Rapid progress in the development of DSSCs and perovskite-based solar cells shows the great potential in the application of hybrid solar cells. Major improvements in the areas of stability, especially for perovskites, and longevity are key to further progress. DSSCs have proved to be low-cost alternatives to traditional inorganic cells with good performance under low-light conditions. The rapid increases in efficiency and their relatively low cost have positioned organometal halide perovskites for success in high-performance applications. If the inherent stability problems can be solved and large-scale manufacturing processes developed, these cells will be viable alternatives to traditional inorganic cells and may have a large impact on the solar cell technology market.

References

1. M. Wright and A. Uddin, *Sol. Energy Mater. Sol. Cells*, 2012, **107**, 87–111.
2. B. O'Regan and M. Grätzel, *Nature*, 1991, **353**, 737–740.
3. P. Clarke, in *EE Times*, 2009.
4. S. Mathew, *et al.*, *Nat. Chem.*, 2014, **6**, 242–247.
5. V. Sugathan, E. John and K. Sudhakar, *Renewable Sustainable Energy Rev.*, 2015, **52**, 54–64.
6. A. Hagfeldt, G. Boschloo, L. Sun, L. Kloo and H. Pettersson, *Chem. Rev.*, 2010, **110**, 6595–6663.
7. C. Longo and M.-A. De Paoli, *J. Braz. Chem. Soc.*, 2003, **14**, 898–901.
8. M. Grätzel, *Nat. Mater.*, 2014, **13**, 838–842.
9. M. Ye, X. Wen, M. Wang, J. Iocozzia, N. Zhang, C. Lin and Z. Lin, *Mater. Today*, 2015, **18**, 155–162.
10. D. O. Scanlon, *et al.*, *Nat. Mater.*, 2013, **12**, 798–801.
11. T. L. Thompson and J. T. Yates, *Chem. Rev.*, 2006, **106**, 4428–4453.
12. D. A. H. Hanaor and C. C. Sorrell, *J. Mater. Sci.*, 2011, **46**, 855–874.
13. N.-G. Park, J. van de Lagemaat and A. J. Frank, *J. Phys. Chem. B*, 2000, **104**, 8989–8994.

14. T. Luttrell, S. Halpegamage, J. Tao, A. Kramer, E. Sutter and M. Batzill, *Sci. Rep.*, 2014, **4**, 4043.
15. G. Li, *et al.*, *Dalton Trans.*, 2009, **353**, 10078.
16. L. Wei, *et al.*, *RSC Adv.*, 2015, **5**, 70512–70521.
17. J. Navas, C. Fernández-Lorenzo, T. Aguilar, R. Alcántara and J. Martín-Calleja, *Phys. Status Solidi*, 2012, **209**, 378–385.
18. M. Stefik, F. J. Heiligtag, M. Niederberger and M. Grätzel, *ACS Nano*, 2013, **7**, 8981–8989.
19. K.-P. Wang, *et al.*, *Phys. Chem. Chem. Phys.*, 2009, **11**, 9489.
20. B. Zhao, *et al.*, *Phys. Chem. Chem. Phys.*, 2015, **17**, 14836–14842.
21. S. Lee, *et al.*, *J. Phys. Chem. C*, 2009, **113**, 6878–6882.
22. H. Tian, *et al.*, *J. Phys. Chem. C*, 2010, **114**, 1627–1632.
23. R. Li, *et al.*, *J. Photochem. Photobiol., A*, 2016, **319**, 62–69.
24. D. M. Antonelli and J. Y. Ying, *Angew. Chem., Int. Ed. Engl.*, 1995, **34**, 2014–2017.
25. A. A. Ismail, *et al.*, *J. Mater. Chem.*, 2011, **21**, 11686.
26. Q. Zhang and G. Cao, *Nano Today*, 2011, **6**, 91–109.
27. X. Feng, K. Shankar, O. K. Varghese, M. Paulose, T. J. Latempa and C. A. Grimes, *Nano Lett.*, 2008, **8**, 3781–3786.
28. W.-Q. Wu, *et al.*, *Sci. Rep.*, 2013, **3**, 737–740.
29. K. Fujihara, *et al.*, *Nanotechnology*, 2007, **18**, 365709.
30. J. M. Macák, H. Tsuchiya, A. Ghicov and P. Schmuki, *Dye-sensitized Anodic TiO_2 Nanotubes*, 2005, vol. 7.
31. J. H. Park, *et al.*, *Chem. Commun.*, 2008, **414**, 2867.
32. F. Shao, J. Sun, L. Gao, S. Yang and J. Luo, *J. Mater. Chem.*, 2012, **22**, 6824.
33. S. S. Mali, *et al.*, *Sci. Rep.*, 2013, **3**, 37–38.
34. A. Janotti and C. G. Van De Walle, *Rep. Prog. Phys.*, 2009, **72**, 126501–126529.
35. R. Vittal and K.-C. Ho, *Renewable Sustainable Energy Rev.*, 2016, 920–935.
36. N. Memarian, I. Concina, A. Braga, S. M. Rozati, A. Vomiero and G. Sberveglieri, *Angew. Chem., Int. Ed.*, 2011, **50**, 12321–12325.
37. L. Luo, *et al.*, *J. Power Sources*, 2011, **196**, 10518–10525.
38. J.-X. Zhao, Y.-Z. Zheng, X.-H. Lu, J.-F. Chen, X. Tao and W. Zhou, *ChemPhysChem*, 2013, **14**, 1977–1984.
39. L. Zhang, *et al.*, *J. Mater. Chem. A*, 2013, **1**, 12066.
40. M. Saito, *et al.*, *Energy Environ. Sci.*, 2008, **1**, 280.
41. H. Minoura and T. Yoshida, *Electrochemistry*, 2008, **76**, 109–117.
42. M. Law, L. E. Greene, J. C. Johnson, R. Saykally and P. Yang, *Nat. Mater.*, 2005, **4**, 455–459.
43. D. Barpuzary, A. S. Patra, J. V. Vaghasiya, B. G. Solanki, S. S. Soni and M. Qureshi, *ACS Appl. Mater. Interfaces*, 2014, **6**, 12629–12639.
44. M. Abd-Ellah, *et al.*, *Nanoscale*, 2016, **8**, 1658–1664.
45. Y.-Z. Zheng, H. Ding, Y. Liu, X. Tao, G. Cao and J.-F. Chen, *J. Power Sources*, 2014, **254**, 153–160.
46. C.-Y. Lin, Y.-H. Lai, H.-W. Chen, J.-G. Chen, C.-W. Kung, R. Vittal and K.-C. Ho, *Energy Environ. Sci.*, 2011, **4**, 3448.
47. A. Birkel, *et al.*, *Energy Environ. Sci.*, 2012, **5**, 5392–5400.

48. Z. Li, Y. Zhou, R. Sun, Y. Xiong, H. Xie and Z. Zou, *Chin. Sci. Bull.*, 2014, **59**, 2122–2134.
49. B. Tan, E. Toman, Y. Li and Y. Wu, *J. Am. Chem. Soc.*, 2007, **129**, 4162–4163.
50. R. Ghosh, *et al.*, *ACS Appl. Mater. Interfaces*, 2011, **3**, 3929–3935.
51. Q. Yu, Y. Wang, Z. Yi, N. Zu, J. Zhang, M. Zhang and P. Wang, *ACS Nano*, 2010, **4**, 6032–6038.
52. S.-W. Wang, *et al.*, *Chem. Sci.*, 2013, **4**, 2423.
53. W. Zeng, *et al.*, *Chem. Mater.*, 2010, **22**, 1915–1925.
54. A. Yella, *et al.*, *Chem. Mater.*, 2013, **25**, 2733–2739.
55. A. Yella, *et al.*, *Science*, 2011, **334**, 629–634.
56. S. Mathew, *et al.*, *Nat. Chem.*, 2014, **6**, 242–247.
57. I. Chung, B. Lee, J. He, R. P. H. Chang and M. G. Kanatzidis, *Nature*, 2012, **485**, 486–489.
58. J. Burschka, N. Pellet, S.-J. Moon, R. Humphry-Baker, P. Gao, M. K. Nazeeruddin and M. Grätzel, *Nature*, 2013, **499**, 316–319.
59. A. Gheno, S. Vedraine, B. Ratier and J. Bouclé, *Metals (Basel, Switz.)*, 2016, **6**, 21.
60. M. Grätzel, *Dyesol*, 2015, 1.
61. T. Wolfram and S. Ellialtioglu, *Electronic and Optical Properties of D-Band Perovskites*, Cambridge University Press, 2006.
62. Q. Chen, *et al.*, *Nano Today*, 2015, **10**, 355–396.
63. C. Li, X. Lu, W. Ding, L. Feng, Y. Gao and Z. Guo, *Acta Crystallogr., Sect. B: Struct. Sci.*, 2008, **64**, 702–707.
64. V. M. Goldschmidt, *Chem. Ber.*, 1927, **60**, 1263–1268.
65. S. D. Stranks and H. J. Snaith, *Nat. Nanotechnol.*, 2015, **10**, 391–402.
66. F. Brivio, A. B. Walker and A. Walsh, *APL Mater.*, 2013, **1**, 42111.
67. Y. H. Chang, C. H. Park and K. Matsuishi, *J. Korean Phys. Soc.*, 2004, **44**, 889–893.
68. C. C. Stoumpos, *et al.*, *Inorg. Chem.*, 2013, **52**, 9019–9038.
69. L. Etgar, *et al.*, *J. Am. Chem. Soc.*, 2012, **134**, 17396–17399.
70. A. Kojima, K. Teshima, Y. Shirai and T. Miyasaka, *J. Am. Chem. Soc.*, 2009, **131**, 6050–6051.
71. S. De Wolf, *et al.*, *J. Phys. Chem. Lett.*, 2014, **5**, 1035–1039.
72. W. Yin, J. Yang, J. Kang, Y. Yan and S.-H. Wei, *J. Mater. Chem. A*, 2015, **3**, 8926–8942.
73. Y. Ogomi, *et al.*, *J. Phys. Chem. Lett.*, 2014, **5**, 1004–1011.
74. N. Pellet, P. Gao, G. Gregori, T. Y. Yang, M. K. Nazeeruddin, J. Maier and M. Grätzel, *Angew. Chem., Int. Ed.*, 2014, **53**, 3151–3157.
75. T. M. Koh, *et al.*, *J. Phys. Chem. C*, 2014, **118**, 16458–16462.
76. R. E. Beal, *et al.*, *J. Phys. Chem. Lett.*, 2016, 746–751.
77. M. Saliba, *et al.*, *Energy Environ. Sci.*, 2016, **9**, 1989–1997.
78. Y. Jiang, M. A. Green, R. Sheng and A. Ho-Baillie, *Sol. Energy Mater. Sol. Cells*, 2015, **137**, 253–257.
79. J. Qiu, Y. Qiu, K. Yan, M. Zhong, C. Mu, H. Yan and S. Yang, *Nanoscale*, 2013, **5**, 3245–3248.
80. G. E. Eperon, S. D. Stranks, C. Menelaou, M. B. Johnston, L. M. Herz and H. J. Snaith, *Energy Environ. Sci.*, 2014, **7**, 982.

81. W. Shockley and H. J. Queisser, *J. Appl. Phys.*, 1961, **32**, 510–519.
82. G. E. Eperon, G. M. Paternò, R. J. Sutton, A. Zampetti, A. A. Haghighirad, F. Cacialli and H. J. Snaith, *J. Mater. Chem. A*, 2015, **3**, 19688–19695.
83. S. D. Stranks and H. J. Snaith, *Nat. Nanotechnol.*, 2015, **10**, 391–402.
84. C. C. Stoumpos, C. D. Malliakas and M. G. Kanatzidis, *Inorg. Chem.*, 2013, **52**, 9019–9038.
85. B. Cheng, Y. Yongbo, F. Yanjun and H. Jinsong, *Adv. Energy Mater.*, 2015, **5**, 1401616.
86. H. Bin Kim, E. Kim, J. Kim and W. I. Lee, *J. Mater. Chem. A*, 2015, **3**, 9264–9270.
87. J. H. Noh, S. H. Im, J. H. Heo, T. N. Mandal and S. Il Seok, *Nano Lett.*, 2013, **13**, 1764–1769.
88. S. A. Kulkarni, T. Baikie, P. P. Boix, N. Yantara, N. Mathews and S. Mhaisalkar, *J. Mater. Chem. A*, 2014, **2**, 9221.
89. F. Deschler, *et al.*, *J. Phys. Chem. Lett.*, 2014, **5**, 1421–1426.
90. C. S. Ponseca, *et al.*, *J. Am. Chem. Soc.*, 2014, **136**, 5189–5192.
91. J. M. Frost, K. T. Butler, F. Brivio, C. H. Hendon, M. Van Schilfgaarde and A. Walsh, *Nano Lett.*, 2014, **14**, 2584–2590.
92. W. Rehman, *et al.*, *Adv. Mater.*, 2015, **27**, 7938–7944.
93. N. K. Noel, *et al.*, *Energy Environ. Sci.*, 2014, **7**, 3061–3068.
94. S. D. Stranks, *et al.*, *Science*, 2013, **342**, 341–344.
95. N. Kedem, *et al.*, *J. Phys. Chem. Lett.*, 2015, **6**, 2469–2476.
96. E. Edri, *et al.*, *Nano Lett.*, 2014, **14**, 1000–1004.
97. Q. Chen, *et al.*, *Nat. Commun.*, 2015, **6**, 1–9.
98. Q. Dong, Y. Fang, Y. Shao, P. Mulligan, J. Qiu, L. Cao and J. Huang, *Science*, 2015, **347**, 967–970.
99. M. B. Johnston and L. M. Herz, *Acc. Chem. Res.*, 2016, **49**, 146–154.
100. I. Nakamura, N. Negishi, S. Kutsuna, T. Ihara, S. Sugihara and K. Takeuchi, *J. Mol. Catal. A: Chem.*, 2000, **161**, 205–212.
101. M. D. McGehee, *Nat. Mater.*, 2014, **13**, 845–846.
102. Y. Shao, Z. Xiao, C. Bi, Y. Yuan and J. Huang, *Nat. Commun.*, 2014, 1–7.
103. H.-S. Kim and N.-G. Park, *J. Phys. Chem. Lett.*, 2014, **5**, 2927–2934.
104. S. H. Im, J.-H. Heo, H. J. Han, D. Kim and T. Ahn, *Energy Environ. Sci.*, 2015, **8**, 1602–1608.
105. R. Gottesman, *et al.*, *J. Phys. Chem. Lett.*, 2014, **5**, 2662–2669.
106. C. Eames, J. M. Frost, P. R. F. Barnes, B. C. O'Regan, A. Walsh and M. S. Islam, *Nat. Commun.*, 2015, **6**, 7497.
107. W. Tress, *et al.*, *Energy Environ. Sci.*, 2015, **8**, 995–1004.
108. A. Kojima, K. Teshima, Y. Shirai and T. Miyasaka, *J. Am. Chem. Soc.*, 2009, **131**, 6050–6051.
109. J.-H. Im, C.-R. Lee, J.-W. Lee, S.-W. Park and N.-G. Park, *Nanoscale*, 2011, **3**, 4088–4093.
110. H.-S. Kim, *et al.*, *Sci. Rep.*, 2012, **2**, 591.
111. M. M. Lee, J. Teuscher, T. Miyasaka, T. N. Murakami and H. J. Snaith, *Science*, 2012, **338**, 643–647.

112. J. M. Ball, M. M. Lee, A. Hey and H. J. Snaith, *Energy Environ. Sci.*, 2013, **6**, 1739.
113. L. Etgar, *et al.*, *J. Am. Chem. Soc.*, 2012, **134**, 17396–17399.
114. T. A. Berhe, *et al.*, *Energy Environ. Sci.*, 2016, **9**, 323–356.
115. N. Aristidou, I. Sanchez-Molina, T. Chotchuangchutchaval, M. Brown, L. Martinez, T. Rath and S. A. Haque, *Angew. Chem., Int. Ed.*, 2015, **54**, 8208–8212.
116. G. Niu, X. Guo and L. Wang, *J. Mater. Chem. A*, 2015, **3**, 8970–8980.
117. C. Müller, *et al.*, *Chem. Mater.*, 2015, **27**, 7835–7841.
118. A. M. A. Leguy, *et al.*, *Chem. Mater.*, 2015, **27**, 3397–3407.
119. B. Hailegnaw, S. Kirmayer, E. Edri, G. Hodes and D. Cahen, *J. Phys. Chem. Lett.*, 2015, **6**, 1543–1547.
120. A. Dualeh, P. Gao, S. Il Seok, M. K. Nazeeruddin and M. Grätzel, *Chem. Mater.*, 2014, **26**, 6160–6164.
121. A. Pisoni, J. Jaćimović, O. S. Barišić, M. Spina, R. Gaál, L. Forró and E. Horváth, *J. Phys. Chem. Lett.*, 2014, **5**, 2488–2492.
122. Y. Zhao and K. Zhu, *Chem. Commun. (Cambridge, U. K.)*, 2014, **50**, 1605–1607.
123. G. Niu, W. Li, F. Meng, L. Wang, H. Dong and Y. Qiu, *J. Mater. Chem. A*, 2014, **2**, 705–710.
124. F. Hao, C. C. Stoumpos, D. H. Cao, R. P. H. Chang and M. G. Kanatzidis, *Nat. Photonics*, 2014, **8**, 489–494.
125. G. E. Eperon, S. D. Stranks, C. Menelaou, M. B. Johnston, L. M. Herz and H. J. Snaith, *Energy Environ. Sci.*, 2014, **7**, 982.
126. J. W. Lee, D. H. Kim, H. S. Kim, S. W. Seo, S. M. Cho and N. G. Park, *Adv. Energy Mater.*, 2015, **5**, 1501310.
127. T. Leijtens, G. E. Eperon, S. Pathak, A. Abate, M. M. Lee and H. J. Snaith, *Nat. Commun.*, 2013, **4**, 2885.
128. A. Mei, *et al.*, *Science*, 2014, **345**, 295–298.
129. J. T. W. Wang, *et al.*, *Nano Lett.*, 2014, **14**, 724–730.
130. X. Zhang, Z. Bao, X. Tao, H. Sun, W. Chen and X. Zhou, *RSC Adv.*, 2014, **4**, 64001–64005.
131. J. Wang, *et al.*, *Appl. Phys. Lett.*, 2015, **106**, 121104.
132. K. Manseki, T. Ikeya, A. Tamura, T. Ban, T. Sugiura and T. Yoshida, *RSC Adv.*, 2014, **4**, 9652.
133. M. Grätzel, *Nature*, 2001, **414**, 338–344.
134. D.-Y. Son, J.-H. Im, H.-S. Kim and N.-G. Park, *J. Phys. Chem. C*, 2014, **118**, 16567–16573.
135. J. Dong, *et al.*, *Chem. Commun. (Cambridge, U. K.)*, 2014, **50**, 13381–13384.
136. S. J. Pearton, D. P. Norton, M. P. Ivill, A. F. Hebard, J. M. Zavada, W. M. Chen and I. A. Buyanova, *IEEE Trans. Electron Devices*, 2007, **54**, 1040–1048.
137. A. Bera, K. Wu, A. Sheikh, E. Alarousu, O. F. Mohammed and T. Wu, *J. Phys. Chem. C*, 2014, **118**, 28494–28501.
138. L. Zhu, Z. Shao, J. Ye, X. Zhang, X. Pan and S. Dai, *Chem. Commun.*, 2016, **52**, 970–973.

139. Z. Liu, *et al.*, *Dalton Trans.*, 2015, **44**, 3967–3973.
140. J. Jang, S. H. Hwang, J. Roh, J. Lee, J. Ryu and J. Yun, *J. Mater. Chem. A*, 2014, **2**, 16429–16433.
141. B.-J. Kim, *et al.*, *Energy Environ. Sci.*, 2015, **8**, 916–921.
142. C.-Y. Hsu, Y.-C. Chen, R. Y.-Y. Lin, K.-C. Ho and J. T. Lin, *Phys. Chem. Chem. Phys.*, 2012, **14**, 14099.
143. D. Liu, J. Yang and T. L. Kelly, *J. Am. Chem. Soc.*, 2014, **136**, 17116–17122.
144. W. H. Nguyen, C. D. Bailie, E. L. Unger and M. D. McGehee, *J. Am. Chem. Soc.*, 2014, **136**, 10996–11001.
145. M. Cheng, *et al.*, *Adv. Energy Mater.*, 2015, **5**, 4101720.
146. J. Dore, *et al.*, *EPJ Photovoltaics*, 2013, **4**, 40301.
147. P. Ganesan, *et al.*, *Energy Environ. Sci.*, 2015, **8**, 1986–1991.
148. Y. S. Kwon, J. Lim, H.-J. Yun, Y.-H. Kim and T. Park, *Energy Environ. Sci.*, 2014, **7**, 1454.
149. N. J. Jeon, J. H. Noh, W. S. Yang, Y. C. Kim, S. Ryu, J. Seo and S. Il Seok, *Nature*, 2015, **517**, 476–480.
150. A. Abrusci, S. D. Stranks, P. Docampo, H. Yip, A. K.-Y. Jen and H. J. Snaith, *Nano Lett.*, 2013, **13**, 3124–3128.
151. J. Xiao, *et al.*, *Adv. Energy Mater.*, 2015, **5**, 1401943.
152. K. Zhao, R. Munir, B. Yan, Y. Yang, T.-S. Kim and A. Amassian, *J. Mater. Chem. A*, 2015, **3**, 20554–20559.
153. S. Ye, *et al.*, *Nano Lett.*, 2015, **15**, 3723–3728.
154. M. Grätzel, *Nature*, 2001, **414**, 338–344.
155. D. Liu and T. L. Kelly, *Nat. Photonics*, 2013, **8**, 133–138.
156. X. Chen, S. Yang, Y. C. Zheng, Y. Chen, Y. Hou, X. H. Yang and H. G. Yang, *Adv. Sci.*, 2015, **2**, 1500105.
157. N. J. Jeon, J. H. Noh, Y. C. Kim, W. S. Yang, S. Ryu and S. Il Seok, *Nat. Mater.*, 2014, **13**, 1–7.
158. Q. Chen, *et al.*, *J. Am. Chem. Soc.*, 2013, 3–6.
159. W. Zhang, *et al.*, *Nat. Commun.*, 2015, **6**, 6142.
160. J. Qing, *et al.*, *Org. Electron.*, 2015, **27**, 12–17.
161. W. Zhang, *et al.*, *Nat. Commun.*, 2015, **6**, 6142.
162. M. Liu, M. B. Johnston and H. J. Snaith, *Nature*, 2013, **501**, 395–398.
163. C.-W. Chen, H.-W. Kang, S.-Y. Hsiao, P.-F. Yang, K.-M. Chiang and H.-W. Lin, *Adv. Mater.*, 2014, **26**, 6647–6652.
164. M. M. Tavakoli, *et al.*, *Sci. Rep.*, 2015, **5**, 14083.
165. S. Chen, J. Briscoe, Y. Shi, K. Chen, R. M. Wilson, S. Dunn and R. Binions, *CrystEngComm*, 2015, **17**, 7486–7489.

CHAPTER 10

Deposition Techniques for Perovskite Solar Cells

DAVID J. LEWIS

School of Materials, University of Manchester, Oxford Road, Manchester, M13 9PL, United Kingdom
*E-mail: david.lewis-4@manchester.ac.uk

10.1 Introduction

Hybrid perovskite solar cells are a promising class of next-generation photovoltaic devices. They incorporate ABX_3 perovskite as the absorber and intrinsic semiconductor layer, where A is an organic cation (*e.g.* $CH_3NH_3^+$), B is a divalent metal ion (*e.g.* Pb^{2+} or Sn^{2+}) and X is a halide anion (*e.g.* iodide, bromide, chloride or mixtures thereof). The perovskite crystal structure is adopted by these so-called inorganic–organic perovskites (IOPs) (Figure 10.1), with the A cations occupying the corners, face-centring by the X anions and the B cation at the centre of the unit cell.

The high efficiencies of perovskite-based solar cells,[1] which have soared to >20% in a short space of time (see Chapter 1, Figure 1.1),[2] have proved attractive to researchers as an alternative to copper indium gallium selenide, silicon and polycrystalline silicon solar cells.[3] Although the application of IOPs in solar devices is now a well-established concept, the materials themselves were known and explored prior to this by Mitzi and co-workers as layered semiconductors with anisotropic transport properties.[4–6] It is only relatively recently that their potential in photovoltaic and other optoelectronic devices,

Nanoscience & Nanotechnology Series No. 45
Nanostructured Materials for Type III Photovoltaics
Edited by Peter Skabara and Mohammad Azad Malik

Published by the Royal Society of Chemistry, www.rsc.org

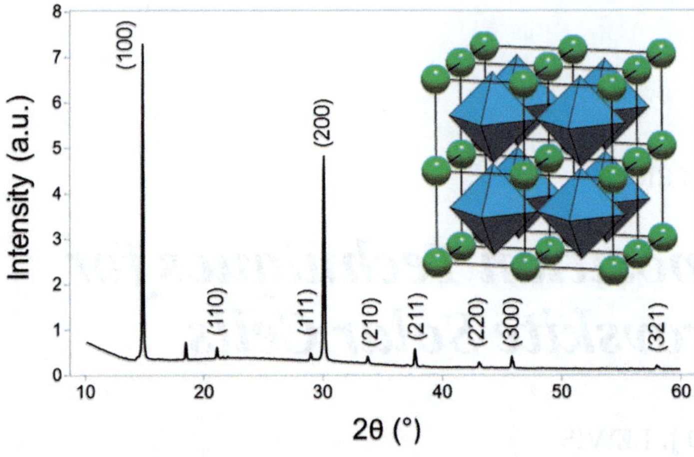

Figure 10.1 Powder XRD pattern of $CH_3NH_3PbBr_3$ (MAPBr) deposited on glass substrates by AACVD and indexed to the cubic perovskite phase. Inset: cubic perovskite structure (eight ABX_3 unit cells). Reproduced from ref. 48 with permission from The Royal Society of Chemistry. Reproduced courtesy of National Renewable Energy Laboratory, Golden, CO, USA.

such as light-emitting diodes, has been explored.[3] This short review discusses the myriad techniques that have evolved for the deposition of active perovskite layers in conventional (*i.e.* non-inverted) solar cell architectures and is limited to inorganic–organic lead halides, although the reader should be aware that fully inorganic lead(II) halide perovskites, such as $CsPbI_3$, have also been used to produce photovoltaic devices,[7] as have tin(II) perovskites, which are isoelectronic with lead(II) perovskites.[8]

10.2 Perovskite Solar Cells

Perovskite solar cells are the direct descendants of the Grätzel-type dye-sensitised solar cell (DSSC) (Figure 10.2). These cells are assembled in a vertical architecture consisting of a bottom anode, usually a transparent conducting oxide such as fluorine-doped tin oxide (FTO) on a glass substrate, and a mesoporous layer of n-type TiO_2 nanocrystals on which inorganic dyes such as ruthenium(II) or cyclometallated iridium(III) complexes are immobilised. The mesoporous dye layer is fully immersed in an iodide/triiodide electrolyte and capped with Pt/FTO on a glass substrate cathode. Light is harvested by the immobilised dye molecules and electrons injected into the TiO_2 nanoparticle layer, while charge transport is effected by the iodide/triiodide redox pair, thus creating a potential difference across the cell. The large surface area of the mesoporous TiO_2 means that there is effectively a large amount of dye adsorbed at the surface (typically a 10 μm layer), ensuring a large absorption cross-section for efficient light harvesting. Typical efficiencies for DSSCs

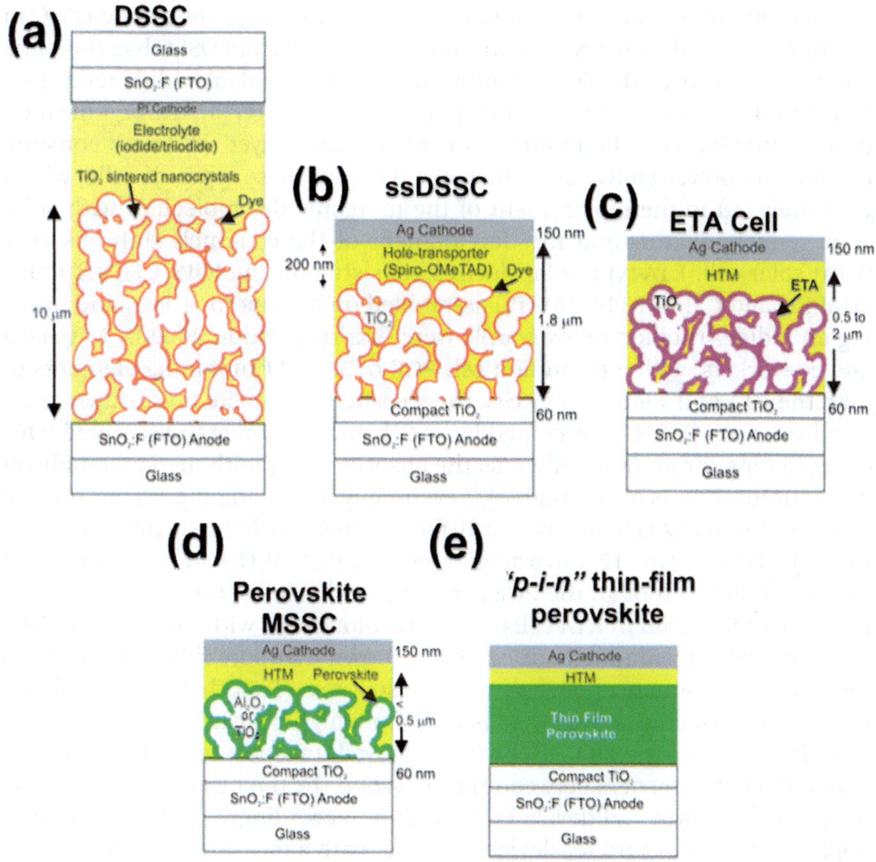

Figure 10.2 Historical evolution of DSSCs to perovskite solar cells (a) The original DSSC, often dubbed the 'Grätzel cell' after its inventor. Inorganic dye molecules such as Ru(II) polypyridyl complexes are adsorbed to TiO_2 and act as light-harvesting centres to effect charge separation. (b) The solid state DSSC, in which the liquid electrolyte was replaced with a solid state organic p-type HTM, spiro-OMeTAD. (c) The ETA cell, which uses a thin layer of semiconducting material on TiO_2 nanoparticles. (d) The MSSC, in which the absorber is a perovskite immobilised on mesoporous alumina or TiO_2. (e) The planar heterojunction architecture, in which a thin film of IOP is sandwiched between n-type compact TiO_2 and typically spiro-OMeTAD. Adapted with permission from Snaith *et al. J. Phys. Chem. Lett.* 2013, **4**, 3623–3630. Copyright (2013) American Chemical Society.

are ~10%, with open circuit voltages (V_{OC}) typically ~0.7 V.[1] The optical bandgap of the complexes used is ~1.4 eV and is thus ideal for the capture of the atmospheric solar photon flux. Although a promising technology, the commercialisation of DSSCs has been hampered by stability and reproducibility issues, the latter especially prominent on scale up of the technology. Leakage of the electrolyte from the cells is also potentially an issue.

The move away from semi-liquid DSSCs gave rise to the solid state DSSC in an attempt to make the DSSC more stable. The solid state DSSC has the same basic architecture as the DSSC, but the semi-liquid iodide/triiodide redox pair is replaced by a solid state hole transporting material (HTM), the organic molecule spiro-OMeTAD. Limitations on the absorber layer thickness conspire to limit the power conversion efficiency (PCE) of these types of cells, which ultimately led to the replacement of the inorganic dye molecules with solid state semiconductors and the development of the extremely thin absorber (ETA) solar cell. ETA-type solar devices consisting of the IOPs $CH_3NH_3PbBr_3$ (MAPBr) and $CH_3NH_3PbI_3$ (MAPI) assembled on mesoporous TiO_2 and utilising an iodide/triiodide redox couple were first reported in 2009 by Miyasaka and co-workers, with a champion cell PCE of 3.5%.[9] Combining elements of both the ETA and solid state DSSC, the so-called perovskite meso-superstructured solar cell (MSSC), developed by Snaith and co-workers, assembled IOPs on insulating mesoporous Al_2O_3 as the absorber layer (although assembly on mesoporous TiO_2 is also possible),[10] eventually increasing the V_{OC} by 300 mV compared with ETA cells and with a fill factor of 0.67; in doing so, they increased the PCE of devices to >12% in a single step change.[11] Al_2O_3 can be co-deposited in a single step, although the PCEs are reduced.[12] Charge transport is notably faster in MSSCs than in ETA cells. In photovoltaic cells with mesoporous TiO_2 as the n-type semiconductor, it has been shown that pore filling is crucial to the device performance; devices with mesoporous TiO_2 that are completely filled with perovskite active layers benefit from higher sustained electron densities injected into the TiO_2 layer with improved transport rates. The recombination of charge carriers between the HTM and the n-type layer is blocked by the perovskite layer and device PCEs improve accordingly.[13] MSSCs currently hold the certified record for device efficiency with a measured PCE of >20%.[2]

The perovskite layer in MSSCs is remarkable in that it fulfils the three primary aspects of an absorber layer: the efficient absorption of light, the generation of charge carriers and the transport of electrons and holes towards their respective electrodes. Thus, the final major evolution of the perovskite solar cell was the development of the planar heterojunction p–i–n perovskite cell.[11] In this type of cell, the mesoporous layer is eliminated completely and the perovskite exists as an intrinsic semiconductor sandwiched between compact n-type TiO_2 and a solid state HTM, initially spiro-OMeTAD, although there are some examples of extremely novel and interesting HTM variants.[14] In some cases, the HTM can be removed completely.[15] There has been work to engineer and improve the interface of the electron-blocking layer with the perovskite material[16] – for instance, band alignment combined with doping of the electron transport layer (*i.e.* TiO_2) has improved electron extraction and can improve device efficiencies to >18%.[17] The planar heterojunction solar cell has been improved and vacuum-processed devices have emerged that increase the PCE of champion cells to >15%.[18] Depending on the work function of the HTM, inverted device architectures are possible[19–30] – for example, an inverted planar heterojunction stack (ITO/NiO/MAPI/PCBM/Al) with a PCE of 7.6% has been reported by Hu *et al.* in which the deep work function NiO acts as the HTM.[31]

One of the most remarkable aspects of perovskites as an absorber is that not only do they absorb photons with a direct bandgap commensurate with the peak solar flux, they also generate and effectively separate charge carriers. Measurements of the carrier distance reveals why the efficiency of MAPI-based solar cells may be so high. Grätzel and co-workers reported the electron–hole carrier distances in MAPI to be >100 nm using transient absorption measurements. Snaith and co-workers have also measured the carrier distance in MAPI and corroborated this report, although the carrier length of $MAPI_{3-x}Cl_x$ was reported to be an order of magnitude greater at >1 μm in length in a planar heterojunction configuration. Because of the lack of radiative recombination in devices with $MAPI_{3-x}Cl_x$ as the active perovskite layer, they display the highest device efficiencies in planar heterojunction architectures,[32] although it is not clearly understood why. The appellation '$MAPI_{3-x}Cl_x$' has courted controversy of late, with Grätzel calling for the nomenclature be retired based on work presented by Erk *et al.* at a conference in Lausanne; the chloride ions are said to become part of an amorphous Pb-containing phase that melts at low temperature.[33] The crystallisation of $MAPI_{3-x}Cl_x$ from $PbCl_2$/PbI_2 precursors has been investigated by Tidhar *et al.* using scanning electron microscopy (SEM), studies on the growth of single crystals and cryo-transmission electron microscopy.[34] There is strong evidence based on these studies that $PbCl_2$ nanoparticles act as heterogeneous nucleation sites (Figure 10.3). Unger *et al.* have suggested, using a combination of X-ray photoelectron spectroscopy (XPS), X-ray fluorescence spectrometry and X-ray diffraction (XRD), that MAI and Cl evaporate during annealing steps, which allows the formation of highly crystalline MAPI. However, even after annealing at 95 °C for 120 min, significant amounts of chloride could be detected in the films by XPS. Thus the argument regarding the nomenclature '$MAPI_{3-x}Cl_x$' seems to be unresolved and will be retained for the purposes of this review, as a general consensus in the solar cell community has not yet emerged.

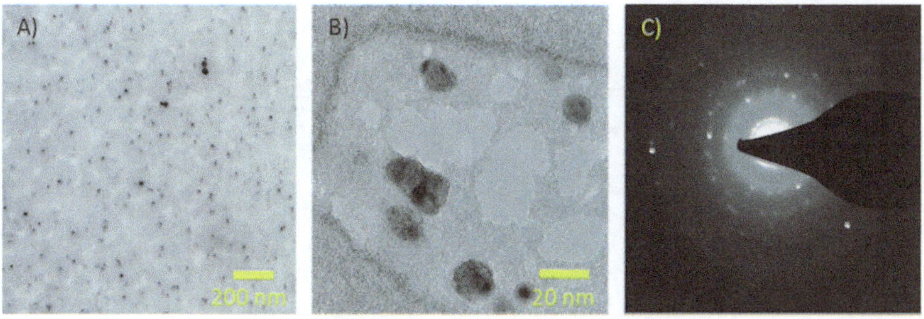

Figure 10.3 (A, B) Bright-field transmission electron microscopy images of $PbCl_2$ nanoparticles observed by Tidhar *et al.* in freeze-dried films of $MAPI_{3-x}Cl_x$, which potentially act as nucleation sites for the formation of high-quality MAPI crystals. (C) Selected-area diffraction pattern of the sample area imaged in (B) Reproduced with permission from ref. 34. Copyright (2014) American Chemical Society.

The most intriguing phenomenon associated with perovskite photovoltaic devices has been the observation of hysteresis in their *J–V* profiles,[35] which is thought to arise from either carrier effects from defects at surfaces causing trap states, ion intercalation/interstitial labile ions or from a ferroelectric-type phenomena[36] (although the latter reason has almost been ruled out completely as the hysteresis effect is remarkably pronounced in devices consisting of $CsPbI_3$, which, due to symmetry considerations, cannot act as a ferroelectric material[7]). It is known that the effect is not observed in the dark and nor is it a consequence of the interface with the HTM.[37] The hysteresis effect is particularly pronounced in planar heterojunction device architectures, as demonstrated by Seok and co-workers in their comparison of planar heterojunction devices *versus* those with MSSC hybrid architectures (Figure 10.4);[38] the latter class of devices seems to not suffer from this effect. However, there is a real implication in the observation of the hysteresis effect in that, especially in earlier reports, this led to an overestimation of the PCE because the forward bias scan usually gives a larger PCE than the reverse bias scan. Thus care must be taken in choosing scanning rates in the forward and reverse directions so that the PCEs are accurate. Snaith *et al.* demonstrated that the rise time to a stable power output is of the order of hundreds of seconds and that a reduction in the scanning rate can exacerbate the hysteresis effect.[36] The cause of this phenomenon remains unresolved and thus questions loom concerning the performance of perovskite devices.

10.3 Deposition Techniques for Perovskite Absorber Layers

For planar heterojunction devices, a smooth layer of perovskite material should be deposited as uniformly as possible. Perovskite film quality, and the presence of deleterious 'pinholes' in particular,[13] has been identified as

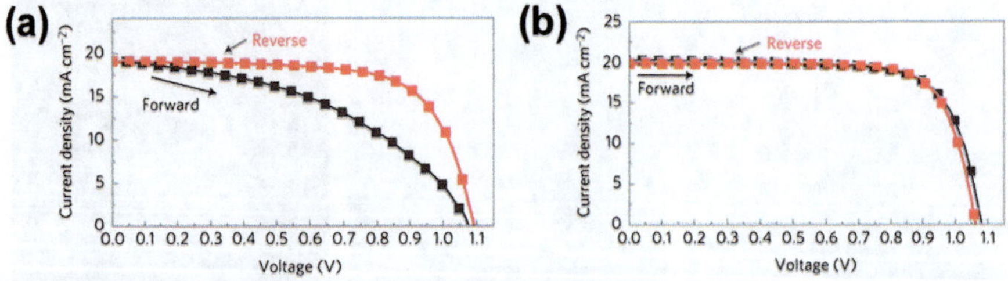

Figure 10.4 Anomalous hysteresis effect observed by Seok and co-workers in type III perovskite solar cells. (a) Significant hysteresis effect observed in a photovoltaic device with a planar heterojunction architecture with $MAP(I_{1-x}Br_x)_3$ as the intrinsic semiconductor layer. (b) No hysteresis is observed in cells with a hybrid MSSC/planar heterojunction architecture. Reprinted by permission from Macmillan Publishers Ltd: *Nature Materials* (ref. 38), copyright 2014.

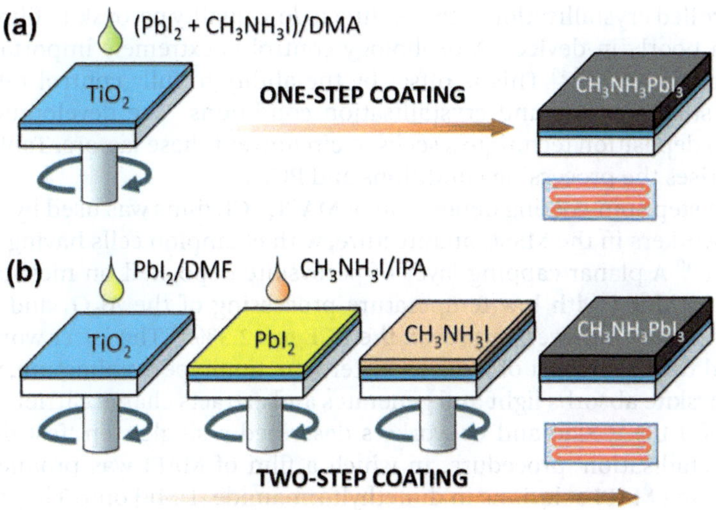

Figure 10.5 Typical work flows based one- and two-step deposition techniques to produce perovskite thin films. (a) One-step techniques generally use a pre-mixed stoichiometric solution of the lead halide and alkylammonium salt in a polar aprotic solvent such as DMF. (b) Two-step deposition procedures typically proceed *via* the deposition of a lead halide film, followed by conversion of the lead halide to the inorganic–organic lead halide perovskite by exposure to the alkylammonium salt either as a liquid, vapour or aerosol. Both procedures may incorporate an annealing step at the end to improve the crystallinity of the perovskite film. Reproduced from Jung and Parl, Perovskite Solar Cells: From Materials to Devices, *Small*, John Wiley and Sons, © 2014 WILEY-VCH Verlag GmbH & Co. KGaA, Weinheim.

being the key to producing champion solar cells with the highest PCEs. The conditions of crystallisation are important to the quality of the generated films. The search for scalable deposition techniques, especially those that can print or deposit perovskite absorber layers over large areas, is ongoing. A number of approaches have emerged to deposit perovskite thin films for photovoltaic devices and these can be grouped into two approaches: one-step and two-step deposition techniques (Figure 10.5). A system for the classification of deposition techniques for perovskite absorber layers has been suggested.[39]

10.3.1 One-step Deposition Techniques

In one-step deposition techniques, the components of the final perovskite, namely PbX_2 (where X = Cl, Br or I) and RNH_2X (where R is typically CH_3) are pre-mixed in a solvent that can solubilise both components. The solution is deposited onto the substrate and the solvent is evaporated to leave the perovskite as a film. The most crucial factor to control in one-step deposition is the precipitation of perovskite species from solution, with fast and uncontrolled crystallisation rates leading to low-quality perovskite films that perform poorly in devices. Morphology control is extremely important for device performance.[40] This is offset by the ability to fully control film elemental stoichiometry and crystallisation conditions. The development of two-step deposition techniques seeks to circumvent these factors. Table 10.1 summarises the processing conditions and PCEs.

Single-step spin-coating deposition of $MAPI_{3-x}Cl_x$ films was used by Snaith and co-workers in the MSSC architecture, with champion cells having a PCE of 10.9%.[10] A planar capping layer of perovskite deposited on mesoporous Al_2O_3, combined with low temperature processing of the Al_2O_3 and thickness-tuning of this layer, improved the PCE to 12.3%.[11] The latter work also included the first report of a planar heterojunction-type architecture, where the perovskite absorbs light and generates and extracts charge carriers,, with a PCE of 9.1%.[11] Xiao and co-workers described a single-step 'fast deposition–crystallisation' procedure, in which a film of MAPI was produced by spin-coating MAPI solutions in dimethylformamide (DMF) onto TiO_2-coated FTO and forcing rapid crystallisation by exposure to chlorobenzene.[41,42] The films were highly uniform and crystallinity was improved compared with spin-coating MAPI alone. Planar heterojunction cells produced by this method with a spiro-OMeTAD HTM yielded an average PCE of 13.9 ± 0.7%. The effect of the annealing temperature on the quality of spin-coated perovskite films was investigated by Dualeh *et al.*[43] They found that the final transformation to the perovskite from the spin-coated solution requires sufficiently high temperatures to ensure the vaporisation of the solvent and thus crystallisation of the material. Increasing the temperature too much, however, leads to the formation of PbI_2 due to volatilisation of the organic element of the perovskite, as evidenced by changes in the powder XRD patterns of films. This effect has also been investigated in detail by the same group.[44] The optimum temperature for annealing under the conditions used in the study was 100 °C, giving champion cells with PCEs of 11.7%. At 60 °C the PCE was 1.8%, while annealing at 200 °C gave poor device PCEs of 0.6%. The optimum temperature range under these conditions was 80–175 °C, giving device PCEs in the range 8.5–11.7%.

Seok and co-workers used solvent engineering to create perovskite layers in photovoltaic devices with MSSC architecture with PCEs >15% and, recently, >20%. In this approach, coordinating solvents other than DMF (including mixtures of solvents) are used for spin-coating of the MAPX film and additional solvent treatment of the deposited MAPX–PbX_2–MAX pseudocrystalline (often still solvated) intermediate is used to encourage the formation of highly crystalline perovskite films.

Jeon *et al.* used a solvent engineering process to produce champion cells with $MAP(I_{1-x}Br_x)_3$ ($0.10 \le x \le 0.15$) as the active layer in an MSSC architecture which displayed PCEs of 16.5%.[38] A mixture of dimethylsulfoxide (DMSO) and γ-butyrolactone was used to solubilise the $MAP(I_{1-x}Br_x)_3$, with a toluene treatment used after spin-coating to produce an intermediate phase film that was then annealed at 100 °C for 10 min (Figure 10.6). *In situ* powder XRD,

Table 10.1 One-step deposition techniques for perovskite thin film formation: processing conditions, architecture and PCE of photovoltaic devices.

Technique	Perovskite absorber	Deposition/ annealing temperature (°C)	Deposition/annealing time (min)	Photovoltaic device architecture	Champion PCE (%)	Ref.
Air-brushing	MAPI	50/100	60	Planar heterojunction	10.2	75
Spin-coating	MAPI	100	10	Planar heterojunction	13.9 ± 0.7	41 and 42
CVD	$MAPI_{3-x}Cl_x$	n/a	n/a	Planar heterojunction	15.4	18
Spin-coating	$MAPI_{3-x}Cl_x$	100	45	MSSC	10.9	10
Spin-coating	$MAPI_{3-x}Cl_x$	100	45	MSSC/planar heterojunction hybrid	12.3	11
Spin-coating	$MAPI_{3-x}Cl_x$	100	45	Planar heterojunction	9.1	11
AACVD	MAPBr	250	–	n/a	n/a	48
AACVD	MAPI	200	–	n/a	n/a	49
Spin-coating	MAPI	100	–	Planar heterojunction	11.7	43
Spray-coating	$MAPI_{3-x}Cl_x$	90	–	Planar heterojunction	11.1	50
Doctor blading	MAPI	125	–	Planar heterojunction	15.1	51
Spin-coating	$MAPI_{3-x}Cl_x$	90	120–180	Planar heterojunction	11.8	45
Spin-coating	MAPI	100	0.5	Planar heterojunction	10.0	46
Spin-coating	$MAP(I_{1-x}Br_x)_3$	100	10	MSSC/planar heterojunction hybrid	16.5	38

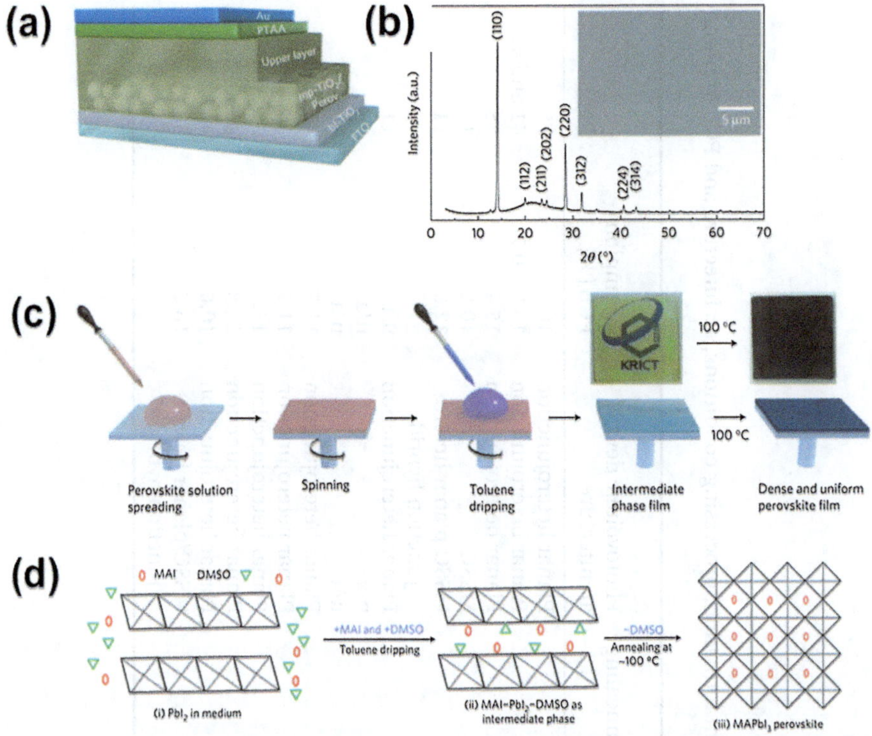

Figure 10.6 Solvent engineering approach used by Seok and co-workers to produce perovskite solar cells with superior efficiencies. (a) The planar heterojunction–MSSC hybrid architecture, in which some of the perovskite sits as a capping layer above that entrained into the mesoporous TiO_2. (b) Representative powder XRD pattern and plan view secondary electron SEM images of the perovskite layer, the former showing the excellent crystallinity and the latter revealing the smooth surface. (c) Solvent engineering workflow, showing the toluene dripping step and subsequent conversion to MAPI films. (d) A cartoon of the proposed crystallisation mechanism involving the intercalation of DMSO and MAI. Reprinted by permission from Macmillan Publishers Ltd: *Nature Materials* (ref. 38), copyright 2014.

elemental analysis and Fourier-transform IR measurements suggested that an intermediate phase was formed of the type MAI–PbI_2–DMSO, with the MAI and DMSO intercalated within the layer structure of the PbI_2, which served to ensure that complete crystallisation occurred when the DMSO was removed in the final toluene washing/annealing step. The toluene washing step is crucial to the formation of the smooth capping layer on top of the mesoscopic TiO_2. An important result in the same study was the comparison of devices with identically synthesised active layers, but with different device architectures: planar heterojunction and MSSC. The MSSC devices, which had a 200 nm layer of mesoporous TiO_2 on top of the compact TiO_2 electron transport layer, had generally higher PCEs than the planar heterojunction devices (~16 *versus* 14%), but the planar heterojunction devices showed marked hysteresis effects in their *J–V* profiles, whereas the MSSC devices did not. This suggests that the planar heterojunction architecture may suffer from slow charge collection and there may indeed be an optimum balance between the thickness of the planar part of the MSSC architecture and the mesoporous part.

Additive-enhanced crystallisation shares many similarities with one-step spin-coating, with a major difference in that additives (other than MAX and PbX_2) are added to the solution to be spin-coated to control the kinetics of crystallisation. Control of the crystallisation of $MAPI_{3-x}Cl_x$ has been reported by Liang *et al.* using the bidentate ligand 1,8-diiodooctane (DIO), which increases the solubility of $PbCl_2$ in precursor solutions by the formation of an heteroleptic octahedral complex of Pb *in situ* (Figure 10.7) and improves the homogeneity of the crystallised film by allowing control of the crystallisation kinetics.[45] The films with the DIO additive are strikingly smoother than those that do not have the additive in the precursor solution. PCEs of planar heterojunction devices formed by this method are 11.8%. Similarly, additive control was used by Jeon *et al.* to produce planar heterojunction solar cells with highly reproducible PCEs of 9.7 ± 0.1%, with champion cells having 10.0% efficiency.[46] *N*-Cyclohexyl-2-pyrollidone (CHP) was used to retard the crystallisation of the MAPI film, leading to smoother active perovskite layers compared with those without the CHP additive in the precursor solution. Solar cells produced without the CHP additive had comparatively lower average PCEs with decreased reproducibility (3.6 ± 1.6%).

Chemical vapour deposition (CVD) techniques rely on the delivery of molecular precursors to a substrate either under a high vacuum (traditional CVD), low pressure (LPCVD, ~10^{-3} Torr) or under ambient conditions, as in the case of aerosol-assisted chemical vapour deposition (AACVD), although many other CVD techniques are known. The substrate may be heated to afford a pyrolytic reaction and crystallise the product. In the latter, volatile precursor molecules can be avoided, thus expanding the range of molecules that can be used in

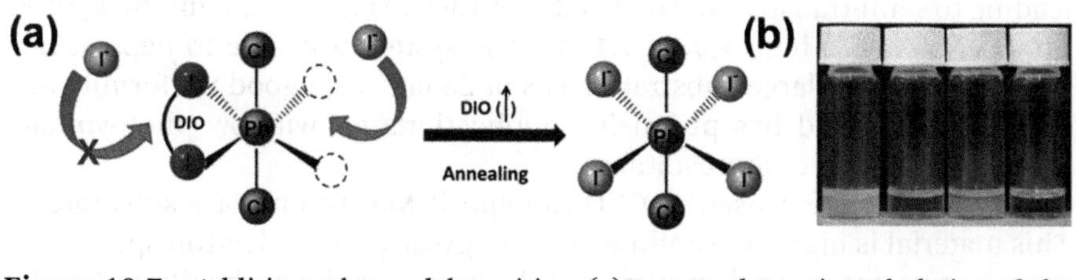

Figure 10.7 Additive-enhanced deposition. (a) Proposed transient chelation of PbI_2 by the bidentate ligand DIO. (b) Solubility of $PbCl_2$ and PbI_2 with and without DIO in DMF. Solutions without DIO are generally heterogeneous at room temperature. Adapted with permission from PW Liang *et al.*, Additive enhanced crystallization of solution-processed perovskite for highly efficient planar-heterojunction solar cells, *Advanced Materials*, John Wiley and Sons, © 2014 WILEY-VCH Verlag GmbH & Co. KGaA, Weinheim.

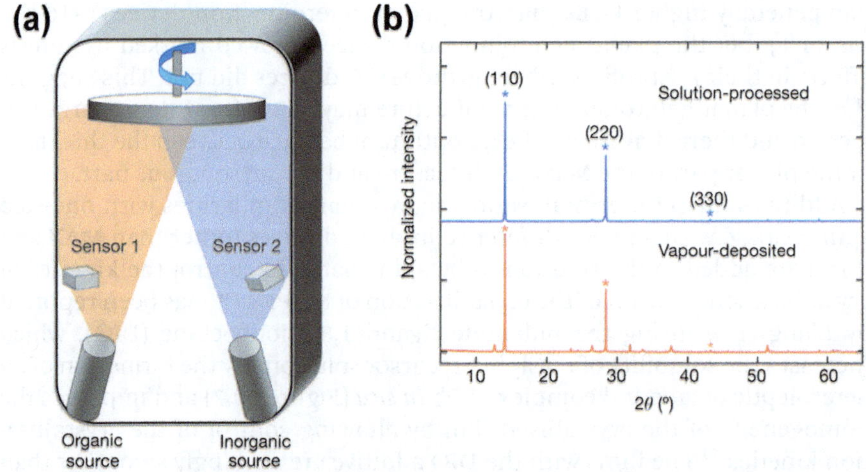

Figure 10.8 CVD system used by Snaith and co-workers to produce high quality $MAPI_{3-x}Cl_x$ films for planar heterojunction devices with high PCEs. (a) Thermal evaporation system. The organic source is MAI and the inorganic source is $PbCl_2$. (b) Powder XRD pattern of $MAPI_{3-x}Cl_x$ films fabricated by CVD compared with solution-deposited material showing comparable crystallinity. Reprinted by permission from Macmillan Publishers Ltd: *Nature* (ref. 18). Copyright (2013).

the feed. CVD techniques have been applied to the formation of hybrid IOP thin films. Snaith and co-workers used a parallel dual-source (PbI_2 and MACl) thermal evaporation CVD method to deposit $CH_3NH_3PbI_{3-x}Cl_x$ ($MAPI_{3-x}Cl_x$) on compact TiO_2 substrates under high vacuum (10^{-5} mbar) (Figure 10.8).[18] The $MAPI_{3-x}Cl_x$ films produced by this method are extremely uniform compared with solution-deposited films of a similar composition. Champion solar cells with a 330 nm thick perovskite layer and spiro-OMeTAD HTM performed excellently, with a PCE of 15.4%. Ono *et al.* described a CVD system that can fabricate MAPI films from the dual deposition of MAI and $PbCl_2$, leading to semi-transparent solar cells with active layers ~125 nm thick, PCEs up to 9.9% and a high V_{OC} of 1.1 V.[47] The system was able to deposit perovskite films over large substrate areas of 25 cm^2 with good uniformity and reproducibility and has promising applications for window photovoltaics (building-integrated photovoltaics).

Lewis and O'Brien used AACVD to deposit MAPBr on glass substrates.[48] This material is highly crystalline and energy-dispersive electron spectrometry spectrum mapping demonstrated the homogeneity of the films. Bhachu *et al.* used AACVD to deposit MAPI on glass and TiO_2 substrates.[49] Films produced by these two attempts at AACVD are not ideal for planar heterojunction architectures and would be better suited to ETA or MSSC solar devices. However, the report by Bhachu *et al.* is striking in that the areas of deposition are up to 40 mm^2; because AACVD operates under ambient pressure it does not share any of scalability issues of the high vacuum used in conventional CVD and circumvents the general impossibility of large-scale spin-coating.

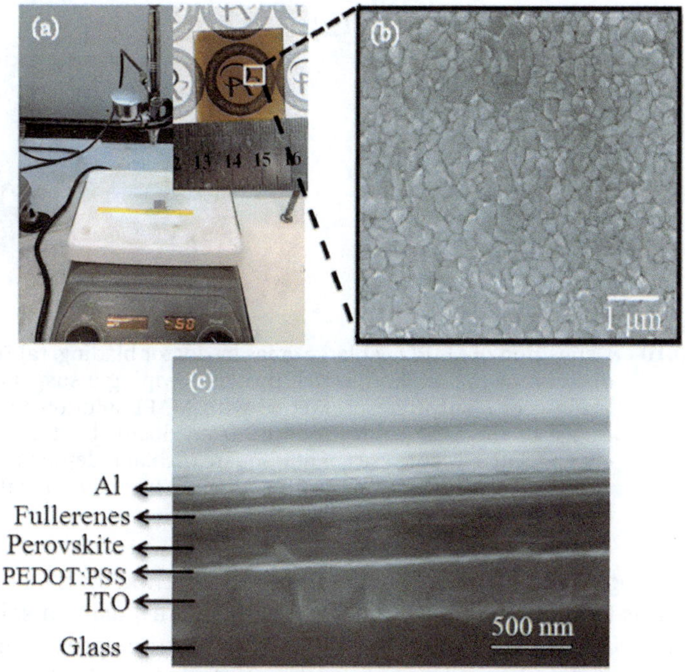

Figure 10.9 Layer-by-layer growth of perovskite thin films using an air-brush pen. (a) The air-brush pen is held over substrates on a hot-plate and the mist produced deposits MAPI onto the substrates. (b) Plan view secondary electron SEM image of the perovskite film produced from air-brush coating. (c) Planar heterojunction photovoltaic device (champion PCE 10.2%) in cross-section with layers annotated to the left. Reproduced with permission from ref. 75. Copyright (2015) American Chemical Society.

Air-brush coating is an inexpensive and facile technique for coating large areas of substrates and consists of a gas-driven pen nozzle generating an aerosol that is sprayed over an area, often manually. Ramesh *et al.* have demonstrated that air-brush coating is viable for the deposition of MAPI over large areas (up to 340 mm^2) of PEDOT:PSS-coated ITO (Figure 10.9). DMF solutions of MAI/PbI_2 were air-brushed onto substrates heated to 50 °C, followed by annealing of the MAPI films formed at 100 °C for 60 min. The perovskite absorber layer produced under these conditions is ~500 nm thick by cross-sectional SEM, with a grain size of ~300 × 300 nm in plan view. Champion devices had measured PCEs of 10.2%, with 85% of the devices having PCEs >8%. The ultrasonic spray deposition of $MAPI_{3-x}Cl_x$ at 35 kHz has been reported by Lidzey and co-workers using a mixture of MAI and $PbCl_2$ sprayed onto PEDOT:PSS/ITO substrates.[50] Solar cells with efficiencies of 11.1% were reported.

Doctor blading is a low-cost method for coating large areas using the swiping motion of a glass or steel blade set at a certain height to thinly coat a solution over a substrate This is followed by evaporation of the solvent to leave a thin crystalline film of active layer on the substrate. It is potentially scalable.

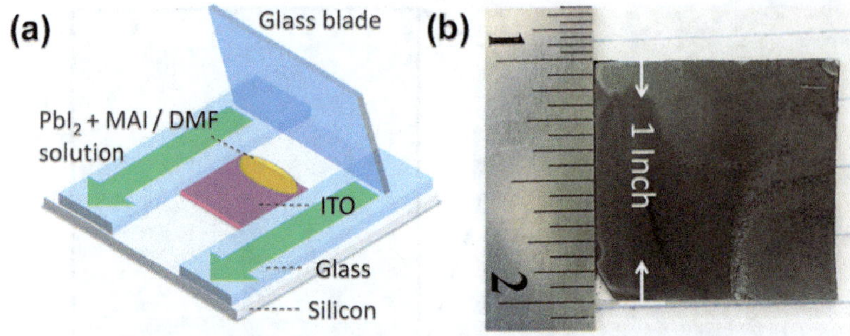

Figure 10.10 Deposition of MAPI over large areas by doctor blading. (a) Schematic diagram of the method, which involves swiping a suspended glass blade over a substrate pre-wetted with MAPI solution in DMF. (b) Image of the MAPI film produced from doctor blading deposited over a relatively large area of one square inch and demonstrating the apparent homogeneity of the film. Adapted from ref. 51 with permission from The Royal Society of Chemistry.

Deng *et al.* reported that doctor blading is a viable method for the deposition of MAPI films over an area of ~1 square inch (~6.5 cm^2) using a solution of MAI and PbI_2 in DMF.[51] The purity of the MAI used in the process is crucial to the quality of the films. Champion cells with fullerene HTM and a novel electron transport polymer layer had PCEs of 15.1% (Figure 10.10).

10.3.2 Two-step Deposition Techniques

Two-step deposition techniques typically involve the initial deposition of PbX_2 films onto the substrate, followed by the *in situ* conversion of the PbX_2 to RNH_3PbX_3 films by the addition of RNH_3X, with the amine salt solution having to fully penetrate the solid state PbX_2 films to afford crystallisation to the perovskite film. Although the most popular deposition methodology, two-step deposition techniques often suffer from uncontrollable crystallisation, stoichiometry and morphology. One of the major problems in the conversion of PbX_2 to RNH_3PbX_3 by the addition of RNH_3X is that the final RNH_3PbX_3 film often contains regions of PbX_2 that remain unconverted, which has a detrimental effect on the carrier mobility and thus the overall PCE of the device. This is deleterious to device reproducibility. The formation of a dense, void-free PbX_2 buffer layer on the compact TiO_2 prior to conversion to perovskite has been reported to be crucial for high charge collection efficiency in planar heterojunction devices made by two-step processes (spin-coat/dip-coat).[52] Therefore the careful characterisation of films produced by this technique is required, usually by XRD, to determine the conversion times and temperatures and thus to ensure that high-quality films are produced. Table 10.2 summarises some of the processing conditions and PCEs obtained.

The first example of the two-step solution-based deposition of perovskite thin films was reported by Mitzi and co-workers in 1998, when lead(II) or

Table 10.2 Two-step deposition techniques for perovskite thin film formation: processing conditions, architecture and PCE of photovoltaic devices.

Technique 1 (PbX_2)	Technique 2 (perovskite)	Perovskite absorber	Deposition/annealing temperature (°C)	Deposition/annealing time (min)	Photovoltaic device architecture	Champion PCE (%)	Ref.
Spin-coating	LPCVD	MAPI	140/145	100/30	Planar heterojunction	12.7	68
Spin-coating	Dip-coating	MAPI	RT/100	10	Planar heterojunction	13.4	56
Spin-coating	Dip-coating	MAPI	90	45	Planar heterojunction	10.5	55
Spin-coating	Spin-coating	$MAPI_{3-x}Cl_x$	95	40	Planar heterojunction	13.1	59
Slot-die coating	Slot-die coating	MAPI	70	–	Planar heterojunction	12.0	71
AACVD	AACVD	MAPI	70/220	–	n/a	n/a	69
CVD	CVD	$MAPI_{3-x}Cl_x$	75/85	–	Planar heterojunction	15.4	70
Spin-coating	Dip-coating	MAPI	RT	–	MSSC	15.0	54
Doctor blading	Vacuum flash evaporation	MAPI	80/125	–	Planar heterojunction	12.2	72
Spin-coating	Dip-coating	$MAPI_{3-x}Cl_x$	90/70	60/30	Planar heterojunction	11.7	60
Spin-coating	Dip-coating	MAPI	70	30	Planar heterojunction	14.3	61
Spin-coating	Dip-coating	MAPI	RT	–	Planar heterojunction	13.5	62
Spin-coating	Dip-coating	MAPI	–	–	Planar heterojunction	10.6 ± 2.0	63
Spin-coating	Sublimation	MAPBr	70/150	30/10	MSSC	9.1	66
Spin-coating	Sublimation	MAPI	110/150	15	Planar heterojunction	12.1	65
Spin-coating	Dip-coating	MAPI	65/150	1	Planar heterojunction	17.2	57
Spin-coating	Inkjet printing	MAPI	100	1	Planar heterojunction	11.6	73
Spin-coating	Spin-coating	MAPI	100	20	MSSC	13.5	58

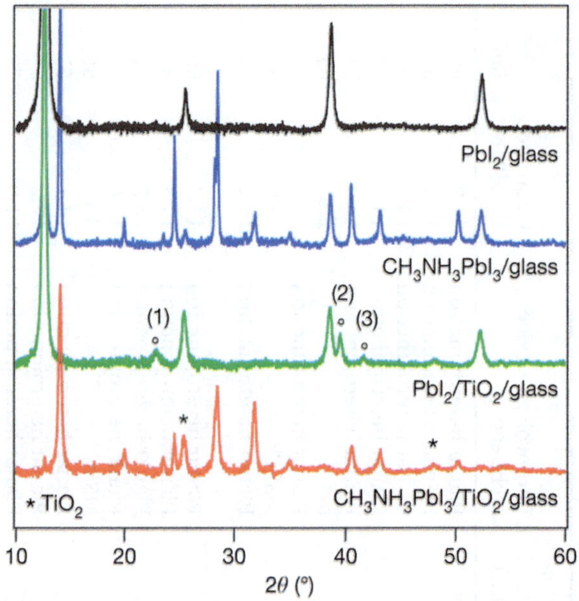

Figure 10.11 Typical powder XRD patterns of intermediates and final MAPI films in two-step spin-coating deposition. Reprinted by permission from Macmillan Publishers Ltd: *Nature* (ref. 54). Copyright 2013.

tin(II) halide thin films were converted to IOPs by dipping in alkylammonium iodide for 1–5 min.[53] Toluene and 2-propanol are the best solvents for this procedure. The application of the two-step dip-coating approach to produce a photovoltaic device was subsequently reported by Grätzel and co-workers in devices with an MSSC architecture (Figure 10.11).[54] Mesopourous TiO_2 films were infiltrated with PbI_2, followed by a wash step with 2-propanol and immersion in MAI to furnish the MAPI perovskite. Champion cells with a spiro-OMeTAD HTM showed PCEs of 15.0%.

Shi *et al.* produced MAPI films by a modified two-step deposition process.[55] A MAPI layer of ~300 nm was deposited on nanoporous TiO_2. The substrate was spin-coated twice with PbI_2 in DMF, producing a relatively thick film of the lead halide, followed by an annealing step at 90 °C for 2 min. The perovskite was generated by immersion of the PbI_2 film in MAI dissolved in 2-propanol for 120 min, followed by annealing at 90 °C for 45 min. HTM-free champion cells produced by this method had PCEs of 10.5%. Xie *et al.* reported an 'optimised' two-step sequential solution-based process for the deposition of MAPI films by first spin-coating DMF solutions of PbI_2 containing a small amount of MAI, followed by immersion in MAI for 25 min to complete the crystallisation process.[56] They claimed that the crystallisation process was enhanced by the addition of a small amount of amine in the lead halide solution. The films were annealed at 100 °C for 10 min. This technique seeks to circumvent the problem of incomplete crystallisation of the lead halide perovskite by 'initiation' of the crystallisation with small amount of amine. The optimum stoichiometry is 20 mol% MAI to PbI_2, leading to less exciton recombination compared with other stoichiometries, determined from photoluminescence emission spectroscopy. SEM shows that, at some stoichiometries, significant pinholes are observed in the MAPI films produced by this method. Champion cells with a planar heterojunction architecture and spiro-OMeTAD as the HTM had PCEs of 13.4%.

Similar work by Zhang *et al.* used mole fractions of MAI in the PbI_2 film precursor solution ($0.1 \leq x \leq 0.3$), followed by immersion of the MAI:PbI_2 film in MAI for 1–3 min to elicit the complete conversion of the films to MAPI, with champion cells fabricated from the precursor PbI_2 to 0.15 MAI solution having measured PCEs of 17.2%.[57] The morphologies of the precursor film are adjustable from the amount of MAI in the PbI_2 solution. The reason for the fast conversion time and eventual homogeneity of the film, as judged by powder XRD, is due to control of the morphology of the deposited PbI_2 film by the MAI 'impurity', which causes the PbI_2 film to be less crystalline, less planar and thus more susceptible to ingress by the MAI solution. This leads to immersion times that are ten-fold shorter than the typical immersion time associated with the sequential deposition of around 10–30 min.

From this report, it seems that a playoff between precursor crystallinity and the eventual roughness of the film could be a crucial aspect of deposition design in terms of chemistry. Modification of the -step process by Bi *et al.*, with the introduction of a drying step and dichloromethane treatment, led to longer photoluminescence emission lifetimes in MAPI thin films. Longer exciton lifetimes, faster electron transport and PCEs up to 13.5% in MSSC devices were reported in the same study.[58] Inverted perovskite solar cells with active layers deposited by the single-step deposition process have recorded PCEs >18% and lack the hysteresis effect.[27]

The mixed-phase perovskite $CH_3NH_3PbI_{3-x}Cl_x$ ($MAPI_{3-x}Cl_x$) has been deposited onto compact TiO_2 by a two-step spin-coating process, first by spin-coating the PbI_2 onto the substrate, followed by the spin-coating of MAI or a mixture of MAI and MACl to form the perovskites.[59] Powder XRD was used to confirm complete film transformation. Champion cells with $MAPI_{3-x}Cl_x$ as the absorber layer and a spiro-OMeTAD HTM had a PCE of 13.1%.

A comparison of the single-step spin-coating deposition of $MAPI_{3-x}Cl_x$ *versus* two-step sequential deposition was reported by Ma *et al.*[60] The two-step deposition involves pre-mixing PbI_2 and $PbCl_2$ for the halide deposition, followed by dip-coating in MAI. The use of PbI_2 and $PbCl_2$ in the halide deposition step, as opposed to pure $PbCl_2$, improves loading in the TiO_2 layer of the planar heterojunction cells and was demonstrated to control the morphology better than single-step deposition. Champion cells from this process with spiro-OMeTAD HTM had measured PCEs of 11.7%; the analogous device fabricated by single-step deposition of the $MAPI_{3-x}Cl_x$ layer reported in this work had a PCE of 4.8%. It was concluded that, in this case, two-step sequential deposition was superior for the fabrication of perovskite solar cells.

Similarly, Grätzel and co-workers compared sequential *versus* one-step deposition and arrived at similar conclusions.[61] Comparison of sequentially deposited MAPI *versus* single-step deposition at different loading concentrations in all cases revealed that planar heterojunction devices with sequentially deposited material had higher a PCE (14.3%) than single-step spin-coated MAPI (7.8–8.9%, depending on the solution concentration used for spin-coating). Sequential mode deposition is cited to be better than single-step deposition because infiltration of the MAPI into the TiO_2 layer is better and the MAPI overlayer, which is in contact with the HTM, is much more flatter and uniform, leading to more efficient charge carrier extraction. Replacement of the usual solvent DMF with DMSO can effectively retard the crystallisation of the PbI_2 during the first spin-coating step and can effect complete conversion of PbI_2 to MAPI after dip-coating, leading to solar cells with both a high PCE (13.5% champion) and good reproducibility (average $12.5 \pm 0.6\%$, $N = 120$), potentially due to the greater solvating power of DMSO for oxophilic PbI_2 compared with DMF.[62]

Bandgap tuning of $MAPb(I_{1-x}Br_x)_3$ ($0 \leq x \leq 1$) films has been demonstrated by Kulkarni *et al.* by spin-coating PbI_2 films, followed by dip-coating into MAI:MABr solutions for 20 min at varying ratios of methylammonium iodide to methylammonium bromide.[63] Visible to near-IR spectrum tuning was demonstrated, with active layers with a higher bromide content displaying absorption onsets higher in energy than those containing more iodide. Champion planar heterojunction solar cells with a PCE of $10.6 \pm 2.0\%$ ($N = 4$) were described that contained solely MAPI with spiro-OMeTAD as the HTM. Increasing the Br content of the thin films eroded the PCE and cells with $MAP(I_{0.05}Br_{0.95})$ had a PCE of $1.0 \pm 0.5\%$, possibly due to the rough morphologies in these films, revealed by SEM. However, bandgap tuning is an important strategy to break the Shockley–Quiessar limit and thus the moderate PCEs and poor reproducibility in this case are not the salient point of the study.

The composition of perovskite thin films can be calculated from the following equation proposed by Seok and co-workers:[64]

$$E_g(x) = 1.57 + 0.39x + 0.33x^2 \tag{10.1}$$

where x is the mole fraction of bromide in the films and E_g is the energy of the optical bandgap in eV (Figure 10.12). Stable PCEs of ~10% have been measured over 22 days with x in the range 0.20–0.29.[64]

Vapour-assisted deposition is a simple technique that relies on the sublimation of MAI or MABr at moderate processing temperatures. The vapour of the alkylammonium salt converts the lead halide deposited onto substrates into the final active perovskite layer (Figure 10.13). The methodology is applicable to the fabrication of photovoltaic devices with planar heterojunction and MSSC architectures. Chen *et al.* used vapour-assisted deposition to produce MAPI-based photovoltaic devices in a planar heterojunction architecture.[65] A layer of spin-coated PbI_2 was converted to MAPI by the sublimation

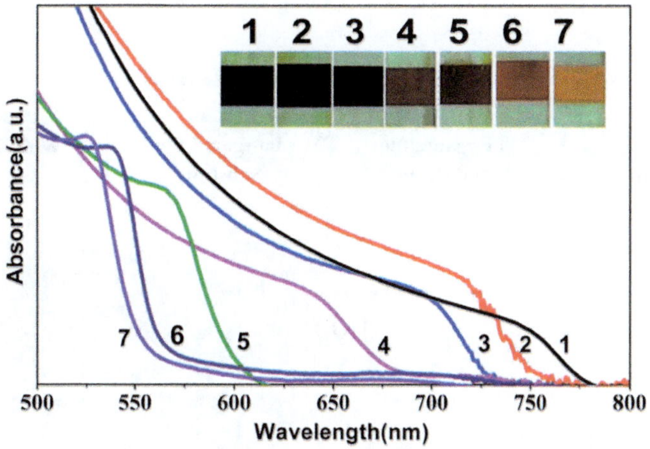

Figure 10.12 Bandgap engineering in $MAPb(I_{1-x}Br_x)_3$ ($0 \le x \le 1$) films by Kulkarni *et al.* by variation of x using a sequential (two-step) deposition process. Film no. (x): 1 (0), 2 (0.12), 3 (0.26), 4 (0.42), 5 (0.59), 6 (0.72) and 7 (0.95). As x increases, the energy of the optical band gap in the $MAPb(I_{1-x}Br_x)_3$ ($0 \le x \le 1$) films also increases, but the overall device efficiency tends to decrease due to reductions in J_{SC}. Reproduced from ref. 63 with permission from The Royal Society of Chemistry.

of MAI in an enclosed vessel at ambient pressure. The crystal grain size of the MAPI was ~$0.5 \times 0.5\ \mu m^2$, with no apparent pinhole in the resulting film and a uniform flatness in cross-section by SEM, crucial to device performance. Champion photovoltaic devices with spiro-OMeTAD as the HTM had PCEs of 12.1%. Vapour-assisted deposition has also been used by Sheng *et al.* to fabricate solar cells consisting of MAPBr on mesopourous TiO_2.[66] A spin-coated layer of $PbBr_2$ on the mesoporous substrates was exposed to sublimed MABr vapour at 150 °C for 10 min in a closed Petri dish. Devices with spiro-OMeTAD as the HTM has champion efficiencies of 9.1%, although the PCEs of the cells were shown to be highly dependent on the scan rate and displayed the hysteretic behaviour synonymous with perovskite devices at higher scan rates of 3 V s^{-1}. To put this effect into perspective, the reverse scan of the champion cell revealed a PCE of 8.3%. Vapour-assisted deposition could be a promising technique to control the crystallisation of the PbI_2 in two-step processing because it is claimed that the conversion of PbI_2 to MAPI goes to completion in short time periods, avoiding the lengthy processing steps associated with dip-coating conversions that can compromise the final film quality. Questions remain about the scalability of the process, which is not limited by the vapour-assisted conversion, but rather by the first spin-coating step. The vapour phase method and the two-step solution deposition method have been directly compared[67] and it was reported that the vapour-processed MAPI layers were more uniform, smooth and thus probably better suited to devices with planar heterojunction architectures. Comparison of the PCEs of

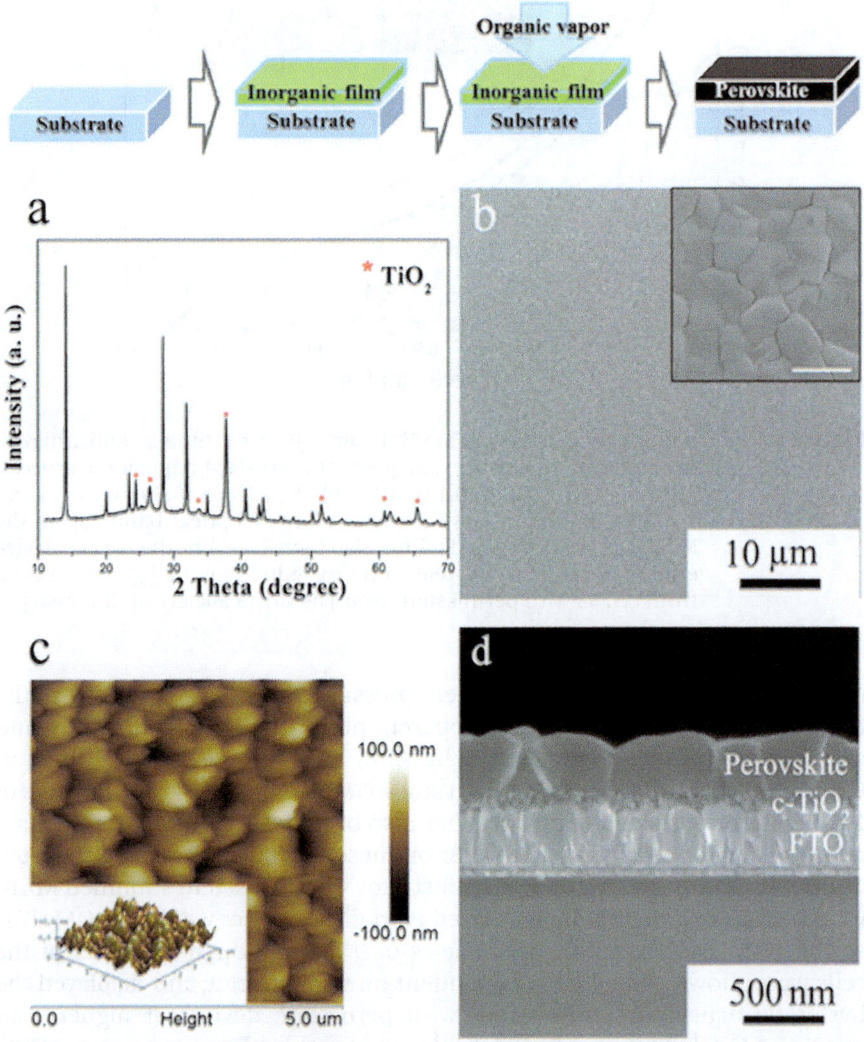

Figure 10.13 Vapour-assisted solution processing (VASP) of perovskite solar cells. Spin-coated PbI_2 is converted to MAPI by exposure to vapour of MAI in an enclosed headspace. (a) Typical powder XRD pattern of MAPI on TiO_2 produced by VASP. (b) Plan view secondary electron SEM image of a MAPI film produced by VASP demonstrating the heterogeneity of the thin film surface. Inset: high magnification image in the same area of the film. (c) Height profile atomic force microscopy image of a MAPI surface produced from VASP. (d) Planar heterojunction half-cell in cross-section, where the active layer is MAPI produced from VASP. Adapted with permission from ref. 65. Copyright (2014) American Chemical Society.

champion cells where both processes were used to fabricate the perovskite layer demonstrated that the vapour phase method was superior under the reported conditions (8.1 *versus* 5.8%).

LPCVD, an industrially mature technique for glass coating, was used by Luo *et al.* for the deposition of MAPI films in the planar heterojunction configuration.[68] A spin-coated PbI_2-coated compact TiO_2 substrate was reacted with MAI vapour at 140 °C. An annealing step of 145 °C in air was then used, optimised at 30 min. The films produced were ~500 nm thick in cross-sectional SEM and uniform in appearance in plan view, with large crystallites sizes of ~300 × 300 nm. Champion photovoltaic cells with spiro-OMeTAD as the HTM displayed PCEs of 12.7%. Binions and co-workers have attempted to improve the initial attempts at AACVD of MAPI[49] and MAPBr[48] by introducing a two-step AACVD process, where PbI_2 films were initially generated and then films reacted with MAI.[69] The films were superior in quality to those generated by one-step AACVD and were highly crystalline with a large grain size of ~2 × 2 μm, with no appreciable voids or pinholes in the films as imaged by SEM. Vacuum-processed cells were prepared by sequential two-step CVD by Chen *et al.*[70] Champion cells with PCE of 15.4% were prepared by the CVD of $PbCl_2$ onto ITO:PEDOT/PSS substrates at high vacuum (10^{-6} Torr) and 75 °C, followed by treatment of this layer by MAI at 85 °C.

Slot-die coating has been reported as a viable two-step scalable method for the production of perovskite solar cells by Hwang *et al.*[71] The slot-die coater was used to put down a layer of PbI_2 on compact TiO_2, then MAI was coated on top of the lead halide layer at 70 °C in the case of the 12.0% efficiency champion cell. Spiro-OMeTAD was slot-die-coated onto the perovskite layer as the HTM. The whole process took place in air with 30–40% relative humidity. The use of slot-die-coating may potentially lead to the roll-to-roll processing of perovskite solar cells assembled on flexible substrates.

Longo *et al.* reported a two-step process where doctor blading was used to initially coat a Ta foil substrate with MAPI from a PbI_2/MAI solution, followed by thermal annealing at 80–125 °C. The final perovskite film was flash-evaporated onto PEDOT on ITO substrates under vacuum.[72] A perovskite layer of 200 nm was produced, which was highly crystalline from XRD measurements and appeared largely defect-free by atomic force microscopy. Devices constructed from this technique had champion efficiencies of 12.2%.

Inkjet printing of perovskites is potentially a facile route to producing large area photoactive layers on flexible substrates. Wei *et al.* spin-coated PbI_2 onto compact TiO_2 glass substrates, then compared two inkjet printing processes for deposition (Figure 10.14). In one process, a carbon electrode was printed on top of the PbI_2 layer, followed by immersion in MAI to furnish the MAPI layer. In the second process, the carbon black and MAI were pre-mixed as an 'ink' and printed on top of the PbI_2.[73] It was found that the second process with the premixed ink gave better planar layers in cross-section by SEM and this was correlated with the device performance, with champion cells from the first process displaying PCEs of 6.5% and the second pre-mixed processing procedure giving cells with PCEs up to 11.6%. Automated patterning of the perovskite active layer is also within the capabilities of the bespoke printer used in the study.

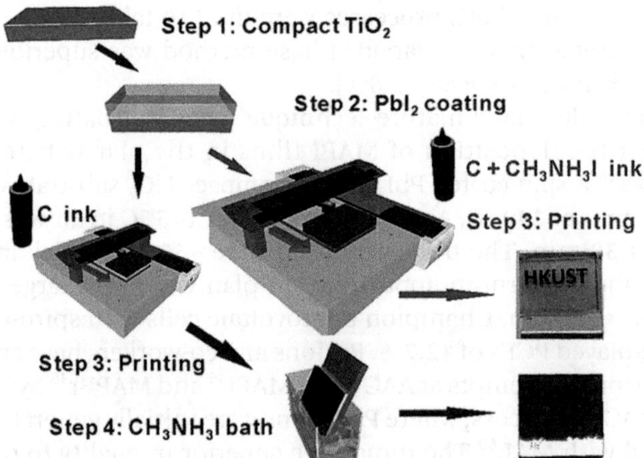

Figure 10.14 Inkjet printing of perovskite layers showing the two possible routes to produce planar heterojunction devices over large areas. Pre-mixing of the C and MAI ink generally leads to devices with superior PCEs compared with the dip-coating route. Adapted with permission from Z. Wei *et al.*, Inkjet printing and instant chemical transformation of a $CH_3NH_3PbI_3$/nanocarbon electrode and interface for planar perovskite solar cells, *Angewandte Chemie International Edition*, John Wiley and Sons, © 2014 WILEY-VCH Verlag GmbH & Co. KGaA, Weinheim.

10.3.3 Three-step Deposition Techniques

Three-step deposition is currently limited to a single example by Sargent and co-workers,[74] who produced MAPI films by atomic layer deposition on glass substrates. Lead(II) sulphide (PbS) was first deposited by the atomic layer deposition of $Pb(tmhd)_2$ and H_2S vapour, followed by iodination of the PbS film to produce PbI_2 by sealing in a chamber containing I_2 chips. Planar MAPI films were produced by immersion of the PbI_2 films in MAI. The films were of extremely high quality and uniform as observed by SEM. No device was reported.

10.4 Conclusions

There has been massive interest in perovskite-sensitised solar cells in the past six years, driven largely by the ease of processing combined with the superior PCEs of such devices, which have made them viable for commercialisation. They are the direct descendants of DSSCs and device architecture permutations have led to the dominance of the MSSC and planar heterojunction architectures as the preferred device configurations. These architectures give the best PCEs, with MSSC devices now reaching >20% certified efficiency.

Many deposition techniques have emerged to produce thin films of IOPs and they can generally be classed into one- or two-step processes. The most popular one-step deposition process is the spin-coating of a pre-mixed MAPX solution onto substrates. Spin-coating of the metal halide followed by dip-deposition of the alkylammonium halide salt, which converts the metal halide film to the inorganic organic perovskite *in situ*, has emerged as the most popular two-step process. Some techniques outlined here (*e.g.* AACVD, slot-die coating, inkjet printing) are scalable and have potential in the eventual commercial exploitation of large area photovoltaic devices based on IOPs. Issues with the role of Cl in $MAPI_{3-x}Cl_x$, the hysteresis effect observed in these photovoltaic devices and the search for an optimised device architecture of the perovskite layer (planar heterojunction *versus* MSSC *versus* hybrids) to ensure stable power output while maximising the PCE make this an intriguing area of research.

References

1. M. A. Green, K. Emery, Y. Hishikawa, W. Warta and E. D. Dunlop, *Prog. Photovoltaics Res. Appl.*, 2015, **23**, 1–9.
2. W. S. Yang, J. H. Noh, N. J. Jeon, Y. C. Kim, S. Ryu, J. Seo and S. I. Seok, *Science*, 2015, **348**, 1234–1237.
3. S. D. Stranks and H. J. Snaith, *Nat. Nanotech.*, 2015, **10**, 391–402.
4. D. B. Mitzi, C. D. Dimitrakopoulos and L. L. Kosbar, *Chem. Mater.*, 2001, **13**, 3728–3740.
5. D. B. Mitzi, C. A. Feild, W. T. A. Harrison and A. M. Guloy, *Nature*, 1994, **369**, 467–469.
6. D. B. Mitzi, S. Wang, C. A. Feild, C. A. Chess and A. M. Guloy, *Science*, 1995, **267**, 1473–1476.
7. G. E. Eperon, G. M. Paterno, R. J. Sutton, A. Zampetti, A. A. Haghighirad, F. Cacialli and H. J. Snaith, *J. Mater. Chem. A*, 2015, **3**, 19688–19695.
8. F. Hao, C. C. Stoumpos, D. H. Cao, R. P. H. Chang and M. G. Kanatzidis, *Nat. Photonics*, 2014, **8**, 489–494.
9. A. Kojima, K. Teshima, Y. Shirai and T. Miyasaka, *J. Am. Chem. Soc.*, 2009, **131**, 6050–6051.
10. M. M. Lee, J. Teuscher, T. Miyasaka, T. N. Murakami and H. J. Snaith, *Science*, 2012, **338**, 643–647.
11. J. M. Ball, M. M. Lee, A. Hey and H. J. Snaith, *Energy Environ. Sci.*, 2013, **6**, 1739–1743.
12. M. J. Carnie, C. Charbonneau, M. L. Davies, J. Troughton, T. M. Watson, K. Wojciechowski, H. Snaith and D. A. Worsley, *Chem. Commun.*, 2013, **49**, 7893–7895.
13. T. Leijtens, B. Lauber, G. E. Eperon, S. D. Stranks and H. J. Snaith, *J. Phys. Chem. Lett.*, 2014, **5**, 1096–1102.
14. Z. Wei, K. Yan, H. Chen, Y. Yi, T. Zhang, X. Long, J. Li, L. Zhang, J. Wang and S. Yang, *Energy Environ. Sci.*, 2014, **7**, 3326–3333.
15. Y. Jin and G. Chumanov, *ACS Appl. Mater. Interfaces*, 2015, **7**, 12015–12021.

16. Q. Hu, J. Wu, C. Jiang, T. Liu, X. Que, R. Zhu and Q. Gong, *ACS Nano*, 2014, **8**, 10161–10167.
17. H. Zhou, Q. Chen, G. Li, S. Luo, T.-b. Song, H.-S. Duan, Z. Hong, J. You, Y. Liu and Y. Yang, *Science*, 2014, **345**, 542–546.
18. M. Liu, M. B. Johnston and H. J. Snaith, *Nature*, 2013, **501**, 395–398.
19. J. Cui, F. Meng, H. Zhang, K. Cao, H. Yuan, Y. Cheng, F. Huang and M. Wang, *ACS Appl. Mater. Interfaces*, 2014, **6**, 22862–22870.
20. V. Trifiletti, V. Roiati, S. Colella, R. Giannuzzi, L. De Marco, A. Rizzo, M. Manca, A. Listorti and G. Gigli, *ACS Appl. Mater. Interfaces*, 2015, **7**, 4283–4289.
21. W. Wang, J. Yuan, G. Shi, X. Zhu, S. Shi, Z. Liu, L. Han, H.-Q. Wang and W. Ma, *ACS Appl. Mater. Interfaces*, 2015, **7**, 3994–3999.
22. F. Xia, Q. Wu, P. Zhou, Y. Li, X. Chen, Q. Liu, J. Zhu, S. Dai, Y. Lu and S. Yang, *ACS Appl. Mater. Interfaces*, 2015, **7**, 13659–13665.
23. Z. Zhu, Y. Bai, T. Zhang, Z. Liu, X. Long, Z. Wei, Z. Wang, L. Zhang, J. Wang, F. Yan and S. Yang, *Angew. Chem., Int. Ed.*, 2014, **53**, 12571–12575.
24. H. Dong, Z. Wu, B. Xia, J. Xi, F. Yuan, S. Ning, L. Xiao and X. Hou, *Chem. Commun.*, 2015, **51**, 8986–8989.
25. P. Cottingham, D. C. Wallace, K. Hu, G. Meyer and T. M. McQueen, *Chem. Commun.*, 2015, **51**, 7309–7312.
26. W. Chen, Y. Wu, J. Liu, C. Qin, X. Yang, A. Islam, Y.-B. Cheng and L. Han, *Energy Environ. Sci.*, 2015, **8**, 629–640.
27. J. H. Heo, H. J. Han, D. Kim, T. K. Ahn and S. H. Im, *Energy Environ. Sci.*, 2015, **8**, 1602–1608.
28. S. Ye, W. Sun, Y. Li, W. Yan, H. Peng, Z. Bian, Z. Liu and C. Huang, *Nano Lett.*, 2015, **15**, 3723–3728.
29. C. Kuang, G. Tang, T. Jiu, H. Yang, H. Liu, B. Li, W. Luo, X. Li, W. Zhang, F. Lu, J. Fang and Y. Li, *Nano Lett.*, 2015, **15**, 2756–2762.
30. H. Choi, C.-K. Mai, H.-B. Kim, J. Jeong, S. Song, G. C. Bazan, J. Y. Kim and A. J. Heeger, *Nat. Commun.*, 2015, **6**, 7348.
31. L. Hu, J. Peng, W. Wang, Z. Xia, J. Yuan, J. Lu, X. Huang, W. Ma, H. Song, W. Chen, Y.-B. Cheng and J. Tang, *ACS Photonics*, 2014, **1**, 547–553.
32. S. D. Stranks, P. K. Nayak, W. Zhang, T. Stergiopoulos and H. J. Snaith, *Angew. Chem., Int. Ed.*, 2015, **54**, 3240–3248.
33. M. Graetzel, *Nat. Mater.*, 2014, **13**, 838–842.
34. Y. Tidhar, E. Edri, H. Weissman, D. Zohar, G. Hodes, D. Cahen, B. Rybtchinski and S. Kirmayer, *J. Am. Chem. Soc.*, 2014, **136**, 13249–13256.
35. E. L. Unger, E. T. Hoke, C. D. Bailie, W. H. Nguyen, A. R. Bowring, T. Heumueller, M. G. Christoforo and M. D. McGehee, *Energy Environ. Sci.*, 2014, **7**, 3690–3698.
36. H. J. Snaith, A. Abate, J. M. Ball, G. E. Eperon, T. Leijtens, N. K. Noel, S. D. Stranks, J. T.-W. Wang, K. Wojciechowski and W. Zhang, *J. Phys. Chem. Lett.*, 2014, **5**, 1511–1515.
37. A. Dualeh, T. Moehl, N. Tetreault, J. Teuscher, P. Gao, M. K. Nazeeruddin and M. Graetzel, *ACS Nano*, 2014, **8**, 362–373.
38. N. J. Jeon, J. H. Noh, Y. C. Kim, W. S. Yang, S. Ryu and S. I. Seok, *Nat. Mater.*, 2014, **13**, 897–903.

39. S. T. Williams, C.-C. Chueh and A. K. Y. Jen, *Small*, 2015, **11**, 3088–3096.
40. L. Zheng, D. Zhang, Y. Ma, Z. Lu, Z. Chen, S. Wang, L. Xiao and Q. Gong, *Dalton Trans.*, 2015, **44**, 10582–10593.
41. M. Xiao, F. Huang, W. Huang, Y. Dkhissi, Y. Zhu, J. Etheridge, A. Gray-Weale, U. Bach, Y.-B. Cheng and L. Spiccia, *Angew. Chem.*, 2014, **126**, 10056–10061.
42. M. Xiao, F. Huang, W. Huang, Y. Dkhissi, Y. Zhu, J. Etheridge, A. Gray-Weale, U. Bach, Y.-B. Cheng and L. Spiccia, *Angew. Chem., Int. Ed.*, 2014, **53**, 9898–9903.
43. A. Dualeh, N. Tetreault, T. Moehl, P. Gao, M. K. Nazeeruddin and M. Graetzel, *Adv. Func. Mater.*, 2014, **24**, 3250–3258.
44. A. Dualeh, P. Gao, S. I. Seok, M. K. Nazeeruddin and M. Graetzel, *Chem. Mater.*, 2014, **26**, 6160–6164.
45. P.-W. Liang, C.-Y. Liao, C.-C. Chueh, F. Zuo, S. T. Williams, X.-K. Xin, J. Lin and A. K. Y. Jen, *Adv. Mater.*, 2014, **26**, 3748–3754.
46. Y.-J. Jeon, S. Lee, R. Kang, J.-E. Kim, J.-S. Yeo, S.-H. Lee, S.-S. Kim, J.-M. Yun and D.-Y. Kim, *Sci. Rep.*, 2014, **4**, 6953.
47. L. K. Ono, S. Wang, Y. Kato, S. R. Raga and Y. Qi, *Energy Environ. Sci.*, 2014, **7**, 3989–3993.
48. D. J. Lewis and P. O'Brien, *Chem. Commun.*, 2014, **50**, 6319–6321.
49. D. S. Bhachu, D. O. Scanlon, E. J. Saban, H. Bronstein, I. P. Parkin, C. J. Carmalt and R. G. Palgrave, *J. Mater. Chem. A*, 2015, **3**, 9071–9073.
50. A. T. Barrows, A. J. Pearson, C. K. Kwak, A. D. F. Dunbar, A. R. Buckley and D. G. Lidzey, *Energy Environ. Sci.*, 2014, **7**, 2944–2950.
51. Y. Deng, E. Peng, Y. Shao, Z. Xiao, Q. Dong and J. Huang, *Energy Environ. Sci.*, 2015, **8**, 1544–1550.
52. Y. H. Lee, J. Luo, R. Humphry-Baker, P. Gao, M. Graetzel and M. K. Nazeeruddin, *Adv. Funct. Mater.*, 2015, **25**, 3925–3933.
53. K. N. Liang, D. B. Mitzi and M. T. Prikas, *Chem. Mater.*, 1998, **10**, 403–411.
54. J. Burschka, N. Pellet, S.-J. Moon, R. Humphry-Baker, P. Gao, M. K. Nazeeruddin and M. Graetzel, *Nature*, 2013, **499**, 316–319.
55. J. Shi, Y. Luo, H. Wei, J. Luo, J. Dong, S. Lv, J. Xiao, Y. Xu, L. Zhu, X. Xu, H. Wu, D. Li and Q. Meng, *ACS Appl. Mater. Interfaces*, 2014, **6**, 9711–9718.
56. Y. Xie, F. Shao, Y. Wang, T. Xu, D. Wang and F. Huang, *ACS Appl. Mater. Interfaces*, 2015, **7**, 12937–12942.
57. T. Zhang, M. Yang, Y. Zhao and K. Zhu, *Nano Lett.*, 2015, **15**, 3959–3963.
58. D. Bi, A. M. El-Zohry, A. Hagfeldt and G. Boschloo, *ACS Appl. Mater. Interfaces*, 2014, **6**, 18751–18757.
59. Y. Xu, L. Zhu, J. Shi, S. Lv, X. Xu, J. Xiao, J. Dong, H. Wu, Y. Luo, D. Li and Q. Meng, *ACS Appl. Mater. Interfaces*, 2015, **7**, 2242–2248.
60. Y. Ma, L. Zheng, Y.-H. Chung, S. Chu, L. Xiao, Z. Chen, S. Wang, B. Qu, Q. Gong, Z. Wu and X. Hou, *Chem. Commun.*, 2014, **50**, 12458–12461.
61. N. Yantara, D. Sabba, F. Yanan, J. M. Kadro, T. Moehl, P. P. Boix, S. Mhaisalkar, M. Grätzel and C. Grätzel, *Chem. Commun.*, 2015, **51**, 4603–4606.
62. Y. Wu, A. Islam, X. Yang, C. Qin, J. Liu, K. Zhang, W. Peng and L. Han, *Energy Environ. Sci.*, 2014, **7**, 2934–2938.

63. S. A. Kulkarni, T. Baikie, P. P. Boix, N. Yantara, N. Mathews and S. Mhaisalkar, *J. Mater. Chem. A*, 2014, **2**, 9221–9225.
64. J. H. Noh, S. H. Im, J. H. Heo, T. N. Mandal and S. I. Seok, *Nano Lett.*, 2013, **13**, 1764–1769.
65. Q. Chen, H. Zhou, Z. Hong, S. Luo, H.-S. Duan, H.-H. Wang, Y. Liu, G. Li and Y. Yang, *J. Am. Chem. Soc.*, 2014, **136**, 622–625.
66. R. Sheng, A. Ho-Baillie, S. Huang, S. Chen, X. Wen, X. Hao and M. A. Green, *J. Phys. Chem. A*, 2015, **119**, 3545–3549.
67. T. Du, N. Wang, H. Chen, H. Lin and H. He, *ACS Appl. Mater. Interfaces*, 2015, **7**, 3382–3388.
68. P. Luo, Z. Liu, W. Xia, C. Yuan, J. Cheng and Y. Lu, *ACS Appl. Mater. Interfaces*, 2015, **7**, 2708–2714.
69. S. Chen, J. Briscoe, Y. Shi, K. Chen, R. Wilson, S. Dunn and R. Binions, *CrystEngComm*, 2015, **17**, 7486–7489.
70. C.-W. Chen, H.-W. Kang, S.-Y. Hsiao, P.-F. Yang, K.-M. Chiang and H.-W. Lin, *Adv. Mater.*, 2014, **26**, 6647–6652.
71. K. Hwang, Y.-S. Jung, Y.-J. Heo, F. H. Scholes, S. E. Watkins, J. Subbiah, D. J. Jones, D.-Y. Kim and D. Vak, *Adv. Mater.*, 2015, **27**, 1241–1247.
72. G. Longo, L. Gil-Escrig, M. J. Degen, M. Sessolo and H. J. Bolink, *Chem. Commun.*, 2015, **51**, 7376–7378.
73. Z. Wei, H. Chen, K. Yan and S. Yang, *Angew. Chem., Int. Ed.*, 2014, **53**, 13239–13243.
74. B. R. Sutherland, S. Hoogland, M. M. Adachi, P. Kanjanaboos, C. T. O. Wong, J. J. McDowell, J. Xu, O. Voznyy, Z. Ning, A. J. Houtepen and E. H. Sargent, *Adv. Mater.*, 2015, **27**, 53–58.
75. M. Ramesh, K. M. Boopathi, T.-Y. Huang, Y.-C. Huang, C.-S. Tsao and C.-W. Chu, *ACS Appl. Mater. Interfaces*, 2015, **7**, 2359–2366.

CHAPTER 11

CIGS and CIS Nanomaterials for Solar Cells

SHIVRAM S. GARJE*[a], JAGRUTI S. SUROSHE[a] AND NEERISH REVAPRASADU*[b]

[a]Department of Chemistry, University of Mumbai, Vidyanagari, Santacruz (East), Mumbai, 400098, India; [b]Department of Chemistry, University of Zululand, Private Bag X1001, Kwadlangezwa, 3880, South Africa
*E-mail: RevaprasaduN@unizulu.ac.za

11.1 Introduction

The use of thin film photovoltaic technology started with the second generation of solar cells. Thin films, ranging from a few nanometres to micrometres in thickness, were deposited on suitable substrates, such as glass, plastic or metal. The first generation of solar cells based on crystalline Si made use of Si wafers with thicknesses up to 200 μm. The emerging third generation of photovoltaics includes copper zinc tin sulfide nanocrystals and organic dye-sensitised solar cells (DSSCs). The small thickness of thin film photovoltaics provides unique advantages over conventional crystalline Si solar cells. These advantages include lower weight, flexibility and low drag.[1–3] Despite these advantages, thin film technology has not completely replaced conventional Si-based technology due to the low efficiency of the cells.

Nanostructured materials are promising candidates for enhancing the performance of solar cells and show improved light trapping and photocarrier collection.[4–7] The fabrication of these materials is relatively cheap, making

Nanoscience & Nanotechnology Series No. 45
Nanostructured Materials for Type III Photovoltaics
Edited by Peter Skabara and Mohammad Azad Malik

Published by the Royal Society of Chemistry, www.rsc.org

their production cost-effective. The properties requiring attention in developing photovoltaic materials include the bandgap, the absorption coefficient, the recombination rate and the method of synthesis. Despite the rapid development of photovoltaic materials, global photovoltaic power generation accounts for <0.1% of the electricity produced worldwide.[8] The obvious reason for this is that photovoltaic power generation is expensive. Efforts are being made to develop materials to enhance photovoltaic power generation. The fabrication of conventional semiconductor materials in nanostructures results in an improvement in the properties of these materials. Three-dimensional arrays of nanostructures enhance photon absorption and photocarrier collection efficiencies. These nanostructures include nanowires,[9,10] nanopillars[11,12] and nanocones.[4]

Nanomaterials are expected to induce new photovoltaic mechanisms involving multiple exciton generation. The performance of photovoltaic devices is mainly controlled by three factors: (1) the absorption of light to generate the charge carriers, *i.e.* holes (p-type) and electrons (n-type); (2) the separation of the charge carriers; and (3) the collection of charge carriers at the respective electrodes to establish the potential difference across the p–n junction. The generation of a difference in voltage at the p–n junction of the cell is a response to visible radiation, which is utilised to do work.[13] Therefore these parameters need to be optimised to enhance device performance. Three-dimensional arrays of nanostructures significantly improve these parameters. Researchers have used various theoretical models to study optical absorption by nanostructures. For example, the effect of diameter, length and filling ratio in nanowires on optical absorption has been investigated using the transfer matrix method.[14]

Semiconductor materials, such as amorphous Si, polycrystalline CdTe and copper indium gallium selenide (CIGS) have been developed for thin film photovoltaic solar cells. Polycrystalline CIGS thin films, which are a member of the I–III–VI group of chalcopyrite semiconductors, have the highest conversion efficiency. The CIGS material has a suitable energy bandgap (1.04–1.68 eV) with a high optical absorption coefficient in the visible spectrum of incident sunlight. The absorption coefficient of CIGS thin films in the visible spectrum is 100 times larger than that of Si materials. Solar cells based on CIGS thin films show excellent outdoor stability, a tunable bandgap and provide specific power up to 919 W kg^{-1}, the highest of all types of solar cell. The highest efficiency of 20.3% with an active area of 0.5 cm^2 for CIGS thin film solar cells was reported by the Centre of Solar Energy and Hydrogen Research, Württemberg, Germany and 19.9% efficiency has been reported by the National Renewable Energy Laboratory in Colorado, USA. To achieve a higher efficiency for CIGS solar cells, the parameters that need to be optimised include the strict standardisation of the process flow, the use of clean materials, procedures and working environments, variation of the selenium rate and fine-tuning of the cell stack.[15–17]

11.2 Solar Cell Generations

Figure 11.1 is a flow diagram showing the development of various generations of solar cells.

11.2.1 First Generation Solar Cells

The first generation solar cells were based on Si wafers. This is the oldest and most popular technology due to the high power efficiencies achieved. Si-based solar cells are further categorised into two types of solar cell: monocrystalline and polycrystalline Si solar cells.[18–22]

11.2.1.1 Monocrystalline Silicon Solar Cells

These consist of Si in which the crystal lattice of the entire solid is continuous, free of any grain boundaries. Monocrystalline Si can be prepared intrinsically and consists mainly of pure Si or Si doped with very small amounts of other elements to change its semiconducting properties. Most Si monocrystals are grown by the Czochralski process into ingots of up to 2 m in length

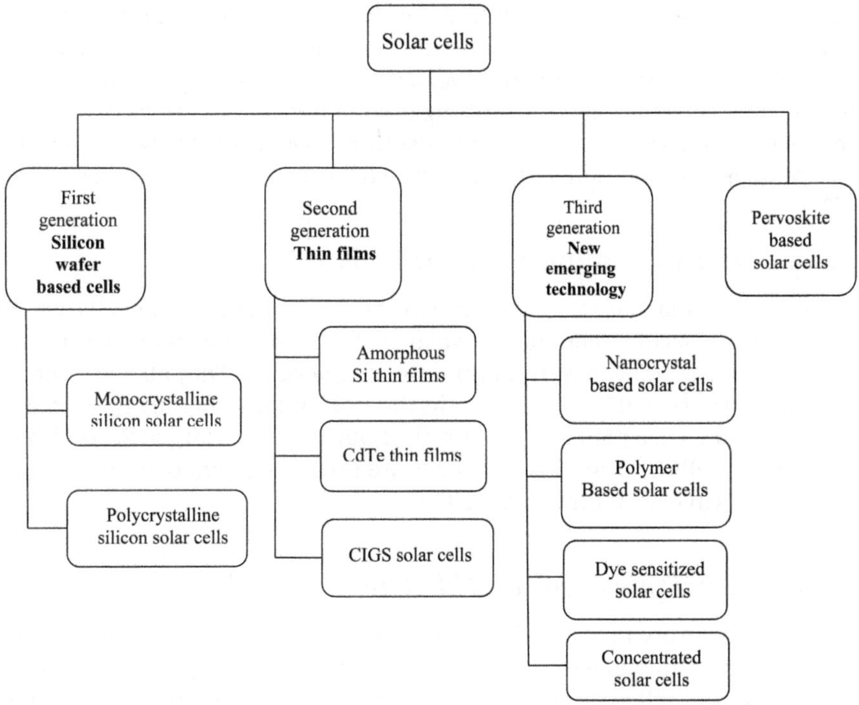

Figure 11.1 Three generations of solar cells.

and weighing several hundred kilograms.[23,24] These cylinders are then sliced into thin wafers, a few hundred micrometres thick, for further processing. Production of the single crystal is a precise process and hence these solar cells are expensive. Their efficiency lies between 17 and 18%.[25]

11.2.1.2 *Polycrystalline Silicon Solar Cells*

Polysilicon is manufactured from metallurgical grade Si by a chemical purification process, known as the Siemens process. This process includes distillation of volatile Si compounds and their decomposition into Si at high temperatures. These are composed of number of different crystals coupled with one another in a single cell. They are produced by cooling a graphite mould filled with molten Si. During solidification of the molten Si, various crystal structures are formed. These cells are known to occupy about 48% of worldwide solar cell production during year 2008.[26] Polycrystalline Si does not need to be deposited on a Si wafer to form a solar cell, it can be deposited on other-cheaper materials, thus reducing the cost. They have low efficiency ~12-14%.[27]

First generation solar cells cannot be operated at high temperature and are two times more expensive compared to thin film solar cells. As this technology has matured, costs have become increasingly dominated by material costs. This trend is expected to continue as the photovoltaic industry continues to mature. A 1997 study shows that the cost of manufacturing is greatly increased. 500 MW/y production volume suggests material costs in such volumes would account for over 70% of total manufacturing costs.[28] Silicon wafer cells have light absorbing layers up to 350 μm thick, while thin film solar cells have a very thin light absorbing layers, generally of the order of 1 μm thickness and hence thin film based solar cells were discovered.[29]

11.2.2 Second Generation Solar Cells

For the last few years, researchers have focused on thin film solar cells, which are known as second generation solar cells. Thin films offer significant reductions in material costs by eliminating the Si wafer. They also offer other advantages, particularly an increase in the unit of manufacturing from a Si wafer (~100 cm^2) to a glass sheet (~1 m^2), about 100 times larger. Second generation solar cells or thin film solar cells are further classified as amorphous Si, CdTe and CIGS thin film solar cells.[28]

11.2.2.1 *Amorphous Silicon Thin Film Solar Cells*

These solar cells are fabricated by coating the doped Si material to the back side of the substrate/glass plate. These solar cells generally are dark brown in colour on the reflecting side and silvery on the conducting side.[30] Amorphous Si photovoltaic modules are primitive solar cells, which were the first to be manufactured commercially. These solar cells are manufactured at a low temperature and are hence relatively cheaper. The main drawback of amorphous Si-based solar cells is their poor and unstable efficiency, ranging between 4 and 8%. The advantage of these cells is that they can be easily operated at elevated temperatures and are suitable for use in changing climatic conditions when the Sun shines for only a few hours.[31]

11.2.2.2 CdTe Thin Film Solar Cells

CdTe is an attractive candidate for the development of cheap, economically viable photovoltaic devices.[25,32,33] CdTe has a direct optimum bandgap of ~1.5 eV and a high optical absorption coefficient >5 × 10^{15} cm^{-1}.[34] The direct bandgap of CdTe makes the absorption of light easier and improves efficiency, which is usually in the range 9–11%.[25,35] The devices are generally fabricated by sandwiching CdTe between layers of CdS to form a p–n junction diode. The manufacturing process consists of two steps. First, the CdTe-based solar cells are synthesised from polycrystalline materials using glass as a substrate. The second process involves deposition: multiple layers of CdTe solar cells are coated onto the substrate using different, economical methods. CdTe solar cells are flexible and can be constructed on polymer substrates. However, there are various environmental issues with the Cd components of these solar cells. Cd is a toxic heavy metal and can accumulate in humans, animals and plants. The disposal and recycling of Cd-based materials is both expensive and damaging to the environment.[21,31] The limited supply of Cd and the associated environmental hazards are the main drawbacks to its use in CdTe technology.[33–36]

11.2.2.3 CIGS Solar Cells

CIGS is a direct bandgap semiconductor and has a higher efficiency (~10–12%) than CdTe thin film solar cells. CIGS solar cells are produced by sputtering, evaporation, electrochemical coating, printing and electron beam deposition techniques.[22,37] The sputtering can be a one-step reactive process or a two- or multi-step process involving deposition and subsequent interaction with Se. The substrates for CIGS materials can be glass plates, polymer substrates, steel, or Al. CIGS thin film solar cells have several advantages, such as a long lifetime without considerable degradation, flexibility, light weight and less expensive than conventional solar cells. They can be operated under both low and high-temperature conditions and have a higher efficiency.[22,23]

In terms of the energy conversion efficiency, this second generation technology might, over time, bridge the gap between the first and third generation products. As this second generation thin films technology matures, the costs will be progressively dominated by those of the constituent materials. There will be a lower limit on these costs (say, US$ 30 m^{-2}) which, when combined with the likely cell efficiency (15% or 150 Wp m^{-2}), determines the lower limit on the costs of photovoltaic modules (US$ 0.20 Wp^{-1}) and electricity generation (US$ 0.02–0.04 kW h), assuming that the module costs are half of the system costs and a range of insolation and cost-of-money scenarios.[28]

11.2.3 Third Generation Solar Cells

Third generation solar cells are new, promising technologies, but have not been investigated in detail commercially.

11.2.3.1 Nanocrystal-based Solar Cells

Nanocrystal-based solar cells are also known as quantum dot solar cells. These solar cells are composed of semiconductors from the transition metal groups – for example, materials such as porous Si or porous TiO_2.[38] Using advanced nanotechnology, semiconducting materials in the bulk state (*e.g.* Si, CdTe or CIGS) are replaced by their nanocrystalline counterparts. In this type of solar cell, the nanocrystals are mixed in a bath and coated onto the Si substrate. These crystals rotate very fast and flow outwards due to centrifugal forces. In conventional semiconductor-based solar cells, a photon will generally excite an electron by creating one electron–hole pair. However, when a photon strikes a quantum dot made of a similar semiconductor material, numerous electron–hole pairs can be formed, usually two or three, although seven have been observed in a few instances.[18,36] They have some advantages, including flexibility, high thermal stability, a short installation time and lower cost than conventional Si solar cells. The only disadvantage is that they require a large space for installation. Figure 11.2 is a schematic representation of a third generation solar cell.

11.2.3.2 Polymer-based Solar Cells

The first polymer solar cells were discovered by the research group of Tang *et al.* at the Kodak Research Laboratory.[18] They were composed of serially connected thin functional layers coated onto a polymer foil. They usually work as a combination of a donor (polymer) and an acceptor (fullerene). Polymer solar cells and other organic solar cells operate on the principle known as the photovoltaic effect, *i.e.* where energy in the form of electromagnetic

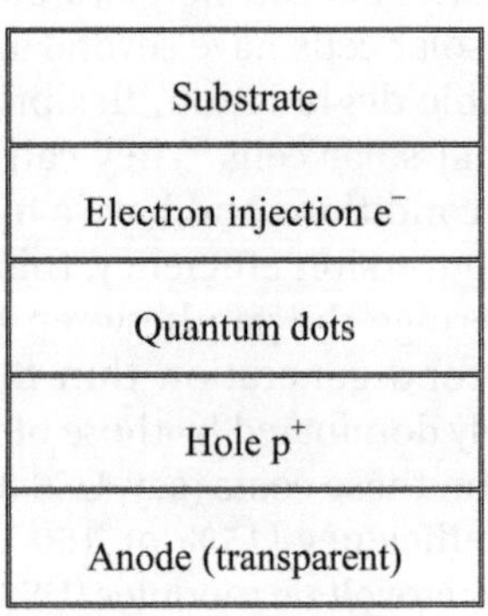

Figure 11.2 Schematic representation of a third generation solar cell.

radiation is transformed into an electrical current. Wudl *et al.* mixed poly[2-methoxy-5-(2′-ethylhexyloxy)-*p*-phenylene vinylene], C60 and other derivatives to develop the first polymer solar cell and obtained a high power conversion efficiency.[39] After the optimisation of various parameters, an efficiency of ~3–10% was achieved for polymer solar cells.[39–41] As a result of their unique properties, they can be used in stretchable solar devices, including textiles and fabrics. To increase the function of liquid crystal displays, a modern recycling concept known as polarising organic photovoltaics has been developed using the same polariser, a photovoltaic device and the correct light conditions/solar panel.[40–42] They have several advantages, including good flexibility due to the polymer substrate, they are less expensive than conventional Si solar cells, they require only a small space for installation and require only short installation times. However, they cannot be operated at high temperatures and have a low efficiency of ~3–10%.

11.2.3.3 Dye-sensitised Solar Cells

Research has recently focused on molecular manipulation to improve the solar efficiency with the help of nanotechnology for harvesting light energy. The first DSSC was introduced by Michel Grätzel at the Swiss Federal Institute of Technology.[21,22] DSSC devices consist of four components: a semiconductor electrode (n-type TiO_2 and p-type NiO); a dye sensitiser; a redox mediator; and a counter electrode (C or Pt).[43] An efficiency >10% is observed for DSSCs due to the photosensitisation of nanograined TiO_2 coatings coupled with dyes optically active in the visible region. DSSCs are attractive as a result of their simple conventional processing methods, such as printing techniques, and are highly flexible, transparent and low cost.[21] However, there are also certain challenges, such as the degradation of dye molecules, which leads to stability issues. The dye molecules generally degrade after exposure to UV and IR radiation, leading to a decrease in the lifetime and stability of the cells.

11.2.3.4 Concentrator Photovoltaics

Concentrator photovoltaics is a technology that has been established since the 1970s.[44,45] The main principle of concentrator cells is to collect a large amount of solar energy onto a tiny region of the photovoltaic solar cell (Figure 11.3).[13]

Large mirrors and lenses are used to focus rays of sunlight onto small regions of the solar cells. In concentrator photovoltaics, solar energy is directly converted into electricity on the basis of ‘photovoltaic effect’. The cells can be classified into low, medium and high concentrator solar cells depending on the power of the lens system.[45] The advantages of concentrator photovoltaics include solar cell efficiencies >40%, the absence of moving parts, no thermal mass, a speedy response time and scalable to a range of sizes.

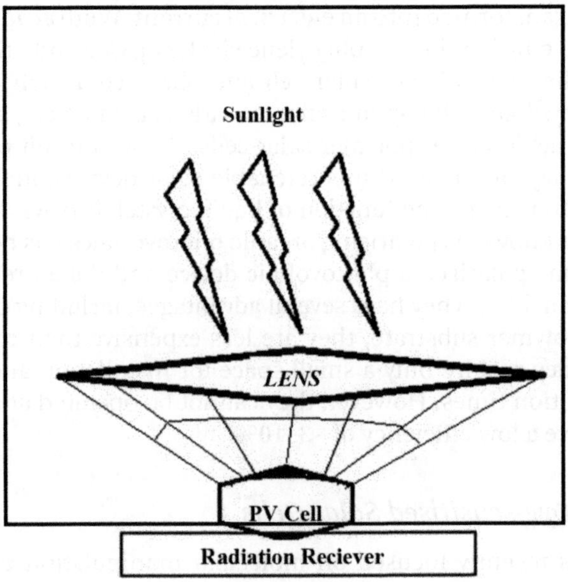

Figure 11.3 Schematic representation of concentrator solar cells. Adapted from *Materials Sciences and Applications*, 6, Sharma, K. K. Jain and A. Sharma, Solar Cells: In Research and Applications—A Review, 1145–1155 (2015) from Scientific Research Publishing (ref. 13).

11.2.4 Perovskite-based Solar Cells

Perovskite solar cells have several advantages over conventional solar cells based on Si and thin film. Perovskites are a class of compounds defined by the formula ABX_3, where X represents a halogen such as I^-, Br^- or Cl^-, and A and B are cations of different sizes. They have many advantages, including a high efficiency (~31%), minimum space requirements and a short installation time. They may also play an important part in the next generation of electric automobile batteries.[46,47] However, the material degrades over time, resulting in a decrease in stability and overall efficiency.

11.3 Third Generation CIGS Solar Cells

11.3.1 Construction of CIGS Solar Cells

Figure 11.4 shows a schematic representation of a third generation CIGS solar cell. A typical CIGS solar cell consists of a soda lime glass (SLG) substrate, an Mo back contact, a wide bandgap CIGS absorber layer, an n-type buffer layer composed of n-CdS (CdS, ZnS, ZnSe) and a window layer made of n-ZnO:Al as the front contact.[48,49] The conventionally used SLG is preferred over borosilicate glass because of its suitable coefficient of thermal expansion and also for the beneficial supply of Na during CIGS growth. Gerthoffer *et al.* demonstrated CIGS-based solar cells grown on flexible ultra-thin glass with a thickness of 100 μm and an efficiency of 11.2%.[50]

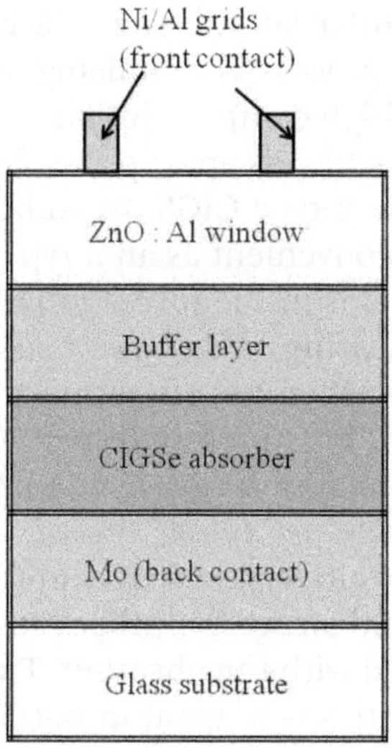

Figure 11.4 Schematic structure of a CIGS solar cell.

Mo is the most commonly used back contact and is preferred to Pt, Au, Au/Be, Al, Ni, Ag or Cu because of its excellent properties, such as not forming an alloy with Cu, a matching thermal coefficient of expansion with the SLG substrate and low resistivity.[51–53] Generally, the back contact consists of a sputtered Mo thin film forming an ohmic contact with the p-type CIGS absorber material.[29] Mo has been reported to form $MoSe_2$ at the interface during selenisation, which acts as an ohmic contact between Mo and the CIGS absorber, improving the adhesion between them.[54] During selenisation, the diffusion of Na from the glass substrates is dependent on the density of the Mo films. The diffusion of Na helps to improve the crystallinity of the CIGS thin films, which, in turn, improves the performance of the solar cell.[55,56]

The most important layer in a CIGS thin film structure is the $Cu(In_{1-x},Ga_x)Se_2$, *i.e.* the absorber layer. It is a chalcopyrite semiconductor with a high absorptivity, high thermal stability, low cost and an energy bandgap of 1.06–1.7 eV. The exact value of the energy bandgap (E_g) depends on x, namely the Ga/(Ga + In) ratio. An increase in the absorber layer leads to an increase in absorption probability and hence increases the total generation rate. It also leads to a decrease in the back contact recombination current density and, consequently, increases the cell efficiency. However, it is necessary to reduce the thickness of this layer due to material costs and because In and Ga resources are limited. Hence there is a need to reduce the duration of the process to achieve a higher output per unit production cost.[57,58]

The buffer layers on CIGS absorbers are used to enhance the lattice match between the CIGS absorber and the ZnO:Al window; the layers preserve the CIGS surface from ion loss during ZnO/ZnO:Al sputtering. CdS, ZnS, ZnSe and In_2S_3 are often used as buffer materials and the best efficiency is observed with a CdS buffer layer.[59] The CdS buffer layer deposited by a chemical bath deposition process is promising as a result of the high solar conversion efficiency and high comparative photoconductivity.[60,61] CdS has a direct bandgap of ~2.4 eV, which is appropriate for accepting a good portion of the incident light by the p-type CIGS. In addition to having high optical transmission, CdS is also convenient as an n-type semiconductor. Therefore it is presumed that CdS helps in the formation of the heterojunction along with the transparent conducting oxide layer.[62] Solar cells with a thinner buffer layer show a higher efficiency. Usually a thickness of the buffer layer optimised in the range 40–50 nm is preferred for CIGS-based solar cells.[63] The front contact consists of a bilayer of i-ZnO and Al:ZnO, which are sputtered at high substrate temperatures.

The fundamental photovoltaic parameters of CIGS devices are influenced by the optical, electrical and structural properties of the buffer layer and by the defect structure created with the absorber. The effect of post-deposition annealing on these properties is of great importance and can influence the chemical composition of the film after annealing.[64]

11.3.2 Crystal Structure of CIGS Tetragonal Unit Cell

The crystal structure of CIGS is important and makes this material special as a solar cell material. Figure 11.5 is a schematic representation of a CIGS crystal in a tetragonal unit cell.[65] Several studies have shown that the [220/204] orientation affects the solar cell performance relative to the [112] orientation of the CIGS device. The (220) and (204) planes are separate planes in the tetragonal CIGS structure.[66–69] Good quality CIGS thin films exhibit [112] and [220/204] orientations and show high power conversion efficiencies in solar cell devices. The preferred orientation generally depends on process parameters such as the substrate temperature and flux ratio of Se during CIGS deposition.[70]

The Cu and In (or Ga) atoms are alternately arranged along both the [110] and [001] directions in the (220) plane and there are Cu–Cu–In–In arrangements in the (204) plane. The In or Ga occupies the same location depending on the elemental composition of In/Ga. The orange and purple planes indicate the (220) and (204) planes, respectively, and the arrows show the crystallographic directions. The bright balls denote atoms placed on the surface plane, whereas the blurred balls indicate atoms located in the lower planes in Figure 11.5b and 11.5c. The physical and chemical properties are different in each plane. The calculated surface energies of the (220) and (204) planes are 1.08 and 1.16 eV a_0^{-2}, respectively, where a_0 is the calculated bulk lattice constant[71] and C_0 is roughly twice a_0. Therefore it is necessary to separately investigate the [220] and [204] oriented textures. In a typical X-ray characterisation, however, it is difficult to discern whether the X-ray reflections result from either the textures of [220] or [204] or the coexistence of [220] and [204]. This is because the spacing of the (220) and (204) planes is nearly identical in the CIGS unit cell.[67,72]

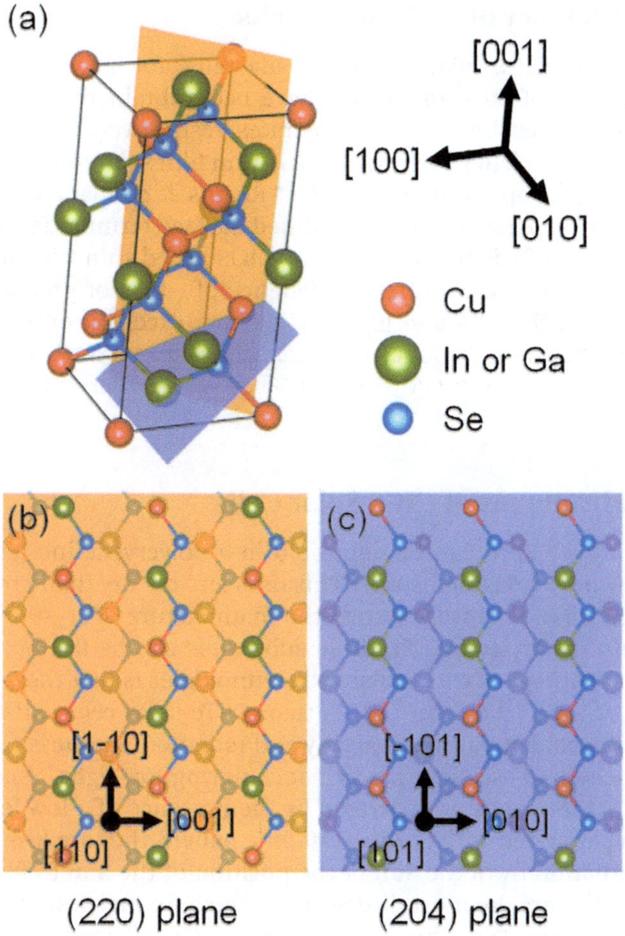

Figure 11.5 (a) Crystal structure of a CIGS tetragonal unit cell. Schematic illustration of the atomic arrangements of (b) the (220) plane and (c) the (204) plane. Reprinted from *Thin Solid Films*, 589, D. H. Cho, J. Kim and Y. D. Chung, Distinction of [220] and [204] textures of Cu(In,Ga)Se_2 film and their growth behaviors depending on substrate nature and Na incorporation, 309–314, Copyright (2015), with permission from Elsevier.[65]

At the condition of $c = 2a$, $d_{(220)}$ and $d_{(204)}$ are exactly the same from the relationship of $d_{(220)} = a\sqrt{2}/4$ and $d_{(204)} = \sqrt{(4a^2 + c^2)}/8$. Even though the X-ray diffraction peak near $2\theta = 44.2°$ was designated as (204) by applying the other peak positions of (112), (103), (101) and (211) in a previous study, the (220) and (204) peaks can coexist in a peak position of $2\theta = 44.2°$. Contreras *et al.* concluded that an X-ray diffraction peak originates from one crystal texture, although it is usually induced by one or more crystal textures that have a similar plane spacing.[73]

11.3.3 Efficiency of CIGS Solar Cells

The efficiency of CIGS solar cells is already >20% for laboratory-scale samples. Efficiencies of 20.3 and 20.4% have been obtained from CIGS solar cells on glass and flexible polymer substrates, respectively.[57] Kawamura *et al.* achieved 15.4% efficiency by growing CIGS thin film solar cells using cracked Se at a cracking temperature of 500 °C.[74] Repins *et al.* made a device using three-stage co-evaporation with a modified surface termination with a new record total area efficiency of 19.9% for CIGS-based thin film solar cells.[75] CIGS solar cells with a conversion efficiency of ~17% or greater are fabricated in Europe, the USA and Japan. They have excellent outdoor stability and radiation hardness.[76] Contreras *et al.* achieved an 18.8% total area conversion efficiency for a ZnO/CdS/Cu(In,Ga)Se_2/Mo polycrystalline thin film solar cell.[72]

11.3.4 Advantages of CIGS Solar Cells

The active layer (CIGS) can be deposited in a polycrystalline form directly onto an Mo-coated glass substrate. This uses less energy than growing large crystals, which is a necessary step in the manufacture of crystalline Si solar cells; also unlike crystalline Si, these substrates can be flexible. One environmental advantage of CIGS solar cell technologies is that they use a much lower level of Cd in the form of CdS than CdTe solar cells. CIGS solar cell panels show excellent outdoor stability and radiation hardness.[76]

CIGS panels have excellent performance during low light, cloudy or hazy conditions. CIGS panels are activated by the presence of a low level of photons. When compared with monocrystalline panels, they are less dependent on direct radiation from the Sun or the position of the Sun to function. For this reason, they are the most diverse and effective panel available for use in low light conditions. Even though the current CIGS efficiency may be slightly lower than that of Si-based panels, CIGS panels convert more sunlight per day due to their shade tolerance, low light performance and ability to tolerance to the build-up of heat. CIGS thin film solar cells have several advantages, such as a long life without considerable degradation, flexibility, light weight, less expensive than conventional solar cells, operation at both low and high temperatures and a higher efficiency.[13,25,35]

11.3.5 Disadvantages of CIGS Solar Cells

Knowledge about the reliability of CIGS photovoltaics is still limited and the lifetime of modules in the field cannot be easily predicted. As warrantees are given that the modules will still yield 80% of their initial power after 20 years of field exposure, more knowledge about the degradation behaviour of CIGS photovoltaic devices is required. Knowledge about this degradation behaviour would also help to lower production costs because increased humidity and temperature have a negative impact on the stability of CIGS photovoltaic modules, which are sold with encapsulation to keep moisture out. Glass is an excellent encapsulation choice for rigid modules, but expensive organic–inorganic multilayer coatings are often used for flexible modules. The demands of these barriers are usually only based on water penetration numbers and, in some cases, on oxygen, although it is not clear which atmospheric gases are actually involved in the degradation of CIGS solar cells.[77,78]

11.3.6 Methods of Preparation of Nanostructured CIGS Solar Cells

CIGS thin films can be grown by both vacuum and non-vacuum methods. A two-stage growth process involves sequential deposition *via* the sputtering of Cu and In layers onto Mo-coated glass and subsequent sulfurisation of this multilayer structure in a vacuum. A non-vacuum process consists of the one-step electrodeposition of these elements. A large number of preparation methods and processes have been reported. These include a colloidal route,[79] a mechanochemical method,[80] green synthesis,[81] precipitation,[82] mechanical alloying,[83] microwave synthesis[84] and a co-evaporation process.[85]

The co-evaporation process is acceptable to inline manufacturing, where In, Ga and Se are co-evaporated first, followed by the co-evaporation of Cu and Se and then the co-evaporation of In, Ga and Se. Co-evaporation is used for the production of monolithically interconnected CIGS photovoltaic modules on 6500 cm^2 glass substrates by Wurth Solar, Germany, achieving a best efficiency of 13%. Global Solar has used co-evaporation processes to prepare CIGS solar cells on stainless-steel foils. They have achieved the best efficiency of 10.1% for a 7085 cm^2 photovoltaic module prepared by connecting cells in series.[85]

Jakhmola *et al.* used a simple liquid phase method, *i.e.* a solvothermal approach using ethanolamine as a solvent with $CuCl_2$, Se powder, indium chloride and gallium chloride precursors.[86] Agrawal and co-workers synthesised CIS and CIGS nanocrystals *via* the hot injection of Se into an oleylamine (OLA) solution containing the corresponding metal chlorides at 285 °C.[87] Wu *et al.* used a simple and convenient modified polyol route to synthesise CIGS nanoparticles with tetraethylene glycol as a green solvent. In this case, polyol acts as a reductant that can reduce the metallic cation to the metallic form in the liquid phase. However, nanoparticles prepared by this method could not be dispersed directly in the solvent.[88]

Non-vacuum processes are attractive for reducing the high initial capital cost of physical vapour deposition processes. Screen printing and spray pyrolysis are fast and simple non-vacuum techniques used to deposit a wide variety of layers. In screen printing, thickness control is difficult at low thicknesses. However, it is difficult to build thickness >1 μm using spray pyrolysis and material usage is low because of losses by overspraying.[89] A novel non-vacuum process of fabricating CIGS solar cells with polymer substrates is used by International Solar Electric Technology. Here, a thin polymer foil

'Upilex' was metallised by the sputter deposition of a 0.4 μm thick layer of Mo and CIGS solar cells were fabricated on this.[90]

11.3.7 CIGS Materials

The optimisation of CIGS-based thin film solar cells crucially depends on minimising current and voltage losses across the surfaces of individual grains, including the grain interfaces with contacts or with other semiconductors. Solar cells based on CIGS thin films have been investigated under different growth conditions – including increased substrate temperatures, the use of an appropriate substrate (SLG) instead of hard glass or alumina and different Na contents – to see the effects on efficiency. Na is often introduced from an SLG substrate, which affects the electrical properties of the CIGS thin films. The incorporation of a minute quantity of Na improves the morphology, the conductivity of CIGS, modifies film growth and can also act at the grain boundaries.[76,90] Canava *et al.* attempted oxidative etching with an aqueous solution of bromine to obtain a specular surface on CIGS thin films.[91]

Cui *et al.* reported a facile and economic route to synthesise chalcopyrite ($CuIn_xGa_{1-x}Se_2$) nanocrystals by heating metal chlorides in an Se solution containing OLA and glycerol.[92] Kobayashi *et al.* investigated the recombination paths of CIGS thin film solar cells with chemical bath deposition ZnS (O,OH) and atomic layer deposition Zn (O,S) buffer layers through temperature-dependent current–voltage and capacitance–frequency measurements on devices before and after heat and light soaking post-treatment. Plasma damage-less metal–organic chemical vapour deposition, which does not use light illumination during the fabrication process, was used for the deposition of the boron-doped ZnO (ZnO:B) window layer.[93]

Fraga *et al.* investigated the properties of Ag and Au enamels as potential back contacts for CIGS solar cells.[94] The enamels were deposited on ceramic tiles by non-vacuum printing techniques. They also explained the CIGS synthesis procedure using the co-precipitation of selenite precursors. A doctor blade method was used to deposit the precursor powders on the substrate. The interface morphology between the ceramic tile, the back contact and the CIGS absorber was studied as a crucial factor in the final solar cell performance. Excellent compatibility was observed between CIGS and the Au layer, with the thickness and chemical composition suitable for photovoltaic applications. Pianezzi *et al.* studied the influence of Ni and Cr impurities on the electronic properties of CIGS thin films.[95] Impurities were introduced into the CIGS layer by diffusion during the CIGS deposition process from a Ni or Cr precursor layer below the Mo electrical back contact. High- and low-temperature CIGS deposition processes were applied to correlate the changes in the photovoltaic parameters with the amount of impurities diffused into the absorber layer. Solar cells with Ni and Cr impurities showed a reduction in device performance and the effect was most pronounced in Ni-containing devices.

11.4 Third Generation CIS Solar Cells

$CuInS_2$ (CIS), a chalcopyrite material, has received considerable interest during the past decade due to its potential as a solar cell material. CIS is a semiconductor with a direct bandgap of 1.5 eV. In particular, its use in quantum dot solar cells has attracted much attention due to its size-dependent properties. The efficiency of solar cells can be increased through design by incorporating various sizes of nanoparticles in a multilayer. CIS has replaced toxic semiconductor materials such as CdS, CdTe and PbS in solar cells. The spectral range of CIS can also be tuned from the visible into the near-IR range.

There has been keen interest in developing routes to micro- and nanostructures of CIS. Castro *et al.* prepared ultrafine CIS nanocrystals by thermally decomposing a Cu–In complex $[(PPh_3)_2CuIn(SEt)_4]$.[96] The $CuInS_2$ nanocrystals were 2–4 nm in diameter. The analogous compound $[(PPh_3)_2CuIn(SePh)_4]$ was used to prepare $CuInSe_2$.[97] Vittal and co-workers also used a single-source precursor, $[(Ph_3P)CuIn(SC\{O\}Ph_4)]$, and dual-source precursors, $[Cu(Sc\{O\}Ph]$ and $[In(bipy)(SC\{O\}Ph)_3]$, to obtain monodispersed wurtzite and zinc blende phases of CIS. They thermolysed the Cu–In complex in dodecanethiol and tri-octylphosphine oxide (TOPO) in the ratio of 50 : 50 V/V.[98] They showed that the ratio of surfactants played a part in the final phase of the particles. When the amount of TOPO was increased, the wurtzite phase was dominant. In the dual-source method, the Cu and In precursors were added to dodecanethiol and TOPO at 175 °C and heated for 16 h. The wurtzite phase was dominant at temperatures <250 °C; however, at 300 °C, less crystalline material was indexed to the zinc blende phase.

A one-pot synthesis of CIS using a metal ethyl xanthate complex as a precursor was also reported.[99] The copper indium xanthate was prepared by reacting appropriate amounts of aqueous potassium ethyl xanthate with a solution containing equimolar concentrations of copper chloride and indium chloride. The copper indium xanthate complex was then thermolysed in OLA and tri-*n*-octylphosphine at a low temperature of 110 °C. The reaction procedure was adapted to overcoat the CIS nanoparticles with CdS and ZnS. The reaction time (3–300 s) was varied to study the effect of the particle size on the optical properties. The CIS nanoparticles under normal and UV light in Figure 11.6 show the variation in the colour of the particles in solution as the particle size changes. The lower reaction decomposition temperature is due to the OLA and tri-*n*-octylphosphine. The CdS/CIS and ZnS/CIS core–shell nanoparticles had a higher quantum yield than the parent CIS particles. Ren and co-workers decomposed a single-source precursor, $(Ph_3P)_2Cu\text{-}(\mu\text{-}SEt)_2In(SEt)_2$, in 3-mercaptopropionic acid *via* a conventional convective heating method.[100] The optimum temperature for the formation of CIS nanocrystals was 150 °C. The NaCl formed as a by-product was used as a heat transfer agent. The sizes of the particles were tuned from 1.8 to 5.2 nm as the reaction temperature was increased from 150 to 190 °C.

The use of dual-source precursors to synthesise $CuInS_2$ nanoparticles is well established. Korgel and co-workers carried out pioneering work

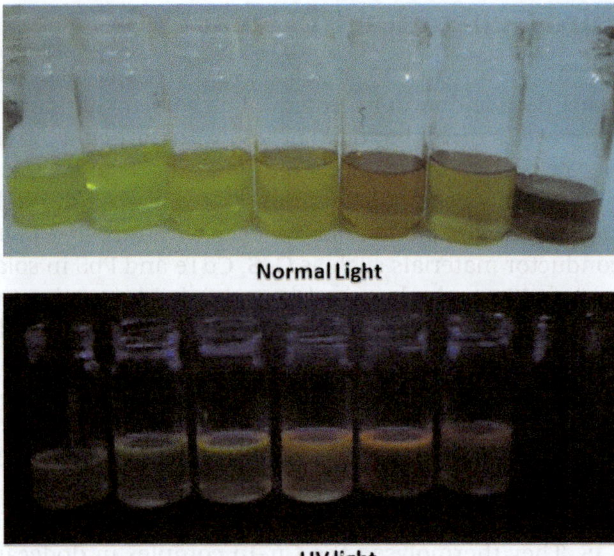

Figure 11.6 Photographs of CIS nanoparticles in normal white light and in UV light. Reprinted from *Colloid and Polymer Science*, One-pot synthesis of $CuInS_2$ and $CuInS_2/MS$ (M = Cd, Zn) core–shell luminescent nanocrystals: a low-temperature and low-cost approach, 292(11), 2014, 2913–2926, (A. Kharkwal, K. Jain, S. B. Tyagi, A. K. Singh, S. N. Sharma, M. Kharkwal, © Springer-Verlag Berlin Heidelberg 2014) With permission from Springer.[99]

$$Cu(acac)_2 + In(acac)_2 + 2S \xrightarrow[\text{oleylamine}]{o\text{-dichlorobezene , } 180\ ^\circ C} CuInS_2 + \text{byproducts} \quad (1)$$

Scheme 11.1

by heating copper chloride, indium chloride and thiourea in OLA to form monodipersed CIS nanodisks.[101] The X-ray diffraction studies revealed the dominant wurtzite phase instead of the chalcopyrite phase. However, the high-resolution transmission electron microscopy results showed polytypism in the nanodisks, whereby the wurtzite phase was interfaced with the chalcopyrite domains. The $CuInS_2$ nanodisks had an average diameter and thickness of 13.4 and 5.7 nm, respectively. The X-ray diffraction patterns of the disks revealed a hexagonal wurtzite phase instead of the expected tetragonal chalcopyrite structure. The same group also reported the synthesis of $CuInS_2$ nanocrystal 'inks' for printable photovoltaic devices.[82] $Cu(acac)_2$, $In(acac)_2$ (where acac = acetyl acetonate) and elemental sulfur dissolved in *o*-dichlorobenzene were reacted in OLA at 180 °C (reaction Scheme 11.1). The nanocrystal size was controlled by varying the OLA to metal ratio. The average size of the particles increased from 6 to 12 nm as the OLA to metal ratio was decreased from 6 : 1 to 3 : 1.

Li and co-workers used similar starting materials in dodecanethiol in a non-coordinating solvent of 1-octadecene at 240 °C.[102] They achieved shape control of the CIS nanoparticles by adjusting reaction parameters such as temperature and time. At 2 h reaction time at 240 °C, nanoparticles with an average diameter of 2.6 nm were observed. As the reaction time increased to 3 and 4 h, elongated rods were observed, with an average diameter of 3.3 nm and an aspect ratio that varied from 1 to 3. Extremely small (~1 nm) particles were obtained at a reaction temperature of 210 °C, whereas larger particles were obtained at 270 °C. The nanoparticles showed tunable absorption, with the absorption edge ranging between 550 and 750 nm and the photoluminescence peaks from 600 to 750 nm. Xie *et al.* synthesised high-quality CIS nanoparticles in the 2–20 nm size range by adjusting the relative reactivity of the copper precursor with an In precursor in a non-coordinating solvent.[103] The lower reaction temperature (<200 °C) meant that the route conformed to some of the precepts of green chemistry. Another simple method to synthesise nearly monodisperse wurtzite $CuInS_2$ nanocrystals and nanorods is the direct reaction of copper and indium nitrate with dodecanethiol under atmospheric conditions.[104] The size and shape of the nanocrystals are controlled by varying the Cu to In ratio and ligands such as OLA and oleic acid. Interesting shapes such as water drop-like nanorods, spindle-like nanorods and straight nanorods were obtained. The anisotropic particles could be assembled into a superlattice on the substrate through solvent evaporation.

Lu *et al.* used $Cu(dedc)_2$ and $In(dedc)_2$, where dedc = diethyldithiocarbamate, as precursors in the hot injection route with oleic acid or dodecanethiol as capping groups.[105] The adjustment of the precursor ratio gave particles with tunable Cu to In composition. The variation in capping group had an influence on the crystalline phase. Hyeon *et al.* reported the synthesis of copper indium sulfide heterostructured nanocrystals with morphologies including acorn, bottle and larva shapes.[106] The reaction procedure involved the thermolysis of metal–oleate complexes in dodecanethiol at 250 °C. The shape of the Cu–In sulfide nanocrystals could be changed by varying the reaction temperature and time. Figure 11.7 shows the transmission electron microscopy and high-resolution transmission electron microscopy images of the nanocrystals in the shape of acorns, bottles and larva. A mixed 1-dodecanethiol and OLA system produced bullet-like wurtzite $CuInS_2$ nanostructures.[107] The OLA had a role of controlling the shape, composition and phase of the nanocrystals. The bullet-shaped particles were 25 nm in diameter and 50 nm in length. Elongated CIS nanoparticles were synthesised by the injection of a mixture of 1-dodecanethiol and *tert*-dodecanethiol in a solution of copper and indium to a solution of copper and indium acetates in OLA and TOPO.[108] The elongated particles had a narrow size distribution. TOPO had a dual role of capping the particles and also regulating the reactivity of the In precursor because it is a Lewis base.

The luminescent properties of $CuInS_2$ were studied in detail by Li *et al.*[109] $CuInS_2$ nanoparticles were synthesised by reacting copper and indium acetate in dodecanethiol at 230 °C. OLA and CTAB were added to the reaction

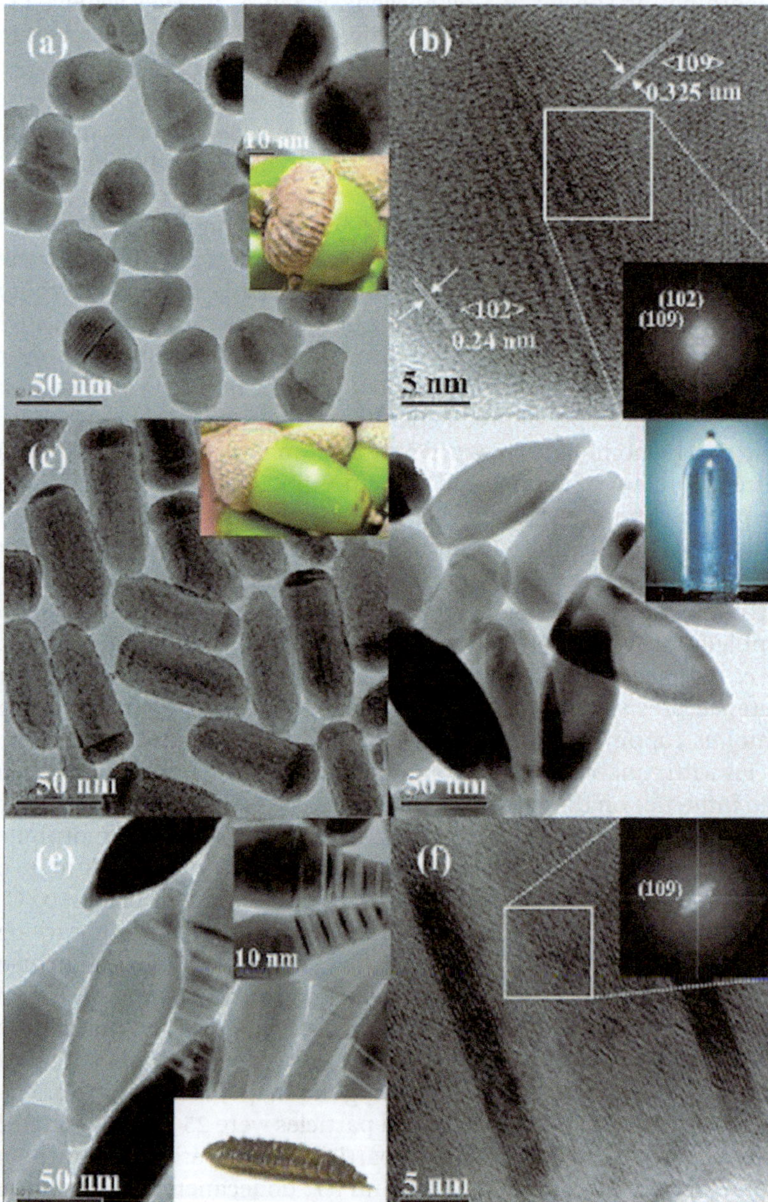

Figure 11.7 (a) TEM image of short nanoacorns (inset, enlarged image). (b) HRTEM image of short nanoacorns (inset, FFT). (c) TEM image of long nanoacorns. (d) TEM image of nanobottles. (e) TEM image of nanolarvae (f) HRTEM image (inset, FFT) of nanolarvae. Reprinted with permission from ref. 106 Copyright (2006) American Chemical Society.

mixture to investigate the effects of ligands. The photoluminescence intensity increased with reaction time. Nanoparticles synthesised at lower reaction temperatures (210 °C) exhibited weak photoluminescence.

The solvothermal route is a well-reported route to CIS nanocrystals. Cu, In and S precursors were reacted in a solvent under mild reaction conditions. Solvents such as ethylenediamine,[110] glycerol,[111] ethylene glycol[112] and benzoyl alcohol[113,114] can be used as reaction media. Thiourea is often used as the S source. Sharma and co-workers synthesised CIS nanoparticles by a solvothermal approach by reacting copper chloride, indium chloride and thiourea in ethanol at 150 °C.[115] The morphology and phase of the products were affected by the reaction time and solvent. They observed porous flower-like nanoparticles as they increased the reaction time. A change of solvent to ethylene diamine gave a plate-like morphology in the form of hexagons. Qi *et al.* prepared wurtzite CIS by a solvothermal route with a mixture of morphologies from nanoplates to microspheres.[116] Qian and co-workers synthesised CIS nanoparticles using a solvothermal route with hexadecylamine as the capping group. Simple precursors, such as metal acetate, chloride or nitrate and carbon disulphide, were used. The $CuInS_2$ nanoparticle self-assembled into a colloidal monolayer with individual particles having a polyhedral morphology in the 13–17 nm size range. They proposed a possible mechanism for crystal growth and suggested that the final morphology and phase were dependent on the reaction temperature and time. A modified surfactant–solvothermal route to hexadecylamine-capped $CuInS_2$ nanoparticles was reported by Gong and co-workers.[117] Other milder solvothermal routes gave good, reasonable quality CIS nanoparticles in the form of cubes and spheres.[114,118]

The use of microwave-assisted methods to synthesise CIS nanoparticles has been extensively reported. Salavati-Niasari and co-workers have carried out extensive work on the synthesis of CIS using microwave heating. They used copper complexes – (1,8-diamino-3,6-dioxaoctan)copper(II) sulfate, $[Cu(DADO)]SO_4$, and *bis*(propylenediamine)copper(II) sulfate, $[Cu(pn)_2]SO_4$, – as the source of copper.[119,120] $InCl_3$, thioacetamide and propylene glycol were used as In, S and the solvent, respectively. The reactants were heated under microwave irradiation in the presence of sodium dodecyl sulfate to form $CuInS_2$ nanoparticles. The effects of the reaction parameters – such as microwave power, irradiation time and precursor type – were studied. It was found that the CIS nanoparticles synthesised from the [Cu(DADO)] SO_4 complex were most effective in solar cell applications. In another study, they used another copper complex – diaquabis(salicylate)copper(II), Cu $(Hsal)_2(H_2O)_2$ – as the copper precursor.[121] The sulfur sources were varied from elemental sulfur to sodium sulfite, thioacetamide, carbon disulfide, thioglycolic acid and thiourea. It was found that the best sulfur source was thiourea. $[Cu(gly)_2]$ (gly = glycine) was also used as the copper source, with $InCl_3$ as the In source and various sulfur-containing compounds as the sulfur source.[122] In a typical reaction $[Cu(gly)_2]$, $InCl_3$ and thiourea was reacted in a 1:1:2 ratio using 0.5 g of sodium dodecyl sulfate under microwave irradiation. The CIS synthesised under all reaction conditions had a tetragonal phase. Other workers have used copper oxide and copper iodide as copper sources.[123,124]

A novel copper precursor, pomegranate marc peel dye–Cu(II), and sodium sulfite (reducing agent) were used to prepare CIS nanoparticles *via* the hydrothermal method.[125] The dyes were extracted by Cu^{2+} ions to form a pomegranate marc peel dye–Cu(II) complex, which was then used in the fabrication of a solar cell. The effect of different dyes and CIS nanoparticles on the solar cell efficiencies was investigated. The hydrothermal route was used with carminic acid–Cu(II) as a novel copper precursor and thioglycolic acid as a capping agent, sulfide source and reducing agent.[126] The carminic acid was extracted from cochineal dye for the synthesis of the Cu precursor. Qian and co-workers also used the hydrothermal route to prepare CIS nanorods 20–25 nm in diameter and 400–500 nm in length.[127] The starting materials – $CuCl_2$, In, CS_2 and NaOH – were reacted at 180 °C for 15 h.

The precursor solution methods to prepare thin films for solar cell applications have the advantages of simplicity, scale-up potential and precise stoichiometric control at the atomic level. A thioacetic acid based solution method to deposit CuIn(S,Se) films was reported by Huang and co-workers.[128] Copper(I) oxide and indium hydroxide were dissolved in a thioacetic acid based solution with the assistance of NH_3. The thioacetic acid acted as the sulfur source. The films were deposited *via* a spin-coating method, followed by sulfurization and selenisation and then annealing at 500 °C. The CISSe thin film solar cells were assembled to give a conversion efficiency of 6.75%. Roggan and co-workers deposited CIS nanoinks as thin precursor layers for $CuIn(S,Se)_2$ solar cells.[129] The CIS nanoparticles were prepared *via* a hot injection route using $[Cu(acac)_2]$, $[In(acac)_3]$ and sulfur in dichlorobenzene followed by the addition of OLA. The OLA was then surface-exchanged to pyridine and allylamine using a previously reported method. The CIS-based inks were then deposited on glass substrates using a doctor blade technique. Zhu and co-workers also reported the deposition of CIS thin films *via* the high-temperature sulfurization of copper indium oxide nanoparticles using a doctor blade technique.[130]

Single-phase $CuInS_2$ nanosheets are synthesised on TiO_2-coated fluorine-doped tin oxide ($F:SnO_2$) by the successive ionic layer adsorption reaction technique, followed by annealing at 500 °C.[131] When used as counter electrodes in a DSSC, the nanosheets had a power conversion efficiency of 6.3% compared with 5% for pristine CIS or 3.5% for CuS. Liu *et al.* reported the use of solvothermally prepared CIS as a counter electrode in a DSSC.[132] They used a novel precursor solution containing Cu_2O, $In(OH)_3$, thioacetic acid and NH_3 to prepare CIS nanopompons and hierarchial nanostructured films. The use of the CIS nanoflake films in a DSSC yielded a power conversion efficiency of 4.8%. Yang *et al.* also used CIS nanosheets synthesised *via* a one-step solvothermal process as the counter electrode in a in DSSC.[133] The catalytic activity of the CIS-based films on different precursor concentrations was investigated using electrochemical methods and a power conversion efficiency of 6.33% was achieved. A inorganic–organic nanocomposite consisting of $CuInS_2$ nanocrystals and poly(styrenesulfonate)-doped pol(3,4-ethylenedioxythiophene) (CIS/PEDOT:PSS) was explored as a counter electrode to increase the efficiency of DSSCs.[134]

11.5 Conclusions

Solar energy is considered to be one of the most promising next generation clean energy resources due to its renewable nature, infinite power, low maintenance and ease of installation. Over the past few decades, the solar cell industry has been dominated by Si-based inorganic solid state devices. However, the many drawbacks of Si-based solar cells has led to an increased activity to find new materials for high-performance, low-cost solar cells and energy storage devices. Second generation solar cells based on thin films is now a mature technology with some commercial success. Third generation solar cells are still laboratory-based, but with huge potential for commercial success; because they contain thin and flexible solar modules, interesting and novel applications in photovoltaics are expected, especially in space, aeronautical and other mobile applications.

Much work has been reported on the development of flexible and lightweight CIS and CIGS modules over the past few years. In addition to their low weight and flexibility, the advantages of a low energy payback time and strong radiation hardness of flexible CIGS solar cells are attractive. The use of nanomaterial thin film technology in solar cells has several advantages compared with the thin film or bulk counterparts. These include significantly improved photon capturing and photocarrier collection capability, potential new photovoltaic mechanisms exceeding current thermodynamic limits, as well as low-cost and scalable processes. However, there are a number of challenges for this industry, including lowering the cost of production, public awareness and developing the best infrastructure. Solar energy is in great demand and research on solar cells has a promising future worldwide.

References

1. F. Kessler, D. Herrmann and M. Powalla, *Thin Solid Films*, 2005, **480–481**, 491.
2. K. Herz, F. Kessler, R. Wachter, M. Powalla, J. Schneider, A. Schulz and U. Schumacher, *Thin Solid Films*, 2002, **403–404**, 384.
3. S. Ishizuka, A. Yamada, K. Matsubara, P. Fons, K. Sakurai and S. Niki, *Curr. Appl. Phys.*, 2010, **10**, S154–S156.
4. J. Zhu, Z. F. Yu, G. F. Burkhard, C. M. Hsu, S. T. Connor, Y. Q. Xu, Q. Wang, M. McGehee, S. H. Fan and Y. Cui, *Nano Lett.*, 2009, **9**, 279.
5. B. M. Kayes, H. A. Atwater and N. S. Lewis, *J. Appl. Phys.*, 2005, **97**, 114302.
6. Z. Y. Fan, R. Kapadia, P. W. Leu, X. B. Zhang, Y. L. Chueh, K. Takei, K. Yu, A. Jamshidi, A. A. Rathore, D. J. Ruebusch, M. Wu and A. Javey, *Nano Lett.*, 2010, **10**, 3823.

7. L. K. Yeh, K. Y. Lai, G. J. Lin, P. H. Fu, H. C. Chang, C. A. Lin and J. H. He, *Adv. Energy Mater.*, 2011, **1**, 506.
8. R. Yu, Q. Lin, S. F. Leung and Z. Fan, *Nano Energy*, 2012, **1**, 57.
9. E. C. Garnett and P. D. Yang, *Nano Lett.*, 2010, **10**, 1082.
10. B. D. Yuhas and P. Yang, *J. Am. Chem. Soc.*, 2009, **131**, 3756.
11. Z. Y. Fan, H. Razavi, J. W. Do, A. Moriwaki, O. Ergen, Y. L. Chueh, P. W. Leu, J. C. Ho, T. Takahashi, L. A. Reichertz, S. Neale, K. Yu, M. Wu, J. W. Ager and A. Javey, *Nat. Mater.*, 2009, **8**, 648.
12. K. Cho, D. J. Ruebusch, M. H. Lee, J. H. Moon, A. C. Ford, R. Kapadia, K. Takei, O. Ergen and A. Javey, *Appl. Phys. Lett.*, 2011, **98**, 203101.
13. S. Sharma, K. K. Jain and A. Sharma, *Mater. Sci. Appl.*, 2015, **6**, 1145.
14. L. Hu and G. Chen, *Nano Lett.*, 2007, **7**, 3249.
15. P. Jackson, D. Hariskos, E. Lotter, S. Paetel, R. Wuerz, R. Menner, W. Wischmann and M. Powalla, *Prog. Photovoltaics*, 2011, **19**, 894.
16. I. Repins, M. Contreras, M. Romero, Y. Yan, W. Metzger, J. Li, S. Johnston, B. Egaas, C. DeHart and J. Scharf, *33rd IEEE Photovoltaic Specialists Conference San Diego*, California, 2008, May 11–16.
17. C. H. Huang, *J. Phys. Chem. Solids*, 2008, **69**, 330.
18. C. W. Tang, *Appl. Phys. Lett.*, 1986, **48**, 183.
19. A. McEvoy, L. Castaner and T. Markvart, *Solar Cells: Materials, Manufacture and Operation*, Elsevier Ltd, Oxford, 2nd edn, 2012, p. 3.
20. A. L. Fahrenbruch and R. H. Bube, *Fundamentals of Solar Cells*, Academic Press Inc., New York, 1983.
21. A. M. Bagher, M. M. A. Vahid and M. Mohsen, *Am. J. Opt. Photonics*, 2015, **3**, 94.
22. B. Srinivas, S. Balaji, M. B. Nagendra and Y. S. Reddy, *Int. J. Eng. Res.*, 2015, **3**, 178–182.
23. P. Wurfel and U. Wurfel, *Physics of Solar Cells: From Basic Principles to Advanced Concepts*, John Wiley & Sons, Hoboken, 2009.
24. S. Dmitrijev, *Principles of Semiconductor Devices*, Oxford University Press, Oxford, 2006.
25. M. Bertolli, *Solar Cell Materials. Course: Solid State II*, Department of Physics, University of Tennessee, Knoxville, 2008.
26. T. Saga, *NPG Asia Mater.*, 2010, **2**, 96.
27. P. Jayakumar, *Solar Energy Resource Assessment Handbook*, Renewable Energy Corporation Network for the Asia Pacific, 2009.
28. M. A. Green, *Physica E*, 2002, **14**, 65.
29. K. L. Chopra, P. D. Paulson and V. Dutt, *Prog. Photovoltaics*, 2004, **12**, 69.
30. Hassan, *Types of Solar Panels,* Grein Energy, Published 22 April 2015, available at http://greinenergy.com/index.php/18-about-solar/types-of-solar-panels/39-types-of-solar-panels.
31. M. A. Maehlum, *Energy Informative the Homeowner's Guide to Solar Panels, Best Thin Film Solar Panels—amorphous, Cadmium Telluride or CIGS?*, 2015.
32. D. Y. Goswami and F. Kreith, *Handbook of Energy Efficiency and Renewable Energy*, CRC Press, Boca Raton, 2007.

33. A. Luque and S. Hegedus, *Handbook of Photovoltaic Science and Engineering*, John Wiley & Sons, Ltd, Hoboken, 2nd edn, 2003.
34. K. M. Elsabawy, W. F. El-Hawary and M. S. Refat, *Int. J. Chem. Sci.*, 2012, **10**, 1869.
35. W. A. Badawy, *J. Adv. Res.*, 2015, **6**, 123.
36. V. K. Sethi, M. Pandey and P. Shukla, *Int. J. Chem. Eng. Appl.*, 2011, **2**, 77.
37. T. M. Razykov, C. S. Ferekides, D. Morel, E. Stefanakos, H. S. Ullal and H. M. Upadhyaya, *Sol. Energy*, 2011, **85**, 1580.
38. H. Hoppe and N. S. Sariciftci, *Adv. Polym. Sci.*, 2008, **214**, 1.
39. F. Wudl and G. Srdanov, Conducting Polymer Formed of Poly (2-Methoxy-5-(2′- Ethylhexyloxy)-P-Phenylene Vinylene), US Patent, 5,189,136, 1993.
40. G. Li, R. Zhu and Y. Yang, *Nat. Photonics*, 2012, **6**, 153.
41. C. J. Brabec, S. E. Shaheen, C. Winder, N. S. Sariciftci and P. Denk, *Appl. Phys. Lett.*, 2002, **80**, 1288.
42. R. Zhu, A. Kumar and Y. Yang, *Adv. Mater.*, 2011, **23**, 4193.
43. S. Suhaimi, M. M. Shahimin, Z. A. Alahmed, J. Chyský and A. H. Reshak, *Int. J. Electrochem. Sci.*, 2015, **10**, 28.
44. P. R. F. Barnes, A. Y. Anderson, S. E. Koops, J. R. Durrant and B. C. O'Regan, *J. Phys. Chem. C*, 2009, **113**, 1126.
45. S. P. Philipps, A. W. Bett, K. Horowitz and S. Kurtz, *Current Status of Concentrator Photovoltaics (CPV) Technology. Report Version 1.2*, Fraunhofer Institute for Solar Energy Systems (NREL), 2015.
46. T. Casey, *An Article on Perovskites Will Power New Low-cost & Highly Efficient Solar Cells*, Clean Technical, 2015.
47. D. Shi, Y. Zeng and W. Shen, *Sci. Rep.*, 2015, **5**, 16504.
48. A. C. Badgujar, S. R. Dhage and S. V. Joshi, *Thin Solid Films*, 2015, **589**, 79.
49. M. Mostefaoui, H. Mazari, S. Khelifi, A. Bouraiou and R. Dabou, *Energy Procedia*, 2015, **74**, 736.
50. A. Gerthoffer, F. Roux, F. Emieux, P. Faucherand, H. Fournier, L. Grenet and S. Perraud, *Thin Solid Films*, 2015, **592**, 99.
51. E. Moons, T. Engelhard and D. Cahen, *J. Electron. Mater.*, 1993, **22**, 275.
52. S. Raud and M. A. Nicolet, *Thin Solid Films*, 1991, **201**, 361.
53. S. Ashour, A. H. Moutinho, R. Matson and F. Abou-Elfotouh, *Thin Solid Films*, 1993, **226**, 129.
54. D. Abou-Ras, G. Kostarz, D. Bremaud, M. Kalin, F. V. Kurdesau, A. N. Tiwari and M. Dobeli, *Thin Solid Films*, 2005, **480–481**, 433.
55. J. H. Scofield, S. Asher, D. Albin, J. Tuttle, M. Contreras and D. Niles, *24th IEEE Photovolt, Specialists Conf*, 1995, p. 164.
56. S. Glynn, L. Mansfield, Y. Yan, M. Contreras, D. Levi and F. Lewis, *37th IEEE Photovolt, Spec. Conf*, PVSC, 2011, p. 002749.
57. N. Khoshsirat, N. A. Md. Yunus, M. N. Hamidon, S. Shafie and N. Amin, *Optik*, 2015, **126**, 681.
58. H. Movla, A. M. Rafi and N. M. Rafi, *Optik*, 2015, **126**, 5088.
59. R. N. Bhattacharya and K. Ramanathan, *Sol. Energy*, 2004, **77**, 679.

60. D. A. Ras, G. Kostorz, A. Romeo, D. Rudmann and A. N. Tiwari, *Thin Solid Films*, 2005, **480–481**, 118.
61. R. O. Landín, J. S. Hernández, O. V. Galan and R. R. Bon, *Sol. Energy*, 2010, **84**, 208.
62. F. R. Ahmad, A. Yakimov, R. J. Davis, J. H. Her, J. R. Cournoyer and N. M. Ayensu, *Thin Solid Films*, 2013, **535**, 166.
63. N. Khoshsirat and N. A. M. Yunus, *IEEE Conference on Sustainable Utilization and Development in Engineering and Technology*, 2013.
64. A. Omelianovych, J. H. Kim, L. Liudmila and B. T. Ahn, *Curr. Appl. Phys.*, 2015, **15**, 1641.
65. D. H. Cho, J. Kim and Y. D. Chung, *Thin Solid Films*, 2015, **589**, 309.
66. M. A. Contreras, M. J. Romero and R. Noufi, *Thin Solid Films*, 2006, **511–512**, 51.
67. N. Ott, G. Hanna, U. Rau, J. H. Werner and H. P. Strunk, *J. Phys.: Condens. Matter.*, 2004, **16**, S85.
68. A. Chirilă, S. Buecheler, F. Pianezzi, P. Bloesch, C. Gretener, A. R. Uhl, C. Fella, L. Kranz, J. Perrenoud, S. Seyrling, R. Verma, S. Nishiwaki, Y. E. Romanyuk, G. Bilger and A. N. Tiwari, *Nat. Mater.*, 2011, **10**, 857.
69. D. H. Shin, Y. M. Shin, J. H. Kim, B. T. Ahn and K. H. Yoon, *J. Electrochem. Soc.*, 2012, **159**, B1.
70. B. Y. Li, Y. Zhang, H. Wang, B. Wang, L. Wu and Y. Sun, *Prog. Photovoltaics*, 2013, **21**, 838.
71. S. B. Zhang and S. H. Wei, *Phys. Rev. B*, 2002, **65**, 081402.
72. M. A. Contreras, B. Egaas, K. Ramanathan, J. Hiltner, A. Swartzlander, F. Hasoon and R. Noufi, *Prog. Photovoltaics*, 1999, **7**, 311.
73. M. A. Contreras, B. Egaas, D. King, A. Swartzlander and T. Dullweber, *Thin Solid Films*, 2000, **361–362**, 167.
74. M. Kawamura, T. Fujita, A. Yamada and M. Konagai, *J. Cryst. Growth*, 2009, **311**, 753.
75. I. Repins, M. A. Contreras, B. Egaas, C. DeHart, J. Scharf, C. L. Perkins, B. To and R. Noufi, *Prog. Photovoltaics*, 2008, **16**, 235.
76. L. Kronik, D. Cahen and H. W. Schock, *Adv. Mater.*, 1998, **10**, 31.
77. D. Coyle, *Prog. Photovoltaics*, 2013, **21**, 156.
78. J. D. Cueto, S. Rummel, B. Kroposki, C. Osterwald and A. Anderberg, *33rd IEEE PVSC*, 2008, pp. 1–6.
79. J. Tang, S. Hinds, S. O. Kelley and E. H. Sargent, *Chem. Mater.*, 2008, **20**, 6906.
80. B. Rehani, J. R. Ray, C. J. Panchal, H. Master, R. R. Desai and P. B. Patel, *J. Nano- Electron. Phys.*, 2013, **5**, 02007.
81. L. A. Juhaiman, L. Scoles, D. Kingston, B. Patarachao, D. Wang and F. Bensebaa, *Green Chem.*, 2010, **12**, 1248.
82. M. G. Panthani, V. Akhavan, B. Goodfellow, J. P. Schmidtke, L. Dunn, A. Dodabalapur, P. F. Barbara and B. A. Gorgel, *J. Am. Chem. Soc.*, 2008, **130**, 16770.
83. B. Vidhya, S. Velumani and R. Asomoza, *J. Nanopart. Res.*, 2011, **13**, 3033.

84. F. Bensebaa, C. Durand, A. Aouadou, L. Scoles, X. Du, D. Wang and Y. L. Page, *J. Nanopart. Res.*, 2010, **12**, 1897.
85. K. Zweibel, H. S. Ullal and B. Roedern, *Photon Int.*, 2004, 48.
86. P. Jakhmola, P. K. Jha and S. P. Bhatnagar, *Appl. Nanosci.*, 2015, **6**, 673.
87. Q. Guo, S. J. Kim, M. Kar, W. N. Shafarman, R. W. Birkmire, E. A. Stach, R. Agrawal and H. W. Hillhouse, *Nano Lett.*, 2008, **8**, 2982.
88. J. D. Wu, L. T. Wang and C. Gau, *Sol. Energy Mater. Sol. Cells*, 2012, **98**, 404.
89. N. G. Dhere, *Sol. Energy Mater. Sol. Cells*, 2006, **90**, 2181.
90. K. Otte, L. Makhova, A. Braun and I. Konovalov, *Thin Solid Films*, 2006, **511–512**, 613.
91. B. Canava, J. F. Guillemoles, J. Vigneron, D. Lincot and A. Etcheberry, *J. Phys. Chem. Solids*, 2003, **64**, 1791.
92. X. Cui, D. Yun, C. Zhong, W. Chen, Q. Cheng, J. Feng and F. Zhang, *J. Sol–Gel Sci. Technol.*, 2015, **76**, 469.
93. T. Kobayashi, Z. J. L. Kao and T. Nakada, *Sol. Energy Mater. Sol. Cells*, 2015, **143**, 159.
94. D. Fraga, T. S. Lyubenova, A. Rey, I. Calvet, R. Marti and J. B. Carda, *Int. J. Appl. Ceram. Technol.*, 2015, **12**, 728.
95. F. Pianezzi, S. Nishiwaki, L. Kranz, C. M. S. Fella, P. Reinhard, B. Bissig, H. Hagendorfer, S. Buecheler and A. N. Tiwari, *Prog. Photovoltaics*, 2015, **23**, 892.
96. S. L. Castro, S. G. Bailey, R. P. Raffaelle, K. K. Banger and A. F. Hepp, *J. Phys. Chem. B*, 2004, **108**, 12429.
97. S. L. Castro, S. G. Bailey, R. P. Raffaelle, K. K. Banger and A. F. Hepp, *Chem. Mater.*, 2003, **15**, 3142.
98. S. K. Batabyal, L. Tian, N. Venkatram, W. Ji and J. J. Vittal, *J. Phys. Chem. C*, 2009, **113**, 15037.
99. A. Kharkwal, K. Jain, S. B. Tyagi, A. K. Singh, S. N. Sharma and M. Kharkwal, *Colloid Polym. Sci.*, 2014, **292**, 2913.
100. C. Sun, Z. Cevher, J. Zhang, B. Gao, K. Shum and Y. Ren, *J. Mater. Chem. A*, 2014, **2**, 10629.
101. B. Koo, R. N. Patel and B. A. Korgel, *Chem. Mater.*, 2009, **21**, 1962.
102. H. Zhong, Y. Zhou, M. Ye, Y. He, J. Ye, C. He, C. Yang and Y. Li, *Chem. Mater.*, 2008, **20**, 6434.
103. R. Xie, M. Rutherford and X. Peng, *J. Am. Chem. Soc.*, 2009, **131**, 5691.
104. X. Lu, Z. Zhuang, Q. Peng and Y. Li, *CrystEngComm*, 2011, **13**, 4039.
105. D. Pan, L. An, Z. Sun, W. Hou, Y. Yang, Z. Yang and Y. Lu, *J. Am. Chem. Soc.*, 2008, **130**, 5620.
106. S. Choi, E. Kim and T. Hyeon, *J. Am. Chem. Soc.*, 2006, **128**, 2520.
107. J. Guo, W. Zhou, M. li, Z. Hou, J. Jiao, Z. Zhou and S. Wu, *J. Cryst. Growth*, 2012, **359**, 72.
108. M. Kruszynska, H. Borchert, J. Parisi and J. Kolny-Olesiak, *J. Am. Chem. Soc.*, 2010, **132**, 15976.
109. D. Li, Y. Zou and D. Yang, *J. Lumin.*, 2012, **112**, 313.

110. Q. Y. Lu, J. Q. Hu, K. B. Tang, Y. T. Qian, G. Zhou and X. M. Liu, *Inorg. Chem.*, 2000, **39**, 1606.
111. M. Li, S. F. Zhou, Y. W. Zhang, G. Q. Chen and Z. L. Hong, *Appl. Surf. Sci.*, 2008, **254**, 3762.
112. L. Wu, J. H. Bi, Z. H. Li, X. X. Wang and X. Z. Fu, *Catal. Today*, 2008, **131**, 15.
113. L. Z. Zhang, I. Djerdj, M. H. Cao, M. Antonietti and M. Niederberger, *Adv. Mater.*, 2007, **19**, 2083.
114. C. Yu, J. C. Yu, H. Wen and C. Zhang, *Mater. Lett.*, 2009, **63**, 1984.
115. A. Kharkwal, S. N. Sharma, K. Jain and A. K. Singh, *Mater. Chem. Phys.*, 2014, **144**, 252.
116. Y. Qi, Q. Liu, K. Tang, Z. Liang, Z. Ren and X. Liu, *J. Phys. Chem. C*, 2009, **113**, 3939.
117. W. Du, X. Qian, J. Yin and Q. Gong, *Chem.–Eur. J.*, 2007, **13**, 8840.
118. S. Han, M. Kong, Y. Guo and M. Wang, *Mater. Lett.*, 2009, **63**, 1192.
119. S. M. Hosseinpour-Mashkani, M. Salavati-Niasari, F. Mohandes and K. Venkateswara-Rao, *Mater. Sci. Semicond. Process.*, 2013, **16**, 390.
120. S. M. Hosseinpour-Mashkani, M. Salavati-Niasari, F. Mohandes and K. Venkateswara-Rao, *Mater. Res. Bull.*, 2012, **47**, 3148.
121. M. Sabet, M. Salavati-Niasari, D. Ghanbari and O. Amiri, *Synth. React. Inorg., Met.-Org., Nano-Met. Chem.*, 2015, **45**, 1025.
122. S. M. Hosseinpour-Mashkani, M. Salavati-Niasari and F. Mohandes, *J. Ind. Eng. Chem.*, 2014, **20**, 3800.
123. M. Suzaki, *J. Appl. Phys.*, 2005, **44**, 866.
124. A. Pein, M. Baghbanzadeh, T. Rath, W. Haas, E. Maier, H. Amenitsch, F. Hofer, C. O. Kappe and G. Trimmel, *Inorg. Chem.*, 2011, **50**, 193.
125. M. Mousavi-Kamazani, M. Salavati-Niasari and M. Sadeghinia, *Mater. Lett.*, 2015, **142**, 145.
126. M. Mousavi-Kamazani, M. Salavati-Niasari, S. M. Hosseinpour-Mashkani and M. Goudarzi, *Mater. Lett.*, 2015, **145**, 99.
127. J. Xiao, Y. Xie, R. Tang and Y. Qian, *J. Solid State Chem.*, 2001, **161**, 179.
128. Y. Xie, Y. Liu, Y. Wang, X. Zhu, A. Li, L. Zhang, M. Qin, X. Lu and F. Huang, *Phys. Chem. Chem. Phys.*, 2014, **16**, 7548.
129. P. Niyamakom, A. Quintilla, K. Kohler, M. Cemernjak, E. Ahlswede and S. Roggan, *J. Mater. Chem. A*, 2015, **3**, 4470.
130. J. Liu, J. Li, G. Jiang, W. Liu and C. Zhu, *Mater. Lett.*, 2015, **156**, 153.
131. R. K. Bhosale, S. A. Agarkar, I. Agrawal, R. A. Naphade and S. Ogale, *RSC Adv.*, 2014, **4**, 21989.
132. Y. Liu, Y. Xie, H. Cui, W. Zhao, C. Yang, Y. Wang, F. Huang and N. Dai, *Phys. Chem. Chem. Phys.*, 2013, **15**, 4496.
133. J. Yang, C. Bao, J. Zhang, T. Yu, H. Huang, Y. Wei, H. Gao, G. Fu, J. Liu and Z. Zou, *Chem. Commun.*, 2013, **49**, 2028.
134. Z. Zhang, X. Zhang, H. Xu, Z. Liu, S. Pang, X. Zhou, S. Dong, X. Chen and G. Cui, *ACS Appl. Mater. Interfaces*, 2012, **4**, 6242.

CHAPTER 12

Copper-based Multinary Materials for Solar Cells

JOHN WATT[a] AND KARTHIK RAMASAMY*[b]

[a]Center for Integrated Nanotechnologies, Sandia National Laboratories, Albuquerque, NM-87185, USA; [b]Los Alamos National Laboratory, Albuquerque, NM-87185, USA
*E-mail: jdwatt@sandia.gov, kramasamy@lanl.gov

12.1 Introduction

Multinary semiconductors derived from chalcopyrite have been widely studied since the 1980s and are currently used extensively in commercially available solar cells. As such, much research has been carried out to optimize their material properties to obtain the highest solar energy conversion efficiencies possible. The material with the best performance for this type of application reported thus far is $CuInGaSe_2$ (CIGS). CIGS is derived from the cubic sphalerite (ZnS) crystal structure in which two Zn^{II} ions are substituted by Cu^{I} and either In^{III} or Ga^{III}, thus preserving the cubic symmetry. Power conversion efficiencies for solar cells derived from this material have reached 19.9% for a 0.5 cm^2 laboratory-scale cell, 16.6% for 16 cm^2 modules and 15.2% for 9703 cm^2 modules. Unfortunately, In and Ga are expensive and rare materials, with future supply being a concern. Therefore finding cheaper and more readily available materials for solar cell applications is a strong driver for current research. An alternative is to substitute with Zn^{II} and Sn^{IV} to give the multinary chalcopyrite Cu_2ZnSnS_4 (CZTS). This semiconductor material has

Nanoscience & Nanotechnology Series No. 45
Nanostructured Materials for Type III Photovoltaics
Edited by Peter Skabara and Mohammad Azad Malik

Published by the Royal Society of Chemistry, www.rsc.org

an I_2–II–IV–VI_4 configuration and has great potential in solar applications due to a bandgap in the useful 1.4–1.5 eV range. It contains exclusively non-toxic and inexpensive elements and has a large absorption coefficient $>10^4$ cm^{-1}.

CZTS has two different crystal structures: stannite and kesterite. Stannite forms in the $I\bar{4}2m$ space group, whereas kesterite forms in the $I\bar{4}$ space group. Both structures can be viewed as an expansion of the zinc blende structure with different ordering in the II–IV sub-cation lattice (Figure 12.1). In the stannite structure, both Cu atoms occupy the 4d position, with the Zn at 2a. In kesterite, one of the Cu atoms occupies the 2a position, with Zn and the remaining Cu occupying the 2d and 2c positions, respectively. In both structures, Sn is located at 2b. The kesterite structure is slightly lower in energy than the stannite structure by ~1.3 meV per atom; however, there is only a small difference in the lattice parameters. As a result of their similarities,

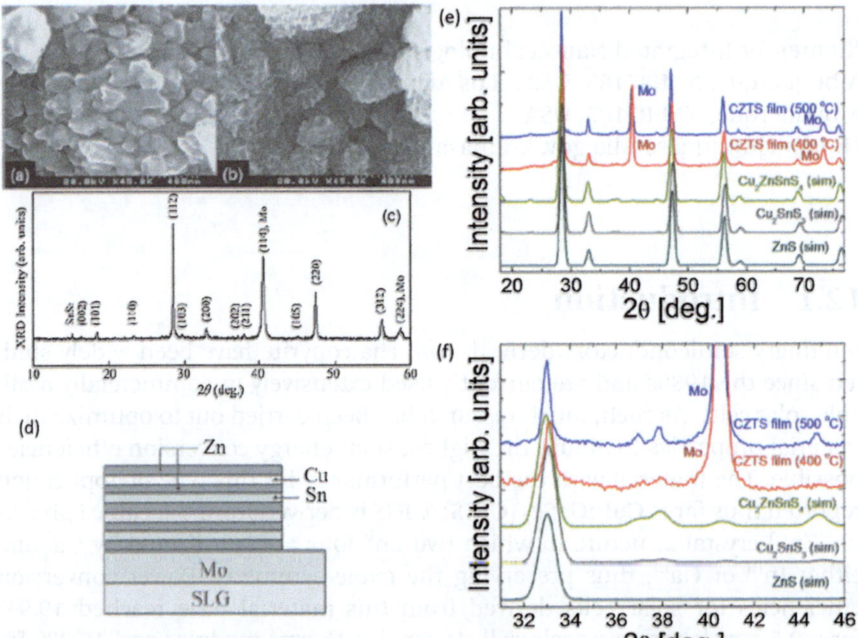

Figure 12.1 Surface morphology of CZTS films synthesized by Htay *et al.*[7] (a) before and (b) after the deposition of a ZnO buffer layer. (c) XRD pattern of the CZTS film. (d) Schematic diagram of a laminated metallic precursor of the CZTS film. Reproduced with permission from ref. 7. Copyright 2011. The Japan Society of Applied Physics. (e) Comparison of experimental XRD patterns from Cheng *et al.*[10] Zinc/copper/tin thin film stacks sulfidized at 400 and 500 °C are compared with the simulated XRD patterns of Cu_2ZnSnS_4, Cu_2SnS_3 and ZnS. (f) The three CZTS diffraction peaks at 37° (202), 37.9° (211) and 44.9° (105) and (213), which are absent in ZnS and very weak in Cu_2SnS_3, can be used to confirm the presence of CZTS. Reproduced with permission from ref. 10. Copyright 2011. American Vacuum Society.

there is often the possibility of both phases occurring in the same material. We present here an in-depth coverage of the most recent advances in the synthesis of Cu_2MSnE_4 (M = Zn or Fe and E = S or Se) materials in the form of nanocrystals and thin films and their fabrication into solar cell devices.

12.2 CZTS Nanocrystals and Thin Films

Ito and Nakazawa were the first (in 1988) to prepare thin films of CZTS by a beam sputtering method.[1] The films were deposited on a glass substrate and had a (112) orientation. Using UV–visible spectrophotometry, they showed that the films had an absorption coefficient >1.2 × 10^4 cm^{-1} at 120 °C, in the photon energy range >1.3 eV. They showed that the resistivity could be reduced to as little as 1.3 Ω cm with an increase in the substrate temperature to 240 °C. A few years later, Ito and Nakayama used spray pyrolysis to form CZTS thin films on glass substrates at elevated temperatures (280–360 °C).[2] The concentration of the solutes in the spray solution were used to control the stoichiometry; however, the as-deposited films were deficient in sulfur. Therefore the structures were thermally annealed at 550 °C in an inert gas containing H_2S to obtain stoichiometric films. The films had a resistivity of 2 × 10^2 Ω cm.

Another approach involved the decomposition of four thiourea–metal complexes, $Cu_2(tu)_6Cl_2{\cdot}2H_2O$, $ZnCl_2(tu)$, $Sn_2Cl_4(tu)_5{\cdot}2H_2O$ and $SnCl_2$ (where tu = thiourea), in a simultaneous thermogravimetric/differential thermoanalytical apparatus.[3] The spray pyrolysis precursors were prepared by dissolving the respective chloride salts in an aqueous solution of thiourea. CZTS films were formed when the substrate temperature was between 225 and 350 °C and required an excess of thiourea beyond the stoichiometric ratio. Kamoun *et al.* also deposited CZTS films from aqueous solutions containing $CuCl_2$, $ZnCl_2$, $SnCl_2$ and thiourea.[4] They found that all the films formed in the kesterite crystal structure with a [112] preferential orientation; however, the film formed at 340 °C for 60 min showed the best crystallinity. By annealing the film at 550 °C for 120 min they were able to improve the optical properties from an estimated bandgap at 1.7 eV to the more optimized 1.5 eV.

Kumar *et al.* carried out a number of studies on the formation of CZTS films deposited onto soda lime glass substrates by spray pyrolysis.[5,6] They investigated the effect of temperature on film deposition and found an optimum temperature range of 370–410 °C for the growth of CZTS films. Outside this temperature range, Cu_2SnS_3 was formed, with Cu_xS or ZnS appearing as a secondary phase. The direct optical bandgap was optimized to lie between 1.40 and 1.45 eV. The films had p-type electrical conductivity and an optical absorption coefficient >10^4 cm^{-1}. The effect of pH was studied by varying the pH of the starting solution from 3.0 to 5.5. Polycrystalline, non-stoichiometric CZTS films were obtained at pH 3.0, with pH 4.5 yielding a mixture of Cu_xS, ZnS and CZTS.

A different approach by Htay *et al.* involved annealing a laminated precursor of Cu, Zn and Sn metal layers (Figure 12.1a–d).[7] Control over the composition was achieved simply by varying the thickness of each metal layer. The film was annealed under a sulfur vapor pressure of 0.27 atm at 550 °C for 15 min. An SnS_2 impurity and voids near the surface of the back electrode were observed. By depositing a ZnO buffer layer *via* ultrasonic spray pyrolysis they were able to reach a solar conversion energy of 4.29% when the thin films were fabricated into a solar cell device. The ZnO layer was in place of the more commonly used CdS layer and gave promising results as an alternative to more toxic materials.

Katagiri *et al.* developed a solar cell device from CZTS films deposited using electron beam evaporation followed by sulfurization.[8] The cell had an open circuit voltage of up to 400 mV; however, the solar energy conversion efficiency was very low (0.66%), which was attributed to a degraded fill factor. The CZTS films were prepared by sequentially depositing layers of Zn, Sn and Cu on a soda lime glass substrate. The precursor films were sulfurized at 500 °C in an H_2S atmosphere to yield stannite CZTS. The group was able to optimize their system to reach a conversion efficiency of 2.62% and an open circuit voltage of 735 mV.[9] To achieve this, they substituted SnS (or SnS_2) in place of pure Sn, which suppressed the volume expansion of the CZTS film and lead to stronger adhesion of the CZTS film to the substrate. They also observed that increasing the substrate temperature from 150 to 400 °C during the evaporation of the precursors was a contributing factor to the increased conversion efficiency.

Cheng *et al.* used confocal Raman spectroscopy to image and identify the phases of CZTS films grown using thermal evaporation.[10] First, Zn, Cu and Sn layers were deposited at ambient temperature in thicknesses that corresponded to the required stoichiometry for CZTS. The substrate was then sealed in an ampule with elemental sulfur and held at a high temperature for sulfurization. They showed that high temperatures (500 °C) lead to predominantly CZTS films (Figure 12.1e and f), whereas sulfurization at lower temperatures (350 °C) led to the coexistence of a secondary Cu_2SnS_3 phase. The formation of the secondary phase was due to the formation of an alloy of Cu and Sn in the initial deposition steps, which was then sulfidized to form Cu_2SnS_3.

A comprehensive study into film formation at low (325 °C) and high (475 °C) temperatures was carried out in 2014 by Du *et al.*[11] This was carried out by growing graded composition Cu–Zn–Sn–S films *via* the co-evaporation of Cu, Zn and Sn using effusion cells and *via* a valve-cracking source for S. This enabled them to visualize the direct effect that film composition has on the crystalline phase content and morphology. Based on X-ray fluorescence spectrometry, scanning electron microscopy, X-ray diffraction (XRD) and Raman spectroscopy, they were able to construct a schematic thin film quasi-ternary phase diagram. They showed that favorable film morphologies were observed in regions of coexisting CZTS and secondary phases of ZnS and Cu_2SnS_3 due to similar unit cells and lattice parameters. These areas were characterized by a composition where Cu/Sn ~2, were strongly [112] textured and showed a columnar growth mode. These regions of structural coherency were postulated to have the potential to increase photovoltaic performance due to a favorable distribution of point defects.

A multi-stage evaporation method was developed by Weber *et al.* in 2009.[12] They investigated and compared kesterite films formed from two different two-stage processes. In the first process, a Mo/SnS_x/CuS stack was deposited at room temperature, followed by the deposition of ZnS at different substrate temperatures. In the second process, ZnS was first deposited at 150 °C, followed by the simultaneous evaporation of Cu, Sn and S. For the first sequence, the layers tended to be Cu-rich due to the re-evaporation of ZnS and CuS. For the second process, they observed a strong decrease in Sn incorporation at temperatures >400 °C. Therefore films with a second stage deposition temperature of 380 °C from both processes were chosen to be processed into photovoltaic cells. They achieved open circuit voltages of 501 and 426 mV and conversion efficiencies of 0.7 and 1.1% for the first and second processes, respectively.

Wang *et al.* developed a method to grow CZTS films by the thermal evaporation of all four elements (Cu, Zn, Sn and S) in a vacuum system.[13] The elements were co-deposited onto 700 nm thick Mo-coated glass substrates held at 110 °C. Interestingly, they were able to achieve incredibly short annealing times (5 min) by annealing the films on a ceramic hot-plate at 540 °C in the presence of S. They then fabricated a device 0.45 cm^2 in size with a conversion efficiency of 6.8% (Figure 12.2). However, the activation energy of the device was 1.05 eV, which is significantly lower than the absorber bandgap of 1.45 eV. This was attributed to a dominant recombination at the buffer–absorber interface, which is a known problem in wide bandgap chalcopyrite solar cells.[14] The same group also carried out a detailed microstructural study of solar cells based on the thermally evaporated CZTS absorber layer.[15]

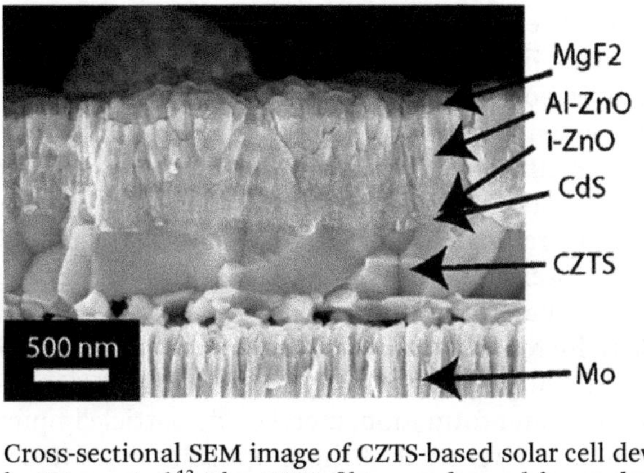

Figure 12.2 Cross-sectional SEM image of CZTS-based solar cell device fabricated by Wang *et al.*[13] The CZTS film was formed by co-deposition of all four elements (Cu, Zn, Zn and S) by thermal evaporation in a vacuum system before annealing. The solar cell device achieved a power conversion efficiency of 6.8%. Reproduced with permission from ref. 13. Copyright 2010. AIP Publishing LLC.

They showed that the interaction between the CZTS layer and the Mo electrode resulted in the formation of an MoS_x layer and phase separation of the CZTS layer. This led to the formation of a lower bandgap Cu_xSnS_y layer and lower than expected open circuit voltages.

The highest efficiency reported for a Cu_2ZnSnS_4 solar cell (*i.e.* not including Se) was 8.4% in a study by Shin *et al.*[16] The film was deposited by a vacuum thermal evaporation process and gave a very thin absorber thickness (~600 nm). The solar cell had an open circuit voltage of 661 mV, with a fill factor of 65.8% and a short circuit current density of 19.5 mA cm^{-2}. The high efficiency was attributed to a Cu-poor and Zn-rich composition, which gave a strong optical performance despite the presence of some secondary phases (*e.g.* ZnS).

A radio frequency magnetron sputtering process was used by Seol *et al.* to form CZTS films on unheated glass substrates.[17] They found that the crystallinity increased as they annealed the films, accompanied by a rapid decrease in sheet resistance to <0.47 Ω cm. The preferred [112] orientation became more pronounced with an increase in annealing time. The absorption coefficient was about 1.0×10^4 cm^{-1} with a bandgap at 1.51 eV. Thus although the absorption coefficient was less than values obtained for films containing In, the bandgap was more appropriate for photovoltaic applications.

Tanaka *et al.* developed a hybrid sputtering method consisting of a deposition chamber with two sputtering sources for Cu and Sn and two effusion cells for Zn and S.[18] A diffusion pumping system was included and the fluxes from all four sources were focused on a quartz glass substrate. The films were grown by the sequential deposition of Sn, Zn and Cu and subsequent annealing in a sulfur atmosphere. They showed that the film thickness decreased as the substrate temperature increased. When the film was fabricated at 400 °C, it had the stoichiometric composition of CZTS and crystallized in the stannite structure without any other phases present. The film had an optical bandgap of 1.5 eV and a resistivity of 0.13 Ω cm.

Good energy conversion efficiencies of 5.33 and 5.74% were obtained from CZTS solar cells by Jimbo *et al.* by a radio frequency co-sputtering method, combined with vapor phase sulfurization.[19] This was performed with an in-line vacuum apparatus to give higher quality CZTS films. The film with 5.33% conversion efficiency was 2.5 μm thick with an open circuit voltage of 646 mV. The film with 5.74% conversion efficiency was Cu-poor and Zn-rich and had an open circuit voltage of 662 mV. Both films had fill factors >0.50. They also found, over a number of synthesized films, that the fill factor was inversely proportional to the composition ratio Cu/(Zn + Sn). The same group was then able to increase the conversion efficiency to 6.7% by introducing an etching step following synthesis.[20] By soaking the CZTS on a Mo substrate in deionized water after formation, metal oxide particle impurities were preferentially etched. The removal of these optically inactive species lead to a greater exposed area of CZTS and hence an increase in the overall conversion efficiency.

Yoo and Kim fabricated CZTS thin films through the sulfurization of stacked metallic films.[21] Three types of films (Cu-rich, Cu-correct and Cu-poor) were sputtered onto Mo-coated glass. After sulfurization at high temperature (570 °C), each film exhibited [112] orientation dominated growth and no presence of secondary phases such as Cu_2SnS_3, ZnS or SnS. However, the Cu-rich and Cu-correct films did show the presence of a $Cu_{2-x}S$ phase. The Cu-poor film was purely kesterite CZTS and showed the smoothest morphology on scanning electron microscopy (SEM) images. These films were postulated to have a larger grain size and fewer voids, which are essential characteristics for high-efficiency CZTS-based photovoltaics.

The deposition of CZTS films has also been achieved in a reactive magnetron co-sputtering system (Figure 12.3a and b).[22] These films had a strong [112] orientation, although secondary minor phases of $Cu_{2-x}S$ and Cu_3SnS_x were observed. Figure 12.3c shows that the film had an optical bandgap of 1.50 eV, p-type conductivity and an optical absorption coefficient >10^5 cm^{-1}. The same group then grew films *via* the same method that had a compact surface morphology and large columnar grains 100 nm in size.[23] The optical bandgap was 1.52 eV and the carrier concentration, resistivity and mobility were 3.889×10^{16} cm^{-1}, 5.4 Ω cm and 29.75 cm^2 V^{-1} s^{-1}, respectively.

CZTS films have also been prepared in a two-step sputtering and annealing process (Figure 12.3d and e).[24] First, Cu–Zn–Sn–S precursor films were deposited on Mo-coated glass substrates in an H_2S atmosphere by reactive pulsed dc magnetron c-sputtering using metallic Cu : Sn (2 : 1) alloy and pure Zn targets. The precursors were then annealed in a tube furnace under a static Ar atmosphere (0.3 atm). XRD analysis showed that, despite the low deposition temperature (120 °C), the sputtered precursor was primarily composed of a CZTS-like phase with a high defect density. Following annealing, the kesterite phase was formed with a remarkable increase in grain size to around 1 μm, comparable with the dimensions of the film thickness (Figure 12.3f and g). The best conversion efficiency (4.6%) for a fabricated device was achieved with a 3 min annealing time at 550 °C.

Wu and coworkers developed an efficient one-step vacuum-based approach to deposit CZTS films without the need to supply excess sulfur or to perform any post-sulfurization treatment.[25] The films were crystalline, phase-pure (kesterite), dense, smooth and continuous. The films were (112) oriented and were processed into devices that had power conversion efficiencies of 5.5% under Cu-poor and Zn-rich conditions. They then grew a ZnO nanorod array onto the films by seeding with an Al-doped ZnO (AZO) layer. This served as an antireflective coating and increased the power conversion efficiency to 6%. This had the added benefit of increasing the solar cell devices short circuit current density by 8.6% without any significant effect on the open circuit voltage or the fill factor.

Shin *et al.* found that the characteristics of the annealed CZTS thin films are strongly dependent on the stacking order of the precursors.[26] By sputtering onto glass substrates at room temperature, they compared three different stacking orders: Cu/SnS_2/ZnS/glass (A), ZnS/Cu/SnS_2/glass (B) and SnS_2/ZnS/Cu/glass (C). The thin films were sulfurized using a tubular rapid thermal annealing system in a mixed N_2 (95%) and H_2S (5%) atmosphere at 550 °C

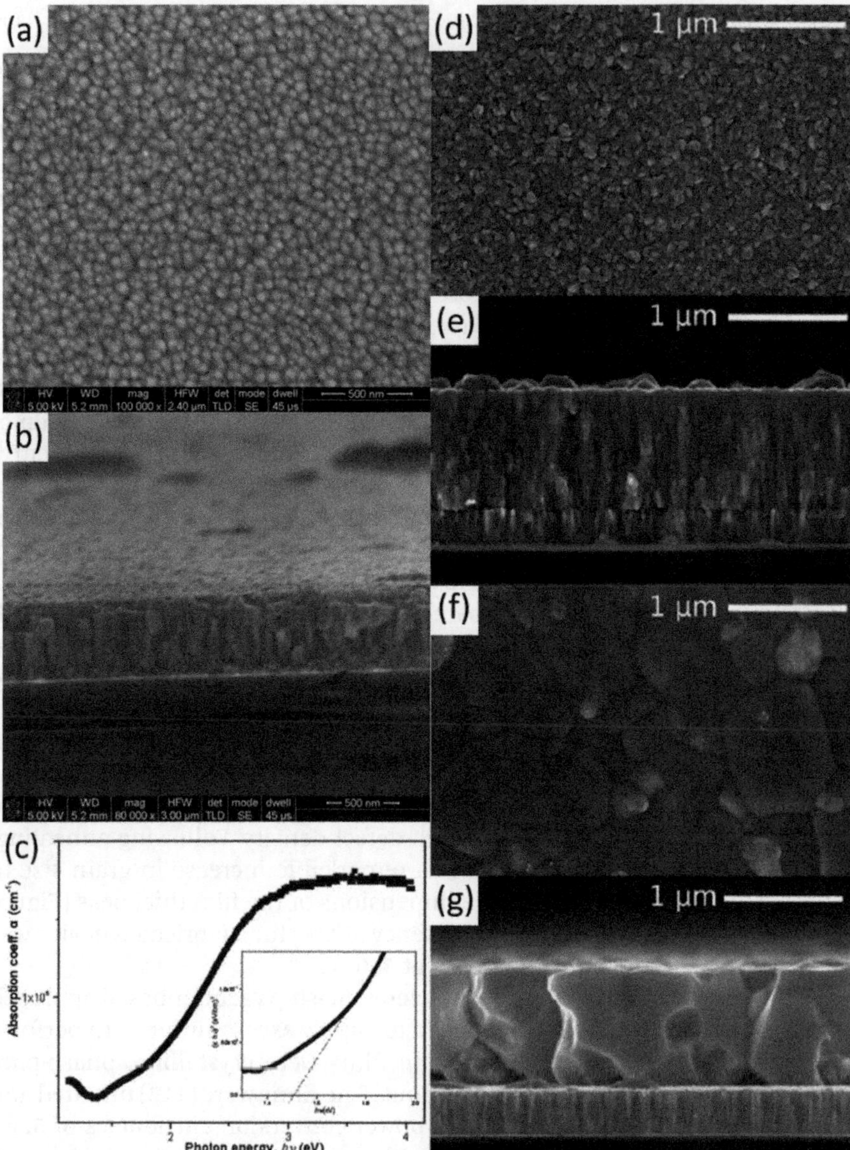

Figure 12.3 SEM images of a typical CZTS thin film deposited by dc reactive magnetron sputtering from a Cu–Zn–Sn alloy target by Liu *et al.*[22] Reproduced with permission from ref. 22. Copyright 2010. The Electrochemical Society. (a) Top view and (b) cross-section. (c) Optical absorption coefficient (α) of the CZTS thin film deposited by dc reactive magnetron sputtering from a Cu–Zn–Sn alloy target. The inset shows the calculation of the estimated bandgap of 1.50 eV by plotting $(\alpha h\nu)^2$ *versus* $h\nu$. (d) Top view and (e) cross-section of the as-sputtered CZTS precursor film of Scragg *et al.*[24] (f) Top view and (g) cross-section of CZTS films after annealing at 550 °C for 3 min. Reproduced with permission from J. J. Scragg *et al.*, Rapid annealing of reactively sputtered precursors for Cu_2ZnSnS_4 solar cells, *Progress in Photovoltaics*, John Wiley and Sons, Copyright © 2012 John Wiley & Sons, Ltd.

for 10 min. They showed that stacking order (A) had a single kesterite crystal structure with no secondary phases. Stacking orders (B) and (C), however, had a kesterite phase alongside secondary phases such as $Cu_{2-x}S$, SnS_2 and SnS. The direct bandgap was also found to depend on the stacking order with stacking orders (A), (B) and (C) having energies of 1.45, 1.35 and 1.1 eV, respectively.

Fairbrother *et al.* developed a selective chemical etching process to remove Zn-rich secondary phases, which are partly responsible for the deterioration of the power conversion efficiency, from the CZTS film surface.[27] The CZTS-based solar cells were fabricated by a two-stage process, which involved the deposition of Sn/Cu/Zn metallic stacks by dc magnetron sputtering, followed by reactive annealing under an atmosphere containing sulfur and tin. The films were then etched in HCl (5% *v*/*v*) at 75 °C. The resulting solar cells had a power conversion efficiency of 5.2%, nearly twice that of devices formed without the etching process.

CZTS thin films have also been grown on sapphire (0001) substrates by radio frequency magnetron sputtering from a single CZTS target. Song *et al.* investigated the influence of working pressure on the composition and the structural, morphological and optical properties of CZTS films.[28] They showed that a working pressure between 20 and 25 mTorr was suitable for the preparation of good quality CZTS films. Films prepared at 20 mTorr were stoichiometric single-phase CZTS, yet showed no photoluminescence. Conversely, films prepared at 25 mTorr were Zn-rich with a broad luminescence from 1.15 to 1.9 eV. The same group showed that ZnS segregation behavior had an influence on device performance in Zn-rich CZTS thin films. They showed that excess Zn diffuses to the surface of the thin films, accumulating in the form of ZnS. Too much ZnS could limit the CZTS grain size and reduce the carrier transportation performance.

Johnson *et al.* reported a systematic study into the formation mechanisms of CZTS films and the factors that control cation stoichiometry.[29] The films were formed *via ex situ* sulfirization of 250 nm thick Cu–Zn–Sn metal films that had been co-sputtered from Cu, Cu/Zn and Cu/Sn targets. Over the course of the study, >30 different precursor film compositions were synthesized and examined. Sulfurization was carried out by sealing the precursor films in an evacuated quartz ampule with solid sulfur and heated to 600 °C for 8 h. They found that the Cu/Sn ratio in the resulting CZTS films was self-regulating, approaching 2 regardless of the initial precursor composition. To achieve phase-pure, void-free films, Sn-deficient precursor films with a Cu/Zn ratio slightly <2 needed to be sulfurized in a closed system that contained excess Sn. This was achieved by including elemental Sn in the sulfurization ampule.

CZTS films have also been prepared by post-annealing films of metal sulfide (Cu_2S, ZnS and SnS_2) precursors deposited on soda lime glass substrates by photochemical deposition.[30] The metal sulfides were formed from an aqueous solution containing $CuSO_4$, $ZnSO_4$, $SnSO_4$ and NaS_2O_3, which was flowed onto the substrate and then illuminated with UV light at a wavelength of 254 nm. The sulfides were only formed in the 10 mm diameter area that experienced irradiation. The films were then annealed in a furnace at high temperature under an atmosphere of N_2 + H_2S (5%). The bandgap of the material was 1.5 eV, although the films were Zn-poor, suggesting that optimization of the technique is required.

Sekiguchi *et al.* grew epitaxial CZTS thin films on GaP substrates using pulsed laser deposition.[31] The optical bandgap was 1.5 eV and the deposition of the thin films was confirmed by powder XRD and SEM analysis. This technique was then applied to the fabrication of a thin film solar cell.[32] The CZTS absorber layer was deposited by pulsed laser deposition and the cell consisted of an Al : ZnO (n-type) window layer and a CdS buffer layer. The bandgap of the Sn-rich absorber layer was 1.5 eV. The device had an open circuit voltage of 546 mV, a fill factor of 0.48 and a power conversion efficiency of 1.74%.

Pawar *et al.* prepared CZTS films by pulsed laser deposition at room temperature using a target made of Cu_2S, ZnS and SnS_2.[33] They then investigated the effect of laser incident energies by varying the energy from 1.0 to 3.0 J cm^{-2}. They showed that the crystalline quality increased with increasing laser incident energy up to 2.5 J cm^{-2}. However, the film formed at 3.0 J cm^{-2} showed a decrease in crystalline quality due to the plasma density being too high. They also showed an increase in grain size with increasing energy. A solar device was then fabricated with a power conversion efficiency of 3.14%.[34] The films were smooth, uniform and homogeneous, with a fill factor of 0.55, despite a relatively large grain size. The quantum efficiency was reported to be 0.34%.

Scragg *et al.* formed CZTS films *via* electroplating to give a p-type absorber.[35] Soda lime glass substrates were coated with a 1 μm radiofrequency magnetron sputtered layer of Mo. Cu, Sn and Zn metallic layers were then deposited using a conventional three-electrode cell with a platinum counter electrode. The depositions were carried out at room temperature. The three-layer stack was annealed in a sulfur atmosphere at 550 °C for 2 h. The films were predominantly CZTS with only trace amounts of other components (mostly SnS_2). However, devices fabricated from these films had a relatively low efficiency of 0.8%, mainly due to a poor open circuit voltage (295 mV) and fill factor (0.32). In contrast with the evaporated films, Zn-rich films prepared by electrodeposition did not display the best performance; a Zn/Sn ratio of 1 : 1 was best. It was also shown that H_2S was more effective than S during the annealing stage.[36] A device was fabricated from this approach which had a 3.2% power conversion efficiency.[37] It had an open circuit voltage of 480 mV, a short circuit density of 15.3 mA cm^{-2} and a fill factor of 45%. Proposed improvements in the device fabrication included better control over the deposition of the Cu layer and the removal of voids at the CZTS–Mo interface.

Ennaoui *et al.* deposited ternary alloys on Mo-coated glass substrates in one step from a cyanide-free alkaline electrolyte containing Cu^{II}, Zn^{II} and Sn^{IV} metal salts.[38] A sulfur reaction was then carried out using a gas mixture of Ar–H_2S (5%) under normal pressure to form CZTS films. A solar cell was fabricated from a chemical bath deposited CdS buffer layer and a sputtered i-ZnO/ZnO:Al bilayer. The estimated bandgap of the optical material was 1.54 eV. The best performing device was Cu-poor and had a power conversion efficiency, open circuit voltage, short circuit density and fill factor of 3.4%, 563 mV, 14.8 mA cm^{-2} and 41%, respectively.

CZTS films deposited from electroplating were sulfurized by annealing with sulfur in the temperature range 300–600 °C by Araki *et al.*[39] The precursors (Cu/Sn/Zn stacked layers) were prepared by sequential electroplating from metal sulfate solutions onto glass/Mo/Pd substrates. The layers were then sulfurized for 2 h under an N_2 atmosphere. They showed that XRD peaks corresponding to CZTS did not appear until 400 °C. Increases of either Zn or Sn in the precursor concentration lead to a matched increase in the final film composition. A photovoltaic cell was prepared that had an open circuit voltage of 262 mV and an efficiency of 0.98%. In a similar process, the performance of the solar device was increased to 540 mV and 3.16% for the open circuit voltage and power conversion efficiency, respectively.[40] They used $CuSO_4{\bullet}5H_2O$, $ZnSO_4{\bullet}7H_2O$ and $SnCl_2$ as the metal precursors, which were electroplated before sulfurization at 580–600 °C. Pillar-shaped grains with a size of 0.5–1.0 μm were observed by SEM and the films had a fill factor of 46.4%.

CZTS films were deposited onto a copper-coated glass substrate using ionic liquids by Chan and coworkers in 2010.[41] The choline-based ionic liquid used was non-volatile, a good solvent for both organic and inorganic substances and was more environmentally friendly than conventional electrolytes. $CuCl_2$, $SnCl_2$ and $ZnCl_2$ were dissolved in a mixture of choline chloride and ethylene glycol (molar ratio 1:2); this was then electroplated onto the substrate using copper as the working electrode. The layers were sulfurized at 450 °C using sulfur. The XRD results showed that the films had a kesterite crystal structure and no oxide impurities, properties which are typically found in CZTS films grown by electrochemical deposition. The films had a bandgap of 1.49–1.5 eV and an absorption coefficient ~10^4 cm^{-1}.

CZTS films have also been grown for use as electrodes for solar fuel production *via* water splitting.[42] The films were prepared by inexpensive and scalable electrodeposition techniques. The CZTS precursor was deposited either sequentially or simultaneously onto Mo foil and then sulfurized using CdS deposited *via* a solution-based chemical bath deposition method. To protect the electrodes, they were covered with over-layers of AZO and TiO_2. The simultaneous deposition route had a better performance and more reproducible results. The best performing electrode (Mo/CZTS/CdS/AZO/TiO_2/Pt) exhibited a photocurrent >1 mA cm^{-2} under standard illumination conditions and showed a significant improvement in stability over unprotected electrodes.

In 2011 Ahmed *et al.* reported a then-record 7.3% power conversion efficiency for electrodeposited Cu_2ZnSnS_4 solar devices.[43] Metal stacks of copper, zinc and tin were sequentially electroplated on a 600 nm sputtered Mo layer before low temperature annealing at 210–350 °C. The samples were then sealed in a quartz tube with 2–5 mg of sulfur and annealed at 585 °C for 12 min under a nitrogen atmosphere. They then formed a series of solar cell devices with the champion device giving a remarkably high open circuit voltage of 567 mV. The fill factor was 58.1% and the photocurrent was 22 mA cm^{-2}. The relatively low fill factor was attributed to a thick MoS_2 layer at the interface between the Mo and the CZTS. This layer contributed to the overall high series resistance of the device.

Uchiki and coworkers developed sol–gel sulfurization methods for the growth of CZTS thin films under non-vacuum conditions.[44,45] The thin films were prepared by sulfurizing glasses coated with copper(II) acetate monohydrate, zinc(II) acetate dehydrate and tin(II) chloride dehydrate. These were mixed with 2-methoxyethanol and monoethanolamine, which acted as the solvent and stabilizer, respectively. This solution was then deposited by spin-coating and dried at 300 °C. The coating procedure was repeated five times to ensure an adequate thickness for sulfurization. This was carried out by annealing in an $N_2 + H_2S$ atmosphere for 1 h at 500 °C. The resulting films were stoichiometric and had a direct optical bandgap of 1.49 eV. This method was applied to the fabrication of solar cells with an Al/ZnO : Al/CdS/CZTS/Mo/ soda lime glass substrate structure.[46] By tuning the thickness of the CdS buffer layer, the highest conversion efficiency of a non-vacuum-prepared CZTS solar cell at the time was achieved. The solar cell had an open circuit voltage of 554 mV, a short current density of 6.70 mA cm^{-2}, a fill factor of 43.4% and a conversion efficiency of 1.61%. Further studies increased the conversion efficiency to 2.03% by controlling the Cu:(Zn + Sn) ratio and thus increasing the overall grain size.[47]

Cho *et al.* formulated a precursor 'paste' from copper nitrate hydrate, zinc nitrate hydrate and tin chloride dehydrate (Figure 12.4).[48] This was prepared by mixing with terpinol and ethyl cellulose in ethanol. After evaporating off the ethanol, they spin-coated the paste onto Mo-coated soda lime glass. The films were then sulfurized in an $N_2 + H_2S$ atmosphere (Figure 12.4a–d). The resulting films were densely packed with grain sizes up to 300 nm; however, some pores or voids were present. Cu, Sn and S were observed by energy-dispersive X-ray spectrometry (EDX) to be homogeneously distributed throughout the films, whereas Zn was inhomogeneously distributed. A solar cell was prepared from the films that showed a conversion efficiency of 3.02%.

Green and coworkers introduced a non-aqueous thiourea–metal–oxygen sol–gel method to form high-quality, phase-controlled CZTS thin films (reaction scheme in Figure 12.4e).[49] Solar cells fabricated from the films reached a power conversion efficiency of 5.10%. The precursor sol solution was a mixture of metal complexes dissolved in 2-methoxyethanol, with excess thiourea added to compensate for the loss of sulfur during the annealing process. To improve the fill factor and open current voltage, NaOH was added as a dopant to avoid cracks forming in the reaction and to improve the crystallinity. This form of doping had the added benefit of reducing the amount of C in the resulting CZTS films.

In 2010, Fischereder and coworkers prepared thin films of CZTS from metal salts and thioacetamide as the sulfur source in another solution-based precursor approach.[50] CuI, $ZnAc_2$, $SnCl_2$ and thioacetamide were dissolved

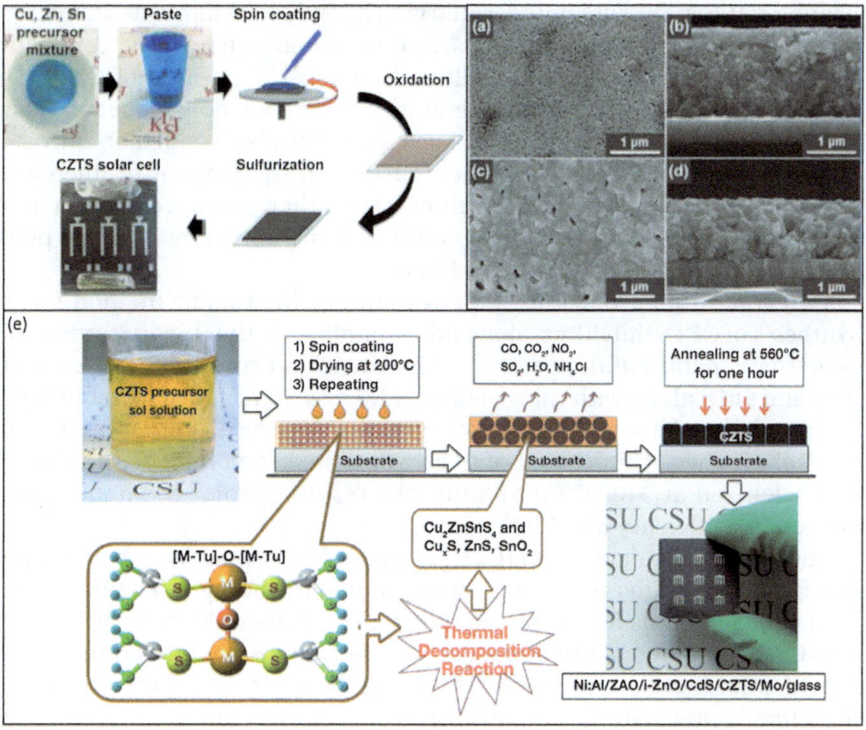

Figure 12.4 Schematic diagram showing the fabrication of CZTS films synthesized by an easily scalable and relatively safe solution-based process developed by Cho *et al.*[48] A precursor gel solution is spin-coated onto Mo-coated soda lime glass followed by a two-step heating stage (oxidation and sulfurization). (a, c) Top view and (b, d) cross-sectional SEM images of the films obtained after (a, b) air annealing and (c, d) sulfurization. Reproduced with permission from ref. 48. Copyright (2013) American Chemical Society. (e) Illustration of the formation of the CZTS thin films *via* thermal decomposition and reaction by the sol–gel route as developed by Su *et al.*[49]

in pyridine and then spin-coated onto the substrates before baking. By using thioacetamide as the sulfur source, the amount of organic impurities was reduced, more volatile decomposition products were formed and more pure CZTS materials were obtained. XRD showed that the films were kesterite CZTS with primary crystallite sizes ranging from 8 to 150 nm depending on the baking temperature. The thin films exhibited a high optical absorption ($>1 \times 10^4$ cm^{-1}) and an optical bandgap between 1.41 and 1.81 eV depending on the heat treatment. The obtained CZTS materials were copper-poor and zinc-rich in nature, the ideal stoichiometry for the use in photovoltaic applications; however, no device or electrochemical measurements were reported.

Knutson *et al.* deposited homogeneous solutions of CZTS precursors directly onto conductive films *via* selective thermolysis by microwave heating.[51] When the substrates coated with a conductive film were submerged in the reaction solution and exposed to microwave energy, the conductive films were selectively heated, resulting in the fast deposition of CZTS. The comparatively low solution temperature meant that homogeneous nucleation was avoided. Heterogeneous nucleation was also avoided on uncoated portions of the substrate due to a reduction in the amount of microwave energy absorbed. The conductive films were either tin-doped indium oxide (ITO) on glass, Mo on glass or Kapton. The films were pure kesterite phase with an average crystallite size of ~12 nm.

Tian *et al.* used a versatile and environmentally friendly method for the synthesis of CZTS thin films, alongside a number of other binary, ternary and quaternary metal sulfide species.[52] Simple oxide precursors were used (CuO, ZnO and SnO) along with thioglycolic acid. These were added to a solution of methoxyethanol and monoethanolamine and dissolved at 65 °C for 30 min. The solution was spin-coated onto an Mo-coated soda lime glass substrate and selenized at 540 °C for 15 min. The resulting solar cell had a power conversion efficiency of 6.83%.

Chemical vapor deposition (CVD) is a popular method for the deposition of high-quality films for various applications, including biomedicine, and electronic materials for sensors and batteries. It was only recently, however, that CVD was demonstrated as a useful approach for the formation of CZTS films for solar cell applications. This could be due to the difficulty in finding suitable precursors for the deposition of such a complicated ternary system. O'Brien and coworkers exploited their expertise in the field of precursor development for CVD to demonstrate that diethyldithiocarbamate complexes of Cu and Zn, and a dibutyl derivate of Sn, are suitable precursors for the formation of CZTS films by aerosol-assisted CVD (AACVD).[53] Powder XRD measurements showed the kesterite phase with no other binary sulfide present. The films were closely packed with granular crystal sizes of 150–200 nm. Elemental mapping by EDX showed that the Cu, Zn, Sn and S were evenly distributed across the surface. By changing the temperature of the AACVD process, the grain size, stoichiometry and film thickness could all be closely controlled.

The synthesis and study of semiconductor nanoparticles has seen a tremendous amount of effort in the past few decades due to the ability to tune a desired material's properties by controlling the dimensions of the nanocrystals. For semiconductor systems such as CZTS, this allows tuning of the optical and electronic properties. This has seen semiconductor nanoparticles being used effectively in various applications, including sensors, light-emitting diodes and photovoltaics.[54–56] In particular, the I–III–VI_2 family of nanoparticles has seen significant research interest due to their effective use in solar energy technologies.[57–61] However, they still suffer from the limited supply and increasing cost of the rare metals, such as Ga and In. Therefore there has recently been much effort to develop methods for the synthesis of CZTS nanoparticles.

The first report of semiconductor CZTS nanoparticles was by Guo *et al.*, who prepared them by a hot injection method and then applied the nanoparticles as an absorber in a solar cell device.[62] A solution of elemental sulfur in oleylamine was injected rapidly into an oleylamine solution containing copper(II) acetylacetonate, zinc acetylacetonate, and tin(IV) *bis*(acetylacetonate) dibromide at 225 °C. The as-synthesized nanoparticles were relatively polydisperse, ranging in size from 15 to 25 nm, with oleylamine acting as the stabilizing agent. Powder XRD measurements showed that the nanoparticles had the kesterite phase, with the estimated crystallite size matching well with transmission electron microscopy (TEM) size analysis. The absorption spectrum of the nanoparticles was measured using UV–visible spectrophotometry and the bandgap was estimated to be 1.5 eV. For the fabrication of solar cells based on nanoparticle ink, the suspended CZTS nanoparticles were applied directly to Mo-coated soda lime glass substrates by drop-casting. The films were then selenized and processed into photovoltaic devices by the chemical bath deposition of CdS, radiofrequency sputtering of ZnO and radiofrequency sputtering of ITO. The fabricated devices had a power conversion efficiency of 0.74%.

By tuning the composition of the CZTS nanoparticles and developing a robust film coating method, the same group was able to increase the total power conversion efficiency of the selenized CZTS nanoparticle film solar cells to 7.2%.[63] The overall composition of the nanoparticles was kept copper-poor and zinc-rich, in agreement with the stoichiometry identified for peak performance in thin film devices. Again, an elemental sulfur solution was injected rapidly into a mixture of Cu, Zn and Sn precursors in oleyamine at 225 °C. They then formed a dense film of nanoparticles on an Mo-coated glass substrate by knife coating and drying at 300 °C in air. The resulting solar cell devices showed a 430 mV open circuit voltage, a 21.3 mA cm^{-2} current density and a 53.9% fill factor. The same group then introduced Ge to form $Cu_2Zn(Sn_{1-x}Ge_x)S_4$ nanocrystals *via* a batch reaction in oleylamine with no additional surfactant present.[64] The nanoparticles were knife-coated onto Mo substrates and then selenized to form a dense layer. The bandgaps of the nanocrystals could then be controlled, and hence the resulting solar cells, by tuning the Ge/(Ge + Sn) ratio. This ratio also affected the overall performance of the devices. When purely Cu_2ZnGeS_4 nanoparticles were used, the power conversion efficiency was a relatively low 0.51%. However, when the Ge/(Ge + Sn) ratio was tuned to 0.7, the devices had a power conversion efficiency of 6.8% when Cu-poor and Zn-rich.

Todorov *et al.* investigated the synthesis of quaternary nanoparticle suspension precursors by a soft chemistry method and their deposition by a printing technique.[65] Metal salts were reacted directly with elemental sulfur in ethyleneglycol to obtain dispersible nanoparticles. Films were then printed on soda lime glass substrates by sliding a glass rod over 50 μm thick spacers before thermal annealing. To obtain crack-free films, a binder was added, which decreased the crystallinity and optical transmission by 30% at lower energies.

In 2009, Riha *et al.* reported a modified approach to the synthesis of CZTS nanoparticles.[66] They injected a cold solution of precursors into a hot solution of surfactant, a method that has been widely successful for the formation of semiconducting chalcogenide systems. First, copper(II) acetylacetonate, zinc acetate and tin(IV) acetate were dissolved in oleylamine at 150 °C. This was added to a solution of sulfur in oleylamine and then rapidly injected into trioctylphosphine oxide (TOPO) at 300 °C. The nanoparticles were 12.8 ± 1.8 nm in size as measured by TEM and had a 2:1:1:4 ratio of Cu, Zn, Sn and S. The measured bandgap was 1.5 eV, which indicates the absence of any quantum confinement effect. They then used this method to form nanoparticles that were fabricated into thin films by drop-casting onto fluorine-doped tin oxide (FTO) substrates.[67] The measured photocurrent efficiency was 1.8% from a cell constructed using stoichiometric nanoparticles and 3.1% for Zn-rich nanoparticles. Both films showed a significant increase in performance after low temperature (350 °C) annealing.

A similar method was used by Steinhagen and coworkers with the results shown in Figure 12.5.[68] However, in their synthesis they first mixed the reagents ($Cu(acac)_2$, $Zn(O_2CCH_3)_2$, $SnCl_2 \cdot 2H_2O$, S and oleylamine) and then heated the reaction to 280 °C under an inert atmosphere. After reacting for 1 h, they formed CZTS nanoparticles that were 10.6 ± 2.9 nm in size (Figure 12.5a–d). EDX mapping (Figure 12.5d) in a scanning tunneling electron microscope showed a stoichiometry of $Cu_{2.08}Zn_{1.01}Sn_{1.20}S_{3.70}$. They also fabricated a device with an open circuit voltage of 321 mV, a short circuit current density of 1.95 mA cm^{-2}, a fill factor of 37% and a power conversion efficiency of 0.23%.

Kameyama *et al.* showed that high temperatures were essential for the formation of a pure CZTS crystal phase in their nanoparticle synthesis.[69] Temperatures <180 °C were observed to give a secondary CuS phase. The synthesis was carried out by thermally treating a mixture of metal acetates and elemental sulfur in an oleylamine solution. This was performed under an inert atmosphere at a range of temperatures. Although they observed a large dependence of the crystal structure on the reaction temperature, there was less variation in the nanoparticle size. TEM analysis gave nanoparticle sizes of 6.6 ± 1.5, 6.6 ± 1.3, 5.6 ± 1.2 and 5.1 ± 1.2 nm at 120, 180, 240 and 300 °C, respectively. Densely packed nanoparticle films were then formed on ITO electrodes *via* layer-by-layer deposition and these showed similar behavior to p-type semiconductors.

Lu and coworkers showed that CZTS nanoparticles could be synthesized in a wurtzite phase (Figure 12.6).[70] In this structure, the sulfur anions are hexagonally close-packed to form the frame of the unit cell (see inset, Figure 12.6g). The cations occupy half of the tetrahedral interstices. The nanoparticles were synthesized using dodecanethiol, which solvates the metal precursors as thiolates and allows their injection into a range of coordinating solvents (*e.g.* oleylamine or oleic acid). Synthesis with oleylamine gave nanoprisms that were 20 nm × 28 nm in size (Figure 12.6a–c). When substituting with oleic acid, nanoplates with a thickness of 14 nm were synthesized (Figure 12.6d–f). High-angle annular dark field (HAADF) STEM EDS mapping showed an even distribution of Cu, Zn, Sn and S within the nanoparticles (Figure 12.6h). The absorption spectra showed

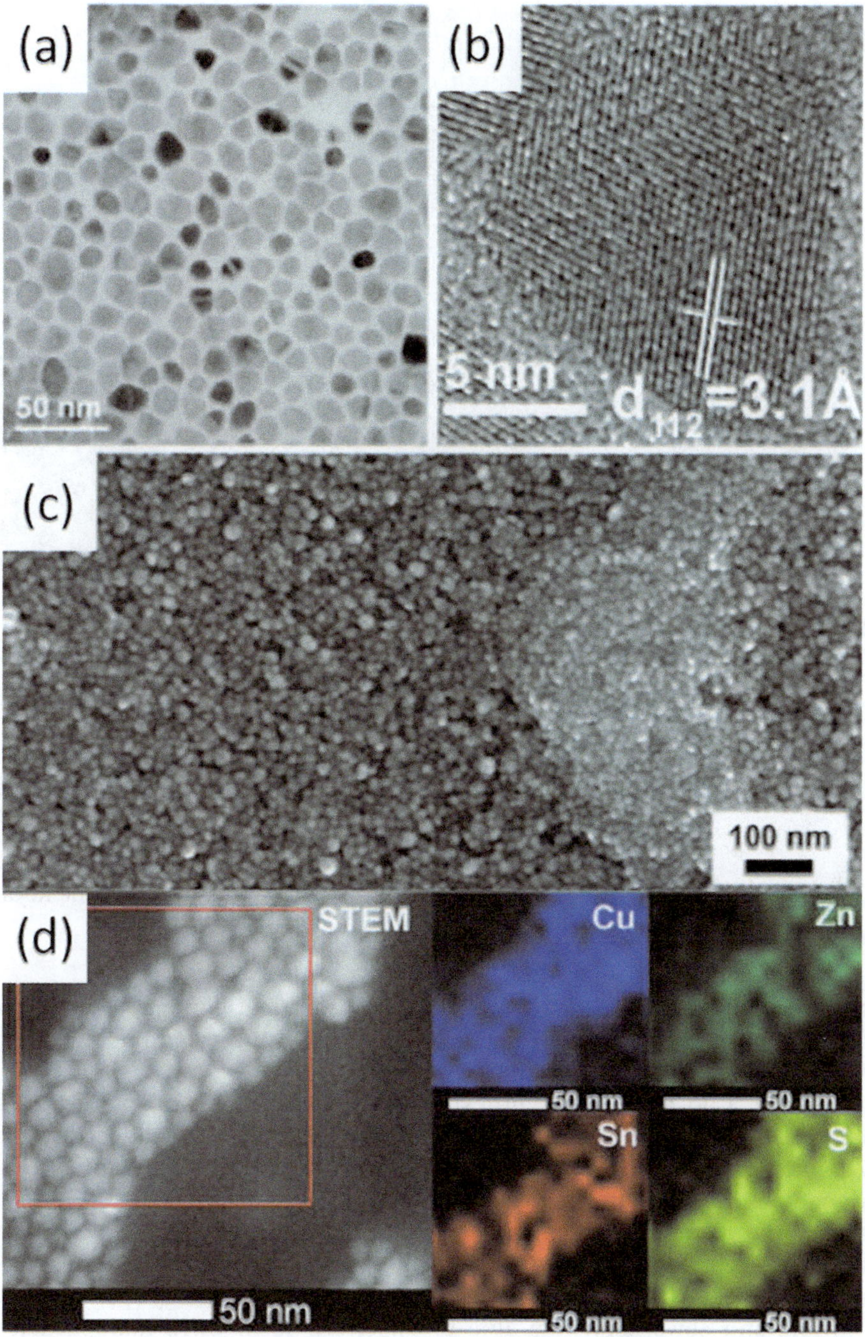

Figure 12.5 (a, b) TEM and (c) SEM images of CZTS nanocrystals synthesized by Steinhagen *et al.*[68] In (b), the nanocrystal is imaged down the [−110] crystallographic zone axis. (d) STEM EDS elemental map of CZTS nanocrystals. Reproduced with permission from ref. 68. Copyright (2009) American Chemical Society.

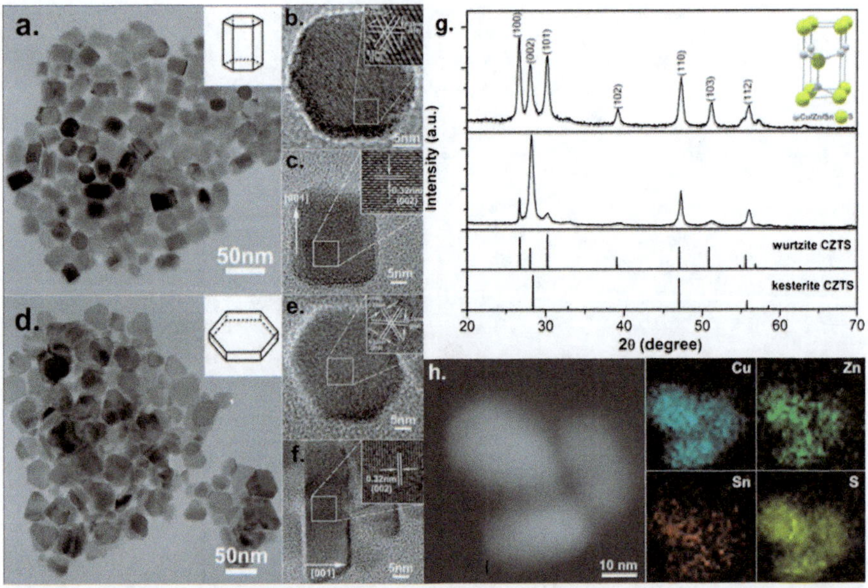

Figure 12.6 Characterization of CZTS nanoparticles synthesized by Lu *et al.* (a) TEM and (b, c) HRTEM images of nanoprisms prepared in DDT and OAm. (d) TEM and (e, f) HRTEM images of nanoplates prepared in DDT and OA. (g) XRD patterns of the CZTS nanoprisms prepared in DDT and OAm and nanoplates prepared in DDT and OA. For reference, the stimulated XRD pattern of wurtzite CZTS and the XRD pattern of kesterite CZTS (JCPDS 26-0575) are also shown. The scheme of the proposed crystal structure of wurtzite CZTS is given in the top right-handcorner. (h) HAADF STEM image of CZTS nanocrystals prepared in DDT and OAm and the corresponding EDS elemental mapping images. Reprinted from ref. 70 with permission from The Royal Society of Chemistry.

that the wurtzite CZTS nanoparticles had an optical bandgap of ~1.4 eV and the resistivities were 0.838 and 5.740 Ω m with oleylamine and oleic acid capping, respectively.

In 2011, Cao and Shen synthesized kesterite CZTS nanoparticles by a relatively simple and convenient solvothermal route.[71] They added $CuCl_2$, $Zn(OAc)_2$, $SnCl_4$ and S to a stainless-steel autoclave filled with 20 mL of ethylenediamine. The autoclave was sealed and maintained at 180 °C for 15 h and 5–10 nm nanoparticles were obtained. EDX analysis showed that the nanoparticles were slightly Sn-rich and S-poor. The bandgap of the CZTS nanoparticles was 1.5 eV, making them suitable for solar energy applications.

Wang and coworkers were able to selectively form CZTS nanoparticles in either the zinc blende or wurtzite crystal structures through a cation-exchange mechanism.[72] Cu_2SnS_3 nanoparticles were synthesized as structural templates by solvothermal methods. The zinc blende structure was synthesized in a hot injection method using elemental sulfur, whereas wurtzite was formed in a heat-up method using dodecanethiol as the source of sulfur. They then used zinc diethyldithiocarbamate as an Zn^+ source to partially substitute cations on the ternary copper tin sulfide (CTS) structure to form CZTS. Cu-poor and Zn-rich thin films were then formed by the deposition of a corresponding concentrated nanoparticle ink. These gave a power conversion efficiency of 2.89% when tested in a solar cell device.

A modified two-phase method was proposed by Wang *et al.*, who demonstrated the versatility of their synthesis by applying it to the formation of other polynary nanoparticle species, *i.e.* CuS, $CuInS_2$ and $Cu_{2-x}Ag_xZnSnS_4$.[73] The CZTS nanoparticles were formed by rapidly injecting sulfur powder dissolved in octadecene and octadecylamine into a reaction solution of metal salts and triethylene glycol with stirring at 220 °C. The solution immediately turned black, indicating the formation of nanoparticles, and was left to react for 1 h. On cooling, the tetraethylene glycol layer separated neatly from the black octadecene layer, allowing easy recovery of the nanoparticle product. The nanoparticles were 6.5 ± 1 nm in size and had a composition of $Cu_{2.73}Zn_{0.6}Sn_{1.14}S_{3.86}$ from EDX analysis. Larger CZTS nanoparticles 15.6 ± 1 nm in size were synthesized by the thermal decomposition of ethylxanthate complexes of Cu and Zn, along with $SnCl_2$ in oleylamine at 280 °C.[74] EDX analysis showed a Cu-poor and Zn-rich composition and an optical bandgap of 1.5 eV.

Zaberca *et al.* reported a surfactant-free synthesis of CZTS nanoparticles, along with a number of other semiconducting species.[75] In a typical experiment, CZTS nanoparticles in ethylene glycol were synthesized using the following molar ratios of metal salts: $2.0(CuCl_2{\cdot}2H_2O):1.3(ZnCl_2{\cdot}6H_2O):1.1(SnCl_2{\cdot}5H_2O:CS(NH_2)_2):2.0(TMaOH)$, where TMaOH is tetramethylammonium hydroxide. The mixture was transferred to a Teflon-lined stainless-steel autoclave and reacted for 16 h at 200 °C. By changing the solvent and reaction mixture composition, they were able to form a number of nanoparticles with differing stoichiometries. The nanoparticle sizes were large, ~250 nm, but could be dispersed in ethanol.

Zhou *et al.* used a similar mixture of metal salts, but substituted thiourea as the sulfur source and added polyvinylpyrolidone.[76] The reagents were mixed in ethylene glycol and reacted in the same reaction set-up, but at 230 °C for 24 h. The resulting nanoparticles were kesterite and displayed a bandgap of 1.48 eV. They then systematically varied the reaction conditions and synthesized a wide range of CZTS nanoparticle shapes and sizes, including hierarchical flower-like structures.[77] They suggested a growth mechanism that included the reduction of the metal salts and nucleation, followed by aggregation, oriented attachment and self-assembly to yield a structural hierarchy. The nanostructures displayed different bandgaps depending on their morphology. The flower-like nanoparticles had a bandgap of 1.47 eV, whereas sphere-like particles and nanocrystals had bandgaps of 1.53 and 1.60 eV, respectively. After forming a device, they observed that the ratio of electrical resistivity of CZTS films was 1 : 20 : 75 for flower-like particles, sphere-like particles and nanocrystals, respectively.

Li and coworkers used templates to direct the formation of single crystalline nanowire arrays of CZTS.[78] Anodic aluminum oxide templates were immersed in a precursor solution, which was then transferred to a stainless-steel Teflon-lined autoclave. The reaction was carried out by heating to 230 °C for 70 h. The morphology of the nanowire bundles was characterized by SEM and the overall crystallinity was measured using XRD. A growth mechanism was developed that suggested the nanowires nucleated homogeneously inside the template before growth occurred through the aggregation of individual nanoparticles. Single crystallinity was observed after a 48 h reaction time. The nanowires had a bandgap of 1.5 eV.

Xin *et al.* used CZTS nanoparticles as counter electrodes in dye-sensitized solar cells.[79] The nanoparticles were formed in a solution phase approach where Cu, Zn and Sn precursors were dissolved in oleylamine and heated to 225 °C under Ar. A sulfur solution was then injected rapidly and the reaction was stirred at 225 °C for 1 h. The nanoparticles were then either spin-coated or drop-casted onto FTO glass and sintered at 540 °C for 1 h in Se vapor. Solar cells were constructed using N719 Ru-based dyes, a 16 μm thick TiO_2 nanoparticle layer on the FTO and ionic liquid electrolytes. The solar cells had an impressive power conversion efficiency of 7.37%, which was slightly higher than a comparable Pt electrode.

CZTS nanoparticles with the kesterite crystal structure were formed by Camara *et al.*[80] Copper, zinc and tin chloride precursors were combined with sodium sulfide in a 2.8 : 1:1 : 6 molar ratio and transferred to a stainless-steel autoclave. This was heated to 230 °C and reacted for 24 h. They then spin-coated the nanoparticles to form a thin film and characterized the films, including photoelectrochemical measurements. They found that this Cu-rich stoichiometry gave a better crystallinity and overall performance.

Starting with dithiocarbamate complexes, Khare and coworkers synthesized colloidal CZTS nanoparticles with sizes ranging from 2 to 7 nm.[81] HRTEM showed that the nanoparticles were crystalline and exhibited lattice spacings consistent with CZTS. This was confirmed by XRD, which was also used to closely match the HRTEM size analysis using the Debye–Scherrer equation. Raman scattering was used to confirm that the nanoparticles were purely CZTS with no secondary phases (*e.g.* Cu_2S, ZnS or SnS_2) present in the system. Optical measurements showed that nanoparticles less than ~3 nm in size exhibited quantum confinement effects.

A flow reactor has been used to continuously form CZTS nanoparticles (Figure 12.7).[82] The nanoparticles were prepared *via* the reaction of metal amino complexes with elemental sulfur at 300 °C. Using this method, gram-scale amounts of spheroid nanoparticles with a controlled composition were obtained in an open-air environment. XRD and Raman scattering showed that the nanoparticles crystallized in the kesterite crystal structure. The different reaction kinetics of Cu, Zn, Sn and S allowed tuning of the nanoparticle composition by adjusting the precursor concentrations, reaction temperature and flow-rate. However, only compositions close to the stoichiometric ratio were stable after annealing.

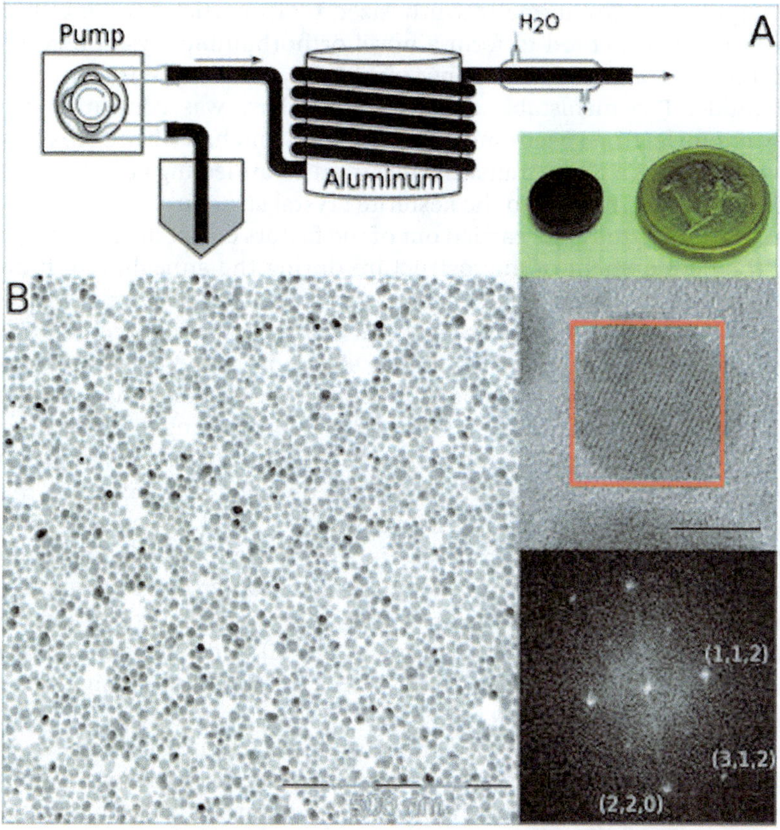

Figure 12.7 (a) Schematic diagram of the flow reactor set-up and image of a 1 g pellet made of CZTS nanoparticles formed by Shavel *et al.*[82] (b) TEM micrograph of cleaned CZTS nanoparticles prepared inside the flow reactor at 300 °C at a flow-rate of 2.0 mL min^{-1}. The inset shows an HRTEM image of a CZTS nanocrystal and the corresponding small-angle electron diffraction pattern. Reproduced with permission from ref. 82. Copyright (2012) American Chemical Society.

CZTS nanoparticles were formed in a nanorod shape and then perpendicularly aligned by controlled evaporation from solution.[83] The nanorods had a stoichiometric composition and a bandgap of 1.43 nm with a defect-free wurtzite crystal structure. The nanorods were highly uniform, which allowed easy self-assembly. To form the low-dispersity nanorods, $Cu(Acac)_2$, $Zn(Ac)_2$ and $Sn(Ac)_2$ were combined with TOPO and 1-octadecene in a round-bottomed flask. This was heated to 240–260 °C and a mixture of dodecanethiol and *tert*-dodecyl mercaptan was rapidly injected at 150–160 °C. After this injection, the reaction was allowed to continue for 15–30 min. Single-phase nanorods were only formed when the thiols were injected at <200 °C because this avoided the formation of unwanted Cu_2S side-products.

Jiang *et al.* hydrothermally synthesized CZTS particles in an ethylenediamine-assisted method to form a novel orthorhombic crystal phase.[84] The metal precursors were $CuCl_2$, $ZnCl_2$ and $SnCl_2$, with a sulfur source of thiocarbamide. The metastable orthorhombic phase was confirmed by XRD, X-ray photoelectron spectrometry and HRTEM and had a bandgap of 1.45 eV. They found that a phase transition occurred on annealing at 500 °C, with the particles transforming into the kesterite crystal structure.

A systematic study was carried out of the factors controlling crystal growth and the development of microstructure during the annealing of films cast from colloidal dispersions of CZTS nanoparticles.[85] The nanoparticles were formed by the decomposition of metal thiocarbamate precursors in octadecene and oleic acid. The nanoparticle solution was then drop-casted onto a range of different substrates, including soda lime glass and quartz substrates, with or without a thin Mo layer. It was shown that varying the annealing temperature, time and sulfur vapor pressure affected the resulting microstructure. It was proposed that two competing mechanisms influence the film microstructure: normal grain growth and abnormal grain growth. During normal grain growth, the grain size increases monomodally; however, at a certain point some fraction of the grains grow larger due to material transport from the nanocrystal layer, resulting in a bimodal distribution of grain sizes. However, at high temperatures (700 °C) and high sulfur pressures (500 Torr), normal grain growth was accelerated and the grain sizes could reach ~500 nm, significantly reducing abnormal grain growth.

Exarhos and coworkers used thiocarbamate precursors in their synthesis of CZTS nanoplatelets.[86] They prepared their precursors by reacting metal chlorides with sodium diethyl-dithiocarbamate in a recipe previously reported by Khare *et al.*[81] They used spray pyrolysis to deposit the precursors directly onto Mo-coated soda lime glass substrates. They showed that films grown at 360 °C were relatively smooth, but increasing the temperature to 460 °C resulted in a dense array of nanoplatelets with an average height of 5 μm. The structures were Cu-rich at the surface, which suggests that copper sulfide species were responsible for the anisotropic growth.

The effect of precursor reactivity on the crystallinity and quality of the final CZTS material has been investigated.[87] First it was investigated how the kinetics of precursor decomposition affected the perfection of CZTS thin films in a solution-processed method. It was shown that although the use of reactive molecular precursors allowed growth under milder reaction conditions, the formation of large crystalline domains, which provide good mobility and carrier diffusion lengths, required less reactive precursors. It was then investigated how the sintering of pre-formed CZTS nanoparticles affected the quality of the resulting films. The nanoparticles were synthesized using a colloidal chemistry route with oleylamine as the surface stabilizer.[68] It was found that slow sintering kinetics inhibited the formation of large CZTS grains; however, the use of wurtzite CZTS nanoparticles significantly increased the mobility of the grain boundaries, enabling the rapid growth of the crystal domains.

Fine control over shape and structure has been reported by Zhang *et al.* in their synthesis of monodisperse CZTS nanocrystals.[88] Nanocrystals with a range of morphologies (*e.g.* spherical, rice-like and rod-like) were synthesized over a wide range of compositions. By tuning the ratio of the precursors, either the kesterite or wurtzite phase could be obtained. To achieve this, metal–oleate complex precursors were formed by reacting the corresponding metal chloride with sodium oleate in water, ethanol and hexane at 60 °C for 4 h. A homogeneous mixture of the metal oleates was then rapidly injected into hot solvents that contained a source of sulfur. The bandgaps for CZTS nanocrystals were 1.56 and 1.45 eV for wurtzite and kesterite, respectively. The CZTS nanocrystals were shown to have a shape-dependent photoresponse behavior, with the rod-like structures performing the best.

In a heating-up method, Collord and Hillhouse showed that controlling the reaction time was crucial in obtaining CZTS nanoparticles with the desired composition.[89] There was a distinct heterogeneity present at short reaction times. Large particles containing primarily Cu and Zn were present alongside small particles of Cu and Sn. This bimodal structural and compositional distribution can be explained by the nucleation of Cu_xS early in the reaction. This was followed by the very rapid addition of Zn or the simultaneous nucleation of ZnS. By extending the reaction, they allowed the nanoparticle population to ripen, decreasing the polydispersity of the sample, but also allowing an average composition closer to that of the starting precursors. This also allowed the synthesis of specific off-stoichiometric composition CZTS nanoparticles.

A similar study by Mkawi *et al.* used a heat-up method to investigate the effect that solvent choice had on the synthesis of CZTS nanoparticles in a heat-up method.[90] Metal chloride precursors were mixed with thiourea as the sulfur source in polyvinylpyrrolidone, ethylene glycol, oleylamine or octadecene. All the solvents gave CZTS nanoparticles with the kesterite crystal structure. They then formed a solar cell device using the CZTS–oleylamine nanoparticles. The device exhibited a short circuit current density of 8.1 mA cm^{-2}, an open circuit voltage of 0.449 V, a fill factor of 0.70 and a power conversion efficiency of 2.5%.

A unique, one-step sonochemical method was reported by Liu *et al.*[91] Metal chloride precursors were mixed with thioacetamide (TAA) and 2-methoxyethanolamine (MEA) in a ratio of Cu : Zn : Sn : TAA : MEA = 2 : 1:1 : 6 : 1. The mixture was then sonicated using a system operating at 59 kHz with a power output of 99 W. The nanoparticles had the kesterite phase with an average nanoparticle size of ~2 nm and a bandgap of ~1.52 eV.

12.3 $Cu_2ZnSnSe_4$ Nanocrystals and Thin Films

$Cu_2ZnSnSe_4$ (CZTSe) is a multinary selenide compound analogous to Cu_2ZnSnS_4, with a nearly identical structural and optical properties. It has also been actively investigated for use as an absorber layer in thin film solar cells in the form of nanocrystals and thin films.

Wei *et al.*[92] reported a hot injection method for the synthesis of CZTSe nanocrystals. In their method of synthesis, Se dissolved in oleylamine was swiftly introduced into a suspension of copper acetate ($Cu(OAc)_2$, $ZnCl_2$ and $SnCl_2 \cdot 2H_2O$) in oleylamine at 100 °C and the temperature was increased to 240 °C and maintained for 2 h. The XRD pattern of the nanocrystals showed peaks from tetragonal phase CZTSe with three distinct diffraction peaks corresponding to the (112), (204) and (312) planes. The average elemental composition of the nanocrystals determined by EDX analysis was $Cu_2Zn_{0.84}Sn_{1.24}Se_{4.08}$, exhibiting a slightly Zn-poor and Sn-rich composition. TEM images of CZTSe nanocrystals showed an average size of 17 ± 2 nm with a mixture of triangular, hexagonal and plate-like shapes. The energy bandgap was determined by extrapolating the plot of $(\alpha h\nu)^2$ *versus* $h\nu$ and was 1.52 eV. Hass *et al.* reported a modified method for $Cu_2ZnSnSe_4$ nanocrystal synthesis using copper(I) acetate, zinc(II) iodide, tin(IV) acetate and elemental Se as the starting materials.[93] This approach gave $Cu_2ZnSnSe_4$ nanocrystals with an average crystalline size of 18 nm.

A number of investigations have been carried out into the growth of thin films of CZTSe using vacuum- or non-vacuum-based techniques, along with their fabrication into solar cell devices. Fella *et al.*[94] deposited CZTSe thin films by the simple, non-vacuum deposition method, in which copper(II) nitrate, zinc(II) nitrate and tin(IV) chloride were dispersed in a mixture of ethanol and 1,2-propanediol. Another viscous solution of ethyl cellulose in 1-pentanol was mixed separately and added into the metal salt mixture. The mixture was then knife-coated (600 nm thick) onto a 1 mm thick soda lime glass with a layer of Mo. The film was first dried under a lamp at 200 °C and then on a hot-plate at 230 °C to evaporate the solvents. It was then selenized in a furnace under a flow of N_2. EDX point measurements confirmed that the decomposed layer consisted of a mixture of binary ZnSe and $Cu_{2-x}Se$ phases due to very high substrate temperatures and an insufficient supply of Se. However, by optimizing the selenization conditions (a chamber pressure of 10 mbar and a substrate temperature of 470 °C), absorber layers with overall metal ratios of Cu/(Zn + Sn) = 0.88 and Zn/Sn = 1.17 were obtained. A solar cell device was fabricated with the CZTSe absorbers deposited under optimum selenization conditions, a CdS buffer layer (~70 nm thick) deposited by chemical bath deposition and an i-ZnO/ZnO:Al window bilayer (~1.2 mm thick) deposited by radiofrequency magnetron sputtering. The device showed a solar energy conversion efficiency of 4.2% with an open circuit voltage of 362 mV, a short circuit current density of 22.2 mA cm^{-2} and a fill factor of 49.6%.

Repin *et al.* fabricated a thin film solar cell by depositing CZTSe films *via* a thermal co-evaporation method.[95] The as-fabricated device showed an energy conversion efficiency of 9.15%, an open circuit voltage of 377 mV, a short circuit current density of 37.4 mA cm^{-2} and a fill factor of 64.9%.

Kim *et al.* developed a thin film solar cell with CZTSe as the photoabsorber.[96] In their method of device fabrication, spray-coated CZTS was selenized to obtain a CZTSe layer and In_2S_3 was used as a buffer layer. For the deposition of the CZTS film, a mixture of $CuCl_2$, $ZnCl_2$, $SnCl_4 \cdot 5H_2O$ and $SC(NH_2)_2$

in water was sprayed onto an Mo-coated glass using N_2 as the carrier gas at a substrate temperature of 450 °C. The CZTS film was then annealed under Se inside a graphite box. The XRD pattern of the CZTSe films showed diffraction peaks for polycrystalline CZTSe with the intensity of the (112) plane dominant. A solar cell with an absorber layer annealed under Se at 530 °C for 15 min (CZTSe-15) showed efficiencies of 0.27% (with an open circuit voltage of 0.08 V and a short circuit current density of 12.48 mA cm^{-2}) and cells using CZTSe annealed for 25 min (CZTSe-25) showed an efficiency of 2.39% (with an open circuit voltage of 0.36 V, and a short circuit current density of 16.91 mA cm^{-2}).

A CZTSe solar cell with an energy conversion efficiency of 7.0% has been fabricated by electrodepositing metal layers, followed by annealing under an Se atmosphere.[97] The CZTSe device was made with the configuration Al : ZnO 450 nm/i-ZnO 80 nm/CdS 60–70 nm/$Cu_2ZnSnSe_4$ 1.4–1.5 mm/$MoSe_2$, along with an additional barrier layer of TiN (15 nm) on the Mo back contact to suppress $MoSe_2$. The solar cell device exhibited the following properties: an open circuit voltage of 369 mV, a short circuit current density of 32.4 mA cm^{-2} and a fill factor of 58.5%.

Brammertz *et al.* fabricated solar cells with a CZTSe absorber layer deposited by the selenization of sputtered Cu, Zn and Sn multilayers.[98] Four different layer structures were studied, including a basic reference layer consisting of a 110 nm Cu/150 nm Zn/170 nm Sn multilayer deposited onto the Mo/glass substrates, with Sn being the bottom layer of the multi-stack, referenced as sample A. A second layer consisting of a similar layer structure, but with only 110 nm of Zn as the middle layer, was referenced as sample B. Sample C consisted of a similar layer structure to sample A, but with only 125 nm of Sn as the bottom layer. Sample D was made of a more complex layer structure with 55 nm Cu/115 nm Zn/255 nm Sn/110 nm Cu/110 nm Zn. These four different layers were then annealed under vacuum for 15 min at 450 °C in a continuous flow of pure H_2Se directed onto the sample surface. EDX measurements showed that all the layers were Cu-poor and Zn-rich, with Cu/(Zn + Sn) ~0.8 and Zn/Sn ranging from 1.14 to 1.46, which can be attributed to the vaporization of species such as SnSe and Zn, which evaporate at higher vapor pressures during the selenization process. Hence the ratio Cu/(Zn + Sn) increases after annealing and Zn/Sn decreases after annealing. Solar devices were fabricated using a standard procedure for CIGS-based solar cells. This consisted of a KCN etch, the chemical bath deposition of a thin n-type CdS buffer layer and ac sputtering of 120 nm of ZnO, 250 nm of highly Al-doped ZnO, 50 nm of Ni and a ~1 μm Al finger contact. The solar cell showed an efficiency of 6.3% with a maximum short circuit current density of 31.3 mA cm^{-2}, an open circuit voltage of 0.39 V and a fill factor of 52%. Cells A and B showed efficiencies slightly >6%, whereas cell D had an efficiency of 4.4%. Cell C had the lowest efficiency of 0.7%. The short circuit current densities of cells A and B were particularly large, with values of 31.3 and 36.1 mA cm^{-2}, respectively. The same research group later fabricated a $Cu_2ZnSnSe_4$ solar cell with a total area efficiency of 9.7%.[99] For this purpose, CZTSe was prepared by the selenization of sputtered $Cu_{10}Sn_{90}$, Zn and Cu multilayers. The device exhibited an open circuit voltage of 408 meV, a short circuit current density of 38.9 mA cm^{-2} and a fill factor of 61.4%.

Kim *et al.* reported the synthesis of large grain size (>1 μm) $Cu_2ZnSnSe_4$ thin films using Cu_2SnSe_3 (CTSe) and ZnSe bilayers by annealing at the relatively low temperature of 500 °C.[100] Two different stacking orders of the precursors CTSe/ZnSe/Mo (precursor 1) and ZnSe/CTSe/Mo (precursor 2) were used and their effect on the chemical composition of CZTSe thin films was studied. The thickness of films deposited using precursors 1 and 2 showed a significant difference and precursor 1 had a lower Zn content than precursor 2. Hence the difference in the thickness was attributed to the loss of Zn; no significant change in atomic composition was observed before and after annealing. The CZTSe solar cells fabricated using the ZnSe/CTSe bilayer precursors showed a power conversion efficiency of 7.14% with an open circuit voltage of 0.331 V, a short circuit current density of 36.7 mA cm^{-2} and a fill factor of 58.8%.

Oueslati *et al.* developed a CZTSe-based solar cell device with a power conversion efficiency of >10%.[101] The metal multilayers were dc sputtered at room temperature from high-purity elemental targets on Mo/soda lime glass. The stacked metal layers were then annealed at a temperature of 460 °C for 15 min with a continuous flow of 10% H_2Se and N_2 (5 mbar) to obtain CZTSe films. A quantitative EDX elemental composition of the absorber after the KCN etching showed the ratios of the elements to be 0.83, 1.13 and 1.2 for [Cu]/[Zn + Sn], [Se]/[Cu + Zn + Sn] and [Zn]/[Sn], respectively. A slightly Zn-rich and Cu-poor layer was noted, which is typical for good performing CZTSe absorber layers. XRD of the CZTSe film revealed several distinct CZTSe peaks, a strong Mo undercoated peak and a low intensity peak from the interfacial $MoSe_2$ layer, which suggests that the selenization reaction continued through the precursor layer, reaching the Mo back contact at a relatively moderate selenization temperature (460 °C). The power conversion efficiency of the device was 10.4% with a maximum short circuit current density of 39.7 mA cm^{-2}, an open circuit voltage of 394 mV and a fill factor of 66.4%.

12.4 $Cu_2ZnSn(SSe)_4$ Nanocrystals and Thin Films

$Cu_2ZnSn(SSe)_4$ (CZTSSe) is a partially selenized Cu_2ZnSnS_4 type structure with a direct energy bandgap suitable for solar energy applications. This material has been synthesized in nanocrystal and thin film form for application in solar cell devices.

Todorov *et al.*[102] used a hybrid hydrazine-based solution–particle approach for the synthesis of a $Cu_2ZnSn(S,Se)_4$ system. Three different types of slurry/ink were used for the deposition of CZTSSe thin films with high Se content (A), an intermediate sulfoselenide content (B) and pure sulfide (C). For the sulfo-selenide films A and B, individual Cu_2S–S (1.2 M) and SnSe–Se (0.57 M) hydrazine solutions were first prepared. The $ZnSe(N_2H_4)$ particles were prepared *in situ* by the stoichiometric addition of Zn powder directly to the

Se solution. For each film, five successive layers were spin-coated at 800 rpm and annealed on a ceramic hot-plate at 540 °C. The sulfur content was adjusted for sample B using an elemental sulfur vapor atmosphere for the final heat treatment. The solar cell device prepared with ink A (high Se content) showed an efficiency of 9.3% (with an open circuit voltage of 412 mV, a short circuit current density of 36.4 mA cm^{-2} and a fill factor of 62%), whereas the cell formed with sample B (mixed sulfoselenide) showed an improved efficiency of 9.6% (with an open circuit voltage of 516 mV, a short circuit current density of 28.6 mA cm^{-2} and a fill factor of 65%).

Fan *et al.*[103] reported a colloidal approach for synthesizing non-branched, but linearly arranged, polytypic non-stoichiometric $Cu_2ZnSnS_xSe_{4-x}$ nanocrystals. A mixture of CuI, $Zn(CH_3COO)_2{\cdot}2H_2O$ and $SnCl_2{\cdot}2H_2O$ was dissolved in oleylamine in the presence of 1-dodecanethiol and heated to 180 °C. At the same time, diphenyl diselenide in oleylamine was injected into the reaction mixture and the temperature was increased from 150 to 280 °C and maintained at 280 °C for 1 h. The polytypic CZTSSe nanocrystals with different ratios of zinc blende and wurtzite were obtained by keeping the reaction temperature at 240, 260, 300 and 320 °C. TEM images of the polytypic nanocrystals showed uniform rugby-ball-like morphology with long (2264 nm) and short diameters (1362 nm). The bandgap calculated from the absorption measurements increased from 1.0 to 1.2 eV with an increase in the synthesis temperature from 240 to 320 °C.

Shin *et al.* adopted a two-step process involving a solution-based precursor and a post-annealing route for the synthesis of CZTSSe nanocrystals.[104] The precursor solution was prepared using $Cu(OAc)_2$, $Zn(OAc)_2$ and $SnCl_2$ in 0.2 M thioacetamide and adjusting the pH to 7 with the addition of ammonia. For the synthesis of CZTSSe nanocrystals, the Se pellets and precursor powders were placed in a furnace under vacuum and then Ar gas was passed through as the temperature was varied from 350 to 550 °C at intervals of 50 °C for 1 h. They were then cooled naturally for 4 h. HRTEM images of CZTSSe nanocrystals formed at an Se vaporization temperature of 550 °C showed the atomic arrangement of the CZTSSe nanocrystal phase with *d*-spacing values of 0.330 and 0.326 nm. EDX and elemental mapping images of CZTSSe nanocrystals showed that Cu, Zn, Sn, S and Se were well distributed. The optical bandgap of the post-annealed CZTSSe formed with an increase in the Se vaporization temperature showed a systemic decrease from 1.46 to 1.14 eV.

Zhang *et al.* used a N_2/H_2 gas mixture to sinter CZTSSe films fabricated by a nanoparticle-based approach.[105] A colloidal dispersion of Cu–Zn–Sn–S–Se in pyridine and methanol was coated on the Mo film with an ultrasonic spray and annealed in an atmosphere containing H_2 as well as Se and Sn vapor. The metal salts of CuI, ZnI_2 and SnI_2 in pyridine were reacted with Na_2S and Na_2Se at 0 °C for 2 min to synthesize the CZTSSe nanocrystals. Two experiments, A and B, were carried out to investigate the effects of N_2 and N_2/H_2 sintering atmospheres at 500 °C for 30 min. The CZTSSe thin films developed by both the experiments were used to produce solar cells with a structure of Al grid/B-doped ZnO/i-ZnO/CdS/CZTSSe/Mo/soda lime glass. The solar cells fabricated using CZTSSe films annealed under the N_2/H_2 atmosphere showed an efficiency of 5.4% with an open circuit voltage of 415 mV, a short circuit current density of 29.6 mA cm^{-2} and a fill factor of 44%, whereas the cell using nanocrystals annealed under N_2 showed an efficiency of 0.677% with an open circuit voltage of 263 mV, a short circuit current density of 8.3 $mAcm^{-2}$ and a fill factor of 31%.

Miskin *et al.* selenized CZTS nanoparticles synthesized using copper(II) acetylacetonate, zinc acetylacetonate hydrate and tin(IV) *bis*(acetylacetonate) dichloride in oleylamine.[106,107] For selenization, CZTS nanoparticles were dispersed in 1-hexanethiol and then spread onto Mo-coated soda lime glass *via* a doctor blading method and subsequently annealed in Se vapor inside a graphite box containing the coated substrates and Se pellets at 500 °C. The device made with these films were formed by a layer of chemical bath deposited CdS (~50 nm), radiofrequency sputtered i-ZnO (~80 nm) and ITO (~220 nm), e-beam evaporated Ni/Al grids and antireflective MgF_2 (~100 nm). The solar cell device showed an efficiency of 9.0% with an open circuit voltage of 404 mV, a short circuit current density of 35.1 mA cm^{-2} and a fill factor of 63.7%.

Cao *et al.* presented a new solution-based method to fabricate CZTSSe thin films using binary and ternary metal chalcogenide nanoparticles as precursors.[108] A key advantage of this approach was the facile control of the metal ratios in the final CZTSSe films. For this purpose, nanoparticles of CTS, ZnS, SnS, CuS and Cu_7S_4 were synthesized by reacting metal salts with sulfur in oleylamine and TOPO as a first step. The CZTSSe thin films were prepared by spin-coating an ink containing an appropriate mixture of binary and ternary nanoparticles, followed by annealing in the presence of Se. XRD showed that the films contained CZTSSe as the dominant phase, along with some secondary phases. The solar cell fabricated using these thin films had a total area efficiency of 8.5% (Figure 12.8).

A hydrazine-based solution processing approach developed by Mitzi and coworkers was optimized further to fabricate more efficient solar cells.[109–112] For instance, the use of In_2S_3/CdS double emitters on CZTSSe solar cells addressed the open circuit voltage deficit and thereby improved the overall performance of the device. In addition, the presence of an In_2S_3 layer enhanced the carrier concentration of both the CZTSSe absorber and the CdS emitter layers. The solar cell device with this configuration showed a maximum efficiency of 12.7%, with an open circuit voltage of 466 mV, a short circuit current density of 38.9 mA cm^{-2} and a fill factor of 69.8%.

Gao *et al.* fabricated solar cells by a screen-printing and high-pressure sintering process using ball-milled CZTSSe.[113] CZTSSe powder was synthesized from elemental powders by planetary ball-milling and post-heating at 200, 300, 400, 500 and 600 °C for 5 h in an N_2 gas atmosphere. The milling was conducted in a planetary ball-mill at a rotational speed of 950 rpm and a milling period of 40 min. Particulate precursor ink was prepared by mixing the obtained Cu-poor and Zn-rich composition of $Cu_{1.9}Zn_{1.25}Sn(S_{0.4}Se_{0.6})_{4.5}$ powders with ethylene glycol monophenyl ether by planetary ball-milling at a rotational speed of 800 rpm for 5 h. The precursor CZTSSe layer was

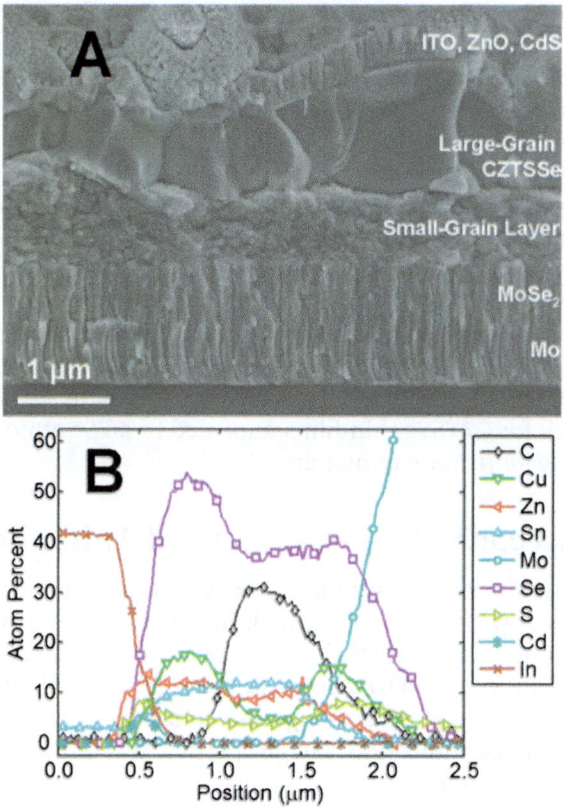

Figure 12.8 (A) Cross-section SEM image of solar cell device fabricated by Cao *et al.*[108] using a nanocrystal approach. (B) Auger depth profile of the solar cell device. Reproduced with permission from ref. 108. Copyright (2012) American Chemical Society.

deposited on a soda lime glass substrate by a screen-printing technique using the precursor ink. The organic solvent was removed from the screen-printed CZTSSe films by heating at ~110 °C in an N_2 atmosphere. The porous precursor layer was sintered into a dense polycrystalline film at 100 °C for 30 min by high-pressure sintering using conventional hot-press equipment. The solar cell device fabricated by this method showed an efficiency of 2.63%, with an open circuit voltage of 372 mV, a short circuit current density of 18.7 mA cm^{-2} and a fill factor of 37.8%.

Bilayered CZTSSe films were deposited using a mixture of binary and ternary chalcogenide nanoparticle solutions as precursors followed by annealing in an Se atmosphere.[114] In this method, a mixture of Cu_2SnS_3, ZnS and SnS was added to a tetrahydrofuran solvent. Layers of ink were then bar-coated onto an Mo-coated soda lime glass substrate and annealed under an Se atmosphere at 550 °C. The bandgap of the CZTSSe absorber was ~1.07 eV. The solar cell device constructed by this method showed an efficiency of 9.02%

with an open circuit voltage of 445 mV, a short circuit current density of 32.01 mA cm^{-2} and a fill factor of 63.3%.

Gurav *et al.* synthesized $Cu_2ZnSn(S_xSe_{1-x})_4$ thin films by selenizing a single-step electrodeposited Cu–Zn–Sn–S precursor.[115] For the single-step electrodeposition, an electrolytic bath with copper sulfate pentahydrate, zinc sulfate heptahydrate, tin(II) sulfate dihydrate and trisodium citrate dihydrate and tartaric acid was prepared in double-distilled water. Deposition was performed potentiostatically (−1.05 V *versus* the saturated calomel electrode) for 45 min without stirring in a three-electrode assembly at room temperature. The Cu–Zn–Sn–S precursor films formed were then annealed in an Ar and Se atmosphere inside a two-zone commercial furnace system at 580 °C for 1 h. The measured bandgap energy for an annealed film without using an Se atmosphere was 1.40 eV and the bandgap systematically decreased from 1.40 to 1.08 eV for CZTSSe thin films annealed in an Se atmosphere with an increasing Se vaporization temperature.

12.5 Cu_2FeSnS_4 Nanocrystals and Thin Films

By substituting the Zn cation of kesterite with Fe, the mineral analog stannite (Cu_2FeSnS_4) (CFTS) can be formed. Just as in the case of kesterite, the cations in stannite are statistically distributed over the cation sub-lattice; however, they are assigned to a different space group.[116,117] Stannite is traditionally used as an ore of Sn, although its economic importance has recently increased due to its semiconducting properties. It is antiferromagnetic with a Neel temperature ~6 K.[118]

Li *et al.* synthesized Cu_2FeSnS_4 nanoparticles by a thermal reaction of metal salts and sulfur in a hot oleylamine solution.[119] The nanoparticles were 15–25 nm in size and were characterized as having the stannite crystal structure. EDX analysis showed a composition close to the ideal 2 : 1 : 1 : 4 ratio and the bandgap was 1.33 eV, which compares well with the theoretical prediction.

A similar method was used by Yan *et al.*, although the precursors were copper(II) acetylacetonate, iron(II) acetylacetonate, and tin(II) chloride.[120] These were mixed with oleylamine and heated to 280 °C before a solution of sulfur and oleylamine was rapidly injected. The stannite structure was confirmed with Raman spectroscopy, which eliminated the possibility of secondary Cu_2SnS_3 and FeS phases. UV–visible spectrophotometry was used to estimate a bandgap of 1.28 eV, which is close to the optimum value for solar cell applications. The bandgap of Cu_2SnS_3 expected at 0.93 eV was not observed, further confirming the presence of a single stannite phase. Photocurrent densities of 11–13 μA cm^{-2} were obtained. A similar bandgap was observed from CFTS microparticles synthesized from metal chloride precursors with thiourea as the sulfur source.[121]

Zhang *et al.* were able to tune the bandgap of CFTS to 1.54 ± 0.04 and 1.46 ± 0.03 eV for oblate spheroid and triangular plate nanostructures (Figure 12.9), respectively.[122] This was the first observation of a wurtzite crystal phase

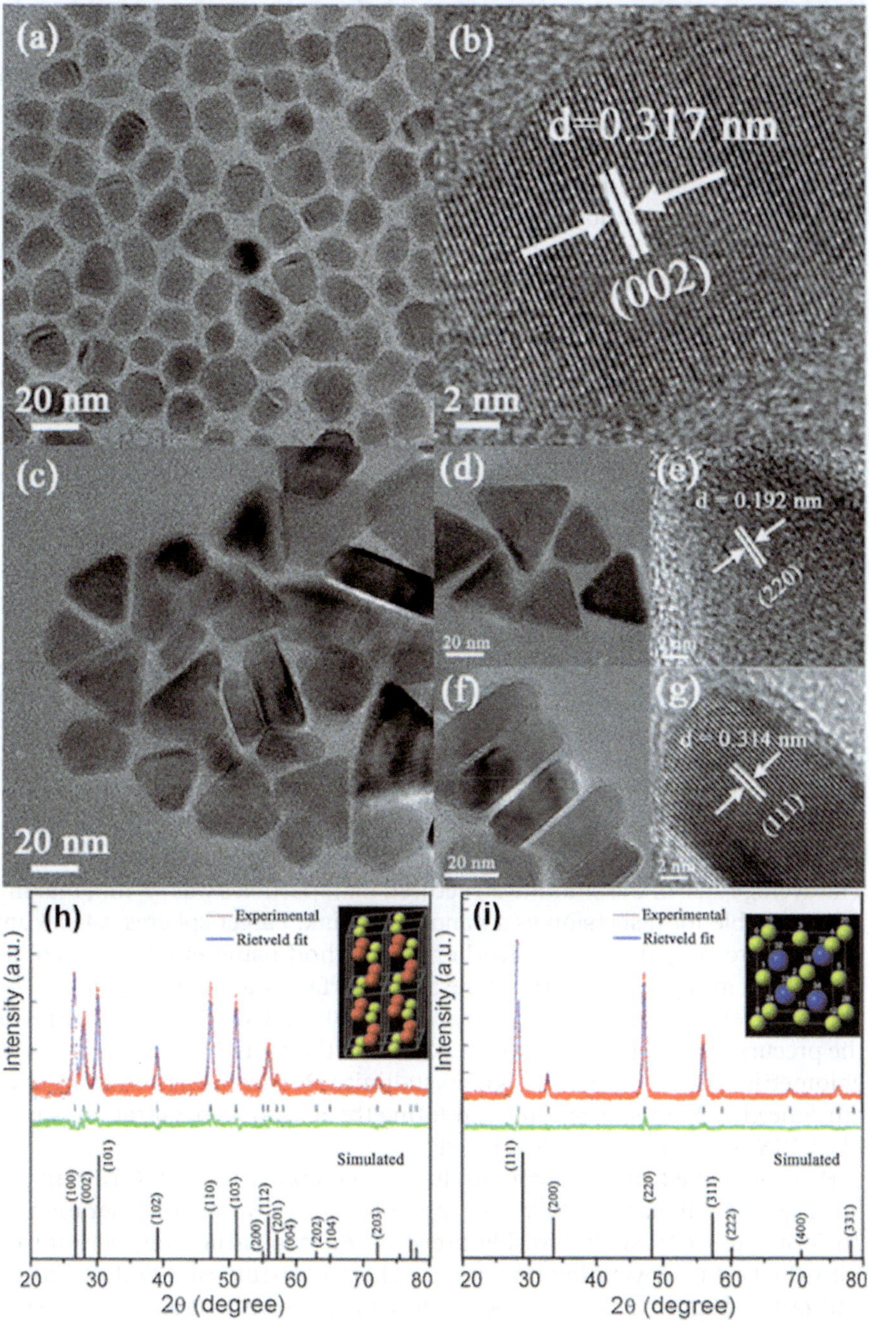

Figure 12.9 (a) TEM and (b) HRTEM images of wurtzite CFTS nanocrystals synthesized by Zhang *et al.*[122] (c) TEM image of zinc blende CFTS nanocrystals. (d) TEM and (e) HRTEM images of zinc blende CFTS from top view; (f) and (g) are the corresponding side-view images. (h) XRD pattern of wurtzite CFTS nanocrystals synthesized in OLA at 210 °C (red dots), the simulated reference pattern (black line) and the Rietveld fit (blue line). (i) XRD pattern of zinc blende CFTS nanocrystals synthesized in ODE and OA at 310 °C (red dots), the simulated reference pattern (black line) and the Rietveld fit (blue line). Insets show corresponding crystal structures. Reprinted from ref. 122 with permission from The Royal Society of Chemistry.

for Cu_2FeSnS_4. Metal salts were heated in oleylamine to 150 °C before a mixture of 1-dodecanethiol and *tert*-dodecanethiol was rapidly injected. This was then heated to 210 °C and reacted for 30 min. Formation of the wurtzite phase was favored at lower temperatures (~210 °C), with the zinc blende phase appearing at higher temperatures (~310 °C). The size and shape of the nanoparticles could be controlled through the choice of solvent and stabilizing agent.

Huang *et al.* tuned the ratio of Fe/(Zn + Fe) across the entire composition range from 0 to 1 by adjusting the Fe/Zn ratio in the metal sources.[123] By doing so, they could tune the bandgap of the $Cu_2(Fe_xZn_{1-x})SnS_4$ nanocrystals from 1.25 to 1.52 eV. The nanocrystals were synthesized by the hot injection method using elemental sulfur. In their system, CZTS and CFTS were formed in the kesterite and stannite crystal structures, respectively, with a phase transition occurring at a Fe/(Zn + Fe) ratio of 0.4. Improvement in the photoelectrical conversion performance was also observed at this ratio.

CFTS nanoparticles were formed by a ball-milling method by Mokurala *et al.* Copper acetate, iron(III) nitrate, tin chloride and thiourea were loaded with ethanol and zirconia balls (3 mm in diameter) into a polypropylene bottle.[124] The bottle was then subjected to wet pot-milling for 1–2 h at 90 rpm. After pot-milling, the homogeneous precursor solution was transferred to an alumina crucible and heated at 400 °C for 1–8 h. The CFTS nanoparticles were 8–15 nm in size and had a bandgap of 1.40 eV.

A similar bandgap of 1.35 eV was observed by Wang *et al.* in their ultrasound-assisted microwave irradiation synthesis of CFTS nanoparticles.[125] The average size of the stannite structured nanoparticles was ~5 nm, making them suitable for dispersion as a nanoparticle ink. Larger spheres 0.4–0.6 μm in size were prepared in a solvothermal method using ethylene glycol as the solvent and polyvinylpyrollidone as the surface-stabilizing ligand.[126] The particles were formed in an autoclave at 180–200 °C using metal chlorides as the precursors. The bandgap was ~1.33 eV and the particles had a nearly stoichiometric composition and high crystallinity. By increasing the amount of ethylene glycol in the reaction, *i.e.* diluting the reagent concentration, sheet-like CFTS structures could be formed.

Ha *et al.* coated spherical and multi-pod nanoparticles with Cu_2FeSnS_4 to form core–shell nanoparticles for photocatalytic hydrogen generation.[127] To form the CFTS shells, freshly prepared Au nanostructures were mixed with Cu–Fe–Sn–oleylamine precursors. Then a mixture of 1-dodecanethiol and *tert*-dodecanethiol was injected into solution with vigorous stirring at 300 °C. Such a high temperature was needed to achieve the stoichiometric CFTS phase. At lower temperatures, secondary phases such as Cu_2S and an Au–Cu alloy were observed. The presence of Au cores led to a 125 and 240% enhancement in the photocatalytic evolution of hydrogen from the spherical and multi-pod cores, respectively.

Cui *et al.* developed a general strategy for the synthesis of Cu_2MSnS_4 nanocrystals, where M = Co^{2+}, Fe^{2+}, Ni^{2+} or Mn^{2+}.[128] The nanoparticles made with

Fe and Co cation substitutions were spherical in shape, whereas the nanoparticles formed with Ni and Mn were nail-like and rod-like in structure. The particles were formed in hexylamine using 3-mercaptopropionic acid as the sulfur source. They showed that Cu_2FeSnS_4 and Cu_2CoSnS_4 were ferromagnetic at room temperature, whereas Cu_2MnSnS_4 and Cu_2NiSnS_4 displayed superparamagnetic behavior. The optical bandgaps ranged from 1.2 to 1.5 eV. A number of other studies have also investigated the use of alternative cations for the formation of ternary chalcopyrites, but this field of research is still in its infancy.[129–134]

Khadka and Kim formed polycrystalline $Cu_2Zn_xFe_{1-x}SnS_4$ thin films (where x was tuned between 0 and 1) using chemical spray pyrolysis and post-sulfurization.[135] Metal chloride salts were deposited on Mo-coated soda lime glass substrates using a laboratory-made ultrasonic spray system. They were then sulfurized under sulfur vapor at ~500 °C for 25 min. By increasing the Zn content, they showed an increase in the bandgap from 1.36 to 1.51 eV. They also observed an expected change in the lattice parameters with increasing Zn content and identified a phase transition from stannite to kesterite at a Zn/(Zn + Fe) ratio of 0.55. By fitting the experimental data on the range of observed bandgaps using the 'bandgap bowing model', they were able to infer a good miscibility of the alloyed constituents.

CFTS thin films have been synthesized as promising cheaper alternatives to replace Pt as the counter electrode in dye-sensitized solar cells. Prabhakar *et al.* reached a power conversion efficiency of 8.03% with a dye-sensitized solar cell fabricated with CFTS thin films as a photocathode in an iodine/iodide electrolyte.[136] The thin films were formed by spray pyrolysis of metal chloride precursors and thiourea, followed by further sulfurization. Stoichiometric CZTS was formed when sulfurization was carried out at 400 °C, with higher temperatures leading to sulfur-rich thin films. XRD and Raman spectroscopy were used to confirm the CFTS phase, which exhibited a bandgap of 1.46 eV.

12.6 $Cu_2FeSnSe_4$ Nanocrystals and Thin Films

Liu *et al.* synthesized $Cu_2FeSnSe_4$ (CFTSe) nanocrystals by a hot injection method using a soluble alkyl ammonium selenide precursor.[137] A transparent solution of Se precursor was prepared by dissolving Se powder in a mixture of oleylamine and dimethyl aminoborane at 110 °C. The as-prepared Se solution was quickly introduced into a mixture of copper(II) acetylacetonate (Cu(acac)$_2$), iron(II) acetylacetonate (Fe(acac)$_2$) and $SnCl_2$ in oleylamine at 270 °C. This method of synthesis yielded irregular and faceted nanocrystals with an average diameter of 15.5 ± 1.9 nm. HRTEM images showed crystalline CFTSe nanocrystals with lattice fringes with an interplanar spacing of 3.27 Å, in good agreement with the (111) plane of the cubic phase of CFTSe (d_{111} = 3.28 Å). The elemental ratio measured by EDX showed Cu : Fe : Sn : Se 2.09 : 0.84 : 0.92 : 4, which is in good agreement with the calculated composition of CFTSe.

The bandgap of the CFTSe nanocrystals was 1.60 ± 0.02 eV. The photoresponse of the films was measured by drop-casting the nanocrystals onto an ITO/glass substrate and mounting in a three-electrode photoelectrochemical cell containing 0.5 M H_2SO_4 and graphite and standard calomel electrodes. The intensity of incident light was 100 mW cm^{-2} and a bias voltage of −0.4 V *versus* the standard calomel electrode was applied. Nanosheets of CFTSe were prepared by a one-pot solvothermal method using copper(II) chloride, iron(II) chloride, tin(IV) chloride and Se powder.[138] The mixture of precursors was sealed in an autoclave and maintained at different temperatures (180, 200 and 220 °C). The XRD patterns of the nanosheets showed diffraction peaks at $2\theta = 27.17, 45.07, 65.65, 72.41$ and 83.22° corresponding to the (111), (220), (311), (400), (331) and (422) planes of the tetragonal structure of CFTSe sheets. HRTEM images of the nanosheets showed an interplanar spacing of 3.27 Å corresponding to the (111) planes of the CFTSe sheets and the measured bandgap was 1.12 eV. The solvothermal method was later extended to the growth of nanoflowers and nanosheets using the same set of precursors.[139] The nanostructures showed absorption bandgap values between 1.19 and 1.25 eV.

Thin films of CFTSe were prepared by radiofrequency magnetron sputtering followed by post-annealing treatment.[140] The Cu/Fe/Sn layers were grown at room temperature on clean and vacuum-dried glass substrates and the selenization process was carried out by rapid thermal process annealing at a furnace temperature of 500 °C. The average chemical composition of the films was estimated to be 27.12 : 11.89 : 11.27 : 49.73 (Cu : Fe : Sn : Se) from EDX and the composition ratio of the thin films was controlled by changing the sputtering time of each layer when growing the precursors. The XRD patterns of CFTSe thin films showed peaks at the (112), (204) and (312) planes of the stannite structure. The lattice parameters calculated from the XRD pattern were 5.689 Å (*a*) and 11.284 Å (*c*). The bandgap of the CFTSe thin films was 1.10 eV.

Khadka *et al.* reported the growth of polycrystalline Cu_2FeSnX_4 (X = S, Se) thin films by chemical spray pyrolysis.[141] For spray pyrolysis, copper(II) acetate, iron(III) chloride hexahydrate, tin(II) chloride and thiourea were dissolved in mixture of dimethyl sulfoxide and ethanol. The solution was sprayed onto Mo-coated soda lime glass at 450 °C and the films were sulfurized or selenized at 500 °C. The Hall effect measurement on post-sulfurized CFTS and post-selenized CFTSe thin films showed p-type conductivity with a carrier concentration of 4.76×10^{21} and 8.01×10^{20} cm^{-3} and a charge carrier mobility of 0.78 and 4.58 $cm^2\ V^{-1}\ s^{-1}$.

A similar method was reported for the growth of CFTSe films using a mixture of $[Cu(PPh_3)[Ph_2P(Se)NP(Se)Ph_2]]$, $[Sn(OAc)_4]$ and $[Fe(acac)_3]$ by an AACVD method.[142] The XRD patterns of films grown at different temperatures showed peaks corresponding to stannite phase CFTSe. Bandgaps determined from UV–visible spectrophotometry were ~1.05 and ~1.18 eV for films at 350 and 400 °C, respectively. Meng *et al.* fabricated solar cells using sputter-grown CFTSe thin films.[143] The solar cell device showed an open circuit voltage of 94 mV, a short circuit current density of 0.79 mA cm^{-2} and a fill factor of 27%.

12.7 Conclusions

Cu-based multinary compounds are exciting candidates as replacements for expensive CIGS solar cells. We have reviewed a large number of different synthetic methods for their formation, including studies showing device fabrication, with the resulting solar cells exhibiting efficiencies up to 12.7% (Table 12.1). Despite a continued increase in the efficiency of the champion materials, the performance of Cu_2MSnE_4 (M = Zn or Fe and E = S or Se) based solar cells is yet to meet efficiency requirements for grid parity.

Table 12.1 Performance of solar cells fabricated using various techniques.

Material	Method	V_{OC} (mV)	J_{SC} (mA)	Fill factor (%)	η (%)	Ref.
Cu_2ZnSnS_4	e-Beam evaporation	400	6.0	27.7	0.66	8
Cu_2ZnSnS_4	e-Beam evaporation	735	14.11	35.54	2.62	9
Cu_2ZnSnS_4	Thermal evaporation	426	6.0	41	1.1	12
Cu_2ZnSnS_4	Thermal evaporation	587	17.8	65	6.8	13
Cu_2ZnSnS_4	Thermal evaporation	661	19.5	65.8	8.4	16
Cu_2ZnSnS_4	Sputtering	662	15.7	55	5.74	19
Cu_2ZnSnS_4	Sputtering	610	17.9	62	6.7	20
Cu_2ZnSnS_4	Sputtering	–	21.7	48	6.0	25
Cu_2ZnSnS_4	Sputtering	656	14.6	53.9	5.2	27
Cu_2ZnSnS_4	Pulsed laser deposition	546	6.78	48.0	1.74	32
Cu_2ZnSnS_4	Pulsed laser deposition	651	8.76	55	3.14	34
Cu_2ZnSnS_4	Electrodeposition	480	15.3	45	3.2	37
Cu_2ZnSnS_4	Electrodeposition	563	14.8	41	3.4	38
Cu_2ZnSnS_4	Electrodeposition	567	22.0	58.1	7.3	43
Cu_2ZnSnS_4	Sol–gel process	554	6.7	43.4	1.61	46
Cu_2ZnSnS_4	Spin-coating	556	13.5	40.3	3.02	48
Cu_2ZnSnS_4	Sol–gel process	650	12.9	56	5.1	49
Cu_2ZnSnS_4	Nano-ink	188	10.5	37.2	0.74	62
Cu_2ZnSnS_4	Nano-ink	430	21.3	53.9	7.2	63
$Cu_2Zn(Sn_{1-x}Ge_xS_4$	Nano-ink	640	21.5	49	6.8	64
Cu_2ZnSnS_4	Nano-ink	–	–	–	2.89	72
$Cu_2ZnSnSe_4$	Nano-ink	362	22.2	49.6	4.2	94
$Cu_2ZnSnSe_4$	Co-evaporation	377	37.4	64.9	9.15	95
$Cu_2ZnSnSe_4$	Spray-coating	360	16.9	–	2.39	96
$Cu_2ZnSnSe_4$	Electrodeposition	369	32.4	58.5	7.0	97
$Cu_2ZnSnSe_4$	Sputtering	390	31.3	52	6.3	98
$Cu_2ZnSnSe_4$	Sputtering	408	38.9	61.4	9.7	99
$Cu_2ZnSnSe_4$		331	36.7	58.8	7.14	100
$Cu_2ZnSnSe_4$	Sputtering	394	39.7	66.4	10.4	101
$Cu_2ZnSn(SSe)_4$	Hydrazine slurry	516	28.6	65	9.6	102
$Cu_2ZnSn(SSe)_4$	Nanocrystals	415	29.6	44	5.4	105

(continued)

Table 12.1 (*continued*)

Material	Method	V_{OC} (mV)	J_{SC} (mA)	Fill factor (%)	η (%)	Ref.
$Cu_2ZnSn(SSe)_4$	Nanocrystals	404	35.1	63.7	9.0	106
$Cu_2ZnSn(SSe)_4$	Hydrazine slurry	466	38.9	69.8	12.7	112
$Cu_2ZnSn(SSe)_4$	Screen-printing	372	18.7	37.8	2.63	113
$Cu_2ZnSn(SSe)_4$	Nanocrystals	445	32.0	63.3	9.02	114
$Cu_2FeSnSe_4$	Sputtering	94	0.79	27.0	-	143

There are a number of challenges in achieving such high-performing CZTS materials. Variations in the stoichiometry, *e.g.* by creating Cu-poor or Zn-rich materials, can lead to significant changes in their optical and electronic properties. These can also be influenced by changes in the macroscopic properties, such as the grain size and surface morphology. Surface effects are also important when fabricating CZTS materials into solar cell devices because voids and surface defects can lead to significant decreases in the resulting power conversion efficiencies. Deposition techniques must be optimized such that the material is deposited in the target phase, stoichiometry and morphology. As a result of the abundance of the constituent elements for CZTS relative to alternative solar cell materials, continued investigation into their synthesis and optimization provides an important opportunity on the path to sustainable energy production.

Acknowledgements

J. W. was supported by the Laboratory Directed Research and Development (LDRD) program at Sandia National Laboratories. K. R. is supported by LANL LDRD funding. This work was performed, in part, at the Center for Integrated Nanotechnologies, an Office of Science User Facility operated for the U.S. Department of Energy (DOE) Office of Science. Los Alamos National Laboratory, an affirmative action equal opportunity employer, is operated by Los Alamos National Security, LLC, for the National Nuclear Security Administration of the U.S. Department of Energy under contract DE-AC52-06NA25396. Sandia National Laboratories is a multimission laboratory managed and operated by National Technology and Engineering Solutions of Sandia, LLC., a wholly owned subsidiary of Honeywell International, Inc., for the U.S. Department of Energy's National Nuclear Security Administration under contract DE-NA-0003525.

References

1. K. Ito and T. Nakazawa, *Jpn. J. Appl. Phys., Part 1*, 1988, **27**, 2094–2097.
2. N. Nakayama and K. Ito, *Appl. Surf. Sci.*, 1996, **92**, 171–175.
3. J. Madarasz, P. Bombicz, M. Okuya and S. Kaneko, *Solid State Ionics*, 2001, **141**, 439–446.

4. N. Kamoun, H. Bouzouita and B. Rezig, *Thin Solid Films*, 2007, **515**, 5949–5952.
5. Y. B. K. Kumar, G. S. Babu, P. U. Bhaskar and V. S. Raja, *Phys. Status Solidi A*, 2009, **206**, 1525–1530.
6. Y. B. K. Kumar, G. S. Babu, P. U. Bhaskar and V. S. Raja, *Sol. Energy Mater. Sol. Cells*, 2009, **93**, 1230–1237.
7. M. T. Htay, Y. Hashimoto, N. Momose, K. Sasaki, H. Ishiguchi, S. Igarashi, K. Sakurai and K. Ito, *Jpn. J. Appl. Phys.*, 2011, **50**, 032301–032304.
8. H. Katagiri, N. Sasaguchi, S. Hando, S. Hoshino, J. Ohashi and T. Yokota, *Sol. Energy Mater. Sol. Cells*, 1997, **49**, 407–414.
9. H. Katagiri, K. Saitoh, T. Washio, H. Shinohara, T. Kurumadani and S. Miyajima, *Sol. Energy Mater. Sol. Cells*, 2001, **65**, 141–148.
10. A. J. Cheng, M. Manno, A. Khare, C. Leighton, S. A. Campbell and E. S. Aydil, *J. Vac. Sci. Technol., A*, 2011, **29**, 051203–051211.
11. H. Du, F. Yan, M. Young, B. To, C.-S. Jiang, P. Dippo, D. Kuciauskas, Z. Chi, E. A. Lund, C. Hancock, W. M. O. O. Hlaing, M. A. Scarpulla and G. Teeter, *J. Appl. Phys.*, 2014, **115**, 173502.
12. A. Weber, H. Krauth, S. Perlt, B. Schubert, I. Kotschau, S. Schorr and H. W. Schock, *Thin Solid Films*, 2009, **517**, 2524–2526.
13. K. Wang, O. Gunawan, T. Todorov, B. Shin, S. J. Chey, N. A. Bojarczuk, D. Mitzi and S. Guha, *Appl. Phys. Lett.*, 2010, **97**, 143508.
14. M. Gloeckler and J. R. Sites, *Thin Solid Films*, 2005, **480–481**, 241–245.
15. K. Wang, B. Shin, K. B. Reuter, T. Todorov, D. B. Mitzi and S. Guha, *Appl. Phys. Lett.*, 2011, **98**, 051912.
16. B. Shin, O. Gunawan, Y. Zhu, N. A. Bojarczuk, S. J. Chey and S. Guha, *Prog. Photovoltaics*, 2013, **21**, 72–76.
17. J. S. Seol, S. Y. Lee, J. C. Lee, H. D. Nam and K. H. Kim, *Sol. Energy Mater. Sol. Cells*, 2003, **75**, 155–162.
18. T. Tanaka, T. Nagatomo, D. Kawasaki, M. Nishio, Q. X. Guo, A. Wakahara, A. Yoshida and H. Ogawa, *J. Phys. Chem. Solids*, 2005, **66**, 1978–1981.
19. K. Jimbo, R. Kimura, T. Kamimura, S. Yamada, W. S. Maw, H. Araki, K. Oishi and H. Katagiri, *Thin Solid Films*, 2007, **515**, 5997–5999.
20. H. Katagiri, K. Jimbo, S. Yamada, T. Kamimura, W. S. Maw, T. Fukano, T. Ito and T. Motohiro, *Appl. Phys. Express*, 2008, **1**, 041201.
21. H. Yoo and J. Kim, *Thin Solid Films*, 2010, **518**, 6567–6572.
22. F. Y. Liu, K. Zhang, Y. Q. Lai, J. Li, Z. A. Zhang and Y. X. Liu, *Electrochem. Solid-State Lett.*, 2010, **13**, H379–H381.
23. F. Y. Liu, Y. Li, K. Zhang, B. Wang, C. Yan, Y. Q. Lai, Z. A. Zhang, J. Li and Y. X. Liu, *Sol. Energy Mater. Sol. Cells*, 2010, **94**, 2431–2434.
24. J. J. Scragg, T. Ericson, X. Fontane, V. Izquierdo-Roca, A. Perez-Rodriguez, T. Kubart, M. Edoff and C. Platzer-Bjorkman, *Prog. Photovoltaics*, 2014, **22**, 10–17.
25. B.-T. Jheng, P.-T. Liu and M.-C. Wu, *Sol. Energy Mater. Sol. Cells*, 2014, **128**, 275–282.
26. S. W. Shin, S. M. Pawar, C. Y. Park, J. H. Yun, J.-H. Moon, J. H. Kim and J. Y. Lee, *Sol. Energy Mater. Sol. Cells*, 2011, **95**, 3202–3206.

27. A. Fairbrother, E. Garcia-Hemme, V. Izquierdo-Roca, X. Fontane, F. A. Pulgarin-Agudelo, O. Vigil-Galan, A. Perez-Rodriguez and E. Saucedo, *J. Am. Chem. Soc.*, 2012, **134**, 8018–8021.
28. N. Song, X. M. Wen and X. J. Hao, *J. Alloys Compd.*, 2015, **632**, 53–58.
29. M. C. Johnson, C. Wrasman, X. Zhang, M. Manno, C. Leighton and E. S. Aydil, *Chem. Mater.*, 2015, **27**, 2507–2514.
30. K. Moriya, J. Watabe, K. Tanaka and H. Uchiki, *Phys. Status Solidi C*, 2006, **3**, 2848–2852.
31. K. Sekiguchi, K. Tanaka, K. Moriya and H. Uchiki, *Phys. Status Solidi C*, 2006, **3**, 2618–2621.
32. K. Moriya, K. Tanaka and H. Uchiki, *Jpn. J. Appl. Phys., Part 1*, 2007, **46**, 5780–5781.
33. S. M. Pawar, A. V. Moholkar, I. K. Kim, S. W. Shin, J. H. Moon, J. I. Rhee and J. H. Kim, *Curr. Appl. Phys.*, 2010, **10**, 565–569.
34. A. V. Moholkar, S. S. Shinde, A. R. Babar, K. U. Sim, H. K. Lee, K. Y. Rajpure, P. S. Patil, C. H. Bhosale and J. H. Kim, *J. Alloys Compd.*, 2011, **509**, 7439–7446.
35. J. J. Scragg, P. J. Dale, L. M. Peter, G. Zoppi and I. Forbes, *Phys. Status Solidi B*, 2008, **245**, 1772–1778.
36. J. J. Scragg, P. J. Dale and L. M. Peter, *Thin Solid Films*, 2009, **517**, 2481–2484.
37. J. J. Scragg, D. M. Berg and P. J. Dale, *J. Electroanal. Chem.*, 2010, **646**, 52–59.
38. A. Ennaoui, M. Lux-Steiner, A. Weber, D. Abou-Ras, I. Kotschau, H. W. Schock, R. Schurr, A. Holzing, S. Jost, R. Hock, T. Voss, J. Schulze and A. Kirbs, *Thin Solid Films*, 2009, **517**, 2511–2514.
39. H. Araki, Y. Kubo, A. Mikaduki, K. Jimbo, W. S. Maw, H. Katagiri, M. Yamazaki, K. Oishi and A. Takeuchi, *Sol. Energy Mater. Sol. Cells*, 2009, **93**, 996–999.
40. H. Araki, Y. Kubo, K. Jimbo, W. S. Maw, H. Katagiri, M. Yamazaki, K. Oishi and A. Takeuchi, *Phys. Status Solidi C*, 2009, **6**, 1266–1268.
41. C. P. Chan, H. Lam and C. Surya, *Sol. Energy Mater. Sol. Cells*, 2010, **94**, 207–211.
42. L. Rovelli, S. D. Tilley and K. Sivula, *ACS Appl. Mater. Interfaces*, 2013, **5**, 8018–8024.
43. S. Ahmed, K. B. Reuter, O. Gunawan, L. Guo, L. T. Romankiw and H. Deligianni, *Adv. Energy Mater.*, 2012, **2**, 253–259.
44. K. Tanaka, N. Moritake and H. Uchiki, *Sol. Energy Mater. Sol. Cells*, 2007, **91**, 1199–1201.
45. Y. Miyamoto, K. Tanaka, M. Oonuki, N. Moritake and H. Uchiki, *Jpn. J. Appl. Phys.*, 2008, **47**, 596–597.
46. N. Moritake, Y. Fukui, M. Oonuki, K. Tanaka and H. Uchiki, *Phys. Status Solidi C*, 2009, **6**, 1233–1236.
47. K. Tanaka, Y. Fukui, N. Moritake and H. Uchiki, *Sol. Energy Mater. Sol. Cells*, 2011, **95**, 838–842.

48. J. W. Cho, A. Ismail, S. J. Park, W. Kim, S. Yoon and B. K. Min, *ACS Appl. Mater. Interfaces*, 2013, **5**, 4162–4165.
49. Z. Su, K. Sun, Z. Han, H. Cui, F. Liu, Y. Lai, J. Li, X. Hao, Y. Liu and M. A. Green, *J. Mater. Chem. A*, 2014, **2**, 500–509.
50. A. Fischereder, T. Rath, W. Haas, H. Amenitsch, J. Albering, D. Meischler, S. Larissegger, M. Edler, R. Saf, F. Hofer and G. Trimmel, *Chem. Mater.*, 2010, **22**, 3399–3406.
51. T. R. Knutson, P. J. Hanson, E. S. Aydil and R. L. Penn, *Chem. Commun.*, 2014, **50**, 5902–5904.
52. Q. W. Tian, G. Wang, W. G. Zhao, Y. Y. Chen, Y. C. Yang, L. J. Huang and D. C. Pan, *Chem. Mater.*, 2014, **26**, 3098–3103.
53. K. Ramasamy, M. A. Malik and P. O'Brien, *Chem. Sci.*, 2011, **2**, 1170–1172.
54. T. Trindade, P. O'Brien and N. L. Pickett, *Chem. Mater.*, 2001, **13**, 3843–3858.
55. N. L. Pickett and P. O'Brien, *Chem. Rec.*, 2001, **1**, 467–479.
56. M. A. Malik, M. Afzaal and P. O'Brien, *Chem. Rev.*, 2010, **110**, 4417–4446.
57. M. G. Panthani, V. Akhavan, B. Goodfellow, J. P. Schmidtke, L. Dunn, A. Dodabalapur, P. F. Barbara and B. A. Korgel, *J. Am. Chem. Soc.*, 2008, **130**, 16770–16777.
58. P. M. Allen and M. G. Bawendi, *J. Am. Chem. Soc.*, 2008, **130**, 9240–9241.
59. E. A. Hernandez-Pagan, W. Wang and T. E. Mallouk, *ACS Nano*, 2011, **5**, 3237–3241.
60. Q. Guo, S. J. Kim, M. Kar, W. N. Shafarman, R. W. Birkmire, E. A. Stach, R. Agrawal and H. W. Hillhouse, *Nano Lett.*, 2008, **8**, 2982–2987.
61. Z. Jiasong, Z. Haijun, Z. Chenglong, M. Xin, P. Lang, L. Xiaojuan and X. Weidong, *J. Alloys Compd.*, 2014, **610**, 392–398.
62. Q. J. Guo, H. W. Hillhouse and R. Agrawal, *J. Am. Chem. Soc.*, 2009, **131**, 11672–11673.
63. Q. Guo, G. M. Ford, W. C. Yang, B. C. Walker, E. A. Stach, H. W. Hillhouse and R. Agrawal, *J. Am. Chem. Soc.*, 2010, **132**, 17384–17386.
64. G. M. Ford, Q. Guo, R. Agrawal and H. W. Hillhouse, *Chem. Mater.*, 2011, **23**, 2626–2629.
65. T. Todorov, M. Kita, J. Carda and P. Escribano, *Thin Solid Films*, 2009, **517**, 2541–2544.
66. S. C. Riha, B. A. Parkinson and A. L. Prieto, *J. Am. Chem. Soc.*, 2009, **131**, 12054–12055.
67. S. C. Riha, S. J. Fredrick, J. B. Sambur, Y. Liu, A. L. Prieto and B. A. Parkinson, *ACS Appl. Mater. Interfaces*, 2011, **3**, 58–66.
68. C. Steinhagen, M. G. Panthani, V. Akhavan, B. Goodfellow, B. Koo and B. A. Korgel, *J. Am. Chem. Soc.*, 2009, **131**, 12554–12555.
69. T. Kameyama, T. Osaki, K. Okazaki, T. Shibayama, A. Kudo, S. Kuwabata and T. Torimoto, *J. Mater. Chem.*, 2010, **20**, 5319–5324.
70. X. T. Lu, Z. B. Zhuang, Q. Peng and Y. D. Li, *Chem. Commun.*, 2011, **47**, 3141–3143.
71. M. Cao and Y. Shen, *J. Cryst. Growth*, 2011, **318**, 1117–1120.

72. Y.-X. Wang, M. Wei, F.-J. Fan, T.-T. Zhuang, L. Wu, S.-H. Yu and C.-F. Zhu, *Chem. Mater.*, 2014, **26**, 5492–5498.
73. X. Y. Wang, Z. C. Sun, C. Shao, D. M. Boye and J. L. Zhao, *Nanotechnology*, 2011, **22**, 245605.
74. Y. F. Liu, M. Y. Ge, Y. Yue, Y. Sun, Y. Z. Wu, X. Chen and N. Dai, *Phys. Status Solidi RRL*, 2011, **5**, 113–115.
75. O. Zaberca, A. Gillorin, B. Durand and J. Y. Chane-Ching, *J. Mater. Chem.*, 2011, **21**, 6483–6486.
76. Y. L. Zhou, W. H. Zhou, Y. F. Du, M. Li and S. X. Wu, *Mater. Lett.*, 2011, **65**, 1535–1537.
77. Y.-L. Zhou, W.-H. Zhou, M. Li, Y.-F. Du and S.-X. Wu, *J. Phys. Chem. C*, 2011, **115**, 19632–19639.
78. L. Shi, C. J. Pei, Y. M. Xu and Q. Li, *J. Am. Chem. Soc.*, 2011, **133**, 10328–10331.
79. X. Xin, M. He, W. Han, J. Jung and Z. Lin, *Angew. Chem., Int. Ed.*, 2011, **50**, 11739–11742.
80. S. M. Camara, L. Wang and X. Zhang, *Nanotechnology*, 2013, **24**, 495401.
81. A. Khare, A. W. Wills, L. M. Ammerman, D. J. Norris and E. S. Aydil, *Chem. Commun.*, 2011, **47**, 11721–11723.
82. A. Shavel, D. Cadavid, M. Ibanez, A. Carrete and A. Cabot, *J. Am. Chem. Soc.*, 2012, **134**, 1438–1441.
83. A. Singh, H. Geaney, F. Laffir and K. M. Ryan, *J. Am. Chem. Soc.*, 2012, **134**, 2910–2913.
84. H. Jiang, P. Dai, Z. Feng, W. Fan and J. Zhan, *J. Mater. Chem.*, 2012, **22**, 7502–7506.
85. B. D. Chernomordik, A. E. Beland, D. D. Deng, L. F. Francis and E. S. Aydil, *Chem. Mater.*, 2014, **26**, 3191–3201.
86. S. Exarhos, K. N. Bozhilov and L. Mangolini, *Chem. Commun.*, 2014, **50**, 11366–11369.
87. C. Jiang, W. Liu and D. V. Talapin, *Chem. Mater.*, 2014, **26**, 4038–4043.
88. X. Zhang, G. Guo, C. Ji, K. Huang, C. Zha, Y. Wang, L. Shen, A. Gupta and N. Bao, *Sci. Rep.*, 2014, **4**, 5086.
89. A. D. Collord and H. W. Hillhouse, *Chem. Mater.*, 2015, **27**, 1855–1862.
90. E. M. Mkawi, K. Ibrahim, M. K. M. Ali, M. A. Farrukh and N. K. Allam, *Chem. Phys. Lett.*, 2014, **608**, 393–397.
91. Y. C. Liu, J. Xu, Z. K. Ni, G. Fang and W. D. Tao, *J. Alloys Compd.*, 2015, **630**, 23–28.
92. H. Wei, W. Guo, Y. Sun, Z. Yang and Y. Zhang, *Mater. Lett.*, 2010, **64**, 1424–1426.
93. W. Haas, T. Rath, A. Pein, J. Rattenberger, G. Trimmel and F. Hofer, *Chem. Commun.*, 2011, **47**, 2050–2052.
94. C. M. Fella, A. R. Uhl, Y. E. Romanyuk and A. N. Tiwari, *Phys. Status Solidi A*, 2012, **209**(6), 1043–1048.
95. I. Repins, C. Beall, N. Vora, C. DeHart, D. Kuciauskas, P. Dippo, B. To, J. Mann, W. Hsu, A. Goodrich and R. Noufi, *Sol. Energy Mater. Sol. Cells*, 2012, **101**, 154–159.

96. S. Y. Kim and J. Kim, *Thin Solid Films*, 2013, **547**, 178–180.
97. L. Guo, Y. Zhu, O. Gunawan, T. Gokmen, V. R. Deline, S. Ahmed, L. T. Romankiw and H. Deligianni, *Prog. Photovoltaics: Res. Appl.*, 2014, **22**, 58–68.
98. G. Brammertz, Y. Ren, M. Buffière, S. Mertens, J. Hendrickx, H. Marko, A. E. Zaghi, N. Lenaers, C. Köble, J. Vleugels, M. Meuris and J. Poortmans, *Thin Solid Films*, 2013, **535**, 348–352.
99. G. Brammertz, M. Buffiè, S. Oueslati, H. ElAnzeery, K. Ben Messaoud, S. Sahayaraj, C. Koble, M. Meuris and J. Poortmans, *Appl. Phys. Lett.*, 2013, **103**, 163904, 1–4.
100. K. M. Kim, K. H. Liao, H. Tampo, H. Shibata and S. Niki, *Appl. Phys. Express*, 2015, **8**, 042301, 1–4.
101. S. Oueslati, G. Brammertz, M. Buffière, H. ElAnzeery, O. Touayar, C. Köble, J. Bekaert, M. Meuris and J. Poortmans, *Thin Solid Films*, 2015, **582**, 224–228.
102. T. K. Todorov, K. B. Reuter and D. B. Mitzi, *Adv. Mater.*, 2010, **22**, E156–E159.
103. F. Fan, L. Wu, M. Gong, S. Y. Chen, G. Y. Liu, H. Yao, H. Liang, Y. Wang and S. Yu, *Sci. Rep.*, 2012, **2**, 952, 1–6.
104. S. W. Shin, J. H. Han, Y. C. Park, G. L. Agawane, C. H. Jeong, J. H. Yun, A. V. Moholkar, J. Y. Lee and J. H. Kim, *J. Mater. Chem.*, 2012, **22**, 21727–21732.
105. Y. Zhang, N. Suyama, M. Goto, M. Yin and A. Yamada, *Appl. Phys. Express*, 2013, **6**, 072302, 1–3.
106. C. K. Miskin, W. Yang, C. J. Hages, N. J. Carter, C. S. Joglekar, E. A. Stach and R. Agrawalin, *Prog. Photovoltaics: Res. Appl.*, 2015, **23**, 654–659.
107. H. Katagiri, K. Jimbo, W. S. Maw, K. Oishi, M. Yamazaki, H. Araki and A. Takeuchi, *Thin Solid Films*, 2009, **517**(7), 2455–2460.
108. Y. Cao, M. S. Denny, J. V. Caspar, W. E. Farneth, Q. Guo, A. S. Ionkin, L. K. Johnson, M. Lu, I. Malajovich, D. Radu, H. D. Rosenfeld, K. R. Choudhury and W. Wu, *J. Am. Chem. Soc.*, 2012, **134**, 15644–15647.
109. D. A. R. Barkhouse, O. Gunawan, T. Gokmen, T. K. Todorov and D. B. Mitzi, *Prog. Photovoltaics: Res. Appl.*, 2012, **20**, 6–11.
110. T. K. Todorov, J. Tang, S. Bag, O. Gunawan, T. Gokmen, Y. Zhu and D. B. Mitzi, *Adv. Energy Mater.*, 2013, **3**, 34–38.
111. W. Wang, M. T. Winkler, O. Gunawan, T. Gokmen, T. K. Todorov, Y. Zhu and D. B. Mitzi, *Adv. Energy Mater.*, 2014, **4**, 13014651–13014655.
112. J. Kim, H. Hiroi, T. K. Todorov, O. Gunawan, M. Kuwahara, T. Gokmen, D. Nair, M. Hopstaken, B. Shin, Y. S. Lee, W. Wang, H. Sugimoto and D. B. Mitzi, *Adv. Mater.*, 2014, **26**, 7427–7431.
113. F. Gao, T. Maeda and T. Wada, *Jpn. J. Appl. Phys.*, 2014, **53**, 04ER11-5.
114. W. Wu, N. G. Tassi, Y. Cao, J. V. Caspar, K. Roy-Choudhury and L. Zhang, *Phys. Status Solidi RRL*, 2015, **9**, 236–240.
115. K. V. Gurav, S. W. Shin, U. M. Patil, M. P. Suryawanshi, S. M. Pawar, M. G. Gang, S. A. Vanalakar, J. H. Yun and J. H. Kim, *J. Alloys Compd.*, 2015, **631**, 178–182.

116. E. R. Infante, J. M. Delgado and S. A. L. Rivera, *Mater. Lett.*, 1997, **33**, 67–70.
117. F. Di Benedetto, G. P. Bernardini, D. Borrini, W. Lottermoser, G. Tippelt and G. Amthauer, *Phys. Chem. Miner.*, 2005, **31**, 683–690.
118. A. Caneschi, C. Cipriani, F. Di Benedetto and R. Sessoli, *Phys. Chem. Miner.*, 2004, **31**, 190–193.
119. L. Li, X. Liu, J. Huang, M. Cao, S. Chen, Y. Shen and L. Wang, *Mater. Chem. Phys.*, 2012, **133**, 688–691.
120. C. Yan, C. Huang, J. Yang, F. Liu, J. Liu, Y. Lai, J. Li and Y. Liu, *Chem. Commun.*, 2012, **48**, 2603–2605.
121. X. Jiang, W. Xu, R. Tan, W. Song and J. Chen, *Mater. Lett.*, 2013, **102**, 39–42.
122. X. Zhang, N. Bao, K. Ramasamy, Y. H. A. Wang, Y. Wang, B. Lin and A. Gupta, *Chem. Commun.*, 2012, **48**, 4956–4958.
123. C. Huang, Y. Chan, F. Liu, D. Tang, J. Yang, Y. Lai, J. Li and Y. Liu, *J. Mater. Chem. A*, 2013, **1**, 5402–5407.
124. K. Mokurala, P. Bhargava and S. Mallick, *Mater. Chem. Phys.*, 2014, **147**, 371–374.
125. W. Wang, H. Shen, H. Yao and J. Li, *Mater. Lett.*, 2014, **125**, 183–186.
126. M. Cao, C. Li, B. Zhang, J. Huang, L. Wang and Y. Shen, *J. Alloys Compd.*, 2015, **622**, 695–702.
127. E. Ha, L. Y. S. Lee, H.-W. Man, S. C. E. Tsang and K.-Y. Wong, *ACS Appl. Mater. Interfaces*, 2015, 7, 9072–9077.
128. Y. Cui, R. Deng, G. Wang and D. Pan, *J. Mater. Chem.*, 2012, **22**, 23136–23140.
129. B. Murali and S. B. Krupanidhi, *J. Appl. Phys.*, 2013, **114**, 144312.
130. B. Murali, M. Madhuri and S. B. Krupanidhi, *Cryst. Growth Des.*, 2014, **14**, 3685–3691.
131. Y. Cui, G. Wang and D. Pan, *J. Mater. Chem.*, 2012, **22**, 12471–12473.
132. M. Cao, L. Li, W. Z. Fan, X. Y. Liu, Y. Sun and Y. Shen, *Chem. Phys. Lett.*, 2012, **534**, 34–37.
133. T.-X. Wang, Y.-G. Li, H.-R. Liu, H. Li and S.-X. Chen, *Mater. Lett.*, 2014, **124**, 148–150.
134. A. Kamble, K. Mokurala, A. Gupta, S. Mallick and P. Bhargava, *Mater. Lett.*, 2014, **137**, 440–443.
135. D. B. Khadka and J. Kim, *J. Phys. Chem. C*, 2014, **118**, 14227–14237.
136. R. R. Prabhakar, L. Nguyen Huu, M. H. Kumar, P. P. Boix, S. Juan, R. A. John, S. K. Batabyal and L. H. Wong, *ACS Appl. Mater. Interfaces*, 2014, **6**, 17661–17667.
137. Y. Liu, M. Hao, J. Yang, L. Jiang, C. Yan, C. Huang, D. Tang, F. Li and Y. Liu, *Mater. Lett.*, 2014, **136**, 306–309.
138. B. Zhang, M. Cao, L. Li, Y. Sun, Y. Shen and L. Wang, *Mater. Lett.*, 2013, **93**, 111–114.
139. M. Cao, B. L. Zhang, J. Huang, Y. Sun, L. J. Wang and Y. Shen, *Chem. Phys. Lett.*, 2014, **604**, 15–21.

140. X. Meng, H. Deng, J. He, L. Zhu, L. Sun, P. Yang and J. Chu, *Mater. Lett.*, 2014, **117**, 1–3.
141. D. B. Khadka and J. Kim, *J. Alloys Compd.*, 2015, **638**, 103–108.
142. P. Kevin, S. N. Malik, M. A. Malik and P. O'Brien, *Mater. Lett.*, 2015, **152**, 60–64.
143. X. Meng, H. Cao, H. Deng, W. Zhou, J. Zhang, L. Huang, L. Sun, P. Yang and J. Chu, *Mater. Sci. Semicond. Process.*, 2015, **39**, 243–250.

CHAPTER 13

Quantum Dots for Type III Photovoltaics

G. KEDARNATH* AND VIMAL K. JAIN*

Chemistry Division, Bhabha Atomic Research Centre, Mumbai, 400 085, India
*E-mail: kedar@barc.gov.in, jainvk@barc.gov.in

13.1 Introduction

The growing worldwide demand for energy is currently of great concern, not only due to the rapid depletion of conventional sources of energy, but also due to the dramatic environmental impact of the combustion of fossil fuels on the Earth's climate. The enormity of the effect of the emission of greenhouse gases on global warming was the subject of the 2015 United Nations Climate Change Conference held in France and attended by 195 countries, where the need to 'go green' with renewable sources of energy such as solar power was emphasised. This has motivated interest in renewable sources of energy such as solar, wind and geothermal power and in fuel cells. Solar energy has immense potential because the supply of energy from the Sun to the Earth (3×10^{24} J yr^{-1}) is about 10 000 times more than is currently consumed by humankind. Thus the conversion of solar energy to electricity is a promising approach to green power generation.

Silicon-based solar cells, which have been known for more than five decades, have dominated the photovoltaic market, with the best values of energy conversion efficiencies reaching 25%, close to the theoretical limit for

Nanoscience & Nanotechnology Series No. 45
Nanostructured Materials for Type III Photovoltaics
Edited by Peter Skabara and Mohammad Azad Malik

Published by the Royal Society of Chemistry, www.rsc.org

single-junction solar cells of 31%. Although the lifetime of these first-generation solar cells is >25 years, their high cost and stringent purity requirements have been limiting factors. Thin film semiconductor-based solar cells (*e.g.* CdTe, efficiency ~16%; $CuInSe_2$, efficiency ~19%), also known as second-generation solar cells, were developed to lower costs. However, their lower conversion efficiencies necessitated the development of other types of solar cell. Nanocrystal-based photovoltaic devices, known as third-generation solar cells, offer several advantages in terms of low-temperature processing, scalability and a lower cost of manufacturing than those based on single crystals and thin films.

13.2 Historical Background of Photovoltaics

The term photovoltaics is a combination of the Greek word 'phos' (means light) and 'volta', a unit of electrical voltage after the Italian physicist A. Volta. The photovoltaic effect refers to the conversion of light into direct current (dc) electricity by photoresponsive semiconductor materials. The photovoltaic effect was first noted by the French physicist Alexandre Edmond Becquerel in 1839 while he was experimenting with metal electrodes placed in an electrolyte solution.[1] This phenomenon was observed in selenium by a number of researchers during the third quarter of the nineteenth century,[2] but the first demonstration of a selenium-based photovoltaic cell (commonly called a solar cell) with an efficiency of ~1% was made by an American scientist, Charles Fritts, in 1883. A better understanding of the photovoltaic effect was augmented by the discovery of the electron in 1898, combined with theoretical work in 1905 by Albert Einstein describing the photovoltaic effect in his landmark paper 'Concerning an heuristic point of view toward the emission and transformation of light'. The paper mentioned the different attributes of light and also a phenomenon called the photoelectric effect.

There have been continuous efforts to develop solar cells based on photovoltaic materials[3,4] since the publication of this seminal work by Einstein, but the major breakthrough occurred in the mid-1950s when the practical utilisation of photovoltaic devices was realised at Bell Laboratories. While working on silicon semiconductors for their possible application in electronics, Gerald Pearson inadvertently discovered silicon's photoelectric properties, which could be used for the development of silicon-based solar cells. The first practical silicon photovoltaic cell was demonstrated by Gerald Pearson, Daryl Chapin and Calvin Fuller of Bell Laboratories with 4.5% efficiency in 1953[5] and 6% in 1954.[6] The prominence of silicon-based photovoltaics came with their incorporation into the Vanguard I satellite in 1958. The development of silicon-based photovoltaics up until the early 1970s was primarily driven by their space applications (*e.g.* Soyuz-I in 1967 and Salyut-I in 1971), where the cost of power was not a major issue.

Solar power initially proved to be costly for terrestrial use. Elliot Berman designed a solar cell using lower grade silicon and a cheaper packaging material, which significantly reduced the cost of solar power.[7] Such solar cells boosted the terrestrial solar cell industry and were used by the gas and oil industries. The efficiency of crystalline silicon cells reached 16.7%[8] by 1976, which further improved to a record efficiency of 24.4% for a single-crystalline silicon cell in 1998.[9] A cheaper alternative to the latter is solar cells based on polycrystalline silicon with efficiencies of ~15%. They became popular in the 1990s and dominated the low-cost panel market by the mid-2000s. The efficiency of polycrystalline silicon improved and reached up to 20.3% with improved material quality, surface passivation and a reduction in the cell thickness to 99 μm.[10]

Although the global photovoltaic market was dominated by solar cells based on crystalline silicon, there was a rapid development of other approaches to improve efficiency and cost reduction, relying on the use of different materials, technologies and methodologies. Such approaches led to the next generation of photovoltaics, referred as second-generation or type II photovoltaics. They were based on thin film technologies and included cells based on amorphous silicon and polycrystalline $CuInGaSe_2$ and CdTe deposited on low-cost substrates such as glass. The quest for cheaper alternatives to crystalline silicon led the way for amorphous silicon based photovoltaics as early as the 1960s[11,12] and the first amorphous silicon solar cell was reported in 1976.[13] Efficiencies in the range 5–15% have been achieved. An efficiency of 10% with a single-junction amorphous silicon cell[14] and 12% in combination with a polycrystalline solar cell have been achieved.[15] The highest efficiency for a single-junction epitaxially grown GaAs cell is 28.3%.[16] The full potential of second-generation photovoltaics will only be seen when large-scale production with the reduced use of the active material is achieved.

Three novel approaches for low-cost photovoltaics emerged during the 1980–1990s: dye-sensitised solar cells (DSSCs) based on sintered TiO_2, organic photovoltaics (OPVs) and quantum dot solar cells. These third-generation or type III photovoltaics eventually exceeded the limits of a single-junction device and resulted in high conversion efficiencies for the same production costs of the first- and second-generation photovoltaics. The modern version of a DSSC, also known as a Grätzel cell, was originally co-invented by O'Regan and Grätzel in 1988,[17] although Deb patented a TiO_2/dye photogalvanic cell in 1978. DSSCs and OPV cells achieved efficiencies of 12 and 8%, respectively.[18,19] With the advent of nanostructures, new strategies evolved, including the replacement of the dyes, electrodes or absorbing material of the solar cells by colloidal quantum dots, self-assembled nanostructures or quantum structures leading to quantum dot solar cells (QDSCs).

13.3 Quantum Dots

The physicochemical properties of bulk materials change drastically as the size and shape of the material approaches the nano-regime as a result of the increase in the surface-to-volume ratio and changes in the motion of the charge carriers (electrons and holes). A pair consisting of an electron and an electron hole (e^-–h^+) is referred to as an exciton. An exciton is bound by a weak Columbic attraction and is an electrically neutral quasi-particle. Excitons have a finite size within the crystal defined by the exciton Bohr radius; this can vary from 2 to 50 nm depending on the material. The Bohr radius can be calculated as $a_B = \varepsilon h^2/\mu e^2$ (where ε is the dielectric constant of the bulk material, h is Planck's constant, μ is the reduced effective mass of the electron and hole and e is the charge of an electron).[20]

When one or more dimensions of a semiconductor material are reduced or become equal to the exciton Bohr radius of the corresponding bulk material, the motion of the charge carriers is spatially confined. Such an effect is referred to as quantum confinement and results in an increase in the bandgap, kinetic energy and density of the carriers. Quantum confinement occurs when one of the dimensions of the nanostructure is comparable with the de Broglie wavelength of the charge carriers of the bulk material. Quantum confined nanocrystals have size-dependent absorption and emission spectra. The quantum confinement in nanocrystals can be in one, two or three dimensions. Accordingly, three types of quantum confined nanocrystals can be isolated. The semiconductor nanocrystals confined in three dimensions are called quantum dots or zero-dimensional nanostructures. Similarly, one- and two-dimensionally confined nanocrystals are called quantum rods (one-dimensional nanostructures) and quantum wells, sheets or plates (two-dimensional nanostructures), respectively.

The fabrication of heterojunction interfaces between materials with different electronic and optical properties opened a new domain in quantum physics. The field grew extensively with investigations of quantum well transport properties,[21,22] energy quantisation, tunnelling effects in GaAs/AlGaAs quantum wells[23] and localisation and interaction effects on electrons and holes confined in one-dimensional quantum wires.[24] Ekimov and Onushchen showed quantum size effects on the absorption spectrum of excitons confined in zero-dimensional quantum boxes in a glass matrix,[25] followed by Brus in colloidal nanocrystals.[26] Reed *et al.*, who coined the term quantum dot, observed a fine structure in resonant tunnelling through three-dimensional quantum wells (quantum dots), providing indirect evidence of the discrete density of states in zero-dimensional systems.[27]

The shape-, size- and composition-dependent electronic and optical properties of quantum dots are mainly due to the difference in potential energies inside and outside the quantum dot experienced by the confined carriers. As a consequence of quantum confinement, the carriers show three important properties: (1) transformation of the continuous band-like structure of the bulk counterpart into discrete energy levels separated by an order of 10^2 meV; (2) the bandgap of a quantum dot increases with the decrease in size; and (3) electrons and holes overlap strongly, resulting in an increase in oscillator strength and hence the probability of radiative recombination. The energy levels for three-dimensional (bulk), two-dimensional (quantum wells), one-dimensional (quantum wires) and zero-dimensional (quantum dots) semiconductor nanostructures are shown in Figure 13.1.[28]

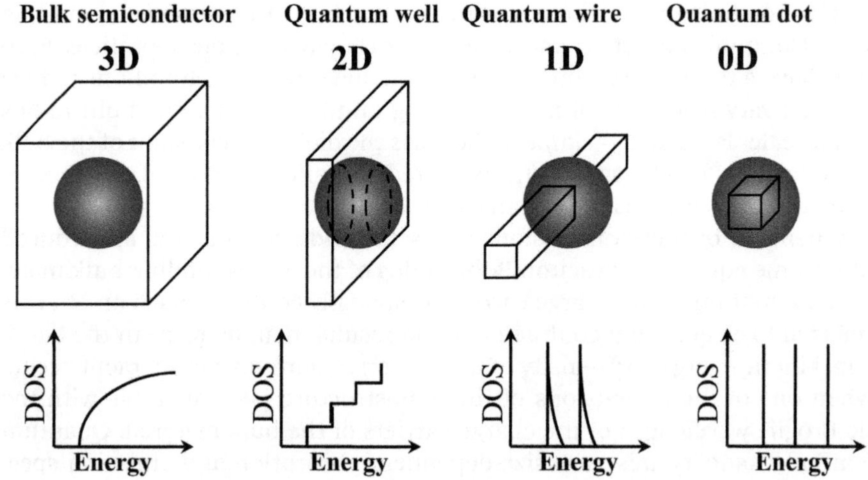

Figure 13.1 Schematic diagram of the energy level structure of bulk, two-, one- and zero-dimensional semiconductor nanostructures. The spheres represent the exciton Bohr diameter of the semiconductor and DOS is the density of states. Adapted with permission from ref. 28. Copyright (1996) American Chemical Society.

Quantum dots can weakly or strongly confine depending on whether their radii are of the order of the exciton Bohr radius or smaller. As a result of the confinement, the bandgap of a quantum dot can be suitably tuned by tailoring its size. The size-dependent bandgap of a quantum dot of radius *r* based on effective mass approximation can be given by eqn (13.1):[29,30]

$$E_{\mathrm{g}}(r) = E_{\mathrm{g}}^{0} + \frac{h^2}{8r^2}\left[\frac{1}{m_{\mathrm{e}}^{*}} + \frac{1}{m_{\mathrm{h}}^{*}}\right] + \frac{1.786e^2}{\varepsilon r} - \frac{0.124e^4}{h^2\varepsilon^2}\left[\frac{1}{m_{\mathrm{e}}^{*}} + \frac{1}{m_{\mathrm{h}}^{*}}\right]^{-1} \quad (13.1)$$

where E_{g}^{0}, m_{e}^{*}, m_{h}^{*}, ε, h and e are the bandgap of the bulk material, the effective mass of an electron, the effective mass of a hole, the dielectric constant of the bulk material, Planck's constant and the charge of an electron, respectively. Eqn (13.1) shows that the theoretical bandgap is inversely related to the square of the radius, which means that the bandgap increases with the decreasing size of the quantum dots.

Quantum confinement is not only influenced by size, but is also dependent on the shape of the nanostructure and hence on the dimensionality. The size of the confinement effect increases with the number of confinement dimensions. For instance, carriers in zero-dimensional nanostructures (quantum dots) confined in three dimensions display a stronger confinement effect than one-dimensional (*e.g.* quantum wires) or two-dimensional (*e.g.* quantum wells) materials. Accordingly, a wider range of bandgap tunability can be attained by increasing the number of confined dimensions, although some exceptions to this type of behaviour are known, such as CdTe nanocrystals.[31]

The size- and shape-dependent surface properties of quantum dots or other confined nanostructures differ greatly from their bulk counterparts due to the increase in the number of atoms or carriers on the crystal surface. As a result, a number of surface dangling bonds and defects are present on the unpassivated surface atoms of quantum dots. These surface atoms trap the charge carriers (electrons and holes) and thereby reduce the charge carrier recombination and increase the probability of non-radiative decay pathways. The surface atoms can be minimised in two ways: (1) passivation of the surface by suitable coordinating organic groups; and (2) inorganic shell passivation. Minimisation of the surface states reduces the non-radiative recombination of carriers and, consequently, significantly increases the chance of fluorescence emission.

The number of surface states is also determined by the shape or geometry of the nanocrystals. For instance, spherical quantum dots have only a small proportion of their total constituent atoms on the surface, whereas quantum rods or wires have the major fraction of their atoms on their surface. Therefore the former have less surface energy and are thermodynamically more stable than quantum rods or wires.

13.4 Classification of Photovoltaics

Since the first commercial production of solar cells in the mid-1950s, research and development in photovoltaics have been growing exponentially. The primary focus has been on boosting solar cell efficiencies (*i.e.* the conversion efficiency from solar to electrical energy), improving the performance and reliability of the photovoltaic components, and lowering the cost of solar cells and modules. Several approaches in terms of material development, processes or cell design have emerged to accomplish these goals. Table 13.1 gives details of the different kinds of materials and their solar cell efficiencies that have evolved.[16] The form of a material results in different efficiencies, *e.g.* silicon in the form of a single crystal, polycrystalline silicon and amorphous silicon have different properties (Table 13.1).

The use of a single material (*e.g.* silicon with a bandgap of 1.1 eV) in a solar cell results in a cell that can only harness solar energy from the energy window in the bandgap of the material, limiting its efficiency. In principle, the efficiency of a cell can be improved by vertically stacking single-junction cells with different bandgaps to harness different regions of solar energy. This approach results in tandem or multi-junction solar cells. A two-layer cell consisting of materials with bandgaps of 1.64 and 0.94 eV should have a theoretical performance of 44%, whereas a three-layer cell consisting of materials with bandgaps of 1.83, 1.16 and 0.71 eV is expected to have an efficiency of 48%.[32] An efficiency of 42.3% for a three-layer InGaAs/GaAs/InGaP solar cell (bandgaps 0.94, 1.42 and 1.89 eV) has been reported.[33] The use of quantum dots for high efficiency solar cell applications was noted by Barnham and Duggan in 1990.[34] As the bandgaps of quantum dots can be tuned over a wide energy range by changing their size, solar cell efficiencies can be improved by the absorption of a large portion of the solar spectrum.

Table 13.1 Solar cells based on different types of materials and their efficiencies (data from ref. 16).

Type of solar cell	Efficiency (%)
Silicon	
Crystalline silicon	25.0 ± 0.5
Polycrystalline silicon	20.4 ± 0.5
Nanocrystalline silicon	10.1 ± 0.2
Amorphous silicon	10.1 ± 0.3
Group III–V materials	
GaAs (thin films)	28.3 ± 0.8
GaAs (polycrystalline)	18.4 ± 0.5
InP (crystalline)	22.1 ± 0.7
Metal chalcogenides	
CdTe	22.1 ± 0.7
$CuInSe_2$	19.0
$CuInGaSe_2$	19.6 ± 0.6
$Cu_2ZnSnS_{4-y}Se_y$ (thin films)	10.1 ± 0.2
Multi-junction	
GaInP/GaInAs/Ge	34.1 ± 1.2
GaInP/GaAs/GaInAs	36.9 ± 1.5
Amorphous-Si/nanocrystalline Si/nanocrystalline Si (thin film)	12.3 ± 0.3
Organic	
Organic thin films	2–10
Photochemical	
Dye-sensitised solar cell	11.0 ± 0.3

Based on manufacturing costs (module and installation costs), the performance and design of photovoltaics fall in three categories: type I or first-generation photovoltaics; type II or second-generation photovoltaics; and type III or third-generation photovoltaics. First-generation photovoltaics include single-junction devices based on single-crystalline or polycrystalline silicon wafers and compound semiconductors such as GaAs. Crystalline silicon photovoltaics currently account for 84% of the world market,[35] with conversion efficiencies of 15–20% for single-crystalline and 10–14% for polycrystalline silicon solar cells. The compound semiconductor GaAs has a higher absorption coefficient, a higher carrier mobility, a record conversion efficiency of 28.8% and requires less material than crystalline silicon,[36] but its higher cost is a limiting factor. GaAs wafers of few micrometres in thicknesses absorb the same amount of sunlight as 200–300 μm thick silicon wafers. The cost of first-generation photovoltaics remains high (~US$ 4 W^{-1}).

To reduce costs, second-generation photovoltaics based on thin film technologies were developed with the aim of reducing the amount of active (absorbing) material and substrate costs while maintaining the efficiency of the first-generation photovoltaics. The former include amorphous silicon and polycrystalline compound semiconductors (*e.g.* CdTe, $CuInSe_2$ and $CuInGaSe_2$). Second-generation photovoltaics use thin films of active material (1–10 μm thick) with a high light absorption deposited on cheaper substrates such as glass, metal and plastic foils. They can be manufactured by roll-to-roll processes, which are cheaper for large-scale production, leading to production costs of US\$ 1–2 W_p^{-1} at peak energy demand and efficiencies in the range 5–12%.

Third-generation photovoltaics may be classified as photovoltaic technologies that can deliver conversion efficiencies greater than the single-junction device limit of 33%, with a lower cost per unit area and production costs of US\$ 0.20 W_p^{-1} or better while maintaining the advantages of second-generation photovoltaics. Third-generation photovoltaics include DSSCs, bulk heterojunction or OPV devices, quantum dot or nanostructured solar cells and hybrid solar cells.

13.4.1 Dye-sensitised Solar Cells

DSSCs are a low-cost photovoltaic technology with a fairly high conversion efficiency and are based on the use of TiO_2 nanocomposites and an inorganic dye.[37,38] Unlike OPV devices, where the organic molecule exhibits both charge generation and carrier transport properties, these two functions are expressed separately in DSSCs. The semiconductor dye generates a charge on the absorption of light and this is transported by the electrolyte and semiconductor.[17,37–41] DSSCs are often referred as Grätzel cells.

A modern DSSC (Figure 13.2[42]) consists of five components. The transparent anode is a glass sheet coated with a conducting oxide layer covered with a thin layer of mesoporous wide bandgap semiconductor (WBSC), such as TiO_2, SnO_2, ZnO or Nb_2O_5. The semiconductor layer is covalently bonded to a dye. An electrolyte containing iodide–triiodide (I^-/I_3^-) or a polysulfide ($S^{2-}/S^{2-}{}_x$) redox couple in an organic solvent is sandwiched between the

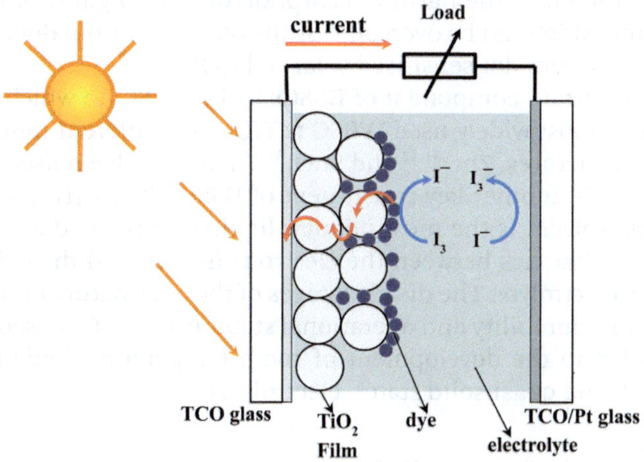

Figure 13.2 Schematic illustration of DSSC. Reprinted from *Renewable and Sustainable Energy Reviews*, 16(8), J. Gong *et al.*, Review on dye-sensitized solar cells (DSSCs): Fundamental concepts and novel materials, 5848–5860, Copyright 2012 with permission from Elsevier.

two electrodes. The cathode is a platinum-coated glass sheet.[37] The function of a typical DSSC involves four important steps. Sunlight is absorbed by the dye layer through the transparent anode. The dye molecules are excited and electrons are injected into the conduction band of the WBSC on an extremely fast timescale (typically 10 ps), while the holes are transported slowly to the counter electrode by the electrolyte.[38] The injected electrons diffuse to the anode through the TiO_2 nanoparticles and are utilised at the external load before being collected by the electrolyte on the cathode surface to complete the cycle. The iodide (I^-) of the electrolyte is oxidised to triiodide (I_3^-) by transferring an electron back to the excited dye to regenerate the dye molecules. The oxidised form of the redox couple (I_3^-) diffuses to the counter electrode and reduces to I^- in the presence of the platinum cathode as a catalyst to complete the circuit.

In DSSCs, the dye generates the photoelectrons, the WBSC transports the charge and charge separation occurs at the surfaces between the dye, the WBSC and the electrolyte. *cis*-[Ru(4,4′-dicarboxy-2,2′-bipyridine)$_2$(NCS)$_2$] (also called N_3)[43] is one of the dyes most commonly used for sensitising TiO_2 because it met the requirements for an ideal sensitiser until the inception of [NHEt$_3$][Ru(4,4′,4″-(COOH)$_3$-terpy)(NCS)$_3$] or 'black dye'.[44] DSSCs using black dye have recorded an incident photon to current conversion efficiency of 11.4%.[18] A zinc porphyrin dye (YD2-o-C8) as the sensitiser co-sensitised with a Y123 dye and a $Co^{II/III}$tris(bipyridyl)-based redox electrolyte showed a high conversion efficiency of 12.3%.[18] Although a number of metal complexes have been developed as sensitisers (Table 13.2 [43–50]), ruthenium polypyridyl complexes are the front runners due to their long-term stability and high photoconversion yield. Several variants of these dyes with a high molar extinction coefficient have also been synthesised.[37] The efficiency of DSSC is, however, limited by the narrow absorption of dyes (organic or inorganic) below 700 nm, which can be overcome by the use of quantum dots instead of dyes to give quantum dot sensitised solar cells (QDSSCs).[38]

Another important component of DSSCs is the WBSC on which the dye is anchored. The most widely used WBSC is TiO_2 with different morphologies and ordered structures. ZnO[51,52] and SnO_2[53,54] have also been used in DSSCs. The electrolyte is another key constituent of DSSCs. The I^-/I_3^- redox couple in an organic solvent is the most popular liquid electrolyte due to the slow recombination kinetics between the electrons in TiO_2 and the oxidised dye and I_3^- in the electrolyte. The disadvantages of the toxic nature of the electrolyte and the low durability and operational stability of DSSCs based on liquid electrolytes led to the development of room temperature ionic liquid,[55,56] solid state[57,58] and quasi-solid state[59] electrolytes.

13.4.2 Organic Photovoltaics

OPV technology is emerging rapidly, with cell efficiencies approaching >8%. In OPV cells an organic molecule or a polymer forms an active absorbing layer that generates an exciton on the absorption of light and produces

Table 13.2 Examples of metal complexes used as dyes in DSSCs.

Dye	Dye structure	Ref.
[Ru(4,4′-dicarboxy-2,2′-bipyridine)$_2$(NCS)$_2$]	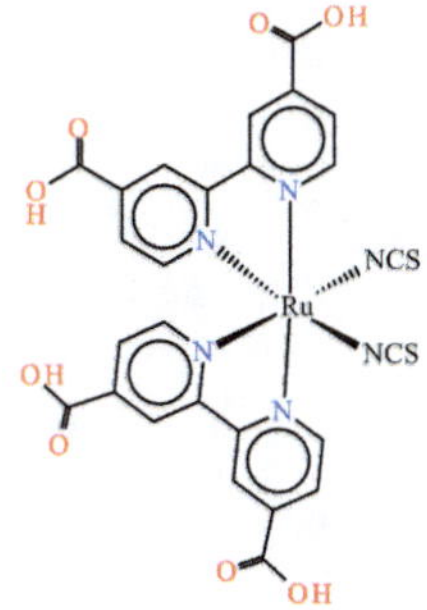	43
[$NHEt_3$][Ru(4,4′,4″–(COOH)$_3$–terpy)(NCS)$_3$]	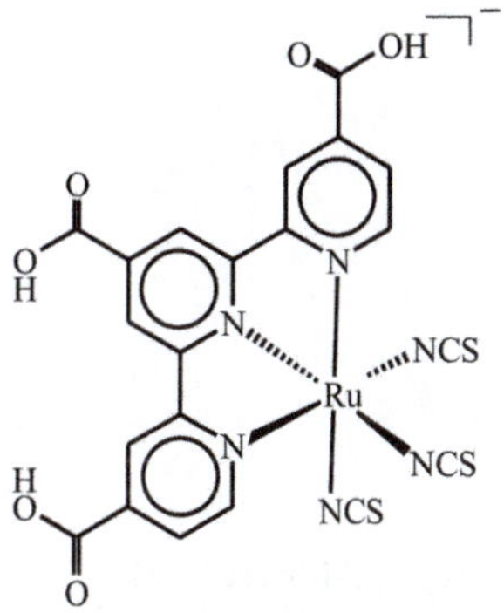	44
CYC-B11	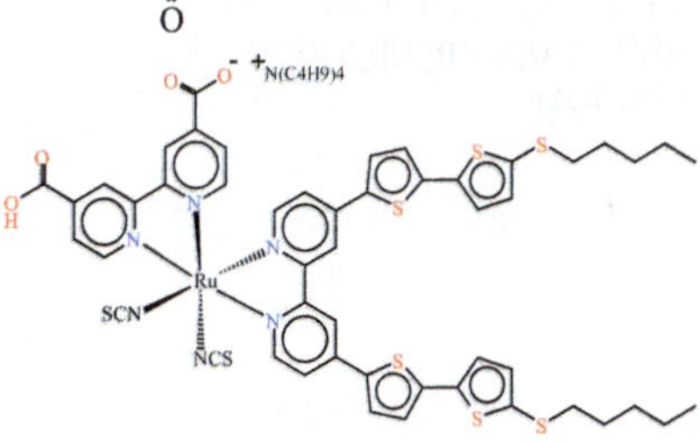	45
Z907	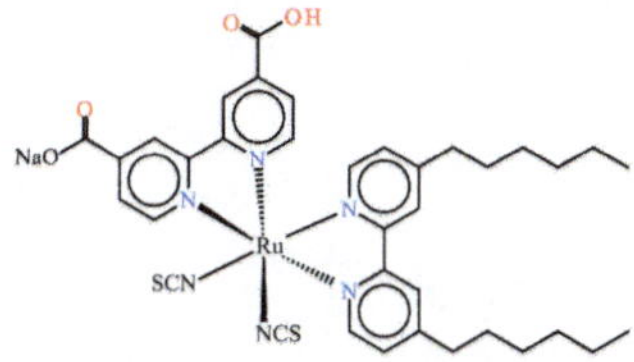	46
Z910	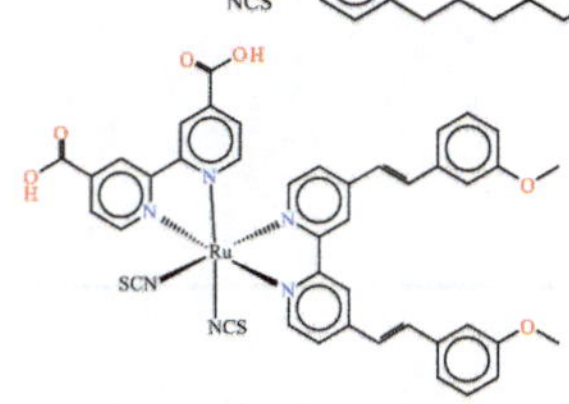	47

(*continued*)

Table 13.2 (*continued*)

Dye	Dye structure	Ref.
CYC-B1	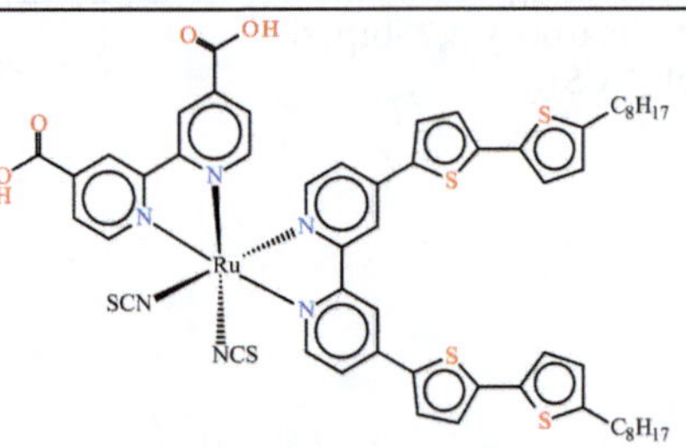	48
N719	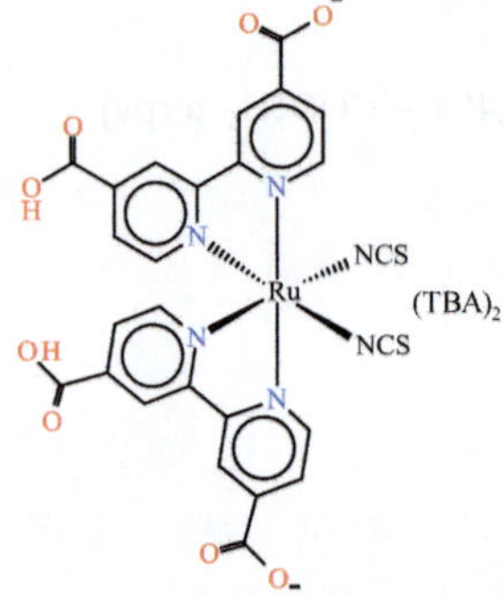	49
YE05 *Bis*(4,4′-dicarboxy-2,2′-bipyridine) 2-(2,4-difluorophenyl)pyridine ruthenium(II)	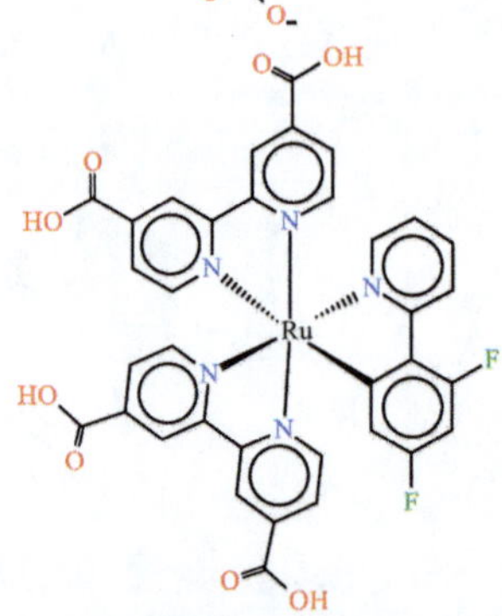	50
YD-2-*o*-C8	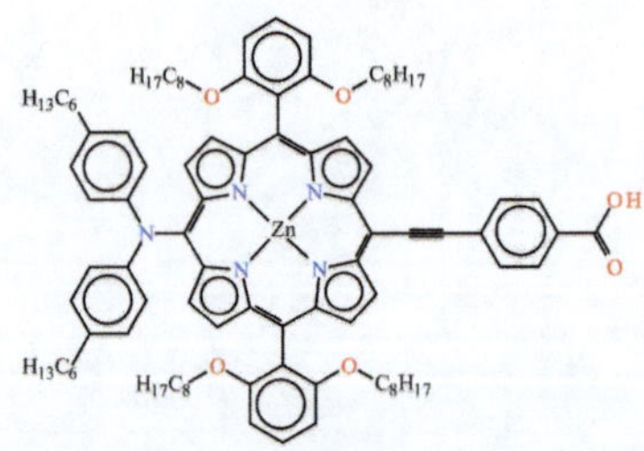	18

electricity due to the separation of charge carriers at an interface by effective fields. They are also called plastic solar cells because low-cost plastics are used as substrates to coat the organic or polymeric active material. The main advantages of organic solar cells are the diversity of organic materials that can be designed and synthesised and also the ease of processing the active material. Large-scale, low-cost roll-to-roll processes, such as printing, spraying and spin-coating on plastic sheets, can be utilised, leading to cost-effective devices. Another important advantage is the high optical absorption coefficient of the organic molecules, resulting in the absorption of large amounts of light with lower amounts of active material. However, there are also disadvantages, such as a poor shelf life and a lower efficiency than solar cells based on inorganic semiconductors.

Depending on the type of junction or design of the solar cell, they can be broadly categorised as single-layer, bilayer and bulk heterojunction cells (Figure 13.3 [60]). Single-layer organic solar cells consist of an active organic layer sandwiched between two metallic conductors with different work functions. The efficiencies are lower for this type of cell because the carriers recombine before they reach the oppositely charged electrodes. Bilayer organic solar cells contain a layer of electron donors and acceptors between the electrodes to generate strong local electric fields that facilitate the efficient separation of charge carriers to enhance the conversion efficiency. They are also called planar donor–acceptor heterojunction solar cells. The thickness of the absorbing medium limits the efficiency of such cells, which led to the development of bulk heterojunction solar cells. These have an active layer of donor and acceptor material blended at the nanoscale. The donor is typically a conjugated polymer, whereas the acceptor is a fullerene derivative.[61]

13.4.3 Quantum Dot Solar Cells

QDSCs are a type of solar cell where the quantum dots act as an absorbing material.[62,63] The size- and shape-dependent tunability of the bandgap has enabled quantum dots to be utilised as solar cell components or photosensitisers in existing solar cell designs to overcome their limitations and also in emerging approaches, such as multi-junction or tandem solar cells, multi-exciton generation, hot electron injection, exciton–plasmonic interaction and nanocrystal assemblies to surpass the theoretical efficiency limits

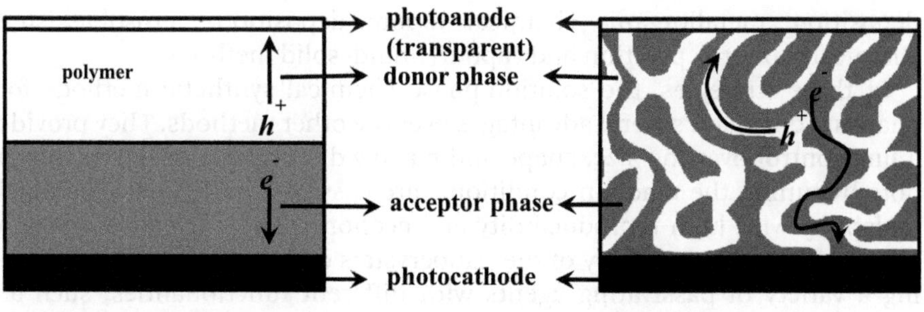

Figure 13.3 Typical architectures of bilayer(left) and bulk (right)heterojunction solar cells. Reprinted from *Advances in Colloid and Interface Science*, B. R. Saunders and M. L. Turner, Nanoparticle-polymer photovoltaic cells, 1–23, Copyright 2008 with permission from Elsevier.

of 31% for single-junction solar cells. In existing cell designs such as DSSCs, the quantum dots are loaded onto mesoporous WBSC anodes by spin-coating, drop-casting, chemical bath deposition, surface ionic layer adsorption and reaction or linker-assisted deposition methods. Quantum dots can also either replace or be integrated with the dyes in DSSCs to fabricate QDSSCs. Emerging strategies have led to the fabrication of various types of QDSC (Figure 13.4 [62,64,65]).

13.4.4 Hybrid Solar Cells

Hybrid solar cells are a mixture of a polymer and inorganic nanocrystals/quantum dots as the active layer sandwiched between two electrodes (Figure 13.5 [62]). The polymer acts as a donor and transports the holes, whereas the inorganic nanocrystals behave as an acceptor and electron transporter. They have the advantages of both organic and inorganic semiconductors.[66,67]

13.5 Synthesis and Characterisation of Quantum Dots

To realise the potential of quantum dots for solar cell applications, the synthesis of phase-pure and monodispersed size/shape quantum dots is essential. There are several methods for the synthesis of quantum dots that can be applied to solar cells.[68–72] These synthetic methods fall into two categories: physical (top-down or fabrication approach) and chemical (bottom-up or synthetic strategy approach) methods. Physical methods involve the gradual breakdown of the bulk material into nanosized particles by either grinding or milling. The chemical approach builds nanostructures atom by atom or molecule by molecule. This approach can further be categorised into solution- (wet chemical) and vapour- (gas) phase methods. Solution-phase processes include sol–gel, solvothermal/hydrothermal, thermal decomposition (hot injection and heating up of either multiple sources or single-source precursors), template, sonochemical, precipitation, polyol and liquid–liquid interface methods. The vapour-phase methods include chemical vapour deposition, metallorganic chemical vapour deposition, aerosol-assisted chemical vapour deposition and vapour–liquid–solid methods.

Of these strategies, the solution-phase chemical synthetic methods for quantum dots have several advantages over the other methods. They provide a fine control over the size, shape and narrow distribution of the quantum dots by tuning the reaction conditions, are easy to handle and have good scalability with high reproducibility and economy. The nucleation, growth rates and surface chemistry of the nanocrystals can be controlled by adopting a variety of passivating agents with different functionalities, such as long-chain amines, acids, esters, phosphonates and thiols. The passivating ligands coordinated on the surface of the quantum dots not only restrict the growth of particles, but also tune the phase,[73] morphology[74] and other

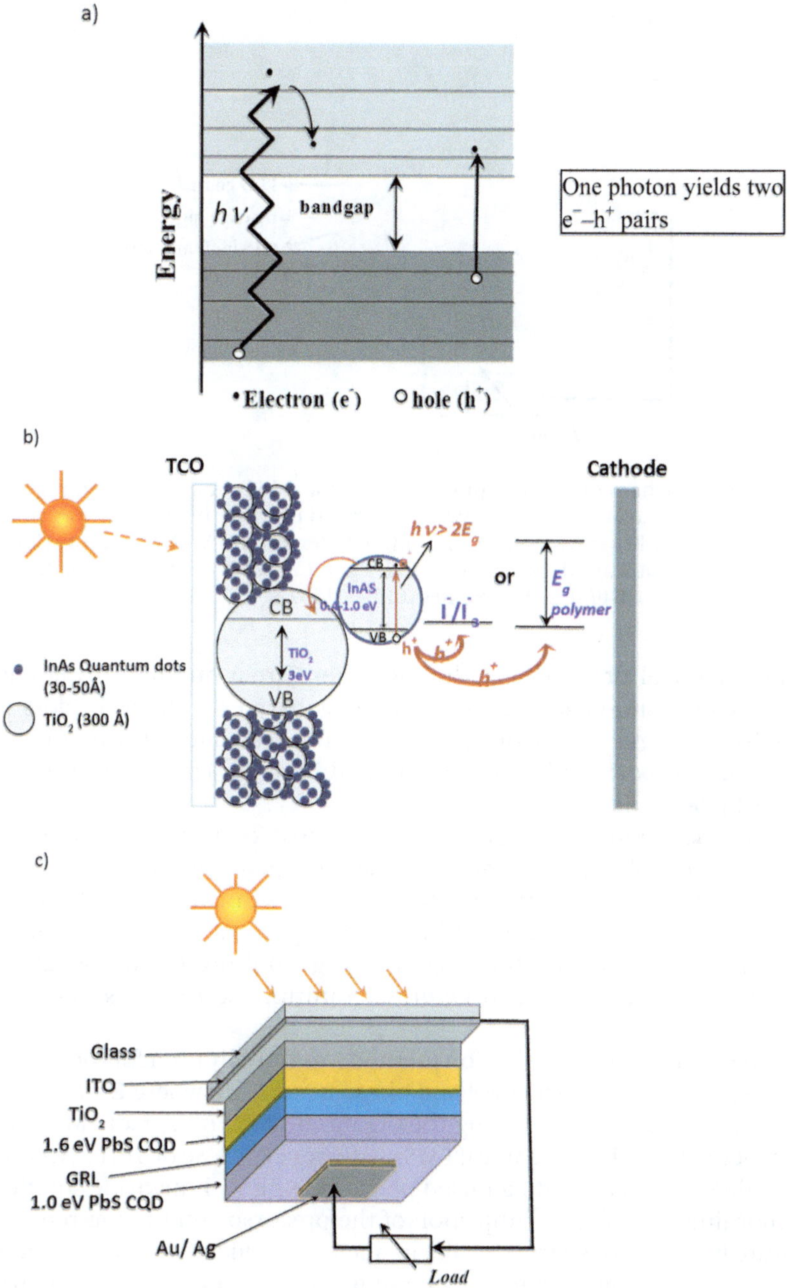

Figure 13.4 Different architectures of quantum dot solar cells: (a) multi-exciton generation solar cell, (b) QDSSC (adapted with permission from ref. 64. Copyright (2010) American Chemical Society) and (c) tandem solar cell (reprinted by permission from Macmillan Publishers: *Nature Photonics* (ref. 65), copyright 2011).

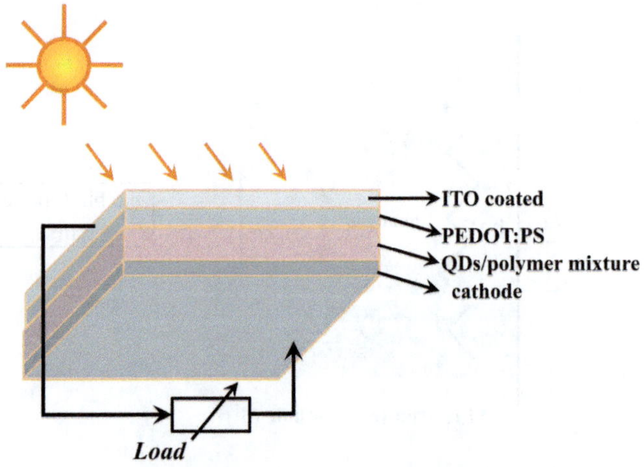

Figure 13.5 Schematic representation of a typical hybrid solar cell configuration based on quantum dots/conjugated polymer heterojunction. Adapted with permission from L. M. Nikolenko and V. F. Razumov, Colloidal quantum dots in solar cells, *Russ. Chem. Rev.*, 2013, **82**(5), 429, DOI: 10.1070/RC2013v082n05ABEH004337.[62]

physicochemical properties.[75] Although there are a number of wet chemical methods, third-generation photovoltaics based on quantum dots need to evolve strategies that can be economically scaled-up. In this direction, thermal decomposition (hot injection[68] and heat-up[72]) methods have shown promising potential.

Thermal decomposition is one of the most cost-effective methods of achieving control over the size and shape of quantum dots in large-scale synthesis by separating the nucleation and growth stages. The former can be realised either by instant hot injection (Figure 13.6) or gradual heating (heat-up method) of the precursors in high boiling point coordinating solvents. The precursors can be either multiple source or single-source molecular precursors.

The path-breaking hot injection method was initially used for the synthesis of high-quality (monodispersed and phase-pure) CdE (where E = S, Se or Te) quantum dots by rapid injecting the organometallic precursors Me_2Cd and *bis*(trimethylsilyl)chalcogenide [$(TMS)_2E$] (where E = S, Se or Te) or tri-*n*-octylphosphine-E at 300 °C into a mixed phosphine/phosphine oxide solution as the coordinating solvent.[76] Injection of the precursor solution at room temperature into the hot solvent results in nucleation due to the supersaturation of the reaction solution and a decrease in the reaction temperature, which quenches nucleation, followed by the start of the growth phase. By gradually heating the precursors, supersaturation is attained by the formation of reactive species within the reaction mixture *via* thermal energy.[77]

The heat-up approach is another important thermal decomposition mechanism for the synthesis of quantum dots. The separation of the nucleation

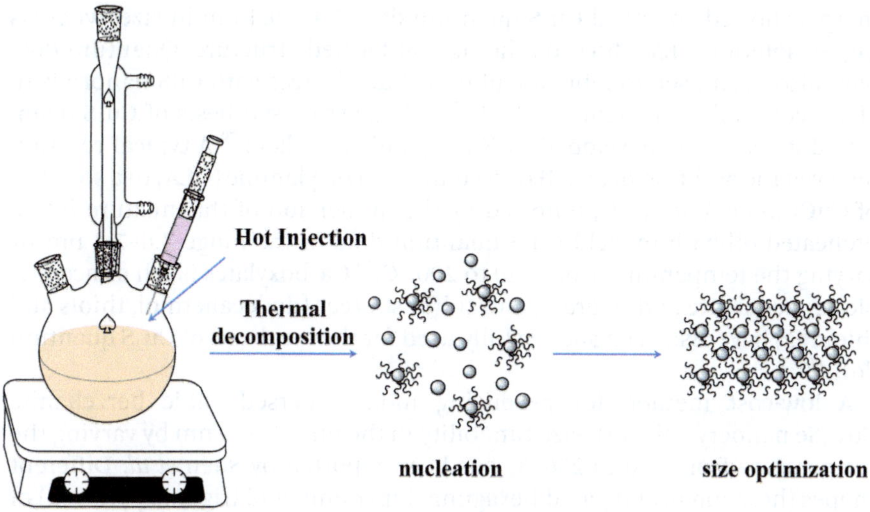

Figure 13.6 Schematic illustration of the hot injection mechanism involving the injection of precursors into surface passivating agents, nucleation, followed by cooling of the reaction leading to a growth phase. Adapted with permission from ref. 63. Copyright (2015) American Chemical Society.

and growth stages can be realised by slow heating of the precursors in suitable coordinating solvents. The practical problems in the hot injection method of the injection step, the reagent mixing time and reaction cooling – which are associated with scalability and consequently affect the reproducibility of batch synthesis – can be avoided in the heat-up method. Hence the heat-up approach seems to be the best method for the industrial synthesis of semiconductor quantum dots/nanocrystals. Some representative examples of the synthesis of semiconductor quantum dots using these methods are discussed in the following sections.

13.5.1 Synthesis of Binary Semiconducting Quantum Dots

13.5.1.1 Group I–VI Quantum Dots

Copper chalcogenide nanocrystals, especially copper-deficient nanocrystals, have received much interest due to their potential applications in third-generation photovoltaics. The growing interest in these quantum dots is due to the increase in the free carrier density caused by the large number of copper vacancies and, as a result, the tunability of their localised surface plasmon resonance in the near- and mid-IR spectral regions.

Alivisatos and coworkers made good-quality Cu_2S nanocrystals by injecting a suspension of $Cu(acac)_2$ in oleic acid into a solution of ammonium diethyl dithiocarbamate in dodecanethiol and oleic acid at 110 °C and quickly heating to 180 °C. Low-resolution transmission electron microscopy (TEM) images showed spherical Cu_2S quantum dots 5.4 ± 0.4 nm in size, whereas high-resolution TEM showed a hexagonal faceted structure. Quantum dots have also been used to fabricate photovoltaic devices with CdS nanorods to give a conversion efficiency of 1.6%.[78] A large-scale synthesis of Cu_2S quantum dots has been developed by Zhang and coworkers.[79] A typical reaction involved the addition of an $(NH_4)_2S$ solution in oleylamine (OLA) to a solution of CuCl in OLA at 50 °C, followed by the immersion of this mixture into a preheated oil-bath to yield Cu_2S quantum dots in the range 4.6–7.4 nm by varying the temperature from 120 to 200 °C.[79] Carboxylates (such as acetate, oleate, octanoate and stearate) and sulfur sources (dodecanethiol, thiols and thiourea) have also been successfully used for the synthesis of Cu_2S quantum dots.[80–83]

A low-cost method for producing monodispersed cubic berzelianite $Cu_{2-x}Se$ nanocrystals with size tunability in the range 4–20 nm by varying the temperature from 120 to 200 °C has been reported by Shen *et al.* Different shapes (hexagonal, elongated hexagonal bipyramid and trigonal pyramid) of $Cu_{2-x}Se$ nanocrystals were achieved by changing the ratio of copper stearate, oleic acid, OLA and Se–1-octadecene (ODE) precursors.[84] Hexagonal copper telluride (Cu_2Te) nanospheres (average diameters 4.6 and 7.4 nm) and nanodiscs were synthesised by injecting different concentrations of TOPTe into a dodecanethiol–oleic acid solution of copper acetylacetonate in the temperature range 160–180 °C.[85]

Single-source precursors have also been used for the synthesis of monodispersed nanostructures by decomposing them in high boiling coordinating solvents such as OLA, dodecanethiol. A variety of copper chalcogenolate complexes derived from dichalcogenocarbamtes,[86–88] dichacogenophosphinates,[89] 2-pyridyl and 2-pyrimidyl chalcogenolates[90,91] have been used for the synthesis of copper chalcogenide nanocrystals. For instance, ultrathin nanowires of Cu_2S with a diameter of 1.7 nm and a length of a few micrometres were fabricated by the thermal decomposition of CuS_2CNEt_2 in a mixed surfactant solvent of dodecanethiol and oleic acid at 160 °C. The nanowires showed bandgaps in the range 3.47–3.69 eV, suggesting a strong quantum confinement effect.[88] Monodispersed and tunable Cu_2S faceted nanocrystals (5–9 nm) and $Cu_{1.75}S$ nanoplates (thickness tuned between 11.7 and 19.3 nm) were synthesised by injecting copper(I) thiobenzoate solution in tributylphosphite or TOP in dodecanethiol. The size of the nanocrystals and the aspect ratio of the nanoplates could be tuned between 2.3 and 4.1 and controlled by varying the temperature and the solvent.[92]

13.5.1.2 Group II–VI Quantum Dots

Among the various binary semiconductor chalcogenide quantum dots and nanocrystals, II–VI nanomaterials, with an emphasis on CdE (where E = S, Se or Te), have been investigated extensively and a wide range of synthetic approaches has been adopted to control their size and morphology. In general, metal oxides/metal carboxylates (usually acetate)/metal chlorides/metal acetylacetonates/Me_2Cd have been used as the source of metals and tertiary phosphine chalcogenides/elemental chalcogens in amine/thiourea have been used as the source of chalcogens to prepare monodispersed II–VI materials in suitable high boiling point solvents. In addition to these two component reactions, single-source precursors, such as metal dithiolates,[93] selenocarboxylates[94] and metal 2-pyridyl chalcogenolates[95,96] have also been used in their synthesis. For example, Hendricks *et al.* demonstrated the controlled growth of CdS nanocrystals through the activated decomposition of $Cd(S_2PPh_2)_2$ in the presence of cadmium tetradecanoate in ODE at 240 °C. Nanocrystals of 3.8–4.1 nm diameter with a size distribution <10% and a photoluminescence quantum yield of 36% have been achieved.[97]

Early investigations after the pioneering work of Murray *et al.*[76] in the preparation of high-quality II–VI nanocrystals involved the use of phosphine chalcogenide precursors. For instance, Peng and Peng reported the synthesis of CdE (where E = S, Se or Te) by injecting stock solutions of chalcogen (tri-*n*-octylphosphine-E) into a hot solution containing CdO, tri-*n*-octylphoshine-oxide (TOPO) and hexylphosphonic acid/tetradecylphosphonic acid. The size of CdTe quantum dots could be tuned from 2 to 8 ± 10% nm depending on the growth time of the quantum dots.[98] Similarly, high-quality wurtzite CdSe quantum dots with a broader tunable size range of 1.5–25 nm have been prepared by injecting TOPSe into a solution of $Cd(OAc)_2$ in different solvents in the temperature range 250–360 °C.[99] Zhang *et al.* demonstrated the large-scale synthesis of different shapes of zinc blende CdS nanocrystals, including spheres, tetrahedral, branched and flower-like structures, by manipulating the amount of TOP.[100] A typical synthesis involved heating a mixture of CdO, sulfur powder, oleic acid and paraffin with or without TOP to 230 °C. The absence of TOP in the reaction mixture gave quantum dots with an average diameter of 3.0 ± 0.2 nm, whereas tetrahedral nanocrystals have been isolated in the presence of TOP.[100]

13.5.1.3 Group III–VI Quantum Dots

Among group III–VI materials, indium chalcogenides have been used as a buffer layer in solar cells.[101,102] Solar cells using β-In_2S_3 as a buffer layer demonstrated a conversion efficiency of 16.4%.[101] Indium chalcogenides have been isolated using multiple sources as well as single-source precursors.

Monodispersed hexagonal β-In_2S_3 nanoplates have been synthesised by heating anhydrous $InCl_3$ and sulfur powder in oleylamine and the thickness of the plates could be tuned from 22 to 63 nm by varying the concentration of the precursor in OLA.[103] Uniform monodispersed In_2S_3 nanorods with an average diameter of 11.5 ± 1.3 nm and a length of 171.7 nm have been isolated by the rapid addition of di-*tert*-butyl disulfide into a solution of indium acetylacetonate in OLA at 180 °C.[104] Indium chaclogenide nanostructures (InS, In_2S_3, InSe, In_2Se_3, InTe and In_2Te_3) have been prepared by injecting indium acetate and the corresponding chalcogen dispersed in dodecylamine into hot hexadecylamine at 300 °C.[105] Single-source precursors, particularly indium tris dithiolate complexes,[106] have also been used successfully for indium sulfide nanocrystals. Nearly monodispersed InS and InSe spherical nanoparticles (5.8 ± 7.0 nm) could be isolated by injecting $[In(E_2CNEt_2)_3]$ (where E = S or Se) in TOP into a hot TOPO solution at 250 °C.[107]

13.5.1.4 Group IV–VI Quantum Dots

Tin and lead chacogenide quantum dots have also been used extensively in solar cells. Tin is less toxic and therefore has advantages over other narrow bandgap semiconducting materials containing lead and cadmium. Tin sulfides (SnS, SnS_2 and Sn_2S_3), selenides (SnSe, $SnSe_2$) and telluride (SnTe) exist in three, two and one forms, respectively. Both hot injection and heat-up methods have been used for their synthesis although the single-source precursor strategy is gaining momentum.[108,109]

Most of the hot injection methods for the preparation of tin chalcogenide nanocrystals involve the use of OLA as a passivating ligand due to its strong affinity for tin, and thioacetamide as the sulfur source. Monodispersed orthorhombic SnS nanocrystals have been synthesised by injecting thioacetamide solution in OLA and TOP into a preheated mixture containing *bis*[bis(trimethylsilyl)amino]tin(II), TOP, oleic acid and ODE at 170 °C. The shape of the SnS nanocrystals can be engineered from spherical to triangular by varying the oleic acid/OLA ratio from 1 : 2 to 1 : 1.[110] High-quality SnS quantum dots have been prepared on a large scale by Zhang *et al.* in a rational synthetic method involving the addition of dried $(NH_4)_2S$ in OLA solution into a solution containing $Sn(OAc)_2$, oleic acid and ODE at elevated temperatures. Monodispersed SnS quantum dots with sizes varying from 6 to 10.5 nm could be isolated by varying the reaction temperature between 70 and 105 °C.[80]

Although significant progress is evident in the syntheses of monodispersed colloidal SnS nanocrystals, relatively less success has been achieved for the synthesis of colloidal tin selenide nanocrystals. Franzman *et al.* reported a hot injection method for the preparation of anisotropic SnSe nanocrystals of width 19.0 ± 5.1 nm, in which di-*tert*-butyl diselenide was injected into a solution of anhydrous $SnCl_2$ in a mixture of dodecylamine and dodecanethiol at 95 °C, followed by heating at 180 °C for 4 min and then cooling. The nanocrystals with a bandgap of 1.71 eV were used to fabricate a hybrid solar cell with an SnSe:PPV active layer.[111] Similarly, a synthesis of tunable irregular-shaped SnSe nanoparticles has been reported using TOPSe in hot OLA.[112,113] The synthesis of a cubic phase of spherical SnTe nanocrystals has been reported by replacing TOPSe by TOPTe in hot OLA; the size of the SnTe could be tuned from 4.5 to 15 nm by altering the reaction temperature and the amount of OLA.[114]

Lead chalcogenide (PbE; where E = S, Se or Te) nanocrystals with narrow bandgaps have attracted significant interest in photovoltaic devices[115,116] and in near-IR communications[117] due to their long carrier lifetimes and multi-exciton generation (MEG) phenomenon.[118] These are generally prepared either by the reaction between $PbO/Pb(OAc)_2/Pb(NO_3)_2/PbCl_2$ and $(TMS)_2E$ (where

E = S or Se), elemental sulfur/R_3PE (where E = Se or Te) in hot coordinating solvents or by the thermolysis of lead(II) chalcogenolate complexes.[119]

Lead sulfide quantum dots have been prepared by injecting a solution of $(TMS)_2S$ in ODE into a preheated (150 °C) mixture of lead oxide (PbO) or lead acetate, oleic acid or TOP in ODE.[120–123] The quantum dots prepared with TOP as the capping agent showed better monodispersity (3.7% for 5.9 nm quantum dots) and exhibited a quantum yield as high as 80% for photoluminescence between 1100 and 1300 nm.[122] A solar cell using PbS quantum dots showed a power conversion efficiency of 3.5%.[122] Photovoltaic devices fabricated using 3.5 nm PbS quantum dots with a bandgap of 1.2 eV had a short circuit current of 1.67 mA cm^{-2} and exhibited the best performance.[124] Low-cost, high-quality PbE (where E = Se or Te) nanospheres and nanocubes scalable to industrial levels have been prepared by heating a mixture of $Pb(acac)_2$, decanoic acid, OLA, Se in ODE or TOPOTe and paraffin oil to 120 or 180 °C for 20 min. The size and shape of the nanocrystals can be varied from spheres to cubes by changing the molar ratios of decanoic acid/OLA (4:1 to 1:3) and the reaction temperature. The size of the nanospheres and nanocubes can be varied between 6.5 and 16 nm by changing the reaction temperature.[125]

13.5.2 Ternary and Quaternary Semiconductor Quantum Dots

A major advantage of multinary semiconducting materials is that their optical properties, such as the bandgap, can be engineered not only by size and shape, but also by composition. However, the synthesis of phase-pure and compositionally uniform materials remains challenging. These challenges can be overcome through the correct selection of the precursors, passivating ligands, solvents, heating conditions and reaction time. Precursors with similar reactivity and the thermodynamic feasibility to yield high-quality nanocrystals of the required materials are always preferable. It is important to characterise the resultant materials with the correct tools to identify secondary phases with nearly the same crystal structure and lattice parameters as the binary material. The simplest ternary I–III–VI_2 (*e.g.* $CuInE_2$, where E = S or Se) (CIS) and quaternary I_2–II–IV–VI_4 (*e.g.* $CuIn_{1-x}Ga_xSe_2$) (CIGS) materials are gaining importance in solar cell applications.

Investigations of ternary and quaternary nanocrystals using the hot injection method are rare compared with the heat-up method due to difficulties such as secondary phase separation and compositional uniformity. Peng and coworkers developed a protocol for the synthesis of tunable, luminescent, monodispersed $CuInS_2$ quantum dots with sizes <2–20 nm. A typical synthesis involved the injection of elemental sulfur in OLA or ODE into a solution containing indium stearate, copper(I)/(II) acetate, oleic acid, ODE and *n*-dodecylthiol at 180 °C.[126] A large-scale synthesis of luminescent, near-stoichiometric colloidal $CuInSe_2$ nanocrystals (3.5 nm) involved the addition of a phosphine selenide (TBP-Se[127] or diphenylphosphine-Se[128]) solution in

ODE into a solution containing CuI, $In(OAc)_3$ in dodecanethiol, oleic acid and ODE at 200 °C, maintaining this temperature for a fixed time. The emission maxima could be tuned between 600 and 850 nm by varying the size of nanocrystals.

A number of ternary and quaternary quantum dots have been prepared by the heat-up method. In a typical synthesis of $CuInE_2$ nanocrystals, copper precursors (CuCl, CuI, Cu(OAc) or $Cu(acac)_2$), indium precursors ($InCl_3$, $In(OAc)_3$ or $In(acac)_3$) and chalcogen sources (dodecanethiol, elemental chalcogen dissolved in ODE or OLA and seleno-urea) were heated in the presence of a suitable capping agent to a temperature at which the crystallisation of the nuclei composed of these elements took place.

Spherical $CuInS_2$ nanocrystals (6–12 nm) have been prepared by refluxing a hot mixture of $Cu(acac)_2$, $In(acac)_3$ in dichlorobenzene, OLA and elemental sulfur dissolved in dichlorobenzene at 180 °C. Irregular $CuInSe_2$ nanocrystals (5–25 nm) have been synthesised by heating a mixture of CuCl, $InCl_3$ and elemental selenium in OLA to 240 °C. The photovoltaic device made from $CuInSe_2$ nanocrystals showed a power conversion efficiency of 0.25%.[129] Yang *et al.* developed a method for the synthesis of $CuInSe_2$ nanocrystals by heating a mixture of CuI, InI_3 and oleylammonium selenocabamate in OLA to 210 °C and continued the reaction for 20 min at this temperature. The size of nanocrystals could be varied from 2.4 to 10 nm by adjusting the growth temperature between 80 and 270 °C. A DSSC fabricated using 4 nm $CuInSe_2$ nanocrystals as a photosensitiser yielded a power conversion efficiency of 4%.[130] Thermolysis of a 1:1 mixture of $Cu\{SeC_5H_3(Me\text{-}3)N\}]_4$ and $[In\{SeC_5H_3(Me\text{-}3)N\}_3]$ in hexadecylamine or TOPO at 330 °C afforded nanocrystals of a tetragonal phase of $CuInSe_2$.[131]

The synthesis of spherical $CuIn_xGa_{1-x}S_2$ nanocrystals with an average diameter of 10 nm has been carried out by a method analogous to the preparation of $CuInS_2$ nanocrystals, except that $GaCl_3$ was also added to the reaction. $CuIn_xGa_{1-x}S_2$ crystallises in a tetragonal crystal structure. The bandgap of the nanocrystals could be modified by changing the In/Ga precursor ratio.[129] The synthesis of Cu_2ZnSnS_4 and related quantum dots using stable and cheap metal precursors was pioneered by Steinhagen *et al.*, who heated of $Cu(acac)_2$, $Zn(OAc)_2$, $SnCl_2.2H_2O$ and elemental sulfur in OLA at 280 °C.[132] The photovoltaic device based on these nanocrystals had a power conversion efficiency of 0.23%.[132] Dithiocarbamate complexes, $[Cu(S_2CNEt_2)_2]$, $[Zn(S_2CNEt_2)_2]$ and $[Sn(S_2CNEt_2)_4]$, have also been used for the synthesis of monodispersed Cu_2ZnSnS_4 nanocrystals (average size 15.1 ± 1.0 nm (length) × 7.6 ± 0.6 nm (base width)) by heating them together in hexadecanethiol and trioctylamine at 250 °C for 30 min.[133]

13.6 Quantum Dot Based Solar Cells

Progress in preparation methods for the synthesis of colloidal quantum dots with tunable sizes and shapes, post-synthetic chemical and physical processing, and developments in the design of devices with emerging new concepts (*e.g.* MEG or carrier multiplication, tandem or multi-junction cells, plasmonic effects) have led to advances in photovoltaic devices based on quantum dots. Some of the design concepts of QDSCs are discussed in the following sections.

13.6.1 Quantum Dot Solar Cells Within the Shockley–Queisser Limit

13.6.1.1 Quantum Dot Sensitised Solar Cells

QDSSCs are, in principle, similar to DSSCs in configuration (Figure 13.7[134]), except that the organic sensitisers (dyes) are replaced by inorganic semiconductor quantum dots. The replacement of organic dyes in DSSCs with quantum dots has, however, not resulted in better efficiencies due to the poor loading of quantum dots on the metal oxide surface.

In QDSSCs, the quantum dots are adsorbed on the surface of meso-porous TiO_2 and absorb solar light to create excitons. The excitons are dissociated into electrons and holes at the quantum dot–TiO_2 interface. The liberated electrons are injected into the porous TiO_2 layer and are subsequently transferred to the anode; the electrons from the electrolyte reinstate the original form of the quantum dots. Cobalt (Co^{2+}/Co^{3+}) and polysulfide (S^{2-}/S_x^{2-}) redox couples are typical electrolytes used in QDSSCs. Quantum dots of indium phosphide (InP) were the first to be used as quantum dot sensitisers in a DSSC architecture with a liquid electrolyte.[135] Since then, a number of inorganic quantum dots with different types of electrolyte have been used. Examples include polymers, such as poly(3-hexylthiophene-2,5-diyl), as a solid electrolyte with Sb_2S_3 nanocrystals as the sensitiser for TiO_2. A conversion

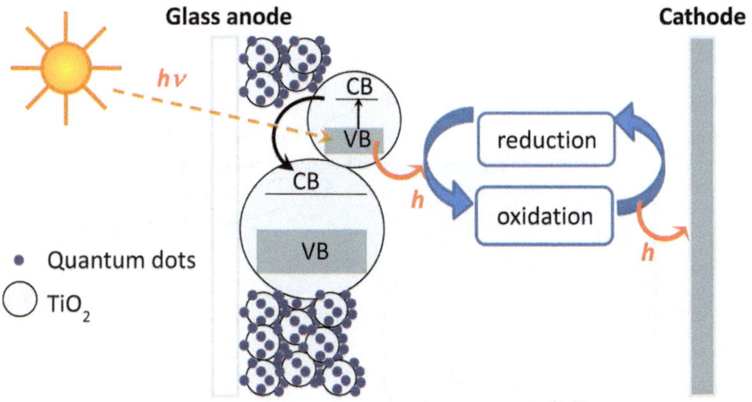

Figure 13.7 Typical configuration of QDSSC and its operating principles. Adapted from *Renewable and Sustainable Energy Reviews*, 37, M. Kouhnavard *et al.*, A review of semiconductor materials as sensitizers for quantum dot-sensitized solar cells, 397–407, Copyright 2014 with permission from Elsevier.

efficiency of 5.13% and good stability was demonstrated for this solar cell.[136] A number of inorganic semiconductor quantum dots, such as Ag_2S,[137] CdS,[138] CdTe,[139] PbS[138] and core–shell quantum dots,[139] have been used as sensitisers and a few investigations have focused on environmentally friendly quantum dots, such as ZnO,[140] $CuInS_2$@ZnS[141] and $(Bi_{1-x}Zn_x)S$.[142]

Quantum dots are deposited on photoanodes in several ways, including chemical bath deposition, successive ionic layer adsorption and reaction[143] and covering with pre-synthesised quantum dots through organic ligands[144] or bifunctional linkers such as thioglycolic.[145] The main drawback of the former two methods is the size control of the quantum dots, whereas the latter faces the problem of inefficient electron injection because the organic ligands separate the quantum dots and photoanode. Despite these drawbacks, their QDSSCs have shown good results. In case of the pre-synthesised quantum dot method, the performance of the QDSSCs can be improved by using passivating ligands or liners with shorter carbon chain lengths to transfer more electrons from the quantum dots to the photoanode.

13.6.1.2 Schottky Junction Solar Cells

The first and most simple photovoltaic device based on quantum dots is the Schottky junction, which consists of an active or absorbing layer of semiconductor quantum dots sandwiched between indium tin oxide (ITO) and metal (*e.g.* Mg or Al) electrodes (Figure 13.8[63]). The ease of fabrication and limited number of interfaces are the prime advantages of this device.

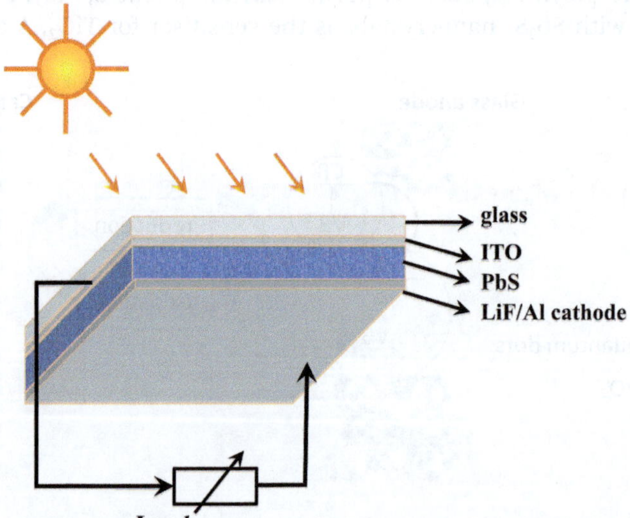

Figure 13.8 Schematic representation of Schottky junction solar cells. Adapted with permission from ref. 63. Copyright (2015) American Chemical Society.

This device is generally used for the evaluation of quantum dot films with different bandgaps.

The active layer in Schottky junction solar cells absorbs light and generates electron–hole pairs. The operating principle of the device involves the creation of charge carriers in the depletion layer, followed by a drift of carriers to the electrodes due to the contact potential difference between the metal and the semiconductor. The efficiency of the device depends on the number of charge carriers generated by the absorption of light by the quantum dots and the extraction of the charge carriers to the electrodes. Both factors depend on the thickness of the active layer. A thicker active layer results in a greater number of carriers, although the sum of the minor charge carrier diffusion length and the width of the depletion layer limit the extraction of carriers. The upper limit on the built-in voltage and the open circuit voltage posed by Fermi level pinning at the Schottky junction are drawbacks of this device.

The Schottky junction uses semiconductor nanocrystals with an optical bandgap in the near-IR region (*e.g.* CdTe, PbS and PbSe) as an active layer. The first Schottky solar cell was based on 6 nm PbS quantum dots as an active layer and aluminium as the metal contact with a conversion efficiency of 1.8% (AM1.5G).[146] The original quantum dots were surface-modified by *n*-butylamine and the quantum dot film was coated with an LiF layer to enhance the operating performance of the device. A record-performing Schottky solar cell with a conversion efficiency of 5.2%, a short circuit current density of 19.3 mA cm^{-2} and an open circuit voltage of 0.46 V was fabricated using a capped oleic acid PbS quantum dot (3 nm) film treated with 1,4-benzenedithiol solution in acetonitrile.[147]

13.6.1.3 Heterojunction Quantum Dot Solar Cells

Heterojunction QDSCs (Figure 13.9[148]) were developed to overcome the drawbacks of Schottky junction solar cells. A heterojunction is the interface between two semiconductors with different bandgaps. A typical device consists of an ITO or fluoride-doped tin oxide (FTO) coated transparent substrate, followed by a layer of doped n-type wide bandgap metal oxide semiconductor (*e.g.* TiO_2 or ZnO). A p-type colloidal quantum dot film acting as an active layer is deposited on the top of this, followed by a final reflective contact of either gold or highly doped MoO_3 and then a silver or aluminium layer. Illumination of the device occurs through the transparent substrate, leading to the generation of carriers close to the depleted junction, thus overcoming the drawback of Schottky devices. The width of the depleted region depends on the free carrier charge density on either side of the junction.

Improvements in the performance of heterojunction QDSCs depend on the use of different semiconductor quantum dots of different sizes, shapes and bandgaps. Tuning of the band alignment between the quantum dot film and the metal oxide to inject electrons into the metal oxide layer without

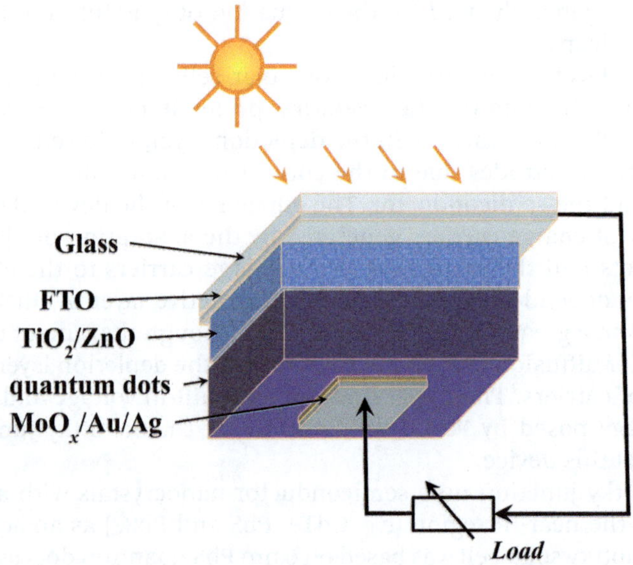

Figure 13.9 Typical configuration of a heterojunction quantum dot solar cell. Reprinted by permission from Macmillan Publishers Ltd: *Nature Nanotechnology* (ref. 148), copyright 2012.

sacrificing the open circuit voltage and optimisation of ohmic contact are two of the challenges in improving the performance of the device. Initial reports on ZnO/PbS heterojunction QDSCs observed that different principles of the operation of these solar cells, *i.e.* excitonic or p–n behaviour, are possible with the same structural design using undoped or doped ZnO.[149]

The rapid advancements in improving the performance of these devices began in 2010 with the introduction of a new heterojunction architecture – the depleted heterojunction quantum dot solar cell – by Pattantyus-Abraham *et al.*[150] In this cell, the charges are separated at a junction between an absorbing layer and a wide bandgap electron acceptor of suitable electron affinity to extract the electrons without sacrificing the open circuit voltage. The new architecture afforded a 5.1% AM1.5 power conversion efficiency. A record power conversion efficiency of 6% has been achieved in an FTO/Zr-doped TiO_2/PbS quantum dots shelled with Br ions/Au cell configuration.[151] A record efficiency of 8.5% has been demonstrated using a new device geometry based on a donor–supply electrode.[152]

13.6.1.4 *Bulk Heterojunction and Bulk Nanoheterojunction Quantum Dot Solar Cells*

This architecture is one of the strategies used to overcome the limitations of planar heterojunction QDSCs by adopting the concept of the bulk heterojunctions from OPVs. A typical configuration of the cell consists of an interpenetrating layer of an n-type WBSC and a quantum dot film formed by structuring them into the electrodes. The structured interface allows the expansion of the depletion region and the accommodation of more absorbing material, thereby improving both the absorption of photons to generate carriers and the extraction of these carriers. Heterojunction architectures containing metal oxide and quantum dot layers demonstrated notable performances by overcoming the drawbacks of earlier structures.

Bulk nanoheterojunction solar cells (Figure 13.10[153]) are a modification of heterojunction solar cells in which the heterojunctions are formed by nanocrystals of different materials, where one material exhibits bulk properties while the other is quantum confined, helping to increase the absorption of light. For instance, Alivisatos *et al.* reported a solar cell with an efficiency of 3% that used CdTe nanocrystals/CdSe quantum dots in which the CdTe acted as a bulk material and CdSe enhanced the absorption of light.[154] A number of combinations with different nanocrystals, such as SnS nanocrystals/PbS nanocrystals,[155] CdSe quantum dots/CdSe nanowires[156] and PbS quantum dots/Si nanoneedles[157] have been reported. A bulk nanoheterojunction device has been fabricated using colloidal Bi_2S_3 nanocrystals and PbS quantum dots as the n- and p-type layers; these were interpenetrated through partial mixing of the two colloidal solutions.[153] The device had a better performance than the bilayer structure in collecting the charge carriers because both materials generated carriers close to the charge-separating junction. It had an efficiency of 4.9%.

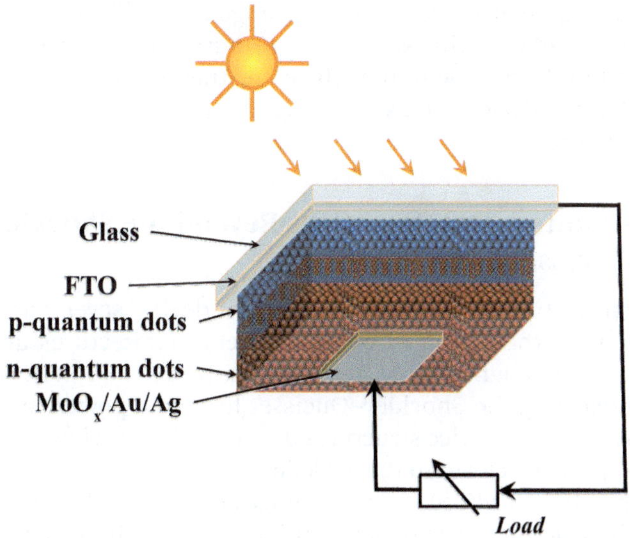

Figure 13.10 Schematic illustration of a bulk nanoheterojunction solar cell. Adapted by permission from Macmillan Publishers Ltd: *Nature Photonics* (ref. 153), Copyright 2012.

13.6.1.5 Quantum Junction Quantum Dot Solar Cells

The capped built-in voltage due to the band offset, the need for band alignment due to the quantum–bulk mismatch of the quantum dot and metal oxide sides, and the interface trap states resulting from imperfect passivation have all limited the utility of earlier structures in multi-junction solar cells. These challenges can be overcome by using a homojunction in which both p- and n-type layers of the same material are used on either side of the junction, facilitating the same degree of quantum tuning and ensuring band alignment for quantum dots of any size. The major hurdle to accomplishing such an architecture is the difficulty of encountering surface oxygen acting as a p-type dopant during the preparation of n-type quantum dots under ambient conditions. The conversion efficiencies of quantum junction solar cells have been improved by using an aluminium zinc oxide electrode for the n-side of the quantum junction, silver doping of the p-layer of PbS[158] or the halide passivation of PbS[159] to the extent of 6.6%.

13.6.1.6 Quantum Funnels

Quantum tuning within a single absorbing layer can enhance the performance of solar cells. The concept of quantum tuning is one approach improving charge carrier transport and extraction through exciton funnelling by generating a gradient bandgap in the conduction band. Quantum dots are well positioned for this concept as their bandgap can be tuned by controlling their size. Quantum funnelling was demonstrated in a heterojunction architecture using five quantum dot layers with bandgaps tuned between 1.06 and 1.35 eV in both graded and antigraded arrangements.[160] An efficiency of 6.6% has been achieved with this structure. The advantage of such a structure is an improved fill factor due to the efficient extraction of minority carriers by ensuring that the device is biased throughout the operation of the cell at its maximum power.

13.6.2 Quantum Dot Solar Cells Beyond the Shockley–Queisser Limit

Since the fabrication of the first quantum dot device, several advances have been made with the introduction of different architectures and concepts in single-junction quantum dot devices to improve their performance with the aim of reaching the Shockley–Queisser limit of a power conversion efficiency of 31%.[161] The device structures discussed in this chapter were mainly focused to approach closer to the 31% limit. Although the potential of quantum dots has been utilised in single-junction devices, the full potential of quantum dots *via* their size-dependent bandgap tunability is yet to be realised and concepts of hot electron transfer and MEG may give devices that reach beyond the Shockley–Queisser limit.

13.6.2.1 Multiple Junction or Tandem Quantum Dot Solar Cells

Multiple junction solar cells (Figure 13.4c[65]) are formed by stacking semiconducting materials of different bandgaps in ideal recombination layers to absorb more photons over the solar spectrum. The maximum theoretical conversion efficiency predicted for solar cells with an infinite number of junctions is 68%.

The challenges associated with the fabrication of such devices are finding materials with the required bandgaps with better lattice matching and developing suitable recombination layers for individual current-matched cells. The best efficiency obtained so far is 44.7% in a wafer-bound four-junction GaInP/GaAs/GaInAsP/GaInAs[162] solar cell. Efficiencies of 43.5 and 40.7% have been realised for three-junction InGaP/GaAs/InGaAs[16] and GaInP/GaInAs/Ge[163] solar cells based on epitaxially grown groups III–V and IV materials. The cost of such systems limits their utility. These problems can be overcome by using quantum dots in the solar cell. The first attempt in this direction was the fabrication of a multi-junction quantum dot cell based on a single quantum dot material (PbS) of different sizes on a graded recombination layer in a two-junction depleted heterojunction architecture. The multi-junction quantum dot cell contained PbS quantum dots with sizes of 2 nm (exciton peak at 1.6 eV) and 4 nm (exciton peak at 1.0 eV) as the absorbing layers in the visible (front cell) region and the IR (back cell) region of the solar spectrum. The graded recombination layer matched the electron and hole currents from the back and front cells, respectively, while adding the open circuit voltages of the two individual junctions to give 1.06 eV and a conversion efficiency of 4.2%.[65]

13.6.2.2 Hot Carrier and Multi-exciton Generation Solar Cells

One of the most important factors that limits the efficiency of single-junction solar cells is the loss of hot carriers through thermal and Auger relaxation. Hot carriers are the charge carriers with an excess kinetic energy equal to the difference between the incident photon energy and the bandgap of the material. The cooling of hot carriers is slowed down in quantum dots due to their discrete energy levels, with the difference between the energy levels being larger than the phonon energies. The cooling of hot carriers can also occur through Auger relaxation as a result of the strong electron–hole Coulombic interaction. Therefore the utilisation of hot carriers, either by capturing using hot carrier solar cells or generating multiple charge carriers (MEG) can boost the efficiency of single-junction solar cells.

Harvesting hot charge carriers in a solar cell requires an increased carrier lifetime, achieved by the slow cooling of the carriers in quantum dots and their transfer to electron and hole acceptors. The former has been proved in CdSe/ZnSe quantum dots,[164] whereas the latter has been demonstrated by withdrawing the hot electrons from PbSe quantum dots to TiO_2 films.[165]

The excess energy of the hot carriers in a solar cell can also be utilised through MEG, when excitons with an energy greater than $2E_g$ created in a semiconductor produce two excitons with an energy equal to E_g each by transferring the excess energy to an electron in the valence band.[166] A QDSC consisting of hydrazine surface-treated PbSe quantum dots was the first to be based on the MEG concept and exhibited an external quantum efficiency of 114% and an internal quantum efficiency of 130%.[115]

13.7 Summary

Growing global demand for energy, rapid depletion of conventional sources of energy and the dramatic environment impact of combustion of fossil fuels have been driving forces to look for renewable sources of energy. Among them, the photovoltaic approach – conversion of solar energy to electricity – has been at the forefront. This chapter covered the historical background and various approaches evolved in photovoltaics since the development of silicon based solar cells in the early 1950s by Bell Labs. Since then there has been rapid progress in terms of improving efficiency and reducing costs as well as in the amount of active (absorbing) material, and use of different materials, technologies and methodologies. As a consequence, second (based on thin film technologies employing CdTe, $CuInSe_2$, *etc.*) and third generation (based on nano-materials) solar cells have emerged. Different designs of third generation solar cells have been developed which include dye-sensitized solar cells (DSSCs), organic photovoltaics (OPVs), quantum dot solar cells (QDSCs), *etc.* Synthesis and characterization of quantum dots of binary, ternary and quaternary semiconducting chalcogenides relevant to third generation photovoltaics have also been included in this chapter.

References

1. A. E. Becquerel, *C. R. Acad. Sci.*, 1839, **9**, 561–567.
2. W. Smith, *Nature*, 1873, 303.
3. (a) R. S. Ohl, US Patent 240252, filed 27 Mar 1941; (b) R. S. Ohl, US Patent 2443542, filed 27 May 1941.
4. E. F. Kingsbury and R. S. Ohl, *Bell Syst. Tech. J.*, 1952, **31**, 8092.
5. G. L. Pearson, *Conference Record, 18th IEEE Photovoltaic Specialists Conference, Las Vegas*, IEEE, New York, 1985.
6. D. M. Chapin, C. S. Fuller and G. L. Pearson, *J. Appl. Phys.*, 1954, **25**, 676–677.
7. N. Williams, *Chasing the Sun: Solar Adventures Around the World*, New Society Publishers, 2005, p. 84.
8. E. S. Rittner and R. A. Arndt, *J. Appl. Phys.*, 1976, **47**, 2999–3002.
9. J. Zhao, A. Wang, M. A. Green and F. Ferrazza, *Appl. Phys. Lett.*, 1998, 73, 1991–1993.
10. O. Schultz, S. W. Glunz and G. P. Willeke, *Prog. Photovoltaics*, 2004, **12**, 553–558.

11. R. C. Chittick, J. H. Alexander and H. F. Sterling, *J. Electrochem. Soc.*, 1969, **116**, 77–81.
12. W. E. Spear and P. G. Le Comber, *J. Non-Cryst. Solids*, 1972, **8–10**, 727–738.
13. D. E. Carlson and C. R. Wronski, *Appl. Phys. Lett.*, 1976, **28**, 671–673.
14. S. Bengali, D. Borrello, E. Vallat-Sauvain, J. Meier, U. Kroll, J. Hötzel, J. Spitznagel, J. Steinhauser, L. Castens and Y. Djeridane, *High Efficiency Amorphous Silicon Device on LPCVD-zno TCO Prepared in Industrial KAI-m R&D Reactor, Presented at 24th European Photovoltaic Solar Energy Conference*, Hamburg, September 2009.
15. A. Banerjee, T. Su, D. Beglau, G. Pietka, F. Liu, G. DeMaggio, S. Almutawalli, B. Yan, G. Yue, J. Yang and S. Guha, *High Efficiency, Multi-junction Nc-SiH Based Solar Cells at High Deposition Rate, Presented at 37th IEEE Photovoltaic Specialists Conferences*, Seattle, June 2011.
16. M. A. Green, K. Emery, Y. Hishikawa, W. Warta and E. D. Dunlop, *Prog. Photovoltaics*, 2012, **20**, 12–20.
17. B. O'Regan and M. Grätzel, *Nature*, 1991, **353**, 737–740.
18. A. Yella, H. W. Lee, H. N. Tsao, C. Y. Yi, A. K. Chandiran, M. K. Nazeeruddin, E. W. G. Diau, C. Y. Yeh, S. M. Zakeeruddin and M. Grätzel, *Science*, 2011, **334**, 629–634.
19. F. He and L. Yu, *J. Phys. Chem. Lett.*, 2011, **2**, 3102–3113.
20. O. E. Semonin, J. M. Luther and M. C. Beard, *Mater. Today*, 2012, **15**, 508–515.
21. L. Esaki and R. Tsu, *IBM J. Res. Dev.*, 1970, **14**, 61–65.
22. R. Tsu and L. Esaki, *Appl. Phys. Lett.*, 1973, **22**, 562–564.
23. R. Dingle, A. C. Gossard and W. Wiegmann, *Phys. Rev. Lett.*, 1975, **34**, 1327–1330.
24. W. J. Skocpol, L. D. Jackel, E. L. Hu, R. E. Howard and L. A. Fetter, *Phys. Rev. Lett.*, 1982, **49**, 951–955.
25. A. I. Ekimov and A. A. Onushchen, *J. Exp. Theor. Phys. Lett.*, 1981, **34**, 345–348.
26. L. E. Brus, *J. Chem. Phys.*, 1983, **79**, 5566–5571.
27. M. A. Reed, J. N. Randall, R. J. Aggarwal, R. J. Matyi, T. M. Moore and A. E. Wetsel, *Phys. Rev. Lett.*, 1988, **60**, 535–537.
28. A. P. Alivisatos, *J. Phys. Chem.*, 1996, **100**, 13226–13239.
29. L. E. Brus, *J. Chem. Phys.*, 1984, **80**, 4403–4409.
30. L. E. Brus, *J. Chem. Phys.*, 1986, **90**, 2555–2560.
31. J. W. Sun, L. W. Wang and W. E. Buhro, *J. Am. Chem. Soc.*, 2008, **130**, 7997–8005.
32. A. S. Brown and M. A. Green, *Phys. E*, 2002, **14**, 96–100.
33. P. Chiu, S. Wojtczuk, X. Zhang, C. Harris, D. Pulver and M. Timmons, *Proc. 37th IEEE Photovolt. Spec. Conf.*, 2011, 771–774.
34. K. W. J. Barnham and G. Duggan, *J. Appl. Phys.*, 1990, **67**, 3490–3493.
35. P. Mints, *Photovoltaic Manufacturer Shipments, Capacity & Competitive Analysis 2010/2011. Palo Alto, CA: Navigant Consulting Photovoltaic Service Program, Report NPS-Supply6*, April 2011.

36. M. A. Green, K. Emery, Y. Hishikawa, W. Warta and E. D. Dunlop, *Prog. Photovoltaics*, 2014, **22**, 701–710.
37. M. K. Nazeeruddin, E. Baranoff and M. Grätzel, *Sol. Energy*, 2011, **85**, 1172–1178.
38. H. M. Upadhyaya, S. Senthilarasu, M. H. Hsu and D. K. Kumar, *Sol. Energy Mater. Sol. Cells*, 2013, **119**, 291–295.
39. B. E. Hardin, H. J. Snaith and M. D. McGehee, *Nat. Photonics*, 2012, **6**, 162–167.
40. M. Grätzel, *Nature*, 2001, **414**, 338–344.
41. M. Grätzel, *Inorg. Chem.*, 2005, **44**, 6841–6851.
42. J. Gong, J. Liang and K. Sumathy, *Renewable Sustainable Energy Rev.*, 2012, **16**, 5848–5860.
43. M. K. Nazeeruddin, A. Kay, I. Rodicio, R. Humphry-Baker, E. Mueller, P. Liska, N. Vlachopoulos and M. Grätzel, *J. Am. Chem. Soc.*, 1993, **115**, 6382–6390.
44. M. K. Nazeeruddin, P. Pechy and M. Grätzel, *Chem. Commun.*, 1997, **18**, 1705–1706.
45. C.-Y. Chen, M. Wang, J.-Y. Li, N. Pootrakulchote, L. Alibabaei, C.-H. Ngoc-le, J.-D. Decoppet, J.-H. Tsai, C. Grätzel, C.-G. Wu, S. M. Zakeeruddin and M. Grätzel, *ACS Nano*, 2009, **3**, 3103–3109.
46. P. Wang, S. M. Zakeeruddin, J. E. Moser, M. K. Nazeeruddin, T. Sekiguchi and M. Grätzel, *Nat. Mater.*, 2003, **2**, 402–407.
47. P. Wang, S. M. Zakeeruddin, J. E. Moser, R. Humphry-Baker, P. Comte, V. Aranyos, A. Hagfeldt, M. K. Nazeeruddin and M. Grätzel, *Adv. Mater.*, 2004, **16**, 1806–1811.
48. C. Y. Chen, S. J. Wu, C. G. Wu, J. G. Chen and K. C. Ho, *Angew. Chem.*, 2006, **45**, 5822–5825.
49. M. K. Nazeeruddin, S. M. Zakeeruddin, R. Humphry-Baker, M. Jirousek, P. Liska, N. Vlachopoulos, V. Shklover, C. H. Fischer and M. Grätzel, *Inorg. Chem.*, 1999, **38**, 6298–6305.
50. T. Bessho, E. Yoneda, J.-H. Yum, M. Guglielmi, I. Tavernelli, H. Imai, U. Rothlisberger, M. K. Nazeeruddin and M. Grätzel, *J. Am. Chem. Soc.*, 2009, **131**, 5930–5934.
51. G. Redmond, D. Fitzmaurice and M. Grätzel, *Chem. Mater.*, 1994, **6**, 686–691.
52. M. Saito and S. Fujihara, *Energy Environ. Sci.*, 2008, **1**, 280–283.
53. Y. Fukai, Y. Kondo, S. Mori and E. Suzuki, *Electrochem. Commun.*, 2007, **9**, 1439–1443.
54. M. K. I. Senevirathna, P. K. D. D. P. Pitigala, E. V. A. Premalal, K. Tennakone, G. R. A. Kumara and A. Konno, *Sol. Energy Mater. Sol. Cells*, 2007, **91**, 544–547.
55. N. Papageorgiou, Y. Athanassov, M. Armand, P. Bonhote, H. Pettersson, A. Azam and M. Grätzel, *J. Electrochem. Soc.*, 1996, **143**, 3099–3108.
56. Y. Bai, Y. Cao, J. Zhang, M. Wang, R. Li, P. Wang, S. M. Zakeeruddin and M. Grätzel, *Nat. Mater.*, 2008, **7**, 626–630.
57. K. Tennakone, G. R. R. A. Kumara, A. R. Kumarasinghe, K. G. U. Wijayantha and P. M. Sirimanne, *Semicond. Sci. Technol.*, 1995, **10**, 1689–1693.

58. L. Heng, X. Wang, N. Yang, J. Zhai, M. Wan and L. Jiang, *Adv. Funct. Mater.*, 2010, **20**, 266–271.
59. P. Wang, S. M. Zakeeruddin, P. Comte, I. Exnar and M. Grätzel, *J. Am. Chem. Soc.*, 2003, **125**, 1166–1167.
60. B. R. Saunders and M. L. Turner, *Adv. Colloid Interface Sci.*, 2008, **138**, 1–23.
61. J. Bisquert and G. Garcia-Belmonte, *J. Phys. Chem. Lett.*, 2011, **2**, 1950–1964.
62. L. M. Nikolenko and V. F. Razumov, *Russ. Chem. Rev.*, 2013, **82**, 429–448.
63. G. H. Carey, A. L. Abdelhady, Z. Ning, S. M. Thon, O. M. Bakr and E. H. Sargent, *Chem. Rev.*, 2015, **115**, 12732–12763.
64. A. J. Nozik, *Nano Lett.*, 2010, **10**, 2735–2741.
65. X. Wang, G. I. Koleilat, J. Tang, H. Liu, I. J. Kramer, R. Debnath, L. Brzozowski, D. A. R. Barkhouse, L. Levina, S. Hoogland and E. H. Sargent, *Nat. Photonics*, 2011, **5**, 480–484.
66. Y. Zhou, M. Eck and M. Kruger, *Energy Environ. Sci.*, 2010, **3**, 1851–1864.
67. T. Xu and Q. Qiao, *Energy Environ. Sci.*, 2011, **4**, 2700–2720.
68. C. de Mello Donega, P. Liljeroth and D. Vanmaekelbergh, *Small*, 2005, **1**, 1152–1162.
69. G. Zhang, S. Finefrock, D. Liang, G. G. Yadav, H. Yang, H. Fang and Y. Wu, *Nanoscale*, 2011, **3**, 2430–2443.
70. S. Suresh, *Nanosci. Nanotechnol.*, 2013, **3**, 62–74.
71. S. Liu and X. Su, *RSC Adv.*, 2014, **4**, 43415–43428.
72. J. van Embden, A. S. R. Chesman and J. J. Jasieniak, *Chem. Mater.*, 2015, **27**, 2246–2285.
73. J. Huang, M. V. Kovalenko and D. V. Talapin, *J. Am. Chem. Soc.*, 2010, **132**, 15866–15868.
74. L. Manna, E. C. Scher and A. P. Alivisatos, *J. Am. Chem. Soc.*, 2000, **122**, 12700–12706.
75. L. Qu and X. Peng, *J. Am. Chem. Soc.*, 2002, **124**, 2049–2055.
76. C. B. Murray, D. J. Norris and M. G. Bawendi, *J. Am. Chem. Soc.*, 1993, **115**, 8706–8715.
77. J. Park, J. Joo, S. G. Kwon, Y. Jang and T. Hyeon, *Angew. Chem., Int. Ed.*, 2007, **46**, 4630–4660.
78. Y. Wu, C. Wadia, W. Ma, B. Sadtler and A. P. Alivisatos, *Nano Lett.*, 2008, **8**, 2551–2555.
79. H. Zhang, B.-R. Hyun, F. W. Wise and R. D. Robinson, *Nano Lett.*, 2012, **12**, 5856–5860.
80. Y. Wang, Y. Hu, Q. Zhang, J. Ge, Z. Lu, Y. Hou and Y. Yin, *Inorg. Chem.*, 2010, **49**, 6601–6608.
81. S.-H. Choi, K. An, E.-G. Kim, J. H. Yu, J. H. Kim and T. Hyeon, *Adv. Funct. Mater.*, 2009, **19**, 1645–1649.
82. T. H. Larsen, M. Sigman, A. Ghezelbash, R. C. Doty and B. A. Korgel, *J. Am. Chem. Soc.*, 2003, **125**, 5638–5639.
83. M. Kanehara, H. Arakawa, T. Honda, M. Saruyama and T. Teranishi, *Chem.–Eur. J.*, 2012, **18**, 9230–9238.

84. H. Shen, H. Wang, H. Yuan, L. Ma and L. S. Li, *CrystEngComm*, 2012, **14**, 555–560.
85. H. Li, R. Brescia, M. Povia, M. Prato, G. Bertoni, L. Manna and I. Moreels, *J. Am. Chem. Soc.*, 2013, **135**, 12270–12278.
86. R. Nomura, K. Kanaya and H. Matsuda, *Ind. Eng. Chem. Res.*, 1989, **28**, 877–880.
87. M. Afzaal, D. Crouch, M. A. Malik, M. Motevalli, P. O'Brien and J.-H. Park, *J. Mater. Chem.*, 2003, **13**, 639–640.
88. Z. Liu, D. Xu, J. Liang, J. Shen, S. Zhang and Y. Qian, *J. Phys. Chem. B*, 2005, **109**, 10699–10704.
89. C. Q. Nguyen, A. Adeogun, M. Afzaal, M. A. Malik and P. O'Brien, *Chem. Commun.*, 2006, 2182–2184.
90. R. K. Sharma, G. Kedarnath, V. K. Jain, A. Wadawale, C. G. S. Pillai, M. Nalliath and B. Vishwanadh, *Dalton Trans.*, 2011, **40**, 9194–9201.
91. R. K. Sharma, A. Wadawale, G. Kedarnath, D. Manna, T. K. Ghanty, B. Vishwanadh and V. K. Jain, *Dalton Trans.*, 2014, **43**, 6525–6535.
92. W. P. Lim, C. T. Wong, S. L. Ang, H. Y. Low and W. S. Chin, *Chem. Mater.*, 2006, **18**, 6170–6177.
93. G. Kedarnath, V. K. Jain, S. Ghoshal, G. K. Dey, C. A. Ellis and E. R. T. Tiekink, *Eur. J. Inorg. Chem.*, 2007, 1566–1575.
94. G. Kedarnath, L. B. Kumbhare, V. K. Jain, P. P. Phadnis and M. Nethaji, *Dalton Trans.*, 2006, 2714–2718.
95. G. Kedarnath, V. K. Jain, A. Wadawale and G. K. Dey, *Dalton Trans.*, 2009, 8378–8385.
96. R. K. Sharma, G. Kedarnath, A. Wadawale, V. K. Jain and B. Vishwanadh, *Inorg. Chim. Acta*, 2011, **365**, 333–339.
97. M. P. Hendricks, B. M. Cossairt and J. S. Owen, *ACS Nano*, 2012, **6**, 10054–10062.
98. Z. A. Peng and X. Peng, *J. Am. Chem. Soc.*, 2001, **123**, 183–184.
99. L. Qu, Z. A. Peng and X. Peng, *Nano Lett.*, 2001, **1**, 333–337.
100. W. Zhang, H. Zhang and X. Zhong, *RSC Adv.*, 2013, **3**, 17477–17484.
101. N. Naghavi, S. Spiering, M. Powalla, B. Cavana and D. Lincot, *Prog. Photovoltaics*, 2003, **11**, 437–443.
102. G. Gordillo and C. Calderón, *Sol. Energy Mater. Sol. Cells*, 2003, **77**, 163–173.
103. K. H. Park, K. Jang and S. U. Son, *Angew. Chem., Int. Ed.*, 2006, **45**, 4608–4612.
104. M. A. Franzman and R. L. Brutchey, *Chem. Mater.*, 2009, **21**, 1790–1792.
105. J. Tabernor, P. Christian and P. O'Brien, *J. Mater. Chem.*, 2006, **16**, 2082–2087.
106. S. Ghoshal and V. K. Jain, *J. Chem. Sci.*, 2007, **119**, 583–591.
107. N. Revaprasadu, M. A. Malik, J. Carstens and P. O'Brien, *J. Mater. Chem.*, 1999, **9**, 2885–2888.
108. R. K. Sharma, G. Kedarnath, A. Wadawale, C. A. Betty, B. Vishwanadh and V. K. Jain, *Dalton Trans.*, 2012, **41**, 12129–12138.

109. A. Tyagi, G. Kedarnath, A. Wadawale, V. K. Jain, M. Kumar and B. Vishwanadh, *RSC Adv.*, 2015, **5**, 62882–62890.
110. S. G. Hickey, C. Waurisch, B. Rellinghaus and A. Eychmüller, *J. Am. Chem. Soc.*, 2008, **130**, 14978–14980.
111. M. A. Franzman, C. W. Schlenker, M. E. Thompson and R. L. Brutchey, *J. Am. Chem. Soc.*, 2010, **132**, 4060–4061.
112. W. J. Baumgardner, J. J. Choi, Y.-F. Lim and T. Hanrath, *J. Am. Chem. Soc.*, 2010, **132**, 9519–9521.
113. D. D. Vaughn, S.-I. In and R. E. Schaak, *ACS Nano*, 2011, **5**, 8852–8860.
114. M. V. Kovalenko, W. Heiss, E. V. Shevchenko, J.-S. Lee, H. Schwinghammer, A. P. Alivisatos and D. V. Talapin, *J. Am. Chem. Soc.*, 2007, **129**, 11354–11355.
115. O. E. Semonin, J. M. Luther, S. Choi, H. Y. Chen, J. B. Gao, A. J. Nozik and M. C. Beard, *Science*, 2011, **334**, 1530–1533.
116. P. D. Cunningham, J. E. Boercker, E. E. Foos, M. P. Lumb, A. R. Smith, J. G. Tischler and J. S. Melinger, *Nano Lett.*, 2011, **11**, 3476–3481.
117. W. W. Scanlon, *Phys. Rev.*, 1958, **109**, 47–50.
118. R. J. Ellingson, M. C. Beard, J. C. Johnson, P. Yu, O. I. Micic, A. J. Nozik, A. Shabaev and A. L. Efros, *Nano Lett.*, 2005, **5**, 865–871.
119. G. Kedarnath, L. B. Kumbhare, S. Dey, V. K. Jain and G. K. Dey, *Polyhedron*, 2009, **28**, 2749–2753.
120. M. A. Hines and G. D. Scholes, *Adv. Mater.*, 2003, **15**, 1844–1849.
121. K. A. Abel, J. Shan, J.-C. Boyer, F. Harris and F. C. J. M. van Veggel, *Chem. Mater.*, 2008, **20**, 3794–3796.
122. L.-Y. Chang, R. R. Lunt, P. R. Brown, V. Bulović and M. G. Bawendi, *Nano Lett.*, 2013, **13**, 994–999.
123. N. Zhao, T. P. Osedach, L.-Y. Chang, S. M. Geyer, D. Wanger, M. T. Binda, A. C. Arango, M. G. Bawendi and V. Bulovic, *ACS Nano*, 2010, **4**, 3743–3752.
124. A. H. Khan, U. Thupakula, A. Dalui, S. Maji, A. Debangshi and S. Acharya, *J. Phys. Chem. C*, 2013, **117**, 7934–7939.
125. J. Niu, H. Shen, X. Li, W. Xu, H. Wang and L. S. Li, *Colloids Surf., A*, 2012, **406**, 38–43.
126. R. Xie, M. Rutherford and X. Peng, *J. Am. Chem. Soc.*, 2009, **131**, 5691–5697.
127. H. Zhong, Z. Wang, E. Bovero, Z. Lu, F. C. J. M. van Veggel and G. D. Scholes, *J. Phys. Chem. C*, 2011, **115**, 12396–12402.
128. W. Li, Z. Pan and X. Zhong, *J. Mater. Chem. A*, 2015, **3**, 1649–1655.
129. M. G. Panthani, V. Akhavan, B. Goodfellow, J. P. Schmidtke, L. Dunn, A. Dodabalapur, P. F. Barbara and B. A. Korgel, *J. Am. Chem. Soc.*, 2008, **130**, 16770–16777.
130. J. Yang, J.-Y. Kim, J. H. Yu, T.-Y. Ahn, H. Lee, T.-S. Choi, Y.-W. Kim, J. Joo, M. J. Ko and T. Hyeon, *Phys. Chem. Chem. Phys.*, 2013, **15**, 20517–20525.
131. R. K. Sharma, G. Kedarnath, N. Kushwah, M. K. Pal, A. Wadawale, B. Bishwanadh, B. Paul and V. K. Jain, *J. Organomet. Chem.*, 2013, **747**, 113–118.

132. C. Steinhagen, M. G. Panthani, V. Akhavan, B. Goodfellow, B. Koo and B. A. Korgel, *J. Am. Chem. Soc.*, 2009, **131**, 12554–12555.
133. M. D. Regulacio, C. Ye, S. H. Lim, M. Bosman, E. Ye, S. Chen, Q.-H. Xu and M.-Y. Han, *Chem.–Eur. J.*, 2012, **18**, 3127–3131.
134. M. Kouhnavard, S. Ikeda, N. A. Ludin, N. B. A. Khairudin, B. V. Ghaffari, M. A. Mat-Teridi, M. A. Ibrahim, S. Sepeai and K. Sopian, *Renewable Sustainable Energy Rev.*, 2014, **37**, 397–407.
135. A. Zaban, O. I. Mićić, B. A. Gregg and A. J. Nozik, *Langmuir*, 1998, **14**, 3153–3156.
136. J. A. Chang, J. H. Rhee, S. H. Im, Y. H. Lee, H.-J. Kim, S. I. Seok, M. K. Nazeeruddin and M. Grätzel, *Nano Lett.*, 2010, **10**, 2609–2612.
137. A. Tubtimtae, K.-L. Wu, H.-Y. Tung, M. W. Lee and G. J. Wang, *Electrochem. Commun.*, 2010, **12**, 1158–1160.
138. H. J. Lee, P. Chen, S.-J. Moon, F. Sauvage, K. Sivula, T. Bessho, D. R. Gamelin, P. Comte, S. M. Zakeeruddin, S. I. Seok, M. Grätzel and M. K. Nazeeruddin, *Langmuir*, 2009, **25**, 7602–7608.
139. X.-Y. Yu, B.-X. Lei, D.-B. Kuang and C.-Y. Su, *Chem. Sci.*, 2011, **2**, 1396–1400.
140. K.-T. Kuo, D.-M. Liu, S.-Y. Chen and C.-C. Lin, *J. Mater. Chem.*, 2009, **19**, 6780–6788.
141. J. Feng, J. Han and X. Zhao, *Prog. Org. Coat.*, 2009, **64**, 268–273.
142. M.-K. Son, H. Seo, I. Shin, I.-K. Kim, J. Choi, K. Prabakar and H.-J. Kim, *Curr. Appl. Phys.*, 2011, **11**, S154.
143. M.-H. Yeh, C.-P. Lee, C.-Y. Chou, L.-Y. Lin, H.-Y. Wei, C.-W. Chu, R. Vittal and K.-C. Ho, *Electrochim. Acta*, 2011, **57**, 277–284.
144. Z. Ning, H. Tian, C. Yuan, Y. Fu, H. Qin, L. Sun and H. Agren, *Chem. Commun.*, 2011, **47**, 1536–1538.
145. T. Lopez-Luke, A. Wolcott, L.-P. Xu, S. Chen, Z. Wen, J. Li, E. D. L. Rosa and J. Z. Zhang, *J. Phys. Chem. C*, 2008, **112**, 1282–1292.
146. J. P. Clifford, K. W. Johnston, L. Levina and E. H. Sargent, *Appl. Phys. Lett.*, 2007, **91**, 253117.
147. C. Piliego, L. Protesescu, S. Z. Bisri, M. V. Kovalenko and M. A. Loi, *Energy Environ. Sci.*, 2013, **6**, 3054–3059.
148. A. H. Ip, S. M. Thon, S. Hoogland, O. Voznyy, D. Zhitomirsky, R. Debnath, L. Levina, L. R. Rollny, G. H. Carey, A. Fischer, K. W. Kemp, I. J. Kramer, Z. Ning, A. J. Labelle, K. W. Chou, A. Amassian and E. H. Sargent, *Nat. Nanotechnol.*, 2012, **7**, 577–582.
149. S. M. Willis, C. Cheng, H. E. Assender and A. A. R. Watt, *Nano Lett.*, 2012, **12**, 1522–1526.
150. A. G. Pattantyus-Abraham, I. J. Kramer, A. R. Barkhouse, X. Wang, G. Konstantatos, R. Debnath, L. Levina, I. Raabe, M. K. Nazeeruddin, M. Grätzel and E. H. Sargent, *ACS Nano*, 2010, **4**, 3374–3380.
151. J. Tang, K. W. Kemp, S. Hoogland, K. S. Jeong, H. Liu, L. Levina, M. Furukawa, X. Wang, R. Debnath, D. Cha, K. W Chou, A. Fischer, A. Amassian, J. B. Asbury and E. H. Sargent, *Nat. Mater.*, 2011, **10**, 765–771.
152. P. Maraghechi, A. J. Labelle, A. R. Kirmani, X. Lan, M. M. Adachi, S. M. Thon, S. Hoogland, A. Lee, Z. Ning, A. Fischer, A. Amassian and E. H. Sargent, *ACS Nano*, 2013, **7**, 6111–6116.

153. A. K. Rath, M. Bernechea, L. Matinez, F. P. G. de Arquer, J. Osmond and G. Konstantatos, *Nat. Photonics*, 2012, **6**, 529–534.
154. I. Gur, N. A. Fromer, M. L. Geier and A. P. Alivisatos, *Science*, 2005, **310**, 462–465.
155. A. Stavrinadis, J. M. Smith, C. A. Cattley, A. G. Cook, P. S. Grant and A. A. R. Watt, *Nanotechnology*, 2010, **21**, 185202.
156. Y. Yu, P. V. Kamat and M. Kuno, *Adv. Funct. Mater.*, 2010, **20**, 1464–1472.
157. C. Y. Huang, D.-Y. Wang, C.-H. Wang, Y.-T. Wang, Y.-T. Jiang, Y.-J. Yang, C.-C. Chen and Y.-F. Chen, *J. Phys. D: Appl. Phys.*, 2011, **44**, 085103.
158. H. Liu, D. Zhitomirsky, S. Hoogland, J. Tang, I. J. Kramer, Z. Ning and E. H. Sargent, *Appl. Phys. Lett.*, 2012, **101**, 151112.
159. Z. Ning, Y. Ren, S. Hoogland, O. Voznyy, L. Levina, P. Stadler, X. Lan, D. Zhitomirsky and E. H. Sargent, *Adv. Mater.*, 2012, **24**, 6295–6299.
160. I. J. Kramer, L. Levina, R. Debnath, D. Zhitomirsky and E. H. Sargent, *Nano Lett.*, 2011, **11**, 3701–3706.
161. W. Shockley and H. J. Queisser, *J. Appl. Phys.*, 1961, **32**, 510–519.
162. F. Dimroth, M. Grave, P. Beutel, U. Fiedeler, C. Karcher, T. N. D. Tibbits, E. Oliva, G. Siefer, M. Schachtner, A. Wekkeli, A. W. Bett, R. Krause, M. Piccin, N. Blanc, C. Drazek, E. Guiot, B. Ghyselen, T. Salvetat, A. Tauzin, T. Signamarcheix, A. Dobrich, T. Hannappel and K. Schwarzburg, *Prog. Photovoltaics*, 2014, **22**, 277–282.
163. R. R. King, D. C. Law, K. M. Edmondson, C. M. Fetzer, G. S. Kinsey, H. Yoon, R. A. Sherif and N. H. Karam, *Appl. Phys. Lett.*, 2007, **90**, 183516.
164. A. Pandey and P. Guyot-Sionnest, *Science*, 2008, **322**, 929–932.
165. W. A. Tisdale, K. J. Williams, B. A. Timp, D. J. Norris, E. S. Aydil and X.-Y. Zhu, *Science*, 2010, **328**, 1543–1547.
166. M. C. Beard, J. M. Luther, O. E. Semonin and A. J. Nozik, *Acc. Chem. Res.*, 2012, **46**, 1252–1260.

CHAPTER 14

Charge Dynamics in Colloidal Quantum Dots: Recombination, Trapping and Multiple Exciton Generation

MARINA A. LEONTIADOU[a], CHARLES T. SMITH[a], CLAIRE LYDON[b] AND DAVID J. BINKS*[a]

[a]School of Physics and Astronomy & Photon Science Institute, University of Manchester, Manchester M13 9PL, UK; [b]School of Chemistry, University of Manchester, Manchester M13 9PL, UK
*E-mail: david.binks@manchester.ac.uk

14.1 Introduction

Colloidal quantum dots (CQDs) are nanometre-scale crystals of semiconductor material synthesised by solution-based techniques.[1] The sizes of these crystals are similar to, or less than, the Bohr radius of one or both of the charge carriers in the bulk material, which produces quantum confinement effects. These include the separation of the conduction and valance bands into sets of discrete quantised states for electrons and holes, respectively. The energy of these states depends inversely on both the size of the CQD and the effective mass of the carrier. The spectral position corresponding to the lowest energy transition between the valance and conduction band

Nanoscience & Nanotechnology Series No. 45
Nanostructured Materials for Type III Photovoltaics
Edited by Peter Skabara and Mohammad Azad Malik

Published by the Royal Society of Chemistry, www.rsc.org

states, *i.e.* the effective bandgap, E_g, can thus be controlled by how large the CQD is allowed to grow, enabling size-tuning of the absorption edge. For electrons promoted across the bandgap, which have a lower effective mass than holes in many of the materials commonly used to produce CQDs, the separation between the lowest energy states can be several hundred meV for small CQDs.[1] This is several times larger than the typical phonon energy for these materials (~30 meV) and therefore the rate of electron cooling by phonon scattering is significantly reduced because it must now involve several phonons simultaneously – an effect that has become known as the 'phonon bottleneck'.[2] Another important result of the nanometre-scale of CQDs is an enhancement in the Coulomb interaction between photogenerated charges due to their close confinement within the restricted volume of a CQD.

These properties of CQDs have the potential to increase solar cell efficiency in a number of ways, which, coupled with the cost-effective and scalable solvent-based methods by they can be produced and processed, promises to improve the economic competitiveness of photovoltaic devices. The detailed balance analysis first described by Shockley and Queisser more than 50 years ago[3] showed that the efficiency of a solar cell can be improved by optimising the E_g of the absorbing species, *i.e.* the absorption edge. For a single-junction device under unconcentrated AM1.5 solar radiation, the ideal value is $E_g \sim 1.3$ eV (which corresponds to an absorption edge at ~950 nm), assuming cooling is the dominant mechanism by which hot carriers lose energy. The absorption edge of a CQD-based solar cell can thus be optimised to this value by controlling the size of the CQD during the growth process. The detailed balance analysis also showed that, for an ideal solar cell with an optimised or near-optimised value of E_g, the most significant source of energy loss is due to relaxation of the photogenerated carriers as they cool towards the band edge. For instance, for a single-junction silicon cell, 47% of the incident solar energy is lost to heat by this mechanism.[4] This was considered to be an effectively unavoidable loss because cooling to the band edge due to phonon scattering is so rapid in bulk semiconductors, occurring on a timescale of a few picoseconds.[2] However, the reduced rate of phonon-mediated cooling in CQDs means that other processes by which hot carriers can lose or transfer their initial energy in excess of the band edge, including those that are non-dissipative, become competitive. The absolute rate of some of these competitor processes depends on the strength of the Coulomb interaction between the carriers and so can be expected to be enhanced by quantum confinement in CQDs. In particular, multiple exciton generation (MEG), sometimes referred to as carrier multiplication,[5–8] was expected to become efficient because of the reduction in the rate of phonon cooling and also the enhancement of the MEG process itself. As will be discussed later, in MEG some or all of the initial energy of a photogenerated carrier in excess of the bandgap is used to excite one or more additional electrons from the valance to the conduction bands, rather than being wasted as heat as the hot carriers cool to the band edges.

The additional carriers thus generated are available to increase the photocurrent and thereby the power output of the cell.

These motivations have prompted a significant amount of work on CQD-based solar cells over the past decade, both on devices and on the underlying materials science of CQDs relevant to their performance in photovoltaic cells. A number of device designs have been investigated, including Schottky and heterojunction cells, some of which utilise CQDs for charge transport as well as light absorption.[9] The initial reported solar power conversion efficiencies were low, <1%,[10] but since 2005 their performance has increased rapidly and, at the time of writing, the record efficiency for a CQD-based solar cell is 10.6%.[11,12] This rapid progress, described and discussed in a number of reviews and perspectives,[9,13–16] has been enabled by a number of different advances. These include an improved understanding and control of the following: the ligands and cross-linkers used to passivate the surface of CQDs and connect them together to form charge transporting matrices with good mobility;[17,18] the surface states produced by unsaturated atoms on the CQD surface;[19] and the fields that drive electrons and holes to their respective contacts.[20]

Measuring and understanding the dynamic behaviour of the charges photogenerated within CQDs by the absorption of solar photons is an important aspect of developing CQDs for photovoltaic applications and of optimising the solar cells that exploit them. Ultimately, good efficiency in a solar cell depends on the rate at which photogenerated charges are produced and extracted to an external circuit significantly exceeding the rate at which they recombine within the cell. MEG enables the more efficient conversion of the energy of an absorbed solar photon into photogenerated carriers; however, the quantum yield of MEG depends on its rate relative to other dissipative processes by which the initial hot carriers relax to the band edge in a CQD. Once at the band edge, geminate charges may recombine radiatively or non-radiatively before they escape from the CQD in which they were photogenerated into the surrounding charge transporting matrix, and hence the efficiency of this escape process will depend on the recombination lifetime of the charges. The charge extraction efficiency to the external circuit also depends on the rate of any recombination in the transport matrix relative to the speed at which the charges are driven to their respective electrodes.

This chapter describes our work on aspects of charge dynamics that are key to the performance of CQD-based solar cells. An overview of CQD synthesis techniques is given, concentrating on those CQDs most commonly studied for application in solar cells, including a description of passivation and ligand-exchange methods. The experimental techniques commonly used to measure charge dynamics in CQDs over a range of timescales are described and the main cooling processes that compete with MEG are discussed, along with the detailed process of MEG itself. The important influence of surface states is described and related to the overall recombination rate within CQDs.

14.2 Synthesis of Colloidal Quantum Dots

14.2.1 Cores

A number of different approaches to the synthesis of CQDs have been extensively covered in books and reviews;[9,21,22] this section gives a brief overview of the most common synthetic methods for the more widely used materials. Desirable properties from CQD syntheses are monodispersity, stability, uniform shape, composition and surface morphology, and good optical properties. The synthesis generally involves nucleation, growth and isolation steps, which can all affect the final properties. The nucleation and growth steps should be separated to give a narrow size distribution and this can be achieved by reducing the reaction temperature or the concentration. The most common synthetic methods include: hot injection, in which the precursors are injected into a hot solvent; non-injection heating, where all the precursors are present at the beginning; and the decomposition of single-source precursors, which are pre-synthesised molecules containing all the main elements of the desired product. A number of groups have also investigated the use of microfluidic methods for the synthesis of nanoparticles.[23,24]

There are a number of factors that ought to be taken into consideration in the synthesis of CQDs. These are temperature, the semiconductor precursors, the organic surfactant, the solvent, the reaction duration, purification (typically by sequential precipitation and redispersion) and any post-synthetic modifications, such as surface passivation or ligand exchange. The choice of precursor can affect the rate of reaction and the overall size of the CQDs (*e.g.* by the use of Cl-, Br- or I-based precursors), as can the temperature and duration of the reaction. The surfactant used plays an important part in solubility, stabilising the nanoparticles and passivating the surface to reduce the number of traps, as well as affecting the kinetics of the nucleation and growth steps.

The kinetics and growth mechanisms of the nanoparticles have also been investigated over the years.[25] Common metal-containing precursors are chlorides, acetates or oxides, although initially metal alkyls were often used. Elemental chalcogenides are often treated with coordinating ligands, such as trioctylphosphine (TOP), prior to injection. The capping agents are often long-chain amines, phosphine oxides or carboxylic acids and, if a non-coordinating solvent is used, it is often octadecene, with the resultant particles being readily dispersed in chloroform, toluene or hexane. The majority of CQD syntheses are now carried out under inert conditions, with dry reagents to exclude air and water, which can lead to poorly crystalline polydispersed particles with low quantum yields.

14.2.1.1 CdX, Where X = S/Te/Se

The II–VI CQDs are among the most common and stable CQDs and there has been a wealth of research into the synthesis and properties of these materials. One of the most important synthetic papers, detailing a solution-based organometallic pyrolysis in coordinating solvents at high temperature, produced monodisperse, crystalline CdS, CdSe and CdTe.[26] Prior to this, papers discussing the luminescence of CdS had been published,[27,28] with syntheses based on aqueous chemistry with inorganic salts. However, the presence of air and water and the lack of high-temperature annealing resulted in polydispersed, poorly crystalline materials. This new synthesis led the way for the syntheses currently in use, which generally use TOP/TOPO (TOP oxide) as a coordinating solvent and are carried out at ~300 °C. These ligands play a key part in allowing controlled growth at high temperatures by coordinating to the crystallite. CdTe is often more air-sensitive than the other Cd chalcogenides and therefore requires a little more consideration during synthesis and storage. The Cd chalcogenides typically form as wurtzite (CdS and CdSe) or zinc blende (CdTe) structures, with the selection of reaction conditions determining which of these is produced.

14.2.1.2 PbX, Where X = S/Te/Se

The first Pb chalcogenides to be synthesised *via* a metal organic route were PbSe and PbS, by decomposition of Pb dithio- and diselenocarbamate single-source precursors.[29] PbTe was later synthesised.[30] The most common synthesis method for these nanoparticles is the hot injection method, with the metal salt being dissolved with a capping agent in solution and the chalcogenide precursor injected after the solution has been heated. This method produced the first absorption spectrum with clear features. This synthesis was carried out at far lower temperatures than the II–VI syntheses.[31]

PbX nanoparticles have also been synthesised using non-coordinating solvents, such as octadecene, by combining PbO with oleic acid and injecting TOPSe.[32] Larger particles can be synthesised at higher temperatures and can favour cubic morphologies. The Pb precursor is also important in determining the shape of the resultant particle. More spherical particles are obtained when the precursor (commonly lead acetate and carboxylic acid) is thoroughly dried; octahedral and star-shaped particles dominate otherwise, due to the presence of acetate salts and acetic acid, which coordinate to the surface. Pb CQDs crystallise in the rock salt structure.

PbSe is relatively air-sensitive and is easily oxidised on the surface to give PbO, $PbSeO_3$ and SeO_2. This initially reduces the emission quantum yield, before enhancing it by providing a passivating layer of oxide, lowering the rate of non-radiative combination. A similar effect is seen with other CQDs, particularly the III–V group, although the emission quantum yield decreases again after a certain thickness of oxide.

14.2.1.3 InAs, InP and GaAs

The III–V CQDs have proved more difficult to synthesise reliably and are particularly air-sensitive. The most successful syntheses have been different to the II–VI and IV–VI strategies. These CQDs are more covalent in character and thus it is more difficult to separate the growth and nucleation steps, requiring more reactive precursors. The materials are often of a lower quality than other CQDs, with lower quantum yields and broader size distributions. The quantum yields are, similarly to Pb chalcogenides, improved on oxidation (although only up to a certain point) and can also be improved by etching the surface with HF (which removes the phosphorus vacancies on the surface). Most syntheses are based on the dehalosilylation reaction.[33,34] Here, $P(SiMe_3)_3$ or $As(SiMe_3)_3$ are injected into a heated solution of $GaCl_3$ or $InCl_3$ to give InP or GaAs. It is also possible to make InAs and GaP using similar procedures. In 2002, Peng *et al.* reported making InP with indium acetate and fatty acids as the capping agent in octadecene, giving monodispersed, crystalline particles with short reaction times. This was followed in 2008 by a synthesis by Nann and coworkers,[35] incorporating Zn undecylenate into the synthesis of the nanoparticles. This gave CQDs with a narrow size distribution and improved quantum yields, and also enabled easy shelling with zinc and sulfur in a one-pot reaction. All these protocols still rely on expensive and air-sensitive $P(SiMe_3)_3$ and, as yet, it is still the main source of phosphorus, although there are a couple of reports of the use of $P(NMe_2)_3$ or its arsenic analogue to make InP, GaAs and InAs.[36–38]

14.2.1.4 $CuInSe_2$ and $CuInS_2$

Ternary copper indium chalcogenides can be formed by hot injection or single-source precursor decomposition routes, with rings, spheres, hexagons and hexagonal rings being reported. $CuInSe_2$ has been prepared using $InCl_3$ and $CuCl_2$, with TOP/TOPO and TOPSe, and also with metal acetates and selenium in oleylamine without TOP. The stoichiometry can be tuned using different metal precursor halides. Varying the temperature of the reaction and the stoichiometries of the CQDs did not result in an excess of trap states. These materials can crystallise in the chalcopyrite, wurtzite and zinc blende phases at room temperature. $CuInS_2$ has been prepared using similar methods and also with non-coordinating solvents. A variety of sulfur sources, such as alkylthiols, dithiocarbamates, CS_2 and elemental sulfur, have been used. To obtain chalcopyrite CQDs, the reactivity of the two cationic precursors must be balanced, which can be achieved using appropriate ligands and reaction temperatures. For the wurzite structure, the ratio of capping agents is instrumental and amines are particularly important and formation takes place *via* a CuE or Cu_2E (where E is S or Se) seed.[39]

14.2.2 Shelling Techniques

Shelling is a procedure for passivating the surface of a particle with an inorganic layer. This can improve the optical properties and remove the dependence on organic capping agents. If the right shelling materials are used, it can also result in a nanoparticle with novel properties. When creating a core/shell nanoparticle, it is important to take into account the lattice mismatch and band edge alignment. If the lattice mismatch is too large, then new defects will be introduced into the material, which could become trapping sites and reduce the emission properties. Depending on which shelling materials are chosen, it is possible to make various types of CQD.[40] Type I quantum dots have a shell with a wider bandgap than the core and the charge carrier is confined to the core, as in CdSe/ZnS and PbS/CdS. Type II quantum dots have staggered bandgaps and the hole and electron are separated between the core and the shell (Figure 14.1) with a significantly

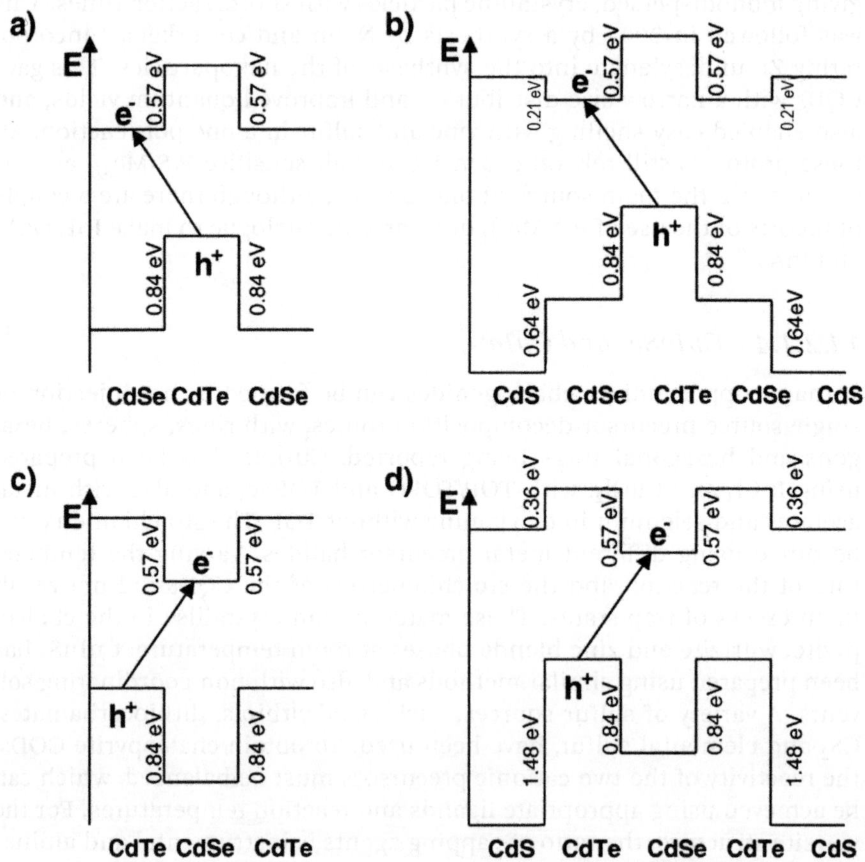

Figure 14.1 Energy level diagrams showing the valence and conduction band positions with for (a) CdTe/CdSe, (b) CdTe/CdSe/CdS, (c) CdSe/CdTe and (d) CdSe/CdTe/CdS. The efficiency of MEG in photovoltaic devices was compared in terms of the variation of cores/shell/shell structures. Reprinted from *Thin Solid Films*, 560, N. McElroy *et al.*, Comparison of solar cells sensitised by CdTe/CdSe and CdSe/CdTe core/shell colloidal quantum dots with and without a CdS outer layer, 65–70, Copyright 2014 with permission from Elsevier.[41]

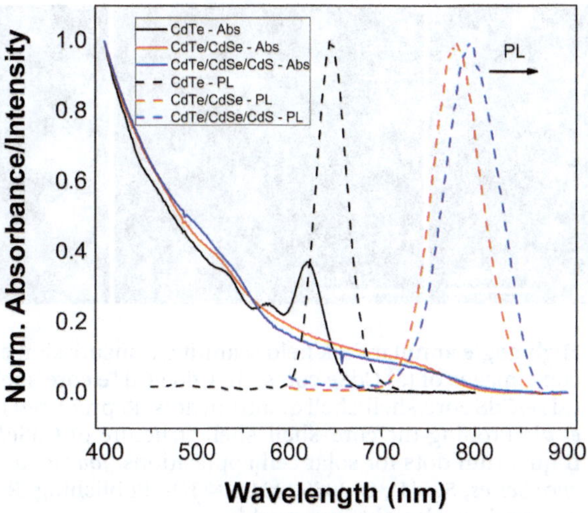

Figure 14.2 Normalised absorbance (solid lines) and photoluminescence (dashed lines) spectra of a CdTe core and after the addition of CdSe/CdS shells. The red shift of the bandgap is evident after the addition of the shells. Reprinted from *Thin Solid Films*, 560, N. McElroy *et al.*, Comparison of solar cells sensitised by CdTe/CdSe and CdSe/CdTe core/shell colloidal quantum dots with and without a CdS outer layer, 65–70, Copyright 2014 with permission from Elsevier.[41]

red-shifted emission, as in CdTe/CdSe (Figure 14.2). This increases the photoluminescence lifetime of the particle because recombination takes longer due to the separation. Less common types of core/shell particle are quasi-Type II particles, where one charge carrier is localised to one part of the quantum dot and the other is delocalised across the whole particle, and inverse Type I quantum dots, which utilise a wide bandgap core and a smaller bandgap shell. In Type I quantum dots, the charge carrier is confined to the shell, in contrast with conventional Type I nanoparticles. An example of scanning transmission electron microscopy images of the growth stages used in the synthesis of CdSe/CdTe/CdS (core/shell/shell) structures is shown in Figure 14.3.

The synthesis of a shell can be carried out in different ways, although successive ion layer adsorption and reaction is a relatively common approach. In this method, the precursors are injected separately, with each injection effectively forming half a monolayer. The temperature of the reaction is kept low enough to prevent nucleation of the injected precursor and the synthesis can be finished with either precursor to give control over the surface of the quantum dot and the thickness of the shell.[43] This can be achieved in a one-pot reaction with the synthesis of the core particle, which works particularly well in the InP/ZnS synthesis proposed by Nann *et al.*, where Zn is used in the initial core synthesis. Single-source precursors have also been used as shelling materials.

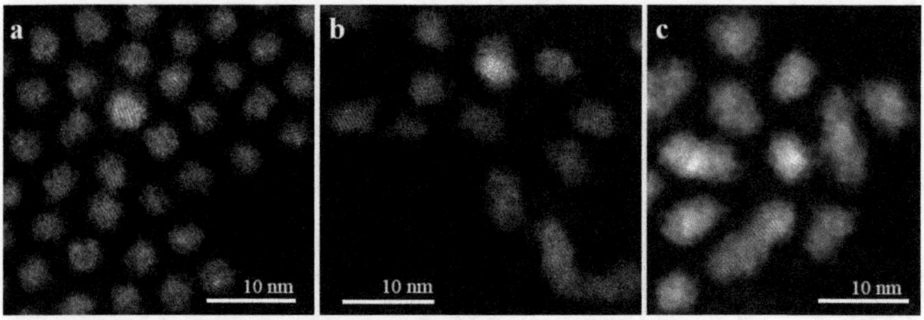

Figure 14.3 High-angle annular dark field scanning transmission electron microscopy images of (a) CdSe cores, (b) CdSe/CdTe core/shell and (c) CdSe/CdTe/CdS core/shell/shell quantum dots. Reproduced from E. A. Lewis *et al.*, Probing the core–shell–shell structure of CdSe/CdTe/CdS type II quantum dots for solar cell applications, *Journal of Physics: Conference Series*, 522(1), 012069, 2014, © IOP Publishing. Reproduced with permission. All rights reserved.[42]

14.2.3 Passivation and Ligands

Passivation can protect a CQD from oxidation, as well as reducing the number of unsaturated dangling bonds (caused by surface atoms having a lower coordination number than the interior atoms) and the associated trap states that decrease the emission quantum yield. Although shelling with an alternative semiconductor is one way to do this, it can also be achieved relatively simply using organic ligands or small inorganic molecules on the surface. The long-chain capping agents used during synthesis interact dynamically with the quantum dot and can be relatively easily exchanged with other organic components. The capping ligands are commonly carboxylic acids, amines or phosphine oxides. Although these capping agents go some way towards preventing trap states, they are often sterically bulky and hence do not passivate the entire surface, allowing oxidation of the less air-stable quantum dots. However, branched ligands may prevent oxygen reaching the surface better than linear ligands. The most common capping ligands used in CQD synthesis are phosphine/phosphine oxide and TOP/TOPO because they allow high-temperature reactions, which facilitate annealing, and can be used in organic solvents, allowing oxygen- and water-free environments. The steric properties of the ligand can also affect the growth and morphology of the particle. Long-chain amines are often used as an alternative in CQD synthesis and can give improved emission quantum yields, possibly due to the closer packing of ligands on the surface or surface reconstruction. Research[44] has indicated that only primary amines promote enhanced emission, although this may be concentration-dependent. Thiols and carboxylic acids are also used relatively frequently, particularly oleic acid, which has a double bond with an associated kink that provides colloidal stability.

Organic ligands have the added benefit of allowing the solubility to be tuned and some functional molecules to induce charge transfer have been investigated. Common ligands for exchange are alkylthiols, alkylamines,

Figure 14.4 CdTe core structure before (bottom left) and after chloride passivation (top). Reproduced with permission from R. C. Page *et al.*, Near-Unity Quantum Yields from Chloride Treated CdTe Colloidal Quantum Dots, *Small*, John Wiley and Sons © 2014 The Authors. Published by WILEY-VCH Verlag GmbH & Co. KGaA, Weinheim.

aromatic thiols and mercaptocarboxylic acids. Most ligand exchanges are carried out in a large excess of the new ligand to drive the process by mass action. There have also been a number of reports considering small inorganic molecules as passivating agents. PbX CQDs have been treated with halides to give well passivated surfaces, resulting in markedly increased quantum yields. This has also been achieved with CdTe (Figure 14.4).[45,46] Talapin and coworkers[47] used molecular metal chalcogenides and metal-free inorganic ligands such as HS^- and elemental chalcogenides as surface ligands.

14.3 Experimental Techniques

To assess the suitability of CQD samples for the enhancement of photovoltaics, a thorough understanding of the potential charge relaxation and recombination processes is required, particularly the circumstances and timescales under which each process can occur and whether the behaviour of the CQDs can be modified or eliminated. Charge dynamics in CQDs have been studied extensively experimentally, and the techniques most commonly used are mature. However, even the most common techniques can be challenging due to the nature of the charge relaxation processes, the potentially strong sample-to-sample variation and the sensitivity of CQD samples to their environment. Charge relaxation processes can occur on timescales spanning several orders of magnitude, from hundreds of femtoseconds for intra-band charge cooling[48] to tens of nanoseconds or even microseconds for radiative recombination,[1] both of which require consideration in photovoltaic devices. Commercially available ultrafast laser systems are capable of producing 100 fs pulses or shorter, which are adequate to probe even the fastest of such relaxation processes. However, the high fluence of such systems can lead to a range of experimental artefacts due to non-linear effects, which can be heavily sample-dependent,[49] necessitating careful consideration of the experimental conditions and subsequent data analysis.

Before charge dynamics are investigated in CQDs, a series of complementary characterisation processes are typically performed. The most important of these are absorption and photoluminescence spectra, which reveal the band edge transition energies and can, in some cases, reveal some higher lying transitions.[50] In addition, confirmation of the CQD composition is often performed through combinations of X-ray diffraction to confirm the crystal structure, X-ray photoelectron spectroscopy to determine the chemical composition, and transmission electron microscopy to determine their size and shape.[1] Another common measurement on CQDs, particularly when non-radiative recombination or trapping is of interest, is the photoluminescence quantum yield (PLQY). The PLQY, usually given as a percentage, is the fraction of absorbed photons (at a given wavelength) that result in photons emitted from the CQD sample as photoluminescence, such that a PLQY of 100% indicates that every absorbed photon creates a hot exciton, which then cools to the band edge and radiatively recombines. The PLQY is usually measured in a continuous wave photoluminescence spectrometer, with the sample housed in an integrating sphere such that every unabsorbed photon and every emitted photon is detected. The PLQY can be a difficult measurement to take due to the strong dependence on sample concentration and excitation wavelength; careful consideration should be taken of the wavelength-dependent efficiency of the detectors and the collection efficiency of the integrating spheres and other components. The PLQY has been used extensively to indicate the degree of surface trapping and passivation of different types of CQD samples[42,46] and PLQY spectra (the variation in PLQY with excitation wavelength) have been used to indicate the presence of hot exciton trapping.[51]

14.3.1 Transient Photoluminescence Spectroscopy

Transient photoluminescence spectroscopy (TPL) is a time-resolved photoemission experiment that can be used to measure the decay of photoluminescence intensity from a CQD sample following excitation by an ultrafast laser pulse. The resulting decay transients can reveal photoluminescence lifetimes and various other relaxation processes that depopulate (and populate) the radiative band edge state. The technique is usually performed with visible or UV light at a high repetition rate of a few MHz. However, several variants of TPL exist depending on the detection system required by the emission range of the sample.

The most common technique is time-correlated single-photon counting. In this technique, the photoluminescence emission from the sample is passed through a monochromator and then into a photomultiplier tube (PMT), which is capable of the efficient detection of single photons. The time delay between the PMT signal and the excitation pulse is then measured to build a histogram of emission delays, a fit to which gives the emission lifetime, and, if present, the lifetime of any non-radiative decay components. Such a system has a time resolution of tens of picoseconds and the sensitivity and fast response of PMTs allow the use of very low fluence excitation, even at high repetition rates, eliminating the chance of multi-excitation of single CQDs. However, this response is insufficient to resolve biexciton decay or Auger-mediated processes, such as MEG or ultrafast surface trapping. Streak cameras are sometimes used instead, which offer a better time resolution and similar sensitivity. PMTs and stream cameras are typically only sensitive to photons of energy ~1.5 eV or greater, which is too high to detect emission from quantum dots with absorption optimum for exploitation of the solar spectrum.

Better time resolution and near-IR detection can be achieved using up-conversion spectroscopy in TPL. In this technique, the photoluminescence is mixed with an ultrafast laser pulse in a non-linear crystal, which can produce light pulses in the sensitivity range of PMTs with the time resolution of the ultrafast laser pulse. However, the signal-to-noise ratio for such systems suffers due to the efficiency of the up-conversion process. This method has been used to measure the MEG efficiency in some CQDs, where an increased photoluminescence intensity and fast recombination of multi-excitons was observed in the photoluminescence transients, and it has also been used to determine the MEG quantum yield.[52,53]

14.3.2 Ultrafast Transient Absorption Spectroscopy

Ultrafast transient absorption (UTA) spectroscopy is a pump–probe experiment, which can produce signals proportional to the occupation of specific energy levels within a CQD on sub-picosecond timescales. A pump beam of sub-picosecond pulse width excites charges in a CQD sample across the bandgap into an excited or 'hot' state. In an ideal CQD (if pumped below the MEG threshold), these charges will cool to the band edge and then recombine radiatively. During this process, a sub-picosecond probe beam passes through the sample at a given delay time after the pump beam. The probe beam will experience a reduction in absorption, or photobleach, if tuned to a transition corresponding to an energy level populated by the pump beam. Figure 14.5 shows a schematic representation of a typical transient absorption system. The reduction in absorption occurs due to CQD energy levels having low degeneracy, typically 2 for the conduction band minimum.

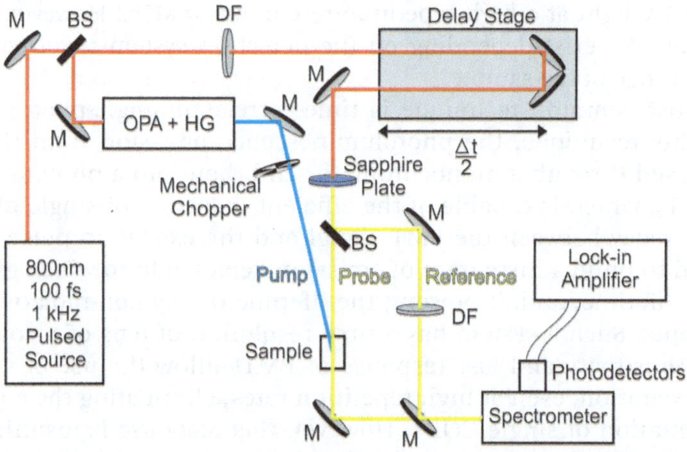

Figure 14.5 Schematic representation of typical transient absorption spectrometer. An 800 nm, 100 fs pulse width source at a repetition of 1000 Hz, 95% of which is sent through an optical parametric amplifier with second harmonic generation allowing tuning of the pump beam, followed by reduction to 500 Hz using a mechanical chopper. The remaining 5% is sent through a delay stage and focused onto a sapphire plate to form a white light continuum. Half of this is focused onto the sample and is coincident with the pump beam. The remaining half forms a reference beam. The reference and probe are passed through a spectrometer and onto two Si photodetectors, the output of which is detected using a lock-in amplifier synchronised to the pump repetition rate. The final output of the system is the time-dependent change in absorption as a function of time and probe wavelength. BS, beam splitter; M, mirror; DF, density filter. More details on the experimental set-up can be found elsewhere.[7,54–58]

An electron occupying such a state would reduce the absorption of the probe beam by half if tuned to a transition involving that level. This reduction in absorption should decay monoexponentially over time if radiative recombination is the only decay path. The probe beam is usually tuned to the first excitonic energy level, the band edge transition; however, other excitonic levels or intra-band transitions are sometimes probed; intra-band transitions are often used when either the laser system or the detection system used are not capable of operating at the band edge transition. This is often true for very narrow bandgap CQDs that have a bandgap in the near-IR, close to the optimum for exploitation of the solar spectrum.

The use of UTA spectroscopy on CQD samples can be challenging due to a range of experimental artefacts caused either by non-linear effects resulting from the short pulse duration of the lasers inducing a response in the sample solvents or system optics,[49] or by potentially misleading non-radiative processes in CQDs. Non-linear effects can usually be corrected post-measurement.[49,59] Non-radiative processes, such as surface trapping, can lead to trion formation in CQDs,[7,60] where the geminate partner of a trapped charge remains in a CQD for longer than the repetition rate of the pump pulse. This enables fast Auger recombination, which was a source of controversy in early MEG measurements when it was mistaken for MEG signals. To avoid this artefact, CQD samples are stirred or flowed to refresh the sample volume inside the pump beam faster than the repetition rate.

As in TPL, the decay of UTA signals can be used to determine the recombination dynamics in CQDs. However, tuning of the probe beam can allow isolation of the signals due to only electron or hole states rather than the radiative exciton state. The probe beam is usually scanned for a given pump–probe delay to measure a transient absorption spectrum, which can contain several bleach peaks corresponding to absorption peaks, and sometimes increases in absorption caused by artefacts or non-radiative processes, often called photoinduced absorption. Other variations of transient absorption spectroscopy exist that allow the further isolation of different transitions in CQDs, such as multiple population period transient spectroscopy,[61–63] which can also isolate exciton and biexciton decay kinetics.

14.4 Multiple Exciton Generation and Other Cooling Processes

MEG was first demonstrated by Schaller and Klimov[5] and is also known as carrier multiplication or impact ionization.[5–7] One of the conditions required for it to occur is that the absorbed photon energy has to be above a threshold value, $h\nu_{th}$ that corresponds to at least $2E_g$. In this way, at least one of the photogenerated charge carriers will have sufficient energy to promote another electron across the bandgap, creating a second exciton. Thus one absorbed photon will create two excitons. The MEG process is shown in Figure 14.6 for a CQD and compared with carrier relaxation in a bulk semiconductor. The average number of excitons created in a quantum dot per absorbed photon defines the quantum yield and depends on the efficiency of the MEG process. By energy conservation, the ideal MEG quantum yield is a step function with the photon energy normalised to the bandgap, *i.e.* it increases by 100% for every additional photon energy equal to the E_g above the threshold value of $2E_g$ (Figure 14.7). However, for most materials the MEG quantum yield is, in practice, found to have a linear dependence beyond $h\nu_{th}$, the value of which can be significantly greater that the predicted value. This deviation from ideal behaviour is due to competition from other cooling processes.[64] The experimentally observed behaviour is characterised by $h\nu_{th}$ and the gradient of this linear increase in quantum yield is termed the MEG efficiency, η. Figure 14.7 compares the ideal and observed relationship between quantum yield and photon energy. Table 14.1 gives some examples of threshold values, along with the reported relevant MEG efficiencies for different materials and nanostructures (*e.g.* CQDs, nanorods and nanosheets). An increase in η, and hence an increase in MEG quantum yield, can be achieved by understanding the processes that compete with MEG and then reducing them by engineering the CQDs.

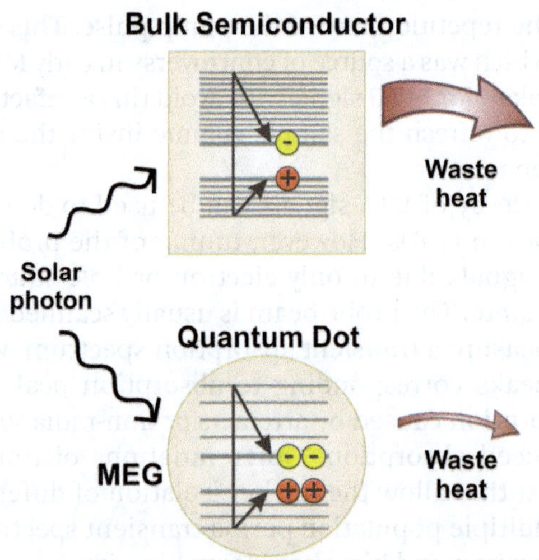

Figure 14.6 Multiple exciton generation in a nanocrystal quantum dot. In contrast to the traditional bulk semiconductor, the excess energy of the generated charge will create another electron–hole pair, reducing the energy wasted as heat. Diagram modified from ref. 7 © 2013 by MDPI.

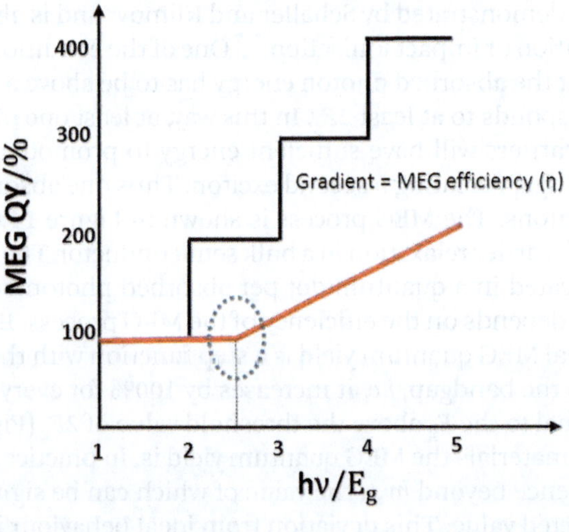

Figure 14.7 MEG efficiency as a function of the photon energy normalised to the bandgap.

Once the hot carriers have been generated (after absorbing a photon), they may cool into the band edge of the conduction band (or valence band), releasing the excess energy, which can then be dissipated as heat *via* electron–phonon interactions, as energy transferred into the surrounding ligand or solvent environment through vibrational modes,[75] or *via* Auger relaxation, in which the electron transfers the energy to the hole, which, in turn, cools

Table 14.1 Experimental values of MEG threshold and efficiencies for a range of CQDs and related nanocrystal structures. NC = glass-embedded nanocrystal.[58,65–74]

Nanocrystal	$h\nu_{th}/E_g$	MEG_{eff} ($\%/E_g$)
PbSe[66,67]	3	40
PbSe[68] nanorods	2.5	26
PbS[59]	3	40
PbS[69] nanosheets	4	90
InP[70]	2.1	30
InAs[71]	2	35
CdTe/CdSe[72]	2.7	<53
PbSe/CdSe[68]	2	52
$CuInSe_2$[73]	2.4	36
Ge NCs[74]	2	50
Si NCs[75]	2	140

with the emission of phonons. Figure 14.8 is a schematic representation of these cooling processes. Also shown is the case were the carriers cool into a surface state. Surface states produced by dangling bonds are known to enable a non-radiative recombination pathway that traps band edge charges on a picosecond timescale.[56,57,76] These surface states can also trap hot charges, introducing an additional cooling pathway and thus increasing the overall cooling rate.[51] These are the main cooling process, the total rate of which competes with MEG.

There are number of ways in which the rate of individual cooling processes may be reduced by control of the CQD structure, thus improving the MEG quantum yield. The separation between energy levels in CQDs can be increased due to the quantum confinement effect, so that for small diameter CQDs the separation of the energy levels is larger than the phonon energy, which reduces any unwanted phonon cooling.[77,78] In Type II CQDs, the reduction in wave function overlap, produced because the electrons and holes are confined in different volumes, increases the Auger relaxation lifetime.[41] Figure 14.1 shows the energy level diagrams for Type II and quasi-Type II CdTe/CdSe CQDs passivated with a CdS shell. This is an example where the existence of the second shell protects the holes from any surface traps. Similarly, to reduce surface-mediated cooling for the electrons, any unwanted surface traps could be passivated by one or more of the different methods described earlier.[46,79] We discuss later how passivation can prevent the photogenerated charges from cooling into the surface states, hence increasing the MEG quantum yield.

The contribution of each of these competitor processes, and hence the efficiency of MEG, depends on the CQD under investigation and on the excitation conditions, which requires thorough study to fully understand. Several different techniques can be used and these can provide not only a typical optical (absorption and photoluminescence spectroscopy) and structural (scanning transmission electron microscopy, X-ray photoelectron spectroscopy)

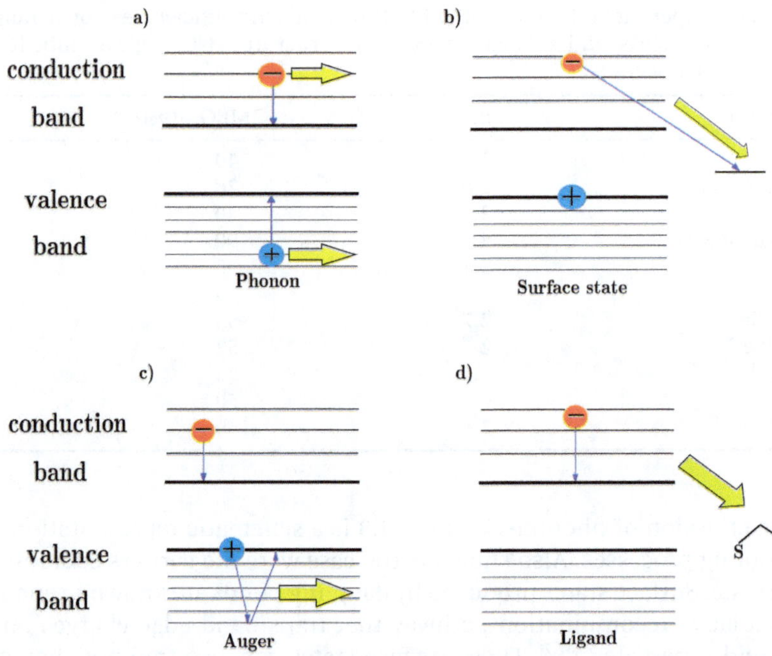

Figure 14.8 Cooling processes competinh with MEG. (a) Phonon emission; (b) electron transfer to a surface state; (c) Auger relaxation; and (d) transfer of energy to surface ligands through vibrational coupling. The yellow arrows represent waste energy. Diagram taken from ref. 7 © 2013 by MDPI.

characterisation of the CQDs, but also a deeper insight into charge dynamics *via* ultrafast spectroscopies (PLTA, UTA) for excitation above and below the threshold value for MEG. Once it is confirmed that MEG produces additional excitons efficiently in the CQD, then if charge extraction from the CQD can be achieved (in less than the typical biexciton lifetime of tens or hundreds of picoseconds), in principle the photocurrent and efficiency of the CQD solar cell will be increased. However, the benefit of MEG to the overall photovoltaic performance is yet to be realised because charge recombination within the CQD and within the charge transport media between the CQDs and the cell electrodes has not yet been sufficiently reduced. For example, at the first demonstration of enhanced photocurrent in a PbS CQD-based device where additional charges were efficiently produced by MEG (with a quantum yield up to 170%), the power conversion of this device remained low due to significant subsequent recombination.[80] However, research since then has led to the optimisation of the device architecture through improved ligand exchange to enhance carrier mobility,[45] extended recombination times *via* passivation of the surface of the CQDs,[81] and optimisation of the quantum dot size and band alignments in oxides.[82,83] This has led to certified power conversions of 9.9[84] and 10.6%.[11,12] Clearly MEG is more efficient in CQDs than in bulk semiconductors, but it still remains to be fully exploited in photovoltaic technologies.

14.5 Trapping

Trap sites can occur on interfaces or dislocations within or on the surface of semiconductors, where dangling bonds, impurities or other discontinuities have the potential to form energetically favourable states for electrons or holes, either hot or cool. The transition into such states, or 'trapping', is a widely studied and debated phenomenon due to a wide range of circumstances leading to trap formation, and their variable energy and efficiency of trapping. For CQDs, the synthesis techniques allow the fine control of nucleation and growth, such that impurities and defects within the dots are rare. As a result, the trap sites are limited to the surface of the CQD or the interfaces between cores and shells. However, the surface chemistry of CQDs is poorly understood and broad in scope, such that many different surface configurations can lead to surface trapping depending on the composition of the CQD, the passivation method and even the position of the trap site relative to the crystal structure. This broad range of traps can affect the CQD charge dynamics and device performance in a number of different ways.

14.5.1 Dangling Bonds and Surface States

Surface effects in CQDs, such as trapping, can allow non-radiative charge relaxation and are a problem for two key reasons. First, they compete with other potentially beneficial processes to reduce the efficiency and performance of CQDs[41,85] – for example, competing with MEG in photovoltaic devices or competing with photoluminescence in emitting devices (Figure 14.8). Second, surface effects significantly complicate the characterisation of, or can be mistaken for, other charge relaxation processes in CQDs.[7,60] This is because they occur on similar timescales and have similar effects on the occupation of energy levels, leading to comparable signals in transient absorption spectroscopy. The surface is thought to activate other potentially unwanted processes, such as CQD blinking.[86–88]

Even if successfully identified, the surface effects themselves are challenging to characterise. This is because they show very dramatic sample-to-sample variation, be this in trapping rates, trap energies or even which carrier is trapped.[51] They are particularly sensitive to the preparation and handling of the sample, such that ostensibly identical syntheses can produce CQDs with different surface properties, which are not readily detected by facile characterisation measurements such as absorption and photoluminescence spectroscopy. This sensitivity is primarily because a large fraction of the constituent atoms of CQDs lie on its surface, where they may have one or more dangling bonds, and because the particular position of these atoms determine the trap properties. For example, a CdTe CQD with a zinc blende structure and a 1.1 nm radius has 76 distinct Te atoms on its surface, although this can be reduced to five inequivalent sites due to the underlying crystal symmetry.[89] Each of these inequivalent sites could form traps with different energies and trapping rates, and each could form a distribution of trapping rates in an ensemble of CQDs due to the variation in alignment with the energy levels of differently sized CQDs.

CQD surface atoms are usually passivated by long-chain organic ligands and any unpassivated dangling bonds can produce surface traps. The binding of these ligands and their interaction with the surface is a widely studied field in itself, where vibrational spectroscopy and density functional theory are often utilised.[90–92] Crucially, the ligands can be loosely bound to the surface and steric hindrance can prevent the often bulky ligands from passivating every surface site, allowing the formation of trap sites and leaving the surface vulnerable to oxidation if not handled carefully. As a result of the large number of surface atoms and the nature of the colloidal growth process, the extent of surface passivation achieved is highly sensitive to the reaction conditions, such as the reagent concentrations, temperature and reaction times; for this reason, reproducibility in synthesis is important for studies of surface effects in CQD.

CQDs are typically synthesised and stored in oxygen-free environments to eliminate or limit the oxidation of the surface. The effects of oxidation of the surface have been studied in many systems.[93] A direct link between the oxidation of the surface of CQDs and the rate of surface trapping has been observed, with a gradual quenching of the PLQY of samples over time on exposure to air.[94] The degree of oxidation of the surface of CQDs can itself have a varied effect on the nature of the trap states and other charge dynamics. A small amount of oxidation can lead to an effective reduction in the size of the charge confinement region of the CQD, which can affect both the transition energies within the dot, the electron–hole wave function overlap — which can change the favourability of Auger processes and recombination — and the relative alignment of CQD energy levels with the trap state levels. A large degree of oxidation can lead to the formation of an oxide shell, which can significantly alter the dielectric environment of the CQD and lead to order of magnitude changes in trapping rates.[89]

14.5.2 Effects of Trap States on Ultrafast Charge Dynamics: Cooling and Recombination

There is a wide range of potential effects on CQD charge dynamics due to surface trapping, many of which can be subtle and often require a combination of several experimental techniques to fully characterise. This is, in part, due to changes in the way that the charge carrier (electron or hole) is trapped, and how the dynamics of each can be resolved separately or together in different techniques.

Sub-nanosecond decay components in transient absorption or TPL signals can be introduced for CQD samples due to the decay of multi-exciton states. These multi-excitons are produced either by the absorption of more than one photon per CQD per pump pulse at high pump fluences, or by MEG if $h\nu \geq h\nu_{th}$. In both of these cases, the fast decay component originates from the Auger recombination of the multi-excitons and terminates when only a single exciton is left in the CQD. In studies which have eliminated these sources of multi-exponential decay, by ensuring low pump fluence and $h\nu < h\nu_{th}$, any remaining multi-exponential decay components are usually attributed to surface trapping. Surface trapping has also been identified to occur on timescales comparable with Auger relaxation, but can also have several more subtle effects on studies of charge dynamics in CQD samples and can obscure measurements of MEG, multi-exciton recombination, optical gain and single dot blinking.[7,51]

One of the first effects of charge trapping that was recognised in CQDs was a consequence of the charges remaining trapped for extended periods. After a charge is trapped, its geminate partner can remain in the CQD core for a long period of time, as long as 20 s,[95] due to the absence of other carriers for charge dissipation or recombination. This process is often referred to as photocharging because the CQD core acquires a net charge.[76,95,96] Although photocharged, if a second electron–hole pair is excited, a trion is formed, where there are three excited charge carriers in the CQD. A trion can recombine by fast Auger recombination in the same way as a multi-exciton, after which a single charge remains, potentially leading to the formation of another trion. An accumulation of many such states over repeated excitation by a pump beam can lead to sub-nanosecond decays in transient absorption studies, even at low pump fluences and below the MEG threshold. This effect was probably mistaken for MEG in some early studies and was later demonstrated to give an apparent MEG yield of 170%.[76,95] Several careful studies[58,60,66] have shown that sufficient stirring or flowing of samples can refresh the sample volume inside the excitation volume of the pump beam faster than the pulse repetition rate, eliminating this misleading signal.

For a transient absorption study on a given CQD sample with surface traps, the effect of charge trapping depends on which charge carrier is trapped and whether that carrier makes a significant contribution to the pump-induced bleach signal at the probe wavelength. For example, in most CQD types, the probed transition is the band edge transition, for which the contributing hole state usually has high degeneracy and a high density of states. It is also common for the band edge transition to correspond to a hole state just below the valance band maximum, meaning that a hole which has cooled to the valance band maximum cannot significantly contribute to the measured photobleach.[97,98]

Isolation of the cooling dynamics of the electron or hole can also be achieved through state-selective pumping and probing, where the optical transitions observable in linear absorption spectra need to be identified as excitonic levels with either relaxed holes or electrons to remove their cooling contribution to the dynamics.[99,100] This can allow the clear determination of hot trapping processes on either carrier. Alternatively, hole traps have been utilised to observe decoupled electron cooling in CdSe CQDs, where the absence of electron Auger cooling by energy transfer to its geminate hole allowed the observation of 3–30 ps electron cooling and phonon bottlenecks.[78,101–103] Such investigations can be combined with TPL studies to identify and extract the radiative lifetimes and other processes.[104]

Transient absorption spectra, where the pump–probe time delay is fixed, but the probe wavelength is scanned, can also yield useful information about non-radiative processes, such as trapping. A typical inter-band transient absorption spectrum will show the development of a single bleach peak in the first tens of picoseconds after the sample is pumped (depending on the time taken for hot charges to relax to the band edge), comparable in width to the absorption peak observed at the edge of absorption spectrum; higher energy transitions are sometimes also observable.[48,105] This large peak corresponds to bleaching of the band edge exciton and is usually labelled B1. If the excited state excitons have a significant binding energy (of the order of 0.1 times the peak width), then a shift in the absorption wavelength is experienced by the probe beam due to the biexciton-induced stark effect[97] for early time spectra while there are still hot excitons in the CQD ensemble. This results in a photoabsorption peak to the side of the main bleach peak (on the low-energy side for negative biexciton binding energy and *vice versa*), which is usually labelled A1. The magnitudes of A1 and B1 can be used to determine the biexciton binding energy.[106]

One signature of trapping in transient absorption spectra has been identified as a broad photoabsorption feature on the low-energy side of the main bleach peak. This was demonstrated by Tyagi and Kambhampati[76] by phototreating the surface of CdSe/ZnS CQDs. The phototreated dots were prepared by irradiating the sample with intense (~2.8 μJ per pulse at 500 Hz) 3.10 eV radiation for ~100 min. This prolonged irradiation altered the surface of the dots such that they exhibited efficient surface trapping.[95,107] The rise of this photoabsorption showed trapping within 3 ps. This broad low-energy photoabsorption is now considered to be a signature of surface effects in transient absorption studies; however, its origin remains elusive[7] and may not appear under low intensity excitation.[108,109]

Determining whether charge trapping of hot or cooled charge carriers is occurring can be difficult to determine, particularly if the trapping process is very fast or does not result in bleach or photoabsorption features in transient absorption spectroscopy, or if they overlap with the B1 signal. In addition to introducing additional time constants in the rise and fall of transient absorption signals,[100,105] surface trapping can significantly reduce the amplitude of the measured signals. Such an effect can have significant implications in a transient absorption study if not recognised, beyond a simple reduction in the signal-to-noise ratio. It has been shown by comparing equivalent samples with and without surface trapping, indicated by a near-unity PLQY in the trap-free samples, that traps can significantly reduce the amplitude of the band edge bleach peak and introduce a decay component of the order of tens of picoseconds (Figure 14.9).[54] Such a reduction in the amplitude of transient absorption signals could have a significant impact on transient absorption experiments, which rely on saturating the band edge bleach to determine

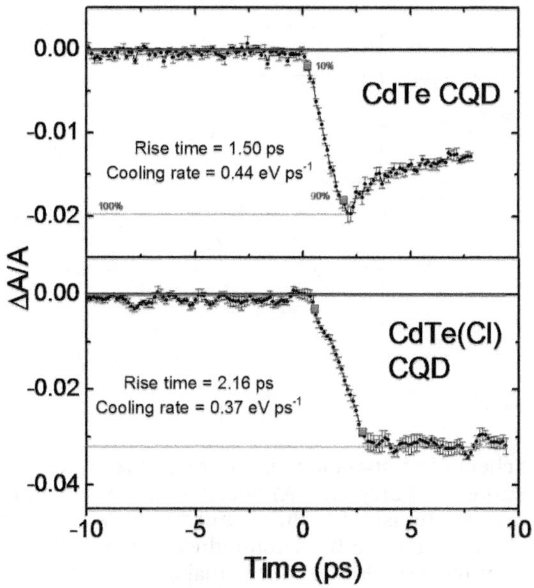

Figure 14.9 Increase in the fractional absorption change, $\Delta A/A$, signal measured on a transient absorption spectrometer for a CdTe CQD sample with and without passivation with chloride ions, which completely eliminate the surface traps.[46,54] The samples were pumped into a hot electron and hole state and probed at their band edge transition. Trap elimination increases the transient absorption response and eliminates a fast decay component. The rise time, defined as the period for the signal to rise from 10 to 90% of the maximum bleach, as indicated by the square symbols, highlights the enhanced rate of cooling to the valence band maximum caused by surface traps.

the CQD absorption cross-sections. This has the potential to lead to the over-excitation of CQD samples, introducing other experimental artefacts.[7]

The dynamics of hot exciton trapping have been reported in some CQD systems and, as with cold trapping, the typical trapping times are of the order of a few picoseconds.[76,86,110] However, because no cooling is required before trapping, the resulting measured dynamics can vary significantly. Kambhampati[51] showed that hot exciton surface trapping can effectively compete with hot exciton relaxation, thereby obfuscating these processes. The subtle effects of hot and cold trapping on the A1 feature depending on whether hot or cold trapping occurs were highlighted; in both cases, the trapped exciton results in a photoabsorption feature due to a biexciton shift (Figure 14.10). The late-time A1 feature in this illustration is a photoabsorption feature (positive), but in different systems the sign will depend on the width of the bleach peak, the magnitude of the biexciton shift and the lifetimes of cooling and trapping.

Studies attempting to determine the rate constants for hot and cold trapping usually find that a single constant for each does not fit the observed decay kinetics. This is usually rectified by assigning several rate constants[111,112]

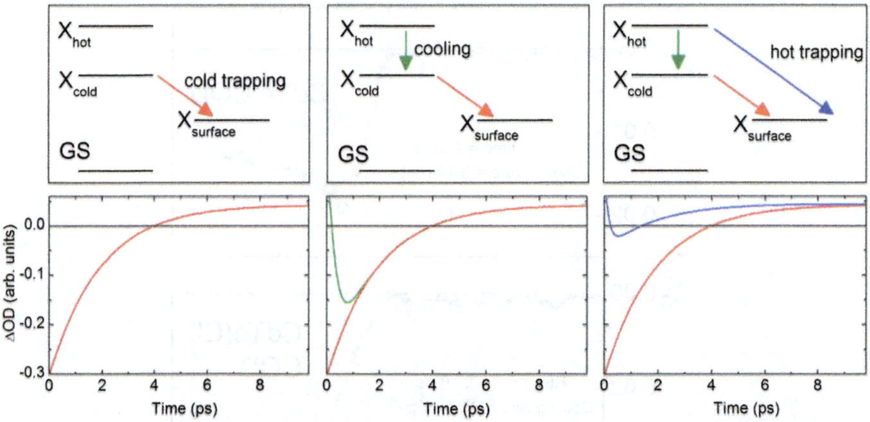

Figure 14.10 Schematic illustration of the real-time observation of hot exciton surface trapping. The A1 spectral feature is monitored in the experimental transients. This feature reflects the excitonic charge distribution *via* the biexciton-induced energy shift. In this illustrative example, we consider a minimal system consisting of a hot exciton (X_{hot}), a cold exciton (X_{cold}) and a surface-trapped exciton ($X_{surface}$). Each exciton has some value for the ΔOD signal in the A1 spectral region. Direct excitation into X_{cold} tracks exciton trapping to the surface (left, red line). Excitation into X_{hot}, followed sequentially by cold exciton trapping (centre, green line) meets the X_{cold} pumping signal after cooling is complete. Excitation into X_{hot}, with direct-to-surface hot exciton trapping (right, blue line), meets the X_{cold} pumping signal only at late times. Reprinted from P. Kambhampati, *J. Phys. Chem. C*, 2011, **115**, 22089–22109. Copyright (2011) American Chemical Society.[51]

or fitting the ensemble kinetics with a slow power law in the form of t−α. This was done by Kern *et al.*[62] when studying the electron trapping kinetics in CdSe nanocrystals using multiple population period transient spectroscopy:[61–63,113] they fitted a power law with $\alpha = 0.16$ over more than three decades in time. Such fitting techniques rely on the fact that there are several inequivalent distributions of trap types on the surface of CQDs, either due to their geometric position or the specific degree of capping/number of dangling bonds.[62,89,104,113]

Signatures of surface trapping have also been identified in photoluminescence spectra. It has been shown that quantum dots can emit broadened and red-shifted photoluminescence, which is attributed to recombination involving an electron trapped on the surface of the nanocrystal. The observed white light emission is temperature-dependent due to an optically bright surface state on very small CdSe CQDs (<2 nm in diameter).[114] Traditionally, this surface radiation had been thought to arise from a broad distribution of mid-bandgap deep trap states.[87,114,115] However, this cannot account for an observed temperature variance in quantum yield. Krause *et al.*[114] reconciled this with a semi-classical electron transfer approach, where a combination of ligand and internal LO phonon interaction is required for transfer to the surface states. There are still some phenomena in photoluminescence studies of trapping that have only been reported in certain systems. Jones *et al.* observed light-activated stabilisation of surface trap states in CdSe CQDs, where an increase in quantum yield and photoluminescence lifetime was observed following prolonged excitation over several hours.[112] Gong *et al.* showed the effect of shell passivation on lifetime and found that the lifetime of luminescence varies across the emission peak of CdSe and CdSe/ZnS CQDs; they attributed this to hole surface trapping.[116]

As a result of the complicated nature of the surface chemistry of passivation arising from the geometric position or specific degree of capping/number of dangling bonds, the modelling of trapping states and their associated rates is particularly challenging. In particular, no explicit calculation of trapping rates using realistic trap state wave functions and reliable coupling matrix elements to core states exists.[63] However, atomistic models considering specific sites on the surface of CQDs can model trapping rates and show good agreement with the experimental results. Califano[89] identified two distinct trapping rates to the CQD surface that are strongly dependent on the dielectric environment of the CQD, allowing the reconciliation of a range of trapping times reported by different groups.

14.5.3 Effect of Passivation on Trapping Dynamics

Organic ligands on the CQD surface can play an important part in the cooling of hot charges to the band edge. The energy of a hot carrier can be transferred to the vibrational modes in the ligand, another form of phonon cooling.[79] This type of coupling was found to favour one carrier over another in some systems[117] and the introduction of specific ligands can lead to the deliberate formation of traps.[51,75,78,97,101,103,118]

Shelling has been utilised to achieve near-unity quantum yields of the core/shell CdSe/CdS CQDs used to enhance the performance of light-emitting diodes[119] and has been successful in improving the quantum yield in other systems.[120,121] It has been shown to remove the photoluminescence originating from defects on the surface of the core before shelling in CdSe/ZnS dots.[122] This enhancement of emission efficiency indicates the suppression of non-radiative recombination and thus the successful passivation of surface states associated with trapping. Although shelling has been shown to almost eliminate the impact of surface trapping in some CQD systems, the shelling process can have some detrimental effects on the potential applications of CQDs. The shell can act as a barrier to one or both carriers for charge extraction in photovoltaic devices. Even in a Type I system, the addition of the shell alters the charge confinement of carriers, such that the energy levels can shift and alter the favourability of other charge relaxation processes. This can make shelling an inappropriate method of characterising surface traps by comparing shelled and unshelled cores.

A recent passivation technique in some types of CQD systems involves the passivation of the surface with halide ions. Chloride ions have primarily been used,[45,123–130] but bromide[45,128] and Iodide[128,131] have also been explored.[128] The halide ions are sufficiently compact to avoid steric hindrance and they passivate the surface sites inaccessible to long-chain organic ligands.[46,129,130] One system that has proved particularly effective is the treatment of CdTe CQDs using chloride ions. The chloride binds to the surface Cd ions and etches and replaces unpassivated surface Te, while leaving the long-chain organic ligands on the surface.[46] A near-unity PLQY after chloride treatment was reported, together with the elimination of two non-radiative components in TPL decays,[94] indicating the almost complete elimination of trap sites. A subsequent report detailed the ultrafast charge dynamics in these dots and found that the chloride treatment eliminated the fast decay components in transient absorption studies, providing evidence that the surface traps exhibit both hot and cold trapping.[54] This form of trap elimination is more suited to a comparison of trap-free and surface trapping cores than Type I core/shell CQDs because passivation is achieved without significant alteration to the confinement of charges in the CQD core.

14.5.4 Auger-mediated Trapping Model

The mechanism for trapping is not completely understood and is still an area of active research. One model that has been used to successfully describe the observed effect on sub-nanosecond timescales is Auger-mediated trapping. In this model, one of the charges in an electron–hole pair relaxes to a lower energy trap state and the energy lost is transferred to its geminate charge partner (Figure 14.11). For example, in a cool electron–hole pair, the electron (hole) at the conduction band minimum (valance band maximum) could relax to a mid-band trap state and transfer its energy to the hole (electron) exciting it deeper (higher) into the valence (conduction) band, where it can then re-cool to the valance band maximum (conduction band minimum) by phonon scattering. The trapping rate constants measured are often comparable with Auger-mediated processes, such as multiple exciton recombination

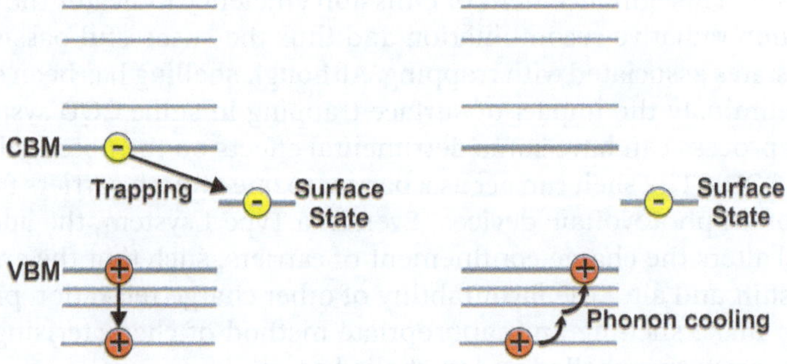

Figure 14.11 Schematic energy level diagram for the mechanism of auger-mediated trapping.

and MEG, and several experimental and theoretical works have found further evidence to support this model in some CQD systems.[63,89,94]

14.6 Recombination Dynamics

Recombination lifetimes for single excitons in CQDs range from a few nanoseconds up to about a microsecond[1] and are typically determined experimentally by studying photoluminescence decay transients. These transients are usually multi-exponential in form, but a mono-exponential or near-mono-exponential decay can be observed if the CQDs are well passivated – examples of both types of transient are shown in Figure 14.12. Passivation saturates dangling bonds on the CQD surface, *i.e.* removes the surface states, results in an increased PLQY and also tends to make the form of the photoluminescence decay closer to mono-exponential. We reported CQDs which, after treating the surface with chloride ions, had a near-unity PLQY,[46] demonstrating that the non-radiative recombination of single excitons in CQDs is caused by surface states and that it is possible to completely passivate these states so that recombination becomes purely radiative. The decay transient for these CQDs changed from a multi-exponential form before treatment to one that was mono-exponential over several decades of intensity afterwards (Figure 14.12), showing that the multi-exponential photoluminescence decay typically seen for CQDs is also due to the influence of surface states. This ability to completely remove the

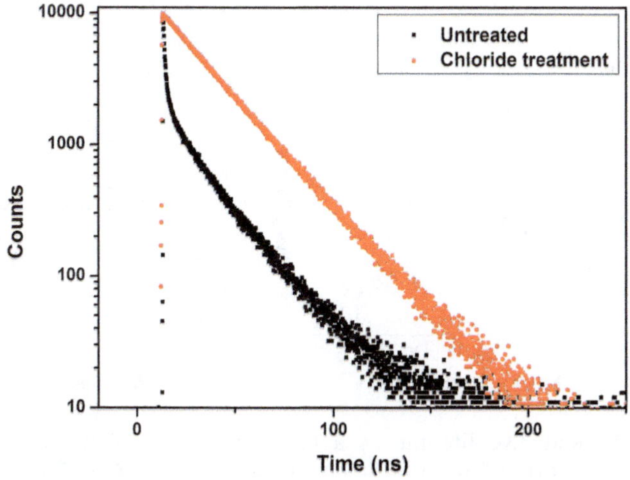

Figure 14.12 photoluminescence transients obtained for CdTe CQDs before and after treatment with chloride ions, which resulted in the complete passivation of surface states. Reproduced with permission from R. C. Page *et al.*, Near-Unity Quantum Yields from Chloride Treated CdTe Colloidal Quantum Dots, Small, John Wiley and Sons © 2014 The Authors. Published by WILEY-VCH Verlag GmbH & Co. KGaA, Weinheim.[46]

surface states, and hence 'switch off' all non-radiative recombination for the first time, enables the radiative recombination process that remains to be studied directly, with the photoluminescence decays now free of any influence of charge trapping by the surface states. Conversely, by comparison of the transients obtained before and after the chloride treatment, the contribution of radiative recombination can be clearly identified in the decay for the untreated CQDs, which allows the influence of surface states to be better understood.

Figure 14.13a shows the radiative lifetimes extracted from the mono-exponential photoluminescence transients obtained for chloride-treated CdTe CQDs as a function of CQD size. The lifetime is observed to increase linearly with the CQD volume, as expected for core-only CQDs.[132] In this case, the CQD volume was selected from the distribution of sizes within a CQD ensemble by observing the decay transients at different wavelengths across the photoluminescence spectrum and using a previously established relationship between CQD size and bandgap.[133,134] Similar results were also obtained when the photoluminescence transients were measured at the photoluminescence peak for different CQD samples with a range of average sizes.[94] Although this simple linear scaling with CQD volume holds for core-only or Type I core/shell CQDs, more complex behaviour is expected for Type II core/shell CQDs. In a Type II CQD, because of the staggered band offset between the two materials, one of the charge carriers will tend to localise in the core and the other in the shell. For instance, for a CdSe/CdTe core/shell CQD, the electron will localise more in the core and the

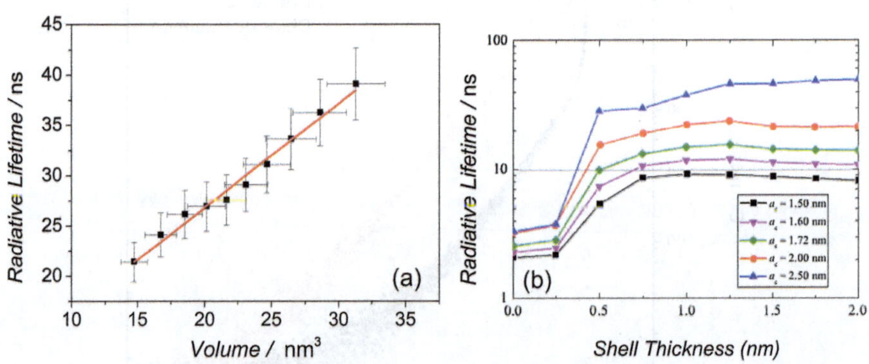

Figure 14.13 Radiative lifetime as a function of (a) volume for CdTe CQDs (reprinted from D. Espinobarro-Velazquez *et al.*, Effect of Chloride Passivation on Recombination Dynamics in CdTe Colloidal Quantum Dots, *ChemPhysChem*, John Wiley and Sons, © 2015 The Authors. Published by Wiley-VCH Verlag GmbH & Co. KGaA[94]) and (b) CdTe shell thickness for several different CdSe core radii, a_C, in a Type II core/shell CQD (reprinted from *Solar Energy Materials and Solar Cells*, 159, M. A. Leontiadou *et al.*, Influence of elevated radiative lifetime on efficiency of CdSe/CdTe Type II colloidal quantum dot based solar cells, 657–663, Copyright 2017 with permission from Elsevier[136]).

hole will tend to move to the shell. This separation of the charges, *i.e.* a reduced overlap of the wave functions of the two carriers, reduces the interaction rate between them, resulting in an increased lifetime for radiative recombination. This effect has been studied theoretically by calculating the radiative lifetimes for a CdSe/CdTe CQD as a function of shell thickness for different core sizes using a detailed model that includes both Coulomb and configuration interactions between the carriers[78] (Figure 14.13b). For thin shells <0.5 nm, there is very little increase in the lifetime compared with CdSe core-only CQDs because the hole is not yet able to localise in the shell.[135] However, for greater shell thicknesses this localisation becomes possible and the consequent reduction in carrier wave function overlap causes the radiative recombination lifetime to increase by an order of magnitude. Thus a Type II CQD design allows the rate of radiative recombination, which competes with charge escape from the CQD, to be decreased by a much greater degree than can be achieved with the linear increase with volume seen for a core-only or Type I core/shell CQD.

The photoluminescence behaviour of CQDs that are not fully passivated allows some features of the process of non-radiative recombination to be deduced. These CQDs typically exhibit a PLQY significantly <100% and a photoluminescence decay transient that is multi-exponential in form. The phenomenon of photoluminescence intermittency, also known 'blinking', is also often seen for such CQDs.[137] Individual CQDs are observed to become non-emissive (*i.e.* 'blink off') for periods of time longer than the duration of a photoluminescence decay experiment, but can then become emissive again at a later time (*i.e.* 'blink on'). These observations can be explained by regarding a typical population of CQDs to consist of three types: trap-free, dark-trapping and bright-trapping. 'Traps' in this instance are mid-gap states created by the imperfect passivation of the surface that can capture photogenerated charge carriers.

The recent demonstration of a unity quantum yield for chloride-passivated CdTe CQDs shows that it is possible for CQDs to be free of such traps and we assume that any population of CQDs will contain some fraction in which recombination is purely radiative. The non-emissive periods observed for individual CQDs during blinking experiments has been well explained by the long-term trapping of charges in surface states and it is this type of CQD that we describe here as dark-trapping. However, although these two types of CQD are sufficient to explain the observed PLQY values and blinking effects, they do not explain the multi-exponential form of the photoluminescence decay transient that is often seen. Dark-trapping CQDs produce blinking and, during a non-emissive period, contribute to the reduction in the PLQY. However, because they do not emit, they do not have an effect on the photoluminescence decay transient, which, if it was solely due to trap-free CQDs, would be mono-exponential.

The multi-exponential form actually observed has been explained by allowing for bright-trapping CQDs.[138] This type of CQD possesses 'shallow traps' – that is, mid-gap surface states that are sufficiently close in energy to the band edge to allow a trapped electron or hole to be thermally liberated from the surface state back to the conduction band or valance band, respectively, as well as being a pathway for non-radiative recombination. A rate equation analysis of this model reveals that shallow trapping results in a photoluminescence decay transient that is bi-exponential in form, with the two associated time constants both depending on the rates of radiative recombination from the band edge, capture into the trap, escape from the trap and non-radiative recombination from the trap.[138] The photoluminescence decay transient from the whole ensemble will thus be tri-exponential, with a mono-exponential contribution from the trap-free CQDs that decays with the radiative recombination lifetime and a bi-exponential contribution from the bright-trapping CQDs with associated lifetimes that are typically different from that of radiative recombination.

Theoretical modelling of the states formed by unsaturated atoms on a CQD surface has been reported,[139] substantiating this description of recombination and allowing the formation of different types of traps to be better understood. This approach calculates the effect on the electronic structure of a CQD produced by individual unsaturated atoms at different points on its surface. For instance, for CdTe CQDs this approach reveals that unsaturated Te atoms produce surface states within the bandgap that can trap holes.[89] The precise energy of these states depends on the position of the atom on the CQD, but is largely determined by the number of dangling bonds per atom. For the surface Te atom with two dangling bonds, these states are clustered about an energy 50 meV above the position of the valence band maximum, whereas the corresponding states for Te atoms with a single unsaturated bond are centred at an energy 450 meV above the valence band maximum. Thus holes trapped in the states produced by two dangling bonds can escape back to the valance band using thermal energy at room temperature, *i.e.* they can act as the shallow traps used to explain the multi-exponential form of the photoluminescence decay transient. By contrast, the states formed when a Te atom has just one dangling bond are too deep within the bandgap for thermal liberation.

The characteristic trapping time from the valence band maximum into these states can be calculated using Auger-mediated trapping.[94] These vary sensitively with the position of the Te atom on the CQD surface leading, as shown in Figure 14.14, to a distribution of times that ranges across three orders of magnitude. The dielectric environment was also found to be an important factor in the magnitude of these times. If the CQD is embedded in a material with a low relative dielectric constant of $\varepsilon = 2.2$, corresponding to common organic solvents, then the trapping times range from 1 ps to 1 ns, but if the ε of the environment is increased to 6.0, a value corresponding to CdO, then the times in this distribution increase by an order of magnitude. This indicates that one of the consequences of the exposure of these CQDs to atmospheric oxygen is a significant slowing of the trapping process.

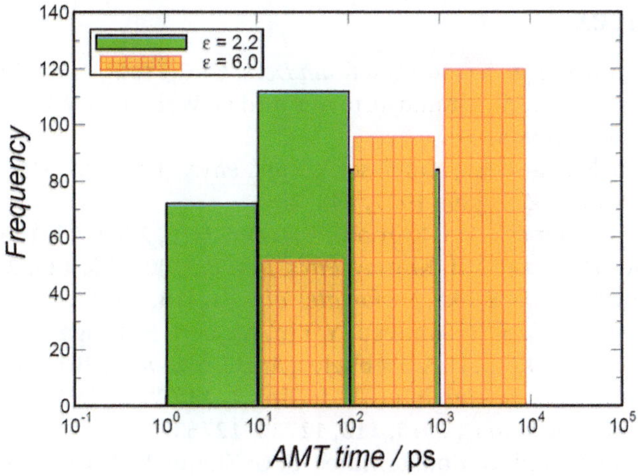

Figure 14.14 Distribution of auger-mediated trapping (AMT) times into states produced by Te atoms with either one or two dangling bonds across the range of possible Te sites on a CdTe CQD surface for surrounding materials with relative dielectric constants of 2.2 (green) and 6 (orange). Reprinted from D. Espinobarro-Velazquez *et al.*, Effect of Chloride Passivation on Recombination Dynamics in CdTe Colloidal Quantum Dots, *ChemPhysChem*, John Wiley and Sons, © 2015 The Authors. Published by Wiley-VCH Verlag GmbH & Co. KGaA.[94]

14.7 Summary

The size-tunable absorption edge and the potential for MEG of CQDs, combined with the facile and scalable solvent-based methods by which they can be made and processed, make them promising as the photoabsorbing species for Type III photovoltaic devices. However, realising these potential benefits depends crucially on the relative rates of the various processes that occur for photogenerated charges. The quantum yield of MEG is determined by its rate relative to the other cooling processes and the overall efficiency of a solar cell depends on the rate of recombination relative to charge extraction to the external circuit. We have described the processes key to charge dynamics in CQDs and how the size, structure and surface passivation of CQDs may be engineered to minimise those processes that reduce the effectiveness of CQDs as the photoabsorbing species in solar cells.

Acknowledgements

The authors thank EPSRC for financial support under grant no. EP/K008544/1 and the following collaborators for useful conversations: Paul O'Brien and Wendy Flavell (University of Manchester), Marco Califano (University of Leeds) and Stanko Tomic (University of Salford).

References

1. V. I. Klimov, *Nanocrystal Quantum Dots*, CRC Press, 2nd edn, 2010.
2. H. Benisty, C. M. Sotomayortorres and C. Weisbuch, *Phys. Rev. B*, 1991, **44**, 10945–10948.
3. W. Shockley and H. J. Queisser, *J. Appl. Phys.*, 1961, **32**, 510–519.
4. H. A. Muser, *Z. Physik*, 1957, **148**, 385.
5. R. D. Schaller and V. I. Klimov, *Phys. Rev. Lett.*, 2004, **92**, 186601.
6. V. I. Rupasov and V. I. Klimov, *Phys. Rev. B.*, 2007, **76**, 125321.
7. C. Smith and D. Binks, *Nanomaterials*, 2014, **4**, 19–45.
8. S. ten Cate, C. S. S. Sandeep, Y. Liu, M. Law, S. Kinge, A. J. Houtepen, J. M. Schins and L. D. A. Siebbeles, *Acc. Chem. Res.*, 2015, **48**, 174–181.
9. G. H. Carey, A. L. Abdelhady, Z. J. Ning, S. M. Thon, O. M. Bakr and E. H. Sargent, *Chem. Rev.*, 2015, **115**, 12732–12763.
10. S. A. McDonald, G. Konstantatos, S. G. Zhang, P. W. Cyr, E. J. D. Klem, L. Levina and E. H. Sargent, *Nat. Mater.*, 2005, **4**, 138–142.
11. *Best Research Cell-efficiencies*, National Renewable Energy Laboratory, 2016, http://www.nrel.gov/ncpv/images/efficiency_chart.jpg, accessed 01.11.2016.
12. G. H. Kim, F. P. G. de Arquer, Y. J. Yoon, X. Z. Lan, M. X. Liu, O. Voznyy, Z. Y. Yang, F. J. Fan, A. H. Ip, P. Kanjanaboos, S. Hoogland, J. Y. Kim and E. H. Sargent, *Nano Lett.*, 2015, **15**, 7691–7696.
13. A. J. Nozik, *Nano Lett.*, 2010, **10**, 2735–2741.
14. R. Debnath, O. Bakr and E. H. Sargent, *Energy Environ. Sci.*, 2011, **4**, 4870–4881.
15. F. Hetsch, X. Q. Xu, H. K. Wang, S. V. Kershaw and A. L. Rogach, *J. Phys. Chem. Lett.*, 2011, **2**, 1879–1887.
16. A. Shabaev, C. S. Hellberg and A. L. Efros, *Acc. Chem. Res.*, 2013, **46**, 1242–1251.
17. Y. Liu, M. Gibbs, J. Puthussery, S. Gaik, R. Ihly, H. W. Hillhouse and M. Law, *Nano Lett.*, 2010, **10**, 1960–1969.
18. J. M. Luther, M. Law, Q. Song, C. L. Perkins, M. C. Beard and A. J. Nozik, *ACS Nano*, 2008, **2**, 271–280.
19. P. Nagpal and V. I. Klimov, *Nat. Commun.*, 2011, **2**, 486.
20. X. Z. Lan, S. Masala and E. H. Sargent, *Nat. Mater.*, 2014, **13**, 233–240.
21. *Semiconductor Quantum Dots: Organometallic and Inorganic Synthesis*, ed. M. Green, RSC, 2014.
22. J. Y. Kim, O. Voznyy, D. Zhitomirsky and E. H. Sargent, *Adv. Mater. (Weinheim, Ger.)*, 2013, **25**, 4986–5010.
23. S. Krishnadasan, R. J. C. Brown, A. J. deMello and J. C. deMello, *Lab Chip*, 2007, **7**, 1434–1441.
24. T. W. Phillips, I. G. Lignos, R. M. Maceiczyk, A. J. deMello and J. C. deMello, *Lab Chip*, 2014, **14**, 3172–3180.
25. X. Peng, J. Wickham and A. P. Alivisatos, *J. Am. Chem. Soc.*, 1998, **120**, 5343–5344.
26. C. B. Murray, D. J. Norris and M. G. Bawendi, *J. Am. Chem. Soc.*, 1993, **115**, 8706–8715.

27. H. Weller, H. M. Schmidt, U. Koch, A. Fojtik, S. Baral, A. Henglein, W. Kunath, K. Weiss and E. Dieman, *Chem. Phys. Lett.*, 1986, **124**, 557–560.
28. L. E. Brus, *J. Chem. Phys.*, 1984, **80**, 4403–4409.
29. T. Trindade, P. O'Brien, X.-M. Zhang and M. Motevalli, *J. Mater. Chem.*, 1997, **7**, 1011–1016.
30. W. G. Lu, J. Y. Fang, K. L. Stokes and J. Lin, *J. Am. Chem. Soc.*, 2004, **126**, 11798–11799.
31. C. B. Murray, S. Sun, W. Gaschler, H. Doyle, T. A. Betley and C. R. Kagan, *IBM J. Res. Dev.*, 2001, **45**, 47–56.
32. W. W. Yu, J. C. Falkner, B. S. Shih and V. L. Colvin, *Chem. Mater.*, 2004, **16**, 3318–3322.
33. R. L. Wells, C. G. Pitt, A. T. Mcphail, A. P. Purdy, S. Shafieezad and R. B. Hallock, *Chem. Mater.*, 1989, 4–6.
34. M. D. Healy, P. E. Laibinis, P. D. Stupik and A. R. Barron, *J. Chem. Soc., Chem. Commun.*, 1989, 359–360.
35. S. Xu, J. Ziegler and T. Nann, *J. Mater. Chem.*, 2008, **18**, 2653–2656.
36. M. A. Malik, P. O'Brien, S. Norager and J. Smith, *J. Mater. Chem.*, 2003, **13**, 2591–2595.
37. M. Green, S. Norager, P. Moriarty, M. Motevalli and P. O'Brien, *J. Mater. Chem.*, 2000, **10**, 1939–1943.
38. M. D. Tessier, D. Dupont, K. De Nolf, J. De Roo and Z. Hens, *Chem. Mater.*, 2015, **27**, 4893–4898.
39. D. Aldakov, A. Lefrancois and P. Reiss, *J. Mater. Chem. C*, 2013, **1**, 3756–3776.
40. P. Reiss, M. Protiere and L. Li, *Small*, 2009, **5**, 154–168.
41. N. McElroy, R. C. Page, D. Espinbarro-Valazquez, E. Lewis, S. Haigh, P. O'Brien and D. J. Binks, *Thin Solid Films*, 2014, **560**, 65–70.
42. E. A. Lewis, R. C. Page, D. J. Binks, T. J. Pennycook, P. O'Brien and S. J. Haigh, *J. Phys.: Conf. Ser.*, 2014, **522**.
43. J. J. Li, Y. A. Wang, W. Guo, J. C. Keay, T. D. Mishima, M. B. Johnson and X. Peng, *J. Am. Chem. Soc.*, 2003, **125**, 12567–12575.
44. C. Bullen and P. Mulvaney, *Langmuir*, 2006, **22**, 3007–3013.
45. J. Tang, K. W. Kemp, S. Hoogland, K. S. Jeong, H. Liu, L. Levina, M. Furukawa, X. H. Wang, R. Debnath, D. K. Cha, K. W. Chou, A. Fischer, A. Amassian, J. B. Asbury and E. H. Sargent, *Nat. Mater.*, 2011, **10**, 765–771.
46. R. C. Page, D. Espinobarro-Velazquez, M. A. Leontiadou, C. Smith, E. A. Lewis, S. J. Haigh, C. Li, H. Radtke, A. Pengpad, F. Bondino, E. Magnano, I. Pis, W. R. Flavell, P. O'Brien and D. J. Binks, *Small*, 2015, **11**, 1548–1554.
47. M. V. Kovalenko, M. Scheele and D. V. Talapin, *Science (Washington, DC, U. S.)*, 2009, **324**, 1417–1420.
48. F. C. M. Spoor, L. T. Kunneman, W. H. Evers, N. Renaud, F. C. Grozema, A. J. Houtepen and L. D. A. Siebbeles, *ACS Nano*, 2016, **10**, 695–703.
49. M. Lorenc, M. Ziolek, R. Naskrecki, J. Karolczak, J. Kubicki and A. Maciejewski, *Appl. Phys. B: Lasers Opt.*, 2002, **74**, 19–27.
50. D. J. Norris and M. G. Bawendi, *Phys. Rev. B*, 1996, **53**, 16338–16346.
51. P. Kambhampati, *J. Phys. Chem. C*, 2011, **115**, 22089–22109.

52. G. Nair and M. G. Bawendi, *Phys. Rev. B*, 2007, **76**, 081304(R).
53. R. D. Schaller, M. Sykora, S. Jeong and V. I. Klimov, *J. Phys. Chem. B*, 2006, **110**, 25332–25338.
54. C. T. Smith, M. A. Leontiadou, R. Page, P. O'Brien and D. J. Binks, *Adv. Sci.*, 2015, **2**, 7.
55. A. Al-Otaify, M. A. Leontiadou, F. V. E. dos Reis, T. C. Damato, P. H. C. Camargo and D. J. Binks, *Phys. Chem. Chem. Phys.*, 2014, **16**, 14189–14194.
56. M. Cadirci, S. K. Stubbs, S. J. O. Hardman, O. Masala, G. Allan, C. Delerue, N. Pickett and D. J. Binks, *Phys. Chem. Chem. Phys.*, 2012, **14**, 15166–15172.
57. M. Cadirci, S. K. Stubbs, S. M. Fairclough, E. J. Tyrrell, A. A. R. Watt, J. M. Smith and D. J. Binks, *Phys. Chem. Chem. Phys.*, 2012, **14**, 13638–13645.
58. S. J. O. Hardman, D. M. Graham, S. K. Stubbs, B. F. Spencer, E. A. Seddon, H. T. Fung, S. Gardonio, F. Sirotti, M. G. Silly, J. Akhtar, P. O'Brien, D. J. Binks and W. R. Flavell, *Phys. Chem. Chem. Phys.*, 2011, **13**, 20275–20283.
59. O. Devos, N. Mouton, M. Sliwa and C. Ruckebusch, *Anal. Chim. Acta*, 2011, **705**, 64–71.
60. J. A. McGuire, J. Joo, J. M. Pietryga, R. D. Schaller and V. I. Klimov, *Acc. Chem. Res.*, 2008, **41**, 1810–1819.
61. H. R. Wu and M. A. Berg, *J. Chem. Phys.*, 2013, **138**, 034201.
62. S. J. Kern, K. Sahu and M. A. Berg, *Nano Lett.*, 2011, **11**, 3493–3498.
63. F. M. Gómez-Campos and M. Califano, *Nano Lett.*, 2012, **12**, 4508–4517.
64. D. J. Binks, *Phys. Chem. Chem. Phys.*, 2011, **13**, 12693–12704.
65. J. A. McGuire, M. Sykora, J. Joo, J. M. Pietryga and V. I. Klimov, *Nano Lett.*, 2010, **10**, 2049–2057.
66. A. G. Midgett, H. W. Hillhouse, B. K. Hughes, A. J. Nozik and M. C. Beard, *J. Phys. Chem. C*, 2010, **114**, 17486–17500.
67. C. M. Cirloganu, L. A. Padilha, Q. L. Lin, N. S. Makarov, K. A. Velizhanin, H. M. Luo, I. Robel, J. M. Pietryga and V. I. Klimov, *Nat. Commun.*, 2014, **5**, 4148.
68. M. Acrts, T. Bielewicz, C. Klinke, F. C. Grozema, A. J. Houtepen, J. M. Schins and L. D. A. Siebbeles, *Nat. Commun.*, 2014, **5**, 3789.
69. S. K. Stubbs, S. J. O. Hardman, D. M. Graham, B. F. Spencer, W. R. Flavell, P. Glarvey, O. Masala, N. L. Pickett and D. J. Binks, *Phys. Rev. B*, 2010, **81**, 081303(R).
70. R. D. Schaller, J. M. Pietryga and V. I. Klimov, *Nano Lett.*, 2007, **7**, 3469–3476.
71. D. Gachet, A. Avidan, I. Pinkas and D. Oron, *Nano Lett.*, 2010, **10**, 164–170.
72. C. J. Stolle, T. B. Harvey, D. R. Pernik, J. I. Hibbert, J. Du, D. J. Rhee, V. A. Akhavan, R. D. Schaller and B. A. Korgel, *J. Phys. Chem. Lett.*, 2014, **5**, 304–309.
73. S. Saeed, C. de Weerd, P. Stallinga, F. C. M. Spoor, A. J. Houtepen, L. D. A. Siebbeles and T. Gregorkiewicz, *Light: Sci. Appl.*, 2015, **4**, e251.
74. M. T. Trinh, R. Limpens, W. D. A. M. de Boer, J. M. Schins, L. D. A. Siebbeles and T. Gregorkiewicz, *Nat. Photonics*, 2012, **6**, 316–321.

75. P. Guyot-Sionnest, B. Wehrenberg and D. Yu, *J. Chem. Phys.*, 2005, **123**, 074709.
76. P. Tyagi and P. Kambhampati, *J. Chem. Phys.*, 2011, **134**, 094706.
77. C. Smith, E. J. Tyrrell, M. A. Leontiadou, J. Miloszewski, T. Walsh, M. Cadirci, R. Page, P. O'Brien, D. Binks and S. Tomić, *Solar Energy Materials and Solar Cells*, Editon edn, 2016.
78. A. Pandey and P. Guyot-Sionnest, *Science*, 2008, **322**, 929–932.
79. M. D. Peterson, L. C. Cass, R. D. Harris, K. Edme, K. Sung and E. A. Weiss, *Annu. Rev. Phys. Chem.*, 2014, **65**, 317–339.
80. J. B. Sambur, T. Novet and B. A. Parkinson, *Science*, 2010, **330**, 63–66.
81. D. C. J. Neo, C. Cheng, S. D. Stranks, S. M. Fairclough, J. S. Kim, A. I. Kirkland, J. M. Smith, H. J. Snaith, H. E. Assender and A. A. R. Watt, *Chem. Mater.*, 2014, **26**, 4004–4013.
82. W. A. Tisdale, K. J. Williams, B. A. Timp, D. J. Norris, E. S. Aydil and X. Y. Zhu, *Science*, 2010, **328**, 1543–1547.
83. C. H. M. Chuang, P. R. Brown, V. Bulovic and M. G. Bawendi, *Nat. Mater.*, 2014, **13**, 796–801.
84. A. J. Labelle, S. M. Thon, S. Masala, M. M. Adachi, H. P. Dong, M. Farahani, A. H. Ip, A. Fratalocchi and E. H. Sargent, *Nano Lett.*, 2015, **15**, 1101–1108.
85. P. V. Kamat, *J. Phys. Chem. Lett.*, 2013, **4**, 908–918.
86. J. Mooney, M. M. Krause and P. Kambhampati, *J. Phys. Chem. C*, 2014, **118**, 7730–7739.
87. J. Mooney, M. M. Krause, J. I. Saari and P. Kambhampati, *Phys. Rev. B*, 2013, **87**, 081201(R).
88. J. Tang and R. A. Marcus, *J. Chem. Phys.*, 2005, **123**, 054704.
89. M. Califano, *ACS Nano*, 2015, **9**, 2960–2967.
90. A. M. Abuelela, T. A. Mohamed and O. V. Prezhdo, *J. Phys. Chem. C*, 2012, **116**, 14674–14681.
91. B. von Holt, S. Kudera, A. Weiss, T. E. Schrader, L. Manna, W. J. Parak and M. Braun, *J. Mater. Chem.*, 2008, **18**, 2728–2732.
92. S. Kilina, S. Ivanov and S. Tretiak, *J. Am. Chem. Soc.*, 2009, **131**, 7717–7726.
93. M. Sykora, A. Y. Koposov, J. A. McGuire, R. K. Schulze, O. Tretiak, J. M. Pietryga and V. I. Klimov, *ACS Nano*, 2010, **4**, 2021–2034.
94. D. Espinobarro-Velazquez, M. A. Leontiadou, R. C. Page, M. Califano, P. O'Brien and D. J. Binks, *ChemPhysChem*, 2015, **16**, 1239–1244.
95. S. L. Sewall, R. R. Cooney, K. E. H. Anderson, E. A. Dias, D. M. Sagar and P. Kambhampati, *J. Chem. Phys.*, 2008, **129**, 084701.
96. P. Tyagi, R. R. Cooney, S. L. Sewall, D. M. Sagar, J. I. Saari and P. Kambhampati, *Nano Lett.*, 2010, **10**, 3062–3067.
97. V. I. Klimov, *Annual Review of Physical Chemistry*, Editon edn, 2007, vol. 58, pp. 635–673.
98. S. C. Boehme, J. M. Azpiroz, Y. V. Aulin, F. C. Grozema, D. Vanmaekelbergh, L. D. A. Siebbeles, I. Infante and A. J. Houtepen, *Nano Lett.*, 2015, **15**, 3056–3066.
99. B. L. Wu, W. Cai and S. K. Gayen, *Multimodal Biomedical Imaging Viii*, 2013, vol. 8574.

100. K. J. Schnitzenbaumer, T. Labrador and G. Dukovic, *J. Phys. Chem. C*, 2015, **119**, 13314–13324.
101. P. Guyot-Sionnest, M. Shim, C. Matranga and M. Hines, *Phys. Rev. B*, 1999, **60**, R2181–R2184.
102. V. I. Klimov, S. A. Ivanov, J. Nanda, M. Achermann, I. Bezel, J. A. McGuire and A. Piryatinski, *Nature*, 2007, **447**, 441–446.
103. J. Urayama, T. B. Norris, J. Singh and P. Bhattacharya, *Phys. Rev. Lett.*, 2001, **86**, 4930–4933.
104. K. E. Knowles, E. A. McArthur and E. A. Weiss, *ACS Nano*, 2011, **5**, 2026–2035.
105. Z. J. Jiang and D. F. Kelley, *J. Phys. Chem. C*, 2010, **114**, 17519–17528.
106. V. I. Klimov, *Appl. Phys. Lett.*, 2006, **89**, 123118.
107. E. J. Tyrrell and J. M. Smith, *Phys. Rev. B*, 2011, **84**, 165328.
108. C. Burda, S. Link, T. C. Green and M. A. El-Sayed, *J. Phys. Chem. B*, 1999, **103**, 10775–10780.
109. D. H. Son, J. S. Wittenberg, U. Banin and A. P. Alivisatos, *J. Phys. Chem. B*, 2006, **110**, 19884–19890.
110. L. A. Padilha, I. Robel, D. C. Lee, P. Nagpal, J. M. Pietryga and V. I. Klimov, *ACS Nano*, 2011, **5**, 5045–5055.
111. M. Jones, S. S. Lo and G. D. Scholes, *Proc. Natl. Acad. Sci. U. S. A.*, 2009, **106**, 3011–3016.
112. M. Jones, J. Nedeljkovic, R. J. Ellingson, A. J. Nozik and G. Rumbles, *J. Phys. Chem. B*, 2003, **107**, 11346–11352.
113. M. A. Berg, *Adv. Chem. Phys.*, 2012, **150**, 1–102.
114. M. M. Krause, J. Mooney and P. Kambhampati, *ACS Nano*, 2013, **7**, 5922–5929.
115. A. Hässelbarth, A. Eychmüller and H. Weller, *Chem. Phys. Lett.*, 1993, **203**, 271–276.
116. H. M. Gong, Z. K. Zhou, H. Song, Z. H. Hao, J. B. Han, Y. Y. Zhai, S. Xiao and Q. Q. Wang, *J. Fluoresc.*, 2007, **17**, 715–720.
117. K. Hyeon-Deuk, A. B. Madrid and O. V. Prezhdo, *Dalton Trans.*, 2009, 10069–10077.
118. M. S. Kang, A. Sahu, D. J. Norris and C. D. Frisbie, *Nano Lett.*, 2011, **11**, 3887–3892.
119. A. Saha, K. V. Chellappan, K. S. Narayan, J. Ghatak, R. Datta and R. Viswanatha, *J. Phys. Chem. Lett.*, 2013, **4**, 3544–3549.
120. X. G. Peng, M. C. Schlamp, A. V. Kadavanich and A. P. Alivisatos, *J. Am. Chem. Soc.*, 1997, **119**, 7019–7029.
121. O. Chen, J. Zhao, V. P. Chauhan, J. Cui, C. Wong, D. K. Harris, H. Wei, H. S. Han, D. Fukumura, R. K. Jain and M. G. Bawendi, *Nat. Mater.*, 2013, **12**, 445–451.
122. H. Yokota, K. Okazaki, K. Shimura, M. Nakayama and D. Kim, *J. Phys. Chem. C*, 2012, **116**, 5456–5459.
123. W. K. Bae, J. Joo, L. A. Padilha, J. Won, D. C. Lee, Q. L. Lin, W. K. Koh, H. M. Luo, V. I. Klimov and J. M. Pietryga, *J. Am. Chem. Soc.*, 2012, **134**, 20160–20168.

124. D. Zhitomirsky, M. Furukawa, J. Tang, P. Stadler, S. Hoogland, O. Voznyy, H. Liu and E. H. Sargent, *Adv. Mater.*, 2012, **24**, 6181–6185.
125. J. S. Owen, J. Park, P. E. Trudeau and A. P. Alivisatos, *J. Am. Chem. Soc.*, 2008, **130**, 12279–12281.
126. M. Zanella, L. Maserati, M. P. Leal, M. Prato, R. Lavieville, M. Povia, R. Krahne and L. Manna, *Chem. Mater.*, 2013, **25**, 1423–1429.
127. L. M. Wheeler, N. R. Neale, T. Chen and U. R. Kortshagen, *Nat. Commun.*, 2013, **4**, 2197.
128. J. B. Zhang, J. B. Gao, E. M. Miller, J. M. Luther and M. C. Beard, *ACS Nano*, 2014, **8**, 614–622.
129. A. H. Ip, S. M. Thon, S. Hoogland, O. Voznyy, D. Zhitomirsky, R. Debnath, L. Levina, L. R. Rollny, G. H. Carey, A. Fischer, K. W. Kemp, I. J. Kramer, Z. J. Ning, A. J. Labelle, K. W. Chou, A. Amassian and E. H. Sargent, *Nat. Nanotechnol.*, 2012, **7**, 577–582.
130. K. Katsiev, A. H. Ip, A. Fischer, I. Tanabe, X. Zhang, A. R. Kirmani, O. Voznyy, L. R. Rollny, K. W. Chou, S. M. Thon, G. H. Carey, X. Y. Cui, A. Amassian, P. Dowben, E. H. Sargent and O. M. Bakr, *Adv. Mater.*, 2014, **26**, 937–942.
131. Z. J. Ning, Y. Ren, S. Hoogland, O. Voznyy, L. Levina, P. Stadler, X. Z. Lan, D. Zhitomirsky and E. H. Sargent, *Adv. Mater.*, 2012, **24**, 6295–6299.
132. A. Javier, D. Magana, T. Jennings and G. F. Strouse, *Appl. Phys. Lett.*, 2003, **83**, 1423–1425.
133. W. W. Yu, L. H. Qu, W. Z. Guo and X. G. Peng, *Chem. Mater.*, 2003, **15**, 2854–2860.
134. W. W. Yu, L. H. Qu, W. Z. Guo and X. G. Peng, *Chem. Mater.*, 2004, **16**, 560.
135. E. J. Tyrrell and S. Tomic, *J. Phys. Chem. C*, 2015, **119**, 12720–12730.
136. M. A. Leontiadou, E. J. Tyrrell, C. T. Smith, D. Espinbarro-Velasquez, J. Miloszewski, T. Walsh, D. Binks and S. Tomic, *Sol. Energy Mater. Sol. Cells*, 2017, **159**, 657–663.
137. C. Galland, Y. Ghosh, A. Steinbruck, M. Sykora, J. A. Hollingsworth, V. I. Klimov and H. Htoon, *Nature*, 2011, **479**, 203–275.
138. H. Y. Wang, S. B. Cai, A. Ernstberger, B. J. Bailey, M. Z. Wang, W. J. Cai, W. S. Goebel, M. B. Czader, C. Crean, A. Suvannasankha, I. Shokolenkoc, G. L. Wilson, A. R. Baluyut, L. D. Mayo and K. E. Pollok, *Clin. Cancer Res.*, 2013, **19**, 2699–2709.
139. M. Califano and F. M. Gómez-Campos, *Nano Lett.*, 2013, **13**, 2047–2052.

Subject Index